Information Systems Engineering and Management 80

The book series "Information Systems Engineering and Management" (ISEM) publishes innovative and original works in the various areas of planning, development, implementation, and management of information systems and technologies by enterprises, citizens, and society for the improvement of the socio-economic environment.

The series is multidisciplinary, focusing on technological, organizational, and social domains of information systems engineering and management. Manuscripts published in this book series focus on relevant problems and research in the planning, analysis, design, implementation, exploration, and management of all types of information systems and technologies. The series contains monographs, lecture notes, edited volumes, pedagogical and technical books as well as proceedings volumes.

Some topics/keywords to be considered in the ISEM book series are, but not limited to: Information Systems Planning; Information Systems Development; Exploration of Information Systems; Management of Information Systems; Blockchain Technology; Cloud Computing; Artificial Intelligence (AI) and Machine Learning; Big Data Analytics; Multimedia Systems; Computer Networks, Mobility and Pervasive Systems; IT Security, Ethics and Privacy; Cybersecurity; Digital Platforms and Services; Requirements Engineering; Software Engineering; Process and Knowledge Engineering; Security and Privacy Engineering, Autonomous Robotics; Human-Computer Interaction; Marketing and Information; Tourism and Information; Finance and Value; Decisions and Risk; Innovation and Projects; Strategy and People.

Indexed by Google Scholar. All books published in the series are submitted for consideration in the Web of Science.

For book or proceedings proposals please contact Alvaro Rocha (amrrocha@gmail.com).

Luís Almeida · Graça Azevedo · Elisabete Vieira ·
Rui Marques
Editors

The Challenges of Sustainability in Accounting, Finance and Innovation

Springer

Editors
Luís Almeida
University of Aveiro, Campus Universitário de Santiago
ISCA-UA
Aveiro, Portugal

Graça Azevedo
University of Aveiro, Campus Universitário de Santiago
ISCA-UA
Aveiro, Portugal

Elisabete Vieira
University of Aveiro, Campus Universitário de Santiago
ISCA-UA
Aveiro, Portugal

Rui Marques
University of Aveiro, Campus Universitário de Santiago
ISCA-UA
Aveiro, Portugal

ISSN 3004-958X ISSN 3004-9598 (electronic)
Information Systems Engineering and Management
ISBN 978-3-032-16267-0 ISBN 978-3-032-16268-7 (eBook)
https://doi.org/10.1007/978-3-032-16268-7

This Springer imprint is published by the registered company Springer Nature Switzerland AG
The registered company address is: Gewerbestrasse 11, 6330 Cham, Switzerland

Preface

Welcome to the Sixth International Conference on Accounting and Finance Innovation (ICAFI 2025), held in Aveiro, Portugal. We are delighted to present this collection of papers under the central theme "The Challenges of Sustainability in Accounting, Finance and Innovation".

In an era marked by global uncertainty and technological disruption, sustainability has become one of the defining challenges for organizations, policymakers, and researchers alike. The intersection between sustainability, accounting, and finance is increasingly critical, not only for ensuring transparency and accountability but also for promoting innovation that aligns with social and environmental responsibility. ICAFI 2025 provided a valuable forum for exchanging ideas and discussing these pressing issues, encouraging dialogue on how these themes can contribute to a more sustainable and equitable future.

This conference continues its tradition of fostering collaboration between academia and practice, offering a space where researchers and professionals exchange insights and explore new frontiers in accounting and finance innovation. The papers in this volume reflect a wide range of perspectives and approaches, addressing themes such as sustainable financial reporting, ESG disclosure, the integration of digital technologies in sustainable business models, and the evolving role of accounting in advancing the sustainability agenda.

We extend our sincere gratitude to all the authors who contributed to this volume. Your dedication to research excellence and your commitment to advancing our understanding of sustainability in accounting and finance are deeply appreciated. We also wish to thank the reviewers for their rigorous evaluations and valuable feedback, which ensured the scientific quality of every paper included in this volume.

We extend our sincere gratitude to all authors and reviewers for their valuable contributions and to the organizing team whose commitment made ICAFI 2025 a success.

As you explore the contributions gathered in this book, we invite you to reflect on how innovation and sustainability can walk hand in hand, reshaping accounting and finance to better serve organizations, society, and the planet.

We hope this volume inspires continued research, collaboration, and meaningful action toward a more sustainable future!

Luís Almeida
Graça Azevedo
Elisabete Vieira
Rui Marques

Organization

Organizing Committee

Luís Almeida	Higher Institute of Accounting and Administration, University of Aveiro, Portuga
Graça Azevedo	Higher Institute of Accounting and Administration, University of Aveiro, Portugal
Elisabete Vieira	Higher Institute of Accounting and Administration, University of Aveiro, Portugal
Rui Pedro Marques	Higher Institute of Accounting and Administration, University of Aveiro, Portugal

Scientific Committee

Adriana Tiron-Tudor	Faculty of Economics and Business Administration, Babes-Bolyai University, Romania
Agnieszka Perepeczo	University of Szczecin, Polónia
Alberto J. Costa	Higher Institute of Accounting and Administration, University of Aveiro, Portugal
Alcina Portugal Dias	Higher Institute of Accounting and Administration of Porto, Portugal
Aldónio Ferreira	Monash University, Australia
Ana Arromba Dinis	Polytechnic Institute of Cávado and Ave, Portugal
Ana Bela Teixeira	Polytechnic Institute of Setúbal, Portugal
Ana Estima	Higher Institute of Accounting and Administration, University of Aveiro, Portugal
Ana Fialho	University of Évora, Portugal
Ana Isabel Morais	University of Lisboa, Portugal
Ana Lucia Fontes de Souza Vasconcelos	Federal University of Pernambuco, Brazil
Ana Maria Bandeira	Higher Institute of Accounting and Administration of Porto, Portugal
Ana Maria Paixão	State University of Paraíba, Brazil
Andreea Ioana Coste	Faculty of Economics and Business Administration, Babes-Bolyai University, Romania
Antonio Manuel López-Henández	University of Granada, Spain

António Martins	University of Coimbra, Portugal
Augusta Ferreira	Higher Institute of Accounting and Administration, University of Aveiro, Portugal
Beatriz Cuadrado-Ballesteros	University of Salamanca, Spain
Carla Carvalho	Higher Institute of Accounting and Administration, University of Aveiro, Portugal
Carla Fernandes	Higher Institute of Accounting and Administration, University of Aveiro, Portugal
Carlos Andrade	Higher Institute of Accounting and Administration, University of Aveiro, Portugal
Carlos Santos	Higher Institute of Accounting and Administration, University of Aveiro, Portugal
Catarina Delgado	School of Economics and Management, University of Porto, Portugal
Cecília Carmo	Higher Institute of Accounting and Administration, University of Aveiro, Portugal
César Bastos	Higher Institute of Accounting and Administration, University of Aveiro, Portugal
Cidália Lopes	Higher Institute of Accounting and Administration of Coimbra, Portugal
Claudio Teodori	University of Brescia, Italy
Cleber Broietti	Federal University of Santa Catarina, Brazil
Cleber Rodrigo Zanetti	Fundação Getúlio Vargas, Brazil
Cristian Carini	University of Brescia, Italy
Cristina Góis	Higher Institute of Accounting and Administration of Coimbra, Portugal
David Perea	Huelva University, Spain
Deborah Agostino	Milano Politechnic, Italy
Desi Arisandi	Rutgers Business School, Rutgers University, USA
Elisabete Fernanda Mendes Duarte	Polytechnic Institute of Leiria, Portugal
Elisabete Neves	Higher Institute of Accounting and Administration of Coimbra, Portugal
Elisabete Vieira	Higher Institute of Accounting and Administration, University of Aveiro, Portugal
Elvettin Akman	Süleyman Demirel Universit, Turkey
Enrique Bonson	Huelva University, Spain
Fátima Borges	Higher Institute of Accounting and Administration, University of Aveiro, Portugal
Fawzi Laswad	Massey University, New Zealand
Fernanda Alberto	Higher Institute of Accounting and Administration of Coimbra, Portugal

Fernando Hanjam	National University of Timor-Leste, Timor-Leste
Fernando Ramos	Universidade de Vale do Rio dos Sinos, Brazil
Fernando Tavares	Portucalense University, Portugal
Flora Lobo	Higher Institute of Accounting and Administration, University of Aveiro, Portugal
Francisco Carreira	Polytechnic Institute of Setúbal, Portugal
Francisco Javier Valencia Duque	Universidad Nacional de Colombia, Colombia
Graça Azevedo	Higher Institute of Accounting and Administration, University of Aveiro, Portugal
Harmeet Kaur Khanuja	Marathwada Mitra Mandal's College of Engineering, SPPU, India
Haslinda Yusoff	Universiti, Teknologi Mara, Malaysia
Helena Inácio	Higher Institute of Accounting and Administration, University of Aveiro, Portugal
Isabel García- Sanchez	University of Salamanca, Spain
Isabel Lourenço	ISCTE Business School, Portugal
Isabel Pedrosa	Higher Institute of Accounting and Administration of Coimbra, Portugal
Jesús Mauricio Flórez-Parra	University ECCI, Colombia
João Teodósio	Santarém Polytechnic Institute
Jonas OliveiraI	SCTE Business School, Portugal
Jorge Mota	University of Aveiro, Portugal
José Aguado-Romero	University of Granada, Spain
Jose Angel Perez Lopez	University of Seville, Spain
José António Moreira	School of Economics and Management, University of Porto, Portugal
José Vale	Higher Institute of Accounting and Administration of Porto, Portugal
Kátia Lemos	Polytechnic Institute of Cávado and Ave, Portugal
Leonor Fernandes Ferreira	Nova University, Portugal
Liang Guo	California State University, United States of America
Liliane Cristina Segura	Mackenzie University, Brazil
Louis Smidt	Tshwane University of Technology, South Africa
Luís Almeida	Higher Institute of Accounting and Administration, University of Aveiro, Portugal
Luís Lima Santos	Polytechnic Institute of Leiria, Portugal
Luís Pacheco	Portucalense University, Portugal
Lurdes Silva	Polytechnic Institute of Cávado and Ave, Portugal
Manuel Castelo Branco	School of Economics and Management, University of Porto, Portugal
Mara Madaleno	University of Aveiro, Portugal

Marcelo de Santana Porte	Federal University of the South and Southeast of Pará, Brazil
Marco Maffei	University of Naples Federico II, Italy
María Antonia García Benau	University of Valencia, Spain
Maria Anunciação Bastos	Higher Institute of Accounting and Administration, University of Aveiro, Portugal
Maria Cláudia Teixeira	Higher Institute of Accounting and Administration of Porto, Portugal
Maria da Conceição Tavares	Higher Institute of Accounting and Administration, University of Aveiro, Portugal
Maria del Carmen Caba-Pérez	University of Almeria, Spain
Maria João Major	ISCTE Business School, Portugal
Maria José Fernandes	Polytechnic Institute of Cávado and Ave, Portugal
Maria José P. Carvalho da Silva	Higher Institute of Accounting and Administration, University of Aveiro, Portugal
Marta Guerreiro	Polytechnic Institute of Viana do Castelo, Portugal
Melina Manochin	Aston University, United Kingdom
Miguel Gonçalves	Higher Institute of Accounting and Administration of Coimbra, Portugal
Miguel Resende	Higher Institute of Accounting and Administration, University of Aveiro, Portugal
Mirko Pecaric	Faculty of Administration, University of Lujbljana, Slovenia
Muganda Munir Manini	Kibabii University, Kenya
Ninel Kiosseva	New Bulgarian University Montevideo, Bulgaria
Patrícia Gomes	Polytechnic Institute of Cávado and Ave, Portugal
Patrícia Quesado	Polytechnic Institute of Cávado and Ave, Portugal
Paul L. Rideru	University of Craiova, Romania
Paula Neto	Higher Institute of Accounting and Administration, University of Aveiro, Portugal
Paulo Alves	Católica Porto Business School, Portugal
Paulo Gama	University of Coimbra, Portugal
Pedro Mendonça da Silva	ISCAP, Instituto Politécnico do Porto, Portugal
Pedro Pavone	University of Naples Federico II, Italy
Radiah Othman	Massey University, New Zealand
Rashid Ameer	IPU Tertiary Institute, New Zealand
Ricardo Costa	Faculty of Applied and Social Sciences of Petrolina, Brazil
Rui Pedro Marques	Higher Institute of Accounting and Administration, University of Aveiro, Portugal
Rui Robalo	Polytechnic Institute of Santarém, Portugal
Rute Abreu	Polytechnic Institute of Guarda, Portugal

Sandra Alves	Higher Institute of Accounting and Administration, University of Aveiro, Portugal
Sandra Caçador	Higher Institute of Accounting and Administration, University of Aveiro, Portugal
Sara Serra	Polytechnic Institute of Cávado and Ave, Portugal
Sérgio Cruz	Higher Institute of Accounting and Administration, University of Aveiro, Portugal
Shital Jayantilal	Portucalense University, Portugal
Sofia Gomes	Portucalense University, Portugal
Sónia Monteiro	Polytechnic Institute of Cávado and Ave, Portugal
Susana Jorge	University of Coimbra, Portugal; CICP-University of Minho
Susana Marques	Higher Institute of Accounting and Administration, University of Aveiro, Portugal
Tarek Rana	RMIT University, Australia
Tatiana Antipova	Institute of Certified Specialists, Perm, Russia
Teresa Eugénio	Polytechnic of Leiria, Portugal
Umulkher Ali Abdillahi	Masinde Muliro University of Science and Technology, Kenya
Verónica Ribeiro	Polytechnic Institute of Cávado and Ave, Portugal

Contents

Disclosure of Intangible Assets: Evidence from Portuguese Football Clubs

Mário Reimão[1](✉) and Carla Carvalho[2]

[1] Higher Institute of Accounting and Administration, University of Aveiro, Aveiro, Portugal
mariovivas@ua.pt

[2] GOVCOPP – Research, Unit on Governance, Competitiveness and Public Policies, Higher Institute of Accounting and Administration, University of Aveiro, Aveiro, Portugal
carla.carvalho@ua.pt

Abstract. This study examines a relatively underexplored topic in accounting literature: the disclosure of intangible assets, particularly those associated with professional football teams. It longitudinally analyzes the evolution of the proportion of such intangibles within the total assets of the largest Portuguese Sports Corporations (*Sociedades Anónimas Desportivas* - SADs) and assesses their compliance with mandatory disclosure requirements under International Accounting Standards (IASs) 36 and 38. Adopting a mixed-method approach that combines qualitative content analysis with quantitative descriptive analysis of the consolidated annual reports from 2017 to 2023, this study reveals that squad values represent a significant share of total assets, ranging from 16% to 36%, with Sporting SAD standing out. Sports passes account for over 97% of intangible assets in most SADs, except Benfica SAD, where they reach a maximum of 74%. Regarding disclosure compliance, the SADs of Benfica, Porto, and Sporting exhibit an average compliance index of 67% in the "Intangible Assets – Squad Value" category, while Braga's SAD reports only 45%. For the "Other Intangible Assets" category, compliance levels are lower. The most frequently disclosed elements include the useful life of assets, amortization rates, and reconciliation of carrying amounts. In contrast, the least reported items relate to impairment losses and the estimation of recoverable amounts. These findings contribute to expanding the empirical evidence on disclosure practices in a unique and under-researched context, but economically significant industry - professional football - and highlighting persistent gaps in transparency, particularly regarding IAS 36. The study also offers practical insights for SADs administrators, managers, regulatory bodies, and stakeholders concerned with the financial transparency of sports entities.

Keywords: Intangible Assets · Mandatory Disclosures · IAS 36 and 38 · Sports Corporations · Football Clubs

1 Introduction

Football possesses the unique ability to attract supporters and participants worldwide. However, it has evolved beyond a mere sport; it is now a benchmark industry capable of generating billions of euros annually (Braga, 2024; Santos, 2020). The economic

L. Almeida et al. (Eds.): ICAFI 2025, ISEM 80, pp. 1–18, 2026.
https://doi.org/10.1007/978-3-032-16268-7_1

growth of the sports market—and football in particular—has been significant, with the transfer of players' sporting rights representing one of the primary means of economic sustainability for football clubs (Alkhanaizi, 2024).

The organization and professionalism of *Sociedades Anónimas Desportivas* (SADs - Sports Corporations, in english) have driven this evolution within the football industry, with the rights over players constituting their main asset (Gazzola et al., 2019; Limba & Sapulette, 2021; Maroun et al., 2022). Nevertheless, the current accounting model has been criticized for not adequately reflecting the value of the football teams' squad, especially concerning internally trained players, as they are not capitalized as club assets (Bengtsson & Wallström, 2014; Maroun et al., 2022; Steenkamp, 2020). Therefore, the appropriate disclosure of SADs' intangible assets to their stakeholders is of particular importance—not only because they represent one of their main assets but also because the squad, whose rights fall within these intangibles, is under great scrutiny from investors, fans, and the general public.

SADs are increasingly important in both the sporting and financial context of professional football clubs (Sobral, 2017). However, limited research addresses the accounting treatment—and even less so the disclosure—of SADs' intangible assets. So, to enrich the existing literature, this study focuses on the intangible assets of Portuguese SADs, a relevant and underexplored context. It aims to analyze the evolution longitudinally, over the 2017–2023 period, of the proportion of intangible assets—particularly those related to football team squads—within the total assets of the largest Portuguese SADs, as well as to assess their compliance with mandatory disclosure requirements under IASs 36 and 38. To define the sample, the SADs of the most relevant clubs in the current Portuguese football landscape were selected (UEFA, 2024). Specifically, *Futebol Clube do Porto* (FCP), *Sport Lisboa e Benfica* (SLB), *Sporting Clube de Portugal* (SCP), and *Sporting Clube de Braga* (SCB). Their consistent participation in European competitions justifies the choice of these four SADs, as well as their representativeness in Portuguese professional football, and the public availability of their financial reports, which was not the case for other SADs. A qualitative methodology was adopted for data collection, using content analysis of quantitative information and narrative disclosures from the consolidated annual reports of the SADs. A quantitative approach was employed for data processing, involving the construction of absolute and relative frequency tables, complemented by graphs, followed by a descriptive analysis of the results.

The study makes theoretical contributions by expanding empirical evidence on disclosure practices of intangibles in a unique and under-researched context, but economically significant industry - professional football - and highlighting persistent gaps in transparency, especially regarding IAS 36. Focusing on a national setting that applies the International Financial Reporting Standards (IFRS) also adds comparative value for future cross-country studies. The study offers practical insights for SAD administrators, managers, regulatory bodies, and stakeholders concerned with the financial transparency of sports entities.

The study is structured into five sections. Following this introduction, section two presents the literature review. Section three is dedicated to the methodology of the empirical study, outlining its objectives, sample, and the process of data collection and analysis. Section four presents the analysis and discussion of the results, and the final section provides the study's conclusions, contributions, limitations, and directions for future research.

2 Literature Review

2.1 The Relevance of Football and SADs in Portugal

Football is a global phenomenon that has evolved from a simple sporting activity into a robust cultural, social, and economic industry. The Fédération Internationale de Football Association (FIFA) recognizes the year 1863 as the official starting point of modern football, marked by the foundation of The Football Association in the United Kingdom (FIFA, 2024). The internationalization of this sport was consolidated with the creation of FIFA in 1904 and its Olympic debut in 1908, culminating in the FIFA World Cup (FIFA, 2024).

In Portugal, football arrived in the 19th century and quickly gained popularity, leading to the establishment of clubs and competitions (Carvalho, 2007). The Portuguese Football Federation organized the first official championship, which evolved into the Portuguese Professional Football League in 1978. Today, this entity serves as the primary governing body of professional football in Portugal (Liga Portuguesa de Futebol Profissional, 2024). On the international stage, two organizations stand out: the Union of European Football Associations (UEFA) and FIFA.

The passion for football leads many people to engage in the sport—whether through youth development, amateur and professional football, or as a leisure activity—as a form of stress relief from daily life (Mourão, 2010). This passion is strongly influenced by each country's culture and population size. There is a strong relationship between a region's economic development and its clubs' performance, as countries with larger populations and regional centrality are more likely to achieve sporting and financial success (Mourão, 2010). Football generates nearly 28 billion euros annually due to its widespread appeal, making it the most profitable sport for sponsors and broadcasters (Santos, 2020). Considering the 2022/23 season at the Portuguese level, professional football contributed approximately 667 million euros to Portugal's Gross Domestic Product (GDP) (Braga, 2024).

The increasingly ambitious goals of football clubs have led to the adoption of new governance policies, with recent decades marked by the transition from amateurism to professional sports and management (Lemos, 2017). Revenues from sponsorship and advertising have become essential for these entities, driving the industrialization of sports (Sobral, 2017). This new governance model led to the creation of SADs, which professionalized and structured clubs differently. This model aims to ensure financial sustainability, professional management, and accountability while promoting trust among investors, sponsors, and fans, thereby contributing to the credibility and stability of the sector.

SADs were legally established in Portugal by Decree-Law No. 19/96 of June 25 and may be formed either from existing clubs or created from scratch. They are governed by specific legislation, namely the Legal Framework for Sports Companies (Law No. 39/2023 of August 4), and must comply with the regulations applicable to public limited companies. From an accounting standpoint, SADs are subject to the Portuguese Accounting Standards System (*Sistema de Normalização Contabilística* – SNC). When listed on a regulated market, they must apply the IFRS issued by the International Accounting Standards Board (IASB), adopted by the European Union (EU). Accounting transparency is ensured through various legal requirements, including independent audits of financial statements, by the Portuguese Commercial Companies Code, the Securities Code, and international auditing standards. Law No. 39/2023 reinforces these requirements, mandating rigorous financial reporting and responsible management.

2.2 Disclosure Requirements for Intangible Assets

Intangible assets have been gaining increasing relevance in the economy, representing one of the primary sources of competitive advantage for entities (Criekingen et al., 2022; Hoss et al., 2010), particularly for SADs. Lev (2001, 2019) highlights intangibles as the primary source of value for organizations, reinforcing the need for greater and more effective disclosure of information about these assets. Adequate disclosure enhances comparability and facilitates investment decisions (Dancaková et al., 2022).

International Accounting Standard (IAS) 38 – *Intangible Assets* defines intangibles as identifiable non-monetary assets without physical substance, emphasizing three essential characteristics for their recognition: identifiability, control, and the expectation of future economic benefits. The Notes constitute an important financial statement and, like the others, must provide relevant and easily understandable information to stakeholders. In this regard, IAS 38 prescribes a broad set of mandatory disclosure requirements for intangible assets in the Notes. First, a distinction must be made between acquired intangible assets and those generated internally, including disclosure of useful lives, the amortization method and rates adopted, the carrying amount and accumulated amortization at the beginning and end of the periods, and finally, a reconciliation of the carrying amount from the beginning to the end of the period (§118 of IAS 38).

Furthermore, following paragraph 118(e) of IAS 38, the reconciliation of the carrying amount between the beginning and end of the period must explain how the asset was acquired, as well as the classification of the assets, the amount of recognized and reversed amortization and impairment losses, and other changes in the carrying amount of the intangible asset. There is also a requirement to disclose research and development costs for intangible assets recognized as expenses (§126 of IAS 38).

It is also important to highlight the disclosure requirements under IAS 36 – *Impairment of Assets*, which apply to intangible assets, particularly paragraphs 126 and 130. Paragraph 126 requires entities to disclose the amount of impairment losses recognized or reversed during the period for each class of assets separately, specifying whether the losses or reversals were recognized directly in other comprehensive income or the profit or loss for the period. According to paragraph 130, an entity must disclose the reasons for recognizing or reversing impairment losses and the nature of the asset whose value has changed. The entity must also explain how the recoverable amount of its assets is calculated and identify the key assumptions used in this calculation.

2.3 Studies on the Disclosure of Intangible Assets by Football Clubs

Bengtsson and Wallström (2014) analyzed, from an investor's perspective, whether the accounting and disclosure of football players comply with the applicable standards and whether the disclosed information allows investors to obtain a realistic view of the clubs' financial position. The study focused on publicly listed clubs participating in European competitions during the 2012/2013 season, specifically Arsenal FC, Manchester United, Borussia Dortmund, Juventus FC, and FCP. The authors concluded several similarities among the clubs in accounting recognition and disclosure of these assets, with general compliance with the mandatory requirements established by IAS 38 and UEFA. However, differences were noted in the type, quantity, and quality of information disclosed by the clubs. It was concluded that FCP and Juventus FC provided more and higher-quality information to their stakeholders. This high level of compliance by Juventus FC with disclosure requirements regarding player registration rights was maintained during the 2018–2021 seasons (Limba & Sapulette, 2021). These differences are explained by the lack of uniform and clear valuation models for football players, which affects their accounting measurement. This conclusion was drawn by Gazzola et al. (2019) in a study examining clubs listed in the STOXX Europe Football Index across various European countries between 2012 and 2016. The study highlights that the application of impairment tests (IAS 36) and the measurement of the recoverable amount of such assets are neither properly carried out nor disclosed, compromising the reliability and relevance of financial information. Gazzola et al. (2019) also document a high degree of heterogeneity in compliance levels with the requirements of IAS 36 and IAS 38, with some clubs demonstrating high compliance and others displaying very low levels of compliance.

Regarding impairment testing specifically, clubs tend to disclose limited information about the application of such tests (Maglio & Rey, 2017), and there are various interpretations of IAS 36 for recognizing and measuring impairments of player rights (Biancone & Solazzi, 2012). Maglio and Rey (2017) report a low level of compliance with IAS 36 disclosure requirements among football clubs in Italy and the United Kingdom, particularly regarding the measurement criteria adopted. They, therefore, argue that UEFA and national federations should strengthen their regulations to improve the quality of such disclosures.

Lozano and Gallego (2011) focused on the valuation and disclosure of player registration rights at Sevilla FC between 1997 and 2009, highlighting shortcomings in accounting standards regarding the non-recognition of internally trained players as assets. This leads to an undervaluation of the intangible assets category. They also observed a significant discrepancy between the book value and market value of players' registration rights, with the book value representing only 25% of the market value. In a more recent study (Lozano & Gallego, 2024), based on interviews with football club auditors, the authors reinforced the notion that the disclosure of measurement criteria for intangible assets is insufficient, suggesting a review of the IASB's accounting standards governing such assets.

Finally, Shareef and Davey (2005) analyzed how the publicly listed companies of 19 English football clubs disclosed their intellectual capital in 2002. More specifically, they examined the extent and quality of information disclosed regarding intellectual capital, which was structured into three components: internal, external, and human capital. These components include, among others, performance analysis policies for players and information regarding the clubs' financial agreements and partnerships. The authors identified a higher level of care and disclosure by managers concerning information on salaries and registration values of players, player negotiations, broadcasting rights, executive compensation, and the organization's management philosophy. The external capital component had the highest disclosure rate (53%), followed by human capital (44%), and lastly, internal capital (33%). The study concluded that financial statements remain very limited in disclosing internal capital-related information, an issue that warrants a review of current disclosure practices by clubs. Moreover, there is substantial room for improvement in the external and human capital disclosures. Shareef and Davey (2005) also found a positive correlation between the size of the club, its performance, and the level of intellectual capital disclosure.

3 Methodology

3.1 Objectives

The economic growth of the sports market—particularly football—has become increasingly evident in recent years, with the transfer of players' registration rights representing one of the primary means of financial and economic sustainability for clubs. These are further complemented by revenues from participation in European competitions, broadcasting rights, and the commercial use of club brands (Alkhanaizi, 2024). Therefore, the registration rights of athletes play a crucial role in the economic performance of SADs, which directly influence their medium- and long-term sporting performance (Mnzava, 2013). Given the relevance of the squad's value within SADs and in the football industry, the present study on the disclosure of SADs' intangible assets is fully justified.

Accordingly, this study aims to longitudinally analyze the evolution of the proportion of intangible assets—particularly those related to football team squads—within the total assets of the largest Portuguese SADs and to assess their compliance with mandatory disclosure requirements under IASs 36 and 38. This longitudinal perspective enables the identification of disclosure patterns and persistence over time, which are crucial to understanding how SADs approach transparency and regulatory compliance regarding intangible assets.

To achieve these objectives, the following research questions are formulated:

- What is the evolution of the proportion of squad value in the total assets of SADs?
- What is the evolution of the proportion of squad value within the total intangible assets of SADs?
- What is the level of compliance by SADs with the mandatory disclosure requirements of IAS 36 and IAS 38 regarding intangible assets?

3.2 Sample

The SADs of the most prominent clubs in the current Portuguese football landscape were selected to address the research questions, specifically FCP, SLB, SCP, and SCB

(Table 1). These clubs consistently occupied the top four positions in the Portuguese professional football league between the 2017/2018 and 2022/2023 seasons and hold the highest UEFA club coefficients among Portuguese teams (UEFA, 2024).

Table 1. Sample.

CLUB
Futebol Clube do Porto - Futebol SAD
Sport Lisboa e Benfica - Futebol SAD
Sporting Clube de Portugal - Futebol SAD
Sporting Clube de Braga - Futebol SAD

All four SADs apply the same accounting framework, namely the IFRS adopted by the EU, which allows for the comparability of the results obtained. It should be noted, however, that the limited access to the annual reports of other Portuguese SADs—due to their unavailability in the public domain—and the lack of response after being contacted via email, constrained the expansion of the sample to other SADs. Table 2 provides a summary characterization of each of the SADs included in the sample.

The data was collected from the consolidated annual reports of the SADs for the financial periods from 2017 to 2023. The selection of this period is justified by the fact that the 2016/2017 season marked the last time one of the sampled teams did not finish in at least fourth place in the Portuguese professional football league. Furthermore, at the beginning of this study, only the consolidated annual reports up to the 2022/2023 season were publicly available.

3.3 Data Collection and Processing

Data were manually collected from the consolidated annual reports of the four SADs, encompassing the analysis of 24 reports. Thus, the data collection methodology is qualitative, consisting of a content analysis of these reports, which include both quantitative and narrative information. A quantitative approach was adopted for data processing, utilizing the construction of relative and absolute frequency tables complemented by graphs and conducting descriptive analysis of the results.

Given that the first two research questions about the proportion that the "Player Squad" item represents within the total assets and intangible assets of the SADs, values were extracted from the balance sheet and notes related to the items "Intangible Assets – Player Squad Value," "Other Intangible Assets," and total assets of the different SADs for all considered financial periods.

A disclosure checklist was constructed, detailing the requirements outlined in IAS 38 and IAS 36, which totaled 58 items to calculate compliance with mandatory disclosure requirements. This checklist was subdivided into two parts to calculate: the disclosure

Table 2. Sample characteristics.

CLUB	Year of foundation (SAD)	Number of members	Business volume (2022/23) (million euros)	Total assets (2022/23) (million euros)	Relevant titles won
SLB	2000	327,000	284.7	557	38 national championships, 26 cups of Portugal, 2 Europeans cups.
SCP	1997	158,000	243.7	315.9	20 national championships, 17 cups of Portugal.
FCP	1997	141,600	192.5	307.7	30 championships, 19 cups of Portugal, 2 UEFA Champions League, 2 Europa League.
SCB	1998	31,200	77.2	118.6	3 cups of Portugal, UEFA European League finalist.

Source: Compiled by the authors based on information gathered from the clubs' official websites (FCP, 2024; SCB, 2024; SCP, 2024; SLB, 2024)

index for the "Intangible Assets – Player Squad Value" and the index for "Other Intangible Assets". For the construction of the disclosure index, each of the 58 items was assigned a value of "1" if the SAD complied with the disclosure requirement, "0" if the SAD did not comply, or "Not Applicable" (N/A) when the requirement did not apply to the SAD under analysis. In cases where an item was not included in the notes on intangible assets and, cumulatively, there was no information indicating that such omitted information was applicable and required for the SAD, it was classified as "Not Applicable". Finally, to calculate the mandatory disclosure compliance index for each SAD and year, the total number of items disclosed by the SAD was divided by the total number of applicable items. This result was multiplied by 100 to obtain a percentage value.

Based on this checklist, it was possible to determine each SAD's overall disclosure compliance index and specific compliance indices for the two classes of intangible assets, namely "Intangible Assets – Player Squad Value" and "Other Intangible Assets". Additionally, it was possible to analyze the evolution of these indices over the years for each SAD, as well as in relation to standards and disclosure requirements.

4 Analysis and Discussion of Results

4.1 Proportion of the Squad Value in SADs

Relative to Total Assets. Before analyzing the proportion of the squad value, identified under the item "Intangible Assets – Player Squad Value", in relation to the total assets of each SAD, it is essential to examine the absolute values of these two items for each club and year within the sample (Table 3).

Table 3. Squad value and total assets of the SADs.

Club	Item	Sports Season					
		2017/18	2018/19	2019/20	2020/21	2021/22	2022/23
SLB	Intangible assets – squad value	113,491	80,426	102,884	146,162	111,866	126,462
	Total assets	485,077	500,780	487,066	523,260	533,732	557,832
FCP	Intangible assets - squad value	82,656	74,990	76,144	110,058	79,384	85,465
	Total assets	399,439	342,029	300,640	407,817	418,449	307,730
SCP	Intangible assets - squad value	78,479	87,863	70,060	60,081	94,190	96,163
	Total assets	221,044	251,928	228,196	260,053	319,595	315,901
SCB	Intangible assets - squad value	17,444	16,416	25,896	24,704	17,444	19,558
	Total assets	57,430	71,522	99,426	93,766	91,461	118,623

(*values in thousands of euros*).

SLB stands out as the club with a substantially higher total asset value and squad valuation than the other SADs. Referencing the figures from the 2022/2023 season, SCP has surpassed FCP in total asset value, positioning itself as the second-largest Portuguese club in balance sheet size. SCB ranks fourth, having consistently increased its asset value over the years.

Figure 1 illustrates the percentage weight of each club's squad value relative to their total assets over the six-year sample period.

It is observed that, across all SADs and economic periods analyzed, the weight of the squad in the total assets ranges between 16% and 36%, with some fluctuations over the years. Notably, Sporting CP (SCP) stands out as the club where the value of the squad has the most significant weight in total assets across all sporting seasons, except for the 2020/21 season. Furthermore, except for the 2020/21 and 2021/22 seasons, this proportion has consistently exceeded 30%, reaching 36% and 35% in the 2017/18 and 2018/19 seasons, respectively. At SL Benfica (SLB), it is evident that the weight of "Intangible Assets – Squad Value" fluctuates over time, experiencing several increases and decreases. However, in 2022/23, this weight is the same as in 2017/18 (23%). The

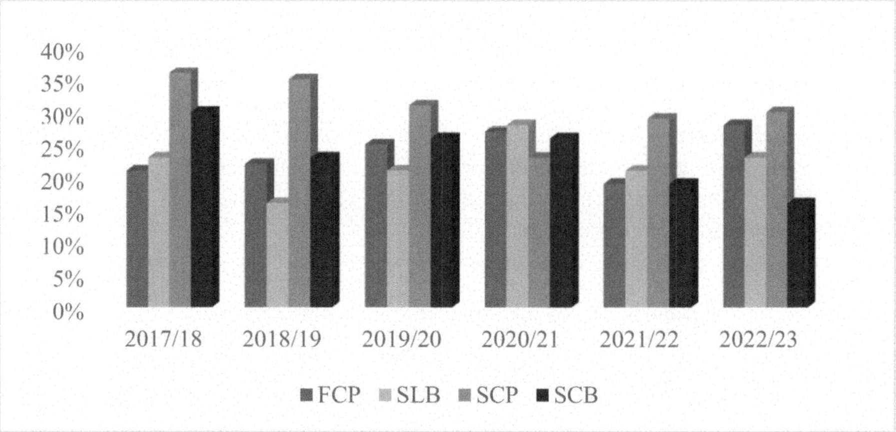

Fig. 1. The proportion of squad value in the SAD's total assets.

2020/21 season marks the highest proportion of squad value in SLB's total assets (28%). In contrast, FC Porto (FCP) shows a growing trend in the proportion of squad value to total assets over the first four periods (from 21% in 2017/18 to 28% in 2022/23). Conversely, in SC Braga (SCB), the proportion of the squad value in total assets has not followed a linear pattern, with a significant decline over the years (from 30% in 2017/18 to 16% in 2022/23).

In summary, it can be concluded that the weight of the squad value in the total assets of SADs ranged from a minimum of 16% to a maximum of 36%, representing a considerable share of the SADs' total assets, particularly in the case of SCP. This high proportion of the squad in total assets demonstrates the economic centrality of sports rights in the patrimonial structure of the clubs and their importance as a driver of value creation. This evidence therefore reinforces the need for rigorous accounting practices and prudent management of these intangible assets. Contrary to the trend indicated in the literature review, there is no notable tendency for the proportion of squad value to increase in the clubs' total assets during the period analyzed.

Relative to Total Intangible Assets. Table 4 presents the relative weight of the squad value and other intangible assets in the total value of intangible assets for each SAD over the six years under analysis.

In the cases of FCP, SCP, and SCB, the squad value accounts for nearly all the intangible assets, consistently representing more than 97%, except SCB in the 2021/22 season (96.46%). It is also worth highlighting the 2022/23 sporting season for both FCP and SCP, as it marked the only instance in which the squad value exceeded 99% of total intangible assets for FCP and simultaneously represented the lowest proportion for SCP (98.95%), although still very high. In the case of SLB, the percentage is not pronounced, with the highest proportion observed in the 2020/21 season (74.34%) and the lowest in the 2018/19 season (59.80%).

Thus, it is evident that the value of players' registration rights, which comprise the various squads, is the most significant item, consistently representing almost the entirety

Table 4. Proportion of squad value in the SAD's total intangible assets.

Club	Item	Sports Season					
		2017/18	2018/19	2019/20	2020/21	2021/22	2022/23
FCP	Intangible assets - squad value	98.76%	98.87%	97.88%	98.62%	98.13%	99.06%
	Other intangible assets	1.24%	1.13%	2.02%	1.38%	1.87%	0.94%
SLB	Intangible assets - squad value	67.10%	59.80%	66.36%	74.34%	69.61%	72.73%
	Other intangible assets	32.90%	40.20%	33.64%	25.66%	30.39%	27.27%
SCP	Intangible assets - squad value	99.60%	99.55%	99.62%	99.22%	99.33%	98.95%
	Other intangible assets	0.40%	0.45%	0.38%	0.78%	0.67%	1.05%
SCB	Intangible assets - squad value	99.92%	99.35%	97.81%	97.10%	96.46%	97.28%
	Other intangible assets	0.08%	0.65%	2.19%	2.90%	3.54%	2.72%

of each SAD's intangible assets. The remaining amount is attributed to "Other Intangible Assets." An exception is Benfica SAD, where the value of players' registration rights accounts for between 60% and 74% of the total value of its intangible assets. In fact, despite SLB SAD having the highest monetary value in terms of the squad, its proportion of intangible assets is lower than the other SADs, although consistently higher than "Other Intangible Assets." This is due to SLB holding a significantly higher amount in "Other Intangible Assets" than the other clubs.

4.2 Compliance with Intangible Asset Disclosures

By Category of Intangible Asset. The transparency of information disclosed to the market is a fundamental pillar for the effective functioning of SADs and for informed decision-making by their stakeholders. This section presents an analysis of the evolution of the compliance index with mandatory disclosure requirements for intangible assets, as prescribed by IAS 36 and IAS 38. The analysis is divided by category of intangible assets (squad value vs. other intangible assets), standard (IAS 38 vs. IAS 36), and disclosure requirement.

Table 5 presents the evolution of compliance levels by each SAD with the disclosure requirements of the two standards concerning intangible assets associated with the squad. It can be concluded that there are a few fluctuations in the compliance index for each SAD. This was already evident during the data collection process, as it became clear that the Note models used by each SAD follow the same disclosure pattern from year to year.

SLB is the most compliant club, presenting an average compliance index of 67.1%. This percentage experienced minor fluctuations over the years, reaching a maximum value (69%) in the 2018/19 and 2020/21 seasons and a minimum value (65.5%) in the

Table 5. Disclosure index of "Intangible Assets – Squad Value".

Club	Sports Season						Mean
	2017/18	2018/19	2019/20	2020/21	2021/22	2022/23	
SLB	66.7%	69.0%	65.5%	69.0%	65.5%	66.7%	67.1%
FCP	66.7%	66.7%	66.7%	66.7%	66.7%	66.7%	66.7%
SCP	66.7%	66.7%	66.7%	66.7%	66.7%	66.7%	66.7%
SCB	48.0%	44.0%	44.0%	44.0%	44.0%	44.0%	44.7%

2019/20 and 2021/22 seasons. On the other hand, SCB is the club with the lowest level of compliance, recording an average value of 44.7% over the six years analyzed. Except for the 2017/18 season, in which the entity reached its maximum value of 48%, the compliance index remained at 44% in all other periods—the lowest value among all years and clubs analyzed. FCP and SCP are the most consistent in compliance with disclosures regarding their squads, presenting precisely the same compliance index (66.7%) in all seasons analyzed.

Concerning the "Other Intangible Assets" category (Table 6), the level of compliance with the disclosure requirements of both standards is substantially lower than that observed for squad-related intangible assets across all SADs and years analyzed. The lowest average value is 26.3% (SCB), and the highest is 37.5% (SLB). However, similarly to what was observed for the intangibles related to the squad, the compliance index for other intangible assets also remained unchanged over the years, reinforcing the adoption of a consistent disclosure pattern by the SADs.

Table 6. Disclosure index of "Other Intangible Assets".

Club	Sports Season						Mean
	2017/18	2018/19	2019/20	2020/21	2021/22	2022/23	
SLB	37.5%	37.5%	37.5%	37.5%	37.5%	37.5%	37.5%
FCP	33.3%	33.3%	33.3%	33.3%	33.3%	33.3%	33.3%
SCP	37.5%	37.5%	37.5%	33.3%	37.5%	37.5%	36.8%
SCB	33.0%	25.0%	25.0%	25.0%	25.0%	25.0%	26.3%

It can be concluded that the "Intangible Assets – Squad Value" category shows a substantially higher disclosure compliance index than the "Other Intangible Assets" category. This indicates an apparent concern on the part of the SADs to disclose more information about the assets most relevant to their balance sheets, specifically those related to their squads, which also receive greater attention from stakeholders. However, the level of compliance with the requirements established by the two standards (IAS 36 and IAS 38) remains low for intangibles related to squads. We are analyzing listed companies, which are subject to high scrutiny by the market, their supporters, and the general public, from whom greater compliance with mandatory information disclosure is required. These results contradict those obtained by Bengtsson and Wallström (2014) and Limba and Sapulette (2021), who found high compliance with the disclosure of intangibles. However, they reinforce the findings of Shareef and Davey (2005) that financial statements still provide minimal information regarding these assets.

By Standard. Regarding the most complied-with standards, Table 7 provides insight into the level of disclosure for each of the standards under analysis.

Table 7. Disclosure Index according to the standard.

Club	IAS 36						IAS 38					
	2017/18	2018/19	2019/20	2020/21	2021/22	2022/23	2017/18	2018/19	2019/20	2020/21	2021/22	2022/23
SLB	40%	45%	45%	45%	45%	40%	58%	61%	58%	61%	58%	58%
FCP	44%	40%	40%	44%	40%	44%	55%	55%	55%	55%	55%	55%
SCP	47%	50%	50%	47%	50%	47%	55%	55%	55%	52%	55%	55%
SCB	13%	13%	13%	13%	13%	13%	55%	45%	45%	45%	45%	45%
Mean	36%	37%	37%	37%	37%	36%	56%	54%	53%	53%	53%	53%

There have been a few changes in the disclosure index over the years across the different SADs and for both standards. The index is higher for IAS 38, with an average disclosure level exceeding 50%; for IAS 36, the average level of compliance consistently remains below 40%. Notably, the average disclosure index for IAS 36 remains the same in both the first and last financial periods (36%), increasing to 37% in the remaining periods. In contrast, the average disclosure level for IAS 38 requirements has continuously declined over time, reaching a peak of 56% in 2017/18 and a low of 53% between 2019/20 and 2022/23.

When analyzing compliance with the disclosure requirements of IAS 36, an apparent discrepancy is observed between the index values for SLB, FCP, and SCP, which are similar (ranging between 40% and 50%), and that of SCB. SCB presents the lowest index (13%), significantly below the others. SCP, on the other hand, has the highest index (50%). The variation in index values between 13% and 50% reveals the limited importance that SADs attribute to compliance with IAS 36 disclosure requirements. The compliance index for IAS 38 requirements shows higher values but remains below the expected and desirable levels, given the relevance of the SADs involved in the study. In the 2018/19 and 2020/21 financial periods, SLB achieved an index of 61%—the highest for IAS 38, never recording a value below 58%. Conversely, SCB had the lowest index (45%), which remained constant throughout all years except for the 2017/18 season, when it reached 55%.

In summary, SADs comply more with the disclosure requirements of IAS 38 than those of IAS 36, although the index values are well below what would be expected for these entities. The most and least compliant SADs for both standards are SLB and SCB, respectively. Gazzola et al. (2019) also state significant heterogeneity in compliance with IAS 36 and 38, with clubs having very different levels of compliance. Maglio and Rey (2017) emphasize that clubs' compliance with the disclosure requirements of IAS 36 is low, particularly regarding the measurement criteria adopted.

By Disclosure Requirement. This section analyzes the compliance index for the items "Intangible Assets – Squad Value" and "Other Intangible Assets" by mandatory disclosure requirement under IAS 36 and IAS 38, allowing for identifying items with the highest and lowest levels of disclosure.

About the "Intangible Assets – Squad Value," of the 58 disclosure items analyzed, 31 do not apply (N/A) to the four SADs during the six periods under review. These requirements mainly relate to changes in accounting policies, intangible assets acquired through government grants, and assets measured at revalued amounts or subject to impairment losses. On the other hand, there are nine requirements for which the average disclosure index was 100%. These requirements are mainly based on paragraph 118 of IAS 38, which includes information on whether the useful life is finite or indefinite and, in the case of a finite life, the useful lives or amortization rates used, the gross carrying amount and accumulated amortization at the beginning and end of the period; and the line items in the income statement in which any amortization of intangible assets is included. A similar pattern was observed for specific IAS 36 paragraph requirements, particularly those concerning the disclosure of impairment losses recognized during the period, the nature of the asset for which impairment was recognized, the line items in the statement of comprehensive income in which those losses were included, and the events that led to their recognition. The average disclosure index for these requirements is 75%, as they are disclosed by all SADs in the sample across all periods except SCB.

SCB distinguishes internally generated intangible assets from other types, describing the remaining amortization period in all its annual reports. Since the other SADs never make this distinction, the average disclosure index for these requirements is only 25%. Another requirement with an average disclosure index of 25% is stated in paragraph 122(b) of IAS 38, where only SLB discloses information regarding impairment losses reversed during the period. Additionally, some items were never disclosed during the six years, mainly related to information about the recoverable amount of intangible assets. Although SADs mention that the value considered is the higher net selling price and value in use, they do not specify which of the two is used on a yearly basis. They also fail to mention the fair value hierarchy level when using the net selling price or the discount rate applied in the previous and current estimates when considering the value in use. These results are consistent with those of Lozano and Gallego (2024), who evidence a lack of disclosure of the criteria for measuring intangible assets.

Concerning the "Other Intangible Assets", out of the 58 disclosure items analyzed, 33 were never applicable (N/A) to any SAD. These requirements correspond to those already identified for the squad-related item. On the other hand, there are requirements where the average disclosure index reaches 100%, particularly whether the useful life is finite or indefinite, and the corresponding line item in the statement of comprehensive income. Among the items applicable but not disclosed in any financial period by any SAD, the most notable are those related to impairment losses, the recoverable amount of assets, and information distinguishing internally generated assets from other types of assets. For the remaining items under IAS 38 and IAS 36, the average disclosure index was 75% or lower.

In conclusion, the items with the highest disclosure indices are those related to the useful life of assets, amortization rates, and the reconciliation of the carrying amount at the beginning and the end of the period. Conversely, those with the highest non-compliance are related to impairment losses—whether regarding their recognition or reversal—as well as the calculation of the recoverable amount. When compared to the findings in the literature, the results are consistent with those of Gazzola et al. (2019) and Lozano and Gallego (2024), particularly in terms of the application of impairment tests under IAS 36 and the assessment of the recoverable amount of assets. Similarly, Maglio and Rey (2017) and Biancone and Solazzi (2012) highlight the limited disclosure of information related to impairment testing and the challenges associated with achieving consistent interpretations of IAS 36.

5 Conclusions

This study aimed to examine longitudinally, over the period 2017–2023, the evolution of the proportion of intangible assets—particularly those related to football team squads—within the total assets of the largest Portuguese SADs, and to assess their compliance with the disclosure requirements established in IAS 36 and 38.

It is concluded that SLB is the club with a substantially higher value of total assets and squad value than the other SADs. In relative terms, however, SCP is the club with the highest proportion of squad value in total assets (30%) in the 2022/23 season, while SCB has the lowest proportion (16%). In response to the research question "What is the evolution of the proportion of squad value in the total assets of SADs?", it is concluded that this proportion ranged from a minimum of 16% to a maximum of 36%, which represents a considerable share of SADs' total assets—especially in the case of SCP. However, this proportion shows no clear upward trend over the analyzed period.

Regarding the second research question, "What is the evolution of the proportion of squad value within the total intangible assets of SADs?" it is concluded that the value of player registration rights, which make up the squads, represents nearly all of the intangible assets of each SAD (above 97%), with the remaining amount attributed to "Other Intangible Assets". An exception is SLB, where the value of player rights accounts for between 60% and 74% of its total intangible assets.

To address the final research question, "What is the level of compliance by SADs with the mandatory disclosure requirements of IAS 36 and IAS 38 regarding intangible assets?", an analysis was conducted by category of intangible asset, by standard, and by disclosure requirement. Regarding the categories, the SADs of SLB, FCP, and SCP show an average disclosure index of 67% for the "Intangible Assets – Squad Value" category, while SCB's SAD shows only 45%. For the "Other Intangible Assets" category, the compliance index is substantially lower, with the highest index also belonging to SLB (37.5%) and the lowest to SCB (26.4%). As for compliance with disclosure requirements by standard, few changes were observed in the index over the years across the various SADs. Higher compliance was found with the requirements of IAS 38 (45% to 61%) compared to IAS 36 (40% to 50%, except for SCB, which had an index of only 13%). The most frequently disclosed requirements by SADs are those concerning the useful life of assets, amortization rates, and reconciliation of the carrying amount at the beginning and the end of the period. The least complied requirements relate to impairment losses—both recognition and reversal—and the estimation of recoverable amounts. The complexity and subjectivity associated with recognizing impairment losses—particularly the estimation of recoverable amounts—and the potential for earnings management may explain why SADs show lower compliance with the disclosure requirements outlined in IAS 36.

This study provides important contributions to both theory and practice. From a theoretical perspective, it expands the scarce existing literature by examining how SADs disclose their intangible assets—particularly those related to player squads—and their level of compliance with relevant accounting standards. More specifically, it provides longitudinal empirical evidence on the importance of squad value within total assets and the disclosure behavior of SADs under IAS 36 and 38, thereby responding to calls for more research in economically significant yet underexplored sectors, such as professional football. The study also highlights persistent gaps in disclosure transparency, especially about impairment tests and the estimation of recoverable amounts. By focusing on a national setting where IFRS is mandatorily applied to listed companies, the findings also contribute to the broader international accounting literature by offering insights that can inform future comparative or cross-country studies on the disclosure of intangible assets. From a practical standpoint, the findings are valuable for SAD managers, supervisory entities, club regulators, fans, and the broader public concerned with the financial transparency of sports entities. This study has limitations, the most significant being the limited sample size. Furthermore, the study analyzed the extent of the disclosure and not its quality. Future research is therefore encouraged to extend this study to the most relevant SADs in European football, given the difficulty in accessing financial and non-financial information from other Portuguese SADs. This would allow for a comparative analysis of results across different jurisdictions.

Acknowledgments. This work was supported by the UID Research Unit on Governance, Competitiveness and Public Policies (GOVCOPP), funded by national funds through FCT - Fundação para a Ciência e a Tecnologia.

References

Alkhanaizi, A.: Beyond the pitch: Unveiling the financial dynamics driving modern football. https://resoinsights.com/insight/beyond-the-pitch-unveiling-the-financial-dynamics-driving-modern-football/. (2024, 19 de julho)

Bengtsson, M., Wallström, J.: Accounting and disclosure of football player registrations: do they present an accurate and fair view of the financial statements? [Master's Thesis, Jönköping University] (2014). https://hj.diva-portal.org/smash/get/diva2:721716/FULLTEXT01.pdf

Biancone, P.P., Solazzi, A.: Financial communication in professional football clubs. Bus. Manag. Sci. Int. Q. Rev. **3**(1), 153–174 (2012)

Braga, N.: Impacto do futebol profissional sobe 8% para 667 milhões de euros. Jornal Económico. https://jornaleconomico.sapo.pt/noticias/impacto-do-futebol-profissional-no-pib-sobe-8-para-667-milhoes-de-euros/. (2024, 27 de fevereiro)

Carvalho, M.J.: Os elementos estruturantes do regime jurídico do desporto profissional em Portugal [Tese de Doutoramento, Faculdade de Desporto da Universidade do Porto] (2007). https://hdl.handle.net/10216/96858

Criekingen, K., Bloch, C., Eklund, C.: Measuring intangible assets: a review of the state of the art. J. Econ. Surv. **36**(5), 1539–1558 (2022)

Dancaková, D., Sopko, J., Glova, J., Andrejovská, A.: The impact of intangible assets on the market value of companies: cross-sector evidence. Mathematics **10**(20), 3819 (2022)

Fédération Internationale de Football Association: How the FIFA World Cup format has evolved. https://www.fifa.com/pt/tournaments/mens/worldcup/canadamexicousa2026/articles/copa-mundo-evolucao-formatos-historia-1930-2026. Accessed 12 Feb 2024

Futebol Clube do Porto – Futebol SAD: História. https://www.fcporto.pt/pt/clube/historia. Accessed 12 Feb 2024

Gazzola, P., Amelio, S., Papagiannis, F., Vatamanescu, E.-M.: Financial reporting in European football teams: a disclosure analysis of player registrations. Int. J. Acad. Res. Account. Financ. Manag. Sci. **9**(4), 182–206 (2019)

Hoss, O., Rojo, C. A., Grappegia, M.: Gestão de ativos intangíveis: Da mensuração à competitividade por cenários. Atlas, São Paulo (2010)

Lemos, P.C.S.: Ativos intangíveis: O caso dos jogadores de futebol nas sociedades desportivas [Dissertação de Mestrado, Instituto Politécnico do Cávado e do Ave] (2017). http://hdl.handle.net/11110/1338

Lev, B.: Intangibles: management, measurement and reporting. Int. J. Account. **36**(4), 501–512 (2001)

Lev, B.: Ending the accounting-for-intangibles status quo. Eur. Account. Rev. **28**(4), 713–736 (2019)

Liga Portuguesa de Futebol Profissional: História da Liga Portugal. https://www.ligaportugal.pt/pt/40anos/. Accessed 29 June 2024

Limba, F.B., Sapulette, S.G.: European football player transfer scheme according to IFRS and IAS: case study on Juventus Football Club S.p.A.. Budapest Int. Res. Crit. Inst.-J. (BIRCI-Journal) **4**(4), 12355–12365 (2021)

Lozano, F.J.M., Gallego, A.C.: Critical discussion regarding the valuation of the most relevant assets in soccer clubs: auditors' insights. Revista de Contabilidad - Spanish Accounting Review **27**(1), 130–145 (2024)

Lozano, F.J.M., Gallego, A.C.: Deficits of accounting in the valuation of rights to exploit the performance of professional players in football clubs: a case study. J. Manag. Control. **22**, 335–357 (2011)

Maglio, R., Rey, A.: The impairment test for football players: the missing link between sports and financial performance? Palgrave Commun. **3**(17055), 1–9 (2017)

Maroun, W., van Zijl, W., Chesaina, R., Garnett, R.: The beautiful game: fair value, accountability and accounting for player registrations. Aust. Account. Rev. **32**(3), 334–251 (2022)

Mnzava, B.: Do intangible investments matter? Evidence from soccer corporations. Sport Bus. Manag. Int. J. **3**(2), 158–168 (2013)

Mourão, P.: Regional determinants of competitiveness: the Case of European soccer teams. Int. J. Sport Financ. **5**(3), 222–234 (2010)

Santos, M.M.: Futebol e Objetivos do Desenvolvimento Sustentável: O caso da Federação Portuguesa de Futebol [Dissertação de Mestrado, Instituto Superior de Economia e Gestão da Universidade de Lisboa] (2020). http://hdl.handle.net/10400.5/21108

Steenkamp, M.: Capturing the value of football players in financial reporting [Master Dissertation, University of Johannesburg] (2020). https://hdl.handle.net/10210/430883

Shareef, F., Davey, H.: Accounting for intellectual capital: evidence from listed English football clubs. J. Appl. Account. Res. **7**(3), 78–116 (2005)

Sobral, F.: Futebol, o estádio global. Fundação Francisco Manuel dos Santos (2017)

Sport Lisboa e Benfica – Futebol SAD: Fundação - Como nasceu o Sport Lisboa e Benfica (2024). https://www.slbenfica.pt/pt-pt/instituicao/clube/historia/fundacao. Accessed 12 Feb 2024

Sporting Clube de Braga – Futebol SAD: História. https://scbraga.pt/historia/. Accessed 12 Feb 2024

Sporting Clube de Portugal – Futebol SAD: História. https://www.sporting.pt/pt/clube/infraestruturas/estadio-jose-alvalade/historia. Accessed 12 Feb 2024

Union of European Football Associations: Financial Sustainability. https://pt.uefa.com/running-competitions/integrity/financial-sustainability/. Accessed 29 June 2024

Impression Management *Strategies* in Companies' Reports - Systematic Literature Review

Maria de Fátima Simões[1(✉)] and Augusta Ferreira[1,2]

[1] University of Aveiro, Aveiro, Portugal
{mfbs,augusta.ferreira,augusta.ferreira}@ua.pt
[2] CICF, Instituto Politécnico do Cávado e do Ave, Barcelos, Portugal

Abstract. Companies need to disclose their information, both financial and non-financial, not only to comply with requirements, standards, and regulations but also to meet stakeholders' expectations. Increasingly competitive markets impact companies by demanding a favorable and acceptable image, which allows them recognition and visibility.

In this context, information preparers' use of Impression Management strategies to hide negative disclosures and highlight positive information in their reports is relevant.

For this study, a systematic literature review was carried out, with scientific articles obtained from the Scopus database, to provide structured knowledge about the investigation in *Impression Management*. The results allowed us to know the evolution of the research on *Impression Management*, verifying that organizations use several IM strategies to highlight their performance, namely, the concealment of negative impacts; the strategic manipulation of reports; the word count; the presentation of images and sentimental expressions; the reference to financial and *benchmarking indicators;* to self-promotion measures and the attribution of positive factors to internal causes and negative aspects to external causes. The results also allowed us to identify the theories that have supported studies on *Impression Management*, as well as to provide clues for future research.

Keywords: Strategies · I*mpression management* · Reports · Theories · Methodologies

1 Introduction

The present work aims to provide structured knowledge about Impression Management (IM) research in the area of accounting, namely in the disclosure and reports of companies, through a systematic literature review (SLR).

According to the behavioral perspective of [1], IM is the process through which people control impressions, that is, the image they cause in others, through the motivation for the impression and the construction of impressions. This image that is intended to be built is related to the business context and the dissemination of the accounting practices

L. Almeida et al. (Eds.): ICAFI 2025, ISEM 80, pp. 19–40, 2026.
https://doi.org/10.1007/978-3-032-16268-7_2

of organizations, as there is a need to present a favorable and acceptable image before the various stakeholders.

The concepts related to IM began in the area of psychology and social behavior, in 1959, by the author Goffman E., stating that any social establishment can be studied in a useful way, from the point of view of IM.

Later, it was possible to relate this theme to the areas of management, finance, economics, and accounting and [2] mentions that managers use consciously and unconsciously, interpretative techniques in their messages, unfavorable disclosures are intentionally more difficult to perceive, resulting in vague information and, consequently, conflicts between managers. On the other hand, information may be disclosed that presents an incomplete picture of the company or that is different from the actual situation.

According to [3], two reasons motivate the use of IM, they are concealment and attribution.

Concealment aims to hide negative information and highlight positive information, while attribution consists of attributing good organizational performance to internal factors and poor performance to external factors of the companies. In other words, company managers relate good performance to their management capacity and blame the external environment for poor performance, deflecting responsibility for poor financial results [4].

Associated with these ratios, seven IM strategies are used [3]: manipulation of readability; manipulation of rhetoric; thematic manipulation; visual and structural manipulation; performance comparisons; choice of performance indicators; and attribution of organizational performances.

The manipulation of legibility is a way of confusing, obfuscating, making information less transparent [5], and hiding the reality of companies through complex language.

Rhetorical manipulation consists of analyzing the way "how" companies say and not about "what they say" [6], with the tendency to hide the truth when performance is negative.

In the case of thematic manipulation, [3] state that it is a strategy that is based on an optimistic perspective, leading managers to present the good performance of organizations through emotion in words, evidencing positive factors, through word count and the use of a greater number of positive words.

Visual and structural manipulation is a strategy that represents the visual domain, that is, the use of various techniques related to repetition; the reinforcement of ideas; the use of adjectives; visual prominence, through graphics, colors, font sizes; location and organization of information, in order to orient the reader's interest towards a certain objective [3, 7].

The strategy of performance comparisons uses benchmarking technique, that is, through performance comparisons, companies' influence the perceptions of stakeholders [7] as the results are compared with leading companies in the market.

In the strategy of choosing performance indicators, the most relevant indicators that present a more favorable image of the company's position are selected, for example: sales, results before taxes, and results per share [4].

The strategy of attributing organizational performance, on the other hand, is based on social psychology, which attributes causes for the events of organizations, relating to the theory of attribution. Managers attribute good results to internal factors and bad results to external factors, trying to look for those responsible [3].

Companies' disclosures in their reports can be manipulated in a variety of ways, compromising their transparency and credibility [8]. The tools for preparing and analyzing financial and non-financial reports are increasingly developed and sophisticated, in such a way that the results of companies can be overshadowed, to benefit their interest and show an image of organizations that does not always correspond to reality.

According to [9], financial performance requirements in the public sector, related to financial statements and statistical indicators, encourage the use of IM strategies.

Considering that IM has the consequence of manipulating information and, therefore, can affect stakeholders' decisions, the study of IM and the strategies used for this purpose has theoretical and practical relevance.

SLR is a methodology that has the potential to allow for a rigorous and objective assessment of the existing literature on a given topic. In addition, SLR can include quantitative analyses and meta-analyses [10].

From the research carried out, no SLR studies on MI were found. For this reason, and considering as previously expressed, the importance of research on MI and the potential of SLR, the present work aims to provide structured knowledge about Impression Management. In this sense, the following research questions (RQ) are posed:

RQ1 - How has research in IM evolved?

RQ2 - What is the impact of IM strategies?

RQ3 - What are the theories that support research in IM?

RQ4 - What are the directions for future research in IM?

To answer the research questions, the work adopts the SLR methodology, as it allows contributing to the perception of scientific development and existing knowledge on a given topic.

After the introduction, in Sect. 2 the research strategy is defined and detailed. The results of the research are presented in Sect. 3. The conclusions are presented in Sect. 4.

2 Research Methodology – Systematic Literature Review

According to [11], the SLR complements the traditional literature review, as it provides perceptions, critiques, and clues for future studies and research questions, and may include quantitative and qualitative approaches, in a structured way. Thus, this methodology develops broadly and transparently each study included in the research, to respond to identified research questions.

In fact, the first step in the development of an SLR is to select the articles that allow the research objective to be achieved.

In this study, the database used was Scopus, since it is one of the largest and most credible scientific databases in several areas of knowledge, including reliable, relevant, and current sources.

The survey, carried out in December 2024, followed the protocol indicated in Fig. 1. The search was carried out with the keywords: *account*; "impression management" and disclosure OR report*. It was decided to search the title of the article using the expression "*impression management*", in the title of the article, abstract, and keywords using the words *account* and disclosure OR report,* and the result of 52 articles was obtained. Then, the research was limited to the areas of "*Business, Management and Accounting*", and "*Economics, Econometrics and Finance*", having obtained a total of 44 articles. The inclusion and exclusion criteria were defined to exclude book chapters, conference papers, letters, and books. After applying these criteria, 37 articles were obtained. Finally, it was decided to select the articles in English and open access, resulting in 20 articles.

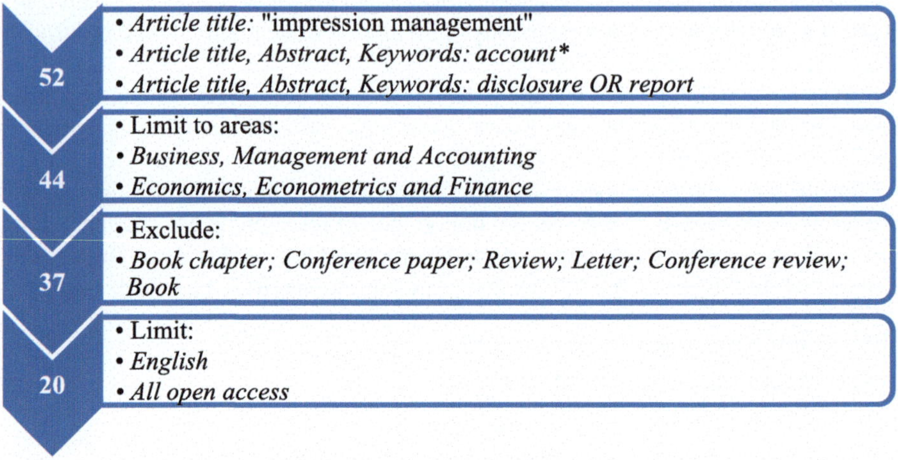

Fig. 1. Research strategy at Scopus.

The second phase of the SLR consists of evaluating the relevance of the articles obtained from the research, according to the analysis of titles, abstracts, and keywords [11].

3 Presentation and Discussion of the Results

The list of articles that resulted from the research for the SLR, as well as the number of citations, are presented in Table 1. From his analysis, it can be seen that the most cited article is "*Impression Management and retrospective sense-making in corporate narratives: A social psychology perspective*" with 195 citations, by the authors Doris M. Merkl-Davies, Niamh M. Brennan, Stuart J. McLeay, published in the journal *Accounting, Auditing, and Accountability Journal,* in the year 2011. This has a great interest in the scientific community, in the area of IM and business reporting.

Table 1. List of articles obtained in the research for the SLR

Title	Authors	Magazine	Year	No. of Citations
Impression management and retrospective sense-making in corporate narratives: A social psychology perspective	Merkl-Davies, D.M., Brennan, N.M., McLeay, S.J	Accounting, Auditing and Accountability Journal	2011	195
Accounting narratives and impression management on social media	Yang, J.H., Liu, S	Accounting and Business Research	2017	71
A sticky chocolate problem: Impression management and counter accounts in the shaping of corporate image	Perkiss, S., Bernardi, C., Dumay, J., Haslam, J	Critical Perspectives on Accounting	2021	63
Reporting practice, impression management and company performance: A longitudinal and comparative analysis of water leakage disclosure	Cooper, S., Slack, R	Accounting and Business Research	2015	53
A longitudinal study of the textual characteristics in the chairman's statements of Guinness: An impression management perspective	Moreno, A., Jones, M.J., Quinn, M	Accounting, Auditing and Accountability Journal	2019	35
Impression management in corporate annual reports during the global financial crisis	Moreno, A., Jones, M.J	European Management Journal	2022	13

(*continued*)

Table 1. (*continued*)

Title	Authors	Magazine	Year	No. of Citations
The effect of cognitive reflection on the efficacy of impression management: An experimental analysis with financial analysts	Cardoso, R.L., Leite, R.O., Aquinas, A.C.B	Accounting, Auditing and Accountability Journal	2018	13
Framing the Magdalen: Sentimental narratives and impression management in charity annual reporting	Evans, L., Pierpoint, J	Accounting and Business Research	2015	13
The influence of audit committee characteristics on impression management in chairman statement: Evidence from Malaysia	Al-Sayani, Y.M., Mohamad Nor, M.N., Amran, N.A	Cogent Business and Management	2020	11
What do audit committees do? Transparency and impression management	Al-Shaer, H., Malik, M.F., Zaman, M	Journal of Management and Governance	2022	9
Do poor and good-performing companies report differently? The readability and impression management in corporate narrative documents: Evidence from Northern Europe	Pasko, O., Minta, S., Rudenko, S., Hordiyenko, M	Business: Theory and Practice	2020	4

(*continued*)

Table 1. (*continued*)

Title	Authors	Magazine	Year	No. of Citations
The content, evolution, and determinants of COVID-19 disclosures in Canadian financial statements and MD&A documents: An impression management perspective	Bujaki, M., Mansurov, A., McConomy, B	Canadian Journal of Administrative Sciences	2024	3
Environmental disclosures in the non-financial reporting of energy companies. Creating a reliable business image or impression management?	Szadziewska, A., Kujawski, J	Zeszyty Teoretyczne Rachunkowosci	2022	3
Impression management and ancient Ceylonese rulers	Kumarasinghe, S., Samkin, G	Accounting History	2020	2
City governance and visual impression management: Visual semiotics and the Biccherna panels of Siena	Davison, J., Giovannoni, E	Human Relations	2023	1
Substantive rationality of impression management: An analysis of corporate reports of JBS	Cruz, J.C.S., Rodrigues, M.D., Araújo, R.A.M., Silva, A.R	Mackenzie Management Journal	2022	1
Impression management in financial reporting: Evidence on management commentary	Jaworska, E., Bucior, G	IBIMA Business Review	2020	1

(*continued*)

Table 1. (*continued*)

Title	Authors	Magazine	Year	No. of Citations
Impression management of biodiversity reporting in the energy and utilities sectors: An assessment of transparency in the disclosure of negative events	Blanco-Zaitegi, G., Álvarez Etxeberria, I., Moneva, J.M	Journal of Behavioral and Experimental Finance	2024	0
The picturesque roots of impression management: framing, Claude glass and "rose"-tinted lens?	McBride, K., Atkins, J.F., Atkins, B.C	Accounting, Auditing and Accountability Journal	2024	0
Impression management in chairmen's letters: An empirical study of banks' annual reports in MENA region	Mahboub, R., Mostapha, N., Hegazy, W	Corporate Board: Role, Duties and Composition	2016	0

3.1 Developments in IM Research

Of the twenty articles that resulted from the research, the authors Jones, M.J. of the United Kingdom and Moreno, A., from Spain, as shown in Fig. 2, and Jones, M.J. has an h-index of 36, which reflects a strong impact on the scientific community.

The journals with the most articles published on this topic are *Accounting Auditing And Accountability Journal* and *Accounting And Business Research,* as shown in Fig. 3, which are positioned in the 97th and 54th percentile of the ranking of journals, with an impact index of 10.1 and 3.4, respectively.

With regard to the years in which the studies were carried out, it is observed that there is an evolution over time, with a greater incidence in the years 2008 and 2012, as observed in Fig. 4, due to the increase in laws and standards, which oblige companies to issue sustainability and social and environmental responsibility reports and due to economic crises, that increase research in the area of financial and non-financial disclosures. It may also be due to *stakeholders* showing more interest in non-financial information and increasing the use of IM strategies in business reporting.

Another aspect may be technological development, which facilitates access to and analysis of financial reports. World events, linked to economic, political, and even pandemic crises, such as the case of COVID-19, also influence companies to justify their decisions and maintain an image of trust.

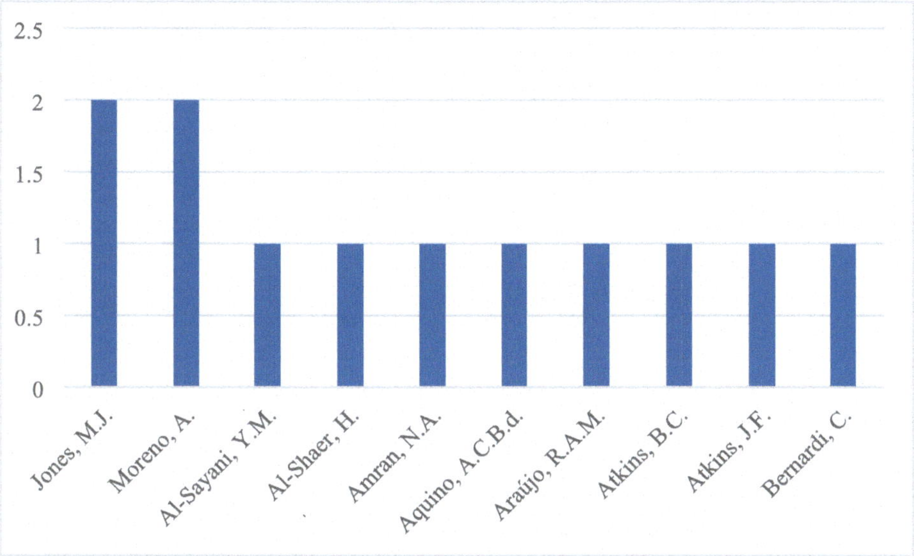

Fig. 2. Publications by author.

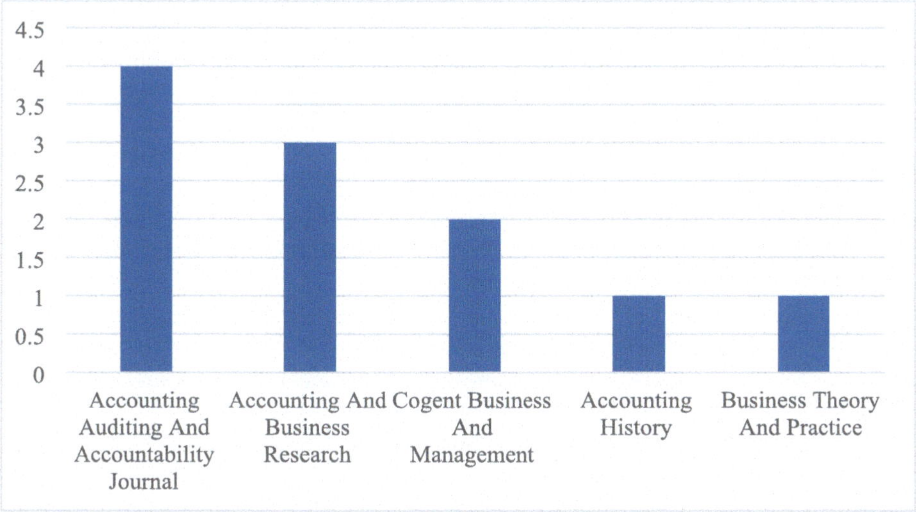

Fig. 3. Publications by journal.

The analysis of the articles allowed us to verify a diversity of research methodologies, as well as contexts (types of companies, sectors of activity, and countries on which the studies were carried out).

In the case of the methodologies used, these were qualitative and quantitative. Regarding studies with a qualitative approach, these focused on the content analysis

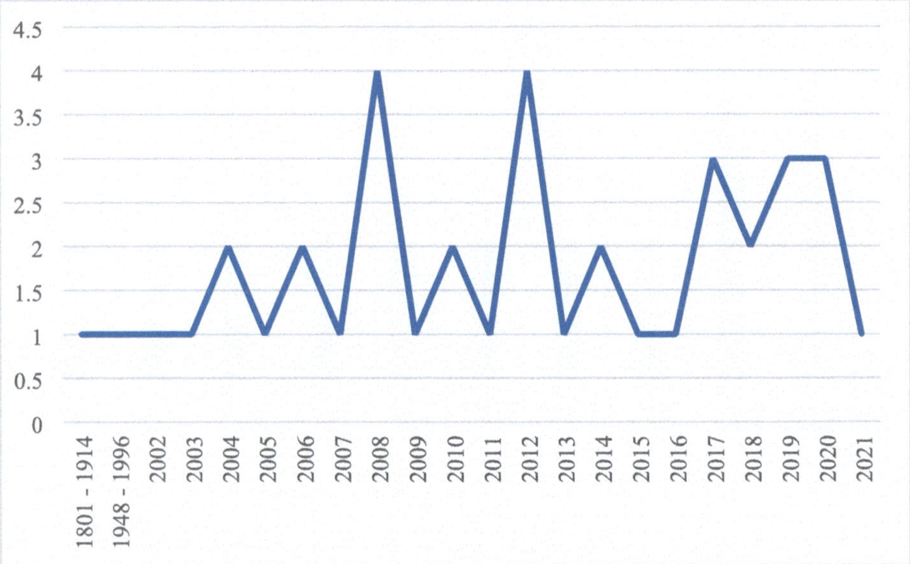

Fig. 4. Number of studies per year.

of the companies' reports and accounts [12, 13, 23]; non-financial [13] and sustainability [14] reports [15]; internal and external audit reports [16]; websites and letters from presidents [17, 18]. Questionnaires were also carried out with accounting professionals [19].

As part of the quantitative approach, it was carried out using statistical models that allow the research questions to be answered [17, 18, 20–24] and that use companies' financial indicators (earnings before taxes, earnings per share, sales, dividends, and cash flows) as variables [25]; financial impacts related to environmental disclosures [15] and on populations [14].

Longitudinal studies [23] also constitute a research methodology on IM strategies, through the analysis of variations in the characteristics of the texts of the presidents' letters, combining qualitative methods (greetings, numerical references, and first-person pronouns) and quantitative methods (returns, numerical references).

Some authors use case studies as a methodology [13]. It should be noted that several studies use both approaches, qualitative and quantitative, to complement the results obtained.

In terms of contexts, studies are observed in companies with values listed on the London Stock Exchanges, in the United Kingdom [18, 24] from Bursa, Malaysia [17]; from Stockholm, Sweden [25]; and Toronto, Canada [20], in different sectors of activity. Studies were found in the materials, consumer goods, services, financial, health, energy, technologies, and telecommunications industry [16]; industrial, telecommunications and state-owned enterprises [13]; water and waste companies [12]; energy sector of European companies [14, 15]; banking sector from different countries in the Middle East and North Africa: Egypt, Jordan, Lebanon, Saudi Arabia, and the United Arab Emirates [22]; food companies, such as Nestlé's Ivory Coast [29]; a group of companies under the Guinness brand [23]; social solidarity institutions [21] and professional accounting workers in Brazil [19].

The analysis of the articles that are the object of the SLR also allows us to conclude that, in the geographical context, the studies focus on several countries, as shown in Table 2.

Table 2. Number of studies by country

Countries	Number of studies
Germany	1
Saudi Arabia	1
Austria	1
Belgium	1
Brazil	2
Canada	1
Côte d'Ivoire	1
Egypt	1
United Arab Emirates	1
Spain	2
Finland	1
France	1
Ireland	1
Italy	2
Jordan	1
Lebanon	1
Malaysia	1
Poland	2
Portugal	1
United Kingdom	4
Sri Lanka	1
Sweden	1

If we group by continent, Fig. 5, we see that there is a greater number of studies carried out on the basis of samples of companies from European countries.

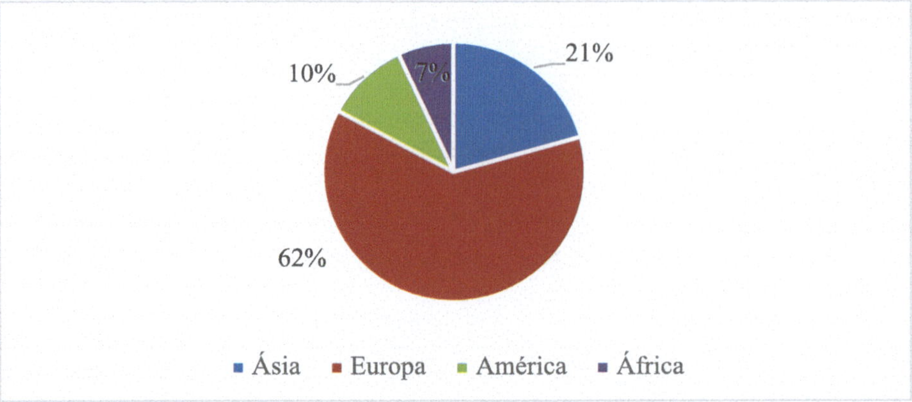

Fig. 5. Distribution of studies by geographical area - Continents.

3.2 IM Strategies Used in Company's Disclosures and Their Impact

As early as the sixteenth century, IM techniques were used in stone inscriptions, to show a favorable image on the part of governments in Asian territories, which at the time were the kings, and who intended to demonstrate their charismatic leadership [26]. Over the years, there has been a process of evolution, in which government officials and CEOs use various IM strategies, namely the manipulation of legibility and rhetoric; visual and structural manipulation; performance comparisons; the choice of performance indicators, and the attribution of organizational performances, to communicate their practices in accounting narratives and account reports.

For [27], the economic vision of the IM is reduced, since it analyzes the annual reports of the companies, looking at their rationality and usefulness. According to the authors, companies operate in social contexts, so accounting research benefits from the use of theories related to social psychology. Managers tend to use IM to influence the various *stakeholders*, through the neutralization of negative information. Accountability and disclosure of information about organizations lead to a disguise regarding explanations of certain decisions and a favorable presentation of companies' financial performance [27].

Later, [25], carried out a study on companies in Sweden, through the analysis of the strategy of readability and thematic manipulation, considering the number of negative words or phrases and personal references in managers' statements. The authors concluded that contrary to other studies, there is no difference in the use of these IM strategies when comparing companies with good and bad performance. The results show evidence that companies with low performance are more focused on the future when compared to companies that present good performance, since they hide unfavorable results, while entities with better performance tend to highlight good results without having the need for future projections.

The analysis of financial reports in the banking sector shows that there are four IM strategies: visual and structural manipulation of information, performance comparisons, *benchmarking,* and performance attribution. It is verified that the narrative of some reports is difficult to read, which can be considered a barrier to effective communication for decision-making. The banking sector highlights good news and puts this information more prominently in the letters of the presidents, blaming external factors for the bad news [22]. The authors refer to the importance of auditing to confirm the reliability of the information provided in the banks' reports.

According to [23], the characteristics of the narratives of the presidents' letters have evolved a lot over the years, as well as the influences of IM, due to the use of strategies for attributing organizational performance and the choice of performance indicators.

Another analysis related to IM and the letter of the president is the link between IM and the characteristics of the audit committee [17]. Due to the financial scandals Enron, WorldCom, and Toshiba, the IM practices in the letters of the presidents of companies listed on the Bursa stock exchange in Malaysia, were used as a way to assess the quality of financial reports and according to [17], meetings with the audit committee have a negative impact with the level of IM in financial disclosures.

The theme of sustainability and non-financial information is increasingly playing a relevant role in the study of IM strategies. Increased regulation of social and environmental disclosures and stakeholder accountability requirements are an incentive for companies to use IM strategies to disclose investor-friendly accounting and financial information [28].

It is therefore important to verify that organizations transparently report negative events related to the environment (biodiversity) [14], using IM strategies. According to [14], there is a lack of commitment to transparently communicate negative environmentally related outcomes, since environmental incidents may not be disclosed, or if they are, IM strategies are used to hide negative situations, through the manipulation of legibility and the manipulation of rhetoric.

Large companies have the power to control the media, which leads to distortion of trust in news, such as company disclosures regarding social, environmental, and economic practices, more specifically, issues of child labor and unsustainable agricultural work, according to [29]. These companies need to restore legitimacy, as well as the interests of investors and the public, to this end, they use IM strategies to disseminate positive initiatives in the media, to control negative news [29].

Related to major world events, with times of uncertainty, such as the Pandemic, COVID-19 [20] and financial crises [23], disclosures have implications at the accounting level, organizations use IM strategies, through sentiment-related information, on the other hand, government characteristics and jurisdictional factors influence decisions to disclose financial performance, as cited by [20], emphasizing, in these situations, the organizational performance strategy.

Still, in the context of financial crises, poor performance is explained by external factors. Companies do not use many IM strategies, they are careful and try to justify the bad results of the crisis, and on the other hand, they use positive attributions through *benchmarking* and favorable practices.

The crisis influences the way in which company reports are presented, and in the financial and real estate sectors, IM techniques are more reactive. Reports prepared with high scrutiny present good news and favorable comparisons, and information users should be aware of the lack of transparency about useful data for decision-making [23].

In fact, the media are currently relevant factors in the presentation and promotion by organizations, which implies that they present favorable images.

According to [24], entities minimize the disclosure of negative information, however, they use various techniques to show positive information, namely regarding performance, presenting themselves opportunistically, to building a positive public image.

According to [19], the use of different IM strategies is effective, and the cognitive capacity of accounting professionals attenuates the manipulation of results, with an improvement in the presentation of results, using the visual and structural manipulation strategy. There is a greater critical spirit in relation to visual presentations, such as colors and graphics, since they are more selective and effective techniques that emphasize the recovery of profits.

The use of visual and structural manipulation strategy was used by [12] to analyze whether the practice of reporting changes according to the performance of companies, considering IM strategies. To this end, text contents, graphics, statements, colors, photographs, and visual images were observed. The authors show evidence that the level, nature, and presentation of information change significantly with the performance of firms. These changes include the use of IM strategies, which leads to concerns about the reliability of disclosures. Thus, the authors suggest that the *International Accounting Standards Board* (IASB) issue guidance on narratives, including the form of presentation, to reduce the scope of application of IM in business reporting.

According to [30], concerns about dealing with conflicts and uncertainties are major challenges for managers. The strategy of visual and structural manipulation (calligraphy, paintings, and images) is an important way to show how art and accounting narratives are intertwined and are relevant management tools in modern times.

[31], formulate the research question: how is the substantive rationality inherent in IM evidenced in companies' reports? The authors observe that the rational actions of the information preparers promote an approximation between the values of the companies and the values of the *stakeholders,* to legitimize the actions of the companies. The discourse used removes illegitimate actions and highlights aspects that generate favorable reactions to *stakeholders,* with the fulfillment of objectives acceptable and desirable by society, thus maintaining its reputation and legitimacy. In this study, the use of the strategy of attribution of organizational performance is verified.

A study conducted by [15] to determine which are the most commonly used IM strategies in listed companies in the Polish energy sector, in their non-financial reporting reports, to create a favorable image, even if it is not true, and concluded that most used are performance comparisons through *benchmarking*; thematic manipulation, through word counting and the attribution of organizational performances.

Taking into account that the study of IM strategies also covers social entities, [21], show evidence that IM strategies related to the choice of performance indicators, applied in the narratives of annual reports of a women's support institution, were used with the aim of removing the responsibility of directors, show their work as socially and economically useful, thus ensuring funding and employment opportunities for women.

The analysis of Table 3 allows us to verify that IM strategies are used, highlighting the manipulation of legibility, rhetoric, and theme, through the omission and concealment of negative factors, excuses and distancing from responsibilities, strategic manipulation of reports, use of personal references, the count of words and sentimental expressions. Other strategies used are visual and structural manipulation, with images, visual impact and positive news, and performance comparisons, through benchmarking.

The strategies for choosing performance indicators and assigning organizational performance stand out. In the case of performance indicators, financial information, numerical references, number of years with financial results available, and data on employment and human resources are used.

Regarding the strategy of attributing organizational performance, this is the most used by the financial and non-financial information preparations of companies, highlighting the reference to positive results, the acceptance of partial responsibility in unfavorable situations, positive attributions to internal factors and negative attributions to external factors, use of assertive measures, related to integration, self-promotion, transmission of values to society; show the example, compensate victims to avoid an image of injustice and disloyalty, carrying out social work and delivering donations, and also the reference to the Covid-19 pandemic as a cause for less favorable performances.

Table 3. Summary of IM strategies used in the companies' reports

IM Strategies		Authors
Handling Readability Manipulation of Rhetoric	Omit and hide Maintain a neutral position	[12]; [14]; [25]
Handling Readability Manipulation of Rhetoric	Denying the significant impacts	[14]; [27]; [15]
Handling Readability Manipulation of Rhetoric	Excuses Distancing from responsibilities Uncertainties	[14]; [20]; [12]; [31]
Thematic Manipulation Manipulation of Rhetoric	Strategic Reporting Manipulation	[13]; [25]; [29]; [15]
Thematic Manipulation	Personal references (first-person pronouns)	[32]; [23]; [25]
Thematic Manipulation	Word Count	[16]; [20]; [21]; [13]; [25]; [15]
Thematic Manipulation	Sentimental words	[21]
Visual and Structural Manipulation	Presentation; visual impact; Using images Positive news	[12]; [30]; [25]; [29]; [15]; [24]
Performance Comparisons	Benchmarking and competition	[12]; [23]; [15]
Choice of Performance Indicators	Financial indicators Numerical references	[32]; [23]; [25]; [24]
Choice of Performance Indicators	Number of years with financial results available	[21]

(continued)

Table 3. (*continued*)

IM Strategies		Authors
Handling Readability Manipulation of Rhetoric	Omit and hide Maintain a neutral position	[12]; [14]; [25]
Choice of Performance Indicators	Employment and Human Resources Information	[21]
Attribution of Organizational Performances	Highlight positive results	[14]; [31]
Attribution of Organizational Performances	Accept partial responsibility	[12]
Attribution of Organizational Performances	Covid-19	[20]
Attribution of Organizational Performances	Positive attributions – internal factors Negative Attributions - External Factors	[31]; [12, 13]; [32]; [23]; [25]; [15]; [24]
Attribution of Organizational Performances	Assertive measures Integration, self-promotion, show example	[12]; [26]; [24]
Attribution of Organizational Performances	Compensating victims to avoid an image of injustice and disloyalty	[12]
Attribution of Organizational Performances	Transmitting values to society	[31]
Attribution of Organizational Performances	Corrective measures Creation of ethics and quality policies	[31]
Attribution of Organizational Performances	Social work and donation deliveries	[21]; [26]

The analysis carried out allowed us to answer the research question 2.

3.3 Theoretical Framework

It is important to highlight the theoretical framework, to explain some reasons that motivate companies to use IM strategies in their financial and non-financial reports.

In the analyzed articles, several explanatory theories were identified, namely, *Impression Management*, the agency theory; the attribution theory; the institutional theory; the legitimacy theory; the signaling theory; the *Stakeholder theory;* the theory of charismatic leadership, and Goffman's theory.

According to studies that address the theme of IM, this can be considered as an explanatory theory in the area of accounting. Despite having its origin in social psychology, IM is analyzed from an organizational point of view, since people, internal and external, are part of companies and human behavior influences the manipulation of certain factors inherent to the disclosure of financial and non-financial information [18]. This information is used to promote communication between organizations and the public, in a transparent way [31].

According to [20] and in the specific case of the COVID-19 pandemic, IM is assumed as a theory to explain that companies disclose more information about external events when their results are negative, thus attributing these results to external factors (attribution theory) and using fewer financial indicators so as not to highlight poor performances.

IM theory explains the disclosure of sustainability information of companies [14] and [15] through the use of IM techniques, namely, a focus on positive aspects; highlighting efforts and measures to protect the environment; minimization of environmental impact; absence of comparisons with previous data, in case of negative data; reports embellished with images, in the case of positive disclosures.

On the other hand, [24] show evidence of the theory of IM as being crucial, due to the use of media that allows companies to reduce information gaps, regardless of their performance, and show a favorable image on social networks.

The agency theory aims to reduce the asymmetrical relationship between managers and holders of capital and their agency costs. According to this theory, there is a tendency that reflects the opportunistic behavior of managers when selecting a style of presentation or the best information they should communicate, to obtain benefits [3]. This theory was used by the authors [17]; [16] and [25], to justify that the independence factor between managers, capital holders, and even auditors, has a strong influence on the use of IM in company reports, as managers of companies with unfavorable performance are less likely to use IM strategies.

The theory of attribution is related to the understanding of the issues involved in certain causes, that is, positive factors are attributed to internal causes, and negative factors are attributed to causes external to companies [33]. In fact, it can be seen that, in crisis situations, companies assign more favorable financial indicators, with the indication of the expression "*despite the context*" ([23], p. 514) and more generally, less based on the values of quotas or shares.

According to institutional theory, it is assumed that companies act in accordance with the norms and expectations adopted by institutions [3]. In response to pressures from customers, suppliers, and regulators, companies use IM strategies to convey their actions and show their reputation [25].

According to [3], the legitimacy theory is used to explain companies' disclosures, more specifically, related to social and environmental issues [29]. This theory is related to the concerns of companies in complying with laws and standards and responding to the demands and influence of the media. The diversity of tools to maintain the legitimacy of companies allows organizations to manage the risks associated with their image through economic, accounting, and social disclosures, in accordance with social norms and beliefs that convey credibility, trust, security, and responsibility [31].

The *stakeholder theory* explains the need to respond to the expectations of the various *stakeholders*, such as employees, customers, governmental and non-governmental entities, pressure groups, and others [25].

According to [3], the signaling theory refers to the signals that the heads of organizations emit in their reports, namely, related to positive factors and good performance, in order to present superiority and transparency [17] and [25].

Another theory is that of Goffman's theory, according to which people manage their impressions in a social context, that is, they have certain behaviors, in different situations, based on the expectations of others. Social interactions are like theatrical performances, where people play the roles of actors to make a good impression on others. Thus, in the context of companies, this theory is applied, with the objective of understanding how organizations build the desired image before stakeholders and explaining the practices of information disclosure in business reports. This theory, used in the study by [29], shows that companies have the power and resources to circumvent the news, to establish their legitimacy, even if it does not correspond to the reality of the actions they perform.

For [26], there is also the theory of charismatic leadership, as a way of explaining that the leader has characteristics and qualities linked to an intrinsically charismatic personality that can motivate their employees to achieve organizational goals.

In view of the above, it is observed that there is a strong connection between the various theories identified in the analyzed studies, and that explains how IM strategies influence the disclosure of company reports, allowing us to answer research question 3.

3.4 Directions for Future Research

Based on the clues for future studies, some areas that may constitute trends for future research can be listed.

Replicate existing studies, for samples from other companies, from other sectors of activity [29, 32]; or from other countries [23].

[24] suggest analyzing the factors that determine disclosure decisions over time, expanding the sample size in order to obtain a greater number of observations, and studying other company scenarios, such as situations of acquisitions and mergers of companies, investments, and new product launches, including other forms of IM study, through graphs, photographs, and tables [22]. Use other data sources with official information, such as interviews, websites, and media publications [14, 23].

Citing [15], companies that also use IM strategies in their non-financial reporting can be studied and the results can be compared in order to identify the most used IM strategies.

Another suggestion is to analyze the direct involvement with *stakeholders* and management bodies in relation to disclosures, as well as specific data from companies, or other sectors of activity [12]; explore the characteristics of the chairman of the board of directors and the directors and the audit committee, such as compensation, ownership and quality [17]; study the results of sector-by-sector disclosures and the impact of governance variables [23].

Analyze the relationship between expected results and actual results and how this relationship motivates managers to use IM tools [25]; use longitudinal research methodology in companies from other sectors of activity that obtained profits and losses [32]. According to [26], other models of communication from ancient times could be explored in depth, in addition to stone engravings.

Some studies suggest examining the work done by third parties to gain a good understanding of corporate governance and outcomes [16].

The implementation of environmentally related standards could also be analysed in European companies to see if transparency in information on ecosystems and biodiversity has improved [14].

Regarding the strategy of thematic manipulation, the authors [21] suggest analyzing the sentimental narrative used in current social solidarity institutions.

Given the importance of world events, [20, 23], consider it relevant to understand the evolution and content of the disclosure about COVID-19 in companies' reports, using IM strategies.

Based on the above, research question 4 was answered.

4 Conclusions

The present work aimed to provide structured knowledge about *Impression Management* (IM) research in the area of accounting, namely in the dissemination and reporting of companies.

It is observed that there has been an evolution of studies over the years, with an increase in the years 2008 and 2012, due to the increase in laws and standards, which oblige companies to issue sustainability and social and environmental responsibility reports and due to economic crises, which increase research in the area of disclosure of financial and non-financial information, by the heads of the organizations. The observation of the studies allows us to verify that the methodologies used are qualitative and quantitative, and focused on the analysis of the content of the reports and accounts of the companies, the sustainability reports, the letters of the presidents, the audit reports and the observation of websites. The quantitative studies were carried out using statistical models. Questionnaires and case studies were also carried out.

By analyzing the articles, it was possible to verify that several IM strategies are used, highlighting the manipulation of readability, rhetoric, and theme, through the omission and concealment of negative factors, excuses and distancing from responsibilities, strategic manipulation of reports, use of personal references; the count of words and sentimental expressions. Other strategies used are visual and structural manipulation, with images, visual impact and positive news, and performance comparisons, through benchmarking.

The strategies for choosing performance indicators and assigning organizational performance stand out. In the case of performance indicators, financial information, numerical references, number of years with financial results available, and data on employment and human resources are used.

Regarding the strategy of attributing organizational performance, this is the most used by the financial and non-financial information preparations of companies, highlighting the reference to positive results, the acceptance of partial responsibility in unfavorable situations, positive attributions to internal factors and negative attributions to external factors, use of assertive measures, related to integration, self-promotion, transmission of values to society; show the example, compensate victims to avoid an image of injustice and disloyalty, carrying out social work and delivering donations, and also the reference to the Covid-19 pandemic as a cause for less favorable performances.

In theoretical terms, the authors refer to the use of several theories to justify the use of IM strategies in organizations. Thus, the theory of IM is highlighted; the agency theory; theory of attribution; institutional theory; theory of legitimacy; theory of signaling; theory of *Stakeholders;* theory of charismatic leadership, and Goffman's theory [34], and strong interconnection between all of them.

In fact, in the geographical context, there is a diversity of countries with studies about IM, however, the European continent stands out, namely in companies with listed values.

According to the articles analyzed, the authors suggest, for future studies, extending the scope of research to other companies and sectors of activity; analyze the factors that determine company disclosures; use other data sources; to verify whether there is an improvement in transparency on non-financial information and to understand the evolution of the content of the disclosures in contexts of crises and uncertainties.

The realization of this work brings to an increase of knowledge in the area of accounting, as the use of IM strategies by organizations is current and pertinent within the scientific community and grounds interests in the context of practical applications of companies.

A limitation of the present work has to do with the fact that it is based only on selected articles in a database of scientific articles.

For future studies within the scope of SLR, on IM strategies and company reports, it is suggested to expand the scope of the investigation of this work through other research strategies and/or use other databases, in order to include in the analyses, other articles with greater relevance and impact on the theme of IM and company disclosures.

References

1. Leary, M.R., Kowalski, R.M.: Impression management: a literature review and two-component model. Psychol. Bull. **107**(1), 34–47 (1990). https://doi.org/10.1037/0033-2909.107.1.34
2. Adelberg, A.H.: Narrative disclosures contained in financial reports: means of communication or manipulation? Account. Bus. Res. (Summer) 179–89 (1979). https://doi.org/10.1080/00014788.1979.9729157
3. Merkl-Davies, D.M., Brennan N.M.: Discretionary disclosure strategies in corporate narratives: incremental information or impression management? J. Account. Literat. **26**, 116–194 (2007). https://ssrn.com/abstract=1089447
4. Clatworthy, M.A., Jones, M.J.: Differential patterns of textual characteristics and company performance in the chairman's statement. Account. Auditing Account. J. **19**(4), 493–511 (2006). https://doi.org/10.1108/09513570610679100
5. Courtis, J.K.: Corporate report obfuscation: artifact or phenomenon? Br. Account. Rev. **36**(3), 291–312 (2004). https://doi.org/10.1016/j.bar.2004.03.005
6. Pennebaker, J.W., Mehl, M.R., Niederhoffer, K.: Psychological aspects of natural language use: our words, ourselves. Annu. Rev. Psychol. **54**, 547–577 (2003). https://doi.org/10.1146/annurev.psych.54.101601.145041
7. Brennan, N.M., Guillamon-Saorin, E., Pierce, A.: Impression management: developing and illustrating a scheme of analysis for narrative disclosures – a methodological note. Account. Auditing Account. J. **22**(5), 789–832 (2009). https://doi.org/10.1108/09513570910966379

8. Albuquerque, F., Stoltzemburg, V., Cariano, A.: A literature review on impression management in accounting through a textual analysis. J. Account. Auditing: Res. Pract. (2023). https://doi.org/10.5171/2023.586125
9. Bisogno, M., Donatella, P.: Earnings management in public-sector organizations: a structured literature review. J. Public Budg. Account. Financ. Manag. **34**(6), 1–25 (2022). https://doi.org/10.1108/JPBAFM-03-2021-0035 (1959)
10. Grant, M.J., Booth, A.: A typology of reviews: an analysis of 14 review types and associated methodologies. Health Info. Libr. J. **26**(2), 91–108 (2009). https://doi.org/10.1111/j.1471-1842.2009.00848.x
11. Massaro, M., Dumay, J., Guthrie, J.: On the shoulders of giants: undertaking a structured literature review in accounting. Account. Auditing Account. J. **29**(5), 767–801 (2016). https://doi.org/10.1108/AAAJ-01-2015-1939
12. Cooper, S., Slack, R.: Reporting practice, impression management and company performance: a longitudinal and comparative analysis of water leakage disclosure. Account. Bus. Res. **45**(6–7), 801–840 (2015). https://doi.org/10.1080/00014788.2015.1081554
13. Jaworska, E., Bucior, G.: Impression management in financial reporting: evidence on management commentary. IBIMA Bus. Rev. 1–13 (2020). https://doi.org/10.5171/2020.693684
14. Blanco-Zaitegi, G., Etxeberria, I.Á., Moneva, J.M.: Impression management of biodiversity reporting in the energy and utilities sectors: an assessment of transparency in the disclosure of negative events. J. Behav. Exp. Financ. **42**, 100942 (2024). https://doi.org/10.1016/j.jbef.2024.100942
15. Szadziewska, A., Kujawski, J.: Environmental disclosures in the non-financial reporting of energy companies. Creating a reliable business image or impression management? Zeszyty Teoretyczne Rachunkowości. **46**(2), 157–194 (2022). https://doi.org/10.5604/01.3001.0015.8814
16. Al-Shaer, H., Malik, M.F., Zaman, M.: What do audit committees do? Transparency and impression management. J. Manag. Gov. **26**, 1443–1468 (2022). https://doi.org/10.1007/s10997-021-09591-9
17. Al-Sayani, Y.M., Mohamad Nor, M.N., Amran, N.A.: The influence of audit committee characteristics on impression management in chairman statement: evidence from Malaysia. Cogent Bus. Manag. **7**(1), 1774250 (2020). https://doi.org/10.1080/23311975.2020.1774250
18. Merkl-Davies, D.M., Brennan, N.M., McLeay, S.J.: Impression management and retrospective sense-making in corporate narratives: a social psychology perspective. Account. Auditing Account. J. **24**(3), 315–344 (2011). https://doi.org/10.1108/09513571111124036
19. Cardoso, R.L., Leite, R.D.O., Aquino, A.C.B.D.: The effect of cognitive reflection on the efficacy of impression management: an experimental analysis with financial analysts. Account. Auditing Account. J. **31**(6), 1668–1690 (2018). https://doi.org/10.1108/AAAJ-10-2016-2731
20. Bujaki, M.L., Mansurov, A., McConomy, B.: The content, evolution, and determinants of COVID-19 disclosures in Canadian financial statements and MD&A documents: An impression management perspective. Canadian Journal of Administrative Sciences / Revue Canadienne des Sciences de l'Administration (2024). https://doi.org/10.1002/cjas.1756
21. Evans, L., Pierpoint, J.: Framing the Magdalen: sentimental narratives and impression management in charity annual reporting. Account. Bus. Res. **45**(6–7), 661–690 (2015). https://doi.org/10.1080/00014788.2015.1039931
22. Mahboub, R., Mostapha, N., Hegazy, W.: Impression management in chairmen's letters: an empirical study of banks' annual reports in MENA region. Corporate Board: Role Duties Compos. **12**(3), 69–80 (2016). https://doi.org/10.22495/cbv12i3art8
23. Moreno, A., Jones, M.J.: Impression management in corporate annual reports during the global financial crisis. Eur. Manag. J. **40**(4), 503–517 (2022). https://doi.org/10.1016/j.emj.2021.08.007

24. Yang, J.H., Liu, S.: Accounting narratives and impression management on social media. Account. Bus. Res. **47**(6), 673–694 (2017). https://doi.org/10.1080/00014788.2017.1322936
25. Pasko, O., Minta, S., Rudenko, S., Hordiyenko, M.: Do poor and good-performing companies report differently? The readability and impression management in corporate narrative documents: evidence from Northern Europe. Bus. Theory Pract. **21**(2), 835–849 (2020). https://doi.org/10.3846/btp.2020.12583
26. Kumarasinghe, S., Samkin, G.: Impression management and ancient Ceylonese rulers. Account. Hist. **25**(1), 5–26 (2020). https://doi.org/10.1177/1032373218802892
27. Merkl-Davies, D.M., Brennan, N.M., McLeay, S.J.: Impression management and retrospective sense-making in corporate narratives: a social psychology perspective. Account. Auditing Account. J. **24**(3), 315–344 (2011)
28. McBride, K., Atkins, J.F., Atkins, B.C.: The picturesque roots of impression management: framing, Claude glass, and "rose"-tinted lens? Account. Auditing Account. J. **37**(4), 1125–1152 (2024). https://doi.org/10.1108/AAAJ-08-2022-6004
29. Perkiss, S., Bernardi, C., Dumay, J., Haslam, J.: A sticky chocolate problem: Impression management and counter accounts in the shaping of corporate image. Crit. Perspect. Account. **81**, 102229 (2021)
30. Davison, J., Giovannoni, E.: City governance and visual impression management: visual semiotics and the Biccherna panels of Siena. Hum. Relat. **76**(11), 1767–1801 (2023).
31. Cruz, J.C., Rodrigues, M.D., Araújo, R.A., Silva, A.R.: Substantive rationality of impression management: an analysis of corporate reports of JBS. Mackenzie Manag. J. **23**(3) (2022). https://doi.org/10.1590/1678-6971/eRAMR220239.en
32. Moreno, A., Jones, M.J., Quinn, M.: A longitudinal study of the textual characteristics in the chairman's statements of Guinness: an impression management perspective. Account. Auditing Account. J. **32**(6), 1714–1741 (2019). https://doi.org/10.1108/AAAJ-01-2018-3308
33. Aerts, W.: Picking up the pieces: impression management in the retrospective attributional framing of accounting outcomes. Acc. Organ. Soc. **30**(6), 493–517 (2005). https://doi.org/10.1016/j.aos.2004.07.001
34. Goffman, E.: The Presentation of Self in Everyday Life. Doubleday Anchor Books, New York (1959)

Institutionalization of the Accounting Profession in Mozambique: OCAM's Role in Light of Institutional Theory

Victorino Branco[1(✉)], Graça Azevedo[1], and Cristina Góis[2]

[1] University of Aveiro, 3810-193 Aveiro, Portugal
{vbranco,graca.azevedo}@ua.pt
[2] Polytechnic University of Coimbra, 3045-093 Coimbra, Portugal
crgois@sapo.pt

Abstract. This research analyses the role played by the Mozambican Association of Accountants and Auditors (OCAM) in the process of institutionalising and professionalising accounting in the Mozambican context. Based on Institutional Theory, the study examines how coercive, normative and mimetic pressures have influenced the evolution of the national accounting field, with emphasis on the impact of the reforms introduced by OCAM between 2012 and 2018. Documentary and historical analysis shows significant progress in the standardisation of practices, professional qualification and integration with international organisations. However, structural limitations are identified, such as institutional centralisation and training disparities among members. It is concluded that OCAM has operated as a strategic institutional actor, promoting the legitimacy and modernisation of the profession, although challenges remain regarding its national reach and regulatory effectiveness. The study contributes to the debate on professional regulation in emerging economies, offering theoretical and practical implications for the consolidation of sustainable professional institutions in post-colonial contexts.

Keywords: Accounting · Institutional Theory · Mozambique · OCAM · Professionalisation · Professional Regulation

1 Introduction

Accounting plays an essential role in promoting transparency, integrity, and efficiency in economic systems, and is particularly relevant in low- and middle-income countries, where institutional challenges are more pronounced. In the African context, the evolution of this discipline is deeply marked by colonial legacies and the adoption of exogenous regulatory models, which are often ill-suited to local realities [2, 5, 7].

L. Almeida et al. (Eds.): ICAFI 2025, ISEM 80, pp. 41–56, 2026.
https://doi.org/10.1007/978-3-032-16268-7_3

In Mozambique, the process of institutionalising accounting as a regulated profession is recent and was consolidated with the creation of the Order of Accountants and Auditors of Mozambique in 2012, under Law No. 8/12 [18]. This measure was the result of a joint effort between the Mozambican government and international organisations, such as the World Bank, in response to the recommendations of the Report on the Observance of Standards and Codes [22], which identified significant deficiencies in the quality of financial information in the country [16].

OCAM was established as an independent entity with regulatory and training powers, aiming to ensure ethical and technical standards in line with international best practices. Since its foundation, it has played a structuring role in the reform of the national accounting field, promoting professional certification, continuing education and regulatory development [16].

This study adopts Institutional Theory as its theoretical framework, highlighting the mechanisms of institutional isomorphism—coercive, normative and mimetic—proposed by DiMaggio and Powell [6] and Meyer and Rowan [14]. In this context, we aim to analyse OCAM's role as a promoter of accounting professionalisation in Mozambique, considering the dynamics of regulatory compliance and the challenges inherent to a post-colonial context.

Thus, we seek to answer the following research question:

How has the Order of Accountants and Auditors of Mozambique (OCAM) contributed to the institutionalisation and professionalisation of accounting in the country, in light of institutional pressures and the post-colonial historical context?

To this end, the objective of this study is to analyse the role of OCAM in the transformation of the Mozambican accounting field, exploring how this entity has promoted institutional modernisation and compliance with international standards, in conjunction with the challenges arising from the historical legacy and contemporary demands.

This study is organised into six sections. Following the Introduction, Sect. 2 presents the theoretical framework. Section 3 addresses the institutional context of accounting in Mozambique. Section 4 describes the methodology adopted, while Sect. 5 presents the results obtained and the respective discussion. Finally, Sect. 6 sets out the main conclusions of the study.

2 Theoretical Framework

2.1 Institutional Theory

Fundamental Concepts and Assumptions. Institutional Theory maintains that the behaviour of organisations is shaped by contextual factors, such as social, cultural and political norms, which influence their practices and structures [6]. In this framework, the objective of organisations goes beyond mere technical efficiency, focusing on institutional legitimacy, achieved through compliance with the expectations of the surrounding environment.

This theory identifies three central mechanisms of institutional isomorphism:

- Coercive, resulting from legal and regulatory impositions;
- Normative, resulting from academic training and professional values;
- Mimetic, associated with the tendency to imitate organisations considered successful, especially in contexts of uncertainty.

Studies such as [1, 9], and [11] present in the analysed database, demonstrate that the theory is particularly useful for understanding the adoption—often symbolic—of governance and accounting practices in countries with weak institutions. In these cases, practices are often implemented ceremonially to satisfy external pressures, without effectively transforming the organisational culture.

2.2 Application of Institutional Theory in the Field of Accounting

The application of Institutional Theory to the field of accounting has allowed us to understand how international standards and practices are adopted in different ways, influenced by cultural, political and professional factors.

For example, [1] points out that, in the public sector of developing countries, the adoption of international accounting standards is determined mainly by external institutional pressures, rather than by efficiency considerations. Subsequently, [11], demonstrate that in countries such as Nigeria, there is often a dissonance between formally adopted standards and actual practice, which is strongly conditioned by local and informal dynamics.

The study by [15], in Uganda shows that the effectiveness of boards of directors depends on institutional acceptance of practices, confirming that governance and accounting are strongly conditioned by local norms and structures.

This evidence shows that theory can explain the divergence between the formal adoption of standards and their effective implementation, a common phenomenon in African contexts and other emerging economies.

2.3 Application of Institutional Theory to the Context of OCAM (Order of Accountants and Auditors of Mozambique)

In the context of the Order of Accountants and Auditors of Mozambique (OCAM), Institutional Theory provides a valuable analytical framework for understanding the challenges associated with the professionalisation of accounting and the legitimisation of international standards in the country.

Analysis of the database shows that OCAM operates in an environment permeated by coercive pressures (e.g., IFAC requirements for the adoption of international standards), but faces local implementation difficulties related to weak technical capacity and cultural resistance to change [9, 19].

[4], in analysing corporate governance in Zambia, emphasise that institutional effectiveness depends on the articulation between formal standards and local practices—a finding that is relevant to Mozambique, where OCAM seeks to reconcile international standards with the local context.

Similarly, [15], conclude that the performance of boards of directors in Uganda is conditioned by social acceptance of standards, a reality comparable to that in Mozambique, where the success of OCAM will depend on its ability to align international standards with local values and capacities.

More recent studies focused specifically on Mozambique deepen this institutional analysis:

- [13], investigate the impact of cultural conservatism on the professional judgement of Mozambican accountants, based on IAS 37. The results indicate that cultural values significantly influence normative interpretation, affecting comparability and consistency in the application of international standards.
- [10], analyse the transition to accrual accounting in the Mozambican public sector. The study identifies a mismatch between the formal adoption of the standard and the institutional capacity for its implementation, attributed to the lack of political impetus and technical preparation of the actors involved. Despite recognition of the benefits of the new standard, its application remains limited, reinforcing the argument of symbolic compliance.

Table 1 presents a summary of the main empirical studies used in this theoretical framework, based on Institutional Theory. These studies were selected because they focus on accounting, governance and institutional legitimacy in emerging economies, with an emphasis on the African continent and, in particular, Mozambique. The systematisation of these works highlights the methodological approaches, the institutional contexts analysed and the main conclusions, reinforcing the relevance and timeliness of the applied theory.

Table 1. Summary of empirical studies based on Institutional Theory

Author(s)/Year	Country of Sample	Article Title	Study Objectives	Theory	Main Findings and Conclusions
Azuruddin (2011)	Bangladesh	Administrative reform and institutions in developing countries	To analyse institutional influences on administrative reforms	Institutional	Adoption of international standards and conduct driven by external pressures, with limited local implementation
Nkunda & Nyamongo (2014)	Uganda	Board role performance in Uganda: A factor analysis and stakeholder approach	To analyse board performance considering the institutional environment	Institutional	Conditional legitimacy based on organisational performance
Hopper et al. (2017)	Developing countries	Globalisation, accounting and developing countries: An institutional perspective	To study globalisation and accounting practices	Institutional	Identifies symbolic use, international norms, and local resistance
Chibuye et al. (2017)	Zambia	The role of corporate governance in enhancing economic development in Zambia	To assess the role of corporate governance in economic development	Institutional	Governance depends on alignment between international norms and local practices
George et al. (2019)	Developing countries	Does integrity improve business performance? Evidence from developing countries	To assess the relationship between governmental integrity and business growth	Institutional	Institutional integrity correlates with business performance
Owolabi et al. (2020)	Nigeria	Accounting and accountability for anti-corruption effectiveness in developing democracies: Evidence from Nigeria	To study accounting and control in the context of corruption	Institutional	Accounting practices shaped by institutional frameworks and political context
Langa & Albuquerque (2023)	Mozambique	Conservatism as cultural values in accounting: An empirical study from Mozambique	To analyse cultural conservatism in professional judgement	Institutional	Cultural values influence normative interpretation and comparability bias
Gomes et al. (2025)	Mozambique	Transition to accrual-based accounting in the public sector: Institutional perceptions	To study the transition to accrual-based accounting in the public sector	Institutional	Institutional recognition, technical expertise, and effective implementation

A comparative analysis of the studies included in Table 1 confirms a recurring pattern in emerging economies: the formal adoption of international standards, often driven by external pressures, occurs without the corresponding institutional internalisation [1, 11]. This situation generates ceremonial compliance [9], where practices are officially implemented but lack effectiveness due to a lack of local legitimacy or technical capacity [4, 19].

In the case of Mozambique, studies by [13] and [10] reveal that factors such as cultural conservatism and insufficient technical skills compromise the effective application of standards. Both studies confirm the usefulness of Institutional Theory in explaining barriers to change, emphasising that legitimacy depends not only on legal compliance but also on social, professional and political acceptance of the standards [15].

3 Institutional Context of Accounting in Mozambique

3.1 History and Structure of OCAM

Colonialism left deep and enduring marks on African countries, extending beyond mere territorial occupation to transform fundamental aspects of their social and cultural structures [2]. This legacy is particularly evident in the adoption of colonisers' languages, the spread of new religions and spiritual doctrines, and the imposition of foreign legal systems [17]. These imported legal models, in turn, directly influenced the development of accounting standards and practices adopted in various African nations, replicating the principles and approaches of the colonial powers, as demonstrated by several academic studies [2, 3, 5, 7, 8, 12, 13, 20, 21].

The Order of Accountants and Auditors of Mozambique (OCAM), established as a public legal entity, holds a unique position within Mozambique's institutional landscape. Its independence from the State and from both public and private entities reinforces its mission to serve the public interest and to regulate the professional practice of accounting and auditing autonomously [12, 13].

OCAM's organisational structure comprises two distinct branches, corresponding to specific areas of professional activity: the College of Certified Accountants and the College of Certified Auditors, both governed by a management board and presided over by their respective leaders. OCAM's highest representative is the Bastonário (President), currently Professor Dr Mário Vicente Sitoe, and in his absence, responsibilities are assumed by the Vice-President of the General Council. For technical matters specific to each College, the respective presidents hold representational authority [16].

The legal recognition of OCAM was consolidated with the approval of its Statute by the Assembly of the Republic on 30 November 2011, which was subsequently promulgated by the President of the Republic, Armando Emílio Guebuza, on 18 January 2012. This statutory foundation provides OCAM with the necessary framework to operate on a non-profit basis and in service of the public interest, in alignment with its internal regulations and relevant Mozambican legislation [18].

Located in Maputo, OCAM holds institutional flexibility to expand its national representation through the establishment of regional branches, thereby enhancing its accessibility and presence across the national territory. This capacity for geographically distributed operation underscores its commitment to inclusion and the standardisation of accounting and auditing practices in Mozambique.

3.2 OCAM's Mission and Vision

OCAM's mission is to collectively promote a professional class that is fully aware of its obligations and rights, through the rigorous definition of technical, ethical, and deontological standards. It aims to equip its members with the necessary competencies, empowering and preparing them to attain a professional level suitable for accreditation and recognition both nationally and internationally, with the required independence and rigour [16].

Its vision is to build a relevant institution in the provision of highly credible and internationally recognised information, by promoting and encouraging its members to engage in training programmes and continuous quality control. This seeks to provide society with principles of ethical and transparent management, fostering the effective promotion and consolidation of an ideal business environment in the country, thereby contributing to the construction of a fairer and more informed society [16].

3.3 Evolution of the Accounting Profession Over Time

The accounting profession in Mozambique has undergone significant transformation over the past decade and a half [12]. From an initial scenario marked by a lack of formal regulation and low-quality financial reporting, the sector has advanced to a robust institutional structure that is ethical and aligned with international standards. This progress has been strongly driven by the establishment of the Order of Accountants and Auditors of Mozambique (OCAM), whose mission has been to ensure the training, certification, supervision, and continuous development of accounting and auditing professionals [16].

Below is a timeline that highlights the key milestones in this evolution, from the initial diagnostic assessment in 2008 to the most recent initiatives aimed at consolidating and internationalising the profession, as shown in Table 2.

In the initial period of its establishment, the Order of Accountants and Auditors of Mozambique (OCAM) faced significant challenges in adhering to international accounting and auditing standards. In 2012, the admission of members was not fully aligned with these global norms, largely due to the need to integrate professionals who were already registered with the General Directorate of Taxes, the body then responsible for licensing such professionals [16].

Nevertheless, the number of registered members grew exponentially from 453 in 2012 to approximately 4,680 in 2017, demonstrating a significant increase in the recognition and regulation of accounting and auditing as professions in Mozambique. The expectation for 2018 was ambitious, with the projected admission of approximately 800 new members, in addition to around 200 trainees. This projected growth was in line with the certification process for the public sector and indicated a growing commitment to professional development and compliance with international standards [16].

Table 2. Timeline of the evolution of the accounting profession in Mozambique

Period	Historical Milestone	Detailed Description
Before 2012	Preliminary Situation of the Profession	The accounting profession operated in a context of institutional disorganisation. There were no clear standards, supervision, ethics, or structured training, and no regulatory body. Most professionals worked based on practical experience, without recognised certification
2008	ROSC Report by the World Bank	The Report on the Observance of Standards and Codes (ROSC) raised concerns about the poor quality of financial reporting in Mozambique. It recommended the creation of an independent body to regulate accounting and auditing based on international standards
2012	Establishment of OCAM	By Law Decree No. 8 of 8 February 2012, the Order of Accountants and Auditors of Mozambique (OCAM) was created, with technical support from Irish Aid and the World Bank. OCAM was born as a regulatory body, certifier, and promoter of professional ethics and quality in accounting
2013–2017	Initial Consolidation of OCAM	- Adoption of IFAC Code of Ethics (2014); - Publication of admission regulations, quality control, and continuing professional development; - Creation of a digital database with more than 4,600 registered members by 2017; - Delivery of 14 national continuous professional training editions
2014	Launch of Continuing Professional Development	Launch of new training editions, focusing on both the private and public sectors. Ongoing training becomes mandatory for certification and maintenance of professionals' technical and ethical quality

(*continued*)

Table 2. (*continued*)

Period	Historical Milestone	Detailed Description
2017	Regional and International Recognition	- OCAM becomes a member of PAFA (Pan-African Federation of Accountants); - Establishes cooperation with orders from Portugal, Brazil, Angola, South Africa, among others; - Participation in IFAC forums to align with SMOs (Statements of Membership Obligations for international adherence)
2018	Strengthening the Regulatory Framework	- Publication of main regulations: - Admission and Exams (11/08/2017); - Quality Control (13/08/2017); - Training (29/08/2018); - Code of Ethics based on IFAC 2016 version
2019	Technical-Scientific Strengthening	- OCAM translates and disseminates 800 pages of IPSAS standards; - Establishes cooperation protocols with CPCC and CPECA for international exam equivalency

3.4 General Requirements for Admission to the Profession (under Article 51, Law No. 8/2012)

The formalisation of criteria for admission as a full member of OCAM reflects an organisational strategy aimed at ensuring professional competence and integrity in the fields of accounting and auditing in Mozambique. This set of requirements not only aligns OCAM with international regulatory practices but also establishes a benchmark of rigour and quality for practising the profession in the country [17].

a) Be of Mozambican nationality or, if a foreign national, have legal permission to work in the country or be integrated into Certified Accountants' Firms or Certified Auditors' Firms established in Mozambique for more than one year; in addition, provide evidence of aptitude under the terms defined by OCAM and possess the technical capacity to practise the profession;
b) In the case of foreign applicants, demonstrate knowledge of the Portuguese language, Mozambican tax and commercial law, and other competencies as defined by OCAM;
c) Not have been convicted of any intentional criminal offence, particularly of a fiscal, economic, or financial nature, unless legally rehabilitated;
d) Not be prohibited or disqualified from practising the profession;
e) Hold the academic qualifications stipulated in Article 68 of these Statutes;
f) Pass the qualifying examination;
g) Complete the internship as defined in these Statutes and in the internship regulations;

h) Fulfil all requirements set out in reciprocity and mutual recognition agreements with foreign counterpart bodies, as well as the conditions outlined in items b), c), and d) above.

Table 3. Specific requirements for registration as certified accountants (under Article 52)

Requirement	Condition	Description
Admission Exam	Mandatory	Passing the admission exam acts as an initial filter to ensure verification of knowledge and competencies, establishing an entry standard aligned with professional expectations
Academic Qualifications (a)	As per Article 53	Confirms the need for a solid academic foundation, ensuring that candidates possess in-depth knowledge of the core areas essential to accounting practice
Professional Internship (b)	Satisfactory completion	The supervised professional internship is crucial for the transition from theoretical knowledge to practical application, thereby enabling adaptation to the realities of the job market
Approval by Mentor (c)	Mentor requirement	The requirement for acceptance by a mentor adds an additional layer of professional validation, whereby an experienced accountant endorses the candidate's potential
Internship Plan Submission	Structured learning	The internship plan must outline a clear learning path and development objectives aligned with OCAM's expectations
Exam Exemption	Special conditions	Exam exemption for foreign degree holders and members from countries with reciprocity agreements acknowledges the equivalence of qualifications and promotes professional integration

This academic Table 2 details the specifications and considerations relating to the registration process for certified accountants with OCAM, including the framework for competency assessment and the integration of international qualification standards. It serves as a valuable resource for understanding the professional and academic requirements of the Order and its alignment with global practices (Table 4).

Table 4. Requirements for Registration as Certified Auditors (under Article 53)

Item	*Requirement*	*Description*
1	Admission Exam	Candidates must pass the admission exam established by the College of Certified Auditors of OCAM.
2	Academic Qualifications	Candidates must hold academic qualifications in accordance with Article 68, which details the acceptable degrees and fields of study.
3a	Registration as Certified Accountant	It is a prerequisite to be registered as a Certified Accountant with OCAM in order to proceed with auditor certification.
3b	Professional Internship (General)	A regulated professional internship must have been completed with satisfactory performance.
3c	Professional Internship (3 years)	A three-year professional internship with satisfactory completion is required, in accordance with the rules of the College.
4	Sponsorship and Internship Plan	Internship candidates must be accepted by a mentor and submit an internship plan approved by the College of Certified Auditors.
5	Specialisation Exams	Candidates must successfully pass specialised exams in areas such as taxation, commercial and civil law to qualify as Certified Auditors.
5a	Exemption for Mozambicans Abroad	Mozambicans with equivalent qualifications and professional licences obtained abroad are exempt from the admission exam.
5b	Exemption for Members of SADC, CPLP or Commonwealth	Nationals from these blocs with the required qualifications may be exempt from the exam, subject to reciprocity.
5c	Exemption for Other Countries	Nationals from other countries may also be exempt from the exam under the same reciprocity conditions.
6	Proficiency in Portuguese Language	Candidates who obtained their qualifications in non-Portuguese-speaking countries must pass a Portuguese language exam.

This Table 3 reflects an academic interpretation of the requirements for auditor certification by OCAM, highlighting the integration of international standards and the emphasis on training and supervised professional practice as key elements in ensuring quality and competence in the auditing sector in Mozambique.

3.5 Internship and Examination (under Articles 53.1/54.1)

1. The internship for certified accountants and auditors shall last three years, with a minimum of 2,080 h per year, two-thirds of which must be spent under the supervision of a qualified mentor. The candidate must have a mentor, who may be a certified accountant, a certified auditor, or a certified auditing firm. In the case of a certified auditing firm, one of the partners shall be appointed as the supervisor responsible for overseeing the internship.
2. For the purposes of the above, mentors must have been registered with the Order for at least three (3) years.
3. During the internship, candidates will be assessed on their knowledge of the subjects included in the internship syllabus for the categories of certified auditor and certified accountant.
4. Successful completion of the internship is a prerequisite for admission to the examination.

3.6 Academic Qualifications Required

(Certified Accountant – under Article 53, paragraph 2(a); Certified Auditor – under Article 54, paragraph 2).

1. The academic qualification required to apply for registration as a certified accountant or certified auditor is:
   a) A degree (bachelor's, master's or doctorate) in accounting, management, economics, auditing, business sciences, or taxation, awarded by a Mozambican higher education institution;
   b) A higher education degree obtained abroad in one of the areas mentioned above, which has been officially recognised as equivalent to a bachelor's, master's or doctoral degree, or recognised as producing the effects of one of these degrees.
2. The amendments relating to registration do not apply to auditors and accountants who are already certified.

4 Methodology

4.1 Methodological Approach

This study follows a qualitative approach, grounded in Institutional Theory—particularly the concept of institutional isomorphism as proposed by [6]—to analyse the process of institutionalisation and professionalisation of accounting in Mozambique through the work of the Order of Accountants and Auditors of Mozambique (OCAM).

A qualitative methodology based on documentary and historical analysis was chosen, with the aim of exploring in depth the institutional mechanisms—coercive, normative, and mimetic—that have shaped the accounting profession in Mozambique from the creation of OCAM in 2012 through to 2018. This approach allows for a detailed understanding of the institutional and historical context, which is essential for capturing the nuances of this process in post-colonial and emerging environments.

The selection of this topic is also justified by the scarcity of previous literature on the institutionalisation of the accounting profession in emerging contexts such as

Mozambique, marked by the recent consolidation of OCAM, its unique historical and socio-economic characteristics, and historical limitations in access to detailed primary data. Thus, this study seeks to fill a significant academic gap and contribute to the expansion of Institutional Theory application in contexts that remain underexplored in the literature.

4.2 Data Sources

Data were obtained from both primary and secondary documentary sources. Official documents published by OCAM were analysed, including statutes, annual reports, strategic plans, and official announcements. Additionally, regulatory documents such as Law No. 8/12 and relevant international reports—particularly the *Report on the Observance of Standards and Codes* [22], produced by the World Bank—were examined. Prior academic studies on accounting development in African contexts, and particularly in Mozambique, were also consulted.

4.3 Analytical Procedures

The qualitative documentary analysis was conducted using an interpretive, chronologically structured approach, identifying critical institutional milestones in the evolution of OCAM and their implications for the Mozambican accounting profession. This analysis was supported by the assumptions of Institutional Theory, with particular attention to identifying the coercive, normative, and mimetic pressures that influenced the institutional configuration of the profession.

Triangulation techniques were used to validate the consistency of information drawn from the various documentary sources, ensuring robustness and credibility in the results obtained.

4.4 Methodological Limitations

Limitations are acknowledged with regard to the documentary methodology adopted, particularly the absence of primary data collected directly from accounting and auditing professionals through interviews or surveys. This limitation suggests that future research could benefit from mixed qualitative and quantitative approaches, thereby enriching the understanding of the perceptions and experiences of professionals involved in the institutional process under study.

5 Results

The work of the Order of Accountants and Auditors of Mozambique (OCAM) reveals significant transformations in the structuring of the accounting profession in the country, in light of the foundations of Institutional Theory, particularly those of [14] and [6]. OCAM has emerged as a relevant institutional actor in promoting compliance with international standards, notably through the adoption of practices aligned with the International Federation of Accountants (IFAC) and the implementation of International Financial Reporting Standards (IFRS).

Between 2012 and 2017, the number of registered members of OCAM increased from 453 to approximately 4,680, indicating not only numerical growth but also the institutionalisation of professional control mechanisms such as entrance examinations, continuous training, and supervised internships. These measures represent normative responses to institutional pressures aimed at strengthening the legitimacy of the accounting profession [6].

OCAM's membership in networks such as the Pan African Federation of Accountants (PAFA) and cooperation agreements with counterpart institutions reflect a process of mimetic isomorphism, reinforcing institutional legitimacy on the international stage. This movement is consistent with trends identified in other developing countries, where the symbolic adoption of international standards seeks to address legitimacy gaps [4, 9, 19].

However, structural challenges remain that limit the effectiveness of the institutionalisation process. The centralisation of operations in Maputo has created regional disparities in access to certification and training, compromising territorial equity. Additionally, the initial integration of professionals without formal training during the post-regulation transition period (from 2012 onwards) resulted in technical qualification asymmetries, which continue to be reflected in accounting practice.

From an analytical perspective, OCAM has responded predominantly through normative and mimetic means to institutional pressures, although its coercive role is evident in the formal regulation imposed by Decree-Law No. 8/12, in alignment with the diagnosis outlined in the *Report on the Observance of Standards and Codes* [22].

In summary, the findings suggest that OCAM has been successful in consolidating a robust institutional infrastructure geared towards the standardisation, credibility, and recognition of the profession in Mozambique. However, the sustainability of this process will depend on concrete measures to:

Expand territorial presence and access to certification in the provinces; Strengthen mechanisms for supervision and continuous evaluation; Structure an ethical and technically demanding training ecosystem, tailored to Mozambique's reality and the demands of a globalised economy.

6 Conclusions

The findings of this study demonstrate that the Order of Accountants and Auditors of Mozambique (OCAM) has played a strategic role in the institutionalisation and professionalisation of accounting in the country, within a context shaped by colonial legacies and growing demands for compliance with international standards. Through the lens of Institutional Theory, it was possible to identify how coercive pressures (stemming from legal and external requirements), normative pressures (arising from educational and professional expectations), and mimetic pressures (linked to the pursuit of international legitimacy) have shaped OCAM's actions since its establishment in 2012.

The consolidation of regulatory, certification, and continuous training mechanisms represents significant progress in building a more structured, ethical, and technically competent profession. OCAM not only responds to the requirements of multilateral bodies and international accounting networks but also acts as a catalyst for internal change, promoting a professional culture oriented towards quality, transparency, and accountability.

Nonetheless, important challenges persist. The geographical centralisation of OCAM's operations, asymmetries in the profile of certified professionals, and the difficulty in ensuring uniform application of standards across the national territory limit the scope of reforms. Furthermore, overcoming tensions between local practices and global normative models continues to require context-sensitive approaches supported by inclusive capacity-building policies.

This study does present certain limitations. Firstly, its documentary nature does not include triangulation with primary empirical data (such as interviews or surveys), which could be addressed in future research. Secondly, the analysis is focused on the period between 2012 and 2018, which may restrict the observation of more recent dynamics in OCAM's activities. This temporal delimitation is based, on one hand, on the founding milestone of OCAM, whose creation in 2012 marks the beginning of a new institutional phase in the Mozambican accounting sector; and on the other hand, on the greater availability, reliability, and systematisation of institutional data up to 2018. Although this study was developed in 2025, it was decided not to include the subsequent period, given that the available data post-2018 remains limited or lacks sufficient updating to ensure analytical robustness. Thus, the selected timeframe allows for a critical analysis of the initial cycle of professional institutionalisation, laying a solid foundation for future longitudinal studies.

This study therefore contributes to the debate on accounting professionalisation in emerging economies and highlights the role of professional orders as agents of institutional transformation. It is recommended that future research include empirical studies using qualitative or quantitative methods to deepen the understanding of professionals' perceptions regarding OCAM's impact on their practices and career trajectories, as well as comparative analyses with other professional bodies in similarly situated African contexts.

References

1. Azizuddin, M.: Administrative reform in Bangladesh: an overview of political dynamics. Nepalese J. Public Policy Gov. **28**(1), 49–68 (2011)
2. Boolaky, P.K.: Accounting development in Africa: a study of the impact of colonisation and the legal systems on accounting standards in Sub-Saharan African countries. Delhi Bus. Rev. **5**(1), 1–13 (2003)
3. Boolaky, P.K.: Measuring de jure harmonisation: a content analysis of the accounting standards of three countries: South Africa, Mauritius and Tanzania and International Financial Reporting Standards. J. Appl. Acc. Res. **8**(2), 110–146 (2006). https://doi.org/10.1108/96754260680001051
4. Chibuye, J.D., Mumba, F., Chishimba, M., Chishimba, C.: The role of corporate governance in enhancing economic development in Zambia. Int. J. Account. Financ. Rep. **7**(2), 80–94 (2017). https://doi.org/10.5296/ijafr.v7i2.12031
5. Degos, J.G., Levant, Y., Touron, P.: The history of accounting standards in French-speaking African countries since independence: the uneasy path toward IFRS. Account. Auditing Account. J. **32**(1), 75–100 (2019). https://doi.org/10.1108/AAAJ-03-2016-2459

6. DiMaggio, P.J., Powell, W.W.: The iron cage revisited: institutional isomorphism and collective rationality in organizational fields. Am. Sociol. Rev. **48**(2), 147–160 (1983)
7. Elad, C.: The development of accounting in the Franc zone countries in Africa. Int. J. Account. **50**(1), 75–100 (2015). https://doi.org/10.1016/j.intacc.2014.12.006
8. Elad, C., Shah, N., Agyeman, C.: Accounting classification in the era of international financial reporting standards: the case of Africa. J. Int. Account. Auditing Taxation (In Press) 1–17 (2023). https://doi.org/10.1016/j.intaccaudtax.2023.100546
9. George, B.A., Walker, R.M., Monster, M.W.L.: Does integrity improve business performance? Evidence from developing countries. Public Integrity **21**(5), 471–490 (2019). https://doi.org/10.1080/10999922.2019.1587401
10. Gomes, P., Salato, R., Ferreira, C.: Transition to accrual accounting in a developing country: the perception of accounting professionals. J. Account. Emerg. Econ. (ahead of print) (2025). https://doi.org/10.1108/JAEE-10-2024-0447
11. Hopper, T., Lassou, P., Soobaroyen, T.: Globalisation, accounting and developing countries. Crit. Perspect. Account. **43**, 125–148 (2017). https://doi.org/10.1016/j.cpa.2016.06.003
12. Juaniha, A.M.: Desafios de implementação das normas internacionais de contabilidade na educação contabilística em Moçambique. Master's thesis, Universidade Federal do Paraná (2016)
13. Langa, E., Albuquerque, F.: Conservatism as a cultural accounting value: an empirical study from the perspective of chartered accountants and auditors in Mozambique. Contaduría y Administración **69**(1), 1–22 (2023). https://doi.org/10.22201/fca.24488410e.2024.4682
14. Meyer, J.W., Rowan, B.: Institutionalized organizations: formal structure as myth and ceremony. Am. J. Sociol. **83**(2), 340–363 (1977). https://doi.org/10.1086/226550
15. Nkundabanyanga, S.K., Augustine, A.: Board role performance in Uganda's services sector firms. J. Public Adm. Policy Res. **4**(5), 115–124 (2014). https://doi.org/10.5897/jpapr12.019
16. OCAM: Plano de atividades. https://ocam.org.mz/wp-content/uploads/2024/06/Proposta-de-Plano-de-Actividades-2018-.pdf. Accessed 13 Mar 2025
17. OCAM: Proposta de revisão dos estatutos da OCAM. https://ocam.org.mz/wp-content/uploads/2024/06/Proposta-de-revisao-dos-estatutos_2.pdf. Accessed 13 Mar 2025
18. República de Moçambique: Decreto-Lei n.º 8/2012, de 8 de Fevereiro. Diário da República, 1.ª Série, n.º 6, p. 1 (2012)
19. Owolabi, S.A., Obialor, M.O., Olayanju, O.M.: Accounting and accountability for anti-corruption effectiveness in developing democracies: evidence from Nigeria. J. Financ. Crime **27**(2), 525–542 (2020). https://doi.org/10.1108/JFC-01-2019-0014
20. Pires, A.M.M., Rodrigues, A.F.J.P., Branco, V.M.C.: O sistema contabilístico angolano: Análise e reflexão com base em fontes e evidência empírica. In: I Encontro Internacional de Língua Portuguesa e Relações Lusófonas (LUSOCONF), Lisboa, Portugal, pp.1–26 (2018)
21. Tawiah, V., Oyewo, B.M., Doorgakunt, L.D.B., Zakari, A.: Colonisation and accounting development in Sub-Saharan Africa. Cogent Bus. Manag. **9**(1), 1–18 (2022). https://doi.org/10.1080/23311975.2022.2087465
22. World Bank: Report on the observance of standards and codes (ROSC): Mozambique. Available at: https://documents1.worldbank.org/curated/en/231131480543149188/pdf/110657-Rosc-Public-rosc-aa-mozambique-2008-jun.pdf. Accessed 13 Mar 2025

Legitimacy Theory as an Explanatory Framework for Sustainability Reporting: A Systematic Literature Review (2014–2024)

Aníbal Magalhães(✉)

University of Aveiro, Aveiro, Portugal
anibalrui@ua.pt

Abstract. The growing demand for transparency and corporate accountability has transformed sustainability reporting into a strategic tool for managing legitimacy, especially considering the emergence of regulations such as Corporate Sustainability Reporting Directive (CSRD) and the standards of the International Sustainability Standards Board (ISSB). This study conducts a Systematic Literature Review (SLR) on the application of Legitimacy Theory to sustainability reporting between 2014 and 2024. Combining the PRISMA protocol with bibliometric and qualitative analyses, 51 scientific articles indexed in the Web of Science were analyzed. The results show a dual theoretical perspective of legitimacy: on the one hand, as institutional compliance and alignment with regulatory requirements; on the other, as a symbolic strategy to manage stakeholder perception, often through selective disclosure practices or greenwashing. The review also highlights a strong concentration of studies in European journals, a limited geographical and sectoral diversity, and a growing interest in digital mechanisms of legitimation—such as blockchain-based reporting, the influence of social media, and algorithmic accountability. This study contributes to mapping the conceptual evolution of Legitimacy Theory in a context marked by digital transformation and regulatory intensification, while also identifying promising directions for future research that articulates sustainability, technological innovation, and the construction of legitimacy.

Keywords: Legitimacy Theory · Social and Environmental Accounting · Sustainability Reporting

1 Introduction

The profound transformation of the corporate reporting landscape, driven by intensifying environmental, social, and governance (ESG) requirements, reached a decisive milestone in 2024 with the entry into force of Corporate Sustainability Reporting Directive (CSRD) in the European Union, which consolidated a process that began in 2014 with the Non-Financial Reporting Directive (NFRD), which already imposed non-financial disclosure obligations on some entities. At the same time, there was a consolidation of global standards promoted by the International Sustainability Standards Board (ISSB), contributing

L. Almeida et al. (Eds.): ICAFI 2025, ISEM 80, pp. 57–84, 2026.
https://doi.org/10.1007/978-3-032-16268-7_4

to greater international harmonization of sustainability reporting. The CSRD represents a substantial reinforcement of the NFRD, expanding the number of companies covered, clarifying materiality criteria, and imposing external audit and certification requirements.

In this new regulatory and social context, organizations face the challenge of continually legitimizing their practices in the face of an increasingly attentive and critical stakeholder ecosystem. Legitimacy Theory, by providing a conceptual framework that explains how companies seek to ensure social and institutional acceptance through compliance or perception management, takes a central position in understanding sustainability reporting. However, the growing digitization of organizational communication—through interactive platforms, social networks, blockchain, or artificial intelligence—introduces new mechanisms of legitimation that have not yet been fully integrated into traditional theory, raising the question: is Legitimacy Theory sufficient to explain the processes of legitimation in the digital age?

Despite its conceptual and empirical richness, recent academic literature continues to present significant gaps: fragmentation of approaches, scarcity of systematic syntheses on the current state of theory, weak articulation with the digital dimensions of legitimation, and underrepresentation of non-European institutional contexts. It is precisely at this point that the present study positions itself.

This article aims to map the conceptual and empirical evolution of Legitimacy Theory applied to sustainability reporting between 2014 and 2024, based on a Systematic Literature Review (SLR), complemented by an in-depth qualitative analysis of the most influential articles. To this end, the PRISMA protocol is adopted and bibliometric and analytical tools such as Bibliometrix and VOSviewer are used.

This work contributes to research in accounting and sustainability on three levels: (i) it offers an integrated view of the main theoretical and empirical currents in the use of Legitimacy Theory; (ii) it identifies emerging theoretical tensions—between substantive and symbolic approaches, between compliance and the instrumentalization of legitimacy; (iii) it proposes new directions for research, focusing on the challenges posed by legitimacy in digital environments and changing regulatory contexts.

The article is structured as follows: Sect. 2 presents the theoretical framework of Legitimacy Theory and its articulation with other relevant approaches. Section 3 describes the RSL methodology and analysis procedures. Section 4 presents the bibliometric and qualitative results, critically analyzing trends in literature. Section 5 proposes lines of future research based on the gaps identified. Finally, Sect. 6 summarizes the main conclusions, limitations of the study, and implications for academic research and professional practice.

1.1 Research Questions and Study Objectives

In recent years, sustainability reporting has become an essential component of corporate management, driven by regulations such as Corporate Sustainability Reporting Directive (CSRD) and the standards of the International Sustainability Standards Board (ISSB). Thus, Legitimacy Theory has been widely used to explain how organizations shape their non-financial information disclosure practices to maintain or reinforce their social and institutional acceptance (Suchman, 1995; Deegan, 2002).

Despite the growing relevance of the topic, academic literature still has significant gaps regarding the conceptual evolution of Legitimacy Theory and its practical applicability in the context of sustainability reporting. Some of the main issues include the adaptation of business strategies to new regulatory requirements, the influence of institutional pressures on disclosure narratives, and how legitimacy is perceived by stakeholders in different sectors of activity (Bebbington & Larrinaga, 2014; Michelon, 2011).

Given this scenario, this study proposes to answer the following main research question:

> "How has the application of Legitimacy Theory evolved in explaining corporate sustainability reporting practices, and what are the main implications for advanced accounting?"

To address this question in a structured manner, the study unfolds into the following sub-research questions:

RQ1 – How has the application of Legitimacy Theory as an explanatory lens for corporate sustainability reporting practices evolved over the last ten years?
RQ2 – How has this theory been applied in advanced accounting, particularly in ESG reporting and sustainability auditing?
RQ3 – What gaps, challenges, and opportunities emerge from recent literature that could guide future research in this field?

Based on these questions, the main objectives of the study are:

- Conduct a systematic literature review (SLR) on the application of Legitimacy Theory to corporate sustainability reporting between 2014 and 2024.
- Analyze the conceptual evolution of the theory and identify its main practical applications in the context of advanced accounting.
- Identify trends, gaps, and potential lines of future research, based on bibliometric and qualitative analyses.

1.2 Study Relevance

The growing demand for transparency and corporate responsibility in the field of sustainability has generated renewed interest in the theoretical foundations that underpin non-financial reporting. Legitimacy Theory, widely applied in this field, has evolved in response to new institutional, technological, and regulatory dynamics, notably with the entry into force of the CSRD and the emergence of global standards such as those of the ISSB. Nevertheless, literature remains fragmented and lacks a critical and systematic synthesis that identifies the main contributions, gaps, and lines of theoretical evolution. This study responds to this need by conducting a Systematic Literature Review (SLR) between 2014 and 2024, combining bibliometric and qualitative analysis.

In addition to offering a detailed mapping of the application of Legitimacy Theory to sustainability reporting, this study identifies the main research trends, under-explored areas, and emerging opportunities—particularly in the use of digital technologies and sustainability auditing. In this way, the article contributes to the theoretical and methodological advancement of research in advanced accounting and sustainability, while also providing relevant insights for policy and regulatory decision-makers and managers who face the challenge of legitimately communicating organizations' ESG commitments.

2 Theoretical Framework

Legitimacy Theory has been widely adopted in accounting and sustainability reporting as a lens for explaining how organizations seek to ensure their survival and social acceptance. For Suchman (1995), Legitimacy represents a widespread perception that an entity's actions are desirable, adequate, or appropriate in the context of a socially constructed system of norms and values. The author proposes a typology that distinguishes three forms of legitimacy: pragmatic (based on stakeholder interests), moral (based on ethical compliance with social expectations), and cognitive (based on the inevitability or naturalness of the organization).

In the field of sustainability reporting, the application of Legitimacy Theory has oscillated between two major approaches: the institutional approach, which emphasizes regulatory compliance and organizational isomorphism (Meyer & Rowan, 1977), and strategic, which sees legitimation as a deliberate action to manage perception, often associated with symbolic practices such as greenwashing (Cho & Patten, 2007). This dichotomy has generated debate about the adequacy of the theory to explain contemporary disclosure phenomena.

To enrich the theoretical framework, several studies have proposed linking Legitimacy Theory with other approaches. Freeman's (1984) Stakeholder Theory introduces the idea that legitimacy depends on multiple audiences, whose interests and levels of influence vary. Agency Theory is often used to explain the incentives or constraints in the disclosure of non-financial information, especially when there are information asymmetries between managers and investors (Mio et al., 2020). Institutional Theory contributes with the concept of coercive, normative, and mimetic pressures that shape organizational behavior and the way companies adopt sustainable reporting practices (Dimaggio & Powell, 1983).

Despite the robustness of these frameworks, recent literature has identified growing tensions and limitations in the application of Legitimacy Theory. Authors such as Eitrem et al. (2024), Cho et al. (2015) and Hahn and Kühnen (2013) highlight that, although the theory continues to be widely used in the analysis of sustainability reporting, its explanatory power proves insufficient in the face of more complex contemporary phenomena, such as the digitization of corporate communication, increased multichannel social pressure, and the risk of greenwashing practices. These approaches warn of the danger of symbolic legitimization, where communicational compliance is not accompanied by substantial changes, thus requiring critical integration with other complementary theoretical lenses. On the one hand, the theory continues to be used as an institutionalized justification for compliance-oriented reporting practices. On the other hand, it is increasingly mobilized to explain symbolic reputation management strategies, which may not reflect real changes in the environmental or social performance of organizations (Bebbington & Larrinaga, 2014; Cho et al., 2015). This theoretical ambiguity requires critical analysis and conceptual updating.

In addition, new digital contexts for organizational communication, including social networks, online reporting platforms, artificial intelligence, and technologies such as blockchain, profoundly alter traditional mechanisms of legitimation. In this environment, organizational legitimacy is constructed in real time through interactive narratives mediated by algorithms. Legitimacy Theory, as conceived in the 1990s, may not fully capture these new dynamics of perception and influence (Hahn & Kühnen, 2013).

Accordingly, this article adopts Legitimacy Theory as its central framework, articulating it with complementary approaches and critically analyzing its evolution and applicability in the field of sustainability reporting. The objective is to understand whether the theory maintains its relevance as an explanation of organizational behavior in the face of institutional, social, and technological pressures in the contemporary context.

3 Research Methodology

The methodological approach is explained and defined as follows. Having proposed the main objective and research questions in the introduction, we begin by defining the identification of the data set. The Web of Science (WoS) database was chosen for the literature search, as it is considered to provide adequate and comprehensive collections of relevant academic literature (Raff et al., 2020). The choice of Web of Science as the exclusive basis for this review is because most articles indexed in Scopus are also present in WoS, in addition to the latter offering greater precision in segmentation by academic impact (Massaro et al., 2016). Secondly, it was decided to use the PRISMA protocol as the systematic review protocol that describes the logic, hypothesis, and planning methods of the review (Shamseer et al., 2015).

The application of this protocol helped to create a dataset of articles. Thirdly, content-based research was conducted to analyze the bibliographic information in each article and its relationship to different topics related to Legitimacy Theory as an explanatory framework for Sustainability Reporting. Finally, research trends and thematic areas are defined through a co-occurrence analysis of keywords using Bibliometrix and VOSviewer software.

Keyword co-occurrence analysis is a technique for analyzing the primary content of selected publications (Guo et al., 2019) with which essential words can be viewed in a search field together with co-occurrences (Secinaro et al., 2022). According to Škare et al. (2022), The co-occurrence analysis of keywords also involved an analysis of article abstracts to identify clusters of terms. A cluster is a set of interconnected terms that do not overlap. The co-occurrence of terms in the abstracts formed several groups. The strength of the links between terms based on co-occurrence frequency was ranked.

VOSviewer software can be used to build relationships and visualize literature based on information from academic literature. It can show the development and emerging publication trends of discipline over time. This software identifies the predominant terms and co-occurrence in all articles during analysis (Biggi & Giuliani, 2021). The result of the process is the grouping of terms into dominant themes.

The following sections provide a detailed description of the analysis and research results.

3.1 Rationale for the Systematic Literature Review

Systematic reviews of the literature are part of modern scientific research practices, providing comprehensive syntheses of available scientific evidence. The development of these synthesis studies makes it possible not only to summarize the main results of the research, but also to highlight the limitations and gaps detected in the research processes under analysis (Weir et al., 2016). Systematic reviews have proven to be the gold standard for synthesizing scientific evidence due to their methodological rigor, and are becoming increasingly common (Moher et al., 2016).

With literature reviews, authors aim to achieve various objectives. According to Petticrew and Roberts (2008), literature reviews can be used to study old theories or provide a basis for interventions, a guide for future research and study, or summaries of a specific issue. Due to the different objectives and increasing complexity, as well as the growing number of publication and research methods, there are several approaches to conducting literature reviews.

In recent years, the increasing digitization of organizational communication has profoundly transformed institutional legitimation processes. The proliferation of digital channels, social networks, online reporting platforms, and real-time feedback mechanisms has changed how organizations build and maintain legitimacy with stakeholders. Legitimacy management now depends not only on institutional compliance or traditional disclosure strategies, but also on digital narratives, visibility algorithms, and digital reputation. This new context poses challenges to the traditional application of Legitimacy Theory, requiring more dynamic theoretical approaches that are integrated with the phenomena of the digital age (Suchman, 1995; Hahn & Kühnen, 2023).

Massaro et al. (2016) identified what they called the "Literature review continuum," which ranges from rapid reviews, characterized by few rules, to systematic literature reviews (SLRs), which must follow very specific rules. While the rapid review aims to summarize and interpret the results of previous studies, the SLR goes much further, providing a clear, transparent, and replicable research methodology to evaluate and classify each study included in the review and propose future lines of research.

3.2 The PRISM Protocol and Selection Process

To understand the relevant trends and developments in studies on Legitimacy Theory as an explanatory framework for sustainability reporting, a Systematic Literature Review (SLR) was conducted using a consolidated approach: the PRISMA protocol (Moher et al., 2016). The SLR was conducted with the purpose of identifying and selecting scientific articles related to the specific research question for evaluation and summary, following rigorous and transparent procedures (Chaudhary et al., 2021), to identify possible future lines of research and highlight the limits of existing knowledge (Tranfield et al., 2003).

Systematic literature reviews, as the name implies, involve careful and comprehensive planning and a pre-designed search strategy to reduce bias (Chaudhary et al., 2021), identifying, evaluating, and summarizing all relevant studies on a given topic.

The PRISMA protocol provides a flowchart that assists in the stages of the SLR process: identification, screening, eligibility, and inclusion. The first objective of this method is to plan, identify, and evaluate studies to extract and summarize data from the literature (Tranfield et al., 2003), ensuring that bibliographic research is objective, transparent, and replicable. The reason for choosing this protocol (PRISMA) among other standards and guidelines (Snyder, 2019) is that its analytical methodological approach is very clear and easy to understand (Moher et al., 2009). The protocol is presented below in Fig. 1.

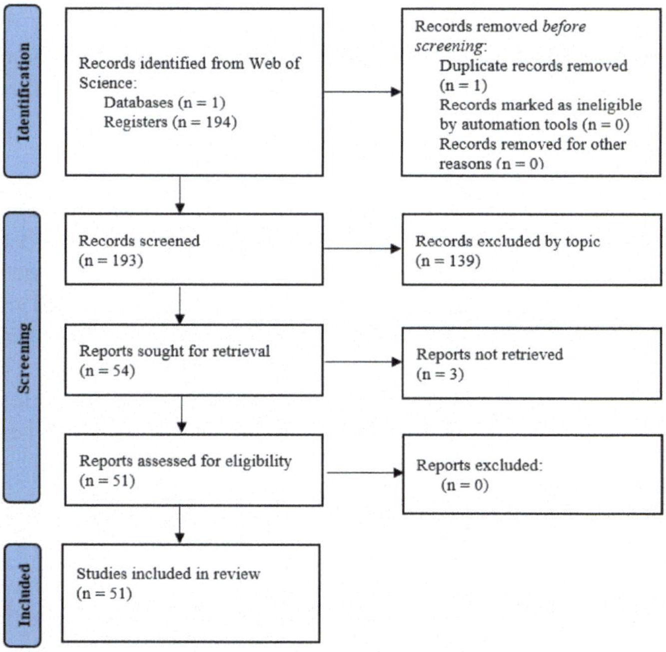

Fig. 1. Prisma Protocol.

3.2.1 Search Criteria

The preliminary research was conducted in two scientific databases: Scopus and Web of Science. After analyzing the initial results, it was found that most of the articles obtained using the same search term were indexed in both databases. Therefore, it was decided to use Web of Science exclusively as the primary source, since, in addition to containing virtually all the articles also indexed in Scopus, it had a higher number of relevant publications. WoS is an internationally recognized database that highlights patterns of activity in global scientific production and has been extensively used as a reference in numerous scientific studies (Thompson & Walker, 2015; Yu & Liao, 2016; Muhuri et al., 2019). The search was conducted in the WoS database on January 6, 8, and 12, 2025.

Given the aim of conducting exhaustive research with a view to systematizing the current state of the literature and scientific knowledge on sustainability reporting, to understand the evolution of Legitimacy Theory as an explanatory lens for sustainability reporting practices, we opted for the database with the largest number of articles available.

The careful definition of keywords and the construction of rigorous search expressions are fundamental and complex tasks in the process of systematic literature review (Kraus et al., 2020). Articles are identified by combining carefully selected keywords, allowing researchers to circumscribe the literature relevant to the object of study. In this specific case, the following search query was used: (TITLE-ABS-KEY (sustainab*) AND TITLE-ABS-KEY ("Legitimacy theory") AND TITLE-ABS-KEY (environmental AND disclosure) AND TITLE-ABS-KEY (corporate AND environmental)) AND (LIMIT-TO (DOCTYPE, "ar")) AND (EXCLUDE (LANGUAGE, "Chinese")).

3.2.2 Eligibility Criteria

Prior to conducting the research, a set of inclusion and exclusion criteria was established. In the preliminary phase, the following inclusion criteria were defined: publication in peer-reviewed scientific journals, document type (research and review articles), and indexing in Web of Science (WoS). The research included the following keywords: sustainab*; "Legitimacy theory"; environmental AND disclosure; corporate AND environmental, considering only documents classified as articles.

Regarding the exclusion criteria, all documentation that did not correspond to the defined thematic scope was eliminated, as well as articles written in Chinese. Finally, a duplicate article was removed, resulting in a final sample of 51 articles.

3.2.3 Inclusion

The next step was to do a qualitative assessment of the publications, which included a detailed analysis of the abstracts and a careful selection of articles, resulting in a final sample of 51 articles. The screening process resulted in the exclusion of 143 articles based on various criteria: one article due to duplication, three due to inaccessibility to the original sources, and 139 which, after analysis of their abstracts, were found not to correspond to the thematic scope of the research.

3.3 Qualitative Analysis

In addition to the bibliometric analysis, an in-depth qualitative analysis was conducted on a subsample of 10 articles drawn from the final dataset. The choice of this subsample was based on criteria of theoretical relevance and academic impact, namely: (i) number of citations (Scopus/WoS); (ii) publication in high impact factor journals (Q1); and (iii) explicit theoretical contribution to the development or application of Legitimacy Theory in sustainability reporting.

The qualitative analysis followed a thematic analysis approach, based on a full reading of the articles and the identification of key concepts, theoretical arguments, methods used, and practical implications. To this end, an analysis grid was developed containing predefined categories (e.g., type of legitimacy addressed, disclosure mechanisms, role of stakeholders, strategic vs. institutional motivations, challenges identified, contributions to future research).

This grid was constructed based on the main dimensions of Legitimacy Theory identified in the literature (e.g., Suchman, 1995; Deegan, 2002) and applied systematically to the selected articles, allowing us to identify patterns, divergences, and gaps in the theoretical and empirical approach to the topic. The result of this analysis enriched the theoretical discussion and provided a basis for the proposals for future research presented in Sect. 5.

The main results of this qualitative analysis are systematized in Table 4, which presents the explanatory theories used, as well as the most relevant conclusions of the selected articles.

4 Results

Starting with the research question, this section presents the main results.

4.1 "How Has the Application of Legitimacy Theory Evolved in Explaining Corporate Sustainability Reporting Practices, and What Are the Main Implications for Advanced Accounting?"

This section presents the main analyses, as described in the previous section, beginning with the distribution of articles by year, scientific journal of publication, and country. These analyses are carried out with the aim of understanding how the application of Legitimacy Theory has evolved in explaining corporate sustainability reporting practices, and what are the main implications for advanced accounting?

In Fig. 2, the graph shows a significant variation in scientific production over the years, with an initial phase of low production (2016–2019), followed by sharp growth from 2020 and a peak in 2022. After this peak, there is a slight reduction in scientific production in 2023 and 2024. The slight reduction in scientific output in 2023–2024 may be associated with an exhaustion of initial exploratory research and a transition to more applied and regulatory studies. In addition, the growing focus on new frameworks such as the ISSB, for example, may have shifted academic interest to other complementary theoretical approaches.

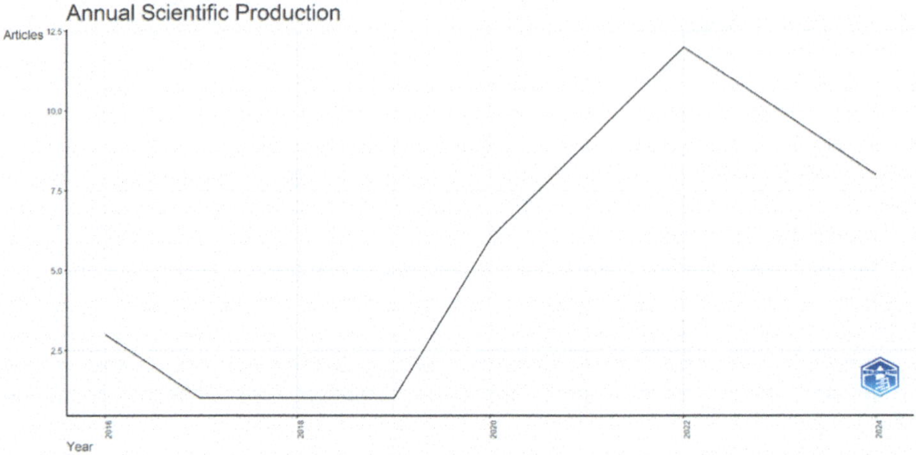

Fig. 2. Annual scientific Production.

Several studies show that this period was characterized by a significant increase in scientific production. During the COVID-19 pandemic, global scientific output grew exponentially in several disciplines, as demonstrated by data extracted from various international scientific bibliographic databases (Mohadab et al., 2020). This may explain the growth observed in the period between 2020 and 2022.

Table 1 shows the ten most relevant articles in the study.

The graph in Fig. 3 shows the main scientific journals that have published relevant articles in the field of Legitimacy Theory and sustainability reporting.

The predominance of the journal Sustainability, known for its focus on empirical studies and organizational practices, highlights the strong alignment between Legitimacy Theory and current concerns with corporate accountability and transparency. The concentration of publications in Q1 journals such as Accounting, Auditing & Accountability Journal and Business Strategy and the Environment reinforces the centrality of legitimacy as a robust theoretical framework for analyzing ESG reporting at a time when regulatory requirements make this type of disclosure not only desirable but also mandatory. This pattern suggests that the theory is being mobilized in both institutional and strategic perspectives, reflecting the tension between regulatory compliance and reputation building.

The predominance of Q1 journals in the field of accounting and sustainability suggests a consolidation of Legitimacy Theory as a robust and widely accepted theoretical framework. However, geographical concentration in European publications reveals an institutional bias that may limit the generalization of the theoretical models used. This observation reinforces the need to extend the application of the theory to less regulated or culturally distinct contexts to test its conceptual flexibility.

Analysis of these sources suggests some important directions. The Accounting, Auditing & Accountability Journal is one of the most respected journals in the field of accounting and auditing and is an important reference for the application of Legitimacy Theory to sustainability reporting. Business Strategy and the Environment and

Table 1. TOP 10 most relevant articles.

Rank	Author(s)	Article	Nº Citations	
			WOS	Scopus
1	Hummel, et al. (Hummel & Schlick, 2016)	The relationship between sustainability performance and sustainability disclosure - Reconciling voluntary disclosure theory and legitimacy theory	297	274
2	Torelli, et al. (Torelli et al., 2020)	Greenwashing and environmental communication: Effects on stakeholders' perceptions	213	149
3	Lai A, et al. (Lai et al., 2016)	Corporate Sustainable Development: is 'Integrated Reporting' a Legitimation Strategy?	163	168
4	Manning B., et al. (Manning et al., 2019)	Corporate governance and sustainable business conduct-Effects of board monitoring effectiveness and stakeholder engagement on corporate sustainability performance and disclosure choices	63	64
5	Simoni L., et al. (Simoni et al., 2020)	Effects of social, environmental, and institutional factors on sustainability report assurance: evidence from European countries	64	58
6	Crossley R, et al.(Crossley et al., 2021)	Sustainability and legitimacy theory: The case of sustainable social and environmental practices of small and medium-sized enterprises	71	50
7	Vourvachis P, et al. (Vourvachis et al., 2016)	CSR disclosure in response to major airline accidents: a legitimacy-based exploration	51	59

(*continued*)

Table 1. (*continued*)

Rank	Author(s)	Article	Nº Citations	
			WOS	Scopus
8	Mio C, et al. (Mio et al., 2020)	The predictive ability of legitimacy and agency theory after the implementation of the EU directive on non-financial information	47	45
9	Aray Y, et al. (Aray et al., 2021)	The hunt for international legitimacy: Examining the relationship between internationalization, state ownership, location and CSR reporting of Russian firms	41	38
10	Di Vaio A, et al. (Di Vaio et al., 2022)	Corporate social performance and non-financial reporting in the cruise industry: Paving the way towards UN Agenda 2030	41	33

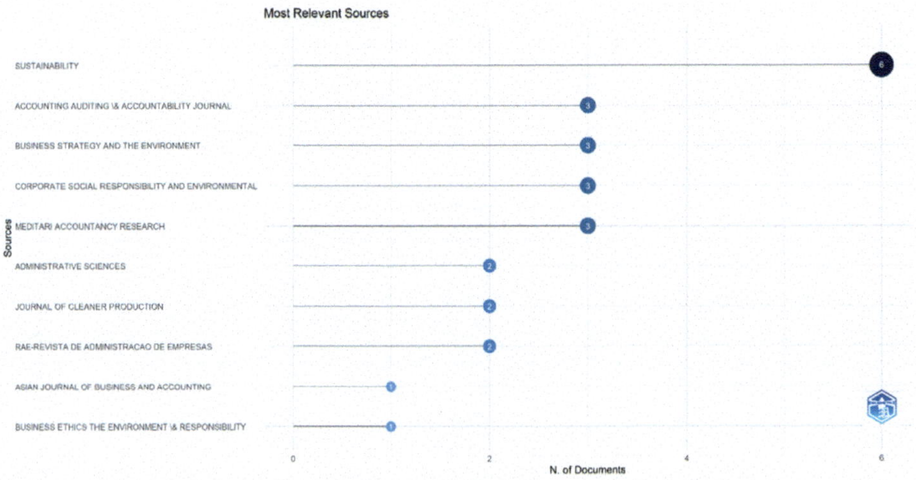

Fig. 3. Most relevant journals (number of articles per publication).

Corporate Social Responsibility and Environmental Management indicate that research on sustainability and organizational legitimacy is often addressed in the literature on corporate strategy and corporate social responsibility. The prominence of the journal Sustainability, with six publications, suggests that the topic has been widely explored in the context of organizational sustainability and ESG (Environmental, Social, and Governance).

The articles selected for the study were published in 35 specialized journals, edited by 20 different publishers, with the three main publishers accounting for about 55% of the journals: Emerald Group Publishing, Ltd. (22%), MDPI (18%), and Wiley (16%). The specialized journals included in the study are based in 10 different countries, nevertheless, three of these countries account for 75% of the selected articles and 66% of the publishers: the United Kingdom (45% of articles and 40% of publishers), Switzerland (20% of articles and 11% of publishers), and the United States (10% of articles and 14% of publishers) (Table 2).

Table 2. TOP 10 journals with the most articles and SJR score 2023.

Rank	Journal	Number of Articles	H-Index	Scimago/SJR 2023	País
1	Sustainability	6	169	0.67/Q1	Switzerland
2	Accounting auditing & accountability journal	3	121	1.85/Q1	United Kingdom
3	Business strategy and the environment	3	147	3.67/Q1	United Kingdom
4	Corporate social responsibility and environmental management	3	113	2.20/Q1	United Kingdom
5	Meditari accountancy research	3	39	0.93/Q1	United Kingdom
6	Administrative sciences	2	35	0.63/Q2	Switzerland
7	Journal of cleaner production	2	309	2.06/Q1	United Kingdom
8	Rae-revista de administracao de empresas	2	23	0.22/Q3	Brazil
9	Journal of business ethics	1	253	2.62/Q1	Netherlands
10	Business ethics, environment & responsibility	1	50	0.80/Q1	United Kingdom

An analysis of the most relevant journals reveals important trends in the dissemination of knowledge about Legitimacy Theory applied to sustainability reporting. Most of the journals belong to the Q1 category in the SJR ranking, indicating high quality and impact in the areas of accounting, sustainability, and business management. The Journal of Cleaner Production (H-Index 309) is one of the most cited journals in the study, focusing on sustainability and environmental practices. Sustainability (H-Index 169) is the journal with the most articles in the sample (6), showing that research on legitimacy and sustainability reporting has been widely published in this journal. The Accounting, Auditing & Accountability Journal (H-Index 121) is a central journal in the field of accounting and auditing, fundamental to studies on Legitimacy Theory in the context of financial and non-financial reporting.

Most of the journals are British (United Kingdom), suggesting that scientific production relevant to the topic has a strong base in the European context. The number of journals from Switzerland, such as Sustainability and Administrative Sciences, is also noteworthy, indicating that research on sustainability and organizational legitimacy has a strong presence in European-based publications.

In short, the predominance of Q1 and Q2 journals demonstrates that research on Legitimacy Theory in sustainability reporting is robust and published in high-impact journals. The strong concentration of publications in European journals suggests that the dominant theoretical approaches may have a perspective more aligned with the European regulatory and institutional context.

Figure 4 shows a map of collaboration between countries, highlighting regional contributions to scientific production in the context of Legitimacy Theory and sustainability reporting. This map provides important information about international collaboration networks and the concentration of publications per country.

Analysis of the map indicates that there is a significant concentration of scientific output in certain countries, particularly in Europe and North America, with notable contributions from emerging countries.

The United Kingdom is the main contributor, accounting for 45% of publications. This dominance reflects a strong British academic tradition in areas such as accounting, sustainability, and corporate social responsibility. Switzerland is the second largest contributor, with 20% of publications, standing out as a hub for sustainability research. This result is in line with the presence of important journals based in the country, such as Sustainability. The United States ranks third, accounting for 10% of publications. Despite being a global research center, its contribution to the Theory of Legitimacy seems to be smaller compared to the United Kingdom.

The concentration of scientific production in the United Kingdom and Switzerland suggests that the most relevant developments in the theory are strongly anchored in European regulatory contexts. This geographical bias may limit the applicability of the theory to emerging economies, where legitimation mechanisms differ substantially. This opens space for future research exploring the validity of the theory in alternative cultural and institutional contexts.

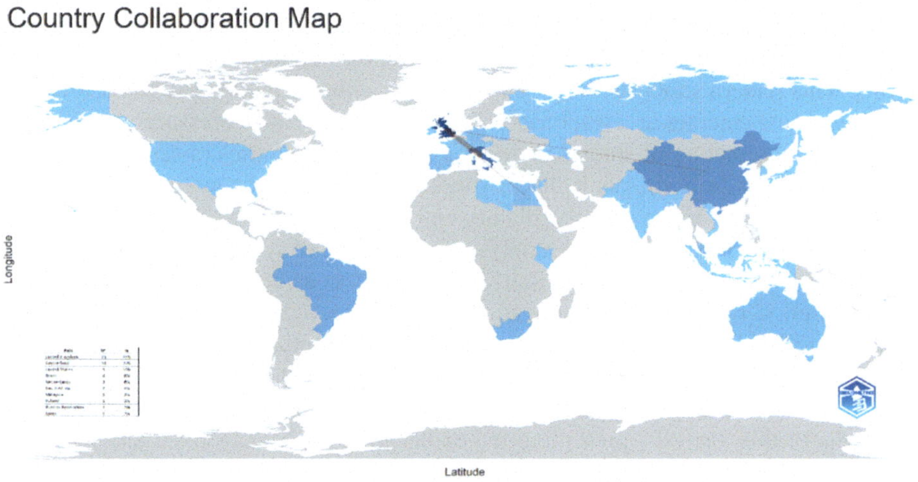

Fig. 4. Collaboration map by country.

Analysis of Table 3 reveals that most of the selected articles come from Europe (37 articles), with the United Kingdom (6 articles) standing out as the country with the highest concentration. America (10 articles) makes a significant contribution, with Brazil and the United States each representing three articles. Africa and Asia have a reduced representation, with only two articles each. These data indicate that research on sustainability reporting is more prevalent in European countries.

Table 3. Analysis by Countries and Continents.

Countries	Nº Articles	Continents	Nº Articles
United Kingdom	6	Europa	37
Switzerland	3	América	10
United States	3	Africa	2
Brazil	3	Asia	2
Netherlands	3		
South Africa	2		
Malaysia	2		
Poland	2		
Russian Federation	1		
Spain	1		

The strong predominance of studies from the United Kingdom and Switzerland reinforces the link between legitimacy and institutional contexts with high normative density. This pattern supports the institutionalist reading of the theory, where sustainability reporting emerges as a response to coercive and normative pressures. In contrast, the weak representation of studies in African and Asian countries raises doubts about the universality of the dominant legitimacy model, suggesting an opportunity for critical research in contexts with weaker regulatory enforcement.

To complement the bibliometric analysis, a subsample of 10 articles with the greatest academic impact was read and qualitatively analyzed, considering the number of citations, theoretical relevance, and publication in highly prestigious journals. This analysis allowed for a more in-depth identification of the theoretical approaches used, the main contributions to the understanding of Legitimacy Theory in the context of sustainability reporting, and the emerging challenges highlighted by the authors. Table 4 summarizes the results of this analysis, presenting the explanatory theories invoked and the main conclusions of each article, highlighting patterns, convergences, and points of theoretical tension that inform the subsequent discussion on gaps and opportunities for future research.

Table 4. Summary of the main conclusions of the most relevant articles.

Rank	Author(s)	Explanatory theories	Main conclusions
1	(Hummel & Schlick, 2016)	• Voluntary Disclosure Theory • Legitimacy Theory	Sustainable companies disclose high-quality information to signal their competitive advantages (Voluntary Disclosure Theory), while those with poorer performance use low-quality information to preserve legitimacy (Legitimacy Theory). The quality, rather than the quantity, of disclosure is essential to understanding this relationship. Standardization and external auditing are fundamental to ensuring greater reliability, reducing opportunistic practices in corporate sustainability disclosure
2	(Torelli et al., 2020)	• Legitimacy Theory • Signaling Theory	Different levels of greenwashing affect stakeholders' perceptions of corporate environmental responsibility. Companies use misleading environmental communications to preserve legitimacy, but reactions vary depending on the degree of greenwashing and the industry involved. The most environmentally sensitive sectors amplify the negative reaction of stakeholders when misleading practices are exposed. The study reinforces that greenwashing can damage corporate reputation and generate mistrust, highlighting the need for greater transparency in environmental communication

(*continued*)

Table 4. (*continued*)

Rank	Author(s)	Explanatory theories	Main conclusions
3	(Lai et al., 2016)	• Legitimacy Theory	The study rejects the hypothesis that companies adopt Integrated Reporting (IR) to regain legitimacy after negative sustainability assessments. Companies with higher ESG ratings are more likely to adopt IR, indicating a prior commitment to transparency. Factors such as size, profitability, and indebtedness do not significantly influence this decision. The findings question the exclusive applicability of Legitimacy Theory, suggesting that other theoretical approaches may better explain the adoption of IR
4	(Manning et al., 2019)	• Legitimacy Theory • Signaling Theory	Companies with better sustainable performance publish high-quality reports to signal credibility (Signaling Theory), while those with poorer performance use regulatory compliance to mitigate risks to legitimacy (Legitimacy Theory). Board monitoring and stakeholder engagement improve reporting quality but do not directly impact sustainable performance in the short term. In the long term, stakeholder engagement can drive more sustainable practices and reduce the use of compliance as a legitimacy strategy
5	(Simoni et al., 2020)	• Legitimacy Theory • Stakeholder Theory • Signaling Theory • Institutional Theory	Companies with greater commitment to stakeholders and better environmental performance are more likely to certify their sustainability reports. Quality corporate governance also influences this decision. Companies in countries with weaker sustainability policies tend to seek certification to gain legitimacy. However, ethical controversies do not significantly affect this choice. Certification strengthens the credibility of disclosure and reduces risks associated with lack of transparency
6	(Crossley et al., 2021)	• Legitimacy Theory	SMEs adopt sustainable social and environmental practices for both substantive and symbolic reasons, balancing real commitment with legitimate strategies. Legitimacy is achieved through community involvement, differentiation within the sector, and reputation building. Companies with limited resources tend to prioritize symbolic strategies, while more structured companies adopt substantive approaches. The study highlights the importance of continuous legitimation to ensure social acceptance and long-term sustainability

(*continued*)

Table 4. (*continued*)

Rank	Author(s)	Explanatory theories	Main conclusions
7	(Vourvachis et al., 2016)	• Legitimacy Theory	Airlines significantly increase CSR disclosure after accidents, reinforcing health and safety practices to mitigate negative impacts on their legitimacy. Nevertheless, the response is not uniform: in some cases, the lack of acceptance of responsibility leads to no significant changes in disclosure. The study confirms the strategic use of CSR disclosure to restore public confidence, highlighting the need for greater transparency and standardization in corporate practices
8	(Mio et al., 2020)	• Legitimacy Theory • Agency Theory	Legitimacy Theory remains relevant in the study and understanding of NFI disclosures even after the implementation of Directive 2014/95/EU. However, Agency Theory has gained greater importance, indicating that mandatory reporting has increased the reliability of non-financial information. The study shows that the Directive has significantly increased the quantity and quality of disclosure, especially in social and environmental aspects, but strategic motivations to ensure legitimacy still influence the reporting decision
9	(Aray et al., 2021)	• Legitimacy Theory	Internationalized Russian companies increase CSR disclosure to improve external legitimacy. The impact is greater in state-owned companies and more evident in markets outside the Commonwealth of Independent States (CIS). State-owned companies prioritize the disclosure of social and community actions over environmental ones to avoid perceptions of hypocrisy. Transparency in CSR communication is essential to mitigate negative stereotypes and strengthen acceptance in foreign markets
10	(Di Vaio et al., 2022)	• Legitimacy Theory	The study shows that companies in the cruise sector use sustainability reporting to reinforce their legitimacy and credibility, promoting their environmental and social responsibility. The report highlights commitments to the SDGs, prioritizing environmental and community issues. Transparency in disclosure improves reputation and facilitates engagement with stakeholders. Nevertheless, the quality and standardization of information remain challenges, suggesting the need for greater regulation and oversight of sustainable reporting

Thematic analysis of the 10 most influential articles reveals a clear theoretical tension between substantive approaches, where reporting is an expression of real commitment to sustainability, and symbolic approaches, where strategic perception management predominates. This theoretical ambiguity, which manifests itself using multiple theories (Legitimacy, Signaling, Stakeholders), challenges Legitimacy Theory to evolve to accommodate emerging practices such as greenwashing or the instrumental use of ESG certifications.

Qualitative analysis reveals a persistent duality in the application of the theory: while some studies highlight legitimacy as a response to regulatory and institutional pressures, others highlight its instrumentalization in greenwashing strategies or selective reporting. This ambivalence reinforces the criticism that Legitimacy Theory, although useful, lacks conceptual updating to cover contemporary phenomena such as digital reputation, algorithmic accountability, or legitimacy based on unaudited ESG metrics.

Legitimacy Theory remains a central framework for explaining corporate sustainability disclosure, especially in response to social, regulatory, and market pressures. Recent literature highlights that companies with superior sustainability performance tend to use reporting to signal credibility, while those with inferior performance resort to reporting to mitigate risks to legitimacy. The quality of disclosure, rather than quantity, is essential to ensure transparency and reliability.

In addition, sectors more exposed to public scrutiny, such as air transport and the cruise industry, demonstrate reporting strategies aligned with reputation management and stakeholder engagement. The introduction of regulations, such as Directive 2014/95/EU, has strengthened the standardization of non-financial disclosure, but the motivation for compliance is still influenced by the pursuit of legitimacy.

In the future, the integration of sustainability into the business model, combined with greater transparency and external certification, will be decisive for the credibility of sustainability reporting.

The graph in Fig. 5 shows that the most cited works belong to leading authors considered to be pioneers of Legitimacy Theory, such as Suchman (1995), Clarkson et al. (2008) and Dowling e Pfeffer (1975). These studies played a fundamental role in the development of Legitimacy Theory.

Suchman (1995) is the most cited reference and provides a solid conceptual basis for Legitimacy Theory, addressing different forms of organizational legitimacy. Dowling and Pfeffer (1975), were the authors of one of the first articles to discuss the need for organizations to maintain legitimacy with stakeholders.

Deegan (2002) is cited multiple times in the graph, standing out as a key author in the application of Legitimacy Theory to accounting and sustainability reporting.

The studies by Clarkson et al. (2008) and Cho and Patten (2007), explore the relationship between sustainability reporting and the perception of legitimacy, analyzing how companies use reporting to respond to social and regulatory pressures.

Milne and Patten (2002) and O'Donovan (2002) investigate the role of sustainability reporting as a legitimacy management strategy.

A more recent study, such as Hahn and Kühnen (2013), suggests the continuity of research on sustainability and legitimacy, adapting to new challenges and regulatory frameworks.

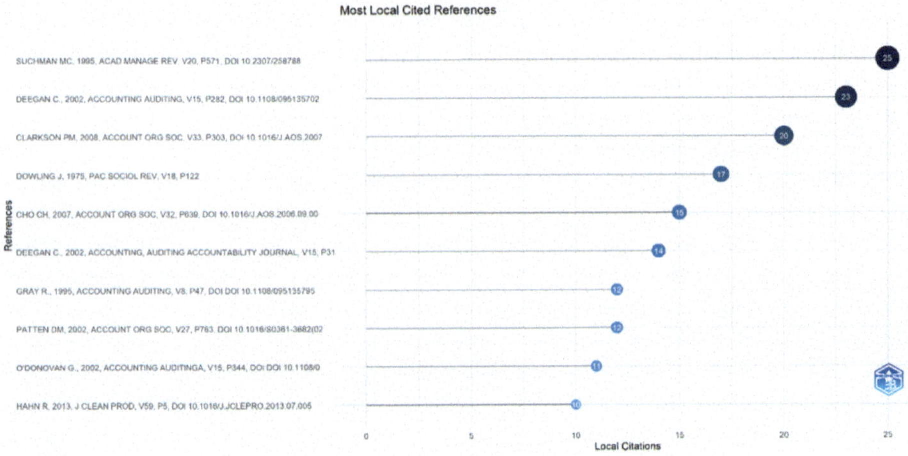

Fig. 5. Most cited authors.

This observation is corroborated by the analysis of the 'Co-citation Network, also obtained from the Bibliometrix software (Fig. 6).

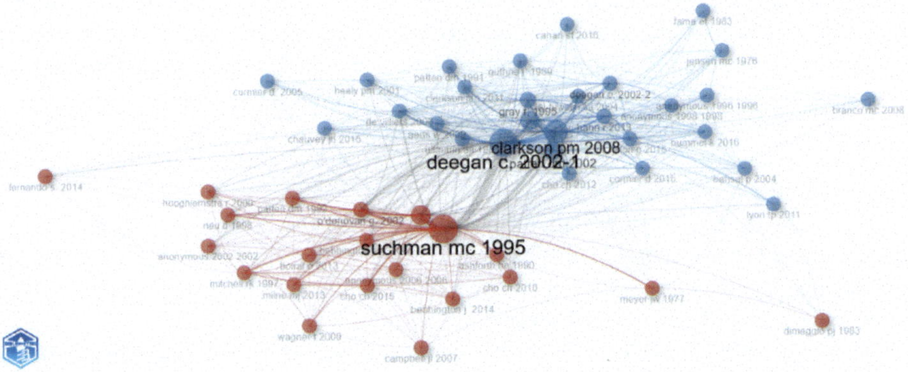

Fig. 6. Co-citation network: links between classical and recent approaches to Legitimacy Theory.

The separation between classical clusters (Suchman, Deegan) and emerging ones (Cho, Clarkson) indicates that the theory is at an inflection point. While traditional literature focuses on legitimacy because of institutional alignment, more recent work shifts the focus to the active role of organizations in the symbolic construction of legitimacy, often without real changes in performance. This theoretical shift requires the academic community to clarify the explanatory limits of the theory and explore its reconceptualization considering contemporary disclosure practices.

Analysis of the co-citation network reveals a theoretical divide between classical approaches (Suchman, 1995; Deegan, 2002) and more recent perspectives that seek to integrate new forms of communication and accountability. The connection between the clusters shows that literature remains rooted in normative foundations but is beginning to explore legitimacy as a strategic construct in digital communication environments. This theoretical tension between tradition and innovation highlights the need to reevaluate the role of theory in a more dynamic, interactive, and decentralized communication ecosystem.

The network displays two main clusters, represented by different colors, in addition to some intermediate nodes that interconnect the two clusters.

The red cluster highlights a more central and prominent node, citing Suchman (1995). This author developed one of the most influential studies on Legitimacy Theory. This article provides a solid conceptual framework, differentiating pragmatic, moral, and cognitive legitimacy, and is widely cited in studies on sustainability reporting.

Other authors in this cluster, such as Meyer and Rowan (1977), are references in Institutional Theory, suggesting that the literature on legitimacy in sustainability reporting has a strong institutionalist influence.

Studies such as O'Donovan (2002) and Patten (2020) are references in Institutional Theory, suggesting that the literature on legitimacy in sustainability reporting has a strong institutionalist influence.

The Blue cluster, which presents the other central node in the network, highlights Deegan (2002), who empirically applies Legitimacy Theory to sustainability reporting, exploring how companies use environmental disclosure to respond to legitimacy crises.

It also points out Clarkson et al. (2008) and Cormier et al. (2005), who focus their studies on the relationship between environmental performance and disclosure, analyzing how companies adjust disclosure to maintain legitimacy with stakeholders.

Gray et al. (1995) and Guthrie e Parker (1989) explore the evolution of corporate reporting practices considering social and institutional theories.

Finally, the connections between the Clusters show the link between the contributions of Suchman (1995) and Deegan (2002) to the current literature on sustainability reporting, which are still strongly based on the theoretical foundations of Suchman (1995), using Legitimacy Theory as an explanatory framework for companies' disclosure strategies.

Studies such as Bebbington and Larrinaga (2014) and Cho et al. (2015) appear in intermediate positions, still suggesting an important role in the integration between theoretical bases and the most recent empirical applications.

The following Fig. 7 reveals a complex network of interconnected terms, indicating how different concepts relate to Legitimacy Theory and sustainability reporting. The colors represent thematic clusters, which group highly correlated topics, allowing for the identification of trends in literature.

The presence of terms like "transparency," "social media," and "disclosure management" in the most recent clusters suggests an evolution of the concept of legitimacy to encompass digital and interactive dimensions. Digital reputation management, mediated by algorithms and online channels, poses new challenges to theory, which has historically favored the analysis of institutionalized legitimacy based on formal documents. These findings reinforce the need to update the theory for contexts in which public perception is volatile, distributed, and shaped in real time.

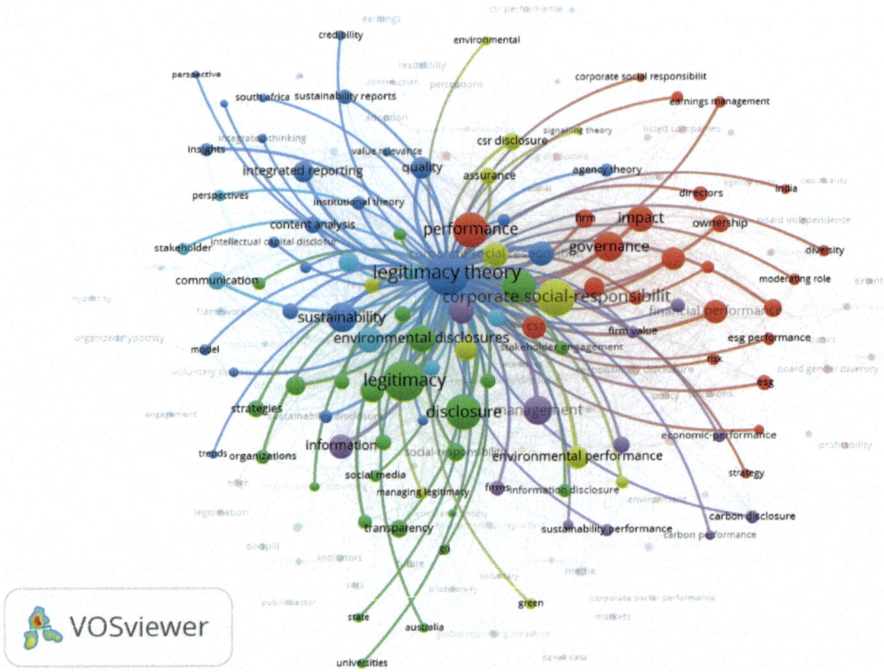

Fig. 7. Digital and traditional clusters in the construction of legitimacy.

The blue cluster highlights "legitimacy theory," which strongly connects to terms such as "sustainability," "integrated reporting," "environmental disclosures," and "communication." This connection suggests that Legitimacy Theory has been widely used to explain environmental disclosure and integrated reporting practices, reinforcing the idea that companies use such reports as a strategy to maintain or obtain legitimacy.

The green cluster includes terms such as "legitimacy," "disclosure management," "information," "transparency," and "social media." It highlights the importance of strategic management of information disclosure in the legitimacy process, especially through transparency and the use of digital media.

The red cluster highlights terms such as "corporate social responsibility (CSR)," "financial performance," "governance," "impact," "ownership," and "risk." It's possible to intuit that the explanatory lens of Legitimacy Theory is often used to explain, or is a privileged vantage point for, observing and attempting to find explanatory narratives for, phenomena related to corporate governance, financial performance, and risks associated with sustainability.

Finally, the purple cluster encompasses concepts such as "environmental performance," "carbon disclosure," "sustainability performance," and "greenhouse gas emissions." This suggests a growing focus on the relationship between companies' environmental performance and sustainability reporting, which is aligned with current regulatory demands and stakeholder expectations.

4.2 Conceptual Evolution of the Theory of Legitimacy in the Context of Sustainability Reporting

Legitimacy Theory has been widely used to explain how organizations adjust their sustainability reporting practices to ensure social acceptance and alignment with institutional expectations. Since the seminal contributions of Suchman (1995) and Deegan (2002), the theory has evolved to incorporate new factors, such as regulatory pressure, the digitalization of corporate information, and the role of artificial intelligence in analyzing legitimacy.

An analysis of the literature between 2014 and 2024 reveals that the application of the theory has followed three distinct phases. First, a period of consolidation (2014–2019), in which studies focused on the relationship between legitimacy and voluntary disclosure of sustainability reports, emphasizing how companies use such practices to mitigate reputational risks. A second period of expansion and integration (2020–2022), in which a pandemic crisis and increased ESG regulation led to more intense scientific production, with a focus on studies on how sustainability reporting influences market value and investor confidence. Third, a period of regulatory transformation (2023–2024), with the implementation of the CSRD and ISSB standards, which redefine companies' approach to sustainability reporting, making legitimacy not just a strategic issue but an essential regulatory requirement.

Bibliometric analysis confirms this evolution, evidenced by the growing co-occurrence of terms such as "disclosure management," "corporate social responsibility," and "governance" in recent publications. The findings indicate that Legitimacy Theory not only remains a solid theoretical framework but is also expanding into new approaches, including computational analysis of organizational legitimacy and the use of big data to monitor reputational risks.

Bibliometric and co-occurrence data indicate that Legitimacy Theory is being reinterpreted in the context of the digitalization of reporting. The emergence of terms such as transparency, social media, disclosure management, and communication in thematic clusters suggests that legitimacy mechanisms are increasingly associated with digital perception management. In this new landscape, organizations must respond to continuous public scrutiny, mediated by digital platforms and social media, which requires real-time legitimacy strategies and greater attention to narrative coherence across multiple channels. This phenomenon points to an expansion in the scope of Legitimacy Theory, which must incorporate emerging technological and communicational dimensions.

4.3 Applicability of Legitimacy Theory in Advanced Accounting

Advanced accounting, driven by the growing relevance of ESG metrics, has incorporated Legitimacy Theory into various aspects of corporate reporting and sustainable auditing. The literature analyzed points to three main practical applications. Legitimacy mechanisms in accounting reporting, in which companies adjust financial and non-financial narratives to demonstrate alignment with social and regulatory standards. The introduction of standards such as CSRD and ISSB reinforced the need for more structured and verifiable reports, reducing the risk of greenwashing practices. The impact on corporate governance and risk assessment, in which Legitimacy Theory has been used to understand how boards of directors and auditors incorporate ESG concerns into decision-making and corporate reporting. The concept of "dynamic materiality" emerges as a new approach to integrating ESG risks into financial statements. Finally, the challenges and opportunities in sustainability auditing. The systematic review identified an increase in scientific literature on ESG auditing, indicating that advanced accounting is expanding beyond traditional financial reporting. Recent studies suggest that the future of auditing will involve the adoption of AI-based technologies to validate the authenticity of sustainability reports and reinforce corporate legitimacy.

In summary, the bibliometric results not only demonstrate the vitality and relevance of Legitimacy Theory in the field of sustainability reporting, but also reveal an important theoretical tension: on the one hand, the theory continues to be used as a solid explanatory lens for institutionalized disclosure practices; on the other, it appears increasingly associated with symbolic strategies, such as greenwashing, calling into question its explanatory sufficiency. This theoretical ambivalence deserves further exploration and is explored in the following section.

5 Possible Line of Future Research

This systematic literature review on Legitimacy Theory as an analytical lens for sustainability reporting provides avenues for future research, which are identified in the map in Fig. 8.

A particularly promising line of research concerns the application of Legitimacy Theory in digital contexts. How do companies use digital platforms to legitimize their sustainability practices? What role do social media, automated feedback systems, blockchain-based reporting, or artificial intelligence play in building organizational legitimacy? These emerging themes, still underexplored in the literature, represent relevant opportunities to deepen our understanding of the new dynamics of disclosure in the digital age.

- Concrete proposals for future research:
- How do digital platform algorithms shape the perceived legitimacy of organizations?
- How does AI-based ESG certification affect stakeholder trust?
- Does Legitimacy Theory remain valid in contexts with low regulatory enforcement (e.g., developing economies)?

Figure 8 provides a thematic map of the literature on sustainability reporting, based on an analysis of centrality (relevance) and density (degree of development). This type of bibliometric analysis helps identify central, emerging, and specialized topics within the field of study.

Analysis of the map allows us to structure the systematic review into three main areas: CSR and Organizational Performance, how legitimacy influences the adoption of corporate social responsibility practices. Transparency and Sustainability Reporting, as companies use disclosure to ensure legitimacy with stakeholders; and finally, Criticism of Organizational Legitimacy, exploring the concept of organized hypocrisy and the limitations of sustainability reporting.

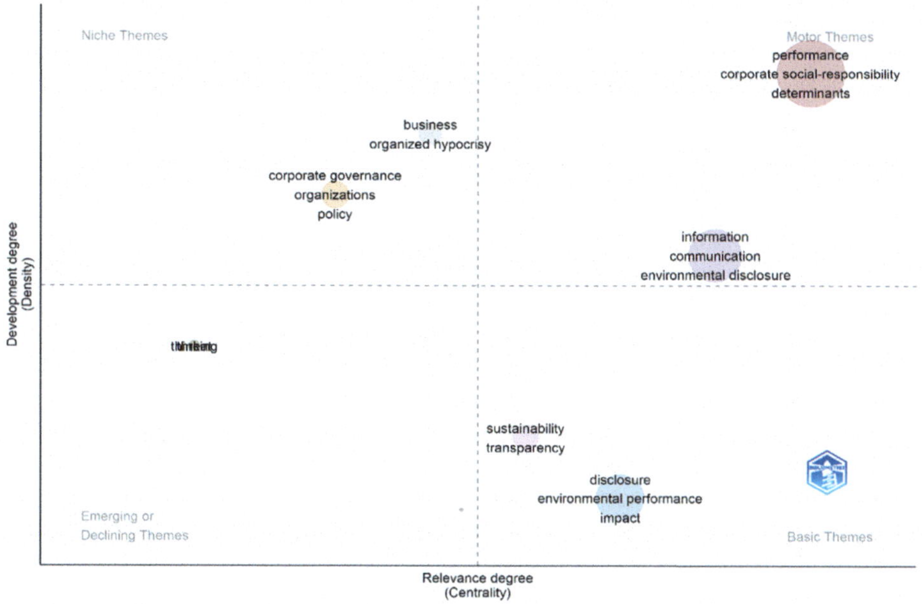

Fig. 8. Thematic map of literature.

6 Main Conclusions

This study conducted a Systematic Literature Review (SLR) on the application of Legitimacy Theory to sustainability reporting between 2014 and 2024, combining bibliometric and qualitative analyses. The results show that, despite its widespread use and persistent influence, Legitimacy Theory is at a critical point in its evolution. On the one hand, it remains the dominant explanatory framework for institutional disclosure practices; on the other, it reveals significant limitations when addressing emerging phenomena such as greenwashing, digital reputation, or the influence of algorithms and artificial intelligence on organizational legitimacy.

The analysis of publications shows significant growth in scientific production, with a predominance of European studies, suggesting an institutional and regulatory bias in theoretical development. The theoretical duality between substantive and symbolic approaches was particularly evident in the qualitative analysis, confirming the ambivalence of the concept of legitimacy. This tension, already identified by authors such as Bebbington e Larrinaga (2014) and Cho et al. (2015), reinforces the need to critically reconceptualize the dominant theoretical model. In this sense, the central question that guided this research—can Legitimacy Theory adequately explain the processes of legitimation in the digital age?—seems to deserve a nuanced answer: the theory remains relevant, but it needs to be updated and articulated with complementary approaches.

In this sense, this article contributes to academic research by mapping, in an integrated manner, the trends and gaps in literature, proposing that Legitimacy Theory be reconfigured to incorporate digital mechanisms of legitimation, hybrid regulatory contexts, and real-time perception dynamics. The emergence of technologies such as artificial intelligence, blockchain, or interactive reporting platforms requires a more dynamic, pluralistic, and interdisciplinary theoretical framework that goes beyond the traditional isomorphic model.

The limitations of this study include its exclusive focus on the Web of Science database and the fact that the qualitative analysis was conducted by a single researcher. Nevertheless, the findings offer valuable clues for future research, namely: (i) the empirical validation of digital legitimacy; (ii) the comparative study of weak vs. strong institutional contexts; (iii) the articulation between legitimacy, financial impact, and verified ESG practices; and (iv) the analysis of the ethical implications associated with algorithmic manipulation of reputation.

In conclusion, this study demonstrates that Legitimacy Theory remains an essential theoretical pillar, but that it needs to evolve to respond to the challenges posed by digital transformation, regulatory complexity, and new demands for transparency and accountability. Its ability to explain—and not just describe—contemporary phenomena will depend on its capacity for critical adaptation, lest it become an anachronistic lens for rapidly changing realities.

References

Aray, Y., Dikova, D., Garanina, T., Veselova, A.: The hunt for international legitimacy: examining the relationship between internationalization, state ownership, location and CSR reporting of Russian firms. Int. Bus. Rev. **30**(5) (2021). https://doi.org/10.1016/j.ibusrev.2021.101858

Bebbington, J., Larrinaga, C.: Accounting and sustainable development: an exploration. Account. Organ. Soc. **39**(6) (2014). https://doi.org/10.1016/j.aos.2014.01.003

Biggi, G., Giuliani, E.: The noxious consequences of innovation: what do we know? Ind. Innov. **28**(1) (2021). https://doi.org/10.1080/13662716.2020.1726729

Chaudhary, S., Dhir, A., Ferraris, A., Bertoldi, B.: Trust and reputation in family businesses: a systematic literature review of past achievements and future promises. J. Bus. Res. **137**, 143–161 (2021). https://doi.org/10.1016/j.jbusres.2021.07.052

Cho, C.H., Michelon, G., Patten, D.M., Roberts, R.W.: CSR disclosure: the more things change …? Account. Auditing Accountab. J. **28**(1), 14–35 (2015). https://doi.org/10.1108/AAAJ-12-2013-1549

Cho, C.H., Patten, D.M.: The role of environmental disclosures as tools of legitimacy: a research note. Account. Organ. Soc. **32**(7–8) (2007). https://doi.org/10.1016/j.aos.2006.09.009

Clarkson, P.M., Li, Y., Richardson, G.D., Vasvari, F.P.: Revisiting the relation between environmental performance and environmental disclosure: an empirical analysis. Account. Organ. Soc. **33**(4–5) (2008). https://doi.org/10.1016/j.aos.2007.05.003

Cormier, D., Magnan, M., Van Velthoven, B.: Environmental disclosure quality in large German companies: economic incentives, public pressures or institutional conditions? Eur. Account. Rev. **14**(1) (2005). https://doi.org/10.1080/0963818042000339617

Crossley, R.M., Elmagrhi, M.H., Ntim, C.G.: Sustainability and legitimacy theory: the case of sustainable social and environmental practices of small and medium-sized enterprises. Bus. Strateg. Environ. **30**(8), 3740–3762 (2021). https://doi.org/10.1002/bse.2837

Deegan, C.: The legitimising effect of social and environmental disclosures. Account. Auditing Account. J. **15**(3) (2002)

Di Vaio, A., Varriale, L., Di Gregorio, A., Adomako, S.: Corporate social performance and non-financial reporting in the cruise industry: paving the way towards UN Agenda 2030. Corp. Soc. Responsib. Environ. Manag. **29**(6), 1931–1953 (2022). https://doi.org/10.1002/csr.2292

Dimaggio, P.J., Powell, W.W.: The iron cage revisited: institutional isomorphism and collective rationality in organizational. Source Am. Sociol. Rev. **48**(2) (1983)

Dowling, J., Pfeffer, J.: Organizational legitimacy: social values and organizational behavior. Sociol. Perspect. **18**(1) (1975). https://doi.org/10.2307/1388226

Eitrem, A., Meidell, A., Modell, S.: The use of institutional theory in social and environmental accounting research: a critical review. Account. Bus. Res. **54**, 775–810 (2024). https://doi.org/10.1080/00014788.2024.2328934

Freeman, E.R.: Strategic Management: A Stakeholder Approach (1984)

Gray, R., Kouhy, R., Lavers, S.: Corporate social and environmental reporting: a review of the literature and a longitudinal study of UK disclosure. Account. Auditing Account. J. **8**(2) (1995). https://doi.org/10.1108/09513579510146996

Guo, Y.M., Huang, Z.L., Guo, J., Li, H., Guo, X.R., Nkeli, M.J.: Bibliometric analysis on smart cities research. Sustainability **11**(13) (2019). https://doi.org/10.3390/su11133606

Guthrie, J., Parker, L.D.: Corporate social reporting: a rebuttal of legitimacy theory. Account. Bus. Res. **19**(76), 343–352 (1989). https://doi.org/10.1080/00014788.1989.9728863

Hahn, R., Kühnen, M.: Determinants of sustainability reporting: a review of results, trends, theory, and opportunities in an expanding field of research. J. Clean. Prod. **59** (2013). https://doi.org/10.1016/j.jclepro.2013.07.005

Hummel, K., Schlick, C.: The relationship between sustainability performance and sustainability disclosure – reconciling voluntary disclosure theory and legitimacy theory. J. Account. Public Policy **35**(5) (2016). https://doi.org/10.1016/j.jaccpubpol.2016.06.001

Kraus, S., Breier, M., Dasí-Rodríguez, S.: The art of crafting a systematic literature review in entrepreneurship research. Int. Entrep. Manag. J. **16**(3) (2020). https://doi.org/10.1007/s11365-020-00635-4

Lai, A., Melloni, G., Stacchezzini, R.: Corporate sustainable development: is "integrated reporting" a legitimation strategy? Bus. Strategy Environ. **25**(3), 165–177 (2016). https://doi.org/10.1002/bse.1863

Manning, B., Braam, G., Reimsbach, D.: Corporate governance and sustainable business conduct-Effects of board monitoring effectiveness and stakeholder engagement on corporate sustainability performance and disclosure choices. Corp. Soc. Responsib. Environ. Manag. **26**(2), 351–366 (2019). https://doi.org/10.1002/csr.1687

Massaro, M., Dumay, J., Guthrie, J.: On the shoulders of giants: undertaking a structured literature review in accounting. Account. Auditing Accountab. J. **29**(5) (2016). https://doi.org/10.1108/AAAJ-01-2015-1939

Meyer, J.W., Rowan, B.: Institutionalized organizations: formal structure as myth and ceremony. New Econ. Sociol. Reader (1977). https://doi.org/10.1086/226550

Michelon, G.: Sustainability disclosure and reputation: a comparative study. Corp. Reput. Rev. **14**(2), 79–96 (2011). https://doi.org/10.1057/crr.2011.10

Milne, M.J., Patten, D.M.: Securing organizational legitimacy: an experimental decision case examining the impact of environmental disclosures. Account. Auditing Account. J. **15**(3) (2002). https://doi.org/10.1108/09513570210435889

Mio, C., Fasan, M., Marcon, C., Panfilo, S.: The predictive ability of legitimacy and agency theory after the implementation of the EU directive on non-financial information. Corp. Soc. Responsib. Environ. Manag. **27**(6), 2465–2476 (2020). https://doi.org/10.1002/csr.1968

Mohadab, M.E., Bouikhalene, B., Safi, S.: Bibliometric method for mapping the state of the art of scientific production in Covid-19. Chaos Solitons Fractals **139**, 110052 (2020). https://doi.org/10.1016/j.chaos.2020.110052

Moher, D., et al.: Preferred reporting items for systematic reviews and meta-analyses: the PRISMA statement. PLoS Med. **6**(7) (2009). https://doi.org/10.1371/journal.pmed.1000097

Moher, D., et al.: Preferred reporting items for systematic review and meta-analysis protocols (PRISMA-P) 2015 statement. Revista Espanola de Nutricion Humana y Dietetica **20**(2) (2016). https://doi.org/10.1186/2046-4053-4-1

Muhuri, P.K., Shukla, A.K., Abraham, A.: Industry 4.0: a bibliometric analysis and detailed overview. Eng. Appl. Artif. Intell. **78** (2019). https://doi.org/10.1016/j.engappai.2018.11.007

O'Donovan, G.: Environmental disclosures in the annual report: extending the applicability and predictive power of legitimacy theory. Account. Auditing Account. J. **15**(3), 344–371 (2002). https://doi.org/10.1108/09513570210435870

Patten, D.M.: Seeking legitimacy. Sustain. Account. Manag. Policy J. **11**(6), 1009–1021 (2020). https://doi.org/10.1108/SAMPJ-12-2018-0332

Petticrew, M., Roberts, H.: Systematic reviews in the social sciences: a practical guide. In Systematic Reviews in the Social Sciences: A Practical Guide (2008). https://doi.org/10.1002/9780470754887

Raff, S., Wentzel, D., Obwegeser, N.: Smart products: conceptual review, synthesis, and research directions. J. Prod. Innov. Manag. **37**(5) (2020). https://doi.org/10.1111/jpim.12544

Secinaro, S., Calandra, D., Lanzalonga, F., Ferraris, A.: Electric vehicles' consumer behaviours: mapping the field and providing a research agenda. J. Bus. Res. **150**, 399–416 (2022). https://doi.org/10.1016/j.jbusres.2022.06.011

Shamseer, L., et al.: Preferred reporting items for systematic review and meta-analysis protocols (prisma-p) 2015: elaboration and explanation. BMJ **349** (2015). https://doi.org/10.1136/bmj.g7647

Simoni, L., Bini, L., Bellucci, M.: Effects of social, environmental, and institutional factors on sustainability report assurance: evidence from European countries. Meditari Account. Res. **28**(6), 1059–1087 (2020). https://doi.org/10.1108/MEDAR-03-2019-0462

Škare, M., Blanco-Gonzalez-Tejero, C., Crecente, F., del Val, M.T.: Scientometric analysis on entrepreneurial skills - creativity, communication, leadership: how strong is the association? Technol. Forecast. Soc. Chang. **182**, 121851 (2022). https://doi.org/10.1016/j.techfore.2022.121851

Snyder, H.: Literature review as a research methodology: an overview and guidelines. J. Bus. Res. **104**, 333–339 (2019). https://doi.org/10.1016/j.jbusres.2019.07.039

Suchman, M.C.: Managing legitimacy: strategic and institutional approaches. Acad. Manag. Rev. **20**(3) (1995). https://doi.org/10.2307/258788

Thompson, D.F., Walker, C.K.: A descriptive and historical review of bibliometrics with applications to medical sciences. Pharmacotherapy **35**(6) (2015). https://doi.org/10.1002/phar.1586

Torelli, R., Balluchi, F., Lazzini, A.: Greenwashing and environmental communication: effects on stakeholders' perceptions. Bus. Strategy Environ. **29**(2), 407–421 (2020). https://doi.org/10.1002/bse.2373

Tranfield, D., Denyer, D., Smart, P.: Towards a methodology for developing evidence-informed management knowledge by means of systematic review. Br. J. Manag. **14**(3) (2003). https://doi.org/10.1111/1467-8551.00375

Vourvachis, P., Woodward, T., Woodward, D.G., Patten, D.M.: CSR disclosure in response to major airline accidents: a legitimacy-based exploration. Sustain. Account. Manag. Policy J. **7**(1), 26–43 (2016). https://doi.org/10.1108/SAMPJ-12-2014-0080

Weir, A., Rabia, S., Ardern, C.: Trusting systematic reviews and meta-analyses: all that glitters is not gold! Br. J. Sports Med. **50**(18) (2016). https://doi.org/10.1136/bjsports-2015-095896

Yu, D., Liao, H.: Visualization and quantitative research on intuitionistic fuzzy studies. J. Intell. Fuzzy Syst. **30**(6) (2016). https://doi.org/10.3233/IFS-162111

Management Accounting and Intellectual Capital: Bibliometric Analysis and Science Mapping

Vânia Dias[1], Patrícia Quesado[2(✉)], and Lurdes Silva[2]

[1] Polytechnic University of Cávado and Ave, Barcelos, Portugal
vdias@ipca.pt

[2] CICF - Research Centre on Accounting and Taxation, Polytechnic University of Cávado and Ave, Barcelos, Portugal
{pquesado,lsilva}@ipca.pt

Abstract. The present study employed bibliometric performance analysis and science mapping techniques to explore scientific production on the relationship between management accounting and intellectual capital, as well as to map the intellectual and social structure. Furthermore, the theoretical approaches employed in the documents developed within the scope of the studies linking the two themes were analyzed. To this end, a sample of 59 publications was extracted from the Scopus and Web of Science (WoS) databases. The analysis was executed using the Biblioshiny application. The results indicate an oscillatory pattern in annual scientific production. The subject is contemporary, as more than half of the publications under analysis were published in recent years. The study further revealed that the paper "The management, measurement and the reporting of intellectual capital," by Guthrie (2001) and published in the Journal of Intellectual Capital, has been identified as the most frequently referenced work in its field. The findings indicate the heterogeneity of institutional roles in delineating the thematic scope of research, with the notable predominance of Portuguese institutions being a salient point. In terms of social structure, the data revealed the presence of five distinct clusters. The theoretical approaches identified in the studies linking the two themes included orchestration theory and the resource-based view. This study contributes to the extant research in this area, which has been under-explored. It is the first bibliometric study to relate management accounting to intellectual capital.

Keywords: bibliometric analysis · intellectual capital · management accounting · systematic literature review

1 Introduction

In the contemporary economy, characterized by the centrality of knowledge, organizations compete essentially through their intangible resources, known as intellectual capital. However, leveraging this strategic resource fully necessitates the implementation of robust management control systems and advanced management accounting practices (Al-Dhubaibi, 2024).

L. Almeida et al. (Eds.): ICAFI 2025, ISEM 80, pp. 85–104, 2026.
https://doi.org/10.1007/978-3-032-16268-7_5

The advent of the digital revolution has precipitated a substantial metamorphosis in the domain of management accounting. This evolution has transcended the confines of a mere system for reporting historical data, thereby transcending into a multifaceted instrument that encompasses the assessment of performance and the facilitation of strategic decision-making (Appelbaum et al., 2017). According to Kalogiannidis et al. (2024), its primary function is to gather, organize, and store data from various organizational subsystems. This enables the provision of quantitative, financial, and non-financial information, which facilitates decision-making and offers a comprehensive and unambiguous perspective on organizational performance (Gonidakis et al., 2020).

The strategic importance of intellectual capital is widely recognized in contemporary business discourse. Organizations are shifting their focus from tangible assets to the management of intangible assets, which are recognized as significant contributors to value creation (Mehralian et al., 2013). This has led to a marked increase in the interest of both public and private organizations, as well as academic institutions, in delving deeper into the intricacies of these assets. This heightened interest has spurred the development of models and tools for their analysis, evaluation, and comprehension. In this sense, intellectual capital is a crucial resource in the process of creating organizational value and strategic innovation (Cabrita, 2009). Therefore, the integration of management accounting and control tools with other organizational functions is imperative to optimize knowledge management and foster the development of intellectual capital (Novas et al., 2017).

In this context, the analysis of scientific production relating to the theme of intellectual capital in the context of management accounting is of pertinence, both within the academic and business contexts. The objective of this study is to facilitate a comprehensive understanding of the specified theme through a systematic review of the extant literature, identifying the main trends, the gaps, the most relevant authors and the leading journals. Our research questions are: what is the current state of research linking management accounting and intellectual capital? What are the main characteristics of the intellectual and social structure of the bibliographic sample on management accounting and intellectual capital? And what theoretical approaches are used in documents developed within the scope of studies linking the two themes? To this end, a sample of relevant documents has been extracted from the Scopus and WoS databases. By examining this sample, the current scientific production will be assessed, and the intellectual and social structure will be mapped.

This study is significant for advancing the understanding of multidisciplinary approaches that encompass intellectual capital and management accounting. Concurrently, it emphasizes the pressing imperative to prioritize research in these domains, given their pivotal function in contemporary society as instruments for generating and disseminating information that facilitates decision-making processes, thereby contributing to the advancement of scientific knowledge. The theoretical implications of this research are twofold. Firstly, it deepens the analysis of the interrelationship between intellectual capital and management accounting. Secondly, it highlights their contribution to value creation and the enhancement of organizational innovation.

According to Quesado et al. (2012), the preponderance of researchers concur that the genesis of sustainable competitive advantage lies in creating, managing, and disseminating knowledge. Consequently, organizations would be wise to prioritize the formulation of knowledge-based strategies. In light of the considerations mentioned above, it is imperative to undertake this study to ascertain the present state of scientific research and to identify prospective avenues for future research within this domain of knowledge. This undertaking is of paramount importance to both the academic and business communities.

To the best of our knowledge, this study is the first to conduct a bibliometric analysis combined with a systematic literature review focusing on the relationship between intellectual capital and management accounting, thus representing a pioneering contribution to the field.

Given the objective, this work is structured as follows: initially, we present a literature review on management accounting and intellectual capital. Then, we present the methodology that guided the investigation, and the main results obtained. Finally, we highlight in the conclusion of the work the final considerations and limitations of the study and present suggestions for future research.

2 Methodology

2.1 Research Strategies

The present study employed systematic literature review, bibliometric analysis, and science mapping protocols to investigate studies on management accounting and intellectual capital indexed in the Scopus and WoS databases. These databases are considered in the literature as the main databases for conducting the type of research proposed (Gaviria-Marin et al., 2019; Liu & Avello, 2021).

According to Siddaway et al. (2019), a systematic review constitutes a distinct literature review that confers additional benefits. This type of review employs explicit methods to identify, select, and critically appraise relevant research.

For Page et al. (2021), systematic reviews synthesize the existing knowledge base in a given field to identify future research priorities and provide insights into the mechanisms and causal factors underlying observed phenomena.

The systematic methodology is designed to minimize subjectivity and bias. The inclusion and exclusion criteria employed in the systematic review are explicitly delineated and consistently implemented (Siddaway et al., 2019).

According to Serra et al. (2018), the value of bibliometric studies lies not only in their capacity to facilitate publication but also in their ability to identify references pertinent to the subject under investigation. Moreover, these studies are instrumental in the development of syllabi and, most crucially, in elucidating the prevailing state of the art within the designated field. Consequently, bibliometrics emphasizes the measurement, dissemination, and utilization of information derived from scientific production (Bujaki & Mcconomy, 2017; Chiu et al., 2019; Saraiva et al., 2016).

Bibliometric analysis is a popular and meticulous technique for investigating and scrutinizing a large amount of bibliographic data (Donthu et al., 2021a). Researchers employ this method for a variety of purposes, including the identification of novel trends in the performance of articles and journals, the discernment of patterns of association, and the analysis of research elements. Additionally, it is utilized for the examination of the cognitive structure of a specific domain within the extant literature (Donthu et al., 2021b; Verma & Gustafsson, 2020).

It is imperative to acknowledge that bibliometric analysis is classified into two primary categories of analytical techniques: performance analysis and science mapping (Aria et al., 2020; Donthu et al., 2021a; Mukherjee et al., 2022). Performance analysis is a process that assesses productivity and impact. Science mapping is a method that identifies clusters of knowledge in a field (Mukherjee et al., 2022). Given the aforementioned points, bibliometric analysis and systematic literature review stand as valuable methods due to their capacity to minimize bias, enhance reliability, and potentially augment the effectiveness of research findings (Sharif et al., 2019; Yao et al., 2024).

Given the importance of integrating these methods to ensure a comprehensive, transparent, and replicable analysis of literature, this investigation applied bibliometric performance analysis and science mapping techniques to analyze the current state of knowledge on this topic.

The research strategies employed encompassed the PRISMA statement (Page et al., 2021), in conjunction with bibliometric analysis and science mapping techniques (Aria & Cuccurullo, 2017; Donthu et al., 2021a; Mukherjee et al., 2022).

These techniques provide a view of scientific production based on bibliometric performance analysis, namely articles, authors, countries, and affiliations. These techniques facilitate comprehension of the subject matter through the application of scientific mapping techniques, encompassing the analysis of historical evolution, and authors' collaboration networks.

The study was guided by the following research questions: what is the current state of research linking management accounting and intellectual capital? What are the main characteristics of the intellectual and social structure of the bibliographic sample on management accounting and intellectual capital? And what theoretical approaches are used in documents developed within the scope of studies linking the two themes?

2.2 Sample Selection and Workflow

To address the defined research questions, this study utilized two primary stages, as illustrated in Fig. 1. In the initial phase of the study, the PRISMA statement was implemented to ensure the research process was systematic. An identical query was used in the Scopus and WoS databases, yielding 49 publications from the former and 32 from the latter. For both databases mentioned above, a linguistic filter was applied (English), and the document type was designated as "all document types." Given the multidisciplinary nature of the discussions in the initial analysis, no filters by field of knowledge were applied. The extraction of metadata was conducted on July 25, 2024.

The current literature search was conducted in the Scopus and WoS databases because they have wide coverage with many editors, are the primary source for indexing scientific articles, and contain an incredible variety of works around the world (Jabeen et al., 2021; Liu & Avello, 2021; Mongeon & Paul-Hus, 2016). In addition, the databases have distinct identities. Scopus is recognized as the largest database of abstracts and

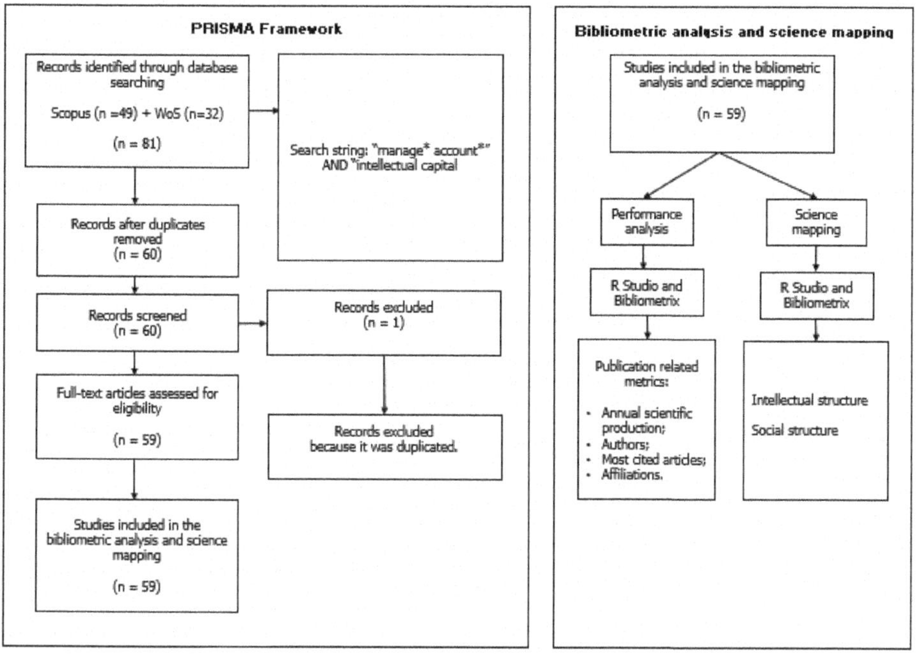

Fig. 1. Methodology workflow

citations in the peer-reviewed literature, providing reliable and high-quality coverage as well as comprehensive data for each reference (Liu & Avello, 2021). The WoS database is similarly recognized as a digital bibliometric platform with high-quality literature, providing metadata for bibliometric analysis and covering a wide range of disciplines (Gaviria-Marin et al., 2019; Liu & Avello, 2021). In addition, it is recommended to utilize both databases, as they are complementary, leveraging each other's strengths while addressing their respective weaknesses (Mongen & Paul-Hus, 2016).

The next phase involved exporting metadata from both databases in BibTeX format, followed by consolidating these records in RStudio using the Bibliometrix package. This process eliminated 21 duplicate records, reducing the sample size to 60 publications for subsequent analysis. After careful examination of the titles and abstracts, it was deemed necessary to exclude one document due to its redundancy. This resulted in a final sample of 59 publications.

Bibliometric analysis and science mapping were performed using Bibliometrix (Aria & Cuccurullo, 2017). RStudio was used for data visualization and interpretation. The Bibliometrix package, which is open source, allowed us to view information on the evolution of scientific production, as well as countries and authors, through the Biblioshiny application.

3 Results

3.1 Bibliometric Performance Analysis

Annual Scientific Production. The analysis of annual scientific output (Fig. 2) from 2001 to 2024 reveals fluctuations in academic interest in the subject of intellectual capital and management accounting. The first contributions date back to 2001, marking the initial phase of research. The year 2019 demonstrated the highest level of productivity, with seven publications. This was followed by 2022, with six. Other significant peaks occurred in 2015 and 2017, with five contributions each, and in 2009, with four publications.

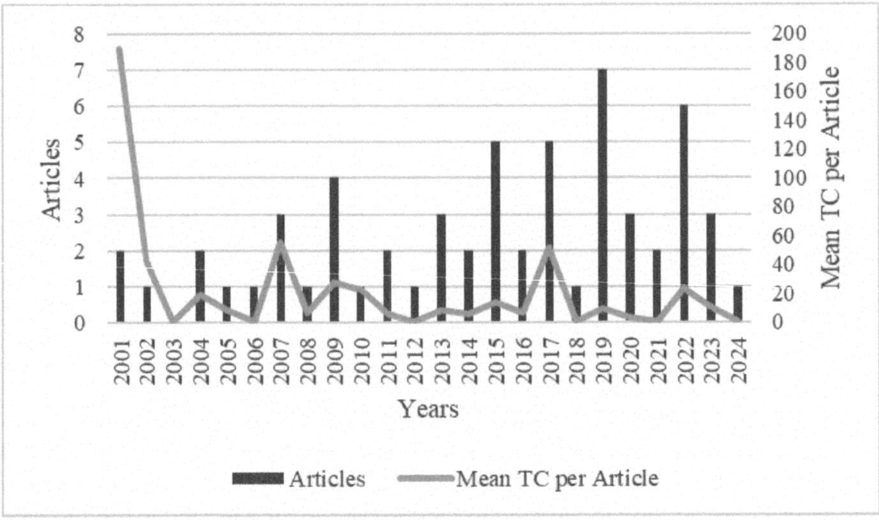

Fig. 2. Annual scientific production.

It is important to note that the periods of greatest publication coincide with key global developments, including the 2008 financial crisis, the growing importance of sustainability, and the emergence of digital technologies in the late 2010s (Asiaei et al., 2022a, 2022b; Švarc et al., 2021).

Regarding the average number of citations per article, there is variability over the period observed. The data indicates a significant increase in citations in the early 2000s, particularly in 2001 when the average number of citations per article exceeded 180. This phenomenon can be attributed to the fact that early contributions are likely to serve as references for later studies. However, in recent years, there has been a general downward trend in the average number of citations per article. The greater importance given to the study of social and environmental accounting, as well as the impact of artificial intelligence in the field of accounting, can explain this decline.

It is considered, however, that this is a global issue whose analysis is constantly evolution, influenced by advances in digital technology, the strategic importance of knowledge assets, and sustainable organizational practices (Asiaei et al., 2022a, 2022b; Švarc et al., 2021).

Authors. As illustrated in Fig. 3, the ten most prolific contributors to the academic discourse on management accounting and intellectual capital are identified. Among the authors, Kaveh Asiaei stands out as the most prolific, having authored two of the four articles published in 2022.

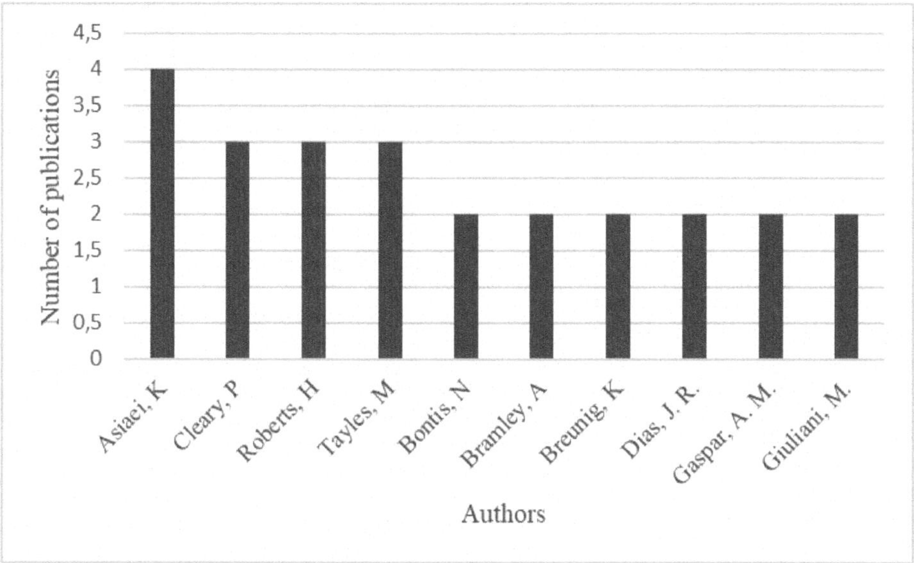

Fig. 3. Most productive authors.

The authors' contributions analyze the intersection of intellectual capital and management accounting through varied lenses, particularly emphasizing sustainability and organizational performance. These studies offer meaningful insights into how strategic management of intangible environmental resources can enhance environmental outcomes and overall organizational effectiveness (Asiaei et al., 2022a, 2022b).

Most Cited Articles. Table 1 presents a comprehensive list of articles that have received the highest number of global citations in the fields of intellectual capital and management accounting. The articles under consideration are noteworthy for their substantial impact, as evidenced by the considerable number of citations received, thereby underscoring their esteemed academic standing.

Table 1. Most cited articles.

Rank	Title	Authors (Year)	Source	Total citations	Total citations per year
1	"The management, measurement and the reporting of intellectual capital"	Guthrie (2001)	Journal of Intellectual Capital	258	10.75
2	"Intellectual capital, management accounting practices and corporate performance Perceptions of managers"	Tayles et al. (2007)	Accounting, Auditing & Accountability Journal	152	8.44
3	"Thinking critically about intellectual capital accounting"	Roslender & Fincham (2001)	Accounting, Auditing & Accountability Journal	122	5.08
4	"Knowledge management and intellectual capital in networks of small and medium-sized enterprises"	Jordão & Novas (2017)	Journal of Intellectual Capital	92	11.50
5	"Green intellectual capital and environmental management accounting: Natural resource orchestration in favor of environmental performance"	Asiaei *et al.* (2022)	Business Strategy and the Environment	86	28.67

(*continued*)

(*continued*)

Rank	Title	Authors (Year)	Source	Total citations	Total citations per year
6	"Wider still and wider? A critical discussion of intellectual capital recognition, measurement and control in a boundary theoretical context"	Gowthorpe (2009)	Critical Perspectives on Accounting	83	5.19
7	"Performance measurement, intellectual capital, and financial Sustainability "	Jordão & Almeida (2017)	Journal of Intellectual Capital	77	9.63
8	"Using a robust performance measurement system to illuminate intellectual capital"	Asiaei & Jusoh (2017)	International Journal of Accounting Information Systems	48	6.00
9	"Dealing with management of intellectual capital. The potential role of strategic management accounting"	Tayles et al. (2002)	Accounting, Auditing & Accountability Journal	43	1.87
10	"The role of management accounting systems in the development of intellectual capital"	Novas et al. (2017)	Journal of Intellectual Capital	41	5.13

A prominent article, entitled "The management, measurement and the reporting of intellectual capital," by Guthrie (2001) and published in the Journal of Intellectual Capital, has been identified as the most frequently referenced work in its field, with

258 citations (Smith, 2019). Conversely, the article entitled "Green intellectual capital and environmental management accounting: Natural resource orchestration in favor of environmental performance" by Asiaei et al. (2022a) and published in Business Strategy and the Environment is distinguished by its noteworthy annual citation rate.

Authors' Affiliation Countries' Frequency Distribution. Figure 4 illustrates the twenty countries that contribute most to research on intellectual capital and management accounting. The columns represent productivity, while the line reflects academic impact.

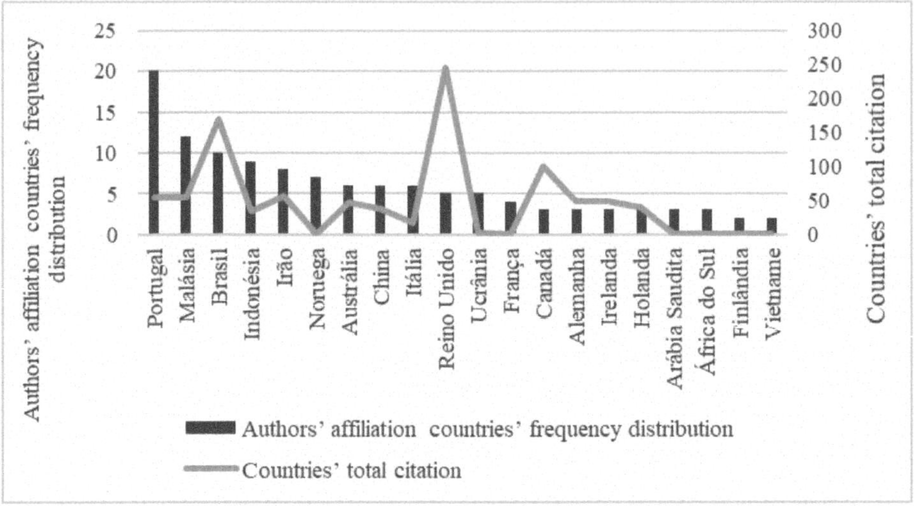

Fig. 4. Authors' affiliation countries' frequency distribution.

Portugal has the largest number of authors, with 20 affiliations and 55 citations. The United Kingdom, on the other hand, although it only has 5 affiliations, leads in citations, with a total of 246, indicating considerable influence in the literature, possibly with authors who are highly relevant in the field.

Brazil and Canada follow Portugal in terms of citations, with 170 and 100 respectively, despite having a smaller number of affiliations compared to Portugal. Countries such as Norway, Saudi Arabia and Finland, however, have no recorded citations, which may suggest that although there is a presence of researchers from these places, their work has not been widely referenced or has not yet achieved significant impact.

On the other hand, countries with a lower representation in terms of number of authors, such as the United Kingdom and Canada, show a significant impact, while some with a greater presence of researchers do not show the same relevance in terms of citations. This data can be useful for identifying trends in academic impact by country and understanding how they contribute to the production and dissemination of knowledge.

Most Relevant Affiliations. The primary affiliations of the organizations are illustrated in Fig. 5, which shows Islamic Azad University as the most prolific, with eight publications in the data set under analysis. The University of Évora has achieved second place with a total of five publications, closely followed by Oslo and Akershus University College of Applied Sciences, the University of Beira Interior, the University of Malaya, and Marche Polytechnic, which have collectively published four publications. The findings indicate the heterogeneity of institutional roles in delineating the thematic scope of research, with the notable predominance of Portuguese institutions being a salient point.

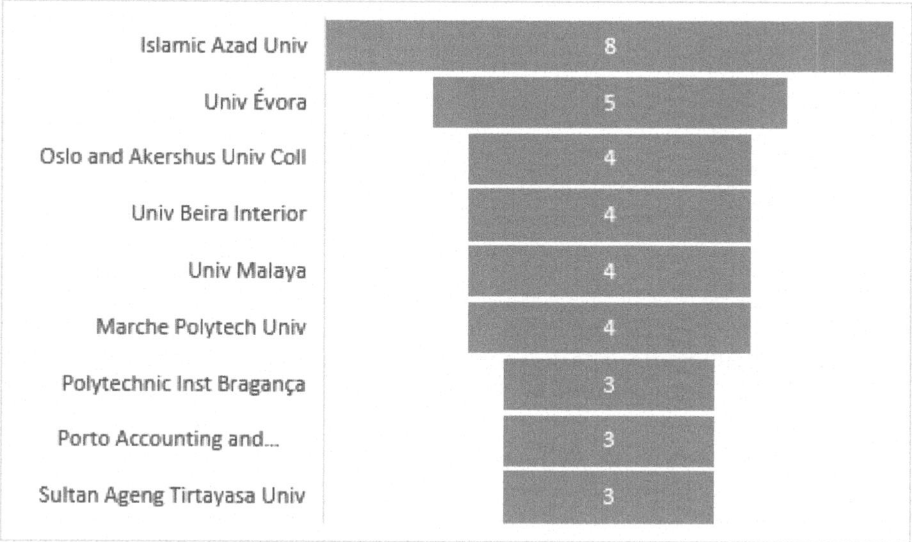

Fig. 5. Most relevant affiliations.

3.2 Science Mapping

Intellectual Structure. The historical evolution of intellectual capital and management accounting is inextricably linked to the emergence of novel concepts and methodologies over time, a phenomenon that is indicative of the shifting demands of organizations and the evolution of economic paradigms.

Figure 6 illustrates the progression of these areas, based on academic production and citation connections between various authors over the years.

The first studies on intellectual capital and management accounting, such as those by Cleary (2009) and Pires and Alves (2011), date back to the beginning of the 21st century, when organizations began to realize the importance of intellectual capital as an intangible asset. The present phase concentrated chiefly on delineating the elements of intellectual capital (human, structural, and relational), resulting in its ramifications for strategic management.

The following section will explore the concepts of expansion and new approaches. The period under consideration extends from 2015 to 2017. In the period between 2015 and 2017, we emphasize the contributions of authors such as Cleary (2015), Novas (2017), and Jordão and Almeida (2017) who pioneered novel methodologies that initiated a more direct linkage between intellectual capital and management accounting practices,

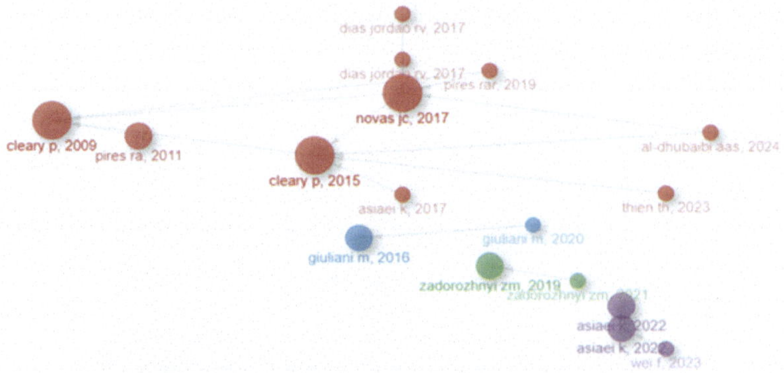

Fig. 6. Citation connections between authors.

as well as the assessment of business performance. Novas et al. (2017) conducted an analysis of the role of management accounting systems in the development of intellectual capital and the resulting effects on organizational performance. The findings substantiate the pivotal function of management accounting systems in fostering human and structural capital development.

From 2019 to 2024, the network underwent significant expansion, with a proliferation of new studies, particularly those conducted by Asiaei et al. (2022a) and Asiaei et al. (2022b). This period is characterized by a substantial advancement, marked by the integration of sustainability principles. At this stage, academic research became increasingly multidisciplinary, integrating conventional practices with technological and social innovations. As demonstrated by Asiaei et al. (2022b), the implementation of environmental performance measurement systems has been shown to effectively mobilize green intellectual capital, thereby exerting a significant influence on organizational performance. Therefore, the extent to which companies prioritize the utilization of environmental performance measurement to translate green intellectual capital into enhanced organizational performance is analyzed. Thien and Hung (2023) conducted a study to ascertain the mediating effect of strategic management accounting practices on the relationship between intellectual capital and investment efficiency. As Al-Dhubaibi (2024) asserts, the effective utilization of intellectual capital necessitates the implementation of robust management control systems and the integration of sound management accounting practices.

Social Structure. The resulting clusters in the author collaboration network illustrated in Fig. 7 were twelve. In this network, each color indicates a single cluster or group of collaborating authors. Five groups stand out, corresponding to clusters that include authors from the top ten of the most prolific (see Fig. 3). Concerning the composition of the collaboration network, cluster 1 demonstrates a propensity for addressing ecological concerns.

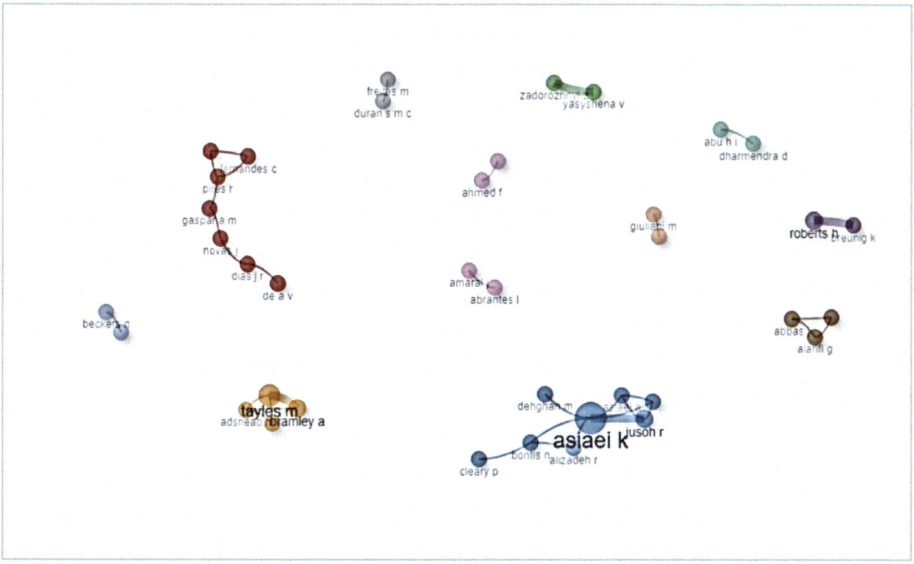

Fig. 7. Author collaboration network.

The authors of cluster 1, which includes the most prolific author, Kaven Asiaei, concentrate on intellectual capital and management accounting from various vantage points, including sustainability and performance. The authors posit that enterprises which implement environmental management accounting practices may attain a substantial strategic advantage.

Conversely, cluster 2, as delineated by Breunig and Roberts (2015), pertains to networked knowledge flows, which are considered antecedents of knowledge-based value creation.

Cluster 3 is associated with the study by Tayles et al. (2005) on the challenge of accounting for intellectual capital and intangible assets. This research pertains to the disclosure of companies' financial results, which has a consequential impact on the balance sheet and management accounting.

The subject of discussion in cluster 4 is the role of strategic management accounting practices in the management and mobilization of intellectual capital (Pires et al., 2019) and the analysis of the role of management accounting systems in the development of intellectual capital (Novas et al., 2017).

Finally, the focus of cluster 5 is on the interplay between the measurement and management of intellectual capital (Giuliani et al., 2016).

Theoretical Approaches. A review of the extant literature reveals a correlation between the two themes, as evidenced by numerous studies that relate them to sustainability. An analysis of the work of Sidik et al. (2019) reveals a reliance on the popular theory of the resource-based view, which explains the status of the company's resources and capabilities to achieve competencies. According to Hart (1995), the development of this theory is predicated on the correlation between environmental challenges and corporate resources, operationalized through the interconnection of strategic capabilities, including pollution prevention, product management, and sustainable development. The theory also analyzes that companies with greater competencies, which are not susceptible to being imitated or copied, have greater potential to improve their performance and maintain sustainable competitive advantages (Barney, 1991). Conversely, the effective utilization of green intellectual capital and environmental management accounting can contribute to enhancing a company's operational efficiency, reducing costs, and improving its image. Consequently, it can assist in gaining competitive advantages and increasing market share, thereby achieving sustainable performance (Sidik et al., 2019).

Asiaei et al. (2022a) developed a theoretical model (Fig. 8) in which the primary assumption is that the implementation of management accounting systems can facilitate the synchronization, measurement, and management of green intellectual capital elements. This, in turn, can result in enhanced environmental performance. To this end, the researchers drew inspiration from the resource-based view of the firm, the resource orchestration theory, and introduced the natural resource orchestration approach to explore the extent to which companies rely on environmental management accounting to translate green intellectual capital into better performance. According to this theory, an organization can optimize its strategic resources and capabilities (e.g., green intellectual capital and environmental management accounting) only when structured and pooled.

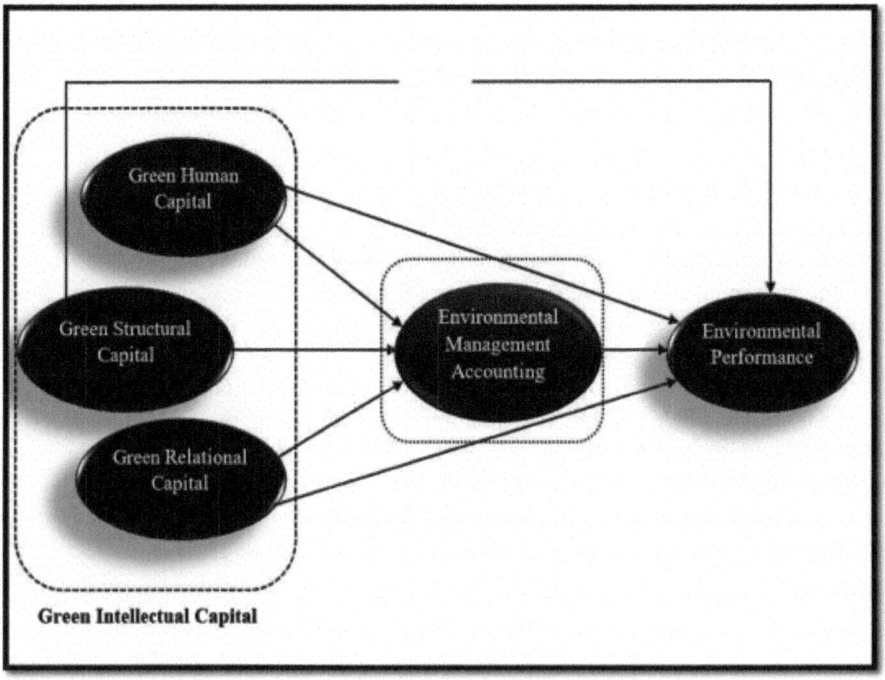

Fig. 8. Theoretical Model: Orchestration Theory.

For these authors, environmental management accounting plays a role in facilitating the implementation of sustainability initiatives by top management, promoting fundamental values, and measuring sustainability performance. It also reduces the strategic risks of sustainability. Therefore, environmental decisions made by managers will be more precise and efficient, thereby minimizing resource waste.

Within the scope of studies linking the two themes, although not related to sustainability, Al-Dhubaib (2024) study draws on the resource-based view to investigate the interaction between management control systems, management accounting practices, and intellectual capital and analyze their combined impact on organizational performance. In the knowledge-based economy, intellectual capital emerges as a critical intangible resource that encompasses knowledge, skills, organizational routines, and other capabilities that are difficult to imitate and contribute significantly to sustained performance (Al-Dhubaib, 2024). This research aligns with the resource-based view by considering intellectual capital as a key strategic resource and exploring how management control systems and management accounting practices facilitate its development and leverage. The study's conclusions align with the notion of resource bundling, as postulated by the resource-based view. This theory posits that valuable resources, such as management control systems and management accounting, complement each other, thereby creating even more valuable resources. An illustration of this phenomenon is the concept of intellectual capital.

According to Cleary et al. (2007), the concept that organizations are inherently heterogeneous in terms of their internal resources, competencies, and capabilities has historically been a central tenet of the field of strategic management. Conversely, it is posited that distinctive or superior resources, when adapted to environmental opportunities in a manner that aligns with their inherent characteristics, have the potential to engender a competitive advantage over rival companies. This assertion underscores the notion of a resource-based view of the company, emphasizing the pivotal role that resources play in shaping competitive advantage.

As stated by Hariyati et al. (2019), intellectual capital is defined as a unique resource that fosters the attainment of competitive advantage. The conclusion drawn from this analysis was that the resource-based theory elucidates how the company's internal resources influence performance, thereby ultimately augmenting the company's value.

4 Conclusion

In the contemporary business landscape, characterized by the demands and increasing intricacy of the market, coupled with a perpetually evolving global environment, management accounting, and intellectual capital have emerged as pivotal components in the decision-making process within organizations. These elements are indispensable for the present and future functioning of organizations.

In addressing the initial research question, which sought to evaluate the contemporary development of research associating management accounting and intellectual capital, the bibliometric analysis revealed an oscillatory pattern in annual scientific production, with a substantial increase observed in the years 2019 and 2022. The year 2019 witnessed the highest number of publications. The digital transformation, propelled by artificial intelligence, and the mounting concern for sustainability have exerted a significant influence in this context. A close examination of the extant literature reveals that the article "The management, measurement and the reporting of intellectual capital," by Guthrie (2001) and published in the Journal of Intellectual Capital, has been identified as the most frequently referenced work in its field. A thorough review of the literature reveals that Asiaei was the author of two of the four articles published in 2022. The analysis of these articles reveals that they address the subjects of intellectual capital and management accounting from different perspectives, namely sustainability and organizational performance. The results highlight the diverse roles played by institutions in defining the thematic focus of research, with a particularly notable predominance of Portuguese institutions.

In addressing the second research question, which sought to ascertain the primary characteristics of the intellectual and social structure, the science mapping revealed for the intellectual structure within the context of historical evolution, the study of intellectual capital and management accounting has demonstrated a discernible progression, reflecting shifts in both the organizational needs and technological capabilities. About the social structure, the authors' collaboration network was comprised of twelve clusters. A total of five clusters were identified and subsequently numbered in accordance with their author productivity, with the clusters being sorted in descending order.

Finally, regarding the third research question, which analyzed the theoretical approaches of the documents developed within the scope of the studies linking the two themes, it was found that the theoretical approaches present in the studies are the orchestration theory and the resource-based view.

The study's limitations include the restriction to only the WoS and Scopus databases, thus restricting a more comprehensive view of knowledge and trends on a given topic (Echchakoui, 2020). Furthermore, the continuous updating of databases has been shown to result in fluctuations in the number of citations and articles (Valenzuela-Fernandez et al., 2019). Given the nature of this bibliometric analysis, which focuses on an emerging topic, it is important to note that substantial changes may occur over the course of the years. A further limitation pertains to the keywords employed in the search, which may have resulted in the exclusion of pertinent studies related to the subject due to their absence from the subjects of the articles.

Notwithstanding, this study is pioneering in its undertaking of a bibliometric analysis and systematic review of the extant literature on the relationship between intellectual capital and management accounting. Consequently, the findings from this study can contribute to the advancement of knowledge in specific domains of this subject, identifying trends, gaps, most influential research, collaboration networks and opportunities for future research. On a practical level, the study can help in understanding how organizational intellectual capital can be quantified, managed and used strategically by accountants and managers.

Considering the identified lacunae, we hereby put forward a series of recommendations for future research endeavors. The initial objective is to investigate the correlation between management accounting and intellectual capital within the framework of environmental sustainability. This investigation will entail an examination of the interplay between environmental management accounting and green intellectual capital. In this context, a detailed analysis of the dimensions of green intellectual capital and the role played by environmental management accounting in each of them is recommended. Furthermore, the present study offers an opportunity to broaden the research by extending it to other databases. The study of intellectual capital in innovative digital contexts is a trend that we cannot fail to mention. This approach would yield a more comprehensive view of the subject and facilitate the integration of publications that were not considered in the present sample.

Despite the recognition of the significance of intellectual capital and its extensive ramifications within the extant literature, there is a paucity of knowledge regarding the nature of green intellectual capital and its consequences in a world where environmental concerns are of paramount importance (Asiaei et al., 2022a, 2022b).

In light of the advancements in digital technologies and artificial intelligence, future research endeavors may encompass the exploration of the role of management accounting and intellectual capital within this emerging technological landscape. Furthermore, the utilization of alternative bibliometric analysis methodologies is advised, offering novel insights into the dissemination of knowledge within this domain.

These recommendations for future research suggest promising avenues for advancing knowledge at the intersection of intellectual capital and management accounting, promoting a deeper understanding and application of these concepts in different business contexts.

Declaration of Conflicting Interests

The author(s) declared no potential conflicts of interest with respect to the research, authorship, and/or publication of this paper.

Funding. This study was developed at the Research Center on Accounting and Taxation (CICF) and was financed by the Foundation for Science and Technology (FCT) through national funds (UIDB/04043: CICF and UIDP/04043:CICF).

References

Al-Dhubaibi, A.A.S.: Unveiling the mediating effect of intellectual capital on the relationship between management control system, management accounting, and business performance. Int. J. Math. Eng. Manag. Sci. **9**(4) (2024)

Appelbaum, D., Kogan, A., Vasarhelyi, M., Yan, Z.: Impact of business analytics and enterprise systems on managerial accounting. Int. J. Account. Inf. Syst. **25**, 29–44 (2017)

Aria, M., Cuccurullo, C.: Bibliometrix: an R-tool for comprehensive science mapping analysis. J. Informet. **11**(4), 959–975 (2017)

Aria, M., Misuraca, M., Spano, M.: Mapping the evolution of social research and data science on 30 years of social indicators research. Soc. Indic. Res. **149**(3), 803–831 (2020)

Asiaei, K., Bontis, N., Alizadeh, R., Yaghoubi, M.: Green intellectual capital and environmental management accounting: natural resource orchestration in favor of environmental performance. Bus. Strateg. Environ. **31**(1), 76–93 (2022a)

Asiaei, K., Jusoh, R.: Using a robust performance measurement system to illuminate intellectual capital. Int. J. Account. Inf. Syst. **26**, 1–19 (2017)

Asiaei, K., Jusoh, R., Barani, O., Asiaei, A.: How does green intellectual capital boost performance? The mediating role of environmental performance measurement systems. Bus. Strateg. Environ. **31**(4), 1587–1606 (2022b)

Barney, J.: Firm resources and sustained competitive advantage. J. Manag. **17**(1), 99–120 (1991)
Breunig, K.J., Roberts, H.: Money talks: the role of communication in monetizing knowledge value. In: Proceedings of the 10th International Forum on Knowledge Asset Dynamics: Culture, Innovation and Entrepreneurship: Connecting the Knowledge Dots, pp. 502–513. IKAM-INST knowledge asset management (2015)
Bujaki, M., Mcconomy, B.: Productivity in Top-10: academic accounting journals by researchers at canadian universities at the Start of the 21st century. Account. Perspect. **16**(4), 269–313 (2017)
Cabrita, M.R.: Capital intelectual e desempenho organizacional. Lidel Editora (2009)
Chiu, V., Liu, Q., Muehlmann, B., Baldwin, A.: A bibliometric analysis of accounting information systems journals and their emerging technologies contribution. Int. J. Account. Inf. Syst. **32**, 24–43 (2019)
Cleary, P., Kennedy, T., O'Donnell, D., O'Regan, P., Bontis, N.: Positioning management accounting on the intellectual capital agenda. Int. J. Account. Audit. Perform. Eval. **4**(4–5), 336–359 (2007)
Cleary, P.: Exploring the relationship between management accounting and structural capital in a knowledge-intensive sector. J. Intellect. Cap. **10**(1), 37–52 (2009)
Cleary, P.: An empirical investigation of the impact of management accounting on structural capital and business performance. J. Intellect. Cap. **16**(3), 566–586 (2015)
Donthu, N., Kumar, S., Mukherjee, D., Pandey, N., Lim, W.M.: How to conduct a bibliometric analysis: an overview and guidelines. J. Bus. Res. **133**, 285–296 (2021a)
Donthu, N., Kumar, S., Pandey, N., Lim, W.M.: Research constituents, intellectual structure, and collaboration patterns in Journal of International Marketing: an analytical retrospective. J. Int. Mark. **29**(2), 1–25 (2021b)
Echchakoui, S.: Why and how to merge Scopus and Web of Science during bibliometric analysis: the case of sales force literature from 1912 to 2019. J. Mark. Anal. **8**, 165–184 (2020)
Gaviria-Marin, M., Merigó, J.M., Baier-Fuentes, H.: Knowledge management: a global examination based on bibliometric analysis. Technol. Forecast. Soc. Chang. **140**, 194–220 (2019)
Giuliani, M., Chiucchi, M.S., Marasca, S.: A history of intellectual capital measurements: from production to consumption. J. Intellect. Cap. **17**(3), 590–606 (2016)
Gonidakis, F.K., Koutoupis, A.G., Tsamis, A.D., Agoraki, M.E.K.: Risk disclosure in listed Greek companies: the effects of the financial crisis. Account. Res. J. **33**(4/5), 615–633 (2020)
Guthrie, J.: The management, measurement and the reporting of intellectual capital. J. Intellect. Cap. **2**(1), 27–41 (2001)
Hart, S.L.: A natural-resource-based view of the firm. Acad. Manag. Rev. **20**(4), 986–1014 (1995)
Hariyati, H., Tjahjadi, B., Soewarno, N.: The mediating effect of intellectual capital, management accounting information systems, internal process performance, and customer performance. Int. J. Product. Perform. Manag. **68**(7), 1250–1271 (2019)
Jabeen, S., Malik, S., Khan, S., Khan, N., Qureshi, M.I., Saad, M.S.M.: A comparative systematic literature review and bibliometric analysis on sustainability of renewable energy sources. Int. J. Energy Econ. Policy **11**(1), 270–280 (2021)
Jordão, R.V.D., Almeida, V.R.D.: Performance measurement, intellectual capital and financial sustainability. J. Intellect. Cap. **18**(3), 643–666 (2017)
Jordão, R.V.D., Novas, J.C.: Knowledge management and intellectual capital in networks of small- and medium-sized enterprises. J. Intellect. Cap. **18**(3), 667–692 (2017)
Kalogiannidis, S., Kontsas, S., Kalfas, D., Chatzitheodoridis, F.: Operational risk management in managerial accounting: a comprehensive examination of strategies and implementation in medium size organizations. Oper. Res. Int. J. **24**(3), 44 (2024)
Liu, Y., Avello, M.: Status of the research in fitness apps: a bibliometric analysis. Telemat. Inform. **57**, 101506 (2021)

Mehralian, G., Rasekh, H.R., Akhavan, P., Ghatari, A.R.: Prioritization of intellectual capital indicators in knowledge-based industries: evidence from, pharmaceutical industry. Int. J. Inf. Manag. **33**(1), 209–216 (2013)
Mongeon, P., Paul-Hus, A.: The journal coverage of web of science and Scopus: a comparative analysis. Scientometrics **106**, 213–228 (2016)
Mukherjee, D., Lim, W.M., Kumar, S., Donthu, N.: Guidelines for advancing theory and practice through bibliometric research. J. Bus. Res. **148**, 101–115 (2022)
Novas, J., Alves, M.C., Sousa, A.: The role of management accounting systems in the development of intellectual capital. J. Intellect. Cap. **18**(2), 286–315 (2017)
Page, M.J., et al.: The PRISMA 2020 statement: an updated guideline for reporting systematic reviews. BMJ **372**(71), 1–9 (2021)
Pires, R.A., Alves, M.C.: Contributions to the measurement and management of intellectual capital: an accounting perspective. In: Proceedings of the 5th European Conference on Information Management and Evaluation, pp. 411–419. Academic Publishing Limited (2011)
Pires, R.A., Alves, M.C., Fernandes, C.: Management and mobilization of intellectual capital using strategic management accounting practices: a theoretical approach. In: 34th International Business Information Management Association Conference, pp. 13914–13924. International Business Information Management Association (2019)
Quesado, P., Aibar-Guzmán, B., Rodrigues, L.: El Cuadro de Mando Integral como Herramienta de Gestión Estratégica del Conocimiento. Revista Perspectivas em Gestão Conhecimento **2**(1), 70–102 (2012)
Roslender, R., Fincham, R.: Thinking critically about intellectual capital accounting. Account. Auditing Account. J. **14**(4), 383–399 (2001)
Saraiva, A., Silva, C., Conceição, I., Lopes, W.: Uma Análise Bibliométrica da Produção Científica sobre Contabilidade. Revista Univap **22**(40), (2016)
Serra, F., Ferreira, M., Guerrazzi, L., Scaciotta, V.: Doing bibliometric reviews for the iberoamerican journal of strategic management. Iberoamerican J. Strateg. Manag. **17**(3), 1–16 (2018)
Sharif, S.P, Mura, P., Wijesinghe, S.N.R.: Systematic reviews in Asia: introducing the "PRISMA" protocol to tourism and hospitality scholars. Quant. Tourism Res. Asia: Curr. Status Futur. Directions 13–33 (2019)
Siddaway, A.P., Wood, A.M., Hedges, L.V.: How to do a systematic review: a best practice guide for conducting and reporting narrative reviews, meta-analyses, and meta-syntheses. Annu. Rev. Psychol. **70**(1), 747–770 (2019)
Sidik, M.H.J., Yadiati, W., Lee, H., Khalid, N.: The dynamic association of energy, environmental management accounting and green intellectual capital with corporate environmental performance and competitive. Int. J. Energy Econ. Policy **9**(5), 379–386 (2019)
Švarc, J., Lažnjak, J., Dabić, M.: The role of national intellectual capital in the digital transformation of EU countries. Another digital divide?. J. Intell. Cap. **22**(4), 768–791 (2021)
Tayles, M., Bramley, A., Adshead, N., Farr, J.: Dealing with the management of intellectual capital: the potential role of strategic management accounting. Account. Auditing Account. J. **15**(2), 251–267 (2002)
Tayles, M., Webster, M., Sugden, D., Bramley, A.: Accounting "gets real" in dealing with virtual manufacturing. J. Intellect. Cap. **6**(3), 322–338 (2005)
Tayles, M., Pike, R.H., Sofian, S.: Intellectual capital, management accounting practices and corporate performance: perceptions of managers. Account. Auditing Account. J. **20**(4), 522–548 (2007)
Thien, T.H., Hung, N.X.: Intellectual capital and investment efficiency: the mediating role of strategic management accounting practices. Cogent Bus. Manag. **10**(2), 2207879 (2023)
Valenzuela-Fernandez, L., Merigó, J.M., Lichtenthal, J.D., Nicolas, C.: A bibliometric analysis of the first 25 years of the Journal of Business-to-Business Marketing. J. Bus. Bus. Mark. **26**(1), 75–94 (2019)

Verma, S., Gustafsson, A.: Investigating the emerging COVID-19 research trends in the field of business and management: a bibliometric analysis approach. J. Bus. Res. **118**, 253–261 (2020)

Yao, X., Xu, Z., Wang, X., Wang, L., Škare, M.: Energy efficiency and COVID-19: a systematic literature review and bibliometric analysis on economic effects. Technol. Econ. Dev. Econ. **30**(1), 287–311 (2024)

Zadorozhnyi, Z.M., Yasyshena, V., Muravskyi, V., Ometsinska, I.: Evolution of accounting for intangible assets and its transformation to international requirements. Financ. Credit Activity: Probl. Theory Pract. **5**(40), 191–203 (2021)

Motivation and Goal Congruence: Evidence in the Portuguese Insurance Sector

Isabel Martins[1(✉)], Maria José Carvalho da Silva[1], Jonas Oliveira[2], and Graça Azevedo[1,3]

[1] Higher Institute for Accounting and Administration of the University of Aveiro, Aveiro, Portugal
icom@ua.pt

[2] University Institute of Lisbon (ISCTE-IUL), Av. das Forças Armadas, 1649-026 Lisboa, Portugal

[3] Research Centre on Accounting and Taxation of the University of Aveiro, Aveiro, Portugal

Abstract. This article investigates the motivational factors that promote goal congruence in commercial structures of the insurance sector, presenting an analysis based on the theory of agency and self-determination.

Based on interviews conducted with commercial managers of seven of the fifteen largest insurance companies operating in Portugal in 2020, the results reveal that 1) information asymmetry and difficulty in aligning interests represent the main agency problems found in commercial insurance structures; 2) the financial reward system, supported by Governance, Risk and *Compliance* mechanisms, proves to be the main way to promote the extrinsic motivation of the commercial and mitigate agency problems; 3) both the financial reward system and the quality of the relationship are forms of incentive, material and immaterial, that increase motivation and contribute to the congruence of goals; and 4) the balanced combination of extrinsic and intrinsic motivational factors seems to be the most effective approach to align goals and interests in this type of commercial structures.

The results also suggest a potential synergy between agency theory and self-determination theory. By proposing an integrated approach that combines economic and motivational fundamentals, this research contributes to deepening the knowledge about the factors that promote the congruence of commercial goals, offering a relevant perspective for a sector little explored so far – the Insurer.

Keywords: Motivation · congruence of objectives · agency theory · self-determination theory · insurance sector

1 Introduction

Commercial motivation can be evaluated by the level of excellence of the respective functions, the pleasure in performing the tasks and the consequent impact on the strategic alignment and results of the company. Motivation is part of the essence of the management of commercial organizations and represents one of the main concerns of the executives who manage these activities (Khusainova et al., 2018; Mallin et al., 2017;

L. Almeida et al. (Eds.): ICAFI 2025, ISEM 80, pp. 105–127, 2026.
https://doi.org/10.1007/978-3-032-16268-7_6

Pullins, 2001). For this reason, it is essential to investigate the factors that influence it, among which rewards stand out, for their ability to promote productivity, retain talent and align strategic objectives with business success.

In the insurance sector, this topic is of special importance due to the social impact of this activity, which not only contributes to mitigating inequalities (Schanz, 2020) and encouraging savings in the context of increased longevity (Aguiar, 2020), but also because of the scarcity of research in the area. Portugal was selected for this study due to its institutional characteristics, which allow for the exploration of the relationships between motivation, rewards, and commercial structure. This market has been characterized by high concentration, with 66 insurance companies active in 2020, of which 38 operated under the direct supervision of the ASF and 28 as branches of the European Union (ASF, 2021; ECO Seguros, 2021).

In fact, the impact of financial rewards on the alignment of interests and on the performance of professionals has been widely investigated (Bhuyan et al., 2022; Menguc & Tansu Barker, 2003; Muthusi, 2016). However, the excessive dependence on this control mechanism has been criticized, due to the latent lack of interest in other relevant motivational factors (Bouillon et al., 2006; Gagné & Forest, 2008). These divergences arise in the context of literature, with authors who defend the motivational power of money (Bonner & Sprinkle, 2002), while others highlight factors that transcend financial compensation (Pullins, 2001). The exclusive use of monetary rewards by disregarding employees' inner comfort levels can even prove counterproductive (Chakrabarty, 2020).

In the face of doubt and lack of consensus on the matter, it is essential to adopt innovative approaches that combine financial and non-financial rewards and work *design* factors that promote the self-determination of professionals (Khusainova et al., 2018).

In this context, this research aims to explore the combination of Agency Theory (TA) and SDT (Self-Determination Theory) constructs. Its main objective is to identify the motivational factors that promote goal congruence in commercial structures in the insurance sector, posing the following research questions:

- What are the main agency conflicts identified in the insurance business structure?
- What mechanisms are used by Insurers to mitigate such conflicts?
- To what extent does the Financial Rewards System (FRS) contribute to reducing agency conflicts and promoting objective congruence?
- What motivational factors are decisive for achieving objective congruence in commercial structures in the insurance sector?

2 Methodology

For the research, an exploratory approach was adopted, using a qualitative methodology involving semi-structured interviews carried out with Directors and Commercial Coordinators of seven companies, positioned among the fifteen largest Insurance Companies and Groups operating in Portugal in 2020, responsible for 69.2% of the total production of direct insurance in the country (ECO Seguros, 2021). This method was adopted because it allows a deeper analysis of the aspects that determine the motivation of the professionals who are part of the commercial structure of this type of institution (e.g., Delpechitre et al., 2020).

The internal commercial structure of insurers is defined by the Board of Directors. It is based on governance criteria that require a clear definition of the hierarchical chain of responsibilities to observe adequate segregation of functions at the level of the entire organizational structure, risk management, and internal control. Based on the business strategy, each Insurance Company defines its internal commercial structure, usually represented by Customer and Agency Assistants, Network Managers/Commercial Technicians, Managers, Coordinators, and Directors.

The sample consisted of seven male participants, with a mean age of 51 years, responsible for the direction/coordination of the underlying commercial structure, whose dimension can be seen in Table 1. In addition to being leaders of the internal commercial structure, these professionals are characterized by bridging the gap between land vendors and the company's management, understanding the commercial structure, knowing the strengths and weaknesses of teams and the organization, and mastering internal communication processes. Aspects that contribute to the enrichment of information and development of the theme in an organizational context.

Table 1. Sample Characterization

Underlying Business Structure	D1	D2	D3	D4	C1	C2	C3
No. of Customer Assistants	13	9	8	14	9	9	8
No. of Sales Managers/Sales Technicians	12	7	7	18	6	8	5
No. of Managers	4	3	3	5	2	2	2
No. of Coordinators	2	1	1	2	0	0	0
Total staff in internal commercial structure	**31**	**20**	**19**	**39**	**17**	**19**	**15**
Professional experience (No.years)	11	9	15	8	12	14	15
No. of Agents/Brokers assisted	530	390	400	600	320	380	310

Note. Created by the author

For legal reasons related to the protection of participants' data and in compliance with the Code of Ethics and Conduct, the identification of the Insurers or the interviewees was not authorized.

The interviews were conducted by a single interviewer, transcribed in full and subjected to a systematic content analysis, following the methodology of Bardin (2016), which involved repeated reading, coding, and categorization. The coding process was iterative and refined through comparison between interviews to ensure consistency and reliability of emerging themes. Selected quotes from participants were used to illustrate the main findings and provide evidence to support the analysis. In this context, each interviewee received an identifier code consisting of a letter representing the position (D = Commercial Director; C = Commercial Coordinator) and a sequential number. This coding process allowed for systematic organization and facilitated comparison between participants, enabling structured processing of information, with the development of an initial content analysis grid to answer the first research question, evolving through the incorporation of indicators linked to empirical data and the Literature Review, culminating in the creation of four analysis grids. Due to the pandemic situation experienced during the investigation, four interviews were conducted by videoconference; a method pointed out by researchers as ideal for the co-creation of knowledge, methodological advancement, and greater relevance of content (Fischer et al., 2017; VanKooten, 2019).

To answer the research questions, we crossed two theories, until then never applied simultaneously to deepen this field of study: TA (Eisenhardt, 1985; Jensen & Meckling, 1976) and SDT (Deci & Ryan, 1985; Ryan & Deci, 2017). Despite the existence of studies based on both AT, which suggest financial rewards as a mechanism to mitigate agency conflicts and promote congruence of objectives (Bhuyan et al., 2022; Saïd Echchakoui & Ghilal, 2019; Eisenhardt, 1985), and in the SDT, which advocate the development of self-determination for the alignment of interests (Aryee et al., 2015; Gagné & Deci, 2005; Güntert, 2014), few studies link the two theories (e.g., Dotter & Johnson, 2020) and no studies have been identified in the insurance sector.

Despite the challenges that integrating theories can bring (Khusainova et al., 2018), we align ourselves with Stathakopoulos' vision (1996) by considering that the joint analysis of different theories can enrich the results by transcending the individual limitations of each one.

3 Literature Review

3.1 Reward System, Motivation, and Goal Congruence

For decades, organizations have used rewards to attract and retain their sales force, ensure the success of their bottom line, and mitigate conflicts of interest, often associated with the nature of their roles (e.g. Abratt & Smythe, 1989; Chonko et al., 1992; Gerhart & Milkovich, 1990; Sprinkle, 2000).

Reward systems are part of the management control systems package (Malmi & Brown, 2008). They are understood as a set of coherent instruments aligned with the organization's strategy, translated into material and immaterial incentives, which the company grants its employees to increase levels of motivation and productivity (Lawler III, 1989; Madeira e Picarelli, 1996).

These incentives correspond to the traditional distinction between intrinsic and extrinsic rewards (Flamholtz et al., 1985; Herzberg, 1987; Mottaz, 1985). Intrinsic rewards derive from the nature of the task, involving challenge, creativity, and *feedback*, while extrinsic rewards refer to financial or tangible compensations, such as salaries and promotions, aimed at improving performance and reducing dissatisfaction (Mottaz, 1985; Flamholtz et al., 1985; Herzberg, 1987).

Several studies suggest financial rewards as a way to achieve goal congruence (Bhuyan et al., 2022; Bonner e Sprinkle, 2002; Menguc e Tansu Barker, 2003; Riedel et al., 1988) and see this type of reward as the main motivational factor of a commercial structure (Abratt & Smythe, 1989; Baker et al., 1998; Chonko et al., 1992; Coughlan e Narasimhan, 1992; Riedel et al., 1988; Sprinkle, 2000). However, some consider them potentially counterproductive and capable of affecting intrinsic motivation (Deci et al., 1999; Gagné & Deci, 2005; Mallin & Pullins, 2009), especially if they are previously hired, perceived as a form of control or when the work itself is already considered attractive (Deci et al., 1999).

Deci et al. (1999) advocate using these rewards in a non-contractual way, adding that if extrinsic rewards are attributed regardless of commitment to a certain level of performance or task, they can be interpreted as a form of recognition and positive *feedback*, increasing intrinsic motivation. The literature also warns of the controversial effects of using goals or budgets to set monetary rewards for performance, emphasizing that this is a form of value destruction and a potential source of dysfunctional behavior (Jensen, 2003; Frey & Osterloh, 2005). Other studies suggest that extrinsic rewards and intrinsic motivation may be complementary (Cerasoli et al., 2014), that performance incentive contracts have variable effects depending on their design (Feichter & Grabner, 2020), and that reward systems must be based on criteria of justice and equity to promote both motivations (Ogunnaike et al., 2014).

In the face of uncertainty and transformations in the business paradigm, driven by digitalization (Feichter & Grabner, 2020; Van den Berg et al., 2020), by redefining the role of sales force management (Mallin et al., 2017) and by the growing appreciation of human resource management (Peccei & Van De Voorde, 2019), It is evident that traditional studies, focused on financial compensation, are insufficient to understand the factors that promote motivation and objective congruence in today's commercial organizations. Thus, it is crucial to consider psychological and behavioral variables that integrate the vision of the commercial manager as a professional and individual.

3.2 Theory of Agency and Theory of Self-determination in Congruence of Objectives

3.2.1 The Agency Theory Perspective

The agency relationship is a contract in which the principal delegates tasks to the agent, giving him or her decision-making authority (Jensen & Meckling, 1976). Agency problems arise when there is a lack of alignment of interests, limited rationality, or information asymmetry, causing costs and opportunistic behaviors such as adverse selection and moral hazard (Eisenhardt, 1985; Broadbent et al., 1996; Kaplan & Atkinson, 1998). These problems are based on aversion to effort, diversion of resources, different time priorities, and risk aversion on the agent's part (Lambert, 2001).

The TA focuses on determining the contract that allows mitigating these conflicts, recommending the implementation of monitoring systems and incentive contracts (Anthony & Govindarajan, 2007; Eisenhardt, 1989; Menguc & Tansu Barker, 2003).

However, resolving agency problems is not limited to implementing monitoring and reward mechanisms, requiring a balance between remuneration levels and the risk transferred to the agent, achieved through structures and systems aligned with the organization's interests (Jensen & Murphy, 1990). In addition, employees must perceive these incentives as fair; otherwise, their motivational potential will be null or even counterproductive (Anthony & Govindarajan, 2007; Aryee et al., 2015). It is also important to link individual objectives to those of the organization and ensure managers' participation in defining them since companies with better performance tend to adopt mechanisms that include negotiation and cooperation with managers (Muthusi, 2016).

AT is seen by several authors as the most relevant and used theory in corporate governance (Bosse & Phillips, 2016; Colina & Crombie, 2010; Solomon & Solomon, 2007; Sundaram & Inkpen, 2004). Several studies on agency conflicts show the contribution of AT to the analysis of aspects such as:

- Conflicts inherent to the separation of ownership and control, where moral hazard and adverse selection stand out (Hill &Crombie, 2010; Marshall, 2009);
- Information asymmetry and difficulty in monitoring (Chen et al., 2007);
- Implementation of measures to control and mitigate the agency's problems, promoting the integration of the manager in the company's strategy (Bhuyan et al., 2022; Hill & Crombie, 2010); and
- Adjustment or implementation of control and compensation mechanisms (Bhuyan et al., 2022; Bouillon et al., 2006; Marshall, 2009; Menguc & Tansu Barker, 2003), considering the resources needed to monitor agent roles (Menguc & Tansu Barker, 2003).

3.2.2 The Self-determination Theory Perspective

SDT studies human motivation by identifying the psychosocial factors influencing intrinsically motivated behavior (Deci & Ryan, 1985; Ryan & Deci, 2000). Pointed out by several authors as a useful model for understanding motivational complexity in organizations, SDT presents a multidimensional approach (Chakrabarty, 2020; Gagné et al., 2020; Gagné & Forest, 2008; Khusainova et al., 2018). According to this theory, three fundamental psychological needs – autonomy, competence, and relationship – promote the individual's psychological growth and well-being, facilitating self-determination, performance, and creativity (Deci & Ryan, 1985). Autonomy reflects the management of one's own decisions in harmony with personal values; competence is related to effectiveness in the face of challenges; and the relationship is linked to the feeling of belonging to the group or organization, driving development (Deci & Ryan, 1985; Ryan & Deci, 2000, 2006).

SDT considers that intrinsic and extrinsic motivation form a *continuum,* mediated by a process of internalization that includes introjection and integration of external rules, both types of motivation being determinant for action and performance (Deci & Ryan, 1985; Ryan & Deci, 2000, 2017; Dotter & Johnson, 2020; Mallin et al., 2017; Miao et al., 2007; Mallin & Pullins, 2009). However, SDT prioritizes the quality of motivation, considering intrinsic motivation as the highest (Ryan & Deci, 2000), especially when associated with high levels of autonomy, competence, and pleasure in the task, which favor the internalization of relevant goals and promote psychological well-being (Ryan & Deci, 2000, 2017).

Several studies highlight intrinsic factors linked to motivation and business performance, such as equity, trust in the organization, connection to the brand, and satisfaction of basic psychological needs (Aryee et al., 2015; Güntert, 2014; Mallin et al., 2017; Khusainova et al., 2018). They also point to the influence of intrinsic rewards, autonomy, and other factors in reducing turnover and promoting goal congruence in different organizational contexts (Pullins, 2001; Mallin & Pullins, 2009; Güntert, 2014; Hohenberg & Homburg, 2016; Kuvaas et al., 2016; Chakrabarty, 2020).

However, in contrast to AT, which focuses on extrinsic motivation and the use of incentives as external stimuli (Ryan & Deci, 2017), Criticism arises from the assumption that the employee is not able to internalize the employer's objectives, resorting to coercive methods, such as linking salary to performance. In addition, several studies indicate that financial incentives do not guarantee the alignment of interests or improve organizational performance (Bouillon et al., 2006; Frey & Osterloh, 2005; Gagné & Forest, 2008).

Still, the practical application of SDT highlights the importance of remuneration systems that consider extrinsic and intrinsic motivation, promoting intrapreneurship and addressing the challenges of the commercial sector (Pullins, 2001; Chakrabarty, 2020). Despite the advances, studies suggest the need to investigate motivation in commercial organizations more deeply and develop new theoretical and practical contributions to assist leaders and companies in emerging challenges (Khusainova et al., 2018).

3.2.3 Contribution of Theories to the Research Gap

Despite the existence of academic studies based on TA (Bhuyan et al., 2022; Kren e Tyson, 2009; Panda & Leepsa, 2017) and in the SDT (Hohenberg & Homburg, 2016; Mallin et al., 2017; Mallin e Ragland, 2017; Shulga & Busser, 2020), few cross both theories to evaluate new contributions to autonomous motivation for work and goal congruence, with emphasis on the Insurance sector, where no contribution was identified.

Nevertheless, some authors, perceiving the controversy based on empirical evidence about the negative impact that financial rewards can have on intrinsic motivation and the potential of SDT in understanding organizational behavior and management, have sought to investigate the topic that is not explored.Kunz and Pfaff (2002) assessed the hidden costs of rewards and the need to incorporate intrinsic motivation into AT. They concluded that these costs exist due to information asymmetry and risk aversion, distancing the agent from cooperative behavior. However, the possible negative effects of rewards on intrinsic motivation do not contradict the principles of AT. Gagné and Forest (2008) point out that SDT has been applied mainly to understand organizational behavior despite its great potential in analyzing processes and consolidating results. In response to Lawler's criticism (1999), which pointed out the lack of clear alternatives to incentive pay, the authors developed a conceptual model that analyzes how compensation systems influence autonomous and controlled motivation. They also argue that studies on compensation are excessively focused on AT, neglecting aspects such as the mental health of employees, which can be explored through motivation theories. Dotter and Johnson (2020) suggest combining AT and SDT can offer valuable insights into factors that encourage the adoption of technology in professional sales environments.

4 Results

4.1 Agency Conflicts Found in Commercial Insurance Structures

As can be seen in Table 2, when asked about the main conflicts that hinder the alignment of interests in insurance commercial structures, the interviewees highlighted problems such as a) information asymmetry, b) difficulties in aligning interests, c) risk aversion, and d) different time priorities.

Table 2. Content Analysis Grid 1 - Agency Conflicts in the Insurance Commercial Structure

QUOTE FROM THE INTERVIEWEE	IDENTIFIED CONFLICT	LITERATURE REVIEW	D1	D2	D3	D4	C1	C2	C3	TOTAL	%
"Because of the pressure to meet objectives, close deals and win commercial campaigns" (D1); "Lack of rigor in the analysis of operations, which can lead to the attraction of high-risk customers" (D2); and "The problems are many and start from the beginning with the selection of people, the lack of aptitude to perform commercial functions. Often, the option for these functions arises, on the one hand, due to circumstances associated with the extinction of other functions and, on the other hand, combined with some attractiveness for their remuneration" (D3)	Information Asymmetry Risk of adverse selection	Arrow (1985); Broadbent et al. (1996); Chen et al. (2007); Lambert (2001); and Marshall (2009)	√	√	√	√	√	√	√	7	100

(*continued*)

Table 2. (*continued*)

QUOTE FROM THE INTERVIEWEE	IDENTIFIED CONFLICT	LITERATURE REVIEW	D1	D2	D3	D4	C1	C2	C3	TOTAL	%
Less ethical behavior "due to excess ambition or necessity" (D1); "Lack of business commitment, individualism and facilitation" (D2); and "Less ethical behavior, often associated with and aligned with the hierarchies that directly or indirectly determine it, due to the pressure to meet objectives. Of course, this can lead, among other corrosive aspects, to the attraction of risky customers to the Insurer" (D3)	Information Asymmetry Moral hazard	Arrow (1985); Broadbent et al. (1996); Lambert (2001); and Marshall (2009)	√	√	√	√	√	√	√	7	100

(*continued*)

Table 2. (*continued*)

QUOTE FROM THE INTERVIEWEE	IDENTIFIED CONFLICT	LITERATURE REVIEW	D1	D2	D3	D4	C1	C2	C3	TOTAL	%
"(…) the strategic objectives of the Insurer do not always coincide with the objectives of the Commercial Managers, who need to sell product A more, in order to fulfill their personal objectives, when the Insurer wants to sell product B" (D1); "There is some competition and opposition of interests between different departments of the same insurer. I am referring to the conflicts of interest between the commercial area and the claims area" (D1); e "(…) the objectives are not always consistent with the macroeconomic reality" (C3)	Difficulty aligning interests/goals	Bhuyan et al. (2022); Bouillon et al. (2006); Broadbent et al. (1996); and Jensen e Meckling (1976)	√	√	√		√	√	√	6	86

(*continued*)

Table 2. *(continued)*

QUOTE FROM THE INTERVIEWEE	IDENTIFIED CONFLICT	LITERATURE REVIEW	D1	D2	D3	D4	C1	C2	C3	TOTAL	%
"Seek immediate results, just to guarantee their performance" (D2); e "(...) when, for example, the salesperson knows that he has reached the minimum necessary for his performance and does not try to go further, which harms the team's overall results" (C3)	Risk aversion	Kunz e Pfaf (2002); and Lambert (2001)		√					√	2	29
"Lack of commercial vision in the medium and long term" (D2); and "The time and strategic objectives of the Insurer do not always coincide with the objectives of the Commercial Department" (D1)	Different time priorities	Lambert (2001)	√	√					√	3	43
---	Difficulty in monitoring									0	0

Note. Created by the author, based on the results of the interviews. *4.1.1 Information asymmetry*

4.1.1 Information Asymmetry

Respondents unanimously highlighted information asymmetry as the main obstacle to aligning interests in the insurance business structure. The principal's dependence on the information provided by the Agent can lead to opportunistic behavior and inappropriate strategic decisions, such as ignoring ordering criteria in the sale or making mistakes in the selection of candidates for the commercial role.

These results are consistent with the literature, which highlights information asymmetry as one of the main challenges in decentralized structures (Arrow, 1985; Broadbent et al., 1996; Lambert, 2001), which can give rise to opportunistic behaviour, described as adverse selection risk and moral hazard (Broadbent et al., 1996; Lambert, 2001; Marshall, 2009).

4.1.2 Difficulty in Aligning Interests

Participants (86%) identified the difficulty in aligning interests as one of the main problems of the agency, justifying it by the high number of objectives defined by the company, conflicts of interest between departments, and the lack of congruence between the interests of the commercial sector and the organization.

This result confirms the literature (Bhuyan et al., 2022; Bouillon et al., 2006; Broadbent et al., 1996; Chen et al., 2007; Jensen & Meckling, 1976), evidencing how divergences of interests influence organizational performance.

4.1.3 Risk Aversion

Problems inherent to risk aversion or effort were also identified by 29% of respondents. This analysis corroborates the studies of Kunz and Pfaff (2002) and Lambert (2001) when they suggest that agency conflicts are based on assumptions such as risk aversion or effort, estimating that the agent will only strive to achieve the objectives if he obtains some personal reward, moving away from cooperative behavior.

4.1.4 Different Time Priorities

Among the respondents, 43% highlighted problems related to time priorities, highlighting discrepancies between the strategic objectives of insurers and commercial ones and the absence of a medium and long-term commercial vision. These results corroborate Lambert (2001), who associates these divergences with conflicts of interest, demotivation, and, in extreme cases, the departure of collaborators.

As for the difficulty of monitoring, none of the interviewees considered it a problem, highlighting that the control systems existing in the companies are sufficient. However, all identified behaviors are incompatible with the organization's values, especially regarding the pressure to meet objectives.

The analysis highlights problems arising from information asymmetry, namely moral hazard and adverse selection (Arrow, 1985; Broadbent et al., 1996; Lambert, 2001; Marshall, 2009). In addition, the lack of alignment of interests emerges as a critical factor in the origin of agency conflicts, directly affecting organizational performance (Bhuyan et al., 2022; Bouillon et al., 2006). In addition, challenges such as risk aversion and divergences in temporal priorities amplify the complexity of conflicts and require effective mitigation mechanisms (Kunz & Pfaff, 2002; Lambert, 2001).

4.2 Mechanisms Used by Insurers to Mitigate Agency Conflicts

Regarding the mechanisms used by Insurers to mitigate agency conflicts (Table 3), it was possible to observe a clear response trend, supported by the existence of Governance, Risk and *Compliance* mechanisms, FRS, and requirements transversal to all Insurers, linked to market conduct (Lei n.º 147/2015, de 09 de Setembro, da Assembleia da República).

Table 3. Content Analysis Grid 2 - Insurers' Mechanisms to Mitigate Agency Conflicts

MECHANISMS	% INCIDENCE
Financial Rewards System (FRS)	100
Code of Ethics and Conduct (CEC)	
Delegation of Powers Manual (DPM)	
Internal Control (IC) and Internal Auditory (IA)	
Internal Regulations (IR) and *Compliance*	

Note. Created by the author

In this context, all interviewees confirmed the implementation of Governance, Risk, and Compliance mechanisms to ensure transparency, corporate responsibility, and ethics, as well as the existence of an FRS integrated into the management control system, structured as a comprehensive package that interacts with other organizational systems and practices (Malmi & Brown, 2008).

4.3 The Reward System and Goal Congruence

Respondents were unanimous in electing the FRS as crucial to mitigating agency problems and promoting alignment of interests (Table 4).

However, they underline the importance of considering the FRS with the Delegation of Powers Manual and with the internal control and internal audit procedures. These results are consistent with AT, which defends control mechanisms and incentive contracts to align interests between agent and principal, both considered egocentric beings with an interest in the process (Eisenhardt, 1985, 1988; Jensen e Meckling, 1976).

Table 4. Content Analysis Grid 3 - Reward System as a Mechanism for Mitigating Agency Conflicts

MECHANISMS	D 1	D 2	D 3	D 4	C1	C2	C3
Panel A: classification of choices 1st choice of respondents 2nd choice of respondents 3rd choice of respondents	FRS DPM IC-IA	FRS IR/*CP* CEC	FRS DPM IC-IA	FRS DPM IC-IA	FRS DPM IC-IA	FRS DPM IR/*CP*	FRS DPM CEC
% INCIDENCE							

Panel B: relative weight of each choice	
FRS	100
DPM	86
IC-IA	57

Note. Created by the author. Classification criteria: selection of three mechanisms, in descending order of importance in the opinion of each interviewee (FRS: Financial Reward System; DPM: Delegation of Powers Manual; IC/IA: Internal Control and Internal Audit; IR/Cp: Internal Regulations and *Compliance*; CEC: Code of Ethics and Conduct)

4.4 Motivational Factors and Goal Congruence

The weighted motivational factors fundamental to the congruence of the objectives are presented in Table 5.

4.4.1 Financial/non-Financial Rewards that Determine Goal Congruence

FRS, as well as the quality of the relationship, stand out as forms of material and immaterial reward that promote motivation and goal congruence. 86% of respondents identified career progression and autonomy as elements that foster trust and accountability. Investment in skills and challenges in projects were mentioned by 71% of participants, while 43% highlighted the impact of work-life balance on motivation and performance.

This analysis highlights the relationship between rewards, motivation, and goal congruence and is consistent with the literature, which classifies rewards as intrinsic and extrinsic (Chiang & Birtch, 2006; Flamholtz, 1983; Herzberg, 1987; Mottaz, 1985). While F represents forms of extrinsic rewards (Flamholtz, 1983; Frey & Osterloh, 2005; Gagné & Deci, 2005; Mallin et al., 2017), Investing in factors such as autonomy, competence, and work-life balance is associated with intrinsic rewards (Ryan & Deci, 2000, 2006). The results also highlight the importance of balancing compensation systems and intrinsic motivation mediating variables to promote goal congruence.

Table 5. Content Analysis Grid 4 - Motivational Factors that determine Objective Congruence

QUESTIONS	SUMMARY - RESPONDENTS' RESPONSES	RESULT
Financial and non-financial rewards that determine motivation, performance, and goal congruence	FRS and Relationship: 100%; Autonomy and Career Progression: 85.7%; Investment in Competence and Challenge in the Project: 71.4%; Work-life balance: 42.9%	Goal congruence requires a balance between extrinsic (FRS: 100%) and intrinsic (relationship quality: 100%) rewards
Motivational factors that can lead the Commercial to leave the company	1st Demotivation (Competence/Relationship Failure); 2nd Remuneration; 3rd Project/Career/Autonomy Challenge: 57.1% 1st Remuneration; 2nd Project/Career/Autonomy Challenge; 3rd demotivation (competence/relationship failure): 42.9%	Low levels of intrinsic motivation enhance the output of the commercial to projects considered more attractive
The importance of autonomy for commercial function	1. 100% consider autonomy fundamental for the proper exercise of commercial functions (increases motivation and responsiveness in a timely manner) 2. 100% consider that the current delegation of powers is not sufficient for the full mastery and management of the function	1. The basic psychological need for autonomy is fundamental for the motivation of the salesperson and for the superior performance of his function. 2. Attention is drawn to the need for greater investment in trade autonomy
The importance of integrating Commercial into the company's strategy	1. 100% consider it important to include the Commercial in the process of negotiating the objectives (it allows a fairer, more realistic and reliable budget in the eyes of the Commercial and increases their levels of motivation and commitment) 2. 100% consider that it is a process to be improved	1. The basic psychological need for relationships is fundamental for the motivation of the salesperson and the good performance of his function. 2. Attention is drawn to the need for greater integration of Commercial in the company's strategy

(continued)

Table 5. *(continued)*

QUESTIONS	SUMMARY - RESPONDENTS' RESPONSES	RESULT
The importance of innovation and creativity in the competence of the Commercial	The nature of the commercial function implies the development of these skills, however, the *modus operandi* does not allow them to be enhanced: 57.1%; Both the function and the modus operandi promote these skills, however, they can be improved in the presence of greater autonomy and integration of the Commercial in the company's strategy: 28.6%	1. The basic psychological need for competence is important for business motivation and professional performance. However, there are external factors that can compromise the satisfaction of this need
Motivational factors that determine the success of the commercial function in favor of the company's objectives	Pleasure in the task, autonomy, competence and relationship: 1st choice (85%); Identifying with management style and work environment/Being well paid: 2nd option (28.6% ex-aequo); Being recognized by superiors and having the possibility of promotion: 3rd option (42.9%)	Pleasure in the task and the satisfaction of basic psychological needs are the most determining motivational factors for the success of the commercial function in favor of the company's objectives
Suggestions from business leaders to increase motivation for goal congruence	Investing in factors that meet basic psychological needs and rewards that encourage intrinsic motivation is a priority (100%); Invest in criteria that align the interests between the commercial sector and the organization, representing (57.1%); Concern about the new challenges in the Insurance sector (28.6%); Expansion of non-financial rewards (14.3%)	1. There is a latent need to invest in the intrinsic motivation of Salespeople. 2. The constructs of both theories allow for a more comprehensive view of how to achieve objective congruence in decentralized commercial organizations

Note. Created by the author

4.4.2 Motivational Factors that Can Lead the Salesperson to Leave the Company

The analysis of Table 5 reveals that 57% of the interviewees point to the lack of identity related to the type of governance, work environment, absence of a relationship between colleagues and superiors, the feeling of stagnation and exclusion from the company's strategy as the main reasons for the departure of the salespeople. Then, there are factors such as remuneration, project challenge, career progression, and increased autonomy.

In addition, the need for competence and relationships was also associated with demotivation, while autonomy emerged as a determining factor in the search for new experiences. These data reinforce the importance of investing in elements that foster extrinsic and intrinsic motivation, ensuring a balance between both. Otherwise, professionals will look for other projects that are considered more attractive.

These results are consistent with the view of Balachandar et al. (2010) and Muthusi (2016), who consider motivation and congruence of objectives as the main concerns of executives linked to the insurance sector.

4.4.3 The Importance of Autonomy for Business Function

The interviewees agree that the current delegation of powers in the commercial structure is insufficient to ensure the integral management of the business, highlighting the need to review this issue, as they consider autonomy essential for the good performance of commercial functions, positively impacting motivation and responsiveness.

This analysis indicates that a salesperson with greater autonomy will adjust their behavior and effort to the company's needs, strengthening the feeling of competence and commitment to results. These results align with the literature, which points out that the employees who add the most value are those who self-manage, perform complex tasks, coordinate their work, and suggest improvements, which require skills, resources, and autonomy (Gagné & Forest, 2008). In addition, supporting leadership's autonomy and understanding of strategy favors performance, job satisfaction, resilience, and acceptance of organizational strategy (Gagne et al., 2000; Güntert, 2014).

4.4.4 The Importance of Integrating Commercials into the Company's Strategy

All interviewees consider it important to include the commercial element in the negotiation process, remembering that, first, the strategy must be communicated, listening to and perceiving the reality of each one, and only then defining the objectives in the incentive plans.

This result suggests the importance of integrating the commercial in the communication process, the implementation of the strategy, and the negotiation of the objectives to be defined. In this sense, the literature highlights the inclusion of the employee in the management of remuneration structures as a form of non-financial reward aimed at increasing the employee's intrinsic motivation and, thus, enabling the achievement of goals (Muthusi, 2016; Aryee et al., 2015). In this domain, the AT argues that the risk of opportunistic behavior can decrease as stakeholders are included in the strategy and/or decision-making process (Bhuyan et al., 2022; Hill & Crombie, 2010).

4.4.5 The Importance of Innovation and Creativity in Business Competence

On this issue, 57% of respondents say that the nature of the commercial function requires the development of innovation and creativity. Although there is no unanimity, everyone highlights the importance of these factors for performance.

The analysis indicates a clear relationship between innovation, creativity, and autonomy, demonstrating that work environments that do not promote autonomy harm these aspects. Studies reinforce that an environment that sustains autonomy improves the satisfaction of basic psychological needs and boosts intrinsic motivation (Chong et al., 2020). More intrinsically motivated employees offer creative and innovative solutions (Gagné & Forest, 2008).

In addition, the results align with the SDT, which suggests that people have motivational tendencies aimed at growth, challenge, achievement, assimilation of knowledge, and development of new skills (Ryan & Deci, 2000).

4.4.6 Motivational Factors that Determine the Success of the Commercial Function in Favor of the Company's Objectives

Most respondents (85%) pointed to pleasure in the task, autonomy, competence, and relationship as the main motivational factors, followed by goals set fairly and reliably. Commercial leaders did not consider the feeling of security at work relevant.

This analysis reinforces the previous conclusions, highlighting the importance of intrinsic motivation for objective congruence in the commercial structures of the insurance sector.

The results are compatible with the study of the relationship between TSS and AT. The SDT considers compensation criteria as a way to stimulate extrinsic motivation, while factors such as trust, equity, and satisfaction of basic psychological needs promote intrinsic motivation, performance, and goal congruence (Chakrabarty, 2020; Deci et al., 1999; Gagné et al., 2020; Gagné & Deci, 2005; Gagné & Forest, 2008).

On the other hand, AT emphasizes financial incentives as a way to mitigate agency conflicts, strengthen satisfaction, increase productivity, and promote goal congruence (Anthony & Govindarajan, 2007; Bhuyan et al., 2022; Echchakoui, 2017; Echchakoui & Ghilal, 2019; Menguc & Tansu Barker, 2003; Panda & Leepsa, 2017; Sprinkle, 2000).

Despite the different approaches, both theories balance several types of motivation, favoring performance and aligning interests.

5 Conclusions

The present study aimed to identify the motivational factors that promote objective congruence in commercial structures in the insurance sector. To this end, two distinct but complementary theoretical approaches were explored by recognizing the possibility of alignment of interests: AT, which emphasizes extrinsic motivation through mechanisms such as FRS, and SDT, which favors intrinsic motivation, promoting the self-determination of sales representatives so that the alignment of interests occurs naturally in the organizational management process.

The results demonstrate that the main agency conflicts include information asymmetry, difficulties in aligning interests, risk aversion, and differences in temporal priorities. Information asymmetry continues to be one of the sector's biggest challenges, associated with moral hazard and adverse selection (Arrow, 1985; Broadbent et al., 1996; Lambert, 2001; Marshall, 2009). However, the lack of alignment of interests contributes to the intensification of these problems, highlighting the need for strategies that minimize the impacts of risk aversion and temporal divergences (Lambert, 2001).

Insurers use specific mechanisms to mitigate agency conflicts, highlighting the role of the FRS as a control tool supported by governance, risk, and *compliance systems,* as well as cross-cutting market conduct requirements.

The FRS proved to be fundamental in reducing agency conflicts and promoting the congruence of objectives. It reinforced its importance as a strategic instrument to ensure the alignment of interests between commercial and organization.

Regarding the motivational factors that favor goal congruence, FRS, the careful definition of objectives, the adaptation of compensation structures, when necessary, and investment in satisfying basic psychological needs stand out. These factors are shown to promote more effective alignment, ensuring that objectives are perceived as fair and achievable (Bouillon et al., 2006; Hill & Crombie, 2010; Aryee et al., 2015), reducing conflicts and boosting performance.

The research reinforces the role of AT in organizational management, positioning FRS as an essential mechanism to increase satisfaction and productivity (Anthony & Govindarajan, 2007; Menguc & Tansu Barker, 2003). On the other hand, SDT complements this perspective by highlighting intrinsic motivation, allowing salespeople to internalize objectives as valuable and relevant, promoting psychological well-being and organizational commitment (Gagné et al., 2020; Gagné & Deci, 2005; Gagné & Forest, 2008; Slemp et al., 2021).

Although distinct, both approaches converge in strengthening motivation, performance, and congruence with objectives in the insurance sector.

Recognizing the limitations of the study, such as the sample size and the specificity of the sector, is the development of new investigations that deepen the intersection of these theories, explore other sectors, and contribute to the evolution of reward and governance models, expanding knowledge about commercial motivation and strategic alignment.

References

Abratt, R., Smythe, M.R.: A survey of sales incentive programs. Ind. Mark. Manage. **18**(3), 209–214 (1989). https://doi.org/10.1016/0019-8501(89)90037-0

de Aguiar, M.C.: O papel da atividade seguradora na recuperação pós pandemia. [Paper apresentado]. 1st Annual ECOSeguros Conference, Portugal. https://www.asf.com.pt/NR/rdonlyres/5B2FE493-CE16-469C-BDE4-5714D87D0BC9/0/IntervençãodaPresidentedaASF1ConferênciaECOseguros27112020.pdf. Accessed 23–27 Nov 2020

Anthony, R., Govindarajan, V.: Management Control Systems (12th ed.). McGraw-Hill/Irwin (2007)

Arrow, K.J.: The economics of agency. In: Pratt, J.W., Zeckhauser, R.J. (eds.) Principals and agents: The structure of business, pp. 37–51. Harvard Business School Press (1985)

Aryee, S., Walumbwa, F.O., Mondejar, R., Chu, C.W.L.: Accounting for the influence of overall justice on job performance: Integrating self-determination and social exchange theories. J. Manage. Stud. **52**(2), 231–252 (2015). https://doi.org/10.1111/joms.12067

Assembleia da República. Lei n.º 147/2015, de 09 de setembro. Diário da República, 1st series, no. 176 (2015). https://dre.pt/application/file/a/70236403

Autoridade de Supervisão de Seguros e Fundos de Pensões.. Prémios de Seguro Direto da Atividade Seguradora-2020 (valores provisórios). In PSDAS (2021). www.asf.com.pt

Baker, G.P., Jensen, M.C., Murphy, K.J.: Compensation and incentives: practice vs. theory. SSRN Electr. J. **43**(3), 593–616 (1998). https://doi.org/10.2139/ssrn.94029

Balachandar, G., Panchanatham, N., Subramanian, K.: Impact of job situation on the motivation of insurance companies officers: a developmental perspective. Int. J. Trade Econ. Financ. **1**(4), 349–353 (2010)

Bardin, L.: Análise de conteúdo (7th ed.). Edições 70 (2016)
Bhuyan, R., Butchey, D., Haar, J., Talukdar, B.: CEO compensation and firm performance in the insurance industry. Manag. Financ. **48**(7), 1086–1115 (2022). https://doi.org/10.1108/MF-04-2019-0154
Bonner, S.E., Sprinkle, G.B.: The effects of monetary incentives on effort and task performance: theories, evidence, and a framework for research. Acc. Organ. Soc. **27**(4–5), 303–345 (2002). https://doi.org/10.1016/S0361-3682(01)00052-6
Bosse, D.A., Phillips, R.A.: Agency theory and bounded self-interest. Acad. Manag. Rev. **41**(2), 276–297 (2016). https://doi.org/10.5465/amr.2013.0420
Bouillon, M.L., Ferrier, G.D., Jr., M.T.S., West, T.D.: The economic benefit of goal congruence and implications for management control systems. J. Account. Publ. Policy **25**(3), 265–298 (2006). https://doi.org/10.1016/j.jaccpubpol.2006.03.003
Broadbent, J., Dietrich, M., Laughlin, R.: The development of principal-agent, contracting and accountability relationships in the public sector: conceptual and cultural problems. Crit. Perspect. Account. **7**(3), 259–284 (1996). https://doi.org/10.1006/cpac.1996.0033
Cerasoli, C.P., Nicklin, J.M., Ford, M.T.: Intrinsic motivation and extrinsic incentives jointly predict performance: a 40-year meta-analysis. Psychol. Bull. **140**(4), 980–1008 (2014). https://doi.org/10.1037/a0035661
Chakrabarty, S.: Intrapreneurship in teams/groups: self-determination theory and compensation. J. Small Bus. Enterp. Dev. (2020). https://doi.org/10.1108/JSBED-10-2019-0344
Chen, X., Yao, T., Yu, T.: Prudent man or agency problem? On the performance of insurance mutual funds. J. Finan. Intermediation **16**(2), 175–203 (2007). https://doi.org/10.1016/j.jfi.2006.10.002
Chiang, F.F.T., Birtch, T.A.: An empirical examination of reward preferences within and across national settings. Manag. Int. Rev. **46**(5), 573–596 (2006). https://doi.org/10.1007/s11575-006-0116-4
Chong, J.X.Y., Beenen, G., Gagné, M., Dunlop, P.D.: Satisfying newcomers' needs: the role of socialization tactics and supervisor autonomy support. J. Bus. Psychol. **36**(2), 315–331 (2020). https://doi.org/10.1007/s10869-019-09678-z
Chonko, L.B., Tanner, J.F., Jr., Weeks, W.A.: Selling and sales management in action: Reward preferences of salespeople. J. Pers. Selling Sales Manag. **12**(3), 67–75 (1992). https://doi.org/10.1080/08853134.1992.10753918
Coughlan, A.T., Narasimhan, C.: An empirical analysis of sales-force compensation plans. J. Bus. **65**(1), 93 (1992). https://doi.org/10.1086/296559
Deci, E.L., Koestner, R., Ryan, R.M.: A meta-analytic review of experiments examining the effects of extrinsic rewards on intrinsic motivation. Psychol. Bull. **125**(6), 627–668 (1999). https://doi.org/10.1037/0033-2909.125.6.627
Deci, E.L., Ryan, R.M.: Intrinsic Motivation and Self-Determination in Human Behavior. Plenum Press, New York (1985)
Delpechitre, D., Gupta, A., Zadeh, A.H., Lim, J.H., Taylor, S.A.: Toward a new perspective on salesperson success and motivation: a trifocal framework. J. Pers. Selling Sales Manag. **40**(4), 267–288 (2020). https://doi.org/10.1080/08853134.2020.1805748
Dotter, T.A., Johnson, V.L.: Interactive visual aid adoption within pharmaceutical sales organizations. Comput. Hum. Behav. Rep. **2**, 100017 (2020). https://doi.org/10.1016/j.chbr.2020.100017
Echchakoui, S.: Sales force commissions in relationship marketing. J. Model. Manag. **12**(1), 53–76 (2017). https://doi.org/10.1108/JM2-02-2015-0003
Echchakoui, S., Ghilal, R.: How sales managers can use salespeople's perceived attributes to monitor and motivate a sales force during relationship marketing. Eur. Res. Manag. Bus. Econ. **25**(2), 99–104 (2019). https://doi.org/10.1016/j.iedeen.2019.02.001

ECO Seguros. Seguros 2020: Fidelidade reforça liderança em mercado recessivo. https://eco.sapo.pt/2021/01/19/seguros-2020-fidelidade-reforca-lideranca-em-mercado-recessivo/. Accessed 9 Jan 2021

Eisenhardt, K.M.: Control: Organizational and economic approaches. Manage. Sci. **31**(2), 134–149 (1985)

Eisenhardt, K.M.: Agency- and institutional-theory explanations: the case of retail sales compensation. Acad. Manag. J. **31**(3), 488–511 (1988). https://doi.org/10.5465/256457

Eisenhardt, K.M.: Agency theory: an assessment and review. Acad. Manag. Rev. **14**(1), 57 (1989). https://doi.org/10.2307/258191

European Insurance and Occupational Pensions Authority. Insurance stress test 2021 (2021). https://www.eiopa.europa.eu/insurance-stress-test-2021_en

Feichter, C., Grabner, I.: Empirische forschung zu management control – ein überblick und neue trends. Schmalenbachs Zeitschrift Für Betriebswirtschaftliche Forschung **72**(2), 149–181 (2020). https://doi.org/10.1007/s41471-020-00092-3

Fischer, A.J., Collier-Meek, M.A., Bloomfield, B., Erchul, W.P., Gresham, F.M.: A comparison of problem identification interviews conducted face-to-face and via videoconferencing using the consultation analysis record. J. Sch. Psychol. **63**, 63–76 (2017). https://doi.org/10.1016/j.jsp.2017.03.009

Flamholtz, E.G.: Accounting, budgeting and control systems in their organizational context: theoretical and empirical perspectives. Acc. Organ. Soc. **8**(2–3), 153–169 (1983). https://doi.org/10.1016/0361-3682(83)90023-5

Flamholtz, E.G., Das, T.K., Tsui, A.S.: Toward an integrative framework of organizational control. Acc. Organ. Soc. **10**(1), 35–50 (1985). https://doi.org/10.1016/0361-3682(85)90030-3

Frey, B.S., Osterloh, M.: Yes, managers should be paid like bureaucrats. J. Manag. Inq. **14**(1), 96–111 (2005). https://doi.org/10.1177/1056492604273757

Gagné, M., Deci, E.L.: Self-determination theory and work motivation. J. Organ. Behav. **26**(4), 331–362 (2005). https://doi.org/10.1002/job.322

Gagné, M., Forest, J.: The study of compensation systems through the lens of self-determination theory: Reconciling 35 years of debate. Can. Psychol. **49**(3), 225–232 (2008). https://doi.org/10.1037/a0012757

Gagne, M., Koestner, R., Zuckerman, M.: Facilitating acceptance of organizational change: the importance of self-determination. J. Appl. Soc. Psychol. **30**(9), 1843–1852 (2000). https://doi.org/10.1111/j.1559-1816.2000.tb02471.x

Gagné, M., Morin, A.J.S., Schabram, K., Wang, Z.N., Chemolli, E., Briand, M.: Uncovering relations between leadership perceptions and motivation under different organizational contexts: a multilevel cross-lagged analysis. J. Bus. Psychol. **35**(6), 713–732 (2020). https://doi.org/10.1007/s10869-019-09649-4

Gerhart, B., Milkovich, G.T.: Organizational differences in managerial compensation and financial performance. Acad. Manag. J. **33**(4), 663–691 (1990). https://doi.org/10.2307/256286

Güntert, S.T.: The impact of work design, autonomy support, and strategy on employee outcomes: a differentiated perspective on self-determination at work. Motiv. Emot. **39**(1), 74–87 (2014). https://doi.org/10.1007/s11031-014-9412-7

Herzberg, F.: One more time: how do you motivate employees?. Harv. Bus. Rev. **65**, 109–120 (1987)

Hill, C., Crombie, N.: Accountability to stakeholders in a student-managed organisation. University of Canterbury. Department of Accounting and Information Systems (2010)

Hohenberg, S., Homburg, C.: Motivating sales reps for innovation selling in different cultures. J. Mark. **80**(2), 101–120 (2016). https://doi.org/10.1509/jm.14.0398

Jensen, M.C.: Paying people to lie: The truth about the budgeting process. Eur. Financ. Manag. **9**(3), 379–406 (2003). https://doi.org/10.1111/1468-036X.00226

Jensen, M.C., Meckling, W.H.: Theory of the firm: managerial behavior, agency costs and ownership structure. J. Financ. Econ. **3**(4), 305–360 (1976). https://doi.org/10.1016/0304-405X(76)90026-X

Jensen, M.C., Murphy, K.J.: Performance pay and top-management incentives. J. Polit. Econ. **98**(2), 225–264 (1990)

Kaplan, R., Atkinson, A.: Advanced Management Accounting. Prentice Hall, Hoboken (1998)

Khusainova, R., de Jong, A., Lee, N., Marshall, G.W., Rudd, J.M.: (Re) defining salesperson motivation: current status, main challenges, and research directions. J. Pers. Selling Sales Manag. **38**(1), 2–29 (2018). https://doi.org/10.1080/08853134.2017.1415761

Kren, L., Tyson, T.: Trade-offs in objective and subjective performance evaluation: a case study examining the validity of agency theory predictions. Manag. Account. Q. **10**(2), 12–23 (2009)

Kunz, A.H., Pfaff, D.: Agency theory, performance evaluation, and the hypothetical construct of intrinsic motivation. Acc. Organ. Soc. **27**(3), 275–295 (2002). https://doi.org/10.1016/S0361-3682(01)00031-9

Kuvaas, B., Buch, R., Gagné, M., Dysvik, A., Forest, J.: Do you get what you pay for? Sales incentives and implications for motivation and changes in turnover intention and work effort. Motiv. Emot. **40**(5), 667–680 (2016). https://doi.org/10.1007/s11031-016-9574-6

Lambert, R.A.: Contracting theory and accounting. SSRN Electr. J. (2001). https://doi.org/10.2139/ssrn.235800

Lawler, E.E.: Rewarding Excellence: Pay Strategies for the New Economy. Pfeiffer (1999)

Lawler, E.E., III.: Pay for performance: making it work. Compens. Benefits Rev. **21**(1), 55–60 (1989). https://doi.org/10.1177/088636878902100108

Mallin, M.L., Gammoh, B.S., Pullins, E.B., Johnson, C.M.: A new perspective of salesperson motivation and salesforce outcomes: the mediating role of salesperson-brand identification. J. Mark. Theory Pract. **25**(4), 357–374 (2017). https://doi.org/10.1080/10696679.2017.1345597

Mallin, M.L., Pullins, E.B.: The moderating effect of control systems on the relationship between commission and salesperson intrinsic motivation in a customer oriented environment. Ind. Mark. Manage. **38**(7), 769–777 (2009). https://doi.org/10.1016/j.indmarman.2008.03.004

Mallin, M.L., Ragland, C.B.: Power-base effects on salesperson motivation and performance: a contingency view. J. Bus. Bus. Mark. **24**(2), 99–121 (2017). https://doi.org/10.1080/1051712X.2017.1313671

Malmi, T., Brown, D.A.: Management control systems as a package: opportunities, challenges and research directions. Manag. Account. Res. **19**(4), 287–300 (2008). https://doi.org/10.1016/j.mar.2008.09.003

Marshall, J.M.: Review of Guillaume Plantin and Jean-Charles Rochet, when insurers go bust: an economic analysis of the role and design of prudential regulation. Int. J. Econ. Bus. **16**(1), 139–146 (2009). https://doi.org/10.1080/13571510802638999

Menguc, B., Tansu Barker, A.: The performance effects of outcome-based incentive pay plans on sales organizations: a contextual analysis. J. Pers. Selling Sales Manag. **23**(4), 341–358 (2003). https://doi.org/10.1080/08853134.2003.10749008

Miao, C.F., Evans, K.R., Shaoming, Z.: The role of salesperson motivation in sales control systems — intrinsic and extrinsic motivation revisited. J. Bus. Res. **60**(5), 417–425 (2007). https://doi.org/10.1016/j.jbusres.2006.12.005

Mottaz, C.J.: The relative importance of intrinsic and extrinsic rewards as determinants of work satisfaction. Sociol. Q. **26**(3), 365–385 (1985). https://doi.org/10.1111/j.1533-8525.1985.tb00233.x

Muthusi, N.C.: The relationship between total compensation and employee performance in the insurance industry: case of Mayfair insurance company limited. Psychol. Behav. Sci. **5**(1), 20 (2016). https://doi.org/10.11648/j.pbs.20160501.14

Ogunnaike, O.O., Akinbola, O.A., Ojo, O.A.: Effect of motivation on job satisfaction of selected sales representatives. J. Educ. Soc. Res. **4**(1) (2014). https://doi.org/10.5901/jesr.2014.v4n1p197

Panda, B., Leepsa, N.M.: Agency theory: review of theory and evidence on problems and perspectives. Indian J. Corp. Governance **10**(1), 74–95 (2017). https://doi.org/10.1177/0974686217701467

Peccei, R., Van De Voorde, K.: Human resource management–well-being–performance research revisited: past, present, and future. Hum. Resour. Manag. J. **29**(4), 539–563 (2019). https://doi.org/10.1111/1748-8583.12254

Pullins, E.B.: An exploratory investigation of the relationship of sales force compensation and intrinsic motivation. Ind. Mark. Manage. **30**(5), 403–413 (2001). https://doi.org/10.1016/S0019-8501(99)00095-4

Riedel, J.A., Nebeker, D.M., Cooper, B.L.: The influence of monetary incentives on goal choice, goal commitment, and task performance. Organ. Behav. Hum. Decis. Process. **42**(2), 155–180 (1988). https://doi.org/10.1016/0749-5978(88)90010-6

Ryan, R.M., Deci, E.L.: Intrinsic and extrinsic motivations: classic definitions and new directions. Contemp. Educ. Psychol. **25**(1), 54–67 (2000). https://doi.org/10.1006/ceps.1999.1020

Ryan, R.M., Deci, E.L.: Self-regulation and the problem of human autonomy: does psychology need choice, self-determination, and will? J. Pers. **74**(6), 1557–1586 (2006). https://doi.org/10.1111/j.1467-6494.2006.00420.x

Ryan, R.M., Deci, E.L.: Self-Determination Theory: Basic Psychological Needs in Motivation, Development, and Wellness. The Guilford Press, New York (2017)

Ryan, R.M., Deci, E.L.: Intrinsic and extrinsic motivation from a self-determination theory perspective: definitions, theory, practices, and future directions. Contemp. Educ. Psychol. **61**, 101860 (2020). https://doi.org/10.1016/j.cedpsych.2020.101860

Schanz, K.-U.: The role of insurance in mitigating social inequality. The Geneva Association - International Association for the Study of Insurance Economics, Zurich. https://www.genevaassociation.org/research-topics/insurance-mitigating-social-inequality-research-brief. Accessed 31 Aug 2020

Shulga, L.V., Busser, J.A.: Customer self-determination in value co-creation. J. Serv. Theory Pract. **31**(1), 83–111 (2020). https://doi.org/10.1108/JSTP-05-2020-0093

Slemp, G.R., Lee, M.A., Mossman, L.H.: Interventions to support autonomy, competence, and relatedness needs in organizations: a systematic review with recommendations for research and practice. J. Occup. Organ. Psychol., joop.12338 (2021). https://doi.org/10.1111/joop.12338

Solomon, J., Solomon, A.: Corporate Governance and Accountability, 2nd edn. John Wiley & Sons Ltd., Hoboken (2007)

Sprinkle, G.B.: The effect of incentive contracts on learning and performance. Account. Rev. **75**(3), 299–326 (2000). https://doi.org/10.2308/accr.2000.75.3.299

Stathakopoulos, V.: Sales force control: a synthesis of three theories. J. Pers. Selling Sales Manag. **16**(2), 1–12 (1996). https://doi.org/10.1080/08853134.1996.10754049

Sundaram, A.K., Inkpen, A.C.: The corporate objective revisited. Organ. Sci. **15**(3), 350–363 (2004). https://doi.org/10.1287/orsc.1040.0068

Van den Berg, M.J., Stander, M.W., Van der Vaart, L.: An exploration of key human resource practitioner competencies in a digitally transformed organisation. SA J. Hum. Resour. Manag. **18**, 1–13 (2020). https://doi.org/10.4102/SAJHRM.V18I0.1404

VanKooten, C.: A research methodology of interdependence through video as method. Comput. Compos. **54**, 102514 (2019). https://doi.org/10.1016/j.compcom.2019.102514

Stakeholder Theory - Bibliometric Analysis

Nina Damas(✉)

University of Aveiro, Aveiro, Portugal
ninadamas@ua.pt

Abstract. This article aims to conduct a bibliometric analysis to study scientific production related to stakeholder theory, quantifying articles published per year, source titles, citations, journal impact, and identifying publications by author, affiliation, country, subject, and type. Additionally, using the VOSviewer software, maps were generated to analyze co-authorship between authors and countries, as well as keyword occurrences. Regarding the research methodology, the investigation criteria were first established using the keywords "stakeholder theory" and "teoria dos stakeholders," followed by the application of certain filters to the results obtained. The main conclusions of this study highlighted that 2024 was the year with the highest scientific research output and the greatest number of citations. The Journal of Business Research stood out as the most prominent publication. The analysis also identified co-authorship relationships among authors. Finally, concerning the documents published by countries, the United States, the United Kingdom, and Australia emerged as the leading contributors, in descending order.

Keywords: Stakeholder Theory · Accounting · Stakeholders · Theory of Stakeholders

1 Introduction

Over the years, stakeholder theory has become a widely studied concept at an international level. The topic has gained depth due to the constant need for information in the fields of accounting and business contexts.

The choice of stakeholder theory for the bibliometric analysis was since it offers a broader view, since it not only deals with the interests of shareholders, but also other groups that are also affected by the organization, such as customers, employees, suppliers and the government.

In summary, stakeholder theory aligns with value maximization and emphasizes that managers must consider all stakeholders who influence the organization. However, this theory delves deeper than the aforementioned definition, as it must guide decision-makers on which interests to prioritize among various stakeholders. For instance, employees seek higher wages, superior working conditions, and benefits, while capital providers aim for low risk and high returns. Communities, on the other hand, expect contributions to charity, community benefits, stable employment, and increased investment (Jensen, 2001).

L. Almeida et al. (Eds.): ICAFI 2025, ISEM 80, pp. 128–149, 2026.
https://doi.org/10.1007/978-3-032-16268-7_7

Consequently, it is evident that any decision to be made should specify how to balance the often inconsistent and conflicting interests of various entities (Jensen, 2001).

Stakeholder theory is regarded as a particularly useful perspective for addressing critical business issues from an international viewpoint. It provides an opportunity to reinterpret a diversity of concepts, models, and phenomena across many different disciplines (Harrison et al., 2015).

This article aims to perform a bibliometric analysis to examine the scientific output on stakeholder theory in accounting over the years and to understand the growth of publications, as well as to identify the main authors, journals, and countries. The analysis also enables mapping co-authorship networks and identifying emerging themes. For this purpose, the Scopus database and VOSviewer software were utilized. At the end of the analysis, sixteen articles deemed most relevant were identified.

From this review, it was possible to understand the relationship of stakeholder theory with three key themes. First, its relationship with corporate social performance in conjunction with corporate financial performance (Barnett & Salomon, 2006; Barnett & Salomon, 2012; Orlitzky et al., 2003; Peloza, 2009; Surroca et al., 2010). Second, corporate social responsibility emerges as another topic supported by stakeholder theory (Barnett & Salomon, 2012; Deng et al., 2013; El Ghoul et al., 2011; Surroca et al., 2010).

Stakeholder groups have the capacity to elicit positive responses from companies and/or transform social responsibility into profit (Eesley & Lenox, 2006; Barnett & Salomon, 2012).

Throughout this paper, the methodology used for developing the bibliometric analysis will be explained, followed by a discussion of the results. Finally, conclusions and proposals for future research will be presented.

In conclusion, following the introduction, Sect. 2 details the research methodology. Section 3 presents the bibliometric analysis, showcasing and analyzing the results. Section 4 provides conclusions and outlines proposals for future research.

2 Research Methodology

Regarding the research methodology, the investigation criteria were initially established. Two keywords related to the topic were defined: "Stakeholders Theory" and "Teoria dos Stakeholders." The choice of English and Portuguese keywords was justified by the fact that Portuguese is the mother tongue of the country where the analysis was carried out. In the case of English, as most of the articles are published in the same language, it was considered essential to use it.

The initial search yielded 5,159[1]documents using the first keyword, with one additional document identified when the second keyword was included.

To refine the results, search filters were applied, such as field of study, document type, and keywords. No temporal limits were imposed, as the intended analysis aimed to encompass the state of the art over the years. The database used for this bibliometric analysis was Scopus, chosen for its reputation as a high-quality international repository of scholarly documents and its broad coverage across various disciplines.

[1] Survey conducted on October 9, 2024.

Subsequently, it was essential to narrow the sample to include only relevant fields of study, specifically Business, Management and Accounting and Economics, Econometrics and Finance. This refinement resulted in 3,849 documents. The analysis was then further limited to document type, selecting only articles, which reduced the sample to 3,131 articles. Next, a keyword filter was applied, retaining only articles containing "Stakeholders Theory," which narrowed the selection to 1,507 articles. Finally, all fields unrelated to the area of study were excluded.

After applying these successive filters, the total number of results was reduced to 657.

In conclusion, the methodology employed in this study is bibliometric analysis, which allows the investigation and measurement of information through a set of methods, providing a general evaluation of the scientific production on the topic under study—stakeholder theory.

Figure 1 presents a summary of the filter application process and the documents obtained at each stage.

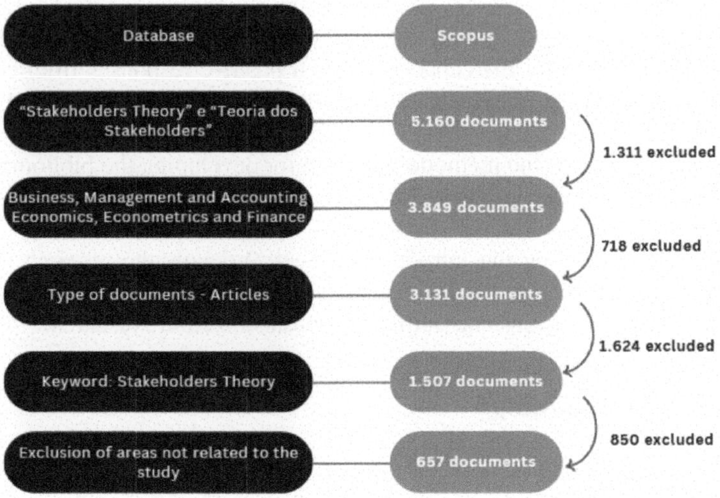

Fig. 1. Summary of methodology, Own elaboration, using Canva

3 Bibliometric Analysis – Presentation and Analysis of Results

This section presents the results and analyzes the information obtained.

A total of 657 articles were identified, from which the total number of citations was analyzed, amounting to 25,962. These citation counts correspond to documents published between 2000 and 2024.

Graphic 1- Citations per year

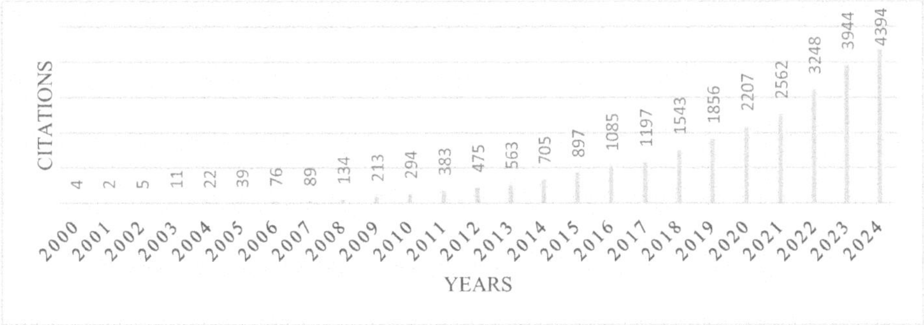

Source - Own elaboration, with data from Scopus

As shown in Graph 1, there has been a steady increase in productivity and citations over the years. This growth can be attributed to the persistent need and interest in research within this area and the use of stakeholder theory as a foundational framework for various studies.

The year that stood out in terms of the number of citations was 2024, with a total of 4,394 citations. It was observed that in 2024, the majority of these citations—exceeding 100—occurred in the Source Titles listed in Table 1.

Table 1. Source Title and no. of citations in 2024

Source Title	No. of citations in 2024	No. of citations out of 25,962
Corporate Social Responsibility And Environmental Management	149	555
Sustainability Switzerland	137	1.291
Business Strategy And The Environment	101	442

Source - Own elaboration, with data from Scopus

The evolution of publications in the sample under study, covering the period from 1998 to 2024, was analyzed. As observed in Graphic 2, there is a consistent increase in publications in this field, with the year 2024 marking the peak of scientific production with a total of 84 articles. A possible explanation for this continuous growth may be the relevance of the topic within the accounting and business context, which also supports other areas of research.

Graphic 2- Documents per year

Source - Own elaboration, with data from Scopus

Following the study of the evolution of publications, interest arose in identifying the sources with the highest number of publications and those with the greatest impact. Based on information from Scopus (Fig. 2), the Journal of Business Research emerges as the source with the highest number of publications—30 articles in our sample over the years, with a peak of seven documents published in 2023.

Fig. 2. Source documents, by year, source by scopus

To complement this, the CiteScore of the leading sources was analyzed (Fig. 3). The Journal of Business Research had the highest citation impact, with a CiteScore of 20.30 in 2023, reflecting 71,115 citations across 3,511 documents from 2020 to 2023. The journal ranks sixth out of 210 in the "Business, Management, and Accounting" category, highlighting its prominence in the field.

Fig. 3. CiteScore of sources, by year, last updated Oct. 9, 2024

An analysis of the SCImago Journal Rank (SJR) for the aforementioned sources was also conducted. This metric measures the prestige of sources based on the average scientific influence of articles published. According to Fig. 4, the Strategic Management Journal stands out with an SJR of 11.035 in 2020, compared to the Journal of Business Research with an SJR of 2.049 in the same year. Despite having a lower CiteScore (13.70) and ranking (29th and 28th in strategy and business management categories, respectively), the Strategic Management Journal is deemed more prestigious by the SJR metric.

Fig. 4. SJR of sources, by year, last updated Oct. 9, 2024

The number of authors per article was examined to understand co-authorship effects (Table 2). Interestingly, articles with two or three authors predominate, indicating a strong prevalence of collaborative research. Articles with one, four, five, six, seven, or eight authors collectively account for fewer publications than those with two or three authors.

Table 2. Number of authors, own elaboration

Authors	Quantity
1	136
2	210
3	162
4	103
5	38
6	5
7	1
8	2

Source - Own elaboration, with data from Scopus

After analyzing the number of authors per article, Table 3 shows the authors who published more than three documents in our sample. This survey served to verify which authors had published the most.

Table 3. Authors who have published more than three articles

Authors	Published Documents
Freeman, R.E	7
San-Jose, L	7
Bosse, D.A	6
Harrison, J.S	6
Retolaza, J.L	6
Velte, P	5
Aguado, R	4
Rashid, A	4
Amran, A	3
Barnett, M.L	3
Civera, C	3
Connolly, C	3
Crilly, D	3
Lee, S	3
Maama, H	3
Phillips, R.A	3
Rundle-Thiele, S	3

Source - Own elaboration, with data from Scopus

For a comprehensive analysis of co-authorship among authors, the VOSviewer software was utilized. Using the sample of 657 documents, a correlation map was generated, capturing the majority of the authors involved. This resulted in Fig. 6, which represents the co-authorship network of authors with at least one document. After creating the network map, it was possible to visualize relationships among fifteen authors, as shown in Fig. 5. Upon analysis, five distinct groups (clusters) with different colors were identified, indicating the proximity among authors within the same cluster. The size of the nodes reflects the number of documents published in the sample, while the connections between authors signify co-authorship relationships. Notably, Freeman, Harrison, Bosse, and Phillips stood out in terms of publications and co-authorship.

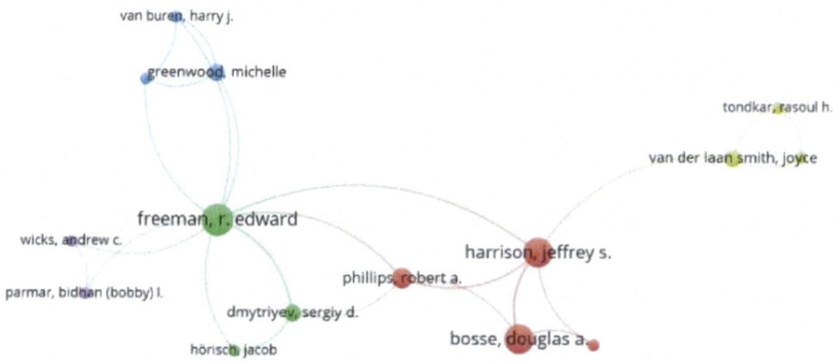

Fig. 5. Co-authorship between authors, Own elaboration, with Vosviewer software

Given the strong co-authorship ties between Freeman, Harrison, Bosse, and Phillips, Table 4 was created to clearly present their relationships across their publications. Notably, the collaboration between Freeman and Harrison resulted in 12 published documents, as did the co-authorship between Freeman and Phillips.

Table 4. Co-authored documents, own elaboration

Co-authored documents	Robert Freeman	Jeffery Harrison	Douglas Bosse	Robert Phillips	University to which they belong
Robert Freeman	---------	12	0	12	University of Virginia, United States
Jeffery Harrison	12	---------	9	7	University of Richmond, United States
Douglas Bosse	0	9	---------	5	University of Richmond, United States

(continued)

Table 4. *(continued)*

Co-authored documents	Robert Freeman	Jeffery Harrison	Douglas Bosse	Robert Phillips	University to which they belong
Robert Phillips	12	7	5	---------	Schulich School of Business, Canada

Source - Own elaboration, with data from Scopus

Subsequently, the initial co-authorship relationships among these authors were analyzed using data obtained from Scopus. It was observed that the first publication by Freeman and Harrison was in 1999 with the article Stakeholders, Social Responsibility, and Performance: Empirical Evidence and Theoretical Perspectives (Harrison & Freeman, 1999). The first joint publication involving Freeman, Harrison, and Phillips occurred in 2019 with the book The Cambridge Handbook of Stakeholder Theory (Harrison et al., 2019). Furthermore, Freeman and Phillips first co-authored an article in 2002 titled Stakeholder Theory: A Libertarian Defense (Freeman & Phillips, 2002). Lastly, Harrison, Bosse, and Phillips collaborated for the first time in 2007, publishing Stakeholder Theory and Competitive Advantage (Harrison et al., 2007). Table 4 also reveals that Harrison and Bosse are affiliated with the same university, which facilitates their co-authorship. In conclusion, these four authors have frequently collaborated on various documents, with the exception of Freeman and Bosse, who lack a direct co-authorship connection.

The co-authorship analysis yielded intriguing findings; however, further investigation was deemed necessary to explore the authors behind the most cited and the most relevant articles in the sample. The most cited article, Corporate Social and Financial Performance: A Meta-Analysis, was authored by Marc Orlitzky, Frank L. Schmidt, and Sara L. Rynes (Orlitzky et al., 2003). These authors represent diverse geographic regions—Australia and the United States—which highlights the cross-regional collaboration in research. According to Scopus, this was their only joint publication. The most relevant article, Prioritization of Organizational Stakeholders: A Managerial Decision-Making Perspective, was authored by Alex Chernyi and Juha Uotila (Chernyi & Uotila, 2024), both affiliated with Warwick Business School, Coventry, UK. Similar to the authors of the most cited article, Chernyi and Uotila collaborated only on this article. Following this analysis, a summary table (Table 5) was created, presenting key metrics for the highlighted authors, such as h-index, citation count, documents cited, and total publications. This table also includes the three prominent authors identified in Fig. 5.

Table 5. Authors' indicators, own elaboration

Indicators[2]	Marc Orlitzky	Frank L. Schmidt	Sara Rynes	Alex Chernyi	Juha Uotila	Robert Freeman	Jeffery Harrison	Douglas Bosse	Robert Phillips
H- index	20	70	42	0	6	47	33	12	21
Quotes	7.547	28.646	13.658	0	819	17.271	10.981	1.896	4.890
Doc. Cited	6.653	21.873	11.834	0	805	12.652	8.900	1.604	3.926
Doc. Published	45	164	81	1	8	124	59	23	42

Source - Own elaboration, with data from Scopus

After studying the authors, it became evident that analyzing the geographical distribution of the published articles was essential. The United States led with 174 publications, followed by the United Kingdom with 85. Table 6 presents additional countries (75 in total) with over ten publications from the sample. Interestingly, the total number of articles from the countries exceeded the sample size (906 articles), suggesting that the same articles might have been published in multiple countries. This underscores the modern reality of research involving co-authors from different countries publishing the same article in their respective regions.

Table 6. Articles published by country

Country	Published documents
United States	174
United Kingdom	85
Australia	77
China	38
Canada	37
Spain	34
Italy	32
France	28
Malaysia	28
Germany	27
India	19
South Africa	19
Netherlands	16
Viet Nam	15

(*continued*)

[2] Data collected on October 24, 2024.

Table 6. (*continued*)

Country	Published documents
Taiwan	14
New Zealand	13
South Korea	13
Denmark	11
Finland	10
Indonesia	10
Portugal	10

Source - Own elaboration, with data from Scopus

Similar to the author analysis, a co-authorship map among countries with more than ten publications was generated using VOSviewer, as shown in Fig. 6. The map highlights the United States, the United Kingdom, and Australia as key contributors, represented by the largest nodes. Six clusters were identified, reflecting the proximity among countries in terms of publications, while the connections denote co-authorship relationships.

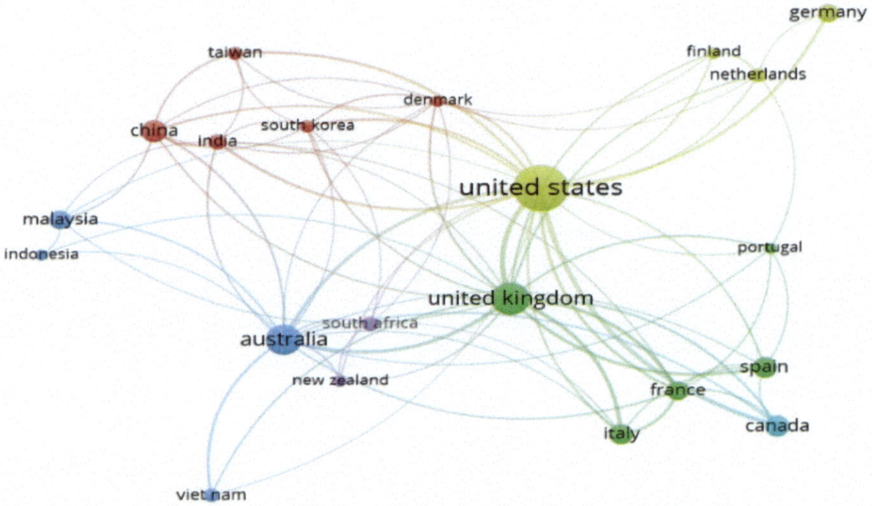

Fig. 6. Co-authorship between countries, Own elaboration, with Vosviewer software

An analysis of institutional affiliations in the sample of 657 articles was conducted and presented in Table 7. To avoid exhaustive rows, only affiliations with more than five articles are included. Summing, the articles by affiliation yielded 569 documents, a figure lower than the sample size, indicating that some articles were published without an institutional affiliation. The top four affiliations represent universities, with the first being located in the United States and the others in Australia. Notably, The University of Queensland Business School specializes in Business, Management, and Accounting, contributing 37.50% of its publications in this area. Further analysis revealed that the top four institutions co-published 34 articles, while their individual contributions totaled 43, indicating joint affiliation for certain articles.

Table 7. Number of articles published by affiliation, plus four articles

Affiliation	Published documents
University of Richmond	11
The University of Queensland	11
Griffith University	11
The University of Queensland Business School	10
University of Virginia	9
Universidad del Pais Vasco	9
Monash University	8
University of Central Florida	8
Griffith Business School	8
Universidad de Deusto	7
University of Southampton	7
University of Pennsylvania	6
Copenhagen Business School	6
Wharton School of the University of Pennsylvania	6
University of Ottawa	6
Leuphana Universität Lüneburg	6
University of Johannesburg	5
Florida State University	5
Arizona State University	5
The Hong Kong Polytechnic University	5
Università degli Studi di Torino	5
University of Birmingham	5
University of Southern Queensland	5
Temple University	5
University of Alberta	5
University of South Australia	5
Rosen College of Hospitality Management	5
Kelley School of Business	5

Source - Own elaboration, with data from Scopus

Following the affiliation analysis, funding sources were examined. Table 8 lists funding organizations that supported more than two articles from the sample. A total of 175 funded articles were identified, a figure lower than the sample size, showing that some articles were published without funding. The two leading funders, both from China, supported 21 articles in total; however, only 18 unique documents were identified, suggesting overlap in funding for some articles.

Table 8. Number of articles published by funding sponsor, with more than two articles

Funding sponsor	Published documents
National Natural Science Foundation of China	13
National Office for Philosophy and Social Sciences	8
European Commission	4
Ministry of Science and Technology, Taiwan	4
Social Sciences and Humanities Research Council of Canada	4
Ministry of Science and Technology of the People's Republic of China	3
Beijing Social Science Fund	2
Coordenação de Aperfeiçoamento de Pessoal de Nível Superior	2
European Regional Development Fund	2
Euskal Herriko Unibertsitatea	2
Fundación Emilio Soldevilla para la Investigación y el Desarrollo en Economía de la Empresa	2
Fundamental Research Funds for Central Universities of the Central South University	2
Imperial College London	2
Japan Society for the Promotion of Science	2
Ministerio de Ciencia e Innovación	2
Ministerio de Economía y Competitividad	2
Ministry of Higher Education, Malaysia	2
Oesterreichische Nationalbank	2
Science and Technology Foundation of State Grid Corporation of China	2
Universitat Politècnica de València	2
Università Bocconi	2

Source - Own elaboration, with data from Scopus

Lastly, a keyword co-occurrence map with at least ten occurrences in the sample was created using VOSviewer (Fig. 7). The map reveals a strong relationship between stakeholder theory and corporate social responsibility. Other related themes include legitimacy theory, institutional theory, agency theory, sustainability, and performance, all closely linked to stakeholder theory.

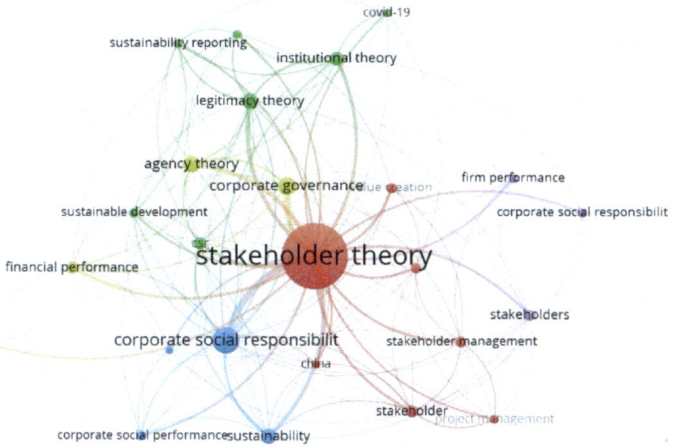

Fig. 7. Occurrence of Keywords, Own elaboration, with Vosviewer software

To deepen the literature review, the ten most cited articles were analyzed and summarized in Table 9, presenting their objectives, methodologies, and conclusions. During this process, it was noted that none of the most cited articles were authored by Edward R. Freeman, considered the "father" of stakeholder theory. Hence, seven documents authored by Freeman from the initial sample were included. After reviewing each, the article The "Business Sucks" Story was excluded as it was not directly aligned with the study's thematic focus despite its relevance to the business world. Consequently, Table 9 presents the top 16 articles for this analysis.

Table 9. Summary of the sixteen relevant articles, in descending order of citations

Authors	Title	Year	No. of citations	Objectives	Methodology	Conclusions	Reference
Orlitzky, M., Schmidt, F.L. e Rynes, S.L	Corporate social and financial performance: A meta-analysis	2003	4745	To show that there is little generalizable knowledge about corporate social/environmental performance (CSP) and corporate financial performance (CFP) and that it is built on unstable foundations	More rigorous than previous efforts, it conducted a meta-analysis of 52 studies (representing the population of the previous quantitative research) producing a total sample size of 33,878 observations	In all the studies, CSP is positively correlated with CFP. The relationship tends to be bidirectional and simultaneous. Reputation seems to be an important mediator of the relationship. The lack of correspondence between stakeholders	(Orlitzky et al., 2003)

(continued)

Table 9. *(continued)*

Authors	Title	Year	No. of citations	Objectives	Methodology	Conclusions	Reference
El Ghoul, S., Guedhami, O., Kwok, C.C.Y. e Mishra, D.R	Does corporate social responsibility affect the cost of capital?	2011	1808	Examines the effect of corporate social responsibility (CSR) on the cost of equity for a large sample of US companies	Using a sample of 12,915 US company-year observations from 1992 to 2007	Various approaches to estimating companies' ex-ante cost of equity have found that companies with better CSR scores have cheaper equity financing. In particular, the findings suggest that investment in improving responsible employee relations, environmental policies and product strategies contributes substantially to reducing companies' cost of equity	(El Ghoul et al., 2011)
Surroca, J., Tribó, J.A. e Waddock, S	Corporate responsibility and financial performance: The role of intangible resources	2010	1430	It examines the effects of a company's intangible resources in mediating the relationship between corporate responsibility and financial performance	It demonstrates theoretical restraint with the use of a database comprising 599 companies from 28 countries	They indicate that there is no direct relationship between corporate responsibility and financial performance, only an indirect relationship that depends on the mediating effect of a company's intangible resources	(Surroca et al., 2010)
Barnett, M.L. e Salomon, R.M	Beyond dichotomy: The curvilinear relationship between social responsibility and financial performance	2006	895	Debate to measure the link between financial and social performance of mutual funds practicing socially responsible investing (SRI). SRI fund managers have a variety of social screening strategies to choose from	Empirical test on a panel of 61 SRI funds from 1972 to 2000	Results show that as the number of social screens used by an SRI fund increases, financial returns decrease at first, but then recover as the number of screens reaches a maximum. In other words, we found a curvilinear relationship, suggesting that two long-competition viewpoints can be complementary	(Barnett & Salomon, 2006)

(continued)

Table 9. (*continued*)

Authors	Title	Year	No. of citations	Objectives	Methodology	Conclusions	Reference
Deng, X., Kang, J.-K. e Low, B.S	Corporate social responsibility and stakeholder value maximization: Evidence from mergers	2013	881	Whether corporate social responsibility (CSR) creates value for the shareholders of acquiring companies	Using a large sample of mergers in the USA	Results suggest that the social performance of acquirers is an important determinant of merger performance and the likelihood of its completion and they support the stakeholder value maximization view of stakeholder theory	(Deng et al., 2013)
Barnett, M.L., Salomon, R.M	Does it pay to be really good? addressing the shape of the relationship between social and financial performance	2012	827	Based on the theoretical argument that a company's ability to profit from social responsibility depends on its ability to influence stakeholders (SIC), they gathered contrasting literature on the relationship between corporate social performance (CSP) and corporate financial performance (CFP) to hypothesize that the CSP-CFP relationship is U-shaped	Test of the hypothetical relationship in an unbalanced panel of 1,214 companies and 4,730 company-year observations during the period from 1998 to 2006. Based on the theoretical conceptualization of CIS	We found that companies with low CSP have higher CFP than companies with moderate CSP, but companies with high CSP have the highest CFP. This supports the theoretical argument that SIC underpins the ability to transform social responsibility into profit	(Barnett & Salomon, 2012)
Harrison, J.S., Bosse, D.A. e Phillips, R.A	Managing for stakeholders, stakeholder utility functions, and competitive advantage	2010	774	A company that manages for stakeholders allocates more resources to satisfying their needs and demands than would be necessary to simply retain their voluntary participation in the company's productive activities. It explains why this type of behavior unlocks additional potential for value creation, as well as the conditions that facilitate or interrupt the value creation process	It integrates some of the central concepts of stakeholder theory with the literature on organizational justice and trust to explain the competitiveness of companies	The competitive advantages resulting from a stakeholder management approach are considered sustainable because they are associated with path dependency and causal ambiguity. These explanations provide a strong justification for including stakeholder theory in the discussion of competitiveness and company performance	(Harrison et al., 2010)

(*continued*)

Table 9. (*continued*)

Authors	Title	Year	No. of citations	Objectives	Methodology	Conclusions	Reference
Eesley, C. e Lenox, M.J	Firm responses to secondary stakeholder action	2006	524	They explore the conditions under which secondary stakeholder groups are likely to provoke positive responses from the company	To test this structure, they built a unique dataset of more than 600 secondary stakeholder actions in the United States, all related to environmental issues in the period 1971–2003	The findings corroborate those of another study that power, legitimacy and urgency are important important salience factors	(Eesley & Lenox, 2006)
Jensen, M	Value maximisation, stakeholder theory, and the corporate objective function	2001	498	They analyze whether companies should have a single value objective and whether that objective should be the maximization of value or another objective such as maintaining employment	Conceptual analysis	Having a single value objective is not enough for organizational success	(Jensen, 2001)
Peloza, J	The challenge of measuring financial impacts from investments in corporate social performance	2009	477	Analyze the business case for CSP and provide recommendations for managers interested in measuring the impacts of CSP investment on financial performance	Conceptual analysis	Positive relationship between CSP and financial performance. The relationship is relatively weak, questions of causality are not answered and the measures used to examine the business are inconsistent. Managers have a vague idea of what CSP pays off financially, but don't know the best way to measure the financial impacts of implementing these initiatives	(Peloza, 2009)

(*continued*)

Table 9. *(continued)*

Authors	Title	Year	No. of citations	Objectives	Methodology	Conclusions	Reference
Freeman, R.E., Dmytriyev, S.D. e Phillips, R.A	Stakeholder Theory and the Resource-Based View of the Firm	2021	239	Analyze how stakeholder theory can inform the resource-based firm (RBV) and suggest a two-part guideline for reconciling stakeholder theory and the RBV	Conceptual analysis	It demonstrates that the RBV, in its current form, is incomplete and can become a more comprehensive theory by incorporating four essential elements of stakeholder theory such as incorporating normativity, recalibrating the idea of sustainability, viewing people in addition to resources and allocating more space to cooperative behaviors	(Freeman et al., 2021)
Dmytriyev, S.D., Freeman, R.E. e Hörisch, J	The Relationship between Stakeholder Theory and Corporate Social Responsibility: Differences, Similarities, and Implications for Social Issues in Management	2021	198	Remove unnecessary barriers in the stakeholder theory and CSR fields; empower academics and organizational practitioners with a more nuanced language for dealing with social issues in management; and enable the creation of a coherent and integrative theoretical base in the area of social issues in management that was previously at a disadvantage in relation to other areas of management	Conceptual analysis	We argue that stakeholder theory and CSR provide distinct but complementary theoretical frameworks, with some overlap. The actual decision to choose a specific framework depends on the problem one wishes to solve and the configurations of that problem	(Dmytriyev et al., 2021)

(continued)

Table 9. (*continued*)

Authors	Title	Year	No. of citations	Objectives	Methodology	Conclusions	Reference
Harrison, J.S., Phillips, R.A. e Freeman, R.E	On the 2019 Business Roundtable "Statement on the Purpose of a Corporation"	2020	140	Briefly describe the main twists in the shareholders vs. stakeholders debate that make this statement so significant for management theory and practice	Statements from 183 top executives	A statement by leading CEOs on the broader purpose of a corporation is an important signal that reinforces a set of ideas that has been gaining traction for some time. This radical shift opens up new research opportunities as academics explore the impact of management actions and strategies on a wider group of stakeholders	(Harrison et al., 2020)
Harrison, J.S., Freeman, R.E. e de Abreu, M.C.S	Stakeholder theory as an ethical approach to effective management: Applying the theory to multiple contexts	2015	124	It provides a brief overview of stakeholder theory, clarifies some widely held misconceptions, explains the importance of examining the theory from a variety of international perspectives and how this kind of research will advance management theory, and introduces the other articles in the special issue	Some of the fundamental ideas of stakeholder theory are discussed, leading to arguments about the importance of the theory for management research, especially in an international context	Stakeholder theory is considered a particularly useful perspective for approaching some of the important issues in business from an international perspective. It offers an opportunity to reinterpret a variety of concepts, models and phenomena in many different disciplines	(Harrison et al., 2015)

(*continued*)

Table 9. *(continued)*

Authors	Title	Year	No. of citations	Objectives	Methodology	Conclusions	Reference
Mitchell, R.K., Van Buren, H.J., Greenwood, M. e Freeman, R.E	Stakeholder Inclusion and Accounting for Stakeholders	2015	75	Returning to the insights of stakeholder theory to better utilize the centrality of risk in understanding the relationships between companies and stakeholders, a theme addressed in each of the articles in the Special Issue	The accounting mechanism we suggest, value creation for stakeholders (VCSPs), is derived from the outline of a practical accounting system focused on stakeholders based on the principles of partnership accounting	Stakeholder theory for value creation: where stakeholder risk is minimized because the four processes of counting, recording, summarizing and reporting support the four premises of value creation: activities, alignment, interaction and reciprocity, respectively. In practice, this step towards transdisciplinary theory suggests the stakeholder theory of value creation (VCSA) which is a superior justification for the inclusion of stakeholders	(Mitchell et al., 2015)
Parmar, B.B.L., Wicks, A.C. e Freeman, R.E	Stakeholder Management & The Value of Human-Centred Corporate Objectives	2022	19	Clarify the role of purpose in stakeholder theory as a mechanism that brings stakeholders together and can help judge their claims, detail the ways in which current scholars recognize that stakeholder management is critical to competitive advantage, and show that competitive advantage needs to be augmented with cooperative advantage	Conceptual analysis	The authors use a straw man conceptualization of stakeholder theory in their argument, detailing the existing strategy literature that connects competitive advantage and stakeholder relationships and arguing that competitive advantage is not enough, it needs to be augmented with a cooperative advantage focused on stakeholders	(Parmar et al., 2022)

Source - Own elaboration

4 Conclusions, Limitations, and Future Research Directions

This study delved into stakeholder theory and associated practices. Its main objective was to identify and analyze articles related to this topic. A bibliometric analysis was conducted, applying necessary filters to refine the sample to 657 articles, 16 of which were selected for a detailed literature review.

The analysis facilitated comparisons among articles across multiple dimensions. Scientifically, 2024 emerged as the year with the highest citations (4,394) and publications (84). The Journal of Business Research stood out as a leading publication venue. Co-authorship and cross-country article publication reinforced the hypothesis of collaborative research and idea exchange among authors. Most works involved two to three authors. Regarding publication distribution by country, the United States, the United Kingdom, and Australia ranked highest, with 75 countries contributing overall. This geographical diversity reflects Scopus's extensive coverage of diverse journals, affirming its status as a premier bibliographic database.

This study has certain limitations, notably its reliance on a single database, Scopus, and the initial focus solely on scientific articles, potentially limiting the scope of the analysis.

Future research directions include:

- Investigating the impact of stakeholder theory on Portuguese companies;
- Utilizing multiple databases for article collection;
- Conducting a bibliometric study to identify over-explored and under-explored themes within stakeholder theory;
- Analyzing the frequency of research related to stakeholder theory and sustainability and assessing whether regulatory changes influence research trends.

These topics could foster intriguing discussions, particularly as stakeholder theory continues to evolve in accounting and the business world.

Declaration. During the preparation of this work, the author utilized ChatGPT for the translation of text from Portuguese to English, ensuring coherence and clarity in English. The author reviewed and edited the content as necessary and assumes full responsibility for the publication's content.

References

Barnett, M.L., Salomon, R.M.: Beyond dichotomy: the curvilinear relationship between social responsibility and financial performance. Strateg. Manag. J. **27**(11), 1101–1122 (2006)

Barnett, M.L., Salomon, R.M.: Does it pay to be really good? Addressing the shape of the relationship between social and financial performance. Strateg. Manag. J. **33**(11), 1304–1320 (2012)

Chernyi, A., Uotila, J.: Prioritization of organizational stakeholders: a managerial decision-making perspective. J. Bus. Res. **185**(1), (2024)

Deng, X., Kang, J.-K., Low, B.S.: Corporate social responsibility and stakeholder value maximization: evidence from mergers. J. Financ. Econ. **110**(1), 87–109 (2013)

Dmytriyev, S.D., Freeman, R.E., Hörisch, J.: The relationship between stakeholder theory and corporate social responsibility: differences, similarities, and implications for social issues in management. J. Manage. Stud. **58**(6), 1441–1470 (2021)

Eesley, C., Lenox, M.J.: Firm responses to secondary stakeholder action. Strateg. Manag. J. **27**(8), 765–781 (2006)

El Ghoul, S., Guedhami, O., Kwok, C.C.Y., Mishra, D.R.: Does corporate social responsibility affect the cost of capital? J. Bank. Finance **35**(9), 2388–2406 (2011)

Freeman, R.E., Dmytriyev, S.D., Phillips, R.A.: Stakeholder theory and the resource-based view of the firm. J. Manag. **47**(7), 1757–1770 (2021)

Freeman, R.E., Phillips, R.A.: Stakeholder theory: a libertarian defense. Bus. Ethics Q. **12**(3), 331–349 (2002)

Harrison, J.S., Barney, J.B., Freeman, R.E., Phillips, R.A.: The Cambridge Handbook of Stakeholder Theory. Cambridge University Press (2019)

Harrison, J.S., Bosse, D., Phillips, R.A.: Stakeholder theory and competitive advantage. In: Taneja, (ed.) Academy of Management Annual Meeting Proceedings, pp. 1–6. Academy of Management (2007)

Harrison, J.S., Bosse, D.A., Phillips, R.A.: Managing for stakeholders, stakeholder utility functions, and competitive advantage. Strateg. Manag. J. **31**(1), 58–74 (2010)

Harrison, J.S., Freeman, R.E.: Stakeholders, social responsibility, and performance: empirical evidence and theoretical perspectives. Acad. Manag. J. **42**(5), 479–485 (1999)

Harrison, J.S., Freeman, R.E., Abreu, M.C.S.: Stakeholder theory as an ethical approach to effective management: applying the theory to multiple contexts. Revista Brasileira de Gestão de Negócios **17**(55), 858–869 (2015)

Harrison, J.S., Phillips, R.A., Freeman, R.E.: On the 2019 business roundtable "statement on the purpose of a corporation." J. Manag. **46**(7), 1223–1237 (2020)

Jensen, M.: Value maximisation, stakeholder theory, and the corporate objective function. Eur. Financ. Manag. **7**(3), 297–317 (2001)

Mitchell, R.K., Van Buren, H.J., Greenwood, M., Freeman, R.E.: Stakeholder inclusion and accounting for stakeholders. J. Manage. Stud. **52**(7), 851–877 (2015)

Orlitzky, M., Schmidt, F.L., Rynes, S.L.: Corporate social and financial performance: a meta-analysis. Organ. Stud. **24**(3), 403–441 (2003)

Parmar, B.B.L., Wicks, A.C., Freeman, R.E.: Stakeholder management & the value of human-centred corporate objectives. J. Manage. Stud. **59**(2), 569–582 (2022)

Peloza, J.: The challenge of measuring financial impacts from investments in corporate social performance. J. Manag. **35**(6), 1518–1541 (2009)

Surroca, J., Tribó, J.A., Waddock, S.: Corporate responsibility and financial performance: The role of intangible resources. Strateg. Manag. J. **31**(5), 463–490 (2010)

Exploratory Study on the Challenges of Forensic Accounting in Portugal: Between the Absence of Standards and the Scarcity of Scientific Research

Djavan de Alcântara Lima(✉)

University of Aveiro, Santiago University Campus, 3810-193 Aveiro, Portugal
djavan.lima@ua.pt

Abstract. Forensic accounting (known in Portugal as *peritagem contabilística* or *perícia contábil*) plays a vital role in supporting legal decision-making by translating complex financial information into technical evidence. However, in Portugal, this professional field faces structural limitations due to the absence of specific regulation and the lack of scientific development. This paper investigates the perceptions of forensic accounting professionals regarding these institutional challenges. A qualitative research design was adopted, based on a structured questionnaire disseminated via LinkedIn to professionals who self-identify as forensic accountants in Portugal. From a targeted sample of 35 individuals, 10 valid responses were obtained. The results indicate strong concerns about the lack of regulatory frameworks, inadequate academic training, limited interdisciplinary integration, and scarce applied research. More than 80% of respondents highlighted the urgent need for legal recognition, formal university education, and the production of context-specific scientific knowledge. Open-ended responses revealed recurring themes related to training, documentation, recognition, and interest in the profession's expansion into extrajudicial arenas, such as arbitration and mediation. The study contributes to the literature by providing empirical evidence from a specialised professional niche and by reinforcing the need for strategic institutionalisation of forensic accounting in underregulated environments. The findings have practical implications for policymakers, academic institutions, and professional associations seeking to strengthen the credibility and development of the field. Due to the small sample size, the findings should be interpreted within the exploratory scope of the study.

Keywords: Forensic Accounting · Regulation · Institutional Challenges

1 Introduction

Forensic accounting emerges as a highly relevant discipline at the intersection between accounting and the judicial system (Cornacchione, 2024; Louro, 2013; Rocha, 2021). It plays a central role in clarifying complex financial issues within both judicial and extrajudicial proceedings (Cornacchione, 2024; Guimarães, 2008; Rocha, 2021).

L. Almeida et al. (Eds.): ICAFI 2025, ISEM 80, pp. 150–163, 2026.
https://doi.org/10.1007/978-3-032-16268-7_8

In Portugal, the practice of forensic accounting - particularly accounting expert witness work - faces serious structural limitations due to the absence of specific regulatory frameworks and the lack of scientific research in the field (Guimarães, 2008; Louro, 2013; Simões, 2022). These gaps affect the development and formal recognition of the profession, compromising the quality and consistency of expert reports and eroding trust in the outcomes (Louro, 2013; Simões, 2022).

As Dias (1926, cited by Guimarães, 2008) noted nearly a century ago, there remains no legal statute in Portugal that clearly defines or protects the designation of accounting expert. Consequently, individuals without adequate qualifications may be appointed by judges as expert witnesses, sometimes without any demonstrated knowledge of accounting.

This regulatory vacuum has led to a scenario in which professionals from diverse backgrounds act as expert accountants without standardised qualifications, duties, or procedures (Guimarães, 2008; Louro, 2013). In contrast, countries like Germany assign forensic experts through formal court procedures and subject them to rigorous professional standards (Cornacchione, 2024).

This article addresses the central research question: How do the absence of standards and the lack of scientific research impact the practice and professional recognition of forensic accounting in Portugal? Despite growing demand for expert accountants, the field remains underregulated and underexplored academically (Guimarães, 2008; Louro, 2013; Russo, 2021).

These challenges result in ambiguity regarding the expert's role, inconsistencies in procedures, and a lack of quality control across expert reports, thereby reducing the reliability of forensic contributions to legal processes (Simões, 2022).

The aim of this study is to analyse the challenges currently faced by forensic accounting in Portugal, particularly focusing on the normative void and the scarcity of applied research (Guimarães, 2008; Louro, 2013; Simões, 2022). It seeks to contribute to a more robust understanding of the field's development potential, based on empirical perceptions collected through a qualitative survey with professional experts.

The practical relevance of this research lies in identifying structural weaknesses and proposing strategies to enhance the technical quality, credibility, and professional standing of forensic accounting. The study also aims to clarify the expert's role within judicial and alternative dispute resolution contexts (Cornacchione, 2024; Louro, 2013).

From an academic perspective, this article addresses a significant research gap by advancing knowledge in a field that remains largely overlooked in Portuguese accounting scholarship. It also encourages further theoretical and applied contributions (Russo, 2021). This limited development is consistent with global trends indicating that forensic accounting research remains unevenly distributed, particularly underdeveloped in Europe and Lusophone contexts (Ozili, 2025).

This paper seeks to contribute to a more robust understanding of the field's development potential, based on empirical perceptions collected through a qualitative survey with professional experts. Therefore, this exploratory research is guided by the following question: *What are the perceptions of forensic accounting professionals regarding the absence of regulation and the lack of scientific development in Portugal?*

The practical relevance of this research lies in identifying structural weaknesses and proposing strategies to enhance the technical quality, credibility, and professional standing of forensic accounting.

This paper is structured as follows: after this introductory section, it presents the theoretical foundations of forensic accounting and its regulatory and scientific context in Portugal. Subsequently, the research methodology is described, followed by an analytical presentation and discussion of the empirical findings. The final section offers concluding remarks and policy recommendations.

Thus, this study does not aim to measure the impact of these limitations but to gather and analyse professionals' perceptions regarding the structural and institutional challenges faced by forensic accounting in Portugal.

2 Theoretical Framework

Forensic accounting - also referred to as accounting expertise or expert witness accounting in various jurisdictions - is a specialist field at the intersection of accounting, law, and litigation support. It encompasses a broad set of activities including the identification, analysis, and communication of financial evidence for use in legal disputes (Crumbley & Fenton, 2020; AICPA, 2014). Its applications span fraud detection, asset tracing, quantification of damages, contract breach assessments, and financial misrepresentation analysis, both in civil and criminal proceedings. This evolving role of forensic accounting reflects a paradigm shift beyond traditional auditing, as it increasingly demands proactive methods to detect and prevent fraud (Afriyie et al., 2023).

According to the American Institute of Certified Public Accountants (AICPA, 2014), forensic accounting involves the application of accounting principles, methodologies, and investigative techniques to matters that may result in legal or administrative action. The Practice Aid 07–1 offers structured guidance on fraud investigations, chain of custody, documentation, and expert reporting. These technical recommendations are widely recognised as professional best practices in jurisdictions with established forensic infrastructure.

A crucial dimension of forensic accounting is its evidentiary value in judicial and quasi-judicial forums. Expert reports are often pivotal in shaping legal outcomes, which raises the bar for accuracy, neutrality, and methodological transparency. However, as Crumbley and Cheng (2014) argue, even technically sound reports can be discredited if they fail to comply with procedural expectations such as those outlined under the Daubert standard in U.S. courts. This evidentiary threshold demands that expert testimony be not only relevant and reliable, but also scientifically validated.

Empirical research has also highlighted the influence of stakeholder perceptions - particularly those of judges and attorneys - on the acceptance and credibility of expert work. Bressler (2012) demonstrates that forensic accountants are evaluated not only by the quality of their analysis, but also by their communication skills, courtroom demeanour, and perceived independence. These dimensions are especially relevant in countries like Portugal, where formal standards and oversight mechanisms are absent, leaving substantial room for discretionary evaluation of expertise.

The regulatory and institutional configuration of forensic accounting varies significantly across jurisdictions. The Council of Europe (CEPEJ, 2014) has issued specific guidelines advocating for minimum standards in the selection, training, and accountability of court-appointed experts in its member states. These recommendations promote transparency, procedural fairness, and public trust in expert systems. Nevertheless, Portugal has yet to incorporate such measures into its legal framework, resulting in a fragmented field where accountants, economists, and even lay individuals may be appointed without consistent criteria or ethical codes (Guimarães, 2008; Simões, 2022).

Furthermore, the absence of state-sponsored oversight bodies or credentialing institutions contributes to the fragmentation of expert witness practice in Portugal. In jurisdictions such as Germany and the United States, forensic accountants must often adhere to specific accreditation systems or court rosters, which impose minimum training and ethical standards (AICPA, 2014; CEPEJ, 2014). The lack of such structures in Portugal reinforces the discretionary nature of expert appointments and allows for significant variance in technical quality. This has also hindered the development of a cohesive professional identity, which DiGabriele et al. (2020) regard as fundamental for the legitimation of forensic roles within adversarial legal systems. In countries where oversight is institutionalised, professional associations often play an active role in the definition and dissemination of standards. Portugal still lacks an association exclusively dedicated to forensic accounting, which would be crucial for consolidating the field, promoting continuing education, and representing practitioners in legal and academic contexts.

The absence of normative structures in Portugal is paralleled by a marked deficiency in academic research on forensic accounting and expert testimony. While countries such as the United States and the United Kingdom have developed postgraduate programmes, certification schemes, and active research communities focused on forensic accounting (DiGabriele, Heitger, & Riley, 2020), Portuguese institutions have produced relatively few empirical or conceptual studies. This research vacuum undermines efforts to build context-specific methodologies and limits opportunities for professional training based on scientific evidence.

From a theoretical standpoint, two perspectives offer insight into this stagnation. First, Institutional Theory, as articulated by DiMaggio and Powell (1983), suggests that professions gain legitimacy through institutional isomorphism - the process by which organisations and professions evolve in response to normative, coercive, and mimetic pressures. The lack of alignment between Portugal and its European peers in terms of forensic accounting standards and practices suggests a failure to undergo such isomorphic transformation.

Second, the Positive Theory of Accounting Regulation, proposed by Watts and Zimmerman (1978), contends that the creation of accounting standards is shaped not only by technical rationale but also by political and economic forces. In this context, the weak institutional lobbying of Portuguese forensic accountants and the limited scholarly attention may explain why the field has yet to achieve regulatory visibility.

Furthermore, the lack of alignment between academia and professional practice restricts the development of curricula and training programmes capable of equipping accountants with the interdisciplinary skills necessary for forensic engagements. As noted by DiGabriele et al. (2020), forensic accounting is inherently multidisciplinary, drawing from auditing, economics, legal reasoning, and investigative methods. Without structured academic foundations, the Portuguese forensic accounting profession remains fragmented, reactive, and institutionally fragile. This intersection is further reinforced by Aksoy and Uzay (2021), who demonstrate how forensic accounting integrates fraud auditing, litigation support, and expert witnessing into a cohesive professional function.

In summary, the literature establishes that forensic accounting requires both a regulatory foundation and a research-based academic infrastructure to function effectively. Countries that have succeeded in institutionalising this profession have done so through the establishment of legal norms, ethical standards, training pathways, and robust scientific output. The Portuguese context illustrates the consequences of neglecting these dimensions and highlights the urgent need for strategic interventions to elevate forensic accounting to the level of a recognised and professionalised field.

3 Methodology

This research adopts a qualitative, exploratory-descriptive approach aimed at investigating the structural challenges faced by the forensic accounting profession in Portugal, particularly in relation to the absence of formal standards and the scarcity of scientific research. Qualitative research is appropriate for contexts where the phenomenon is not yet sufficiently systematised in the literature and where the objective is to explore meanings, perceptions, and complex institutional dynamics (Gil, 2008; Lakatos & Marconi, 2021; Martins & Theóphilo, 2016).

The study employed a structured questionnaire as the primary instrument for data collection. The questionnaire was designed using *Google Forms* and made available to selected forensic accounting professionals operating in Portugal. The choice of a digital instrument enabled broad access, flexibility in response time, and ensured participant anonymity. The questionnaire was composed of both closed and open-ended questions, allowing for the collection of statistical frequencies as well as qualitative narratives that reflect the perceptions, criticisms, and suggestions of professionals directly engaged in forensic engagements.

The questionnaire was organised into five thematic blocks: (1) demographic and professional characterisation of the respondents, (2) academic background and training, (3) perceptions about the regulatory and institutional framework, (4) extrajudicial application potential, and (5) challenges and future needs in the profession. Open-ended responses were analysed through thematic content analysis, seeking recurring keywords, expressions, and patterns. This technique enabled the extraction of core concerns and the contextual interpretation of the professional discourse (Martins & Theóphilo, 2016).

Sampling was intentional and non-probabilistic, targeting respondents who are actively engaged in accounting expertise or have academic or professional experience in forensic accounting. The link to the questionnaire was disseminated exclusively through LinkedIn posts and messages directed to professionals who publicly self-identify as forensic accountants in Portugal. This method is aligned with the objective of reaching a qualified audience capable of contributing relevant and grounded insights (Gil, 2008).

The data were exported into Excel spreadsheets and subsequently processed for tabulation and analysis. Descriptive statistics were used to examine the demographic profile and categorical responses, while qualitative insights were extracted through inductive coding of open responses. The integration of both statistical and textual data supports a more comprehensive understanding of the perceptions and expectations regarding forensic accounting in Portugal (Lakatos & Marconi, 2021).

The research adheres to ethical principles, including the informed consent of participants, data anonymisation, and the exclusive use of collected information for scientific purposes. No personal or identifying information was collected. The study also complies with the standards of transparency and methodological rigour required in the field of applied social sciences research (Martins & Theóphilo, 2016).

While the methodological design centred on a qualitative survey was appropriate for the study's exploratory goals, future research would benefit from mixed-methods approaches. Combining interviews, document analysis, and expert panel techniques (Delphi method, for example) could enhance the triangulation of data and increase analytical depth. Moreover, involving other stakeholders - such as judges, lawyers, or university faculty - could enrich the interpretive scope and provide external perspectives on the role, value, and limitations of forensic accounting in legal proceedings. As noted by Gil (2008), scientific generalisation in qualitative research depends not on sample size, but on the coherence of methodological procedures and the robustness of analytical interpretation.

The questionnaire was available for responses over a four-week period, from April to May 2025. It consisted of 23 questions, organised into four thematic blocks: (1) professional and demographic profile, (2) academic background, (3) perceptions on institutional challenges, and (4) future prospects and needs. Of the 35 professionals identified via LinkedIn and contacted directly, 10 valid and complete responses were obtained. The questionnaire included 22 closed-ended questions and one open-ended question. Although the sample was small, it reflected the views of professionals who actively self-identify with forensic accounting. For future studies, the outreach could be extended to include professionals listed in judicial directories, academic institutions, and professional associations to enhance sample diversity and analytical robustness.

4 Description and Discussion of Results

The findings of this study are based on the analysis of responses from forensic accounting professionals operating in Portugal. The results offer a multifaceted view of the current state of the profession, highlighting institutional gaps, regulatory deficiencies, training needs, and emergent themes related to professional identity and extrajudicial relevance.

4.1 Respondent Profile and Professional Background

The sample consisted of professionals with varying degrees of academic and practical experience in forensic accounting. The majority of respondents were over 40 years old and held degrees in accounting, with a significant proportion possessing postgraduate qualifications. Many reported having over a decade of experience in forensic engagements, reinforcing the credibility and maturity of the sample (Table 1).

The data reveal that most respondents self-identify as male and have a background in accounting or economics. Notably, nearly all participants indicated having practical

Table 1. Characterisation of the Sample

Category	Response	Count
Age Group	26–30 years	1
	31–35 years	1
	41 years or older	8
Gender Identity	Man	6
	Woman	3
Academic Background	Accounting	6
	Economics	1
	Management	1
Highest Academic Degree	Bachelor's Degree	6
	Postgraduate Degree	2
Experience in Forensic Accounting	Less than 5 years	2
	More than 20 years	5
	11 to 20 years	2

experience in preparing expert reports for judicial purposes, although few had received formal academic training in forensic accounting during their university education.

The limited visibility of the title *forensic accountant* (*perito contabilístico*) in Portugal further reinforces the fragmented nature of the field. Despite most respondents having over ten years of experience in expert roles, only a few formally adopt this designation in professional directories or digital platforms. As noted by Marinho (2016) and Rocha (2021), many accounting experts operate under general titles such as "accountant" or "auditor," which obscures the boundaries and scope of forensic practice. This semantic ambiguity affects both public perception and institutional recognition, making it difficult to quantify or regulate the profession. The lack of consensus on nomenclature also limits professional cohesion, reducing opportunities for standardisation, certification, and organised knowledge exchange.

4.2 Training and Professional Education Gaps

One of the key findings relates to the insufficiency of formal academic pathways in forensic accounting, as confirmed by 90% of respondents who reported never having received university-level training specific to this field. This highlights a systemic gap that reinforces the need for curriculum reform in undergraduate and postgraduate programmes.

This confirms what has been previously noted in the literature: that forensic accounting remains underrepresented in Portuguese academic curricula (Simões, 2022; Guimarães, 2008). The lack of formal education and structured competencies hinders the development of a professional identity and reduces opportunities for skill enhancement through evidence-based training (Martins & Theóphilo, 2016).

Additionally, the lack of academic engagement with forensic accounting inhibits the construction of research-based curricula. As Martins and Theóphilo (2016) emphasise, education in applied social sciences must be anchored in empirical reality and supported by methodological training. In Portugal, however, existing programmes rarely offer modules on expert reporting, judicial procedures, or interdisciplinary evidence evaluation. This not only compromises the technical proficiency of future practitioners but also limits the potential for generating context-sensitive research that could inform policy and practice. A sustained effort is needed to integrate forensic accounting into the broader pedagogical frameworks of accounting education. This aligns with international findings that advocate for experiential and case-based methods in forensic accounting education to replicate real-world challenges (Alshurafat et al., 2020).

4.3 Perceptions on Regulatory and Scientific Development

A significant contribution of this study lies in its evaluation of the perceived urgency for institutional reform. Respondents were asked to rate, on a four-point scale, the necessity for improvements in four critical areas: legal and normative frameworks, university education, interdisciplinary integration, and applied scientific research.

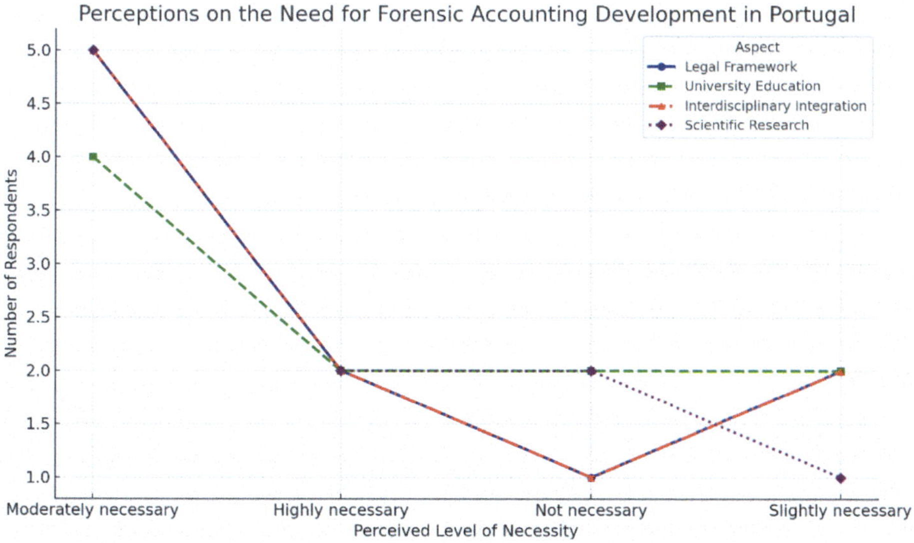

Fig. 1. Perceptions on the need for forensic accounting development in Portugal across four key areas: Legal Framework, University Education, Interdisciplinary Integration, and Scientific Research. The distribution reflects the level of urgency indicated by respondents

The results are striking: over 80% of participants considered the need for legal regulation and university-level training to be "highly necessary." Similarly, the need for greater integration between accounting, legal, and economic knowledge domains, as well as the production of academic research, was emphasised by a majority of respondents.

These concerns are corroborated by recent bibliometric evidence highlighting the role of ethical standards and interdisciplinary collaboration in advancing forensic accounting practices (Al-Raggad et al., 2025).

These results resonate with observations by Louro (2013), who highlighted that the absence of unified criteria in the selection and evaluation of accounting experts has created an uneven and discretionary environment. In practical terms, this translates into a landscape where the admissibility and reliability of expert reports are often challenged not based on their technical merits, but on procedural inconsistencies and lack of formal recognition. Guimarães (2008) similarly argued that without regulatory clarity, the field remains exposed to contestation and undervaluation within the judicial system. This empirical finding confirms that the absence of a regulatory framework weakens not only the profession's legitimacy but also the confidence of judges and attorneys in expert financial evidence - an issue also noted in comparative European studies by the CEPEJ (2014).

These findings strongly align with institutional and regulatory gaps already identified in the literature. The absence of clear rules, ethical codes, and technical procedures in Portugal, as documented by Guimarães (2008), Louro (2013), and the CEPEJ (2014), continues to constrain the quality and legitimacy of forensic accounting practice Although based on a limited sample, the responses collected in this study offer preliminary empirical support for the concerns identified in the literature and provide initial insights into how these deficiencies are perceived by a segment of the professional community. This regulatory vacuum not only undermines the credibility of expert reports in court but also limits the credibility and enforceability of opinions issued in alternative dispute resolution contexts, where evidentiary rigour is equally critical.

4.4 Technical Vocabulary and Professional Identity

The open-ended question at the end of the survey allowed participants to freely express their views on the principal challenges and future directions of the profession. A content analysis of these responses reveals the prominence of certain technical terms - such as "training," "expert report," "documentation," "regulation," and "recognition" - highlighting key themes in the collective discourse of the field.

This interpretation is supported by Aquino and Imoniana (2017), who identified, through a phenomenographic approach, how different professional experiences shape the conceptual understanding of forensic accounting and expert witness work in practice. Notably, many professionals explicitly referenced the need for continuing education, better remuneration, and stronger institutional support. Such concerns echo the findings of Bressler (2012), who emphasised the importance of credibility and perception in expert witness effectiveness, and the AICPA (2014), which advocates for continuous professional development.

The recurrence of terms such as "valuation," "training," and "recognition" in open-ended responses also suggests that practitioners view forensic accounting as a distinct field requiring its own language, epistemology, and institutional space. This supports the view of Simões (2022), who argued that forensic accounting must not be subsumed under generic auditing or consulting categories. Rather, it demands specialised competencies, such as judicial reasoning, legal terminology, and narrative construction of financial facts.

Table 2. Frequency and Context of Forensic Accounting Technical Terms

Technical Term	Frequency	Category	Example of Context
Training	3	Education and Qualification	Need for continuous professional development
Forensic Expertise	2	Field of Practice	Role of the expert in judicial proceedings
Expert Accounting	2	Field of Practice	Regulatory vacuum and expert appointment
Documents	1	Sources of Evidence	Document review during forensic examinations

The development of this professional identity depends not only on legal codification but also on sustained academic production and the formation of communities of practice - none of which can be achieved without targeted investment from universities and professional councils.

4.5 Innovative Contributions and Emerging Issues

Among the more innovative aspects of this study is the documentation of the profession's readiness to expand into extrajudicial domains, including arbitration, mediation, and out-of-court settlements. A large proportion of respondents (9 out of 10) expressed support for this expansion, suggesting a paradigm shift in the scope of accounting expertise in Portugal.

This expansion aligns with international trends toward alternative dispute resolution (ADR), where forensic accountants are increasingly called upon to act as neutral evaluators or technical advisors outside the courtroom (Crumbley & Fenton, 2020; DiGabriele et al., 2020). However, this also raises questions about the adequacy of current ethical, procedural, and pedagogical standards to support such diversification - a theme not yet extensively covered in Portuguese academic literature, thus representing a novel contribution of this study.

Despite the increasing awareness of emerging technologies among practitioners, the literature in Portugal remains largely silent on their implementation and regulatory implications. For instance, Silva et al. (2022) highlight that blockchain technology can enhance transparency and traceability in audit trails, while artificial intelligence tools can assist in anomaly detection, fraud pattern analysis, and predictive modelling. However, these technologies also raise questions about data protection, algorithmic accountability, and the role of expert judgement - issues that demand academic scrutiny and legal adaptation. Integrating these technologies into forensic accounting practice would require not only technical training but also updated regulatory frameworks to ensure admissibility, reliability, and procedural fairness.

Another noteworthy insight concerns the frequent mention of technological themes, including artificial intelligence (AI), blockchain, and advanced data analytics. These responses resonate with the international literature (Grass, 2024; Silva, 2022; Simões,

2022), which has increasingly pointed to the transformative potential of these technologies in forensic practice (DiGabriele et al., 2020), especially in evidence verification, automated anomaly detection, and chain-of-custody control. Recent studies also reinforce the importance of integrating forensic accounting with big data technologies, particularly through machine learning frameworks capable of enhancing fraud detection mechanisms in complex financial systems (Akinbowale, Mashigo, & Zerihun, 2023).

5 Conclusion

This exploratory study offered an initial empirical examination of the state of forensic accounting in Portugal, based on the perceptions of ten professionals with relevant experience in the field who actively engage in expert witness roles. By focusing on the lack of regulation and the scarcity of scientific research, the study revealed how these structural gaps affect professional recognition, practice consistency, and knowledge development in the field.

The continued institutional fragility of the field may also be interpreted through the lens of mimetic and normative isomorphism. As DiMaggio and Powell (1983) suggest, when formal structures are lacking, professions may fail to adopt best practices present in more institutionalised environments. The Portuguese case appears to reflect a stalled process of professionalisation, where neither regulatory coercion nor normative academic production has been sufficient to establish legitimacy. This calls attention to the role of universities, accounting bodies, and judicial authorities in collectively shaping the identity and credibility of forensic experts.

The findings clearly show that forensic accounting in Portugal remains underdeveloped and marginalised. This confirms the persistence of institutional fragility, as framed by DiMaggio and Powell's (1983) concept of institutional isomorphism. The lack of coercive and normative pressure has prevented the field from evolving into a structured and legitimised professional domain. A notable 90% of respondents reported never having had access to specific academic training in forensic accounting during their university education. At the same time, 100% of respondents considered it important or highly important to include a mandatory course on accounting expertise in undergraduate curricula. These results strongly reinforce the literature's argument regarding the disconnection between academic institutions and the technical-professional demands of forensic engagements (Simões, 2022; Guimarães, 2008).

In relation to institutional challenges, over 80% of participants rated as *highly necessary* the development of a legal and normative framework to guide expert appointments, procedures, and ethical standards. Similarly, the production of scientific research and the integration between accounting, legal, and economic domains were also considered urgent by the vast majority of respondents.

As illustrated in Fig. 1, these four dimensions - legal regulation, academic training, interdisciplinary integration, and research - form a shared agenda among professionals who view the current system as fragmented and vulnerable to discretionary practices. These perceptions are consistent with the concerns outlined by CEPEJ (2014) regarding the lack of guarantees and consistency in expert appointment mechanisms across Europe.

Moreover, the open-ended responses revealed recurrent references to *training*, *documentation*, *valuation*, and *professional recognition*, as documented in Table 2. These terms reflect the profession's desire to be seen not only as a technical support to the judiciary but as a specialised and autonomous field, deserving of institutional investment and academic recognition.

In addition, an expressive 86% of participants indicated that forensic accounting should expand beyond judicial proceedings into alternative dispute resolution (ADR) environments - such as arbitration and mediation - suggesting a more proactive and strategic role for expert accountants in conflict resolution. This finding, while still incipient in Portuguese literature, is aligned with international trends that emphasise the utility of forensic accountants in multiparty, non-adversarial contexts (Crumbley & Fenton, 2020; DiGabriele et al., 2020).

Importantly, several participants also referenced technological innovation - highlighting emerging themes such as artificial intelligence, digital platforms, and blockchain as tools for improving the quality and traceability of expert analyses. Although peripheral to the central research question, these signals point to the potential for modernisation and innovation in the profession, especially in terms of evidence analysis and digital documentation.

The main contributions of this exploratory study include: (i) the preliminary identification of priority areas for institutional reform, based on the perceptions of a small group of active professionals; (ii) the empirical confirmation of gaps previously suggested in the literature; and (iii) the introduction of new dimensions for debate, particularly regarding extrajudicial roles and technological advances. One potential avenue for institutional advancement lies in the adoption of conceptual models that integrate forensic accounting into organisational control structures, as proposed by Akinbowale, Klingelhöfer, and Zerihun (2020).

The sample, though analytically relevant, was restricted to 35 professionals, of whom 10 responded. This constraint reflects the niche status of the profession in Portugal. However, it may also introduce a bias of self-representation, as only individuals who publicly identify with the field and are active on professional platforms were included. Such bias may skew perceptions toward more engaged or reform-oriented professionals.

Future studies should consider triangulating data through interviews, institutional document analysis, or comparative research with other legal and academic systems. It would also be valuable to explore the views of other stakeholders - such as judges, attorneys, and regulators - regarding the credibility, utility, and expectations placed upon forensic accountants in dispute resolution contexts.

In conclusion, forensic accounting in Portugal stands at a critical juncture. To advance as a credible, professionalised, and scientifically grounded field, it must be supported by coherent regulation, academic legitimacy, and professional development strategies that reflect the complexity and relevance of expert financial analysis in legal and extrajudicial settings.

Addressing these institutional gaps requires a coordinated agenda involving policymakers, universities, and professional accounting bodies. Academic institutions must recognise forensic accounting as a strategic area of research and training, while professional associations should advocate for the creation of national registers, continuing education standards, and procedural guidelines. As Martins and Theóphilo (2016) affirm, scientific disciplines only achieve autonomy when they are supported by institutional legitimacy and knowledge accumulation. This study thus offers both a diagnosis and a call to action: to advance forensic accounting in Portugal, it is necessary to move beyond isolated professional practice toward a structured, regulated, and academically grounded field.

As with any exploratory study based on a small and purposive sample, the findings presented here should be interpreted with caution and understood as an initial contribution to the broader academic and professional debate on forensic accounting in Portugal.

References

Afriyie, S.O., Akomeah, M.O., Amoakohene, G., Ampimah, B.C., Ocloo, C.E., Kyei, M.O.: Forensic accounting: a novel paradigm and relevant knowledge in fraud detection and prevention. Int. J. Public Adm. **46**(9), 615–624 (2023). https://doi.org/10.1080/01900692.2021.2009855

Akinbowale, O.E., Klingelhöfer, H.E., Zerihun, M.F.: An innovative approach in combating economic crime using forensic accounting techniques. J. Finan. Crime **27**(4), 1253–1271 (2020). https://doi.org/10.1108/JFC-04-2020-0053

Akinbowale, O.E., Mashigo, P., Zerihun, M.F.: The integration of forensic accounting and big data technology frameworks for internal fraud mitigation in the banking industry. Cogent Bus. Manag. **10**(1), 2163560 (2023). https://doi.org/10.1080/23311975.2022.2163560

Aksoy, T., Uzay, S.: Relationship between fraud auditing and forensic accounting. In: Auditing Ecosystem and Strategic Accounting in the Digital Era: Global Approaches and New Opportunities, pp. 127–146. Springer (2021). https://doi.org/10.1007/978-3-030-72628-7_6

Al-Raggad, M., Albalawee, N., Al-Mahasneh, A., Abu Huson, Y., Albajaly, A.: Unveiling financial crimes: advancing forensic accounting practices and ethical integrity through bibliometric insights. Safer Communities **24**(3), 244–264 (2025). https://doi.org/10.1108/SC-11-2024-0069

Alshurafat, H., Beattie, C., Jones, G., Sands, J.: Perceptions of the usefulness of various teaching methods in forensic accounting education. Acc. Educ. **29**(2), 177–204 (2020). https://doi.org/10.1080/09639284.2020.1719425

American Institute of Certified Public Accountants (AICPA): Forensic Accounting: Fraud Investigations (Practice Aid 07–1, Revised). AICPA, New York (2014). https://www.grassiadvisors.com/wp-content/uploads/2021/09/Forensic-Accounting-Fraud-Investigations.pdf

Aquino, A.A., Imoniana, J.O.: Contabilidade forense e perícia contábil: um estudo fenomenográfico. Organizações em Contexto - Revista do Programa de Pós-Graduação em Administração **13**(26), 101–126 (2017)

Bressler, L.: The role of forensic accountants in fraud investigations: importance of attorney and judge's perceptions. J. Finan. Accountancy **9**, 1–11 (2012). https://papers.ssrn.com/sol3/papers.cfm?abstract_id=1788822

CEPEJ – Council of Europe: Guidelines on the Role of Court-Appointed Experts in Judicial Proceedings of Council of Europe's Member States. Council of Europe, Strasbourg (2014). https://rm.coe.int/168074827a

Cornacchione Jr., E.B.: Atuação nacional e internacional em perícia contábil. In: Peleias, I.R. et al. (eds.) Vivências e Evolução da Perícia Contábil, Cap. 12, Academia Paulista de Contabilidade, São Paulo (2024)

Crumbley, D.L., Cheng, C.C.: Avoid losing a Daubert challenge: some best practices for expert witnesses. ATA J. Legal Tax Res. **12**(1), 41–53 (2014). https://doi.org/10.2308/jltr-50765

Crumbley, D.L., Fenton, E.D.: Forensic and Investigative Accounting. 10th edn. CCH, Riverwoods (2020)

DiGabriele, J.A., Heitger, L.E., Riley, R.: A synthesis of non-fraud forensic accounting research. J. Forensic Account. Res. **5**(1), 257–277 (2020). https://doi.org/10.2308/JFAR-19-034
DiMaggio, P.J., Powell, W.W.: The iron cage revisited: institutional isomorphism and collective rationality in organizational fields. Am. Sociol. Rev. **48**(2), 147–160 (1983). https://doi.org/10.2307/2095101
Grass, I.B.P.: Pesquisa científica e avanços metodológicos na perícia. Presentation at Simpósio Perícia e Advocacia: Instrumentos de Promoção da Justiça, Ordem dos Advogados do Brasil - Seção São Paulo (2024). unpublished material
Gil, A.C.: Como Elaborar Projetos de Pesquisa. 5th edn. Atlas, São Paulo (2008)
Guimarães, J.F.C.: A profissão de contabilista em Portugal. Revista Electrónica INFOCONTAB **33**, 38 (2008). https://contabilista.com.pt/anasr/biblioteca/artigo1.pdf
Louro, J.A.G.: Relatório de atividade profissional conducente ao grau de mestre em Gestão. Universidade de Trás-os-Montes e Alto Douro, Vila Real (2013). https://repositorio.utad.pt/handle/10348/2520
Marinho, J.I.S.: A Forensic Accounting e o Ensino Superior em Portugal: Análise Empírica. Master's thesis, Universidade do Porto, Porto (2016). https://hdl.handle.net/10216/87605
Martins, G.A., Theóphilo, C.R.: Metodologia da Investigação Científica para Ciências Sociais Aplicadas. 3rd edn. Atlas, São Paulo (2016)
Moreira, N.R.D.O.: A Forensic Accounting em Portugal: Evidências Empíricas. PhD thesis, Universidade do Minho, Braga (2009). https://hdl.handle.net/1822/10948
Ozili, P.K.: Forensic accounting research around the world. J. Finan. Reporting Account. **23**(1), 128–153 (2025). https://doi.org/10.1108/JFRA-02-2023-0106
Rocha, L.M.G.D.: Contributo da Auditoria Forense na Quantificação de Danos Económicos para a Atividade Seguradora. Master's thesis, Instituto Superior de Contabilidade e Administração do Porto, Politécnico do Porto (2021). https://recipp.ipp.pt/handle/10400.22/19955
Russo, P.T.: Relevância da pesquisa contábil. Revista Fipecafi de Contabilidade, Controladoria e Finanças **2**(2), 1–12 (2021). https://doi.org/10.53826/2763-7069.v2n2.2021.id50
Silva, R., Inácio, H., Marques, R.P.: Blockchain implications for auditing: a systematic literature review and bibliometric analysis. Int. J. Digit. Account. Res. **22** (2022)
Simões, P.A.L.: Contabilidade Forense: Situação Atual e as Perspetivas Futuras em Portugal. PhD thesis, Instituto Superior de Contabilidade e Administração de Lisboa, Lisboa (2022). http://hdl.handle.net/10400.21/15366
Watts, R.L., Zimmerman, J.L.: Towards a positive theory of the determination of accounting standards. Account. Rev. **53**(1), 112–134 (1978). https://www.jstor.org/stable/245729
Lakatos, E.M., Marconi, M.A.: Fundamentos de Metodologia Científica. 9th edn. Atlas, São Paulo (2021)

University Social Responsibility: Advancing the Sustainable Development Goals in Higher Education

Graça Azevedo[1,2] and Maria C Tavares[1,3](✉)

[1] Higher Institute for Accounting and Administration of the University of Aveiro (ISCA-UA), 3810-902 Aveiro, Portugal
{graca.azevedo,mariatavares}@ua.pt
[2] Research Centre on Accounting and Taxation—CICF, 4750-821 Barcelos, Portugal
[3] Research Unit On Governance, Competitiveness and Public Policies (GOVCOPP), 3810-193 Aveiro, Portugal

Abstract. University Social Responsibility (USR) has increasingly consolidated itself as a strategic dimension within higher education, particularly in the era of the United Nations 2030 Agenda. This article aims to systematize the existing knowledge on how higher education institutions integrate the Sustainable Development Goals (SDGs) into their institutional practices, with particular emphasis on sustainability communication and reporting. Through a narrative literature review, the paper analyzes the conceptual evolution of USR, the mechanisms for aligning with the SDGs, and the persistent challenges in their implementation, drawing on both national and international experiences. The analysis highlights disparities between public and private universities, variations in the emphasis placed on social, environmental, and economic dimensions, and gaps in the use of performance indicators. Finally, the transformative potential of artificial intelligence in measuring, managing, and communicating sustainability in higher education is underscored. This study contributes to the academic and institutional debate on the strategic role of universities in promoting sustainable development, identifying trends and opportunities for future research.

Keywords: Social Responsibility · Universities · SDGs · Sustainability · Financial Reporting

1 Introduction

The growing attention to sustainability and social responsibility has led higher education institutions (HEI) to rethink their societal role. Beyond teaching and research, universities are now called upon to play strategic roles in advancing the SDGs, taking on ethical, social, environmental, and economic commitments that shape their internal and external operations. This repositioning requires the adoption of sustainable practices and transparent communication through institutional reports, such as social responsibility or sustainability reports.

L. Almeida et al. (Eds.): ICAFI 2025, ISEM 80, pp. 164–176, 2026.
https://doi.org/10.1007/978-3-032-16268-7_9

Higher Education Institutions (HEI) are recognized as strategic actors in achieving the Sustainable Development Goals (SDGs), and the incorporation of the SDGs is recommended in four core areas: teaching, research, community engagement, and institutional management. However, the study systematizes various—internal and external—obstacles to the implementation of the SDGs in HEI, including budget constraints, governance failures, insufficient faculty training, and institutional resistance (Serafini et al., 2022).

USR should be viewed as a comprehensive concept encompassing various areas of HEI activities to promote the SDGs (Carrillo Valdez et al., 2025). Sawani et al. (2025), Setó-Pamies, D., (2011) discuss the importance of universities in promoting social responsibility and fulfilling the SDGs. They emphasize that HEIs play a crucial role in integrating the SDGs into their operations and curricula, fostering a culture of sustainability and social responsibility. Ebekozien et al. (2025) highlight the need to raise awareness of the importance of social responsibility in HEI infrastructure, involving key stakeholders, and establishing a legal framework to support this infrastructure. In this context, the present study addresses the following research question: What contributions emerge from the literature to understand how universities integrate the SDGs into their social responsibility practices?

This approach is justified because social responsibility in higher education remains an underexplored area within the Portuguese context. The SDGs represent an international commitment that demands concrete adaptation by HEIs, and technology, particularly artificial intelligence, emerges as a catalyst for a new generation of sustainable institutional practices whose impact must be examined.

This article is based exclusively on a narrative and exploratory literature review, aiming to critically describe and understand the state of the art concerning the integration of USR and the SDGs in higher education. It seeks to identify trends, gaps, and best communication practices adopted by higher education institutions at both national and international levels to reflect on the role of universities in fulfilling the 2030 Agenda. The work of Leal Filho et al. (2021) confirms that, due to the still-emerging nature of the field, qualitative and narrative analysis remain relevant for understanding and interpreting institutional practices, trends, and gaps, challenging the predominance of quantitative approaches. Leal Filho et al. (2023) highlight the lack of studies directly linking University Social Responsibility (USR) and the Sustainable Development Goals (SDGs), emphasizing the value and relevance of their scientific contribution.

The literature selection followed a qualitative approach, focusing on empirical and theoretical studies published over the past two decades, emphasizing indexed academic articles, institutional reports, and strategic documents from international organizations (such as the UN and UNESCO). The main inclusion criteria were: (i) direct relevance to the topic of USR or the SDGs in higher education, both in public and private universities, and (ii) contributions to critical reflection on practices of reporting, performance evaluation, or the social and environmental impact of HEIs.

The literature analysis was structured around three principal axes: (i) Social Responsibility in Universities—with a focus on its theoretical framework, historical evolution, and educational and social role; (ii) Integration of the SDGs in HEIs—including international experiences and reporting mechanisms aligned with the 2030 Agenda; and (iii) The role of technological innovation, particularly Artificial Intelligence, as an enabler of sustainable practices in higher education. This critical systematization of literature thus enables the development of a consolidated theoretical framework for future studies. It provides input for policy discussions, reinforcing the strategic role of universities in promoting sustainable development. The conclusion summarizes the key findings and outlines directions for future research.

2 Methodology

This study draws inspiration from the work developed by Costa et al. (2021), conducted within the context of Italian universities. The primary objective of that research was to analyze the interconnections between University Social Responsibility (USR) and the Sustainable Development Goals (SDGs), as framed by the 2030 Agenda. Building upon this foundation, the present research aims to assess whether the findings observed in the Italian context are also applicable to Portuguese higher education institutions, while considering their organizational, cultural, and strategic particularities. To this end, methodological adaptations were made to align the study with the Portuguese context.

As in the original study, the SDGs served as a guiding framework for analyzing the implementation of social responsibility practices within universities. The analytical model focuses on the three core dimensions of sustainable development—social, environmental, and economic responsibility—which are mapped against the 17 SDGs and the 169 targets established by the United Nations. This structure enables the identification of a broad range of Key Performance Indicators (KPIs), relevant for evaluating the degree of institutional alignment with the principles of sustainability and their strategic plans.

In contrast to the empirical and quantitative approach employed by Costa et al. (2021), which was based on direct data collection, the present study adopts an exploratory and narrative methodology. This methodological choice is justified by the emerging and multifaceted nature of SDG integration within Portuguese higher education, allowing for a critical and interpretative analysis of the current state of knowledge.

The selection of scientific literature followed criteria of thematic relevance, timeliness, and disciplinary and geographical diversity, incorporating both national and international contributions from recent years on the themes of university social responsibility and the SDGs. The narrative review approach has been recognized for its ability to foster a comprehensive and interpretative understanding of complex and evolving phenomena such as sustainability in higher education (Leal Filho et al., 2019; Leal Filho et al., 2021; Leal Filho et al., 2023). Although the review process was conducted with methodological rigor, it is acknowledged that a systematic protocol or quantitative bibliometric analysis was not employed. This constitutes a methodological limitation, potentially affecting the breadth and representativeness of the findings.

The analysis is based on academic publications, institutional and regulatory reports, and case studies, with the aim of identifying patterns, trends, and challenges related to the integration of sustainability into the university mission. This qualitative approach enables the capture of the complexity and diversity of ongoing practices, while also highlighting the opportunities and constraints faced by Portuguese institutions in operationalizing the SDGs through social responsibility.

3 Literature Review

3.1 Social Responsibility in Universities

Both scholars and practitioners have long been interested in how organizations respond to increasing stakeholder pressure to go beyond mere profit maximization. These responses are generally framed within Corporate Social Responsibility (CSR), which encompasses organizations' economic, legal, ethical, and philanthropic responsibilities (Carroll, 1999).

The first formal notion of social responsibility emerged with Bowen (1953), who defined it as the obligation of business leaders to make decisions and pursue actions that align with the objectives and values of society (Carroll, 1999; Aggarwal & Singh, 2019). CSR initiatives enable organizations to develop critical success factors and sustainable competitive advantages, positively influencing stakeholders' perceptions and promoting a socially committed image (Alon et al., 2010). According to Yoon et al. (2006), socially responsible behavior may be influenced by the image that companies aim to project to consumers. Thus, CSR is a corporate obligation to protect and enhance social well-being by ensuring equitable and sustainable benefits for a broad range of stakeholders (Alon et al., 2010). It represents a form of self-regulation that may become an integral part of the business model, holding the organization accountable to itself, its stakeholders, and the public (Ali et al., 2022).

Over the years, social responsibility has been extensively debated in business and educational contexts (Setó-Pamies et al., 2011). However, the absence of a unified definition complicates its application in the university setting, where profit is not a primary objective. USR promotes civic engagement and active citizenship, encouraging students and faculty to provide social services to their communities and adopt ecological and environmental practices in support of sustainable development (Vasilescu et al., 2010). Through the WSSSP-HEI survey, involving 103 researchers from over 40 countries, the study by Leal Filho et al. (2023) reveals that the most researched areas in recent years include governance, research, curriculum, and community outreach—all central elements of University Social Responsibility (USR). It highlights that many researchers address both local issues, such as community engagement, and global goals defined by the SDGs. This focus includes practices such as living labs and collaborative initiatives that connect the university with its surrounding territory.

Serafini et al. (2022) emphasize that a genuine commitment by HEI to the SDGs requires their cross-cutting integration—from curriculum and research to institutional management—aligning with the concept of USR. The research shows that USR practices go beyond theory: they translate into behavioral changes, student skill development, and community projects that strengthen the link between HEIs and society (Gallardo Vázquez et al., 2024).

Nejati et al. (2011) emphasize that universities, given their educational and formative nature, play a critical role in addressing societal challenges and contributing meaningfully to a more sustainable future. As institutions deepen their understanding of social responsibility, it becomes crucial to have instruments in place to monitor and assess their impact. To this end, defining key performance indicators (KPIs) is essential to measure outcomes and assess the achievement of goals defined in their institutional strategies (Costa et al., 2021).

According to Meseguer-Sánchez et al. (2020) and Adhikariparajuli (2021), there is a growing interest in USR and its economic and environmental impacts, highlighting the need for more in-depth studies. Nevertheless, Jorge et al. (2012), in their analysis of Spanish public universities, noted that USR is still not perceived as a differentiating factor in strategic planning, reflecting a low level of institutional commitment.

Adams (2013) found that universities lag significantly behind other sectors in sustainability reporting and performance management – an especially critical gap considering their role in generating knowledge and educating socially responsible citizens.

Despite the growing interest, many of the existing studies have focused primarily on teaching and research activities related to social responsibility, sustainability, and sustainable development, as shown by Setó-Pamies et al. (2011), Matten & Moon (2004), and Nejati et al. (2011). Nevertheless, Costa et al. (2021) argue that there is limited evidence on how universities effectively integrate social responsibility principles into their strategic plans or monitor their implementation. The study by Ebekozien et al. (2025) explores how an expanded framework of social responsibility can enhance the infrastructure of HEIs to achieve SDG 4, which aims to ensure quality education. The study by Chigbu and Makapela (2025), utilizing a systematic literature review, identified that integrating data-driven decision-making with the SDGs, particularly SDG 4 (quality education) and SDG 10 (reduced inequalities), can improve institutional effectiveness, align curricula with market demands, strengthen student outcomes, and foster an inclusive and sustainable academic environment. The study by Zhang, Chen, and Xu (2025) develops and validates a scale to measure sustainable leadership development among teachers in Chinese HEI, using tools such as the 'Assessment Model for Sustainable Leadership Development in Higher Education Institutions' (AMDSL-HEI), based on the 'Principles for Responsible Management Education. Therefore, revisiting and updating these analyses in light of current challenges and institutional realities is essential. USR, operationalized through local initiatives and partnerships, is an effective tool for advancing the SDGs in higher education. This connection to the regional context strengthens the transformative role of universities in society, driving the practical implementation of the Sustainable Development Goals (Leal Filho et al., 2019).

3.2 From Social Commitment to the SDGs in Universities

The concept of sustainable development gained international prominence with the publication of the Brundtland Report by the United Nations World Commission on Environment and Development in 1987, titled Our Common Future (United Nations, 1987). This report formally introduced sustainable development as "development that meets the needs of the present without compromising the ability of future generations to meet their own needs" (OECD, 1987; Brundtland Commission, 1987). The concept emphasizes the equitable fulfillment of fundamental human aspirations while recognizing the limits of environmental resources and the biosphere's regenerative capacity (Brown et al., 1987; Paravizo et al., 2018).

Over the following decades, sustainable development was gradually incorporated into economic, political, social, and environmental agendas. Adopting the 2030 Agenda for Sustainable Development by the United Nations in 2015 marked a significant milestone in this process by introducing 17 SDGs with a universal and cross-sectoral approach. The SDGs represent a global call to action, urging countries to implement policies that promote economic growth, enhance human well-being, reduce inequalities, strengthen health and education systems, protect the environment, and foster peace (Russell-Bennett et al., 2024; Saini et al., 2023).

Adopting the 2030 Agenda signaled a significant shift by engaging all sectors of society – public and private alike – through structural reforms and the assignment of specific responsibilities (Bexell & Jönsson, 2017; Erin et al., 2022). Embracing or neglecting the SDGs is no longer a matter of choice: monitoring and evaluating progress have become governmental priorities, directly impacting public policy and institutional action (Firoiu et al., 2022).

For the first time, universities were formally recognized in international agreements as active agents in implementing the SDGs, receiving acknowledgment and incentives from governments and international organizations. This recognition highlights the critical role of education in shaping a sustainable future, raising youth awareness about the 2030 Agenda and positioning higher education institutions at the forefront of social transformation (Borsatto et al., 2024; Angelaki et al., 2024). Within this context, universities assume a broader social commitment – not only through teaching and research, but also through community engagement, the integration of sustainable practices in institutional governance, and the promotion of human development. As Ortiz-Martínez et al. (2023) emphasize, the practical implementation of sustainable development requires a balance among its three core dimensions: social, economic, and environmental, focusing on sustainable and inclusive growth. The study by Cubas et al. (2025) examines the practices and strategies of the QualEnv consortium in promoting sustainable development in HEIs. Using an interdisciplinary approach, the authors identified several effective practices that institutions can adopt to integrate sustainability into their operations and curricula: Education for Sustainability (implementation of educational programs that incorporate sustainability principles); Sustainable Management (adoption of management practices that promote energy efficiency and waste reduction); Community Engagement (partnerships with the local community to promote sustainability initiatives); and Research and Innovation (encouragement of research that seeks innovative solutions to environmental challenges). The article highlights the importance of a collaborative and integrated approach to achieving the SDGs in HEIs. The study by Wajid (2025) examines the role of Saudi Arabian HEIs in sustainable development, focusing on participation, framework alignment, and strategic insights. Emerging technologies are also driving the integration of the SDGs within universities. AI, for example, presents opportunities to enhance sustainability in higher education by improving resource management and supporting teaching and research in strategic areas (Fernandes et al., 2024). Leal Filho et al. (2019) emphasize that universities play a vital role in local ecosystems by promoting sustainable development through research, teaching, and services aimed at communities. The local engagement of universities gives concrete expression to the global goals of the SDGs, reinforcing their social and educational mission. Universities should make their knowledge accessible by creating 'living labs' and outreach programs that actively involve students, faculty, and the broader population.

In this regard, a historic UN resolution adopted on March 21, 2024, stands out. This resolution aims to promote a safer, more secure, and trustworthy use of AI. It encourages its deployment in tackling the world's most pressing challenges – including poverty, health, food security, climate, energy, and education – while addressing digital exclusion in developing countries and ensuring equitable access to the benefits of this technology. Lan (2025) highlights that digital technologies are transforming universities into generators of new knowledge with an impact on quality of life and sustainability, strengthening their connection with the SDGs.

Thus, within the framework of the 2030 Agenda, universities are now called upon to play a strategic role in advancing the SDGs, assuming a global social commitment to sustainability, inclusion, and innovation. This commitment, rooted in their educational mission, now extends to their capacity to transform society to pursue a more just and resilient development path.

3.3 Sustainable Commitment in Higher Education: National and International Evidence

Higher education is central in promoting sustainable development and disseminating social responsibility practices. However, how universities incorporate these practices varies depending on institutional and geographical contexts. Public and private universities, despite sharing the mission of generating and transmitting knowledge, often differ in how they communicate their commitments to social responsibility and sustainability – some responding to legal requirements, others acting voluntarily. The study by Sawani et al. (2025), Setó-Pamies, D., (2011 analyzes the existing literature on the factors that motivate HEIs to disclose their initiatives related to the SDGs online. Using an interdisciplinary systematic review, the authors identified five main clusters related to the key motivating factors for SDG disclosure in higher education institutions: curriculum (integration of the SDGs into educational programs); research activities (focus on research that promotes the SDGs); human resource management (engagement and training of faculty and staff); and research and operational collaborations (partnerships and collaborations that drive sustainability).

Gallardo Vázquez et al. (2024) highlight that the implementation of courses, training programs, and curricula focused on CSR and sustainability has multiple impacts: on institutional performance, academic awareness, and society at large, contributing concretely to the sustainable commitment of universities. The study by Carrillo Valdez et al. (2025) examines the contributions of three Mexican HEIs to sustainable development. Using comparative analysis, the authors explore how these institutions implement sustainable practices and promote the SDGs (Curriculum Integration, Research and Innovation; Institutional Management, Community Engagement). For instance, in Egypt, Ismail (2019) found that public universities disclose more information related to governance, sustainability, and community engagement, while private institutions focus on human resources. This illustrates differing stakeholder communication approaches. Serafini et al. (2022) point to the need for the systematic integration of the SDGs into the strategic processes of universities, emphasizing that raising awareness and providing training for all actors (students, faculty, and staff) are essential to ensure that the commitment goes beyond mere formal discourse.

Several international studies have explored sustainability reporting in universities. In New Zealand, An et al. (2017) noted a narrow environmental focus in sustainability reports and a lack of alignment with frameworks such as the Global Reporting Initiative (GRI). Similar patterns were observed in Canada (Fonseca et al., 2011; Sassen & Azizi, 2018b) and the US (Sassen & Azizi, 2018a), where environmental dimensions dominate reporting, while social and economic aspects remain underrepresented. In the UK, Saha et al. (2021) found limited references to SDGs in carbon emissions disclosures, and in Thailand, Rahman et al. (2019) highlighted a marketing-oriented use of social responsibility with weak conceptual understanding. In Portugal, Silva et al. (2019) noted that public institutions increasingly embed social responsibility in their strategic planning, recognizing teaching, research, and community engagement as key pillars.

These findings align with Behringer et al. (2016), who argued that sustainability and social responsibility have evolved separately for many years but increasingly converged due to their shared goals. Social responsibility contributes directly to sustainable development; without it, sustainability cannot be fully achieved. As Mishra (in Costa et al., 2021) notes, social responsibility encompasses broader, voluntary commitments—often involving reinvestment of profits in social and environmental causes—whereas the SDGs offer specific, measurable targets.

Lu et al. (2021) identified several barriers to integrating social responsibility with the SDGs, including greenwashing, selective SDGs implementation, lack of institutional incorporation, and cross-national disparities. In a study of Italian universities, Costa et al. (2021) found a low degree of institutionalization of social responsibility, with these issues rarely addressed in strategic planning. The study by Lan (2025) highlights the need to guide higher education towards the development of students' moral aspirations, particularly at the master's level, aiming for a conscious regulation of their academic activities and future professional conduct, in line with the SDGs. The study reinforces that digitalization is not merely a pedagogical aspect but a means to enhance universities' commitment to the SDGs by actively and morally engaging students.

Despite these challenges, aligning university strategies with the SDGs through integrated communication and stakeholder engagement is essential. As Chen et al. (2015) argue, social responsibility should be embedded in the institution's core values and principles. Pinto (2009) similarly emphasizes that universities, as open systems, must engage with their environments to foster societal sustainability. As Costa et al. (2021) highlight, social responsibility performance is increasingly relevant to stakeholders' decision-making and must be reflected in sustainability reports using reliable indicators to track institutional progress. However, the study by Kaloutsa et al. (2025) proposes a structured approach to evaluate the progress of HEIs in relation to the SDGs using Multi-Criteria Decision Making (MCDM) through the Analytic Hierarchy Process (AHP), highlighting the importance of a clear and objective framework to assess the progress of HEIs in adopting the SDGs, thereby facilitating future evaluations and improvements.

4 Final Considerations

This literature review enabled the collection, organization, and critical analysis of the existing academic knowledge regarding integrating the SDGs into the social responsibility practices of higher education institutions. Leal Filho et al. (2021) reinforce the recommendation to combine international literature with case studies (such as Costa et al., 2021 and their Portuguese case) to better map experiences and challenges in the operationalization of the SDGs in universities. The article by Gallardo Vázquez et al. (2024) confirms that sustainable commitment in higher education is gaining practical momentum, distinguished by consistent effects on training, institutional culture, and community engagement. It also complements the methodological justification for mixed approaches and local configurations, emphasizing the value and necessity of similar research within the Portuguese context. Lan (2025) deepens the discussion on how HEIs can structure policies, programs, and training to strengthen sustainable commitment at different academic levels. It was found that, although there is growing recognition of the strategic importance of USR, its practical implementation varies significantly across institutional and geographical contexts. By highlighting and systematizing good practices and empirical experiences, Serafini et al. (2022) demonstrate that sustainable commitment must be translated into concrete actions, precisely the focus of their article on advancing the SDGs.

Public universities tend to adopt more formal and comprehensive commitments to sustainability, while private institutions often follow more selective or reputation-driven approaches. Regarding reporting, environmental aspects tend to predominate over the social and economic dimensions, and the use of frameworks such as the GRI or the SDGs indicators remains inconsistent and underdeveloped.

Additionally, promising trends have emerged with the incorporation of technological innovation, particularly AI, as a catalyst for more sustainable, efficient, and transparent management practices. AI is positioned as a driver of institutional transformation with the potential to enhance the alignment between educational mission, social responsibility, and commitment to the 2030 Agenda. We concur with Chigbu and Makapela (2025) in asserting that innovative, inclusive, and data-driven leadership is essential for transforming higher education and contributing to global sustainability. Data-driven leadership can advance the SDGs and promote inclusive transformation in HEIs.

Nevertheless, this study presents limitations inherent to the exploratory and narrative nature of the adopted methodology. Although the selection of literature was rigorous, it did not follow a systematic or quantitative bibliometric analysis protocol, which may have constrained the scope or representativeness of the findings. Furthermore, the predominant focus on public universities may limit the generalizability of the conclusions to the broader spectrum of higher education institutions, particularly private or community-based ones.

Therefore, it is recommended that future research adopt complementary empirical approaches, including the analysis of institutional reports, interviews with university administrators, or surveys targeting various university stakeholders. It would also be valuable to explore the relationship between the use of emerging technologies, such as artificial intelligence, and the sustainability reporting models adopted by higher education institutions. Finally, comparative studies across countries and regions with varying levels of SDGs engagement could offer additional insights into the enabling factors and barriers to implementing USR in alignment with the 2030 Agenda.

References

Adams, C.A.: Sustainability reporting and performance management in universities: challenges and benefits. Sustain. Account., Manag. Policy J. **4**(3), 384–392 (2013). https://doi.org/10.1108/SAMPJ-12-2012-0044/FULL/PDF

Adhikariparajuli, M., Hassan, A., Siboni, B.: CSR implication and disclosure in higher education: uncovered points. results from a systematic literature review and agenda for future research. Sustainability **13**(2), 525 (2021). https://doi.org/10.3390/SU13020525

Aggarwal, P., Singh, A.K.: CSR and sustainability reporting practices in India: an in-depth content analysis of top-listed companies. Soc. Responsib. J. **15**(8), 1033–1053 (2019). https://doi.org/10.1108/SRJ-03-2018-0078

Ali, W., Wilson, J., Husnain, M.: Determinants/motivations of corporate social responsibility disclosure in developing economies: a survey of the extant literature. Sustainability (Switzerland) **14**(6) (2022). https://doi.org/10.3390/su14063474

Alon, I., Lattemann, C., Fetscherin, M., li, S., Schneider, A.M.: Usage of public corporate communications of social responsibility in Brazil, Russia, India and China (BRIC); Int. J. Emerg. Mark. **5**(1), 6–22 (2010). https://doi.org/10.1108/17468801011018248

An, Y., Davey, H., Harun, H.: Sustainability reporting at a New Zealand Public University: a longitudinal analysis. Sustainability **9**(9), 1529 (2017). https://doi.org/10.3390/SU9091529

Angelaki, M.E., Bersimis, F., Karvounidis, T., Douligeris, C.: Towards more sustainable higher education institutions: implementing the sustainable development goals and embedding sustainability into the information and computer technology curricula. Educ. Inf. Technol. **29**(4), 5079–5113 (2024). https://doi.org/10.1007/s10639-023-12025-8

Bartolacci, F., Caputo, A., Soverchia, M.: Sustainability and financial performance of small and medium sized enterprises: a bibliometric and systematic literature review. Bus. Strateg. Environ. **29**(3), 1297–1309 (2020). https://doi.org/10.1002/bse.2434

Behringer, K., Szegedi, K., Professor, A.: The role Of CSR in achieving sustainable development – theoretical approach. Eur. Sci. J. **12**(22), 10–10 (2016). https://doi.org/10.19044/ESJ.2016.V12N22P10

Bexell, M., Jönsson, K.: Responsibility and the United Nations' sustainable development goals. Forum Dev. Stud. **44**(1), 13–29 (2017). https://doi.org/10.1080/08039410.2016.1252424

Borsatto, J.M.L.S., Marcolin, C.B., Abdalla, E.C., Amaral, F.D.: Aligning community outreach initiatives with SDGs in a higher education institution with artificial intelligence. Cleaner Responsible Consumption **12**, 100160 (2024). https://doi.org/10.1016/j.clrc.2023.100160

Brown, B.J., Hanson, M.E., Liverman, D.M., Merideth, R.W.: Global sustainability: toward definition (1987)

Brundtland Comission. Report of the World Commission on Environment and Development (1987). https://sustainabledevelopment.un.org/content/documents/5987our-common-future.pdf

Carrillo Valdez, L., Villanueva Gutiérrez, L., Álvarez-Mendiola, G.: Contributions of three Mexican higher education institutions to sustainable development: a comparative analysis. J. Adult Continuing Educ. **31**(1), 74–95 (2025)

Carroll, A.B.: Corporate social responsibility. Bus. Soc. **38**(3), 268–295 (1999). https://doi.org/10.1177/000765039903800303

Chen, S.-H., Nasongkhla, J., Donaldson, J.A.: University Social Responsibility (USR): identifying an ethical foundation within higher education institutions. Turk. Online J. Educ. Technol. - TOJET **14**(4), 165–172 (2015)

Chigbu, B.I., Makapela, S.L.: Data-driven leadership in higher education: advancing sustainable development goals and inclusive transformation. Sustainability **17**(7), 3116 (2025)

Comissão das Comunidades Europeias. Livro Verde. Promover Um Quadro Europeu Para a Responsabilidade Social Das Empresas, 4–23 (2001). https://www.europarl.europa.eu/meetdocs/committees/empl/20020416/doc05a_pt.pdf

Costa, A., Tafuro, A., Benvenuto, M., Viola, C.: Corporate social responsibility through SDGs: preliminary results from a pilot study in Italian Universities. Adm. Sci. **11**(4), 117 (2021). https://doi.org/10.3390/ADMSCI11040117

Cubas, A.L.V., Provin, A.P., Dutra, A.R.D.A., Andrade Guerra, J.B.S.O.D., Mussi, C.C.: Forging a sustainable path: the QualEnv consortium's practices and strategies in promoting sustainable development in higher education institutions. Int. J. Sustain. High. Educ. **26**(1), 63–82 (2025)

Ebekozien, A., Aigbavboa, C.O., Samsurijan, M.S., Radin Firdaus, R.B., Rohayati, M.I.: Expanded corporate social responsibility framework: companies' role in improving higher education institutions infrastructure to sustainable development goal 4. J. Facil. Manag. **23**(1), 1–18 (2025). https://doi.org/10.1108/JFM-11-2022-0120

Erin, O.A., Bamigboye, O.A., Oyewo, B.: Sustainable development goals (SDG) reporting: an analysis of disclosure. J. Account. Emerg. Econ. **12**(5), 761–789 (2022). https://doi.org/10.1108/JAEE-02-2020-0037

Fernandes, A.B., et al.: A ética no uso de inteligência artificial na educação: implicações para professores e estudantes. Revista Ibero-Americana de Humanidades, Ciências e Educação **10**(3), 346–361 (2024). https://doi.org/10.51891/rease.v10i3.13056

Firoiu, D., Ionescu, G.H., Pîrvu, R., Bădîrcea, R., Patrichi, I.C.: Achievement of the sustainable development goals (SDG) in Portugal and forecast of key indicators until 2030. Technol. Econ. Dev. Econ. **28**(6), 1649–1683 (2022). https://doi.org/10.3846/tede.2022.17645

Fonseca, A., Macdonald, A., Dandy, E., Valenti, P.: The state of sustainability reporting at Canadian universities. Int. J. Sustain. High. Educ. **12**(1), 22–40 (2011). https://doi.org/10.1108/14676371111098285/FULL/PDF

Gallardo-Vázquez, D., De La Cruz Sánchez-Domínguez, J., Valdez-Juárez, L.: Corporate social responsibility and sustainable development goals in higher education: a literature review from 2014 to 2023. Bottom Line (2024). https://doi.org/10.1108/bl-08-2023-0232

Ismail, T.H.: Does Egyptian universities' disclosure on social responsibility enhance sustainable development? J. Humanit. Appl. Soc. Sci. **2**(2), 81–99 (2019). https://doi.org/10.1108/JHASS-08-2019-0042

Jorge, M., Hernández, A., Madueño, J., Peña, F.: Do Spanish Public Universities use corporate social responsibility as a strategic and differentiating factor? Int. J. Humanit. Soc. Sci. **2**(11) (2012)

Kaloutsa, M., Kabassi, K., Martinis, A.: Evaluating the sustainable development goals in higher education institutions using multi-criteria decision making/analysis: calculating the weights of criteria with the analytic hierarchy process. Sustain.: Sci., Pract. Policy **21**(1), 2475592 (2025)

Lan, H.: Sustainable development of social responsibility in universities in the digital era: perceptions and advances. Inf. Dev. (2025). https://doi.org/10.1177/02666669241303124

Leal Filho, W., et al.: The role of higher education institutions in sustainability initiatives at the local level. J. Clean. Prod. **233**, 1004–1015 (2019). https://doi.org/10.1016/j.jclepro.2019.06.059

Leal Filho, W., et al.: Trends in scientific publishing on sustainability in higher education. J. Clean. Prod. **296**, 126569 (2021). https://doi.org/10.1016/j.jclepro.2021.126569

Leal Filho, W., et al.: An overview of research trends on sustainability in higher education–an exploratory study. Int. J. Sustain. High. Educ. **24**(5), 1161–1175 (2023). https://doi.org/10.1108/IJSHE-08-2022-0252

Lu, J., et al.: Assessment of corporate social responsibility by addressing sustainable development goals. Corp. Soc. Responsib. Environ. Manag. **28**(2), 686–703 (2021). https://doi.org/10.1002/CSR.2081

Matten, D., Moon, J.: Corporate social responsibility education in Europe. J. Bus. Ethics **54**(4), 323–337 (2004). https://doi.org/10.1007/S10551-004-1822-0/METRICS

Meseguer-Sánchez, V., Abad-Segura, E., Belmonte-Ureña, L.J., Molina-Moreno, V.: Examining the research evolution on the socio-economic and environmental dimensions on University social responsibility. Int. J. Environ. Res. Public Health **17**(13), 4729 (2020). https://doi.org/10.3390/IJERPH17134729

Nejati, M., Shafaei, A., Salamzadeh, Y., Daraei, M.: Corporate social responsibility and Universities: a study of top 10 World Universities' websites. Afr. J. Bus. Manag. **5**(2), 440–447 (2011). https://ssrn.com/abstract=1868688

OECD. Financing SMEs for sustainability (OECD SME and Entrepreneurship Papers, Vol. 35) (2022). https://doi.org/10.1787/a5e94d92-en

Ortiz-Martínez, E., Marín-Hernández, S., Santos-Jaén, J.M.: Sustainability, corporate social responsibility, non-financial reporting and company performance: relationships and mediating effects in Spanish small and medium sized enterprises. Sustain. Prod. Consumption **35**, 349–364 (2023). https://doi.org/10.1016/j.spc.2022.11.015

Paravizo, E., Chaim, O.C., Braatz, D., Muschard, B., Rozenfeld, H.: Exploring gamification to support manufacturing education on industry 4.0 as an enabler for innovation and sustainability. Procedia Manuf. **21**, 438–445 (2018). https://doi.org/10.1016/j.promfg.2018.02.142

Pinto, M.M.: Responsabilidade social em universidade comunitária: novos rumos para educação superior [Tese de Doutoramento, Faculdade de Educação da Pontifícia Universidade Católica do Rio Grande do Sul, Pontifícia Universidade Católica do Rio Grande do Sul] (2009). https://tede2.pucrs.br/tede2/handle/tede/3612

Rahman, A.A., Castka, P., Love, T.: Corporate social responsibility in higher education. Corp. Soc. Responsib. Environ. Manag. **26**(4), 916–928 (2019). https://doi.org/10.1002/CSR.1731

Ramos-Monge, E.L., Audet, X.L., Martínez, J.B.: Universities as corporate entities: the role of social responsibility in their strategic management. Corp. Governance Strateg. Decis. Mak. (2017). https://doi.org/10.5772/INTECHOPEN.69931

Russell-Bennett, R., Rosenbaum, M.S., Fisk, R.P., Raciti, M.M.: SDG editorial: improving life on planet earth–a call to action for service research to achieve the sustainable development goals (SDGs). J. Serv. Mark. **38**(2), 145–152 (2024). https://doi.org/10.1108/JSM-11-2023-0425

Saha, A.K., Al-Shaer, H., Dixon, R., Demirag, I.: Determinants of carbon emission disclosures and UN sustainable development goals: the case of UK higher education institutions. Aust. Account. Rev. **31**(2), 79–107 (2021). https://doi.org/10.1111/auar.12324

Saini, M., Sengupta, E., Singh, M., Singh, H., Singh, J.: Sustainable development goal for quality education (SDG 4): a study on SDG 4 to extract the pattern of association among the indicators of SDG 4 employing a genetic algorithm. Educ. Inf. Technol. **28**(2), 2031–2069 (2023). https://doi.org/10.1007/s10639-022-11265-4

Sassen, R., Azizi, L.: Assessing sustainability reports of US universities. Int. J. Sustain. High. Educ. **19**(7), 1158–1184 (2018). https://doi.org/10.1108/IJSHE-06-2016-0114/FULL/XML

Sassen, R., Azizi, L.: Voluntary disclosure of sustainability reports by Canadian universities. J. Bus. Econ. **88**(1), 97–137 (2018). https://doi.org/10.1007/S11573-017-0869-1/TABLES/8

Sawani, Y., Joseph, C., Leong, S.H.: Motivating factors of higher education institutions sustainable development goals online disclosure: a systematic literature review. Int. J. Sustain. High. Educ. **26**(3), 651-671 (2025)

Setó-Pamies, D., Domingo-Vernis, M., Rabassa-Figueras, N.: Corporate social responsibility in management education: current status in Spanish universities. J. Manag. Organ., 604–620 (2011). https://doi.org/10.5172/JMO.2011.604

Serafini, P., De Moura, J., Almeida, M., Rezende, J.: Sustainable development goals in higher education institutions: a systematic literature review. J. Clean. Prod. (2022).https://doi.org/10.1016/j.jclepro.2022.133473

Silva, S., Ruão, T., Gonçalves, G.: A responsabilidade social nas universidades: ensino, investigação e extensão. Cadernos de Comunicação e Linguagem **23**(2), 2–18 (2019). http://hdl.handle.net/10400.6/8980

United Nations. Transforming our world: the 2030 Agenda for Sustainable Development - Resolution adopted by the General Assembly on 25 September 2015 (2015). https://documents-dds-ny.un.org/doc/UNDOC/GEN/N15/291/89/PDF/N1529189.pdf?OpenElement

Vasilescu, R., Barna, C., Epure, M., Baicu, C.: Developing university social responsibility: a model for the challenges of the new civil society. Procedia Soc. Behav. Sci. **2**(2), 4177–4182 (2010). https://doi.org/10.1016/J.SBSPRO.2010.03.660

Wajid, H.A.: The Role of Saudi Arabian higher education institutions in sustainable development: participation, framework alignment, and strategic insights. Sustainability **17**(6), 2530 (2025)

Yoon, Y., Giirhan-Canli, Z., Schwarz, N.: The effect of corporate social responsibility (CSR) activities on companies with bad reputations. J. Consum. Psychol. **16**(4), 377–390 (2006). https://doi.org/10.1207/s15327663jcp1604_9

Zhang, G., Chen, P., Xu, S.: Developing and validating a scale for measuring sustainable leadership development among teachers in Chinese higher education institutions. J. Clean. Prod. **486**, 144403 (2025)

Are Firms' Performance Influenced by Corporate Governance?

Luís Almeida[1,2(✉)] and Elisabete Vieira[1,2]

[1] ISCA – Universidade de Aveiro Portugal, Aveiro, Portugal
{gomesalmeida,elisabete.vieira}@ua.pt
[2] GOVCOPP Unit Research, ISCA – Universidade de Aveiro (ISCA-UA), Aveiro, Portugal

Abstract. This paper analyzes the impact of corporate governance on family business performance. Moreover, the study tries to verify whether the relationship between corporate governance and performance varies in periods of instability. We consider a sample of all Portuguese-listed firms on Euronext Lisbon, excluding financial companies and sports clubs, for the period between 2010 and 2023, using the generalized method of moments estimation to account for endogeneity in panel data analysis. The results show that gender diversity positively impacts firms' performance in the cases of accounting performance measures and that smaller boards are associated with higher levels of accounting and market performance measures. Concerning the impact of CEO duality on performance, our results suggest conflicting effects according to the performance measures, since it shows a negative relationship between duality and the return on assets, but a positive relationship in the case of the return on equity and Tobin's Q measure. Moreover, periods of instability have a significant adverse effect on performance. Firms' age, leverage, and economic growth impact on companies' performance.

The results show evidence that corporate governance characteristics are determinants of firm performance, and highlight the conflict between agency and stewardship theories, suggesting the need for context-sensitive governance choices.

Keywords: Performance · Corporate Governance · Instability Periods · Panel Data analysis

1 Introduction

The strategic decisions made by the board of directors can significantly influence firm performance since this group of people, who represent the interests of a firm's shareholders, not only implements governance practices but also influences the strategic direction of firms, which affects the outcomes of companies [1].

The literature review, in general, confirms that the corporate governance characteristics significantly influence the firm's performance. However, the results are not consistent, since some studies find a positive impact of corporate governance on performance, such as [2–6], while other studies find a negative impact [7–9]. There is also empirical

L. Almeida et al. (Eds.): ICAFI 2025, ISEM 80, pp. 177–193, 2026.
https://doi.org/10.1007/978-3-032-16268-7_10

evidence that the effect is nuanced, depending on factors such as external economic conditions and regulatory environments [1, 10, 11].

To overcome the problem of mixed results, we reinforce three corporate governance attributes that tend to present mixed evidence (gender diversity, board size, and CEO duality).

In addition, despite the vast research on the relationship between corporate governance and companies' performance, the performance metrics vary considerably, such as the Return on Assets (ROA), the Return on Equity (ROE), Tobin's Q (TQ), the Return on Capital Employed, the Market Value Added, Economic Value Added, Earnings per Share, Price-to-Earnings, and Market-to-Book ratios, which difficult the results comparison and the conclusions about this relation-ship. To cover this gap, we will use the most common performance metrics: the ROA (an accounting measure) and TQ (a market measure), such as [12].

In this context, we analyze how corporate governance characteristics affect the performance of companies and explore the impact of instability periods on performance.

This study adds several contributions to the literature. First, it tries to cover the gap in what concerns nonconsensual conclusions about the contribution of corporate governance to companies' outputs, as well as the different results according to the performance measure used to proxy the firm's performance. Second, it analyzes a European market that needs research: Portugal. This country is characterized by its very small stock market and a bank-based system. Finally, it explores whether instability periods affect the performance of companies.

Our results show that gender diversity positively impacts firms' performance in the cases of accounting performance measures and that smaller boards are associated with higher levels of accounting and market performance measures. Concerning the impact of CEO duality on performance, our results suggest conflicting effects according to the performance measures, since it shows a negative relationship between duality and the return on assets, but a positive relationship in the case of the return on equity and Tobin's Q measure.

This study is organized as follows. Section 2 presents the literature review and formulates the hypotheses. Section 3 presents the variables, the data, and the methodology. The next section shows and analyzes the empirical results, and finally, Sect. 5 concludes the study.

2 Literature Review and Hypotheses

2.1 Corporate Governance and Performance

Corporate governance refers to a set of practices and mechanisms aimed at protecting the interests of shareholders and improving the efficiency and transparency of business operations.

The main theories related to corporate governance are the agency theory [13, 14], the stewardship theory [15], and the stakeholder theory [16].

Agency theory argues that shareholders (the principal) and managers (the agent) have a working relationship, but their interests may not always match. Because of this, systems are needed to make sure managers act in the best interests of shareholders. Thus, good corporate governance can reduce agency problems by aligning the interests of managers and shareholders, enhancing the firm's performance.

The stewardship theory suggests that managers see themselves as wardens of the company and feel responsible for its success. This theory highlights trust, teamwork, and cooperation between managers and shareholders, which reduces the conflicts between shareholders and managers.

Finally, the stakeholder theory argues that firms should consider not only their shareholders but also employees, customers, suppliers, and the community. Thus, strong corporate governance should balance the interests of all these groups.

The relationship between corporate governance and firm performance has been extensively studied, highlighting the vital role that governance mechanisms play in influencing companies' outcomes since corporate governance helps to ensure that shareholder interests, as well as those of other stakeholders, are prioritized, which can improve the performance of firms [2].

Although a significant part of the empirical studies documents a positive relationship between corporate governance practices and firms' performance, there is mixed evidence. Several studies highlight a positive relationship between corporate governance practices and organizational performance [2].

However, the evidence of a positive impact of corporate governance on performance is not consensual. For example, [17] note that while corporate governance is generally beneficial, it may also correlate negatively with performance in specific contexts, indicating a nuanced relationship. Similarly, [10] conclude that while governance practices can lead to favorable performance, the impact may vary depending on external economic conditions and regulatory environments. [1] have already warned that excessive regulation could inadvertently limit the benefits of corporate governance, suggesting that it is needed to promote balanced and effective governance, without being overly restrictive.

Overall, the literature highlights a complex but significantly positive relationship between corporate governance and firm performance. Most studies confirm that strong governance structures enhance financial outcomes, though the impact varies based on context and specific practices.

Now, we will consider some corporate governance characteristics, according to the availability of information in the database to which we have access.

2.2 Gender Diversity and Performance

Research on gender diversity and firms' performance has increased in recent decades. Gender diversity improves team effectiveness by accessing broader talent pools [18] and enhancing decision-making [19]. Women, typically more risk-averse, may promote sustainable investment strategies [20]. Additionally, diverse teams benefit from broader networks, improved market understanding, and increased innovation [21]. Diversity also aligns with socially responsible management, increasing the firm's reputation and stakeholder relationships [22]. Some authors also argue that the presence of women on the board of directors introduces a greater heterogeneity of experiences and knowledge, skills, which tends to favor more sustainable decisions, strengthening the performance and effectiveness of companies [23].

Most empirical studies find a positive relationship between gender diversity and performance [3–6, 24–27], although some find a negative relationship between these two variables [7, 8], namely because diversity can slow decision-making due to increased opinions and conflicts. In addition, other studies fail to find evidence of a significant impact of gender diversity on companies' performance [17]. [10] find mixed results. Although they find a positive impact of gender diversity on the ROE and TQ, they cannot find a significant relationship when the performance measure is the ROA. There is also evidence of a nonlinear relationship between the two variables [28].

Considering the above-mentioned, we formulate the first hypothesis:

H1: Gender diversity on the board of directors is associated with higher firm performance.

2.3 Board Size and Performance

One of the most relevant governance mechanisms is board size because it is associated with the participation of the board of directors in the company's decisions, activities, and, consequently, future success. It is a critical element of governance since it reflects the balance between diversity of competencies and efficiency in decision-making.

Economic theory views the board of directors as a key factor of corporate governance [29] since board size is associated with directors' capacity to monitor managers. However, others argue that smaller boards perform better because larger groups struggle with communication and decision-making, reducing efficiency [30, 31].

Considering empirical evidence, board size generally shows a negative impact on performance because they are more effective in fulfilling their duties [3, 30], have better coordination, communication and are more agile in the decision-making process [32–35], though some studies find a positive relationship between the two variables, suggesting that larger boards improve firms' performance and enhance monitoring [36, 37], as they offer more specialized skills, greater diversity of knowledge, opinion, and experience in problem-solving [37]. However, they can introduce complexity and slowness into deliberative processes.

Some authors find that the impact of board size on performance depends on the proxies used to measure performance. [38] find that board size negatively impacts firm performance when measured by ROA but positively influences performance when measured by TQ. [39] findings show that board size influences performance differently. Whereas it is negatively related to TQ and ROE, it has no significant impact on the ROA, and [40] conclude that board size negatively (positively) impacts book-based (market-based) performance. [10] also conclude that the impact of board size on a firm's performance depends on the metrics used to measure performance, since they find a negative effect on ROE, a positive impact on TQ, and no significant effect on ROA.

Considering the arguments of [30] and [31], as well as a major part of the empirical findings, we consider the following hypothesis:

H2: Smaller boards of directors are associated with higher firm performance.

2.4 CEO Duality and Performance

The CEO duality happens when the chief executive officer also assumes the position of chairman of the board. Corporate governance suggests that combining the CEO and chair of the board of directors concentrates power, raising concerns about oversight and control, so a relevant characteristic of board independence is the separation of the CEO and the board chair roles.

From the perspective of the agency theory, without effective control, management may prioritize its own interests over shareholders', compromising the supervisory role of the board of directors, and weakening internal control mechanisms [29], suggesting that CEO duality will decrease a company's performance. Several authors, such as [41], argue that CEO duality risks board dominance, allowing the CEO to control meeting agendas, pre-meeting information distribution, and discussions, and [42] consider that duality weakens board independence, reducing governance effectiveness. A significant number of empirical studies prove that CEO duality results in worse firm performance [3, 35, 43–47].

However, stewardship theory offers a different perspective. It suggests that executives aim to act in the company's best interests rather than for personal gain [14]. From this view, CEO duality improves performance, namely by providing clear leadership for strategy development and implementation, and a quicker response to strategic decisions [24, 48]. Concentrating authority in one person can also reduce uncertainty about decision-making responsibility, contributing to improving firms' performance [49]. Some authors find mixed results. For example, [38] find evidence that CEO duality shows conflicting effects on the performance measure (ROA and TQ). The authors find a negative relationship between duality and ROA, suggesting that ROA is higher when the two roles are separated, but find a positive relation in the case of TQ, and [50] finds mixed and inconclusive results regarding the relationship between CEO duality and firm performance. Differently, [39] find that duality is not significantly related to firm performance when performance is measured by TQ and ROE. However, duality is significantly negatively related to ROA. In addition, [51] find no evidence of a significant impact of CEO duality on the performance of companies.

Based on the agency theory (stewardship theory), we expect that duality contributes to decreasing (increasing) a company's performance.

According to the agency theory and most of the empirical results, we formulate the following hypothesis:

H3: CEO duality is associated with lower firm performance.

2.5 Corporate Governance Characteristics and Performance in Periods of Instability

Periods of instability induce social and economic uncertainty and increase volatility in the corporate sector and financial markets, which can affect firm performance. In the short term, instability can lead to leadership changes, board of directors' conflicts, and regulatory uncertainty, reducing investor confidence and organizational efficiency. However, if instability leads to governance reforms, improved oversight, or strategic realignment, it can improve long-term performance. The impact depends, for example, on whether corporate governance characteristics promote better decision-making.

[52] and [53] provide empirical evidence that economic and political instability periods significantly impact firm performance. Their findings suggest that moderate levels of uncertainty may foster innovation, whereas elevated levels of uncertainty tend to hinder it. [54] finds evidence that perceptions of economic policy uncertainty are negatively associated with firm value, indicating that heightened uncertainty prompts more cautious investment decisions, ultimately leading to a decline in firm valuation.

Consequently, we hypothesized that periods of instability negatively affect the performance of companies, formulating the last hypothesis:

H4: Periods of instability are associated with lower firm performance.

3 Methodology

3.1 Variables

3.1.1 Dependent Variables

Based on the literature review, we realize that the most common performance metrics used are the ROA (accounting measure) and TQ (market measure), which agree with the opinion of [11]. Consequently, we use the ROA and the TQ, as [10, 11, 38, 39]. For robustness reasons, we also consider the ROE.

3.1.2 Independent Variables

The independent variables used in this study fall into a set of dimensions associated with corporate governance, namely gender diversity on the board of directors (GENDER), the size of the board (BSIZE), and the dual functions performed by the CEO (DUAL).

We also consider periods of instability (INST), since instability can impact business performance because it creates social and economic uncertainty and increases volatility in the financial markets. Short-term instability can result in conflicts on the board of directors, changes in leadership, and regulatory ambiguity, which lowers organizational effectiveness and investor trust. On the other hand, volatility can enhance long-term performance if it results in better oversight, strategic realignment, or governance reforms. To identify the INST, we consider a dummy variable that takes the value 1 if the firm operates during a financial, economic, or pandemic crisis period and 0 otherwise.

Firm-specific and macroeconomic conditions are tested as control variables.

According to previous research, the firm's age (AGE) is expected to positively influence the performance of companies [22, 55], since their experience and stability are more likely to lead to better decision-making, a stronger reputation, and more resources, which help improve performance. However, [35] finds a negative relation between firms' age and high returns on invested capital of companies.

Financial leverage (LEV) can influence the company's performance. The results concerning the impact of leverage on performance are mixed. Whereas some authors find a negative relationship [56, 57], others find a positive impact of leverage on performance [58]. Moreover, the results of [38] show a negative impact of leverage on ROA measure but a positive association with TQ, suggesting that the market places a premium on companies that use more leverage.

Since macroeconomic factors affect businesses' financial decisions, we also control them [59]. We consider that Gross Domestic Product (GDP) is an adequate proxy for macroeconomic conditions since it can influence performance.

Table 1 shows the description of the variables and the expected signal.

Table 1. Variables description

Variables	Acronym	Expected Signal	Description
Dependent variables			
Tobin's Q	TQ		The ratio of market value to total assets
Return on Assets	ROA		Ratio of earnings before interest and taxes to the book value of total assets
Return on Equity	ROE		The ratio of net earnings to equity
Independent variables			
Gender on board	GENDER	+	The ratio of female directors to the total number of directors
Board Size	BSIZE	+	Number of members of the Board of Directors
CEO Duality	DUAL	-	Dummy variable that takes the value 1 if CEO is chairman, and 0 otherwise
Years of instability	INST	-	The dummy variable takes the value 1 if the firm operates during a financial, economic, or pandemic crisis period and 0 otherwise
Control variables			
Age	AGE	+	The logarithm of the book value of total assets
Leverage	LEV	-	The ratio of current assets to current liabilities
Economic growth	GDP	+	Real GDP growth (%) compared to the previous year

4 Methodology

The methodology adopted in this study is based on the Generalized Method of Moments (GMM). This method is particularly suitable for dealing with endogeneity problems, heteroscedasticity, and unobservable individual effects, guaranteeing consistent and efficient estimates. In this study, we opted for the dynamic GMM of [60, 61], which allow the use of internal instruments based on past values of the explanatory variables, mitigating potential simultaneity. GMM solves the problem of the correlation between $\alpha\gamma_{it-1}$

and the fixed effects μ_i, by applying first-order differences and using internal instruments based on past values of the variables. The dynamic model to be estimated can be generically represented as follows:

$$\Upsilon it = \alpha \gamma it - 1 + X_{it}\beta + \mu_i + \varepsilon_{it} \quad (1)$$

where Υ_{it} represents the dependent variable i in period t. γ_{it-1} is the lagged dependent variable, capturing dynamic effects. X_{it} represents the vector of explanatory variables, μ_i corresponds to the unobservable individual effects, and ε_{it} represents the idiosyncratic error term. To achieve the objectives of the study, the following model was estimated:

$$PERF_{i,t} = \alpha + \beta_1 Gender_{i,t} + \beta_2 Bsize_{i,t} + \beta_3 Dual_{t,i} + \beta_4 INST_{i,t} + \beta_5 AGE_{i,t} + \beta_6 LEV_{i,t} + \beta_7 GDP_{i,t} + \varepsilon_{i,t} \quad (2)$$

Performance (PERF) consists of business performance variables (TQ, ROA, and ROE), which have already been mentioned. The other variables are explained in Table 1.

To ensure the robustness of the estimates and the validity of the instruments used in the GMM model, specific statistical tests are carried out to assess the homogeneity of the instruments, the presence of autocorrelation in the residuals, and the joint significance of the coefficients. The Sargan test checks that the instruments are exogenous, i.e., not correlated with the error term. The Sargan test is based on chi-squared statistics and assumes homoscedasticity.

The Arellano-Bond test assesses the presence of serial autocorrelation in the residuals of the GMM model in differences since the differentiation applied in the estimation can generate time dependence on the errors. The test checks for the existence of first-order autocorrelation [AR(1)] and second-order autocorrelation [AR(2)]. AR(1) is expected due to the transformation applied in the model, so a p-value of less than 0.05 does not invalidate the estimation. However, the absence of AR(2) is essential to guarantee the validity of the instruments, so a p-value greater than 0.05 for this test confirms the adequate specification of the model. Otherwise, the existence of second-order autocorrelation indicates that the instruments may not be valid, which could compromise the reliability of the estimates.

The Wald test assesses the joint significance of the explanatory variables in the model. This test checks whether all the coefficients of the independent variables are simultaneously equal to zero, which would indicate that the explanatory variables have no significant effect on the dependent variable. A p-value of less than 0.05 rejects this hypothesis, confirming that at least one of the explanatory variables contributes significantly to the model. Otherwise, high p-value suggests that the explanatory variables may not have a statistically significant impact, which may indicate problems with the model specification.

Applying these tests ensures the validity of the instruments, the correct specification of the model, and the significance of the explanatory variables, guaranteeing that the GMM method provides robust and reliable estimates. Therefore, the joint interpretation of the results obtained in the Sargan/Hansen, Arellano-Bond, and Wald tests is essential for the correct validation of the model and for obtaining consistent econometric inferences.

4.1 Sample

The sample for this study is made up of all the Portuguese companies listed on Euronext Lisbon between 2010 and 2023, excluding financial and sports clubs due to their different economic and financial characteristics. The final sample consisted of 29 companies, with a total of 406 observations, which clearly shows that Portugal has a small stock market. The purpose of this selection was to ensure the homogeneity of the analysis. The period under study was characterized by profound changes in the market, resulting from economic crises and the pandemic, which significantly influenced business dynamics in Portugal. Financial data was extracted from the Bureau Van Dijk, Orbis database, while governance information was collected manually from the companies' annual financial statements. In addition, the GDP growth rate was obtained from the Pordata database. The panel models were estimated using the GMM method and the Stata 18 software.

5 Results

5.1 Descriptive Statistics

Table 2 shows the descriptive statistics of the variables used in the study.

Table 2. Descriptive Statistic

Variable	Mean	Std. Dev	Min	Max
TQ	1.0091	0.3925	0.0544	2.5700
ROA	2.7685	6.0692	−29.1190	37.3910
ROE	9.4088	35.2600	−434.3910	138.5350
GENDER	0.1727	0.1329	0.0000	0.5000
BSIZE	9.7562	4.8815	3.0000	26.0000
DUAL	0.6601	0.4743	0.0000	1.0000
INST	0.5714	0.4955	0.0000	1.0000
AGE	3.5755	0.5512	1.7918	4.6347
LEV	0.6217	0.2047	0.0273	1.1555
GDP	1.0786	3.7396	−8.2000	7.0000

The variable TQ shows a mean of 1.009 and a standard deviation of 0.393, suggesting a relatively concentrated distribution around the mean. The accounting measures of performance (ROA and ROE) report mean values of 2.769 and 9.409, respectively, showing a higher mean profitability for shareholders than the performance resulting from the investment in assets. However, the ROE is more volatile than the ROA. Corporate governance characteristics show that, on average, 17.3% of board members are women, reflecting a modest female representation among the Portuguese firms under analysis. However, some boards of directors have no women, and others have 50% females in their composition. CEO duality (DUAL) occurs in 66% of the companies, indicating a prevalent practice in Portuguese companies. The variable INST, which captures crisis periods, presents a mean of 0.571, suggesting that approximately 57% of the observed period was characterized by an economic downturn. Concerning the control variables, the average firm age (AGE) is 3.58 years, and the LEV ratio has a mean of 62.17%, showing that the sample companies are somewhat indebted. Finally, the GDP growth exhibits considerable variation, ranging from −8.2% to 7%, indicating substantial economic instability throughout the study period, which is consistent with the results of INST variable.

Table 3 displays the correlation coefficients among the study variables.

Table 3. Correlation matrix

	TQ	ROA	ROE	AGE	BSIZE	GENDER	DUAL	LEV	GDP	INST
TQ	1.000									
ROA	0.232	1.000								
ROE	0.201	0.495	1.000							
AGE	−0.031	0.105	0.020	1.000						
BSIZE	0.269	0.035	0.095	−0.292	1.000					
GENDER	0.034	0.158	0.162	0.035	0.028	1.000				
DUAL	−0.042	0.095	0.034	0.275	−0.542	−0.045	1.000			
LEV	0.323	−0.1486	0.198	0.031	0.252	−0.084	−0.115	1.000		
GDP	−0.037	0.198	0.089	−0.005	−0.024	0.246	0.010	0.013	1.000	
INST	−0.147	−0.0266	−0.041	0.029	0.046	0.107	0.009	0.063	−0.263	1.000

Analyzing the correlation matrix is crucial in assessing the strength and direction of the relationships between variables. According to the specialized literature, correlation coefficients below 0.3 indicate weak relationships, values between 0.3 and 0.8 suggest moderate relationships, and coefficients above 0.8 are generally interpreted as strong [62]. [63] identify 0.8 as a critical threshold, suggesting that correlations above this value may indicate the presence of multicollinearity among explanatory variables, while values below this threshold are typically not associated with such statistical issues.

In our sample, the highest correlation was observed between the DUAL and BSIZE (−0.5425) variables, suggesting that the higher boards have less CEO that is, simultaneously, the chairman of the board of directors. All remaining correlations were below this value, indicating the absence of significant multicollinearity among the variables under analysis.

6 Model Results

Table 4 presents the results of the models estimated using the GMM, considering the dependent variables TQ, ROA, and ROE.

Table 4. Regression results

	Mod. 1 - QTobin	Mod. 2 – ROA	Mod. 3 - ROE
Variable	Coefficient	Coefficient	Coefficient
L1	0.0940	0.0138	0.1026
GENDER	0.0940	5.7711**	30.1706**
BSIZE	−0.0375*	0.4034*	−0.8810**
DUAL	0.2758*	−2.7574***	13.5248*
INST	−0.1589**	−0.1385**	−0.1589**
AGE	0.1680	5.6318	21.9047*
LEV	−0.7548**	−14.549**	12.0450***
GDP	−0.4720**	−0.0097**	−0.0123**
cons	−0.1541	−13.4708	60.0515
Wald	135.40***	44.26***	54.37***
Ar (1)	−1.07**	−2.48**	−1.10**
AR (2)	0.64	0.01	0.13
Sargan	22.64	15.35	13.83

The validation and suitability of the models were checked using the instrument specification and validity tests described above. The Wald test confirms the joint significance of the explanatory variables, indicating that the adjusted models are statistically relevant. The Sargan test suggests that the instruments used are valid since the statistics do not reject the null hypothesis of exogeneity. Finally, the Arellano-Bond test for autocorrelation shows that there is first-order correlation (AR1), but not second-order correlation (AR2), which validates the choice of the dynamic model and confirms the absence of autocorrelation problems in the second-order residuals.

The lag variable (L1.) shows positive coefficients in all models. Still, it is not statistically significant, suggesting that while past performance may influence current performance, this relationship is not strong enough to be considered a significant determinant.

The GENDER variable shows a significant positive signal in the three models. However, it is statistically significant only in the cases of accounting performance measures (ROA and ROE), suggesting that the higher female representation on the board, the better the economic outputs, corroborating the previous empirical literature which shows that gender diversity contributes positively to the performance of companies, such as [3–6, 23–26]. Consequently, the results support the hypothesis that gender diversity is associated with higher performance (H1) for the cases of accounting performance measures. The reasons associated with the fact that the GENDER is only statistically significant for accounting and not for market measures can be related to the decision-making process of females [18], their promotion of sustainable investment strategies [19], as well as their contribution to stakeholder relationships [21], which improves the accounting measures of performance. However, the market might be skeptical and does not realize that this contributes to the enhancement of the firm's value, which does not significantly affect market-based metrics (QT).

BSIZE has a significant and negative effect on performance for all three models, being stronger in the ROE measure, suggesting that smaller boards may perform better because of better coordination, communication, a more agile decision-making process, and because of more effectiveness in fulfilling their duties, which is corroborated by previous studies, such as the ones of [3, 30–35]. Given this result, we corroborate the hypotheses that smaller boards are associated with higher performance (H2).

The CEO duality has distinct effects on the different models. The DUAL variable has a positive and significant impact on performance in the QT and ROE measures, which is in accordance with the stewardship theory arguments [14], and suggests that CEO duality improves performance, specifically by providing clear leadership, a quicker response to strategic decisions, and greater leadership capacity, which is in line with the conclusions of [24, 48, 49]. However, for the ROA dependent variable, DUAL has a significant negative impact, which supports the agency theory [29], and agrees with previous studies [3, 35, 43–47]. This result also suggests that the concentration of power can jeopardize the company's operating profitability due to a lack of oversight and internal control [38, 50]. The evidence of mixed results according to the performance measures used was also found in previous studies [38, 39, 50, 51]. For example, [38] find a negative relationship between duality and ROA and a positive relation in the case of TQ, which agrees with our evidence. Concluding, we find evidence for the hypothesis that CEO duality is associated with lower performance (H3), but only for the ROA performance measure. The mixed results need a more careful analysis of the benefits and disadvantages of accumulating these functions, especially when aiming to maximize operational efficiency [39]. In addition, supporting both agency and stewardship theories suggests that different governance mechanisms align with different theoretical frameworks depending on the performance measure. This duality suggests that future research should avoid framing agency and stewardship theories as mutually exclusive.

Periods of instability, assessed through the INST variable, have a significant adverse effect on performance, considering all the performance measures (models 1–3), which reflect the destructive nature of economic crises on business performance [52–54]. Consequently, we give support to the hypothesis that periods of instability are associated with lower levels of performance (H4).

Regarding the control variables, the results show that AGE has positive effects on performance but is only statistically significant for ROE, suggesting that older companies with more experience tend to generate higher returns for shareholders, consistent with the evidence of [22, 55], among others. This may be an indication that mature companies have greater management capacity and a solid reputation in the market, which favors profitability. However, the evidence suggests that for the market, the AGE is not a key factor for firm performance. The LEV negatively impacts the QT and ROA but has a positive effect on ROE, which reinforces the evidence of mixed results [56–58]. LEV can improve ROE, maybe by the financial leverage effect. On the other hand, the market seems to react negatively to the leverage increase because more debt increases risk, or because the market creates the perception that firms have liquidity problems and need money quickly.

Finally, contrary to our expectations, GDP shows a significant negative effect on all three models, suggesting that economic growth was unfavorable to company performance during the period analyzed. This result may be related to specific economic shocks or macroeconomic instability that negatively affected company performance, despite the economy's aggregate growth. For example, wages and raw materials become more expensive when the economy grows, and higher expenses can damage firms' profits and, consequently, decrease performance.

7 Conclusions

This study investigates the relationship between corporate governance characteristics and firm performance, considering a sample of Portuguese-listed companies for the period between 2010 and 2023, considering both accounting and market-based performance measures, as well as the influence of economic instability on performance.

The findings show that corporate governance plays a significant role in determining firm performance, although the effects vary depending on the governance dimension and the performance metric used. Gender diversity on the board is positively associated with performance, particularly when measured through accounting indicators (ROA and ROE), which may reflect the positive impact of female leadership on internal decision-making and operational efficiency. Smaller boards are consistently linked to better performance across all metrics, supporting the idea that streamlined governance structures enhance agility and effectiveness. CEO duality presents mixed effects. While it negatively impacts ROA, it shows a positive relationship with ROE and Tobin's Q, suggesting that its influence depends on the type of performance being assessed. These results highlight the conflict between agency and stewardship theories and suggest the need for context-sensitive governance choices.

Interestingly, periods of economic instability are shown to have a detrimental effect on firm performance, regardless of the measure used. This underlines the importance of resilient governance structures that can withstand external shocks. Additionally, firm age, leverage, and GDP growth influence performance, with unexpected findings such as a negative association between GDP growth and firm outcomes, possibly due to rising operational costs or resource constraints during periods of economic expansion.

Overall, this study contributes to the literature by providing empirical evidence from a small, underexplored European market and reinforces the importance of tailored governance mechanisms that account for internal and external dynamics.

This study has some implications for firms and policymakers. First, it promotes gender diversity. Since gender diversity improves performance, firms should focus on an inclusive board of directors both for social reasons and performance benefits, which can support initiatives like gender quotas or board diversity policies.

Second, it helps to review the board structure, since the consistent negative effect of larger boards on performance suggests that firms might benefit from leaner and more agile boards, especially in smaller markets like Portugal. Consequently, policymakers and firms may need to rethink governance codes that encourage unnecessarily large boards. Third, given the mixed results of CEO duality depending on the performance measure, firms should assess whether unifying leadership enhances strategic clarity or compromises accountability. Finally, the negative effect of economic instability on performance reinforces the need for resilient governance practices. Boards must adopt proactive risk management strategies to remain effective in unstable contexts.

In what concerns the implications for investors and given the distinct results according to the performance metrics used, they should look beyond market metrics and evaluate accounting measures of performance to assess the impact of governance quality on performance. Investors can determine whether governance characteristics are important for a firm's performance. For example, gender diversity is a sign of good governance and indicative of greater return potential. Investors can identify periods of instability, knowing that they can reduce profits and the market value of companies, pointing to a greater propensity to invest.

By focusing on a small and under-researched European market (Portugal), the study highlights the value of contextual governance research, which contributes to a wider understanding of the relationship between corporate governance and performance in a distinct context of large or Anglo-Saxon markets.

Our study is not free of limitations. First, we have a small sample because Euronext Lisbon is a small stock market. Second, we consider a small number of corporate governance mechanisms, since we do not have much information in the database. Finally, as we ignore industry specificity, it is a limitation. The findings that the conclusions can be different according to the performance metrics used indicate that future research should continue to distinguish between accounting-based and market-based performance to avoid misleading generalizations. Moreover, the evidence that different governance mechanisms align with different theoretical frameworks depending on the performance measure suggests that future studies should avoid framing agency and stewardship theories as mutually exclusive.

For future research directions, we suggest exploring new contexts, external factors, and long-term impacts on performance. It will also be interesting to explore the difference between family and non-family firms concerning the relationship between corporate governance and companies' performance since family-owned businesses have distinctive governance structures, hierarchies, and decision-making processes compared to their counterparts. It will also be interesting to analyze whether corporate governance crosses with corporate social responsibility initiatives, contributing to sustainable business practices, and explore the interaction between corporate governance and human resources practices. In addition, it is important to explore how industry context and periods of instability moderate the relationship between corporate governance and performance.

Acknowledgments. This work was supported by the Research Unit on Governance, Competitiveness and Public Policies (GOVCOPP), through the Portuguese Foundation for Science and Technology (FCT – Fundação para a Ciência e a Tecnologia), references (UIDB/04058/2020) + (UIDP/04058/2020).

References

1. Kamau, G., Aosa, E., Machuki, V., Ganesh, P.: Corporate governance, strategic choices and performance of financial institutions in Kenya. Int. J. Bus. Manage. **13**(7), 169–178 (2018)

2. Arora, A., Sharma, C.: Corporate governance and firm performance in developing countries: evidence from India. Corp. Gov. **16**(2), 420–436 (2016)
3. Berhe, A.G.: Board structure and bank performance: evidence from Ethiopia. Cogent Bus. Manage. **10**(1), 2163559 (2023)
4. Ferrary, M., Déo, S.: Gender diversity and firm performance: when diversity at middle management and staff levels matter. The Int. J. Human Res. Manage. **34**(14), 2797–2831 (2023)
5. Alawi, S.: The role of corporate social responsibility, ownership structure, and gender diversity in firm performance. Int. J. Econ. Financ. Issues **14**(2), 97–110 (2024)
6. Laskar, N., Sahu, J.P., Choudhury, K.S.: Impact of gender diversity on firm performance: Empirical evidence from India. Manag. Financ. **50**(3), 614–633 (2024)
7. Williams, K.Y., O'Reilly, C. A.III: Demography and diversity in organizations: A review of 40 years of research. In: Staw, B.M., Cummings, L.L. (eds.), Research in Organizational Behavior, Vol. 20, pp. 77–140. Jai Press (1998)
8. Wang, K., Ma, J., Xue, C., Zhang, J.: Board gender diversity and firm performance: recent evidence from Japan. J. Risk and Fin. Manage. **17**(1), 20–47 (2024)
9. Raza, M., Bilal, M., Khan, A., Taqi, M.: Adapting corporate governance for sustainable organizational performance: a case study of the banking sector in Pakistan amid the COVID-19 pandemic. J. Innov. Res. Manage. Sci. **4**(2), 53–68 (2023)
10. Khan, I., Saha, A.K., Kamal, Y., Islam, M.S.: Do ownership structure and board composition matter in firm performance? regulatory influence in developing economy. Int. J. Discl. Gov. **22**(1), 263–283 (2025)
11. Elsayed, K., Paton, D.: The impact of environmental performance on firm performance: static and dynamic panel data evidence. Struct. Chang. Econ. Dyn. **16**(3), 395–412 (2005)
12. Jensen, M.C., Meckling, W.H.: Theory of firm: Managerial behavior, agency costs and ownership structure. J. Financ. Econ. **3**(4), 305–360 (1976)
13. Jensen, M.C.: Agency costs of free cash flow, corporate finance, and takeovers. Am. Econ. Rev. **76**(2), 323–329 (1986)
14. Donaldson, L.: The ethereal hand: Organizational economics and management theory. Acad. Manag. Rev. **15**(3), 369–381 (1990)
15. Freeman, R.E.: Strategic management: A stakeholder approach. Pitman, Boston, MA (1984)
16. Hutomo, P.T., Pudjiarti, E.S.: Corporate governance and HRM practice on consumption product sector listed in Indonesia stock exchange. European Res. Stud. J. **XXI**(3), 132–142 (2018)
17. Richard, O., Ford, D., Ismail, K.: Exploring the performance effects of visible attribute diversity: The moderating role of the span of control and organizational lifecycle. The Int. J. Human Res. Manage. **17**(12), 2091–2109 (2006)
18. Dezsö, C.L., Ross, D.G.: Does female representation in top management improve firm performance? a panel data investigation. Strateg. Manag. J. **33**(9), 1072–1089 (2012)
19. Charness, G., Gneezy, U.: Strong evidence for gender differences in risk-taking. J. Econ. Behav. Organ. **83**(1), 50–58 (2012)
20. Ali, M., Kulik, C., Metz, I.: The gender diversity-performance relationship in services and manufacturing organizations. The Int. J. Human Res. Manage. **22**(7), 1464–1485 (2011)
21. Ali, M., Konrad, A.M.: Antecedents and consequences of diversity and equality management systems: the importance of gender diversity in the TMT and lower to middle management. Eur. Manag. J. **35**(4), 440–453 (2017)
22. Azeem, N., Ullah, M., Ullah, F.: Board gender diversity and firms' financial resilience during the Covid-19 pandemic. Financ. Res. Lett. **58**, 104332 (2023)
23. Richard, O.C., Barnett, T., Dwyer, S., Chadwick, K.: Cultural diversity in management, firm performance, and the moderating role of entrepreneurial orientation dimensions. Acad. Manag. J. **47**(2), 255–266 (2004)

24. Pucheta-Martínez, M.C., Gallego-Álvarez, I.: Do board characteristics drive firm performance? An international perspective. RMS **14**(6), 1251–1297 (2020)
25. Ali, M., Grabarski, M., Ahmed, S., Imtiaz, N.: Does leadership gender diversity drive corporate social responsibility and organizational outcomes? The role of organization size. Aust. J. Manag. **49**(3), 319–339 (2023)
26. Wei, Y.: Gender matters: Board gender diversity and firms' export resilience. Humanit. Soc. Sci. Comm. **11**(1), 766 (2024)
27. Jayaraman, J.D., Smita, R., Nilakantan, N.: The impact of board gender diversity on firm performance: does critical mass matter? J. Appl. Acc. Res. **26**(1), 249–277 (2025)
28. Xie, F., Guo, Y., Daniel, S.J., Liu, Y.: The dynamic relation between board gender diversity and firm performance: the moderating role of shareholder activism. Rev. Quant. Financ. Acc. **62**(1), 225–246 (2024)
29. Fama, E.F., Jensen, M.: Separation of ownership and control. J. Law Econ. **26**(2), 301–325 (1983)
30. Jensen, M.: The modern industrial revolution, exit, and the failure of internal control systems. J. Fin. **48**(3), 831–880 (1993)
31. Lipton, M., Lorsch, J.: A modest proposal for improved corporate governance. The Bus. Lawyer **48**(1), 59–77 (1992)
32. Florackis, C.: Agency costs and corporate governance mechanisms: evidence for UK firms. Int. J. Manag. Fin. **4**(1), 37–59 (2008)
33. Saidat, Z., Silva, M., Seaman, C.: The relationship between corporate governance and financial performance. J. Family Bus. Manage. **9**(1), 54–78 (2018)
34. Yan, C., Hui, Y., Xin, L.: The relationship between board size and firm performance. E3S Web of Conferences **257**, 02079 (2021)
35. Danilov, G.: The impact of corporate governance on firm performance: panel data evidence from S&P 500 information technology. Future Bus. J. **10**(1), 86–111 (2024)
36. Anderson, R.C., Mansi, S.A., Reeb, D.M.: Board characteristics, accounting report integrity, and the cost of debt. J. Account. Econ. **37**(3), 315–342 (2004)
37. Williams, R.J., Fadil, P.A., Armstrong, R.W.: Top management team tenure and corporate illegal activity: the moderating influence of board size. J. Manag. Issues **17**(4), 479–493 (2005)
38. Christensen, J., Kent, P., Stewart, J.: Corporate governance and company performance in Australia. Aust. Account. Rev. **20**(4), 372–386 (2010)
39. Ibrahim, H., Samad, F.A.: Corporate governance mechanisms and performance of public-listed family-ownership in Malaysia. Int. J. Econ. Financ. **3**(1), 105–115 (2011)
40. Karim, S., Vigne, S.A., Lucey, B.M., Naeem, M.A.: Discretionary impacts of the risk management committee attributes on firm performance: do board size matter? Int. J. Emerg. Mark. **19**(8), 2222–2240 (2024)
41. Mace, M.L.: Directors: Myth and reality. Harvard Business School Press, Boston MA (1986)
42. Lorsch, J.W., MacIver, E.: Pawns or potentates: The reality of America's corporate boards. Harvard Business School Press, Boston MA (1989)
43. Rechner, P.L., Dalton, D.R.: CEO duality and organizational performance: a longitudinal analysis. Strateg. Manag. J. **12**(2), 155–160 (1991)
44. Balatbat, M.C.A., Taylor, S.L., Walter, T.S.: Corporate governance, insider ownership, and operating performance of Australian initial public offerings. Acc. Fin. **44**(3), 299–328 (2004)
45. Tang, J.: CEO duality and firm performance: The moderating roles of other executives and blockholding outside directors. Eur. Manag. J. **35**, 362–372 (2017)
46. Ali, S., Naseem, M., Jiang, J., Rehman, R., Malik, F., Ahmad, M.: "How" and "When" CEO Duality matter? case of a developing economy. SAGE Open **12**(3), 21582440221116110 (2022)

47. Zahid, Z., Zhang, J., Ali, F., Shahzad, F.: Board diversity and firm performance in Chinese manufacturing firms: Moderating role of CEO duality. Business Research Quarterly, 1–23 (2024)
48. Wicaksono, A.P.N.: Gender diversity, financial expertise, CEO duality and firm performance: Evidence from Indonesian public listed companies. Jurnal Ilmiah Bisnis Dan Ekonomi Asia **16**(1), 1–14 (2022)
49. Wijethilake, C., Ekanayake, A.: CEO duality and firm performance: the moderating roles of CEO informal power and board involvements. Soc. Responsib. J. **16**, 1453–1474 (2019)
50. Yu. M.: CEO duality and firm performance: A systematic review and research agenda. European Management Review **20**(2), 346–358 (2022)
51. Din, S.U., Khan, W., Saleem, F., Khan, M.Y.: CEO duality, board independence and firm financial performance: Evidence from an Asian emerging market. Sarhad J. Manage. Sci. **10**(1), 59–78 (2024)
52. Liang, C., Lee, P.K.C., Zhu, M., Yeung, A.C.L., Cheng, T.C.E., Zhou, H.: The bright side of being uncertain: the impact of economic policy uncertainty on corporate innovation. Int. J. Oper. Prod. Manag. **44**(11), 1886–1913 (2024)
53. Zhang, J., Rodrigues, S.B., Jiang, J., Zhou, C.: Political turnover, power and performance of foreign firms. Multinat. Bus. Rev. (2025)
54. Chao, G.: The impact of perceived economic policy uncertainty on firm value. Trans. Econ. Bus. Manage. Res. **14**, 624–629 (2024)
55. Croci, E., Hertig, G., Khoja, L., Lan, L.L.: Board characteristics and firm resilience: evidence from disruptive events. Corp. Govern. Int. Rev. **32**(1), 2–32 (2024)
56. Tuffour, K., Barnor, C.: The effect of financial leverage on profitability of manufacturing companies listed on the Ghana stock exchange. J. Bus. Res. **8**, 56–65 (2014)
57. Arhinful, R., Radmehr, M.: The impact of financial leverage on the financial performance of the firms listed on the Tokyo stock exchange. SAGE Open **13**(4), 21582440231204100 (2023)
58. Nguyen, T., Van Nguyen, C.: The determinants of profitability in listed enterprises: a study from Vietnamese stock exchange. J. Asian Fin. Econ. Bus. **7**, 47–58 (2020)
59. Mahmood, F., Ahmed, Z., Hussain, N., Zaied, Y.B.: Macroeconomic factors and financing strategies in working capital: evidence from China. Int. J. Financ. Econ. **29**(1), 35–57 (2024)
60. Arellano, M., Bond, S.: Some tests of specification for panel data: monte Carlo evidence and an application to employment equations. Rev. Econ. Stud. **58**(2), 277–297 (1991)
61. Arellano, M., Bover, O.: Another look at the instrumental variable estimation of error-components models. Journal of Econometrics **68**(1), 29–51 (1995)
62. Hair, J.F., Black, W.C., Babin, B.J., Anderson, R.E.: Multivariate Data Analysis, 8th ed. Cengage Learning (2019)
63. Gujarati, D.N.: Essentials of econometrics. Sage Publications (2021)

An Evaluation of Esg Reporting Practices in Banks Holding Automotive Financing Portfolios

Leonardo Bucsan Emrich(✉) and Liliane Cristina Segura

Universidade Presbiteriana Mackenzie, São Paulo, Brasil
leonardo.emrich@machenzista.com.br, liliane.segura@mackenzie.br

Abstract. This study conducts an analysis of the disclosure of information related to ESG (Environmental, Social, and Governance) by banks focused on automotive financing in Brazil. The growing importance of corporate sustainability, regulated by initiatives such as the International Sustainability Standards Board (ISSB) and other international frameworks, highlights the need for banks to adopt transparent practices aligned with the global market. The research evaluates how major Brazilian financial institutions, as well as banks dedicated to the automotive sector, are implementing, and disclosing their ESG practices in official reports. Qualitative and deductive analytical methods are employed to examine ESG data disclosures, with particular attention to the quality of information provided. The study reveals a substantial difference between large banks classified under the Central Bank of Brazil's systemic risk framework (S1 and S2) and smaller banks (S3), in terms of the maturity of ESG disclosures. Emphasis is placed on actions related to climate risk, gender equity, and sustainability. The research also discusses the impact of these practices on governance and the financial performance of the institutions and suggests that the adoption of international standards such as IFRS S1 and S2, and CBPS S1 and S2, may contribute to greater transparency and comparability in disclosed data.

Keywords: ESG (Environmental · Social · and Governance) · banks · automotive financing · sustainability · governance · IFRS (International Financial Reporting Standards) · sustainability disclosure

1 Introduction

Companies around the world are increasingly required to report on their Environmental, Social, and Governance (ESG) practices as a means of communicating with stakeholders and attracting sustainable investment. This demand for disclosure emerged in the 20th century, driven by concerns over climate change, corporate scandals, and, significantly, the societal impact of corporate actions. The momentum for ESG reporting intensified following a statement by BlackRock—the world's largest institutional investor—emphasising its preference for companies demonstrating strong ESG performance. Consequently, ESG has become a prevalent concept across professional, institutional, and societal domains.

L. Almeida et al. (Eds.): ICAFI 2025, ISEM 80, pp. 194–211, 2026.
https://doi.org/10.1007/978-3-032-16268-7_11

According to [37], global debates on climate change, carbon emissions, and other environmental issues have placed sustainability at the forefront of the international agenda. These concerns, in turn, have underscored the relevance of Corporate Social Responsibility (CSR), a principle that resonates strongly with corporate stakeholders. Supporting this view, [13] argue that sustainability is a complex and multidimensional concept, incorporating various theoretical frameworks. CSR, in this context, demands that companies consider the interests of all their stakeholders.

The integration of responsible enterprises into investment portfolios has become a widespread global trend. In this regard, the accounting profession plays a central role in consolidating ESG disclosure practices. This responsibility was institutionalised through the IFRS Foundation, which oversees the International Accounting Standards Board (IASB), and has established the International Sustainability Standards Board (ISSB) to develop global sustainability reporting standards.

In the Brazilian capital markets, regulatory innovation has positioned the country as a pioneer. The Brazilian Securities and Exchange Commission (CVM) introduced Resolution No. 193/2023, enabling listed companies, investment funds, and securitisation entities to voluntarily prepare and disclose sustainability-related financial information in accordance with international standards—specifically, IFRS S1 and IFRS S2, issued by the ISSB.

Further progress was made on 12 September 2024 with the approval of CBPS 01 and CBPS 02 by the Brazilian Committee for Sustainability Pronouncements (CBPS). CBPS 01, aligned with IFRS S1, sets out the General Requirements for Disclosure of Sustainability-Related Financial Information, while CBPS 02, equivalent to IFRS S2, addresses Climate-Related Disclosures. Both standards were formally adopted under Resolution CFC No. 1.670/22 (CBPS, n.d.).

Following the CVM's initiative, the Central Bank of Brazil (Bacen) extended ESG disclosure requirements to regulated financial institutions. Through Resolution CMN No. 5.185, issued by the National Monetary Council (CMN) on 21 November 2024, significant amendments were introduced to Resolution No. 4.818 of 29 May 2020. These modifications establish general guidelines for the preparation and disclosure of financial statements by financial institutions authorised by Bacen. The revised regulation mandates that institutions include a Sustainability-Related Financial Information Report as part of their annual consolidated financial statements. This report must conform to CBPS pronouncements—particularly CBPS 01 and CBPS 02.

The implementation timeline is structured as follows: from 2026, compliance will be mandatory for institutions listed as publicly traded companies or prudential conglomerate leaders in segments S1 and S2; by 2028, the requirements will extend to all other financial institutions [4].

As noted by [32], ESG disclosure has become a growing concern for companies worldwide. Businesses are increasingly aligning themselves with ESG policies, recognising that the financial market places significant value on the disclosure of non-financial information. Within a globalised context, organisations have adopted a post-modern perspective, emphasising social, environmental, and sustainability initiatives.

This study focuses on a group of institutions regulated by the Central Bank of Brazil. It is noteworthy that the Brazilian banking sector has emerged as an active participant in ESG-related disclosures. Given that banks provide critical financial support to the national and global economy, their role in financing sustainable activities is particularly impactful.

Recently, the Central Bank convened a consultation to propose amendments to Resolution CMN No. 4.818/2020, seeking to incorporate the ISSB's international standards—IFRS S1 and IFRS S2—into the national financial system.

At this juncture, it is important to clarify the structure of the Brazilian financial market and the classification of financial institutions under Central Bank of Brazil regulations, as outlined on the Central Bank's official website [4].

> "Prudential regulation is a type of financial regulation that sets requirements for financial institutions, focusing on risk management and minimum capital requirements to address the risks arising from their activities, as well as operational limits. Risk management and minimum capital requirements help ensure that the potential failure of a financial institution does not trigger a domino effect in the financial system and, ultimately, losses for society as a whole. This domino effect is known as systemic risk."

You may see that Table 1 presents the classification of banks, in accordance with CMN Resolution No. 4,557/2017 (updated by CMN Resolution No. 4,943/2021).

Table 1. Classification of Financial Institutions According to CMN Resolution No. 4,557/2017 (as updated by CMN Resolution No. 4,943/2021)

Segment	Composition	Size* and International Activity	Application of Proportionality in Prudential Regulation**
S1	Banks***	Equal to or greater than 10% of GDP (or significant international activity)	Full alignment with Basel recommendations
S2	Institutions and conglomerates	Between 1% and 10% of GDP	Alignment with Basel recommendations, with limited exceptions (no requirement for LCR and NSFR liquidity metrics or full disclosure under Pillar 3). Use of simplified Internal Capital Adequacy Assessment Process (Icaap-simplified)

(*continued*)

Table 1. *(continued)*

Segment	Composition	Size* and International Activity	Application of Proportionality in Prudential Regulation**
S3	Institutions and conglomerates	Between 0.1% and 1% of GDP	Simplified rules for market risk and interest rate risk in the banking book (IRRBB); simplified risk management structure and Pillar 3 requirements
S4	Institutions and conglomerates	Below 0.1% of GDP	Further simplification in prudential requirements, risk management structure, and Pillar 3 disclosures
S5	Non-banking institutions and conglomerates with simplified risk profiles	Below 0.1% of GDP	Optional simplified methodology for calculating minimum prudential requirements. Simplified risk management structure

Source: Bacen chart available at: https://www.bcb.gov.br/estabilidadefinanceira/regprudencialsegmentacao

As part of its ESG disclosure efforts, the Central Bank of Brazil has issued regulations concerning the disclosure and monitoring of climate-related risks by institutions authorised to operate under its supervision. The most prominent of these is CMN Resolution No. 4,557/2017 (updated by CMN Resolution No. 4,943/2021), which establishes the risk management framework for financial institutions, including socio-environmental risks. This regulation has been revised to explicitly incorporate climate-related risks, in addition to environmental and social risks, and must be followed by institutions classified in segments S1, S2, and S3.

It is also worth noting that banks and financial institutions are typically represented by industry associations. The most prominent of these is Febraban, which maintains dedicated committees aimed at aligning market practices among member institutions.

According to [8], the financial sector contributes to sustainable development through the provision of loans and credit to companies that demonstrate greater transparency regarding their ESG practices. At the same time, the financial sector has increasingly focused on environmental disclosures, particularly by linking environmental risk to credit risk. In this regard, Central Bank regulations—most notably, CMN Resolution No. 4,327 of 25 April 2014—emphasise the need to enhance the assessment of socio-environmental risks within Brazil's financial sector.

The aim of this article is to evaluate ESG disclosure among Brazilian banks holding automotive loan portfolios. To address this objective, a deductive analytical study with a qualitative approach was conducted, based on the examination of sustainability reports and integrated disclosures published by banks and financial institutions engaged in automotive financing.

2 Theoretical Framework

To assess the stage of development of this topic, we conducted a literature search using the Web of Science database, applying the keywords: “ESG, Disclosure and Brazilian Banks”. This search yielded a single result: “Evolution and Challenges of Environmental, Social, and Governance Practices: An Analysis of the Brazilian Stock Exchange’s Corporate Sustainability Index”, an article published in August 2024, although its primary focus is not specifically on Brazilian banks [36].

The search was conducted on 28 November 2024. At this point, we revised the search criteria to include the broader keywords: “ESG and Disclosure and Banks”. This returned 113 published articles dating from 2020 onwards. These results indicate a growing global academic interest in the topic; however, there is still a notable gap in the literature specifically focused on the Brazilian banking sector and financial institutions, particularly within high-impact publications.

The findings can be summarised as extracted directly from the Web of Science database (see below).

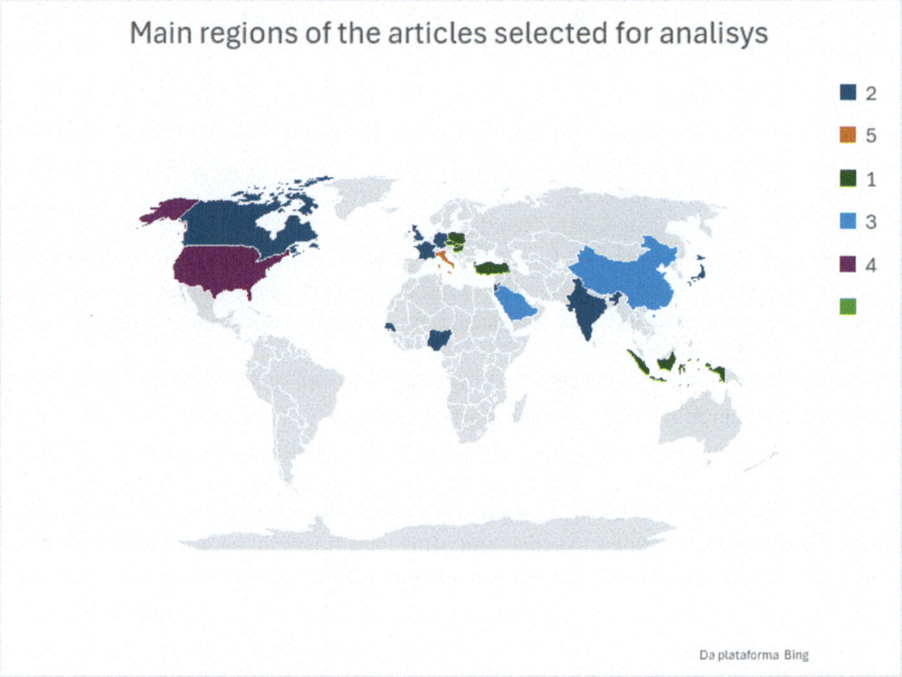

Source: Prepared by the author in Excel, based on the distribution of articles selected among the found on the Web of Science.

The review of the articles identified through the Web of Science database reveals a wide range of perspectives on ESG disclosure and its impact on banking and corporate performance. The influence of ESG practices in the financial sector has been widely examined, particularly regarding their relationship with financial performance, corporate governance, and regulatory transparency. According to [8], banks play a dual role in promoting sustainability: internally, through the management of their operations, and externally, by integrating ESG risks into credit and investment decisions. In emerging markets, ESG governance and regulation are considered essential mechanisms for mitigating risks and ensuring the long-term viability of financial institutions [15].

Notably, the scandals and controversies linked to the role of banks in the 2008 financial crisis revealed failures across various dimensions of corporate social responsibility and control mechanisms [16]. These events have heightened the responsibility of banks to strengthen internal processes and enhance disclosure, thereby reducing information asymmetry for stakeholders [15].

Another frequently discussed theme across the identified articles concerns how ESG disclosure contributes to the increased valuation of banks. The reliability of corporate disclosures is grounded in both disclosure theory and stakeholder theory, both of which stress the critical importance of information for market transparency and trust. Disclosure theory argues that compliance with regulatory standards is vital to maintaining the integrity and legitimacy of the information transmitted by institutions, helping to avoid information asymmetries that can harm stakeholders [12].

To address agency conflicts and meet stakeholder expectations, banks may adopt effective governance mechanisms. The establishment of ESG committees within boards of directors offers closer oversight of managerial decisions, ensuring alignment with the long-term interests of all stakeholders. These committees, by monitoring ESG policies and supervising disclosure practices, help ensure that stakeholder expectations are properly met. Recent studies indicate that integrating ESG factors into banking practices not only promotes financial stability but also enhances the competitiveness of financial institutions, as evidenced by the positive relationship between ESG integration and banking stability [2]. Moreover, research shows that banks' financial performance is influenced by ESG investment, highlighting the importance of governance in strategic decision-making that incorporates sustainability and social responsibility [29]. In this way, ESG committees contribute to transparency and accountability in banking operations, while simultaneously fostering long-term sustainable economic development.

An additional effective measure is the linkage of executive compensation to ESG performance metrics. This approach incentivizes managers to prioritize sustainability objectives and enhance transparency in ESG practices, thereby aligning short-term managerial interests with the long-term goals of shareholders and society at large [14].

Furthermore, stakeholder theory emphasizes that all interested parties—from investors to the wider community—depend on transparent and reliable information to make informed decisions. Recent research on governmental disclosure supports the idea that increased transparency not only corrects information asymmetry but also strengthens public rationality and oversight of governmental decisions [1, 12]. Accurate and transparent disclosure, therefore, serves not only to fulfil legal obligations but also to foster stakeholder trust and engagement, ensuring an organization's long-term competitiveness and sustainability.

Stakeholder theory further asserts that a responsible bank addresses the demands of social actors while disclosing environmental values across its value chain [7]. To be sustainable, a company must minimize its use of non-renewable resources and reduce emissions—criteria that align with the strongest interpretation of sustainability [9].

Banks that adopt a transparent ESG disclosure approach are more likely to reduce principal–agent conflicts, as shareholders gain access to detailed information on risks and the bank's stance regarding social and environmental issues. In this way, ESG disclosure becomes a tool for interest alignment and improved market trust.

However, managers may sometimes be reluctant to disclose ESG-related information due to perceptions that these practices do not yield immediate financial returns, creating a tension between short-term and long-term goals. Agency theory suggests that such conflicts can be mitigated through appropriate governance mechanisms, such as the implementation of ESG committees, which help ensure that all parties' interests are considered [27].

Disclosure theory and stakeholder theory offer complementary perspectives on ESG reporting in the banking sector. The former focuses on conflicts between shareholders and managers, while the latter broadens the scope to include a diverse range of stakeholders whose interests must also be addressed. By adopting governance mechanisms that align the interests of all parties involved, banks can enhance their ESG performance, thereby contributing to the achievement of sustainable development goals. The establishment of ESG committees and the linking of executive remuneration to sustainable performance metrics are examples of practices that can reduce agency conflicts and improve transparency in ESG practices.

3 Method

The method adopted in this study was deductive analytical with a qualitative approach, conducted through the observation of sustainability reports and integrated disclosures published by banks and financial institutions. The analysis was also supported by an extensive literature review, including academic references such as previously conducted studies, theses, dissertations, and peer-reviewed articles published in journals rated under CAPES' Qualis system.

Qualitative research is a widely adopted methodological approach across various academic disciplines, aimed at gaining an in-depth understanding of complex phenomena. Unlike quantitative research, which relies on numerical data and statistical analysis, qualitative research seeks to explore subjective experiences, perspectives, and meanings attributed by individuals or groups. This approach enables a rich and context-sensitive interpretation of behaviors or events, offering insights that are often beyond the reach of quantitative methods. Recent studies highlight the growing importance of qualitative research in the social sciences, particularly for producing detailed understandings of human phenomena and translating them into practical applications [11].

Such investigations demonstrate that the application of qualitative methods enables deeper analyses of social and behavioral issues, thereby contributing to the development of more effective and contextually grounded solutions. Qualitative research is frequently used in the social sciences, humanities, and education to explore topics that are difficult to quantify or measure, such as emotions, attitudes, and cultural practices. In qualitative studies, process-oriented questions are often employed to explore a range of phenomena, their characteristics, and participants' interpretations of events, situations, and experiences [38].

In general, the measurement of both sustainability disclosure and sustainability performance is essential for reconciling disclosure theory and stakeholder theory. This study, therefore, adopts enhanced measurement approaches to assess sustainability disclosure and performance. Regarding sustainability disclosure, the focus is placed on the quality—rather than the quantity—of the information disclosed [22].

Despite its strengths, qualitative research is not without limitations. These include the potential for bias in data collection and analysis, limited generalizability of findings, and the time-intensive nature of the analytical process. However, when conducted rigorously and ethically, qualitative research can generate valuable insights and contribute significantly to the advancement of knowledge across a range of academic fields.

This study adopts a deductive analytical method with a qualitative approach, based on document analysis. The core data sources include ESG and sustainability reports, integrated reports, and publicly available disclosures from financial institutions operating in Brazil. The analysis is also supported by academic literature and institutional frameworks relevant to ESG disclosure and financial sector governance.

The reference date for data collection was June 2024, and the type of institutions analysed were Prudential Conglomerates and Independent Institutions, as classified by the Central Bank of Brazil (Bacen) on its official platform (IF.Data – https://www3.bcb.gov.br/ifdata/#).

The selection criteria comprised two distinct groups:

- The four largest banks classified under Segment S1, representing systemically important financial institutions according to CMN Resolution No. 4,557/2017 (updated by Resolution No. 4,943/2021). These banks serve as benchmarks of excellence due to their significant market share, extensive stakeholder base, and high-quality ESG disclosures.
- Banks operating in the automotive financing sector, classified under Segments S2 and S3. These institutions were selected to enable comparative analysis between systemically important banks and specialised financial entities with narrower operational scope.
- Segment S4 institutions were excluded from the sample due to their relatively limited public disclosures and reduced scale of operations, which would hinder comparability across the dataset.

By comparing large, diversified banks (Segment S1) with niche financial institutions (Segments S2 and S3), the methodology aims to assess the degree of maturity, quality, and comparability of ESG disclosure practices across different types of financial entities. The use of qualitative content analysis enables the evaluation of frameworks adopted (e.g., GRI, SASB, TCFD), independent assurance practices, materiality reporting, climate action disclosures, and gender equity initiatives. Results are summarized in comparative tables and discussed in light of theoretical frameworks such as Legitimacy Theory and Stakeholder Theory.

Institutions

Institutions	Group
Banco Itaú S.A.	Major market institutions
Banco do Brasil S.A	
Banco Bradesco S.A	
Banco Santander	
Banco Safra	Banks in the automotive financing sector – Segment S2
Banco BV	
Banco VW	Banks in the automotive financing sector – Segment S3
Banco GMAC	
Banco Mercedes	
Banco Toyota	
Banco Volvo	
Banco Stellants	Banks in the automotive financing sector – Segment S4
Honda	
BMVW	
Yamaha	
Banco Scania	

The purpose of selecting the largest banks in the market is to establish a benchmark of excellence in ESG disclosure, given the quality and detail of the reports published by these institutions, which—due to their significantly larger size—are classified under Segment S1.

We also included two banks from Segment S2, as they are sizeable institutions and were deemed relevant in terms of scale and influence. The remaining selected banks belong to Segment S3. Banks from Segment S4 were excluded from the comparative analysis.

The theme of sustainability is embedded within a context of complex interactions among various stakeholders. This complexity underscores the need for integrating diverse forms of knowledge through an interdisciplinary perspective—thus moving beyond epistemological paradigms that view the world as homogeneous. The field of sustainability research demands methodological approaches that address social, environmental, economic, political, ethical, and cultural challenges, aiming to understand the intrinsic relationships between human beings, nature, and society.

For this reason, the analysis follows a qualitative approach, based on a questionnaire developed around the key themes outlined in Appendix 1.

4 Results

All reports were analysed, and the data summaries—as well as the author's analysis—are presented in the table on the following page.

It is clearly evident that the larger institutions (Segments S1 and S2) are more advanced in ESG data disclosure. Their reports are complex and contain a wealth of information. The volume of data used for disclosure purposes is substantial, and several frameworks are employed, as illustrated in the figure below:

Acronym	Framework / Initiative
GRI	Global Reporting Initiative (GRI)
SASB	Sustainability Accounting Standards Board (SASB)
IASB	International Accounting Standards Board (IASB)
ISSB	International Sustainability Standards Board (ISSB)
TCFD	Task Force on Climate-Related Financial Disclosures (TCFD)
PRB	Principles for Responsible Banking (PRB)
GHG Protocol	Greenhouse Gas Protocol
TCU	Federal Court of Accounts (Tribunal de Contas da União – Brazil)
SDGs (ODS)	Sustainable Development Goals (SDGs)
UNGC	United Nations Global Compact
PRI	Principles for Responsible Investment (PRI)

All of these frameworks have been employed by the institutions as reference points for reporting and disclosing ESG-related information to the market. However, this multiplicity of frameworks makes reporting more complex and comparability more difficult. As noted by [28]: "The availability of detailed and accurate sustainability data, along with a clear understanding of its format and content, is essential for the quality of ESG reporting."

The 2021 report by the Financial Education & Research Foundation (FERF) on ESG—entitled "Financial Statement Preparers and the ESG Journey: Charting the Course"—examined the role of finance departments in ESG reporting, the degree of integration with financial reporting, and the challenges faced by finance professionals. The research found that confusion regarding disclosure frameworks, methodologies, and stakeholder expectations is causing delays in financial departments' engagement with ESG efforts. It revealed that the integration of ESG and financial reporting is still in its early stages, with 53% of respondents having not yet begun the integration process [28].

Data related to greenhouse gas inventories, in particular Scope 3 emissions, depend heavily on interpretative factors and are not easily comparable. Therefore, it is crucial to harmonise regulatory expectations. The data and information disclosed in reports must be meaningful, comprehensive, comparable, and consistent across time and countries. Moreover, due to the absence of a clear relationship between ESG disclosure and bank profitability (e.g., ROA and ROE), regulatory requirements become a critical driver of sustainable development. As [26] also emphasise, more advanced and sustained regulatory efforts are needed. According to [24], this is essential to maintain the integrity and legitimacy of the information disclosed by institutions and to avoid information asymmetry that may harm stakeholders.

Among vehicle financing banks, except for Volkswagen Bank, there is generally a lack of detailed ESG reporting to the market. This may be due to the close connection between automaker banks and their parent manufacturing companies, which often handle ESG reporting centrally at the corporate level.

5 Discussion and Final Remarks

The analysis of the reports and available information revealed a significant disparity between Brazil's largest banks and smaller institutions, where institutional size clearly impacts financial capacity and the breadth of their stakeholder engagement.

Another conclusion is that the automotive financing sector still requires considerable progress in order to achieve more detailed ESG data disclosure. Numerous frameworks are currently in use, but there is no unified standard. With the adoption of IFRS S1 and S2 by the Central Bank of Brazil—and their consequent mandatory application to institutions authorised by the regulator—it is hoped that greater consistency and comparability in ESG reporting will be achieved.

Future research could extend this analysis to a broader sample of financial institutions and compare ESG disclosure metrics, especially with regard to greenhouse gas inventories, where discrepancies between entities—particularly in Scope 3 emissions—remain evident and challenging to interpret.

The analysis presented in the summary table highlights significant differences among financial institutions in terms of transparency and commitment to the ESG agenda. Banks classified under Segment S1 generally exhibit a high level of disclosure, publishing dedicated ESG reports, adopting internationally recognised frameworks such as GRI, SASB, and TCFD, and engaging independent assurance providers such as PwC and KPMG. In contrast, smaller banks in Segments S2 and S3 typically incorporate ESG reporting into their management reports—an approach that reflects stakeholder-focused priorities aligned with Legitimacy Theory and Stakeholder Theory.

It was also noted that certain S3 banks did not publish ESG-specific reports or relevant disclosures at all, underscoring the role of institutional scale in shaping reporting capabilities and financial resource allocation.

When it comes to climate-related actions and risk disclosures, substantial variation exists among the institutions. Once again, banks classified by Bacen as systemically important (S1 and S2) scored higher in terms of reporting climate initiatives, risk exposure, and sustainable business practices. In contrast, S3 banks (with few exceptions) offered more limited disclosures, and in some cases, provided no detailed information. Moreover, clear targets for reducing carbon emissions across Scopes 1, 2, and 3 were not consistently disclosed, raising concerns about transparency regarding environmental impact. The automotive financing segment, in particular, still requires further development to ensure more comprehensive and comparable ESG reporting.

Gender equity was also examined, although it is not a mandatory requirement under IFRS S1 and S2. Its inclusion was justified by the literature that identifies gender diversity on boards as a key dimension in evaluating institutional governance. Nonetheless, it is difficult to compare such data effectively due to variation in institutional size and the inconsistency in disclosure formats. The data suggest that female representation in leadership and the publication of related targets vary widely. Some banks reported higher percentages and structured policies, while others provided no detailed information on goals or actual participation. Transparency in this area is crucial to ensuring meaningful progress in diversity within the financial sector.

In addition, the variety of frameworks adopted for ESG disclosure continues to impede comparability. The lack of a single reporting standard complicates analysis across institutions. However, with the mandatory adoption of IFRS S1 and S2 by the Central Bank of Brazil, it is anticipated that future disclosures will become more standardised, comparable, and accessible.

From a financial perspective, the study also incorporated available financial data to illustrate the discrepancies in institutional size, which—although not a requirement of ESG frameworks—helps assess materiality and relevance. ESG reports are typically prepared for stakeholder use and tend to present a detailed and optimistic view of the institutions' achievements. It is, therefore, recommended that these reports be read in conjunction with financial statements for a more complete understanding.

In summary, the data analysis reveals varying levels of ESG maturity across Brazilian banks. While some institutions are aligned with global best practices, others still need to improve in areas such as transparency, target setting, and the disclosure of environmental and social impacts. Market trends and increasing pressure from investors and stakeholders suggest that banks which adopt more robust and transparent ESG practices are likely to gain competitive advantages in the long term, both in terms of reputation and attractiveness to investors and clients. A more consolidated regulatory framework may contribute to reducing these disparities and enhancing the overall quality of information provided by the financial sector.

Appendix 1

Reference	Field Name 1	Field Name 2	Description	Possible Response
[1]	Institution		Name of the institution	Name provided
[2]	Total Assets		Total assets in millions of BRL for reference and scale	Value based on available data
[3]	Equity		Total equity in millions of BRL for reference and scale	Value based on available data
[4]	Operating Result		Total Operating Result (EBIT) in millions of BRL for reference and scale	Value based on available data

(continued)

(*continued*)

Reference	Field Name 1	Field Name 2	Description	Possible Response
[5]	Net Profit		Total Net Income (NI) in millions of BRL for reference and scale	Value based on available data
[6]	Dedicated Sustainability Website?		Existence of a dedicated website for sustainability data	Yes / No / N/A / exception noted
[7]	ESG Report Published? *		Existence of an ESG report	Yes / No / N/A / exception noted
[8]	ESG Information Included with Management Report?		Publication of ESG information along with the management report	Yes / No / N/A / exception noted
[9]	Independent Assurance (Yes/No)		Was independent assurance performed?	Yes / No
[10]	Independent Assurance Report Issued By:		Company responsible for conducting the assurance and issuing the report	Company name / N/A
[11]	Framework Used		Description of frameworks used in ESG reporting	Must include data such as GRI, SASB, etc
[12]	Materiality Disclosure?		Was materiality disclosed?	Yes / No (Single materiality / Double materiality)
[13]	Climate Action		Were climate actions disclosed?	Score from 1 to 10 based on disclosure quality, detail, and examples (author's assessment)
[14]	Climate Risk Disclosure		Were climate risks disclosed – RSAC?	Score from 1 to 10 based on disclosure quality, detail, and examples (author's assessment)
[15]	Sustainable Business or Business Ethics		Were disclosures made about sustainable business practices with clients?	Score from 1 to 10 based on disclosure quality, detail, and examples (author's assessment)

(*continued*)

(*continued*)

Reference	Field Name 1	Field Name 2	Description	Possible Response
[16]	Gender Equity		Were gender equity data disclosed?	Score from 1 to 10 based on disclosure quality, detail, and examples (author's assessment)
[17]	Social Actions		Were social action initiatives disclosed?	Score from 1 to 10 based on disclosure quality, detail, and examples (author's assessment)
[18]	Goal Disclosure?		Were the entity's goals clearly disclosed?	Score from 1 to 10 based on disclosure quality, detail, and examples (author's assessment)
[19]	Scope 1 GHG Emissions		Scope 1 emissions disclosed in tCO_2e	Reported value in tCO_2e for Scope 1 GHG emissions
[20]	Scope 2 GHG Emissions	(tCO_2e)	Scope 2 emissions disclosed in tCO_2e	Reported value in tCO_2e for Scope 2 GHG emissions
[21]	Scope 3 GHG Emissions	(tCO_2e)	Scope 3 emissions disclosed in tCO_2e	Reported value in tCO_2e for Scope 3 GHG emissions
[22]	% Women in Leadership	Target	Disclosed target for women in leadership	Target percentage
[22][a]	% Women in Leadership	Actual	Actual percentage of women in leadership	Disclosed actual percentage
[23]	% Women Employed / Hired	Target	Disclosed target for hiring or proportion of women in total staff	Target or reported percentage
[23][a]	% Women Employed / Hired	Actual	Actual percentage of women hired or in workforce	Disclosed actual percentage
[24]	Notes		Relevant notes that highlight specific aspects of the institution	Descriptive comments

Appendix 2

Entidade		ITAUUNIBANCO	BRASIL	BRADESCO	SANTANDER BR	Banco BV	Safra
Total de ativos	Valores em milhões de reais (*)	2.543.100	2.208.054	1.927.523	1.159.076	142.657	312.142
Patrimônio Líquido		199.050	163.638	167.014	85.914	13.980	26.948
Resultado Operacional		39.700	39.531	18.235	8.585	1.443	9.417
Resultado líquido		33.877	33.615	14.503	8.974	1.334	3.335
Possui um site específico para sustentabilidade?		SIM	SIM	SIM	SIM	SIM	SIM
Publicou relatório de ESG?		SIM	SIM	SIM	SIM	SIM	SIM
Publicou informações de ESG juntamente com relatório da administrção?		Sim, possui mais de um relatório.	Sim, possui relatórios auxiliarespara o efetito estufa e para compromisso com mudanças climáticas	Sim, possui mais de um relatório	Sim, possui mais de um relatório	Relatório integrado ao relatório da administração	Sim
Asseguração independente,		SIM	SIM	SIM	Sim	Sim	SIM
Relatório de asseguração indepentende emitido por :		PwC (relatório de asseguraçaõ)	DTT (relatório de asseguração) e Inventário de gases de efeito estufa Instituto Totum	KPMG	PwC (relatório de asseguraçaõ) GHG protocol elaborado por Fundação Vanzolini	KPMG	Não identificado
Framework utilizado		GRI, SASB, TCFD, IASB, PRB	GRI, SASB, TCFD, TCU	GRI, SASB, TCFD, ODS	GRI, ODS, ISSB, TCFD, SASB	GRI, ODS, SASB	GRI, SASB, ONU, ODS
Divulgação de Materialidade?		Sim (dupla materialidade)	Sim (dupla materialidade)	Sim (dupla materialidade)	Sim (dupla materialidade)	Materialidade simples	Materialidade simples
Ações Climaticas		A	A	A	A	A -	B
Divulgação de riscos		A	A	A	A	A -	A-
Negócios sustentáveis ou		A	A	A	A	A-	B
Equidade de genero		A	A	A	A	A	B
Ações sociais		A	A	A	A	A	B
Divulgaçãod e metas?		A	A	A	A	A	C
Relatório inclusivo?		OK	Ok	OK	Ok	OK	Ok
Registro do passivo escopo		19.208	18.494,35	16.540	4.750,81	11	Não divulga
Registro do passivo escopo		14.468	19.710,18	13.933	13,052	53,82	Não divulga
Registro do passivo escopo 3 (tCO2 e)	Total	22.515.204	44.084	122.780	58.260	4.312	Não divulga
	Outras emissões	63.490					
	Emissões Financiadas	22.451.714	44.083,98	122.780	58.259,98	4312,37	
% de mulheres na liderança	Meta	35% a 40%	30%	15% a 22%	40%	50%	não localizado
	Destaque	35,40%	25,33%	16,62%	36,30%	39,70%	26,40%
% de contratação de mulheres, % de mulheres na	Meta	50%	37,17%	35%	não localizado	não localizado	não localizado
	Destaque	53,00%	não identificado	49,47%	20,70%	47,69%	no relatório de ESG 40,3%, relatório da administração 45,2
Observação							

Entidade		Banco VW	Banco GMAC	Banco Mercedes	Banco Toyota	Banco Volvo
Total de ativos	Valores em milhões de reais (*)	42.899	13.762	20.621	11.831	16.499
Patrimônio Líquido		4.098	2.755	148	1.263	2.723
Resultado Operacional		722	250	270	102	380
Resultado líquido		353	226	164	68	322
Possui um site específico para sustentabilidade?		NÃO	Somente um link para o RSAC	SIM	Somente um site dedicado a Letras financeiras verdes	N/A
Publicou relatório de ESG?		SIM	NÃO	NÃO	NÃO	N/A
Publicou informações de ESG juntamente com relatório da administrção?		Relatório integrado ao relatório da administração	Não	Não	Não	N/A
Asseguração independente,		Sim	Não	Não	Não	N/A
Relatório de asseguração		BUREAU VERITAS	Não	Não	Não	N/A
Framework utilizado		GRI, SASB, TCFD	Somante divulgações do bacen	Somante divulgações do bacen	Somante divulgações do bacen	N/A
Divulgação de Materialidade?		Materialidade simples	Não	Não	Não	N/A
Ações Climaticas		Não divulga ações climáticas específicas	N/A	N/A	N/A	N/A
Divulgação de riscos		A	N/A	N/A	N/A	N/A
Negócios sustentáveis ou		B	N/A	N/A	N/A	N/A
Equidade de genero		A	N/A	N/A	N/A	N/A
Ações sociais		A-	N/A	N/A	N/A	N/A
Divulgaçãod e metas?		A	N/A	N/A	N/A	N/A
Relatório inclusivo?		OK	N/A	N/A	N/A	N/A
Registro do passivo escopo		62,07	N/A	N/A	N/A	N/A
Registro do passivo escopo		48,58	N/A	N/A	N/A	N/A
Registro do passivo escopo 3 (tCO2 e)	Total	1.608.339,83	N/A	N/A	N/A	N/A
	Outras emissões					
	Emissões Financiadas					
% de mulheres na liderança	Meta	26%	N/A	N/A	N/A	N/A
	Destaque	38%	N/A	N/A	N/A	N/A
% de contratação de	Meta	50%	N/A	N/A	N/A	N/A
	Destaque	47%	N/A	N/A	N/A	N/A
Observação		O relatório não é assinado por contador				

References

1. Aladwey, L.M.A., Alsudays, R.A.: Unveiling the Significance: Sustainability and Environmental Disclosure in Saudi Arabia through Stakeholders' Theory. Sustainability (Switzerland) **16**(9) (2024). https://doi.org/10.3390/su16093689
2. Arnone, M., et al.: Banking stability in the ESG framework across Italian regions (n.d.). https://hal.science/hal-04647121
3. Banco Central do Brasil: Segmentação prudencial regulatória (n.d.). https://www.bcb.gov.br/estabilidadefinanceira/regprudencialsegmentacao
4. Banco Central do Brasil: IF.Data (n.d.). https://www3.bcb.gov.br/ifdata/#

5. Banco Toyota: Institucional (n.d.). https://institucional.bancotoyota.com.br/
6. Bradesco: Relatório Integrado (n.d.). https://www.bradescori.com.br/o-bradesco/relatorio-integrado/
7. Bătae, O.M., Dragomir, V.D., Feleagă, L.: The relationship between environmental, social, and financial performance in the banking sector: a European study. J. Clean. Prod. **290**, 125791 (2021). https://doi.org/10.1016/j.jclepro.2021.125791
8. Buallay, A., Fadel, S.M., Alajmi, J., Saudagaran, S.: Sustainability reporting and bank performance after financial crisis: evidence from developed and developing countries. Compet. Rev. **31**(4), 747–770 (2020). https://doi.org/10.1108/CR-04-2019-0040
9. Celestino, É.J.M., Lima Pereira, M. de, Camara, R.P. de B.: Assessment of ESG practices in banks listed in [B]3. Revista Catarinense Da Ciência Contábil **22**, e3380 (2023). https://doi.org/10.16930/2237-7662202333802
10. Chevrolet Serviços Financeiros: Sustentabilidade (n.d.). https://www.chevroletsf.com.br/pt-br/inicio/sustentabilidade.html
11. Cordeiro, E.M., et al.: BREVES CONSIDERAÇÕES A RESPEITO DA ANÁLISE DE CONTEÚDO E DA PESQUISA QUALITATIVA (n.d.)
12. da Silva, M.C., Oliveira, E.J. de, Celestino, M. dos S.: Ensaio teórico sobre a teoria da divulgação aplicada ao setor público brasileiro. Revista de Gestão e Secretariado (Management and Administrative Professional Review) **14**(10), 18889–18906 (2023). https://doi.org/10.7769/gesec.v14i10.2921
13. de Souza, J.A.S., Flach, L., Borba, J.A., Broietti, C.: Financial reporting quality and sustainability information disclosure in Brazil. Brazilian Business Review **16**(6), 555–575 (2019). https://doi.org/10.15728/BBR.2019.16.6.2
14. Dell'Erba, M., Gomtsyan, S.: Regulatory and investor demands to use ESG performance metrics in executive compensation: Right instrument, wrong method. J. Corpor. Law Stud. **24**(1), 1–40 (2024). https://doi.org/10.1080/14735970.2024.2350139. Diwan, H., Sreeraman, B.A.: From financial reporting to ESG reporting: A bibliometric analysis of the evolution in corporate sustainability disclosures. Environment, Development and Sustainability **26**, 13769–13805 (2024)
15. el Khoury, R., Nasrallah, N., Alareeni, B.: ESG and financial performance of banks in the MENAT region: concavity–convexity patterns. J. Sustain. Fin. Invest. **13**(1), 406–430 (2023). https://doi.org/10.1080/20430795.2021.1929807
16. Esteban-Sanchez, P., de la Cuesta-Gonzalez, M., Paredes-Gazquez, J.D.: Corporate social performance and its relation with corporate financial performance: International evidence in the banking industry. J. Clean. Prod. **162**, 1102–1110 (2017). https://doi.org/10.1016/j.jclepro.2017.06.127
17. Hummel, K., Schlick, C.: The relationship between sustainability performance and sustainability disclosure – Reconciling voluntary disclosure theory and legitimacy theory. J. Account. Public Policy **35**(5), 455–476 (2016). https://doi.org/10.1016/j.jaccpubpol.2016.06.001
18. IFRS Foundation: International Sustainability Standards Board (ISSB) (n.d.). https://www.ifrs.org/groups/international-sustainability-standards-board/
19. IFRS Foundation: IFRS S1 – General requirements for sustainability-related disclosures (2023). https://www.ifrs.org/issued-standards/ifrs-sustainability-standards-navigator/ifrs-s1-general-requirements.html/content/dam/ifrs/publications/html-standards-issb/english/2023/issued/issbs1/
20. IFRS Foundation: IFRS S2 – Climate-related disclosures (2023). https://www.ifrs.org/issued-standards/ifrs-sustainability-standards-navigator/ifrs-s2-climate-related-disclosures.html/content/dam/ifrs/publications/html-standards-issb/english/2023/issued/issbs2/
21. Itaú: Sustentabilidade – Produtos ESG (n.d.). https://www.itau.com.br/sustentabilidade/produtos-esg/

22. Hummel, K., Schlick, C.: The relationship between sustainability performance and sustainability disclosure – Reconciling voluntary disclosure theory and legitimacy theory. J. Acc. Pub. Policy **35**(5), 455–476 (2016). Disponível em: https://doi.org/10.1016/j.jaccpubpol.2016.06.001
23. Lahiri, S.: A qualitative research approach is an inevitable part of research methodology: an overview. Int. J. Fin. Manage. Res. **5**(3) (2023). www.ijfmr.com
24. Lamanda, G., Tamásné Vőneki, Z.: Is ESG disclosure associated with bank performance? evidence from the visegrad four countries. Manage. Environ. Q. Int. J. **35**(1), 201–219 (2024). https://doi.org/10.1108/MEQ-02-2023-0064
25. Lee, J.J., Ramjan, L., Nuampa, S., Lai, C.K.Y.: Where is the interpretation? Important considerations for the philosophical and disciplinary lens in qualitative health research. Nurs. Health Sci. **25**(4), 493–496 (2023). https://doi.org/10.1111/nhs.13068
26. Mohammed Auwal, B., Ibrahim, M., Sadiq Balarabe, A., Abba Isa, I.: Sustainability reporting frameworks a comparative analysis of reporting standards and their implications for accounting and reporting. In: International Journal of Accounting, Finance, and Administrative Research, Vol. 1 (2024)
27. Park, J.D., Nishitani, K., Kokubu, K., Freedman, M., Weng, Y.: Revisiting sustainability disclosure theories: evidence from corporate climate change disclosure in the United States and Japan. J. Clean. Prod. **382**, 135203 (2023). https://doi.org/10.1016/j.jclepro.2022.135203
28. Raghavan, K.: ESG reporting impact on accounting and finance. The J. Global Awar. **3**(1), 1–16 (2022). https://doi.org/10.24073/jga/3/01/09
29. Salem, M.R.M., Shahimi, S., Alma'amun, S.: Does mediation matter in explaining the relationship between ESG and bank financial performance? a scoping review. J. Risk and Fin. Manage. **17**(8) (2024). https://doi.org/10.3390/jrfm17080350
30. Santander: Sustentabilidade – Resultados (n.d.). https://www.santander.com.br/hotsite/sustentabilidade/#Resultados
31. Santos, V., Bacci, D.: Proposta para governança ambiental ante os dilemas socioambientais urbanos. Estudos Avançados **31**(89), 199–212 (2017). https://doi.org/10.1590/s0103-40142017.31890017
32. Segura, L.C., Naser, A., Abreu, R., Perez-Lopez, J.A.: ESG dimensions and corporate value: insights for sustainable investments. Sustainability **16**(17), 7376 (2024). https://doi.org/10.3390/su16177376
33. Spierings, M.: Linking executive compensation to ESG performance. Harvard Law School Forum on Corporate Governance (2022)
34. Volkswagen Financial Services: Relatório anual – Relacionamento com investidores (n.d.). https://www.vwfs.com.br/volkswagen-financial-services/relacionamento-investidor/relatorio-anual.html
35. Volvo Financial Services: Sustentabilidade (n.d.). https://www.volvofinancialservices.com/br/
36. WEB OF SCIENCE. Resultados de busca para divulgação ESG em bancos. Disponível em: https://www-webofscience-com.ez347.periodicos.capes.gov.br/wos/woscc/summary/93563998-e5eb-4c66-a395-71158ed0a370-012d99a41a/relevance/1. Acesso em: 28 novembro 2024
37. Yip, E., Staden, C. van, Cahan, S.: Corporate Social Responsibility Reporting and Earnings Management: The Role of Political Costs. In: Australasian Accounting Business and Finance Journal Article 3 Australasian Accounting, Vol. 5, Issue 3 (2011). http://ro.uow.edu.au/aabfj
38. Younas, A., Durante, A., Fàbregues, S.: Understanding the nature of and identifying and formulating "research problems" in mixed methods research. J. Mixed Methods Res. (2023). https://doi.org/10.1177/15586898231191441

The Relationship Between ESG and Capital Structure: A Systematic Review of Trade-Off and Performance Feedback Theories

Fernando Ramos dos Reis[1,2,3,4](✉), Mara Teresa da Silva Madaleno[1,2,3], and Johan Hendrik Poker Junior[4]

[1] University of Aveiro, Aveiro, Portugal
{Fernando.ramosr,maramadaleno}@ua.pt
[2] Department of Economics, Management, Industrial Engineering and Tourism – DEGEIT, Aveiro, Portugal
[3] Research Unit on Governance, Competitiveness and Public Policy – GOVCOPP, Aveiro, Portugal
[4] University of Campinas – Unicamp, Campinas, Brazil
johanpkr@unicamp.br

Abstract. This study systematically reviews how ESG (Environmental, Social, and Governance) factors influence corporate capital structure, integrating the Trade-off and Performance Feedback theories. The objective is to propose an integrated framework that explains the Preferred Reporting Items for Systematic Reviews and Meta-Analyses (PRISMA) standards. Using the SPIDER tool, we selected 33 articles from Scopus and Web of Science (2020–2025) that applied Intrinsic Data Quality (IDQ) criteria. Results show that ESG affects financial risk perception, leverage adjustments, and cost of capital, intensifying the structure of capital volatility described by DeAngelo (2015) in the Trade-off Theory. Furthermore, the Performance Feedback Theory suggests that ESG pressures influence managerial aspirations, leading to dynamic changes in a firm's capital structure. The findings suggest that ESG serves as both an external pressure and a strategic driver, challenging traditional assumptions about the stability of capital structure. We conclude that combining Trade-off and Performance Feedback Theories offers a more robust explanation for corporate financial behavior in ESG-driven environments, suggesting important future empirical research directions.

Keywords: ESG · Trade-off Theory · Performance Feedback Theory

1 Introduction

Sustainability has become a central theme in contemporary financial debates, often perceived as challenging stability and economic growth (Akhtar and Rashid, 2024). In response, the United Nations (UN) established the 17 Sustainable Development Goals (SDGs) in 2015, encouraging fairer and more environmentally responsible economic practices (Howard-Grenville et al., 2019). In this scenario, companies and governments have begun to look for ways to balance financial stability and sustainable commitments (Battiston et al., 2021).

L. Almeida et al. (Eds.): ICAFI 2025, ISEM 80, pp. 212–231, 2026.
https://doi.org/10.1007/978-3-032-16268-7_12

Despite this progress, relevant obstacles remain to integrating ESG criteria (Environmental, Social, and Governance) into strategic decisions, especially about capital structure (Adeneye et al., 2023). A recurring challenge has been the absence of a consolidated theoretical basis on how and why these factors should be efficiently incorporated into financial decisions. This scenario has led researchers to investigate the foundations of this integration, seeking to elucidate the mechanisms by which ESG influences crucial corporate decisions.

The growing concern for the environment, as reflected in the adherence of several countries to the Kyoto Protocol and the Paris Agreement (Muchiri et al., 2022; Rajamani and Werksman, 2018), underscores the urgency of understanding how environmental requirements influence business decisions. External pressures refer to regulatory changes, social expectations, and investor demands associated with ESG performance. However, many managers still treat ESG issues as reactive responses (corporate piracy) rather than as integral components of a long-term strategic approach (DeAngelo, 2025), one that considers the needs of institutional investors or society, rather than as part of an integrated, long-term strategy. This limited view compromises the effectiveness of ESG incorporation. It also contributes to the instability of the capital structure.

In this context, we propose that ESG factors act as dynamic drivers of external pressure and strategic aspiration, influencing adjustments in capital structure (Li et al., 2024a; Xiao, 2022; Zahid et al., 2023). This view contrasts with the traditional Trade-off Theory, which suggests that firms maintain an optimal capital structure by balancing the benefits and costs of debt. However, recent studies have challenged the assumption of capital structure stability, arguing that external ESG-related forces disrupt managerial decision-making processes—a phenomenon better explained through the lens of the Performance Feedback Theory (DeAngelo, 2025; DeAngelo and Roll, 2015).

Behavioral finance provides a significant insight regarding this matter. This theory suggests that managers modify their strategies based on the difference between actual performance and organizational aspirations (Berchicci and Tarakci, 2022; Wang and Sun, 2024), enabling a more flexible and responsive approach to volatile environments. Therefore, ESG can be seen as an external factor impacting capital structure.

However, recent studies indicate that Performance Feedback Theory remains almost entirely limited to innovation, internationalisation, and corporate conduct. Virtually no work examines how ESG-driven performance gaps shape firms' capital-structure adjustments. Closing this blind spot is the central contribution of the present study.

Despite the topic's relevance, the literature regarding the relationship between ESG and capital structure remains scarce from the integrated perspective of these two theories—the Trade-off and Performance Feedback theories. In databases such as Scopus and Web of Science, we identified 154 systematic reviews related to ESG and the Trade-off Theory, but none that included the Performance Feedback Theory. This justifies the need for a review that maps the theoretical foundations, variables, and methods used to measure such relationships.

Some early studies suggest that companies with good ESG practices tend to have greater access to equity financing and less reliance on debt (Duguleană et al., 2024; Qiu and Yu, 2024; Xiao, 2022), while others point out that debt can be used strategically to finance sustainable initiatives, serving as a sign of commitment to the ESG agenda (Khan et al., 2024). This ambiguity highlights the complexity of applying the Trade-off Theory to the ESG context.

Recent literature has shown that risks arising from climate and carbon policies pose direct challenges to the definition of capital structure, particularly by intensifying perceptions of bankruptcy risk and financing constraints. Several studies reveal that companies with high carbon emissions reduce their financial leverage after regulatory events (Nguyen and Phan, 2020; Shu et al., 2023) such as the Kyoto Protocol or carbon neutrality policies, precisely because they anticipate higher capital costs and deterioration of intangible assets. This strategic change can be interpreted in the light of the Performance Feedback Theory as a response to external institutional pressures (imposed ESG aspirations) and the fear of poor future performance, which also corroborates the criticism of DeAngelo and Roll (2015) on the structural instability of leverage in volatile environments. Therefore, climate and environmental risks are catalysts for the gap between externally imposed aspirations and the internal capacity to perform, intensifying the need to adapt financial capital structure.

In this context, this study begins with the following problem, structured using the qualitative methodology SPIDER, proposed by Cooke et al. (2012): How does integrating ESG criteria influence capital structure decisions in light of the Trade-off and performance feedback theories? This systematic review focuses on literature published between 2020 and 2025, aiming to understand how these theories explain companies' adjustments in response to ESG pressures, particularly in environments characterised by volatility and uncertainty. Research is guided by a structured design, anchored in the PRISMA protocol (Page et al., 2021), with methodological quality analysis of the selected studies. The main evaluation focuses on the variables, theoretical foundations, and applied methods related to ESG and capital structure decisions. Finally, the type of research is qualitative and descriptive, centered on the conceptual and categorical analysis of scientific literature. This review aims to map the existing literature on ESG and capital structure and propose an integrated theoretical framework that combines the Trade-off Theory and the Performance Feedback Theory. This approach seeks to explain the observed instability in corporate capital structure in response to ESG pressures.

To this end, the following specific objectives will be explored: (1) To identify studies that relate the Trade-off Theory with ESG; (2) Identify behavioral finance studies, especially the Performance Feedback Theory, applied to ESG; (3) To survey the variables used to analyze the Performance Feedback Theory; (4) Map the variables traditionally used to measure ESG; (5) To investigate which other theories are frequently combined with Trade-off Theory and Performance Feedback Theory in the recent literature.

Based on these objectives, it seeks to answer the following research questions:

1. How does the Trade-off Theory incorporate ESG variables into the capital structure?
2. How can the Performance Feedback Theory explain companies' adjustments to ESG requirements?
3. What variables are frequently used in literature to measure the relationship between ESG and capital structure?
4. How does ESG act as a factor of volatility and adjustment in capital structure decisions?

The organization of this article follows the following logical structure: (1) introduction; (2) materials and methods; (3) analysis of the results, subdivided into (3.1) ESG results with the Trade-off Theory, (3.2) results with the Performance Feedback Theory, (3.3) ESG and performance feedback variables, and (3.4) primary statistical methods used; (4) conclusion, limitations, and suggestions for future research.

2 Materials and Methods

In this review, the PRISMA standards were followed (Page et al., 2021), according to the flowchart presented in Figs. 1 and 2, for the search and selection of studies, with the stages of eligibility, ineligibility, selection, sampling, and data analysis as following Ramalho et al. (2025).

Eligibility Criteria: Peer-reviewed English-language articles were searched for in titles, abstracts, and keywords between 2020 and 2025, a period that, according to Palese et al. (2022), includes some recent years and can contribute to the findings for current and future practice. Studies that addressed the theory, trade-off, and, subsequently, the Performance Feedback Theory in the context of sustainability (ESG) were selected.

Ineligibility Criteria: After reading the abstracts, titles, and keywords, studies that did not fit the initial objective of this review were disregarded; that is, when the Trade-off Theory, Performance Feedback Theory, and ESG were not explicit.

Selection of databases: SCOPUS and Web of Science, as they are comprehensive and provide access to diverse literature across various subject areas (Harzing, 2019; Martín-Martín et al., 2021; Stechemesser and Guenther, 2012).

Sampling: Initial sampling was carried out using the following keywords: Trade-off Theory: ("Trade-off" or "tradeoff" or "trade off" or leverage or "capital structure" or optimal) AND ESG. Performance Feedback Theory: ("feedback theory" OR "performance feedback") AND ("firm*" OR "corporate" OR "business*" OR "financ*" OR "companie*") AND ("esg" OR "carbon" OR "co2" OR "environmental" OR "sustainab*" OR "ods").

Data analysis: The data analysis was based on the Intrinsic Data Quality (IDQ) methodology proposed by Rameder et al. (2022). This methodology aims to evaluate the accuracy, quality, credibility, and reputation of the articles, assigning a score from 0 to 100, with weights of 33% for each of the following criteria: SCImago Journal Ranking, Scopus CiteScore, and Scopus quartiles, replacing the discontinued CORE. The scoring criteria were followed as per Rameder et al. (2022) (Table 1).

Table 1. Intrinsic Data Quality (IDQ)

	Criteria	Punctuation
Scopus Quartile	Q1	100
	Q2	75
	Q3	50
	Q4	25
	No quartile	0
SJR	1,000 a 17654	100
	0.754 vs. 0.999	75
	0.500 vs. 0.753	50
	0.499 vs. 0.100	25
	Less than 0.100	0
CiteScore	16 a 30	100
	10 a 15,99	75
	6,00 a 9,99	50
	0,10 a 5,99	25
	Less than 0.10	0

Source: Authors

Article selection process: Article selection process: The initial search resulted in 487 articles in SCOPUS and 384 in Web of Science. After removing 325 duplicate articles, the initial database resulted in 546 articles. A total of 130 articles were excluded from quartiles 3 and 4, as follows Clément et al. (2025), subsequently 24 articles to which we did not have access were excluded, resulting in 392 articles. Based on the IDQ analysis, 244 articles with scores below 75 were excluded, prioritising the most relevant articles. Finally, each article was analysed to determine whether it fell within the scope of the research, and 130 articles were excluded, resulting in a final sample of 18 articles. The selection and exclusion process followed the flowchart illustrated in Fig. 1.

Results of the Performance Feedback Theory: Initially, 35 articles were collected in SCOPUS and 40 in Web of Science. After removing 14 duplicate articles and applying the ineligibility criteria, 36 articles were excluded, resulting in a final sample of 25 articles, as illustrated in the flowchart in Fig. 2.

Following the methodological design, the next section presents the main results, focusing on how the Trade-off and Performance Feedback Theories relate to ESG and capital structure.

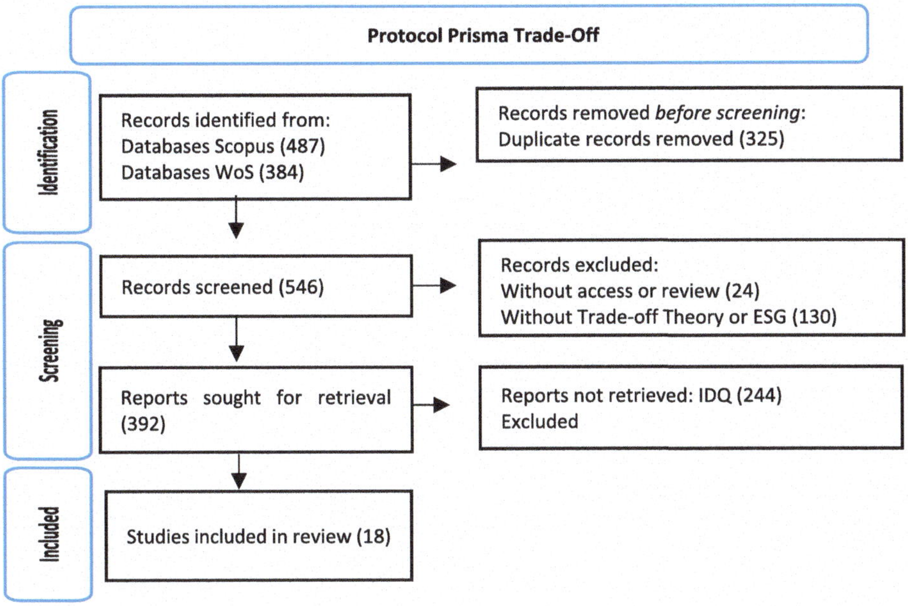

Fig. 1. Flow diagram illustrating the PRISMA protocol, complemented by the IQD protocol, used to select the articles on Trade-off Theory

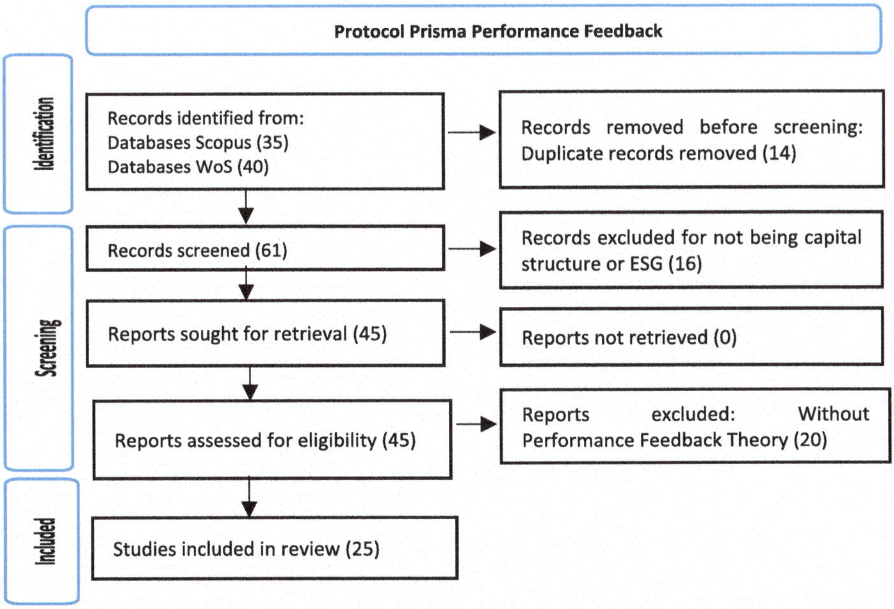

Fig. 2. Flow diagram illustrating the PRISMA protocol, used to select the articles on Performance Feedback Theory.

3 Analysis of the Results

3.1 Trade-Off Theory and ESG

The Trade-off Theory, a traditional approach in finance, seeks to balance the fiscal advantages of debt with the costs associated with financial hardship (Miller, 1977). Our review identified four broad groups of conclusions from the Trade-off Theory: risks (political, climate, environmental), spreads/financing costs, leverage, and profitability.

The first group of conclusions refers to equilibrium risks. ESG performance modifies the perception of financial risk, reducing the cost of capital. However, when significant environmental risks are present, financing costs rise, discouraging leverage (Xiao, 2022; Zahid et al., 2023). Recent literature suggests that uncertainty surrounding carbon policies serves as an external governance mechanism, as it increases the perception of environmental risk among creditors and investors, thereby increasing the cost of debt and inducing companies to reduce leverage or shorten their financing terms (Dang et al., 2023; Elnahas et al., 2018).

The second group of conclusions directly involves leverage. Evidence suggests that companies with greater exposure to ESG risks tend to reduce their leverage, thereby compromising their optimal capital structure (Qiu and Yu, 2024). These risks, particularly those related to carbon, negatively impact company performance, reduce tax benefits, increase financial difficulties, and lead to more conservative financing decisions (Nguyen and Phan, 2020).

The third group of findings refers to spreads and financing costs. Highly polluting firms have higher lending spreads, especially after stricter environmental regulations (Kleimeier and Viehs, 2021), which causes them to pay higher debt financing costs than other firms (Caragnano et al., 2020).

The fourth dimension observed involves profitability. Contrary to the Trade-off Theory, where a company's primary responsibility is to maximise shareholder profits, Qureshi and Ahsan (2022) suggest that spending on environmental, social, and governance (ESG) initiatives is an unnecessary cost that reduces profitability.

An innovative aspect identified in the literature is the concept of leverage targets structured around the golden ratio. Csapi et al. (2024) demonstrate that companies with better ESG performance adjust their leverage targets to levels close to the golden ratio (approximately 61.8% debt and 38.2% equity), suggesting a search for structural harmony. This finding offers an alternative interpretation of the Trade-off Theory, proposing that ESG serves as a regulatory mechanism for symmetry and structural stability. These findings reflect the instability patterns discussed by DeAngelo and Roll (2015).

These results reinforce the existence of non-linear relationships, in which the impact of ESG on debt structure varies according to the stage of involvement (Li et al., 2024b). Additionally, several complementary theories to the Trade-off Theory have been identified. According to Fama and French (2002), the Hierarchy Theory suggests that internal financing is generally preferred over external debt due to issues of information asymmetry. However, when there is adequate ESG disclosure, there is a reduction in informational asymmetry, which facilitates access to credit (Cariola et al., 2020; Restrepo et al., 2020).

From the perspective of Trade-off Theory, Agency Theory suggests that ESG practices mitigate agency conflicts, reduce associated costs, and foster the creation of long-term value. Stakeholder theory, on the other hand, extends this view by demonstrating that ESG strategies strengthen the bond with lenders, investors, and regulators (Duguleană et al., 2024; Khan et al., 2024). However, according to Li et al. (2024b), there are also concerns that some managers may overemphasise ESG investments for personal reasons, which can negatively impact long-term debt. According to the Stakeholder Theory, a company's operations are a social contract with stakeholders, requiring ongoing engagement through ESG practices to enhance long-term corporate success (Qureshi and Ahsan, 2022).

The Signalling Theory complements this scenario by interpreting that high ESG scores signal trust and commitment to stakeholders, which can lead to lower debt costs and greater long-term financing opportunities (Li et al., 2024b). Following Zhang et al. (2023), which is based on Weick (1969), companies are like a system of perception-evaluation feedback, not acting as passive receptors when faced with signs of future stringent carbon regulation, but actively incorporating them into their strategic feedback processes.

Finally, studies such as the one by Li et al. (2024b) highlight that the effects of ESG on the cost of financing are not homogeneous, being mediated by the profile of creditors. Skilled institutional investors and interbank market participants tend to accept lower returns in exchange for good ESG performance, while ordinary individual or institutional investors demonstrate lower sensitivity to these factors. This heterogeneity reveals a dispersion in the preferences of financiers, which is still little explored by the traditional Trade-off Theory but is highly relevant to understanding the instability of the capital structure in contexts of socio-environmental pressure.

Such evidence suggests that ESG practices influence risks, leverage, and financing costs and affect capital structure adjustment speed. As demonstrated by Xiao (2022), policies such as China's "clean air" policy reduce the responsiveness of highly polluting companies due to increased transaction costs and informational asymmetry. These results align with Deangelo and Roll (2015) criticism, indicating that external environmental variables function as structural barriers to optimal capital adjustment, requiring more dynamic analytical models, such as the Performance Feedback Theory, to explain the volatility observed in contemporary financial decisions.

3.2 Performance Feedback Theory and ESG

According to Berchicci and Tarakci (2022), the Performance Feedback Theory, the Behavioral Theory of the Firm (BTOF), proposed by Cyert and March (1963), is characterized by managers' aspirations related to organizational performance. These aspirations represent a gap between actual and expected performance, directly influencing managerial decisions (Lu et al., 2024).

Our review corroborates other research (Sun et al., 2024) in which the Performance Feedback Theory has been mainly focused on areas such as research and innovation (Lu and Wong, 2019; Sun and Qiu, 2022), mergers and acquisitions (Iyer and Miller, 2008), venture capital (Gaba and Bhattacharya, 2012; Wang et al., 2022), internationalization (Chen et al., 2024; Xie et al., 2016), corporate misconduct measured overall by corporate social responsibility (Gao et al., 2021; Ren and Lin, 2024), investments under environmental regulation (Li et al., 2018), financial performance of family businesses (Wang and Sun, 2024), greenwashing (Fiechter et al., 2022), and environmental disclosure (DasGupta and Roy, 2023). However, there is a significant gap in studies that apply this theory to analyzing the capital structure.

The literature distinguishes Performance Feedback Theory into two categories: positive and negative, which according to Wang et al. (2023), the positive Performance Feedback Theory usually brings confidence and satisfaction to decision-makers. However, there is no consensus on whether the Negative Performance Feedback Theory influences corporate decision-making, which suggests the need for further investigation into its impacts.

Several empirical results in corporate sustainability reveal different behaviors in the face of the Performance Feedback Theory. Sun and Qiu (2022) show that positive Performance Feedback is negatively associated with breadth research behavior, indicating that companies with good performance tend to avoid exploring new areas. On the other hand, this same type of feedback has a positive relationship with in-depth research behavior, suggesting a greater propensity to specialize and deepen existing skills.

Zhang et al. (2023) identified that companies with environmental performance below aspiration tend to remodel more slowly, while those with performance above aspiration demonstrate greater agility in strategic adaptations. This is consistent with the Prospective Theory, in which companies more sensitive to losses would not easily increase their ESG performance (Chari et al., 2019; Das Gupta, 2017; Tversky and Kahneman, 1974).

Similarly, Meng et al. (2024) demonstrate that companies whose environmental performance is slightly below that of their peers are more likely to pursue green innovation, evidencing a negative relationship between performance deficiencies and innovation intention.

A notable finding is that of Pan et al. (2023), which observed symbolic management behavior on the part of the companies. The authors identified that negative and positive feedback about environmental performance is associated with publishing less legible Corporate Social Responsibility (CSR) reports to control impressions and postpone possible regulatory sanctions.

We have identified only a few other complement theories, one of which is the Attention-Based View (ABV) of the firm, in which Ocasio (1997, p. 189) summarized, "the cognition and action of individuals are... derived from the specific organizational context and situations that individual decision-makers find themselves in." Hence, the outside environment and internal structure are core to organizational attention.

We also identified the Institutional Theory as strongly related to the variables of government pressure and regulation, formal institutions, and the beliefs and values of other parties, such as the public, consumers, and non-governmental organizations, and influence the metrics of business feedback.

When environmental performance falls short of aspirations, the divergence of interests between managers and shareholders widens. While green innovation can reduce long-term costs and maximize shareholder wealth, it generates immediate expenses that erode managers' earnings. In this situation, equity incentives and intensified supervision act as frameworks that align objectives and encourage strategic responses capable of closing the performance gap.

A theory that corroborates DeAngelo (2022) study is the Social Pressure Theory (Desai, 2014), in which managers experience pressure from the discrepancy between aspiration and performance. We also identified the Resource-Based Theory, where managers look for sufficient resources to carry out innovation and ESG activities, and the Threat-Rigidity Theory, where managers are not prone to significant risks when aspirations are negative.

Despite the growing number of studies that apply the Performance Feedback Theory in areas such as innovation, internationalization, and corporate conduct, there is a significant absence of research that explores this theory in the context of financial adjustments in the face of ESG pressures or capital structure.

3.3 Variables

Tables 2 and 3 summarize the main variables used to explore the relationship between ESG and those related to the Trade-Off and Performance Feedback theories. Table 2 shows that the variables used to study ESG cover several dimensions, such as environmental regulations, scores, performance, climate, political uncertainty, risk, and carbon emissions. By deepening the study of these variables, it is possible to quantify the impacts of environmental, social, and governance factors and theoretically support the incorporation of ESG into financial decisions. These help construct more robust models to explain the volatility observed in capital structure.

The Trade-Off Theory mainly concerns financial variables, such as leverage. However, we identified one variable that analyzes the theory (Optimized supply under trade-off model, Su and Liu, 2024). Variables such as CO_2 emissions and environmental regulation capture the impacts on the cost of debt and the balance between tax benefits and financial risks, which can directly influence the volatility of the capital structure, as the Trade-Off Theory suggests.

As shown in Table 3, we did not identify significant variations in the variables that analyze the relationship between ESG and the Performance Feedback Theory. These variables are mostly aspiration analyses or media disclosures. Most are binary variables or variables that compare the performance of similar companies with the company's results.

The financial variables followed the traditional literature, with variables to control size, profitability, and revenue growth, among other things.

DeAngelo and Roll (2015) pointed out that the instability of the capital structure can be explained in light of ESG as a factor that generates an aspirational gap. On the one hand, ESG creates new organizational aspirations that were not foreseen in traditional models of optimal capital structure. DeAngelo (2025) states that informal ESG pressures can induce firms into reactive strategic adjustments, which he describes as 'corporate piracy'. This dissonance between conventional goals and new socio-environmental demands generates a strategic tension (Agency Theory), which, according to the Performance Feedback Theory, is perceived by managers as a deviation between current and desired performance, causing adjustments that explain, in part, the volatility of the capital structure over time.

Table 2. Trade-Off Theory and ESG Variables

Classification	Variable(s)	Formula	Authors
Financial	ROA, LEV, GROWTH, Firm Age, Firm Size, Cash Flow, Tangibility	Standard accounting calculations	–
Environmental Policy	Environmental Performance Index (EPI)	Country's EPI score	Mohammadi et al. (2015)
	Government environmental pressure	Count of environmental terms / penalties	Wang et al. (2024)
Environmental Regulation	Policy implementation status	Dummy: 0 (pre-policy) / 1 (post-policy)	Qiu and Yu (2024)
	Environmental regulation intensity	ln(SO_2 emissions + 1) of firm (2013)	Xiao (2022)
	Regional regulatory strength	Regional environmental-regulation index	Qiu and Yu (2024)
	Policy toolkit on ES supply	Intervention frequency (e.g., sludge clean-ups)	Su and Liu (2024)
Environmental Risk	Exposure to carbon regulation	High vs. low carbon firm classification	Bharath and Shumway (2008)
	Carbon policy risk	ln(CO_2 emitted) per county-year	Shu et al. (2023)
	Climate policy uncertainty	CPU Index	Gavriilidis (2021)
	Heavy emitter flag	NGER heavy-emitter classification	Nguyen and Phan (2020)
	Market attention to carbon risk	Media, analyst, and investor attention metrics	Wang et al. (2024)
Environmental Score	ESG Rating	Firm's ESG score in year t	Csapi et al. (2024)
	ESG Rating (Huazheng)	Huazheng ESG score	Fang and Hu (2023); Wang et al. (2022); Wei et al. (2023)
	Bloomberg ESG Score	Bloomberg composite ESG	Eccles et al. (2011, 2014)
	Sustainable-development level	Composite of environmental, social, human-capital metrics	Cheng et al. (2023)
Governance	Corporate Governance	Corporate-governance index	Chang et al. (2014); Jiang and Yuan (2018)
Stakeholder	Stakeholder type	Dummies: entrepreneurs, environmentalists, tourists, farmers, government	Su and Liu (2024)
Corporate Social Sustainability	Corporate Sustainability Performance	CSP proxies: environmental, social, governance scores	Ho et al. (2021)
Trade-off	Ecosystem-service efficiency	Optimized supply under trade-off model	Su and Liu (2024)

Source: Authors

Table 3. Performance Feedback Theory and ESG Variables

Classification	Variable(s)	Formula	Authors
Financial	ROA, LEV, GROWTH, Firm Age, Firm Size, Cash Flow, Current Ratio, Operational Turnover, Slack Resources, State-Ownership Dummy	Standard accounting calculations	–
Environmental	Breadth Search	Count of projects/collaborations in new or external fields	Sun and Qiu (2022)
	Environmental Disclosure Strategy	CSR visuals (photos + tables); fewer ⇒ concealment	Zhang (2024)
	Depth Search	Count of projects in known fields	Sun and Qiu (2022)
	EI (Environmental Investment)	Total spending on pollution control and green projects	Cao et al. (2023)
	ESG Score	Aggregate ESG rating from secondary data	Shi and Veenstra (2021)
	Green Innovation	ln (1 + green invention patents)	Liao (2020)
	IW (Innovation Willingness)	R&D expense / Operating income (0 if no R&D)	Wu and Zhang (2021)
	SP (Sustainable Production)	Mean of three 5-point Likert items (energy, waste, etc.)	Geng et al. (2017); Liu et al. (2018); Zhu and Sarkis (2004)
Aspiration	Aspiration Level	Weighted mix of historical & social targets	Greve (2003a)
	Social Aspiration	Competitors' average performance	Greve (2003a)
	Historical Aspiration	Target based on firm's own history	Greve (2003a)
	APG, P_APG, N_APG	Performance − Aspiration (positive / negative)	Cyert and March (1963); Greve (2003a, 2003b)
	CG (Corporate Greenwashing)	Token / substantive CSR disclosures ratio	Fiechter et al. (2022)
Corporate Disclosure	Environmental Strategy	Text analysis of CSR environmental content	Ren and Lin (2024)
	FV (Firm Visibility)	Advertising expenses / Operating income	Luo and Ying (2012); Vesal et al. (2021); Wang et al. (2015)
Environmental Pressure	Gov (Government Regulation)	ln (1 + provincial environmental penalties)	Liu et al. (2022)
	Environmental Pressure	Likert scale of pressure from customers, government, society	Porter and Kramer (2002); Sarkis et al. (2010)
	GS (Government Supervision)	Dummy = 1 if under strict supervision	Pan and Guo (2018)

(*continued*)

Table 3. *(continued)*

Classification	Variable(s)	Formula	Authors
Environmental Regulation	Home-country Environmental Stringency	National regulatory-rigidity index (OECD)	Shou et al. (2020)
	Mandatory Regulation	Local strictness index (standards & penalties)	Li et al. (2018)
	Voluntary Regulation	ISO 14001 or similar certifications	–
	EPBLA	Count of administrative environmental penalties	Zhang (2024)
Media / Reputation	Media Attention	ln (number of ESG-related news)	DasGupta and Roy (2023)
	Media Coverage	ln (number of negative environmental news)	Baron and Kenny (1986)
	Reputation Concern	Index of negative-media history + compliance programs	Zhang (2024)
	Public Environmental Concern	Baidu search index for keyword	Du et al. (2019)

Source: Authors

4 Conclusion, Limitations, and Future Prospects

We have identified several risks involved in companies adopting sustainable practices, such as Shu et al. (2023), which identified carbon policy risk as a significant external risk influencing capital structure decisions. The authors emphasize the importance of addressing carbon policy risks to long-term corporate governance. Companies involved in high-risk sectors significantly reduce debt defaults, i.e., a negative correlation, improve pollution control, and adjust financial structures in response to stricter environmental regulations (Wang et al., 2024).

We identified that these effects are more pronounced in non-state companies with low institutional investor participation, with low agency conflicts, and low corporate social responsibility performance. Most studies suggest that environmental regulation can reconcile environmental protection with economic development, supporting the notion of regulatory gains.

Regarding the disclosure of financial results, we identified studies on the impact of environmental regulations, which significantly influence the capital structure because they are more stringent. Qiu and Yu (2024) and Xiao (2022) state that highly polluting firms, especially in emerging markets, face constraints that limit financial flexibility and that green credit policies in regions such as China compel these firms to reduce leverage, improving their default risk profiles, but possibly stifling growth. This shows that regulation is an indirect tool of macroprudential policy that interferes with the capital structure, causing a dynamic capital adjustment through environmental pressure. This process reinforces the idea of ESG as an element of a gap between the static optimal structure proposed by traditional theory and the structure observed in regulatory contexts, implying adjustments through Performance Feedback Theory to preserve solvency and market value.

Li et al. (2024a) and Li et al. (2024b) report other ESG performance and capital costs findings. According to the authors, companies with strong ESG scores benefit from reduced financing costs. In other words, these companies attract institutional investors due to greater transparency and lower default risks, especially in the bond markets. The authors highlight a non-linear relationship between ESG scores and debt costs. Although costs may be initially high, they tend to stabilize over the long term, confirming the Trade-off Theory.

Zahid et al. (2023) relate ESG performance to audit quality and state that there is no significant relationship between ESG performance and audit-based equity financing decisions. Other studies have examined various factors, such as climate policy uncertainty (CPU), which negatively affects corporate investments, with corporate social responsibility (CSR) playing a complex moderating role (Chang et al., 2024). Zhang (2024) concludes that corporate social responsibility (CSR) does not significantly affect the sources of financing for working capital in companies of all sizes, which contradicts studies of the relationship between CSR and CPU.

Factors such as the specificity of the sector in which the company is inserted or the country's culture also had a role. Cariola et al. (2020) and Csapi et al. (2024) conclude that integrating ESG factors affects sectors differently. In the manufacturing and energy sectors, ESG drives long-term debt strategies aligned with sustainable investments. European companies employ golden index-based targets to align capital structure with ESG targets, while emerging market SMEs struggle with cost-benefit trade-offs due to limited financial resources. Regarding the country's culture, Qureshi and Ahsan (2022) conclude that there is a strong influence between corporate environmental performance (CEP) and investors' assessment of CEP in the market. Investors in cultures with a high power distance, low individualism, high masculinity, and high uncertainty avoidance tend to penalize companies with better environmental performance (Qureshi and Ahsan, 2022). The authors advocate multidimensional policy frameworks prioritizing corporate efforts to reduce CO2 emissions and improve waste recycling.

As for the Performance Feedback Theory, we have identified that performance below short-term aspirations is often perceived as random noise, and executives often fail to take action. In corporate terms, greenwashing affects companies' willingness to innovate and is highly influenced by external factors such as company visibility and government oversight. Second, Lu et al. (2024) suggest that companies should actively avoid greenwashing, although it can bring short-term benefits to a lesser extent if left unexposed.

The relationship between managerial aspirations and actual performance advocated by Performance Feedback Theory, as discussed by Berchicci and Tarakci (2022), highlights the complexity of organizational decisions, especially when performance exceeds or falls short of expectations. Our findings revealed a trend for companies to utilise Performance Feedback Theory, particularly negative feedback, to manage their public image, often through symbolic actions.

We identified several limitations in both theories, such as geographic bias, as studies are mainly concentrated in China, Europe, and the USA; data variability, as we found significant differences in ESG scoring methodologies, which makes direct comparability between studies difficult; time gaps and sectoral scope, as there is a focus on specific sectors such as manufacturing and energy. Future research should address these gaps by exploring diverse geographic contexts, refining ESG measurement methodologies, and examining dynamic interactions between capital structure and sustainability over long periods. It is also possible to explore the nonlinearities of the results, such as the relationship between ESG scores and company profitability. We also suggest studies incorporating behavioral finance to analyze external risks related to social pressure on ESG standards. Studies on the Performance Feedback Theory suggest the need for an in-depth examination of how different types of performance feedback impact corporate strategies and capital structure, as well as the development of a theoretical framework to support this research.

The results of this review show that the relationship between ESG and capital structure cannot be analyzed in isolation within the Trade-off Theory or Performance Feedback Theory. Our results indicate that firms adjust their financial strategies to optimize leverage according to Trade-off Theory principles and respond to ESG pressures described in the Performance Feedback Theory.

Table 4 summarizes the objectives of this review, we identify that (1) the Trade-off Theory, when incorporating ESG variables, needs to consider the volatility generated by the new regulatory and social requirements; the incorporation of ESG impacts the entire capital structure (2) the Performance Feedback Theory can offer support to understand how managers adjust their financial decisions in response to ESG, often caused by external pressures; (3) the ESG variables mainly used in the literature highlight financial and behavioral aspects in isolation, reinforcing the need for hybrid approaches; (4) ESG acts as a factor of volatility and adjustment in capital decisions, challenging the stability assumptions of the traditional capital structure and (5) Performance Feedback Theory is not related in any study to the Trade-off Theory or capital structure, while the Trade-off Theory studies follow the more traditional theories such as Agency, Hierarchy and Stakeholders.

Thus, by combining the Trade-off and Performance Feedback Theories under the lens of ESG, this study offers an alternative theoretical framework capable of explaining the instability of capital structure as a function of managerial aspirations shaped by ESG pressures. Future research should empirically test these hypotheses by measuring the relationship between changes in ESG scores and leverage adjustments under different perceived performance levels.

Table 4. Key results of the objectives

Specific Objective	Main results	Conclusion	Authors
1. Identify studies that relate the Trade-off Theory to ESG	ESG influences risks, leverage, funding spreads, and profitability, with a nonlinear relationship between ESG and capital structure	ESG impacts capital structure	Zahid et al. (2023); Xiao (2022); Qiu and Yu (2024); Nguyen and Phan (2020); Kleimeier and Viehs (2021); Caragnano et al. (2020); Li et al. (2024a)
2. Identify behavioral finance studies, especially the Performance Feedback Theory, applied to ESG	Performance Feedback Theory is little explored in ESG. Companies adjust sustainable actions according to perceived performance, and greenwashing responds to negative feedback	ESG acts as a feedback trigger	Zhang et al. (2023); Sun and Qiu (2022); Meng et al. (2024); Pan et al. (2023); Lu et al. (2024)
3. Survey the variables used to analyze the Performance Feedback Theory	Variables such as CG, FV, ER, and EPBLA measure the mismatch between aspiration and reality	Performance gap influences adjustments	Li, Zheng and Zhang (2018); Fiechter et al. (2022); Wu and Zhang (2021); Xue et al. (2022); Schumacher et al. (2020)
4. Map the variables most used to measure ESG	Broad ESG variables: ratings, CO_2, environmental regulation, carbon risk, climate policy uncertainty, and financial segmentation	High ESG reduces the cost of capital,	Zahid et al. (2023); Qureshi and Ahsan (2022); Xiao (2022); Shu et al. (2023); Huang and Sol (2023); Csapi et al. (2024)
5.Investigate what other theories are often combined with Trade-off Theory and Performance Feedback Theory	Trade-off Theory is combined with: Hierarchy theory, agency, stakeholders, signaling; Feedback combines with resource-based theory, institutional, agency, social pressure, threat-rigidity	Proposed hybrid model	Fama and French (2002); Jiraporn and Gleason (2007); Duguleană et al. (2024); Wang et al. (2024); Berchicci and Tarakci (2022)

Source: Authors

References

Adeneye, Y.B., Kammoun, I., Ab Wahab, S.N.A.: Capital structure and speed of adjustment: The impact of environmental, social and governance (ESG) performance. Sustain. Acc. Manage. Policy J. **14**(5), 945–977 (2023). https://doi.org/10.1108/SAMPJ-01-2022-0060

Akhtar, N., Rashid, A.: Financial development and sustainable development: a review of literature. Sustainable Development, sd.3068 (2024). https://doi.org/10.1002/sd.3068

Battiston, S., Dafermos, Y., Monasterolo, I.: Climate risks and financial stability. J. Financ. Stab. **54**, 100867 (2021). https://doi.org/10.1016/j.jfs.2021.100867

Berchicci, L., Tarakci, M.: Aspiration formation and attention rules. Strateg. Manag. J. **43**(8), 1575–1601 (2022). https://doi.org/10.1002/smj.3371

Cyert, RM., March, J.G.: A behavioral theory of the firm. J. Bus. Comm. **6**(2), 64–67 (1969). Prentice-Hall, Inc., Englewood Cliffs, N.J., 1963, 332. https://doi.org/10.1177/002194366900600211

Caragnano, A., Mariani, M., Pizzutilo, F., Zito, M.: Is it worth reducing GHG emissions? Exploring the effect on the cost of debt financing. J. Environ. Manage. **270**, 110860 (2020). https://doi.org/10.1016/j.jenvman.2020.110860

Cariola, A., Fasano, F., La Rocca, M., Skatova, E.: Environmental sustainability policies and the value of debt in EU SMEs: Empirical evidence from the energy sector. J. Clean. Prod. **275**, 123133 (2020). https://doi.org/10.1016/j.jclepro.2020.123133

Chang, C.-L., Zhang, J., Lin, Y.-E.: Climate policy uncertainty, corporate social responsibility and corporate investments of the energy firms. Energy Economics **140**, 107968 (2024). https://doi.org/10.1016/j.eneco.2024.107968

Chari, M.D.R., David, P., Duru, A., Zhao, Y.: Bowman's risk-return paradox: An agency theory perspective. J. Bus. Res. **95**, 357–375 (2019). https://doi.org/10.1016/j.jbusres.2018.08.010

Chen, W., Zhong, X., Lan, H.: Does CEO openness matter when explaining firm internationalization decisions: Evidence from China. Int. J. Emerg. Mark. **19**(8), 2203–2221 (2024). https://doi.org/10.1108/IJOEM-01-2022-0031

Clément, A., Robinot, É., Trespeuch, L.: The use of ESG scores in academic literature: a systematic literature review. Journal of Enterprising Communities: People and Places in the Global Economy **19**(1), 92–110 (2025). https://doi.org/10.1108/JEC-10-2022-0147

Cooke, A., Smith, D., Booth, A.: Beyond PICO: The SPIDER tool for qualitative evidence synthesis. Qual. Health Res. **22**(10), 1435–1443 (2012). https://doi.org/10.1177/1049732312452938

Csapi, V., Ulbert, J., Tóth-Pajor, Á.: Golden ratio-based leverage targeting and the ESG performance of US and European listed firms. Res. Int. Bus. Financ. **71**, 102469 (2024). https://doi.org/10.1016/j.ribaf.2024.102469

Dang, V.A., Gao, N., Yu, T.: Climate policy risk and corporate financial decisions: evidence from the NO_X budget trading program. Manage. Sci. **69**(12), 7517–7539 (2023). https://doi.org/10.1287/mnsc.2022.4617

Das Gupta, R.: Risk-attitudes of the NSE 500 firms—Bowman's paradox and prospect theory perspectives. IIMB Manag. Rev. **29**(2), 76–89 (2017). https://doi.org/10.1016/j.iimb.2017.04.001

DasGupta, R., Roy, A.: Environmental, Social, Governance Performance and Negative Performance Feedback: Firm Moderators in a Cross-country Context. American Business Review **26**(2), 519–550 (2023). https://doi.org/10.37625/abr.26.2.519-550

DeAngelo, H.: The capital structure puzzle: what are we missing? J. Fin. Quantitat. Anal. **57**(2), 413–454 (2022). https://doi.org/10.1017/S002210902100079X

DeAngelo, H.: ESG, corporate piracy and coasian contracting efficiency. Eur. Financ. Manag. **31**(1), 3–25 (2025). https://doi.org/10.1111/eufm.12522

DeAngelo, H., Roll, R.: How stable are corporate capital structures? J. Financ. **70**(1), 373–418 (2015). https://doi.org/10.1111/jofi.12163

Desai, V.: Learning to behave badly: Performance feedback and illegal organizational action. Ind. Corp. Chang. **23**(5), 1327–1355 (2014). https://doi.org/10.1093/icc/dtt043

Duguleană, C., Duguleană, L., Deszke, K.-D.: Financial performance and capital structure – an econometric approach for Romanian e-commerce companies during the COVID-19 pandemic. Econ. Anal. Policy **83**, 786–812 (2024). https://doi.org/10.1016/j.eap.2024.05.024

Elnahas, A., Kim, D., Kim, I.: Corporate location and the financing decision: evidence form natural disasters. SSRN Electron. J. (2018). https://doi.org/10.2139/ssrn.3123468

Fama, E.F., French, K.R.: Testing trade-off and pecking order predictions about dividends and debt. The Rev. Fin. Stud. **15**(1), 1–33. JSTOR (2002)

Fiechter, P., Hitz, J., Lehmann, N.: Real effects of a widespread CSR reporting mandate: evidence from the european union's CSR directive. J. Account. Res. **60**(4), 1499–1549 (2022). https://doi.org/10.1111/1475-679X.12424

Gaba, V., Bhattacharya, S.: Aspirations, innovation, and corporate venture capital: a behavioral perspective. Strateg. Entrep. J. **6**(2), 178–199 (2012). https://doi.org/10.1002/sej.1133

Gao, Y., Yang, H., Zhang, M.: Too bad to fear, too good to dare? Performance feedback and corporate misconduct. J. Bus. Res. **131**, 1–11 (2021). https://doi.org/10.1016/j.jbusres.2021.03.041

Harzing, A.-W.: Two new kids on the block: How do Crossref and Dimensions compare with Google Scholar, Microsoft Academic, Scopus and the Web of Science? Scientometrics **120**(1), 341–349 (2019). https://doi.org/10.1007/s11192-019-03114-y

Howard-Grenville, J., Davis, G.F., Dyllick, T., Miller, C.C., Thau, S., Tsui, A.S.: Sustainable development for a better world: contributions of leadership, management, and organizations. Acad. Manage. Discov. **5**(4), 355–366 (2019). https://doi.org/10.5465/amd.2019.0275

Iyer, D.N., Miller, K.D.: Performance feedback, slack, and the timing of acquisitions. Acad. Manag. J. **51**(4), 808–822 (2008). https://doi.org/10.5465/amr.2008.33666024

Khan, M.A., Hassan, M.K., Maraghini, M.P., Paolo, B., Valentinuz, G.: Valuation effect of ESG and its impact on capital structure: Evidence from Europe. Int. Rev. Econ. Financ. **91**, 19–35 (2024). https://doi.org/10.1016/j.iref.2024.01.002

Kleimeier, S., Viehs, M.: Pricing carbon risk: investor preferences or risk mitigation? Econ. Lett. **205**, 109936 (2021). https://doi.org/10.1016/j.econlet.2021.109936

Li, D., Huang, M., Ren, S., Chen, X., Ning, L.: Environmental legitimacy, green innovation, and corporate carbon disclosure: evidence from CDP China 100. J. Bus. Ethics **150**(4), 1089–1104 (2018). https://doi.org/10.1007/s10551-016-3187-6

Li, Q., Li, T., Zhang, Y.: ESG performance, heterogeneous creditors, and bond financing costs: Firm-level evidence. Financ. Res. Lett. **66**, 105527 (2024). https://doi.org/10.1016/j.frl.2024.105527

Li, W., Padmanabhan, P., Huang, C.-H.: ESG and debt structure: Is the nature of this relationship nonlinear? Int. Rev. Financ. Anal. **91**, 103027 (2024). https://doi.org/10.1016/j.irfa.2023.103027

Lu, J., et al.: Effective environmental strategy or illusory tactics? corporate greenwashing and innovation willingness. Bus. Strat. Environ. bse.4047 (2024). https://doi.org/10.1002/bse.4047

Lu, L.-H., Wong, P.-K.: Performance feedback, financial slack and the innovation behavior of firms. Asia Pacific J. Manage. **36**(4), 1079–1109 (2019). https://doi.org/10.1007/s10490-018-9634-4

Martín-Martín, A., Thelwall, M., Orduna-Malea, E., Delgado López-Cózar, E.: Google Scholar, Microsoft Academic, Scopus, Dimensions, Web of Science, and OpenCitations' COCI: A multidisciplinary comparison of coverage via citations. Scientometrics **126**(1), 871–906 (2021). https://doi.org/10.1007/s11192-020-03690-4

Meng, L., Su, T., Zhang, J.: Environmental performance shortfalls and corporate green innovation: the role of green credit regulation. Int. Rev. Econ. Financ. **96**, 103624 (2024). https://doi.org/10.1016/j.iref.2024.103624

Miller, M.H.: Debt And Taxes*. J. Financ. **32**(2), 261–275 (1977). https://doi.org/10.1111/j.1540-6261.1977.tb03267.x

Muchiri, M.K., Erdei-Gally, S., Fekete-Farkas, M., Lakner, Z.: Bibliometric analysis of green finance and climate change in post-paris agreement era. J. Risk and Fin. Manage. **15**(12), 561 (2022). https://doi.org/10.3390/jrfm15120561

Nguyen, J.H., Phan, H.V.: Carbon risk and corporate capital structure. J. Corp. Finan. **64**, 101713 (2020). https://doi.org/10.1016/j.jcorpfin.2020.101713

Ocasio, W.: Towards an attention-based view of the firm: attention-based view of the firm. Strateg. Manag. J. **18**(S1), 187–206 (1997). https://doi.org/10.1002/(SICI)1097-0266(199707)18:1+%3c187::AID-SMJ936%3e3.0.CO;2-K

Page, M.J., et al.: The PRISMA 2020 statement: an updated guideline for reporting systematic reviews. BMJ, n71 (2021). https://doi.org/10.1136/bmj.n71

Palese, A., et al.: Framing the time while designing and conducting reviews: a focused mapping review and synthesis. J. Clin. Nurs. **31**(23–24), 3523–3534 (2022). https://doi.org/10.1111/jocn.16180

Pan, X., Chen, X., Sinha, P.: Navigating the haze: environmental performance feedback and CSR report readability. J. Bus. Res. **166**, 114116 (2023). https://doi.org/10.1016/j.jbusres.2023.114116

Qiu, Q., Yu, J.: Green credit policy and default risk of the heavy polluting corporations. J. Clean. Prod. **455**, 142291 (2024). https://doi.org/10.1016/j.jclepro.2024.142291

Qureshi, M.A., Ahsan, T.: Do investors penalize the firms disclosing higher environmental performance? a cross cultural evidence. Environ. Sci. Pollut. Res. **29**(36), 55147–55160 (2022). https://doi.org/10.1007/s11356-022-19716-8

Rajamani, L., Werksman, J.: The legal character and operational relevance of the Paris Agreement's temperature goal. Philosophical Transactions of the Royal Society A: Mathematical, Physical and Engineering Sciences **376**(2119), 20160458 (2018). https://doi.org/10.1098/rsta.2016.0458

Ramalho, E., et al.: Understanding wind energy economic externalities impacts: a systematic literature review. Ren. Sustain. Energy Rev. **209**, 115120 (2025). https://doi.org/10.1016/j.rser.2024.115120

Rameder, H., Di Angelo, M., Salzer, G.: Review of automated vulnerability analysis of smart contracts on Ethereum. Frontiers in Blockchain **5**, 814977 (2022). https://doi.org/10.3389/fbloc.2022.814977

Ren, C., Lin, X.: The impact of performance feedback on corporate ESG performance: mediating role of environmental strategy. PLoS ONE **19**(3), e0298471 (2024). https://doi.org/10.1371/journal.pone.0298471

Restrepo, N., Uribe, J.M., Manotas, D.F.: Dynamic capital structure under changing market conditions in the oil industry: An empirical investigation. Resour. Policy **69**, 101808 (2020). https://doi.org/10.1016/j.resourpol.2020.101808

Shu, H., Tan, W., Wei, P.: Carbon policy risk and corporate capital structure decision. Int. Rev. Financ. Anal. **86**, 102523 (2023). https://doi.org/10.1016/j.irfa.2023.102523

Stechemesser, K., Guenther, E.: Carbon accounting: a systematic literature review. J. Clean. Prod. **36**, 17–38 (2012). https://doi.org/10.1016/j.jclepro.2012.02.021

Su, B., Liu, M.: How to manage the ecosystem services effectively and fairly? J. Clean. Prod. **458**, 142477 (2024). https://doi.org/10.1016/j.jclepro.2024.142477

Sun, Y., Qiu, Z.: Positive performance feedback and innovation search: new ideas for sustainable business development. Sustainability **14**(4), 2086 (2022). https://doi.org/10.3390/su14042086

Sun, Z., Sun, X., Dong, Y.: Does negative environmental performance feedback induce substantive green innovation? The moderating roles of external regulations and internal incentive. Corp. Soc. Responsib. Environ. Manag. **31**(4), 2953–2976 (2024). https://doi.org/10.1002/csr.2722

Tversky, A., Kahneman, D.: Judgment under Uncertainty: Heuristics and Biases: Biases in judgments reveal some heuristics of thinking under uncertainty. Science **185**(4157), 1124–1131 (1974). https://doi.org/10.1126/science.185.4157.1124

Wang, J., Qiang, H., Liang, Y., Huang, X., Zhong, W.: How carbon risk affects corporate debt defaults: Evidence from Paris agreement. Energy Economics **129**, 107275 (2024). https://doi.org/10.1016/j.eneco.2023.107275

Wang, S., Wang, G., Xiao, Y.: How environmental policies affect personal willingness to pay for environmental protection: An investigation of interpretative and resource effects. Environ. Dev. Sustain. **26**(1), 1591–1613 (2022). https://doi.org/10.1007/s10668-022-02775-9

Wang, X., Sun, W.: Aspiration performance gap, second generation involvement in management, and family Firms' environmental responsibility. Bus. Strat. Environ. bse.3903 (2024). https://doi.org/10.1002/bse.3903

Wang, Y., Wang, X., Xu, W.: How does negative performance feedback affect a firm's openness in its innovation search behaviour? The moderating role of organisational slack. Technol. Anal. Strat. Manage. **35**(4), 380–393 (2023). https://doi.org/10.1080/09537325.2021.1976405

Xiao, H.: Environmental regulation and firm capital structure dynamics. Econ. Anal. Policy **76**, 770–787 (2022). https://doi.org/10.1016/j.eap.2022.09.020

Xie, E., Huang, Y., Peng, M.W., Zhuang, G.: Resources, aspirations, and emerging multinationals. J. Leader. Organizat. Stud. **23**(2), 144–161 (2016). https://doi.org/10.1177/1548051816633068

Zahid, R.M.A., Saleem, A., Maqsood, U.S.: ESG performance, capital financing decisions, and audit quality: Empirical evidence from Chinese state-owned enterprises. Environ. Sci. Pollut. Res. **30**(15), 44086–44099 (2023). https://doi.org/10.1007/s11356-023-25345-6

Zhang, D.: Corporate environmental responsibility and financing patterns: Does firm size matter? J. Small Bus. Manage. **62**(5), 2482–2522 (2024). https://doi.org/10.1080/00472778.2023.2241128

Zhang, M., Shui, X., Smart, P., Wang, X., Chen, J.: Environmental performance feedback and timing of reshoring: perspectives from the behavioural theory of the firm. Br. J. Manag. **34**(3), 1238–1258 (2023). https://doi.org/10.1111/1467-8551.12677

Unlocking Factors Influencing Rural Banks' Operational Efficiency in Ghana: Evidence from Panel Models Robust to Cross-Sectional Dependence

Louis David Junior Annor[1](✉), Bright Ferguson Laing[1], John Agyekum Addae[1], Barbara Naa Amanuah Tetteh[2], and Eric Kweku Attefah[3]

[1] Department of Accounting, Banking and Finance, Ghana Communication Technology University, Accra, Ghana
lannor@ua.pt
[2] Department of Business Administration, University of Professional Studies, Madina, Ghana
[3] Department of Accountancy and Accounting Information Systems, Kumasi Technical University, Kumasi, Ghana

Abstract. This study aimed to investigate the drivers of operational efficiency in rural banks within the Ghanaian context, taking into account factors such as size, financial performance, liquidity, ICT investment, and income diversification. Using financial data from 122 rural banks from 2014 to 2020, this study employed the Prais-Winsten panel corrected standard errors (PCSE), feasible generalized least squares (FGLS), and pooled panel OLS techniques to examine the interactions among these variables after testing for multicollinearity, cross-sectional dependence, stationarity, and Westerlund cointegration. This study reveals that, ceteris paribus, larger rural banks are not operationally efficient across all estimations. The quadratic term of bank size indicates that increased size amplifies the decline, resulting in sharper drop in efficiency. Furthermore, income diversification significantly reduces the operational efficiency of banks in Ghana. Interestingly, net interest margin, liquidity, information, communication and technology investments significantly improve the operational efficiency of banks, ceteris paribus. Furthermore, ICT investment negatively moderates the relationship between bank size and efficiency. Previous literature has explored operational efficiency drivers in varied jurisdictions; however, emphasis has been on commercial banks rather than rural banks. Considering the enormous contribution of rural banks to Ghana's financial system, these results are timely in aiding policy directions on banks' efficiency, performance, and ICT diffusion.

Keywords: Financial Performance · Operational Efficiency · Rural Banks · Panels Corrected Standard Errors · Ghana

L. Almeida et al. (Eds.): ICAFI 2025, ISEM 80, pp. 232–251, 2026.
https://doi.org/10.1007/978-3-032-16268-7_13

1 Introduction

The role of financial institutions, particularly banks, is indispensable in a country's socioeconomic growth, whether it is developing or developed. Banks, as depository entities, play a significant role in financial intermediation. They are the principal source of money for businesses because they accept household deposits, which they lend to others. As a result, banks align credit supply and demand in the market [1]. Banks play a crucial role in allocating capital across all productive sectors of the economy, as well as in spurring a country's economic development. Bank operational efficiency is a significant concern for industry players, as it can impact a country's economic growth rate due to internal and external factors related to the financial services sector. The operational effectiveness of each bank is determined by its unique traits and positioning. Efficiency is the capacity to maximise output from a given quantity of inputs, minimize costs related to a given output, or maximize profit for a given output. Banks' efficiency involves reducing costs, increasing profitability, directing more liquidity through the financial system, providing customers with better services and prices, and increasing security through better liquidity and capital buffers to absorb risk [2].

To maximize shareholders' wealth, meet market demands, and adhere to regulations, it is imperative that rural banks increase their operating efficiency. Rural banks serve as financial intermediaries, and their efficient operations essentially contribute to the expansion of the Ghanaian rural economy. Due to the 2017 Bank of Ghana disruption of the banking system and the COVID-19 pandemic, which have impacted enterprises and industry, there is a greater need for both financial and strategic ICT investments to boost productivity and ensure the survival of banks [3]. Notwithstanding their strategic importance, the operational efficiency of rural banks remains a critical subject of interest for scholarly inquiry.

The first RCB in Ghana was established in 1976, in a farming community in the Central Region of Ghana. By the end of 2008, 127 RCBs were operating with 584 service outlets [4]. The Bank of Ghana is the main regulatory body of rural banks in Ghana. The ARB Apex Bank is the central processing centre for the country's RCBs. RCBs are financial services providers in rural areas and represent approximately half of the total banking outlets in Ghana [4]. Today, there are 147 RCBs in Ghana, and more than eight hundred and fifty (850) branch networks providing banking services throughout the country [5]. [6] opines that RCBs are unit banks owned by rural community members who purchase shares and are licensed to provide financial intermediation. These banks are also owned by individual shareholders who reside in the communities in which they operate [7]. RCBs were also established to provide banking services, mobilise funds, and extend credit to salaried workers, farmers, cottage industry operators, and fishermen [8]. RCBs have demonstrated that they play important roles in supporting development and reducing poverty [3, 9].

However, [10] posit that, though Rural banks in Ghana were originally established to bring banking services to the doorstep of rural communities and provide financing support to rural industries, most appear to have relocated to urban areas, thus defeating their original mandate. Part of the relocation is to mobilize funds to increase their capital base, provide credit to the urban poor, and mobilize enough profit to support their activities (social responsibility projects) in their catchment area.

By contrast, apart from chasing profitability and serving the socioeconomic needs of the communities, these RCBs are expected to engage in corporate social responsibility activities, be competitive, maximise shareholders' profit and wealth, and survive in the banking space in Ghana [9]. Their research findings suggest that there is availability of CSR, which has a significant negative effect on return on assets (ROA), return on equity and bank stability. On the other hand, further results showed that CSR has a positive influence on the net interest margin. [9] indicated that ICT investment is an ideal model for a firm and business strategy and growth. The study suggests that ROA and bank size improve rural banks' ICT investment. Furthermore, the development of telecommunication and government effectiveness have significant positive effects on ICT investment among rural banks in the Ghanaian economy.

Sometimes, commercial banks are unwilling to lend to rural communities due to the associated risks and costs. Rural Banks, therefore, represent the only hope of the rural community that they will have access to any financing whatsoever. Gradually, rural banks are becoming increasingly essential in combating poverty, rural-urban migration, and social vices in Ghana. In the Rural Banking Special Report, [11] asserts that "community banks are doing extremely well in driving financial inclusion to support the government to achieve the National Financial Inclusion Strategy, with the target of increasing financial inclusion from 58% to 85% by 2023." The rural banking sector has over 5 million active customers, according to a dataset from ARB Apex Plc, which means that the sector has enabled the vast majority of Ghanaians to access financial services. The rural banking sector has the largest branch network, with over 850 branches spread across Ghana's 16 regions, and a total deposit of GH¢11.98 billion [11].

This study reveals the factors that influence the operational efficiency of rural banks in Ghana. The study identifies a piece of evidence from panel models robust to cross-sectional dependence and heterogeneity. Although operational efficiency has been explored in commercial and universal banks, there is a limited study on the drivers of efficiency in rural banks, particularly in Ghana, despite their significant role in promoting financial inclusion and contributing to local economic growth. Furthermore, whereas previous studies overlook the non-linear effects and interaction dynamics of determinants of operational efficiency, this study examines the moderating effect of the ICT investment nexus on the relationship between bank size and efficiency among rural banks in Ghana.

In recent years, the operational landscape of rural banks in Ghana has been shaped by macroeconomic volatility, regulatory reforms, and technological advancements. Previous studies have explored the relationship between ICT and financial performance [9]; however, the impact of ICT investment on efficiency in rural banks remains underexplored, particularly in terms of its long-term effects using PCSE and FGLS estimators. Insights for multiple stakeholders, including regulatory bodies, investment communities, and policymakers, are provided by delineating the governance variables that can enhance or impede banking efficiency. It also identifies actionable avenues for improvement, specifically in the realms of risk management and board composition, with the potential to catalyse a transformation in Ghana's rural banking landscape.

2 Literature

2.1 Bank Size, Financial Performance and Operational Efficiency

Several studies with an interest in the banking sector have focused on factors influencing bank efficiency [12–15]. Prevalent among them is bank size, which has been argued to have a significant positive effect on operational efficiency [16]. [17] based their argument on the economics of scale, finding that larger banks demonstrate greater profitability. Likewise, [18] revealed the importance of size on bank performance. Moreover, the Basel II and Basel III regulatory changes have diverse effects on performance, depending on bank size [19]. Profitability increases with larger banks, as they can adapt to changing regulations, while small banks are burdened with administrative costs that negatively affect their efficiency. Furthermore, [20] in examining listed banking industries on the Indonesia Stock Exchange between 2018 and 2022, found that bank size and ownership by management positively influence operational efficiency. Again, [21] also observed that the benefits of technology adoption and liquidity on financial performance are amplified by bank size. Larger banks are better equipped to leverage fintech innovations and efficiently manage liquidity due to their greater resources and well-established infrastructures.

Conversely, the works of [15, 17] have opposing results regarding the positive effect of bank size on financial performance and operational efficiency. [15] found a negative moderating effect of bank size on financial performance. Similarly, [22] concluded that bank size beyond a certain level can result in operational inefficiency. Likewise, the X-Efficiency Theory [23] suggests that size does not necessarily lead to efficiency, as other factors, such as managerial competence, can positively impact bank size. Furthermore, [24] in examining the impact of bank size on quoted deposit money banks in sub-Saharan Africa, concluded that smaller banks outperformed their larger counterparts. The results demonstrated a strong and adverse relationship between bank size and financial performance. Additionally, [25] cautioned that an increase in bank size may reduce efficiency, leading to lower Returns on Assets (ROA) and Equity (ROE). This study, therefore, hypothesizes that.

H_1: Bank size has a negative effect on operational efficiency.

2.2 Liquidity and Operational Efficiency

Banks play a crucial role as liquidity providers in the financial system. Banks' top priority is managing liquidity risk, as their capacity to meet short-term obligations and finance operations is closely related to their liquidity levels [26]. Recent research has investigated the relationship between liquidity risk and operational efficiency in banks, revealing a more nuanced and indirect link. For instance, [27] found that decreasing liquidity generation boosts bank stability, whereas increasing bank capital reduces liquidity creation. Therefore, higher liquidity risk is associated with lower operational efficiency and through the mediating function of liquidity creation, bank capital has a favourable effect on stability. Similarly, in [28], liquidity creation is examined as a way for banks to formulate how they can meet consumers' needs while remaining operationally stable. According to their findings, banks with better liquidity management exhibit higher operational efficiency, as they can optimise their asset-liability management systems.

Similarly, [29] investigated the financial performance of Ghana's rural banks, focusing on factors that influence long-term viability and expansion. Their findings demonstrate that proficient liquidity management plays a crucial role in improving the financial outcomes of rural banks. Maintaining adequate liquidity enables these banks to fulfil their obligations and fund lending activities, which are crucial to their profitability and viability.

Other studies have also found an indirect link between liquidity and operational efficiency. [30] investigated the association between liquidity risk, operational efficiency, and financing-to-deposit ratio (FDR) as a liquidity risk measurement indicator and found that liquidity risk does not have a direct influence on operational efficiency. Rather, its effect is indirectly exhibited through credit risk, which can affect banks' overall efficiency. Similarly, [31] found that liquidity risk maximization does not result in a direct and observable relationship with operating efficiency. These findings imply that internally oriented management practices and externally determined economic conditions are disproportionately important drivers of operational efficiency in banking institutions relative to liquidity risk indicators. Nevertheless, as we mentioned, liquidity risk indirectly impacts operational efficiency due to its impact on credit risk and the bank's overall stability. That principle effectively illustrates the phenomenon of effective liquidity management in mitigating credit risk and, consequently, improving efficiency. Furthermore, our research indicates that operational efficiency plays a vital role in helping stabilize banks by being a product of efficient liquidity risk optimization. In the end, however, enough efficiency can cancel out the destabilizing impact that liquidity constraints impose on overall stability.

$\boldsymbol{H_{2:}}$ Optimal liquidity levels enhance the operational efficiency of rural banks.

2.3 Banks' ICT Investment, Performance, and Operational Efficiency

The value of information and communication technology (ICT) investments in enhancing banks' performance and operational efficiency has gained widespread recognition [3, 9]. By implementing advanced technological solutions, financial institutions can streamline their processes, reduce operational costs, and improve customer service [32]. ICT investments enable process automation, leading to efficient service delivery which enhances customer satisfaction by reducing transaction time, contributing to better performance for both customers and banks [9]. Results from the [32] study show a long-term equilibrium link between digital banking technology and the efficiency of commercial banking organisations. This confirms a persistent and connected relationship between the use of digital banking technology and the productivity of commercial banking organisations. Consistent with this study, [33] investigated the relationship between bank performance and financial technology (FinTech) in Pakistan's banking industry from 2018 to 2022. Findings indicate that Capital Adequacy Ratio (CAR), Gross Non-Performing Loans (NPL), ROA, Return on Equity (ROE), and Loan-to-Deposit Ratio (LDR) all showed positive and significant associations with FinTech adoption.

[3] investigated the relationship between information technology investment and rural bank performance and found positive effects of ICT investment on bank performance subject to the indirect influence of ICT diffusion. Moreover, [34] examined the

role of banking digitalization in improving business capacity among rural communities by leveraging digital banking services such as mobile banking, electronic wallets, and online loan platforms. The findings reveal that digital banking reduces barriers to credit access and offers operational efficiencies through round-the-clock, near-real-time payment capabilities, thereby enhancing transaction speed and accessibility for rural businesses and increasing the operational efficiency of banks. Likewise, [35] examined how ICT implementation affects service quality delivery and time savings in Nigerian banks. Based on socio-technical systems theory, the study's findings reveal a statistically significant positive association between the adoption of ICT and the quality of service provided by QDMBs in Nigeria. The study concludes that the deployment of ICT enhances operational efficiency and customer convenience.

With data from audited annual reports of seventeen commercial banks for twelve fiscal years between 2011 and 2022, [36] examined the effects of financial technology on the cost efficiency of commercial banks in Ethiopia. Using a stochastic frontier approach, the study noted that financial technology, including card banking, mobile banking, and internet banking, played a significant positive role in improving the cost efficiency of commercial banks in Ethiopia. Additionally, [37] employed data envelopment analysis to examine the causal relationships between the efficiency of information technology applications and bank performance. The study found a bi-causal relationship between IT efficiency and cost efficiency, arguing that the operational efficiency of ICT applications does not rest solely on their adoption, but rather on how efficiently these applications are deployed within the banking system to deliver the expected output levels. [38] found that there is no relationship between financial performance and operational efficiency in Nigeria.

Contrary to these findings, [10] cautions that ICT investments are costly, and in certain instances, rural banks may not observe an immediate return on investment. Furthermore, [39] and [40] indicate that in remote rural areas, low population density and poor infrastructure exacerbate the challenges of establishing bank branches, leading to higher transaction costs and limited business volume; therefore, investment in ICT may benefit urban banks more than rural banks, where infrastructure is limited. [41] emphasizes that without proper management adaptation, financial institutions, including banks, may face challenges in fully leveraging technological advancements, potentially affecting operational efficiency and competitive positioning. The findings of [42] also suggest that ICT investments without proper support systems often fail to yield the expected operational improvements. In consideration of the relevant literature, we hypothesize that:

***H_3*:** Banks' ICT investment has a negative effect on performance and operational efficiency.

2.4 Income Diversification and Operational Efficiency

Recent empirical research worldwide has yielded conflicting findings, leading to more extensive scholarly discussions regarding the relationship between income diversity and bank efficiency in both developed and emerging markets. Empirically, diversification has generally resulted in operational efficiency [43], though there have been varied opinions.

According to proponents of bank diversification, banks can increase their competitiveness in the market by offering a range of financial products and services [44, 45]. Other studies [46] reveal that income diversification serves as a buffer, ensuring liquidity creation even in competitive environments. [47] suggest that bank management should diversify income streams, particularly through fee-based services, trading operations, and foreign currency transactions, to enhance financial performance and stability in the face of exogenous shocks. Banks' cost efficiency is positively impacted by income diversity, which has a more pronounced effect for unlisted banks [48]. Similarly, [49] posits that income diversity improves banks' performance and lowers the danger of insolvency. According to the portfolio theory, income diversification boosts profitability and lowers income volatility. In line with this, [50] found, through an analysis of 310 years of data from 31 Kenyan commercial banks between 2008 and 2017, that income diversification had a significant impact on bank performance. Conversely, [51] noted that more sectoral loan portfolio diversification lowers bank returns. Nonetheless, [52] showed an inverted U-shaped connection between income diversification and projected efficiency parameters; diversification initially improves efficiency, but beyond a certain point, further diversification reduces efficiency.

[53] employed the system generalized method of moments to investigate the moderating effect of revenue diversification on the relationship between intellectual capital and bank performance in Vietnamese commercial banks from 2007 to 2020. The results demonstrate that income diversification reduces the overall effect of value-added intellectual capital efficiency on bank performance. Also, [54] concluded that financial performance was inversely correlated with income diversification. The findings support the idea [55] that commercial banks tend to stick to traditional business lines rather than branching out into non-traditional ventures, which could result in lower levels of liquidity creation and profitability. However, banks with more market power may benefit more from revenue variety than banks with less market power because they are better equipped to capitalize on their diversified income sources and gain a competitive advantage [50]. With such inconclusive findings in the literature, we hypothesize that:

***H_4*:** Income diversification significantly reduces the operational efficiency of rural banks in Ghana.

3 Methodology

The methods used to conduct empirical estimates are discussed in the methodology section, along with the sources and description of the data. With two primary sections, we first describe the data and their sources that serve as the foundation for the empirical investigation. The second section then presents the empirical models and estimation methods used to analyze the data.

3.1 Data Sources, Variable Definitions and Functional Models

Based on data availability, data for the study were gathered annually from relevant sources for 122 of Ghana's 146 rural banks between 2014 and 2020, specifically from the ARB Apex Bank database. Industry-specific variables, including operating efficiency (RBOE), bank size (BSIZE), the quadratic term of size (SQBSIZE), financial performance (NIM), liquidity (LIQ1), ICT investment (ICTINV), and income diversification (IDIV), were used. RBOE is measured as operating income and operating cost ratio, BIZE is the natural logarithm of total assets, NIM is calculated as interest income – interest expenses/total assets, and LIQ1 is measured as gross loans/total deposits. Again, ICTINV is measured as the natural logarithm of banks' ICT investment, and IDIV is defined as non-interest income/total income.

3.2 Functional Models

The study's objective is to explore the drivers of operational efficiency in rural banks within the Ghanaian financial sector. Thus, based on our argument, the following models are specified to test the relationships under study. Therefore, this study conceptually argues that operational efficiency is a function of bank size, net interest margin, liquidity, ICT investment, and income diversification.

The hypothesis of this study is formulated in Eq. (1) below.

$$RBOE_{it} = f(bsize, nim, Ctrl \tag{1}$$

According to Eq. (1), RBOE represents operating efficiency as the dependent variable, while Ctrls is a vector of control variables comprising liquidity, ICT investment, and income diversification.

$$RBOE_{k,it} = f(bsize, nim, liq1, ictinv, idiv) \tag{2}$$

$$RBOE_{it} = \alpha + \beta_1 BSIZE_{it} + \beta_2 NIM_{it} + \beta_3 LIQ1_{it} + \beta_4 ICTINV_{it} + \beta_5 IDIV_{it} + \mu_{it} \tag{3}$$

Equation 4 examines the quadratic effect of size on the operational efficiency of banks, while Eq. 5 investigates the moderating effect of ICTINV on the relationship between size and efficiency.

$$RBOE_{it} = \alpha + \beta_1 BSIZE_{it} + \beta_2 BSIZE_{it}^2 + \beta_3 NIM_{it} + \beta_4 LIQ1_{it} + \beta_5 ICTINV_{it} + \beta_6 IDIV_{it} + \mu_{it} \tag{4}$$

$$RBOE_{it} = \alpha + \beta_1 BSIZE_{it} + \beta_2 NIM_{it} + \beta_3 LIQ1_{it} + \beta_4 ICTINV_{it} + \beta_5 IDIV_{it} + \beta_6 BSIZE * ICTINV_{it} + \mu_{it} \tag{5}$$

According to Eqs. (1–5), *i* and *t* represent rural banks and years, respectively. The coefficients of the explanatory variables are represented by *β0 through β6, and composite error terms, consisting of residual, time-specific, and bank-specific components,* are indicated by μ_{it}.

3.3 Estimation Strategy and Technique

Multicollinearity tests (correlation test and variance inflation factor), cross-sectional dependence tests, unit root tests, cointegration tests, and panel-corrected standard errors (PCSE) estimates comprise the estimation strategy. In particular, the data is subjected to second-generation unit root tests to avoid inaccurate results when cross-sectional dependence is present. Our findings demonstrate that the variable is stationary at the first-order difference. We fail to accept the null hypothesis because the cross-sectional dependence evidence (see Table 4) indicates that the panels are correlated across panels. If cross-sectional dependence (CSD) is present in the data, estimates and conclusions will be skewed [56]. To avoid potentially erroneous estimates, the study employs the [56, 57] test for cross-sectional dependency (CD), which can be applied to both small and large panels [58].

If a long-term equilibrium relationship exists between variables in panel data, it can be identified using the Westerlund cointegration test. The Westerlund test's variance ratio ($p\text{-}value = 0.000$) indicates that the variables are closely related in the long run. It determines whether the long-term equilibrium between variables across panel units is linked. In contrast to conventional tests, it provides reliable results in the presence of heterogeneity and accounts for cross-sectional dependence within the dataset. Westerlund's test differs from conventional cointegration tests in that it is based on error-correction dynamics. Its robustness and capacity to manage real-world complexities, such as mixed stationarity and interdependence across entities, make it a favourite in econometrics.

Equations 1 and 2 are regression models estimated using the PCSE estimation technique [59]. The PCSE estimator accounts for panel correlation, heteroscedasticity, and finite sample bias, making it a reliable and effective estimator [60]. The estimator assumes that the disturbances are not independent and not equally distributed when calculating the variance, covariance, and standard errors, accounting for unit-level heteroscedasticity and contemporaneous correlation [61]. The PCSE method involves contemporaneous effect controls and panel data heterogeneity in this context [59, 62]. We enumerate the salient characteristics of the PCSE that have been demonstrated to be robust and effective in the literature, such as divergence from spherical errors, handling heterogeneity, autocorrelation, cross-sectional dependency, and unobserved effects [59, 61, 63–66]. Furthermore, to ensure the robustness of the models, this study employed two additional techniques: FGLS and pooled panel OLS, to confirm the validity of the results.

3.4 Analysis and Discussion

Table 1 shows a description of the variables employed in this study. RBOE has an average of 0.174 and a standard deviation of 6.5%, implying that most efficiency scores are relatively close. BSIZE has the second-highest standard deviation of 55.2%, following ICTINV of 80.4%. BSIZE has a mean score of 7.221, while ICTINV averages 5.218. NIM has a mean score of 0.136 and a standard deviation of 8.1%. The banks' liquidity positions are moderately dispersed, considering a deviation of 26.8% from the mean of 0.483. Moreover, IDIV accounts for 27.1% of the income composition of rural banks, with a standard deviation of 28.8%.

Table 1. Descriptive Statistics

Variable	Obs	Mean	Std. Dev	Min	Max
RBOE	854	.174	.065	.002	.621
BSIZE	854	7.221	.552	-.568	8.679
NIM	854	.136	.081	-.816	.495
LIQ1	854	.483	.268	.058	3.409
ICTINV	854	5.218	.804	0	6.624
IDIV	854	.271	.288	.001	4.38

Authors' estimate, 2025

To confirm and ensure that regression results presented in this study are not spurious due to multicollinearity, this study employs a correlation test and variance inflation factor (VIF) to assess the presence of multicollinearity [9]. Highly correlated variables in regression estimates result in inflated standard errors and unstable coefficients, reduce predictive power and make interpretation difficult. According to Table 2, the highest correlation between NIM and RBOE is 0.721, which is lower than the threshold of 0.85 specified in the literature [3, 9, 66]. Aside from IDIV having a negative correlation with efficiency, all other variables have a positive and significant relationship with RBOE. Furthermore, Table 3 displays an average VIF score of 1.184, with the highest VIF score being 1.293. This result is compared to the threshold of 10. Based on the correlation and VIF results in Tables 2 and 3, we conclude that there is no multicollinearity among the variables used in this study and, therefore, the regression results are reliable.

Table 2. Matrix of correlations

Variables	(1)	(2)	(3)	(4)	(5)	(6)
(1) RBOE	1.000					
(2) BSIZE	0.119	1.000				
(3) NIM	0.721	0.236	1.000			
(4) LIQ1	0.152	-0.122	0.088	1.000		
(5) ICTINV	0.093	0.398	0.074	-0.055	1.000	
(6) IDIV	-0.345	-0.209	-0.305	0.188	-0.061	1.000

Authors' estimate, 2025

Table 3. Variance inflation factor

	VIF	1/VIF
BSIZE	1.293	.773
ICTINV	1.19	.84
IDIV	1.18	.848
NIM	1.178	.849
LIQ1	1.078	.928
Mean VIF	***1.184***	

Authors' estimate, 2025.

This study further conducts cross-sectional dependence tests to ascertain the interdependence or correlation across the cross-sectional units (i.e. rural banks). The results shown in Table 4 indicate that the error terms across units are correlated. Thus, with p-values of 0.000 across all variables, we fail to accept the null hypothesis of cross-sectional independence. Subsequent to the evidence of cross-sectional dependence, we further estimated the stationarity using a second-generation technique (Im–Pesaran–Shin unit-root test). From Table 5, whereas some variables are not stationary at their levels, all variables are non-stationary at their first difference.

Table 4. Cross-sectional dependence test

Variables	CD-test	p-value
RBOE	76.462	0.000
BSIZE	143.649	0.000
NIM	83.686	0.000
LIQ1	69.64	0.000
ICTINV	103.503	0.000
IDIV	18.099	0.000

Notes: H_0*: cross-section independence, CD – N (0,1)*
H_1*: Cross-section dependence.*
Authors estimate, 2024.

Moreover, the Westerlund cointegration test was conducted after the first-difference 1*(1)* stationarity test. The t-statistics of 15.5956 and the p-value of 0.000 suggest that the panels are significantly cointegrated. This implies a stable long-term relationship between variables, suggesting an equilibrium relationship.

Table 6 employs a statistical model using PCSEs to illuminate the factors influencing operational efficiency, which is represented by RBOE (Bank Operating Efficiency). We found that, at a 10% significance level, a percentage increase in BSIZE leads to

Table 5. Im–Pesaran–Shin unit-root test and Westerlund test for cointegration

Variables	Z-t-tilde-bar	p-value	Z-t-tilde-bar	p-value
	Levels		*1st difference*	
RBOE	-3.2301	0.0006	-7.6940	0.0000
BSIZE	1.6828	0.9538	-6.5410	0.0000
NIM	-4.3520	0.0000	-8.7008	0.0000
LIQ1	-1.9968	0.0229	-5.7118	0.0000
ICTINV	-1.9611	0.0249	-5.8579	0.0000
IDIV	-6.6034	0.0000	-10.6490	0.0000
Cointegration test [Variance ratio] *H0: No cointegration* *Ha: Some panels are cointegrated*			15.5956	0.0000

Authors' estimate, 2025.

a -0.005 unit decline in RBOE, ceteris paribus. Larger banks may have worse operational efficiency, as indicated by a negative connection. This could be due to size-related inefficiencies, such as management complexity or bureaucratic layers. This finding is corroborated by [20], who state that an increase in bank size may reduce efficiency, resulting in lower returns on assets and equity. Furthermore, [24] found that smaller banks are more efficient, establishing an adverse nexus between bank size and performance. However, [20] found that bank size and management ownership have a positive influence on operational efficiency. This study further finds that NIM has a positive and increasing effect on efficiency. Specifically, a unit increase in profitability leads to a 0.554 unit increase in RBOE at a 1% significance level. This highly substantial and positive coefficient indicates that profitability from core loan and deposit activities is reflected in larger net interest margins, which strongly contribute to bank operational efficiency. Contrarily, [38] found no relationship between operational efficiency and financial performance.

Similarly, this study reveals that a unit increase in LIQ1 results in a 0.031-unit increase in RBOE at a 5% significance level, ceteris paribus. Operational efficiency positively correlates with higher liquidity, suggesting that banks with sufficient liquidity are better equipped to handle operations effectively. Banks with better liquidity management tend to have higher operational efficiency, as they can optimise their asset-liability management systems [28]. Moreover, [29] revealed that proficient liquidity management plays a crucial role in improving the financial outcomes of rural banks. [27] found that decreasing liquidity generation boosts bank stability, whereas increasing bank capital reduces liquidity creation. [30] revealed that liquidity risk does not directly influence operational efficiency. Rather, its effect is indirectly exhibited through credit risk, which can affect banks' overall efficiency.

Additionally, this study examines how ICT investments address the efficiency needs of rural banks in Ghana's financial sector. Our results indicate that ICT investments

Table 6. Determinants of operational efficiency - correlated panels corrected standard errors (PCSE)

	(1)	(2)	(3)
VARIABLES	RBOE	RBOE	RBOE
BSIZE	-0.00533*	0.0218***	0.0402**
	(0.00322)	(0.00735)	(0.0195)
NIM	0.554***	0.559***	0.563***
	(0.0801)	(0.0791)	(0.0741)
LIQ1	0.0309**	0.0258**	0.0334***
	(0.0125)	(0.0116)	(0.00970)
ICTINV	0.00282*	0.00587***	0.0640**
	(0.00154)	(0.00160)	(0.0267)
IDIV	-0.0233***	-0.0236***	-0.0243***
	(0.00846)	(0.00834)	(0.00583)
SQBSIZE		-0.00294***	
		(0.00101)	
ICTINV*BSIZE			-0.00832**
			(0.00355)
Constant	0.113***	0.0570***	-0.222
	(0.0256)	(0.0143)	(0.140)
Obs	854	854	854
R-squared	0.653	0.659	0.872
No. of banks	122	122	122
Wald chi2	127.18	154.71	357.14
Prob > chi2	0.0000	0.0000	0.0000

Standard errors in parentheses *** p < .01, ** p < .05, * p < .1
Authors' estimate (2025).

significantly improve the income-generating capacity of these banks. Specifically, a percentage increase in ICT leads to a 0.003 unit increase in RBOE. Although the effect is weak, investment in information and communication technology favours operational efficiency. This suggests potential improvements in efficiency resulting from technological advancements. Financial institutions can enhance customer service, reduce operating expenses, and streamline their processes by implementing cutting-edge technological solutions [32]. [33] found that adopting fintech helped improve the capital adequacy ratio, return on assets, return on equity and loan-to-deposit ratio in Pakistan. Moreover, [3] noted that the positive effects of ICT investment on bank performance in Ghana are contingent upon the indirect influence of ICT diffusion. In Nigeria, [67] found that the adoption of ICT significantly enhances operational efficiency and customer convenience. By way of caution, [68] indicated that ICT investments are costly, and in certain instances, rural banks may not observe an immediate return on investment. Income diversification is found to have a decreasing effect on operational efficiency. At a 1% significance level, this study reveals that a unit increase in IDIV leads to a 0.113 unit increase in RBOE, ceteris paribus. The significant and negative coefficient suggests that income stream diversification may reduce operational efficiency due to increased expenses or inefficiencies in managing diverse operations.

This study further explored the non-linear relationship between size and operational efficiency as captured in the equation. We found that whereas the linear effect of BSIZE on RBOE is positive at a 10% significance level, the quadratic term of BSIZE in Table 6 (denoted as SQBSIZE) shows that at a 1% significance level, a percentage increase in SQBSIZE aggravates the deteriorating impact of size on efficiency levels, ceteris paribus. Emphasizing both BSIZE and SQBSIZE in Table 6, we can conclude that the initial growth in size potentially increases efficiency, probably due to their ability to attract more customers and negotiate better terms with relevant stakeholders. However, the negative coefficient of 0.003 indicates diminishing returns as bank size continues to grow. Thus, at a certain threshold of size increase, banks experience inefficiencies, likely due to complexity, bureaucracy, and higher regulatory and compliance costs. The interactive effect (i.e. ICTINV*BSIZE) shows a significant negative coefficient of -.008. This implies that by increasing ICT investment, its interaction with BSIZE substantially adds to the cost compared to the income for these banks, thereby significantly reducing efficiency at a 10% alpha level.

3.5 Model Diagnostics

With an average level of operational efficiency of 0.174 and a variability of 6.5%, our PCSE model, as shown in Table 6, predicts that 65.3% of the variability in rural banks' operational efficiency can be attributed to the independent variables used, indicating a good fit. The model as a whole is statistically significant, with a chi-square value of 127.182 (p-value = 0.000), indicating that the predictors collectively have a significant impact on operational efficiency.

3.6 Robustness Checks

To assess the robustness of the results as presented by the PCSE technique, this study further estimates two results using FGLS (see Table 7) and pooled panel OLS (see Table 7) techniques. These results compare the earlier outputs to assess validity and consistency, enabling the formulation of necessary policy recommendations. Consistent with the PCSE results in Table 6, BSIZE and IDIV are found to have a significant adverse effect on the operational efficiency of the banks. Although these signs are the same, the significance level for BSIZE in the FGLS estimate is 1%, whereas the significance level in the PSCE results is 10%. Moreover, at a 1% significance level, NIM, LIQ1, and ICTINV have a positive impact on operational efficiency, ceteris paribus. These results are consistent with PSCE estimates, thereby validating the earlier results. As shown in Table 7, the panel-pooled OLS provided significant coefficients and relationships similar to those of the PSCE and FGLS, as indicated above. The strongest positive correlation with RBOE is found in NIM. Enhancing net interest margins through policy could significantly increase the banks' operational efficiency. RBOE is significantly impacted negatively by BSIZE and IDIV, indicating that higher IDIV and larger banks may impair financial performance. LIQ1 and ICTINV have beneficial and meaningful results despite their relatively minor effects. The robust and well-defined model exhibits high overall explanatory power and includes all statistically significant factors. A better-fitting model is indicated by a lower AIC of -2928.796; this value implies an excellent fit for the data.

Table 7. Determinants of operational efficiency **- FGLS and pooled panel OLS** (Robustness estimates)

	(1)	(2)	(3)	(4)	(5)	(6)
VARIABLES	RBOE *Pooled Panel OLS*	RBOE *Pooled Panel OLS*	RBOE *Pooled Panel OLS*	RBOE ***FGLS***	RBOE ***FGLS***	RBOE ***FGLS***
BSIZE	-0.0108***	0.0123	0.0114	-0.0108***	0.0227***	0.0359
	(0.00306)	(0.0108)	(0.0251)	(0.00305)	(0.00788)	(0.0235)
NIM	0.544***	0.545***	0.541***	0.544***	0.560***	0.579***
	(0.0200)	(0.0199)	(0.0202)	(0.0199)	(0.0206)	(0.0205)
LIQ1	0.0289***	0.0274***	0.0289***	0.0289***	0.0255***	0.0290***
	(0.00576)	(0.00578)	(0.00576)	(0.00574)	(0.00771)	(0.00690)
ICTINV	0.00609***	0.00740***	0.0364	0.00609***	0.00572**	0.0577*
	(0.00202)	(0.00209)	(0.0340)	(0.00201)	(0.00256)	(0.0325)
IDIV	-0.0398***	-0.0404***	-0.0398***	-0.0398***	-0.0229***	-0.0263***
	(0.00560)	(0.00559)	(0.00560)	(0.00558)	(0.00393)	(0.00302)
SQBSIZE		-0.00192**			-0.00306***	
		(0.000856)			(0.000814)	
ICTINV*BSIZE			-0.00417			-0.00748*
			(0.00468)			(0.00442)
Constant	0.144***	0.0712*	-0.0164	0.144***	0.0572**	-0.189
	(0.0209)	(0.0385)	(0.181)	(0.0208)	(0.0262)	(0.171)
Obs	854	854	854	854	854	854
R-squared	0.560	0.563	0.561			
Adj R-squared	0.5577	0.5598	0.5576			
No. of banks	122	122	122	122	122	122
Wald chi2	-	-	-	1088.05	896.47	1097.81
Prob > chi2	-	-	-	0.0000	0.0000	0.0000

Standard errors in parentheses *** $p < .01$, ** $p < .05$, * $p < .1$
Authors' estimate (2025).

4 Conclusion

The study aimed to investigate the determinants of operational efficiency in rural banks within the Ghanaian financial sector, utilising the ARB Apex Bank's financial data from 2014 to 2020. Specifically, the study focuses on how bank size, financial performance, liquidity, investment in ICT, and income diversification potentially drive the efficiency of banks in Ghana, considering the existing literature and the significant contribution of rural banks in championing financial inclusion, business support, and overall economic growth. Operational efficiency (RBOE) was measured as the bank's ability to cover its expenses with its income. We conducted empirical estimates primarily using the Prais-Winsten PCSE method, as it enables the control for cross-sectional dependence, accommodates different forms of autocorrelation in the panel, and accounts for heteroscedasticity. Furthermore, two additional empirical estimates were generated and analysed for robustness using the FGLS and pooled panel OLS methods.

The study's findings reveal that the banks' operational efficiency is adversely affected by large sizes. This outcome suggests that larger rural banks may lose efficiency due to size-related inefficiencies resulting from increased direct and indirect costs. Since smaller banks could be more agile and effective, bigger organizations should consider reducing scale-related inefficiencies. Furthermore, we found that NIM has a positive impact on banks' efficiency. This result suggests that policies that increase net interest margins can enhance operational effectiveness by highlighting the importance of core banking profitability. Again, this study finds that a high liquidity position helps improve the efficiency of banks. This result suggests that resilience and operational efficiency depend on maintaining sufficient cash. We found that an increase in ICT investment leads to improved operational efficiency. Thus, whereas the benefit of technology investment on operational efficiency is minimal, it is likely because ICT returns are of a longer-term nature. This study further demonstrates that IDIV has a negative impact on operational efficiency. Over-diversification into non-core operations (such as branching out into fields lacking economies of scale or expertise) may increase an institution's operational complexity, expenses, or resource misalignment, ultimately lowering operational efficiency. Regulators and management of these banks should establish criteria to guarantee that diversification tactics do not jeopardize financial stability or operational effectiveness. To help institutions better manage a variety of revenue streams, policymakers may encourage the use of tools and technologies (such as artificial intelligence for portfolio management).

Considering the findings of this study, the need for a more in-depth investigation, such as exploring interaction effects and non-linear relationships, was identified. This study reveals that the squared term of BSIZE yields a significant negative effect on banks' efficiency. Interestingly, the ICT investment negatively moderates the negative nexus between BSIZE and efficiency. While ICT investment positively drives up operational efficiency, the negative coefficient from the moderation suggests that additional ICT investment could lead to implementation challenges, increased costs due to initial infrastructure, maintenance, training, and development. Again, larger rural banks may struggle to derive efficiency gains from ICT and thus become competitively disadvantaged when competing with smaller rural banks, which are more adept at adapting to technological changes.

Moreover, banks must focus on developing mechanisms to assess how diversification policies and ICT investments impact operational efficiency over the long term. To respond to external influences that can impact NIM, liquidity, and income diversification, such as economic downturns, organisations should develop adaptable solutions. The conclusions may not be applicable to the entire banking industry or even to the rural and overall banking sectors of other economies worldwide, as the study's scope is limited to Ghanaian rural banks. Therefore, it is proposed that future research focus on the global banking sector, as well as other economies in Africa.

References

1. Khan, S.: Determinants of operational efficiency: the case of Saudi banks. Fin. Int. Q. **18**(3) (2022)
2. Berger, A.N., Hunter, W.C., Timme, S.G.: The efficiency of financial institutions: a review and preview of research past, present, and future. J. Bank. Financ. **17**(2–3), 221–249 (1993)
3. Annor, L.D.J., Agyapong, E.K., Robaina, M., Vieira, E., Anarfo, E.B.: Information technology investment and rural bank performance in Ghana: the moderating role of ICT diffusion and financial development. J. Econ. Administr. Sci. (2024). Vol. ahead-of-print No. ahead-of-print. https://doi.org/10.1108/JEAS-07-2023-0171
4. IFAD (International Fund for Agricultural Development).: The Republic of Ghana Rural and Agricultural Finance Program (RAFiP). Rome (2008)
5. Agyapong, E.K., Annor, L.D.J., Ohemeng, W.: Corporate social responsibility and performance among rural banks in Ghana: the moderating role of governance structures. Int. J. Soc. Econ. (2024). https://doi.org/10.1108/IJSE-02-2023-0116
6. Obeng, S.K.: Rural banking in Ghana: Its impact on rural farmers. Modern Ghana (2008). Retrieved from https://cdn.modernghana.com/images/content/report_content/RURAL%20BANKING%20IN%20GHANA%20ITS%20IMPACT%20ON%20RURAL%20FARMERS.pdf. Accessed on 02 March 2025
7. Asunka, S.: The politics of targeting: State capacity and the implementation of social protection in Ghana. University of California, Los Angeles (2015)
8. Mensah, S., Domfeh, K.A., Osei-Tutu, E.: The role of rural banks in Ghana's rural development: a case study of selected rural banks in Ashanti Region. J. Econ. Sustain. Develop. **8**(12), 1–10 (2015)
9. Agyapong, E.K., Annor, L.D.J., Anarfo, E.B.: Bank performance, governance structures and ICT investment among rural banks in an emerging economy. J. Money and Bus. (2024). https://doi.org/10.1108/JMB-12-2023-0073
10. Akorsu, A.D., Agyapong, D., Osei-Tutu, E.: Rural banking and poverty reduction in Ghana: the case of selected rural banks in the Ashanti Region. J. Bus. Econ. Develop. **3**(1), 1–10 (2015)
11. Akosey, J.: https://thebftonline.com/2024/11/25/rural-banking-special-report-2024-how-rural-banks-are-driving-financial-inclusion/ Rural Banking Special Report 2024: How rural banks are driving financial inclusion Accessed on 3/04/25
12. Al-Zoubi, M., Sha'ban, M.: Bank performance, capital and size: A comparative analysis in MENA and EU. Eurasian Econ. Rev. **13**(4), 677–702 (2023). https://doi.org/10.1007/s40822-023-00248-6
13. Muntaqiah, K., Akhmadi, A.: Banking financial performance as influenced by credit risk, liquidity and bank size in conventional banks listed on the indonesian stock exchange for the period 2014 - 2023. Indonesian J. Innov. Multidiscipl. Res. **2**(4), 149–160 (2024). https://doi.org/10.69693/ijim.v2i4.221

14. Ngumo, S.: Determinants of Financial Performance of Microfinance Banks in **8**(16), 1–8 (2017)
15. Kirimi, P.N., Kariuki, S.N., Ocharo, K.N.: Moderating effect of bank size on the relationship between financial soundness and financial performance. Afr. J. Econ. Manag. Stud. **13**(1), 62–75 (2022). https://doi.org/10.1108/AJEMS-07-2021-0316
16. Siagian, P.: Determinants of banking operational efficiency and the relationship between the factors to market price: evidence from Indonesia. ECONOMICS-Innovative and Econ. Res. J. **11**(2), 153–168 (2023)
17. Gupta, N., Mahakud, J.: Ownership, bank size, capitalization and bank performance: Evidence from India. Cogent Econ. Fin. **8**(1), 1808282 (2020). https://doi.org/10.1080/23322039.2020.1808282
18. Jiménez-Hernández, I., Palazzo, G., Sáez-Fernández, F.J.: Determinants of bank efficiency: evidence from the Latin American banking industry. Appl. Econ. Anal. **27**(81), 184–206 (2019). https://doi.org/10.1108/AEA-09-2019-0027
19. Gržeta, I., Žiković, S., Tomas Žiković, I.: Size matters: analyzing bank profitability and efficiency under the Basel III framework. Financial innovation **9**(1), 43 (2023)
20. Alfariz, R., Asmara, R.Y.: Company size, managerial ownership, sustainability report on financial performance with operational efficiency as an intervening. Asian J. Soc. Humanit. **2**(12), 3125–3135 (2024)
21. Hermuningsih, S., Sari, P.P., Rahmawati, A.D.: The moderating role of bank size: influence of fintech, liquidity on financial performance. Jurnal Siasat Bisnis, 106–117 (2023)
22. Blatter, M., Fuster, A.: Scale effects on efficiency and profitability in the Swiss banking sector. Swiss J. Econ. Statist. **158**(1), 12 (2022)
23. Leibenstein, H.: Allocative efficiency vs. X-efficiency. The American Economic Review **56**(3), 392–415 (1966)
24. Yahaya, K.A., Lawal, A.I., Sulaimon, A.A.: Bank size and performance of quoted deposit money banks in sub-Saharan Africa: Evidence from panel data analysis. African J. Econ. Policy **29**(1), 45–62 (2022)
25. Alfadhli, A., Alali, F.: Bank size and performance: Evidence from GCC countries. Int. J. Fin. Stud. **9**(2), 25 (2021). https://doi.org/10.3390/ijfs9020025
26. Golubeva, O., Duljic, M., Keminen, R.: The impact of liquidity risk on bank profitability: some empirical evidence from the European banks following the introduction of Basel III regulations. J. Acc. Manage. Info. Sys. **18**(4), 455–485 (2019)
27. El Mahdy, A.H.A.M.: Liquidity risk and bank financial performance. Commerce and Finance **43**(2), 41–80 (2023). https://doi.org/10.21608/caf.2023.303697
28. Berger, A.N., Bouwman, C.H.: How does capital affect bank performance during financial crises? J. Financ. Econ. **109**(1), 146–176 (2013)
29. Awo, J.P., Akotey, J.O.: The financial performance of rural banks in Ghana: The generalized method of moments approach. World J. Entrepren. Manage. Sustain. Develop. **15**(1), 2–18 (2019). https://doi.org/10.1108/WJEMSD-02-2018-0012
30. Kaharuddin, K., Yusuf, M.: The Impact of Liquidity Risk Optimization on the Stability of Islamic Commercial Banks in Indonesia. In: Proceeding of The International Conference on Economics and Business, Vol. 1, No. 2, pp. 671–688 (2022)
31. Yulianti, M.L., Pakata, R.: The Impact of liquidity risk optimization on stability. Journal Ekonomi **12**(01), 630–639 (2023)
32. Iwedi, M.: Digital finance infrastructure and growth of commercial banking firms in Nigeria. Discover Analytics **2**(1), 16 (2024)
33. Ashraf, A., Jamal, S., Khan, M.A., Shah, S.A.: Advancing in the change journey towards FinTech: the nexus between fintech and financial performance of commercial banks in Pakistan. J. Soc. Sci. Archiv. **3**(1), 43–55 (2025)

34. Samara, F.N., Prawoto, B.R., Maulana, A., Firdaus, A.R., Dewi, N.G.: The role of banking digitalization in enhancing business capacity among rural communities in Indonesia. Jurnal Akuntansi dan Bisnis **10**(02), 144–153 (2024)
35. Amechi, F.A.: Information and communication technology (ICT) and operational efficiency of quoted deposit money banks In Nigeria. American Int. J. Bus. Manage. (AIJBM) **5**(09), 88–95 (2022)
36. Tariku, T., Id, E., Mersha, D., Id, L.: Cost efficiency of commercial banks in Ethiopia: Does financial technology matter ? 1–18 (2025). https://doi.org/10.1371/journal.pone.0317226
37. Owusu Kwateng, K., Agyei, J., Amanor, K.: Examining the efficiency of IT applications and bank performance. Ind. Manag. Data Syst. **119**(9), 2072–2090 (2019). https://doi.org/10.1108/IMDS-03-2019-0129
38. Umoh, E.B.: Assessing the Influence of Operating Efficiency on the Financial Performance of Industrial Firms in Nigeria (December 24, 2024) (2024). Available at: SSRN: https://ssrn.com/abstract=5076812 or https://doi.org/10.2139/ssrn.5076812
39. Asian Development Bank (ADB).: Asian Development Outlook 2020: What drives innovation in Asia? Asian Development Bank (2020). https://doi.org/10.22617/FLS200317-2
40. Dhar, U.: Asian Development Bank (ADB), Asian Development Outlook 2020: What Drives Innovation in Asia? (2021)
41. OECD.: OECD compendium of productivity indicators 2024. OECD Publishing (2024). https://doi.org/10.1787/b96cd88a-en
42. World Bank Group.: From crisis to green, resilient, and inclusive recovery. World Bank (2021). https://openknowledge.worldbank.org/handle/10986/36067
43. Najam, H., Abbas, J., Álvarez-Otero, S., Dogan., Sial, M.S.: Towards green recovery: can banks achieve financial sustainability through income diversification in ASEAN countries? Economic Analysis and Policy **76**, 522–533 (2022)
44. Ndungu, J.G., Muturi, P.W.: Effect of Diversification on Financial Performance of Commercial Banks in Kenya **3**(V), 267–285 (2019)
45. Sang, N.M.: Income Diversification and Bank Efficiency in Vietnam. **19**(3), 52–67 (2017)
46. Kinini, D.M., Ocharo, K.N., Kariuki, P.W.O.: Do income diversification and capital adequacy affect liquidity creation? a case study of commercial banks in Kenya. Cogent Bus. Manage. **10**(2), 2240082 (2023)
47. Ho, T.H., Nguyen, D.T., Luu, T.B., Le, T.D., Ngo, T.D.: Bank performance during the COVID-19 pandemic: does income diversification help? J. Appl. Econ. **26**(1), 2222964 (2023)
48. Nguyen, P.H., Thi, D., Pham, B.: Income diversification and cost- efficiency of Vietnamese banks **16**(5), 623–643 (2020). https://doi.org/10.1108/IJMF-06-2019-0230
49. Nguyen, D.: Benefit from income diversification of Vietnam commercial banks. VNU J. Sci. Policy and Manage. Stud. **33**(2), 1–11 (2017). https://doi.org/10.25073/2588-1116/vnupam.4088
50. Githaiga, P.N.: Income diversification, market power and performance. J. Econ. Fin. Anal. **3**(2), 1–21 (2020)
51. Huynh, J., Dang, V.D.: Loan portfolio diversification and bank returns: Do business models and market power matter? Cogent Econ. Fin. **9**(1), 1891709 (2021)
52. Vidyarthi, H.: Dynamics of income diversification and bank performance in India. J. Fin. Econ. Policy **12**(3), 383–407 (2019). https://doi.org/10.1108/jfep-05-2019-0084
53. Nguyen, D.T., Tran, S.H., Le, T.D. The moderating role of income diversification on the relationship between intellectual capital and bank performance evidence from Viet Nam. Cogent Business & Manage. **10**(1) (2023). https://doi.org/10.1080/23311975.2023.2182621
54. Kollie, J.: Effects of income diversification on the financial performance of commercial banks listed in Sierra Leone. European J. Manage. Market. Stud. **9**(1) (2024). https://doi.org/10.46827/ejmms.v9i1.1702

55. Hoang, C., Nguyen, L., Tran, M., Hoang, T.: The impact of income diversification on liquidity creation and financial performance of Vietnamese commercial banks. Accounting **6**(4), 553–568 (2020)
56. Pesaran, M.H.: General diagnostic tests for cross section dependence in panels. University of Cambridge, faculty of economics, Cambridge Working Papers in Economics, No. 0435 (2004)
57. Pesaran, M.H.: Testing weak cross-sectional dependence in large panels. Economet. Rev. **34**(6–10), 1089–1117 (2015). https://doi.org/10.1080/07474938.2014.956623
58. Adeleye, B.N., Akam, D., Inuwa, N., James, H.T., Basila, D.: Does globalization and energy usage influence carbon emissions in South Asia? An empirical revisit of the debate. Environ. Sci. Pollut. Res. (2022). https://doi.org/10.1007/s11356-022-24457-9
59. Beck, N., Katz, J.N.: What to do (and not to do) with time-series cross-section data. American Polit. Sci. Rev. **89**(3), 634–647 (1995)
60. Cameron, A.C., Trivedi, P.K.: Microeconometrics: Methods and Applications. Cambridge University Press (2009)
61. Sundjo, F., Aziseh, F.: An empirical investigation into the key drivers of economic performance in the CEMAC zone: A panel corrected standard errors approach. Int. J. Bus. Manage. **5**(6), 189–200 (2018)
62. Greene, W.H.: Econometric analysis, 4 International. Prentice Hall (2000)
63. Wooldridge, J.M.: Econometric analysis of cross-section and panel data. The MIT Press. Cambridge, Massachusetts. London, England (2010)
64. Bailey, D., Katz, J.N.: Implementing panel corrected standard errors in R: The PCSE Package. J. Stat. Softw. **42**(CS1), 1–11 (2011)
65. Annor, L.D.J., Robaina, M., Vieira, E.: Financial development, inclusive growth, and environmental quality: emerging markets perspective. Environment, Development and Sustainability (2023) https://doi.org/10.1007/s10668-023-04198-6
66. Annor, L.D.J., Kyei-Mensah, T., Addae, J.A., Robaina, M., Vieira, E.: Fostering energy transition in Africa: Financial systems, inclusive growth, corruption control and ICT as catalysts. Sustain. Environm. **11**(1) (2025). https://doi.org/10.1080/27658511.2025.2518791
67. Fabian, A.A.: Information and Communication Technology (ICT) and Operational Efficiency of Quoted Deposit Money Banks in Nigeria. American Int. J. Bus. Manage. **5**(9), 88–95 (2022). Retrieved from https://www.aijbm.com/wp-content/uploads/2022/09/J598895.pdf
68. KPMG: Accelerating digital finance: The new reality. EMA Financial Services Regulatory Insight Centre (2021). https://home.kpmg/regulatorychallenges

Artificial Intelligence in Financial Distress Management: Advancing Beyond Portuguese Municipalities

Filipe Caetano[1] (✉), Rute Abreu[2], and Maria Victoria López-Pérez[3]

[1] Instituto Politécnico da Guarda, IPCA-CICF, Guarda, Portugal and Universidad de Granada, Granada, Spain
caetano@correo.ugr.es

[2] Instituto Politécnico da Guarda, IPCA-CICF, IPV-CISeD, CITUR, Guarda, Portugal
ra@ipg.pt

[3] Universidad de Granada, Granada, Spain
mvlopez@ugr.es

Abstract. This paper assesses the financial distress management in the public sector and explores the application of Artificial Intelligence (AI) techniques to the Portuguese municipalities (n = 308). The paper aims to analyse the financial structure of these municipalities by identifying latent factors that characterise their economic and financial performance. The research question addresses which accounting variables provide the most robust explanatory power for managing financial distress. A quantitative methodological approach is adopted, employing Exploratory Factor Analysis (EFA) on variables extracted from municipal financial statements, namely the balance sheet and the statement of income by nature. Data is sourced from the Portuguese Accounting Standardisation System for Public Administrations and covers the most recent available fiscal year. EFA enables the reduction of data dimensionality, revealing principal components that group correlated financial indicators into interpretable factors. These factors are expected to distinguish between financial robustness and exposure to financial risk. By demonstrating the potential of data-driven models for public financial management, this paper sets the groundwork for extending AI-based financial distress detection frameworks to other institutional contexts beyond local government, including corporate and non-profit sectors. The findings aim to contribute to enhanced financial governance, transparency, and early warning systems supported by AI technologies.

Keywords: Artificial Intelligence · Financial Distress Management · Municipalities

1 Introduction

Financial distress in municipalities can disrupt public services and undermine economic stability [1]. Recent advancements in financial analytics and AI have significantly improved management financial distress, but these tools have been applied mostly in the

L. Almeida et al. (Eds.): ICAFI 2025, ISEM 80, pp. 252–269, 2026.
https://doi.org/10.1007/978-3-032-16268-7_14

private sector [2]. In the public sector, especially at the municipal level, proactive financial distress management remains a pressing challenge. So, this paper aims to answer the research question: *Which accounting variables and latent financial factors provide the most robust explanatory power for managing financial distress in Portuguese municipalities?* Indeed, to answer this question, the authors draw objectives: (1) Identify the key accounting variables and latent factors that characterize the economic and financial condition of Portugal's municipalities; (2) Determine which of these factors are most strongly associated with financial distress; and (3) Demonstrate the relevance of AI-based analysis (e.g. exploratory factor analysis) in improving early warning systems for municipal financial distress.

The empirical study focuses on the population of 308 Portuguese municipalities in 2021. Thus, the research develops advanced analytical techniques to assist in municipal financial distress in Portugal is scant. By concentrating on this context, the research, not only aids policymakers, but also develops AI methods that leveraged the public accounting more broadly. Recent advancements in AI analytics and accounting have significantly improved the ability to manage financial distress and subsequent crises [3]. Furthermore, existing tools often overlook behavioral and organizational dynamics, focusing solely on metrics. The absence of interdisciplinary integration limits the applicability and effectiveness of these tools in real-world scenarios.

To address these gaps, this paper starts with this introduction where the research question and specific objectives of the research are identified. The second section proposes a literature review of financial distress. The third section presents the methodology used. The fourth section is dedicated to results analysis, and the last section will provide strategic recommendations, bridging the gap between theoretical models and practical applications in the financial distress of municipalities.

2 Literature Review

The prediction of financial distress has been a focus of research for decades. Altman's seminal Z-Score model [4] demonstrates that quantitative models could significantly outperform expert judgment in identifying distressed firms. Since then, the field has evolved through logistic regression models (e.g., Ohlson [5]) and into various machine learning (ML) techniques. Barboza, Kimura, and Altman [6] tested multiple ML algorithms (support vector machines, bagging, boosting, random forest) and found they achieved higher accuracy in bankruptcy prediction than classic approaches like discriminant analysis or logistic regression. Jones et al. [7] reported that advanced non-linear models can provide better classification of distressed *versus* healthy companies. In summary, the corporate finance literature has firmly established that AI techniques can serve as effective early warning systems for financial distress in firms. This provides a foundation and contrast for the public-sector applications discussed next.

In local governments and the public sector, financial condition analysis has traditionally relied on financial ratio diagnostics and composite indices. Pioneering work by Groves et al. [8] introduced a set of accounting variables to assess municipal distress. Over time, researchers developed more sophisticated frameworks for evaluating municipal financial health. For instance, Cabaleiro et al. [9] proposed a method to quantitatively assess municipal financial health by integrating multiple indicators. Turley et al. [10] defined a comprehensive performance measurement framework for local governments in Irland. Indeed, some studies created early-warning systems, such as Zafra-Gómez et al. [11] developed an alert system for Spanish local governments to predict financial crises, which was later empirically validated with real municipal data (distinguishing financially "healthy" and "at-risk" municipalities). Similarly, Jorge et al. [12] in Portugal examined composite fiscal stress indexes to signal distress. These traditional methods, however, generally did not employ AI or machine learning techniques.

Recent years have seen initial forays of AI techniques into public-sector financial distress forecasting. Jarmulska [13] provides evidence that ML can improve early warnings of fiscal stress in local governments in Polish municipal finances. Chen [14] applied neural networks and decision tree models to subnational governments in China, achieving over 85% accuracy in predicting fiscal stress up to one year in advance. These results are promising, suggesting that AI can handle the complex interactions of economic and fiscal variables that lead to distress in the public arena. Despite these developments, there remain clear gaps in literature related to the scarcity of studies focusing on municipal financial distress with AI methods.

The authors note that factor analysis results depend on the dataset and period. Since the authors are focusing on the singular year 2021, the factor structure will capture the relationships among financial indicators in that year. During an atypical year like 2021, some correlations might be accentuated (for example, perhaps high capital spending correlating with high transfers due to EU COVID funds, forming a distinct factor). Past studies that looked at "normal" years or multi-year averages might have found slightly different structures.

For instance, Navarro-Galera et al. [15] identified revenue, debt, and service dimensions in Spanish data over multiple years. Also relevant are the composition of revenues in 2021 was altered by pandemic-related inflows: state budget transfers to municipalities increased beyond normal patterns (partly due to COVID support programs and the ongoing decentralization of new competencies to local governments).

Also, EU structural funds contributed significantly to capital investments, as projects delayed in 2020 were executed in 2021. These factors likely created a high correlation between capital expenditures and transfers in that year, which may emerge as a factor in our analysis (indicating a distinct "investment with external funding" dimension). On the expenditure side, one notable pandemic-driven cost was the increase in current transfers and subsidies provided by municipalities. Many local governments launched or expanded grant programs to support households, NGOs, and businesses during the crisis (e.g. waiving local fees, subsidizing utility bills, funding protective equipment for schools).

Additionally, personnel costs rose, as some municipalities hired extra temporary staff (such as healthcare workers or sanitization crews) or paid overtime to address COVID-related needs. These expenditure increases were not uniform across all municipalities; those with more fiscal space or higher infection rates spent more on such measures.

The pandemic also brought to the fore the importance of financial flexibility–the ability to reallocate funds or access emergency resources. Local governments with more diversified revenue streams (including local taxes that bounced back, like property taxes which remained robust as real estate boomed in 2021) found themselves in a better position than those largely dependent on flat state transfers. This underscores a known theme in municipal finance: revenue diversification can enhance stability [16].

In summary, the COVID-19 era presents both a stress-test and a learning opportunity for municipal finance. It has accentuated the critical factors of fiscal health identified by prior literature–e.g. prudent debt levels, maintaining some surplus or reserves, and having flexible budgeting. For our study, focusing on 2021 allows us to incorporate these pandemic effects into the analysis. It ensures that the latent factors the authors extract are reflective of a scenario that tested the limits of local financial management. The findings can thus inform not only of peacetime fiscal management but also preparedness for future shocks. In the following section, the authors detail the methodology used to empirically analyse the Portugal's municipalities in 2021.

3 Methodology

The paper focuses on assess the financial distress in the public sector, using the population of Portuguese municipalities (308) as a starting point. Initially, the research focuses on conducting a comprehensive literature review to identify gaps and opportunities in financial distress management, while concurrently collecting quantitative data through webpages of each municipality. These activities have provided critical aspects due to requirements of the accounting information system. Simultaneously, the authors developed and trained predictive algorithms using historical datasets.

Following this, the research starts with the design and prototyping phase. Quantitative validation has been conducted using cross-municipalities case studies to ensure the research's robustness and adaptability across diverse economic contexts, dimensions and several locations. This quantitative methodological approach is adopted, employing EFA on variables extracted from municipal financial statements, namely the balance sheet and the statement of income by nature. Data is sourced from the Portuguese Accounting Standardisation System for Public Administration and covers the most recent available fiscal year. The database has 32.340 observations, based on 308 municipalities and 105 accounting and macroeconomic variables of each municipality of the year 2021. The year 2022 and 2023 has several missing data which does not allowed to assess the full sample of municipalities in Portugal [17].

In this study, authors presented Fig. 1 an innovative computational script designed to ensure data integrity, operational efficiency, and auditability in the automated ingestion of financial data from Excel spreadsheets into a relational database management system (RDBMS). This mechanism follows principles of robust automation, transactional control, and syntactic and semantic argument validation, fully aligned with software engineering best practices for quality assurance and process governance. The process begins with the capture and validation of input arguments, specifically: the file path to the Excel document containing accounting or financial data, and the fiscal year to which the data pertains.

This preliminary validation is critical to ensure that the script is executed with the minimum required parameters (two arguments), preventing faulty or incomplete execution. If validation fails, the script terminates immediately, issuing a detailed error message that ensures transparency of failure while preventing any risk of data corruption in the target database. Upon successful validation, the system proceeds to establish the database connection, configuring key parameters such as the server address, user credentials, and authentication settings. This configuration is designed to be modular and environment-agnostic, enabling the portability and reusability of the solution across different deployment scenarios.

Leveraging the panda's library, a widely recognized framework for structured data manipulation, the script loads the Excel file into memory as a DataFrame. A systematic cleaning process is then applied, which removes empty rows and resets indexing, ensuring that the data structure is consistent and reliable before being committed to the database.

The data ingestion process operates within a controlled transactional context. If the process is completed successfully, changes are made to the database, ensuring permanent data persistence. In the event of any exception or failure, the system triggers a transactional rollback, effectively cancelling all changes and preserving the consistency and integrity of the database. This transactional control reflects a robust and innovative governance model, fully compliant with ACID-Atomicity, Consistency, Isolation, Durability principles.

Regardless of the execution outcome, the system ensures the safe and reliable closure of the database connection, releasing all allocated computational resources. This practice enhances the stability and performance of the execution environment, preventing potential resource leaks or deadlocks.

This methodological framework represents technological advancement by integrating defensive programming practices, data governance principles, and automated audit mechanisms in ETL-Extract, Transform, Load processes. The rigorous control of input parameters, the use of specialized data processing libraries, robust transactional management, and secure connection handling constitute best practices that can be replicated across diverse organizational contexts. This approach promotes the technological and financial sustainability of large-scale data processing operations, contributing to the advancement of data-driven decision-making in both academic research and professional practice.

In Fig. 2 the methodological pipeline is presented with the preliminary treatment of the dataset. This step ensures that the input data is properly formatted and ready for subsequent analytical procedures. Particular attention is paid to the cleaning of variables (columns) that contain excessive missing values, which, if left untreated, could compromise the reliability of the factor solution. Following data cleaning, the process applies standardization procedures to scale the data. This ensures that all variables contribute equally to the analysis, eliminating biases caused by differing measurement units or magnitude discrepancies.

Before proceeding to factor analysis, the Kaiser-Meyer-Olkin (KMO) as Measure of Sampling Adequacy and the Bartlett's Test of Sphericity are employed to evaluate the suitability of the data for factor extraction: KMO < 0.859 or p-value ≥ 0.05 suggests that the data is not factorable, prompting the termination of the process due to inadequacy [18]. Conversely, satisfactory values allow the process to continue toward factor extraction.

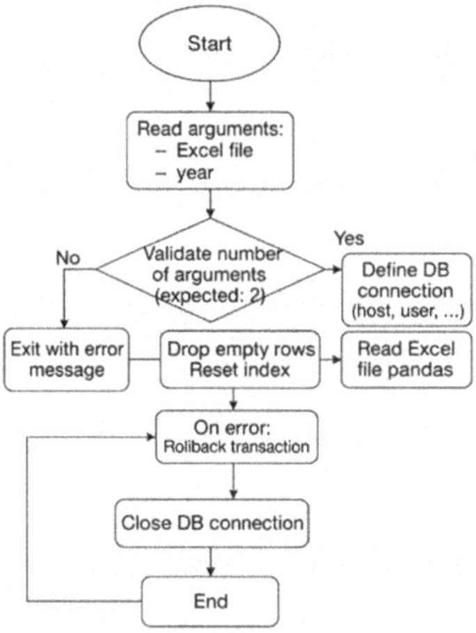

Source: Own elaboration

Fig. 1. Script of the Automated Data Collection. Source: Own elaboration

The next step involves conducting EFA, supported by Screen Plot analysis to determine the optimal number of factors to retain. This graphical method helps identify the 'elbow' point where the explained variance begins to level off, guiding the model specification. Upon determining the number of factors, the model is fitted, and the total variance explained by the retained factors is analyzed. A robust solution typically explains a substantial proportion of the variance, reflecting the latent structure underlying the data.

Lastly, the identified factors are interpreted and named based on the observed loadings and theoretical relevance. This stage ensures the semantic and practical coherence of the extracted factors. If the resulting factor structure fails to meet adequacy criteria—such as poor variance explanation or interpretability—the dataset is classified as inadequate, and the analysis is halted. Otherwise, resulting factors are validated and reported as the outcome of the methodological pipeline.

To enhance the transparency and reproducibility of our analysis, the authors integrated Python-based analytical scripts throughout the research process. These scripts handled data retrieval, preprocessing (as outlined above), and the execution of the factor analysis (using libraries such as Factor Analyzer for Python). The flowcharts in Figs. 1 and 2 were derived from these scripts and depict their logical sequence in a human-readable form. In summary, the methodology is characterized by comprehensive data coverage (all municipalities, wide array of variables), careful preprocessing to ensure robust factor extraction, rigorous statistical validation, and the use of automated scripts and flowcharts to document the process. This sets a solid foundation for the subsequent presentation of results and their analysis in the context of municipal financial distress and the extraordinary conditions of 2021.

The paper uses accounting data from the municipal annual report for the fiscal year 2021. The authors obtained these data from the Portuguese local government finance information system maintained by the Directorate-General for Local Authorities [17], which is the repository for standardized municipal financial statements under

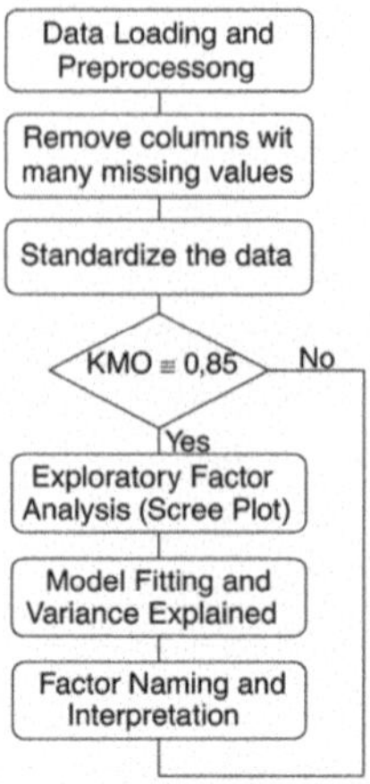

Source: Own elaboration

Fig. 2. Exploratory Factor Analysis Script. Source: Own elaboration

the Accounting Standardisation System for Public Administrations. The dataset covers all municipalities, thereby representing the entire population of Portuguese mainland and autonomous region local governments. Using a single-year, cross-sectional dataset (2021) allows us to analyze the immediate post-COVID financial condition across municipalities under uniform economic conditions.

From the municipal accounts, the authors extracted a broad set of variables encompassing both balance sheet and income statement items, as well as a few derived financial ratios. The goal was to be comprehensive in capturing different aspects of financial health and distress, such as:

- Liquidity and Short-term Solvency: cash and bank balances, short-term accounts payable (creditors), the average payment period (days), and working capital (difference between current assets and current liabilities).
- Debt and Long-term Obligations: total debt outstanding (sum of loans/financial debt), debt per capita, debt as a multiple of current revenue, and interest expenditures.
- Budgetary Performance: annual operating revenue, annual operating expenditure, and the resulting operating balance or surplus/deficit. The authors also include capital revenue and expenditure to capture investment activities, as well as the net result of the year.
- Revenue Structure: breakdown of revenues into own source (municipal taxes, fees), state transfers (the Municipal Fund and other current transfers), and capital transfers (EU funds or national capital grants). Ratios such as "own revenue as % of total revenue" and "transfer dependency ratio" were computed.
- Expenditure Structure: key categories like personnel costs, procurement of goods and services, current transfers (subsidies/grants given), and capital investments. The authors consider "rigid expenditures" (personnel + interest + current mandatory transfers) as a percentage of total spending, as high rigid costs can constrain financial flexibility.

- Socioeconomic and Scale Indicators: population (inhabitants) for 2021, and perhaps GDP per capita or similar if available on the municipal level (though for Portugal, GDP per municipality is not officially provided, the authors proxied economic size by local taxable income or number of businesses where possible). These provide context for financial figures (e.g. debt per capita).
- Financial Sustainability Indicators: such as the debt limit compliance gap (i.e. difference between actual debt and the legal debt limit – a positive value indicates over-indebtedness) and any use of the Municipal Support Fund (an indicator if the municipality was under a recovery plan).

All monetary values were standardized where appropriate to make them comparable across municipalities of different sizes. For instance, expenditure and revenue variables were considered on a per capita basis in some analyses, and likewise debt was examined relative to revenue capacity. This was done to ensure that the factor analysis does not simply pick up "municipality size" as the dominant factor (since large municipalities naturally have bigger absolute numbers). Indeed, without normalization, a factor analysis on raw financial data across municipalities often yields a first factor representing overall size or scale (bigger *versus* smaller). By using ratios and per capita figures, the authors attenuate pure size effects and focus on structural differences in financial profiles.

In this section, the authors described the data sources and preparation steps, the EFA procedure and statistical tests used, and the integration of analytical scripts and flowcharts to ensure a transparent and reproducible analysis.

4 Results Analysis

The results of EFA provide a multifaceted picture of municipal financial distress risk in Portugal, especially in the wake of the COVID-19 shock of 2021. Indeed, the authors discuss the implications of the identified factors in a broader context, relate them back to our research questions, and highlight how the findings contribute to both theory and practice. The authors also address the uniqueness of the year 2021 and why analyzing it yields valuable impacts, consider policy and management implications of our findings, and acknowledge limitations and avenues for future research.

The authors computed the overall KMO measure for the dataset after preprocessing. The KMO for our data turned out to be 0.85, indicating a "meritorious" level of sampling adequacy. In other words, the financial variables have sufficient common variance to justify proceeding with factor analysis – an outcome further evidenced by each variable's individual KMO metric being well above the 0.5 minimum acceptable level. The authors performed Bartlett's test of sphericity, which was highly significant (χ^2 test p-value $<$ 0.001), rejecting the null hypothesis that the variables' correlation matrix is an identity matrix. This implies that there are significant inter-correlations to explain via latent factor analysis [19]. Together, these tests confirmed that the dataset was factorable.

The factor solution is both statistically robust and interpretable. Each factor's composition aligns with distinct conceptual categories of municipal finance, as the authors elaborate next. Table 1 presents the rotated factor loadings allow relatively clean interpretation, with minimal cross-loading conflicts.

Table 1. Rotating Component Matrix. Source: Own elaboration

Rotating component matrix a	Component				
	1	2	3	4	5
Total Liabilities	0,916	0,336	0,034	0,127	0,042
Non-current liabilities	0,911	0,300	0,035	0,093	0,059
Transfers and subsidies granted	0,876	0,394	-0,004	-0,050	-0,134
Services rendered and concessions	0,870	0,348	0,044	-0,038	0,000
Current liabilities	0,861	0,404	0,030	0,205	-0,003
Financing obtained (non-current)	0,859	0,201	0,093	0,181	0,092
Imputed income/expenses from subsidiaries, associates and joint ventures	0,855	0,174	-0,079	-0,066	-0,201
Financing obtained (current)	0,839	0,270	0,101	0,258	0,133
Other accounts payable	0,818	0,304	0,003	-0,052	-0,001
Interest and similar costs incurred	0,755	0,085	0,328	0,206	0,215
Net profit for the period	-0,748	0,279	0,402	0,210	-0,115
Cash and deposits	0,291	0,842	0,079	-0,094	0,003
Current assets	0,522	0,804	0,133	-0,008	0,010
Other changes in net assets	0,488	0,787	0,063	0,107	0,073
State and other public entities (current)	0,298	0,783	0,046	0,190	0,074
Income before depreciation and financial results	-0,139	0,778	0,446	0,191	-0,121
Other income	0,395	0,778	0,183	0,090	0,019
Shareholders' equity	0,579	0,765	0,046	0,031	0,067
Retained earnings	-0,018	0,763	-0,124	0,164	0,080
Current transfers and subsidies obtained	0,401	0,759	0,091	0,192	0,029
Total Assets	0,656	0,715	0,045	0,049	0,065
State and other public entities (non-current)	0,033	0,008	0,775	-0,019	0,122
Investment suppliers (non-current)	0,026	0,063	0,732	0,182	-0,121
Investment property	0,104	0,510	0,544	0,004	0,145
Investment suppliers (current)	0,132	0,098	0,032	0,847	-0,044
Suppliers (current)	0,126	0,264	0,270	0,701	0,329
Impairment of receivables (losses/reversals)	-0,128	0,230	0,117	0,230	0,785
Provisions (increases/decreases)	0,460	-0,170	-0,151	-0,235	0,544

Extraction method: Principal Component Analysis.
a. Rotation converged in 6 iterations.

One of the primary motivations of this paper was to understand how the COVID-19 pandemic's fiscal impact manifested in municipal finances. The factor analysis effectively distilled the complex 2021 financial data into core dimensions, and each of these has a story to tell about the pandemic's effects:

Factor 1 (Debt Burden): The pandemic itself did not directly increase long-term debt for most municipalities in 2020–2021, thanks to emergency transfers and relief measures that allowed local governments to avoid massive borrowing. However, for those already in debt trouble pre-pandemic, 2021 did not offer much respite. Our analysis showed that the same municipalities known for high debt remained outliers in this factor. In fact, a few increased their debt marginally to cover cash shortfalls (reflected in negative liquidity on this factor) [20]. The persistence of this factor as the dominant one implies that pre-existing structural problems (like excessive debt) outweighed COVID-specific issues for many municipalities [21]. This is an important insight: while a global crisis loomed, the municipalities that struggled were largely those that had been struggling before, due to legacy debt and mismanaged finances. The pandemic was more of a stress test that confirmed weaknesses rather than a completely new source of distress. That said, 2021 being the first aggregate deficit year since 2013 hints that if the pandemic response had not included substantial central support, the authors might have seen a wider swath of cities start accumulating debt [22]. Essentially, Factor 1 underscores the importance of entering crises with a strong balance sheet – those that did (low debt) came out without needing bailouts, those that didn't were kept afloat by state support but remain vulnerable ahead.

Factor 2 (Budgetary Performance): This factor was very much influenced by COVID-related financial management. The authors observed a bifurcation: some municipalities tightened belts or got sufficient aid and managed a surplus, while others ran deficits due to pandemic spending or revenue shortfalls. High Factor 2 scorers proved agile or fortunate in 2021, indicating resilience. For example, some municipalities froze hiring and cut non-essential spending anticipating revenue drops, which paid off in maintaining balance. Others perhaps were too optimistic or had unavoidable expenses (like running local vaccination centers, supporting local hospitals) and fell into deficit. The significance of Factor 2 in our results aligns with literature emphasizing early fiscal adjustments during crises (as recommended by IMF or OECD guidelines). Municipalities that adapted quickly to the pandemic (financially speaking) by controlling expenditures or tapping reserves appear healthier in our analysis. Thus, Factor 2 is a barometer of how municipalities handled the shock operationally. It's a call to action that even amid unprecedented crises, maintaining fiscal discipline where possible is key to long-term health. It validates prior findings that prudent budget management is at the core of sustainabilityresearchgate.net, with the pandemic providing a dramatic case study [23].

Factor 3 (Investment & External Funding): The presence of this factor highlights an interesting pandemic side effect – a sort of infrastructure stimulus at the local level. Many municipalities took advantage of recovery programs to invest in capital projects, effectively turning the crisis into an opportunity to improve local infrastructure (roads, digital infrastructure for remote work and others). For those municipalities, our results show a spike in activity (high Factor 3) which could be beneficial long-term: improved infrastructure can enhance economic prospects and future revenues [24]. However, the authors must consider sustainability: if these projects were largely grant-funded, there's minimal immediate financial risk, but if any portion was borrowed or will incur future operating costs, did those municipalities account for it? The factor doesn't directly measure subsequent burdens, but from a discussion standpoint the authors raise caution that a sudden influx of investment can sometimes overshoot, leading to maintenance costs or even white elephants. Nonetheless, seeing Factor 3 clearly separated suggests that some municipalities succeeded in isolating capital expenditures from their operating budget issues (Factor 2) that implies good financial practice: ideally, capital and operating finances should be managed separately to an extent (the golden rule: don't finance operations with debt, only invest with debt if you must). Our data hints that in 2021, many followed that rule – operational health (Factor 2) did not necessarily deteriorate in those that had high capital outlays, since those were covered by capital revenues [25]. This is a positive takeaway: it suggests that extraordinary funds were used as intended to boost investment without derailing day-to-day finances. It also reinforces research like Navarro-Galera et al. [15] who note the importance of capital investments in assessing sustainability, as they tie into service provision and long-run solvency. In future years, if these investments spur economic growth, the municipalities might see improvements in Factor 4 metrics (more own revenue) or maintain Factor 2 surpluses due to efficiencies gained (for example, a new water treatment plant might reduce operating costs of old, inefficient systems).

Factor 4 (Size & Capacity): While not directly about COVID, Factor 4's effects were amplified by the pandemic. Larger cities had to manage public health and economic fallout on a bigger scale; many launched sizeable emergency programs (e.g., Porto's support to cultural sector, Lisbon's free test centers) which could have wrecked their budgets, but their strong revenue rebound compensated. Smaller municipalities, while dealing with pandemic issues, often leaned on national coordination (like nationwide vaccination campaigns) and their finances were more shielded by stable transfer income. The discussion point here is that resilience can depend on scale: bigger municipalities have more diversified economies – a dip in tourism might be offset by a boom in real estate, whereas a small town's economy might hinge on one sector or employer. For instance, a small municipality reliant on a single factory that shut down due to COVID measures might have seen a drastic local recession and drop in local taxes, with little recourse. Factor 4 reminds us that in interpreting factors 1–3, the authors should consider the scale context. For example, a large city with a moderate deficit might be less concerning than a small town with a moderate deficit, because the city can more easily correct it next year with a slight revenue uptick whereas the town might lack means. On the other hand, a large city with very high debt (like historically Lisbon or Oporto had) is a bigger systemic risk (because of absolute amounts) than a small town with high debt – the latter might get a quick bailout from central gov because it's small money, whereas a big one can shake national finances. So, size (Factor 4) plays into risk significance. The uniqueness of 2021 showcased how large *versus* small each had different challenges and advantages, reinforcing that support measures should be tailored: e.g., Portugal's national government perhaps rightly gave more per capita emergency funds to big urban areas hit by COVID outbreaks, while continuing regular support to small ones [14].

Factor 5 (Risks and Accounting Expectations): this factor analysis revealed a fifth, residual factor characterized by moderate loadings on impairment of receivables and provisions. The authors interpret these variables reflecting how municipalities recognize potential future losses and uncertain obligations in their financial statements. Although this factor explains a smaller share of the total variance compared to the four main factors, it captures an important dimension of financial prudence and risk management. Impairment of receivables represents the estimated uncollectible revenues—typically overdue taxes or service fees—thus signaling exposure to revenue realization risk. Provisions, on the other hand, reflect anticipated liabilities such as legal claims or contractual disputes. Together, these variables indicate a municipality's accounting recognition of contingent risks. While these items do not reflect immediate cash outflows, their presence signal underlying structural issues, such as ongoing legal vulnerabilities. Nevertheless, Factor 5 highlights the importance of transparent risk reporting and accounting conservatism as part of sound financial management practices, reinforcing the need for municipalities to anticipate and disclose potential financial pressures.

The pandemic acted as a lens revealing underlying financial structures and testing them. Some municipalities passed the test (surpluses, controlled debt) – often the ones already following good practices. Others struggled and needed support – often those with structural issues (debt, rigid costs). This suggests a reinforcing narrative: crises amplify existing conditions more than create entirely new ones.

From a theoretical standpoint, our findings support the view that municipal financial condition is multidimensional and that these dimensions remain largely consistent even under stress. This adds weight to frameworks of local financial health that include solvency, budgetary, service and investment dimensions [25]. It also demonstrates that factor analysis is a useful tool to empirically validate those conceptual dimensions. From a practical standpoint, focusing on 2021 provided guidance on where to pay attention if similar shocks happen. For instance, if a future crisis hits, oversight authorities might quickly assess which municipalities are likely in trouble by looking at these factor indicators:

Who had high debt going in (Factor 1)?

Who was in deficit (Factor 2)?

Who has ongoing large projects that might be jeopardized by a crisis or conversely could be accelerated as stimulus (Factor 3)?

Who is small and might lack resilience (Factor 4)?

Who captures underlying structural risks to anticipate financial distress (Factor 5)?

The comprehensive analysis of all Portuguese municipalities yields several policy implications:

1. Strengthen Debt Monitoring and Management: Given that Debt Burden (Factor 1) remained the strongest factor, policymakers should continue to rigorously enforce and possibly refine the existing debt limit rules. The fact that over-indebted municipalities stand out so clearly means tools like the Financial Recovery Plans and the Municipal Support Fund should be maintained and perhaps made more proactive. The analysis also showed liquidity issues tied to debt (since Factor 1 included low cash), hence ensuring municipalities maintain a minimum cash buffer (like requiring a certain cash-to-expense ratio) could be beneficial. The Office of Auditor General of Canada [26] recommends indicators of financial condition including debt and liquidity, which could be adapted into Portuguese local oversight in a more formal way – e.g., an annual "financial health report" for each municipality scoring them on factors akin to ours.
2. Promote Balanced Budgets and Flexibility: Our findings reinforce that balanced or surplus budgets (Factor 2) are crucial. The policy here is to strengthen adherence to the "golden rule" – municipalities should not budget persistent operational deficits. Portugal already has rules requiring balanced budgets ex-ante, but ex-post deficits in 2021 happened due to extraordinary reasons. The recommendation is to implement a framework for pandemic or crisis budgeting: essentially, when extraordinary deficits are allowed (like 2020–21), have a medium-term correction plan mandated for municipalities. Some might do so naturally but formalizing it ensures accountability. Also, enhancing budget flexibility by controlling rigid costs is key: policies might include incentivizing shared services or consolidating certain functions among small municipalities to avoid each having high fixed personnel costs (which affect Factor

2). Since Factor 2 negatively correlated with rigidity, the Ministry of Finance could set benchmarks or caps on admin spending relative to total spending, encouraging especially smaller municipalities to keep that ratio in check by pooling resources. Another recommendation: training and resources for modern financial management (e.g., higher education) to improve spending efficiency.

3. Leverage External Funds Without Overreliance: Factor 3 showed the benefit of external capital funding. National authorities should continue to facilitate municipalities in accessing EU funds and other grants, as it allows investment without local fiscal strain. However, planning is needed to ensure these investments are sustainable. A policy idea is to require a sustainability impact assessment before large projects: e.g., if you build a new cultural center with EU money, can you afford its staffing and maintenance later? This could be a criterion in approving projects for co-financing. Also, to avoid communities becoming too reliant on cyclical funds, perhaps set aside a portion of windfall revenues in a capital reserve. For instance, the 37.7% jump in IMT in 2021cfp.pt might not last; municipalities could be encouraged (or required) to reserve a part of such volatile revenue for capital investment or debt reduction, rather than basing recurring spending on it. This builds resilience for when those revenues normalize (which by 2023, with interest rates up, they likely have slowed). The results also suggest that targeted investment in infrastructure was prevalent; ensuring oversight on these projects (to prevent cost overruns which could lead to debt) is a necessity – perhaps through regional coordination by CCDR (regional development coordination commissions) in Portugal, which already help channel EU funds.
4. Tailor Support by Municipality Size: Recognizing Factor 4, policy should not be one-size-fits-all. Large municipalities might benefit from more autonomy in certain financial decisions given their capacity (some argue for losing certain controls for well-performing large cities, allowing them to issue municipal bonds or PPPs responsibly to finance big projects). Our findings show big cities have higher own revenue; maybe incentivize them to use that revenue for local investment, while central funds focus on smaller municipalities. Small municipalities, on the other hand, might need continued equalization transfers and even increased technical support. For example, a small municipality might lack an in-house financial analyst to navigate complex funding, so expanding technical advisory services (perhaps via inter-municipal communities or the DGAL) could help them optimize finances (like better procurement, which can reduce costs and help Factor 2). The negative correlation with transfer dependency means, if possible, helping small towns develop some local revenue (maybe tourism or some niche) can improve their independence. That might involve economic development policies targeted to those areas, aligning with broader regional policy. However, realistically many will remain dependent, so ensuring the transfer system remains predictable and fair is key. The OECD's Government immediately 2021 [27] highlights that predictable intergovernmental frameworks help local planning, something Portugal's Local Finance Law attempts to do. Our results justify maintaining stability, especially for small entities.
5. Early Warning System Integration: The authors could use our factor model as a basis for an early warning system for municipal financial distress. Building on works like Zafra-Gómez et al. [1] who developed alert modelseconpapers.repec.org, Portuguese authorities could establish a scoreboard where each municipality is scored on factors

like 1 and 2 (and perhaps a combination including 4 to adjust for size). A composite "distress risk index" could flag municipalities that exceed certain thresholds. Indeed, Zafra-Gómez et al. [1] model was simple for audit bodieseconpapers.repec.org; our approach could be simplified to a few key ratios under each factor as indicators. For example: Debt per capita, Interest-to-revenue, Operating balance, Rigid cost ratio, all compared to benchmarks. Those consistently in the red on these would get flagged for review. This aligns with recommendations from international practices (e.g., some countries have fiscal rules and also monitoring committees for local finance).

6. Capacity Building and Governance Improvements: The link between factors and governance, while indirect here, should not be forgotten. To the extent that some municipalities struggle due to poor financial management or political incentives (as found by Lobo et al.) [22], continuing to push for transparency and citizen engagement might improve accountability. For instance, if a municipality must publicly explain its deficit (Factor 2 issue) to citizens, there might be pressure to correct it. Tools like participatory budgeting (already used in some Portuguese cities) can ensure spending aligns with citizen priorities and perhaps curb fewer effective expenditures. Also, given that Factor 2 can be influenced by inefficiencies, anti-corruption measures and audits focusing on local spending can indirectly improve financial outcomes. Jorge et al. [12] found issues with over-optimistic revenue budgeting in Portuguese municipalities; addressing such practices by requiring more realistic budget forecasts or penalizing large deviations might help align budgets with actual performance (hence improving Factor 2 outcomes).

In essence, our analysis provides evidence to support a mix of fiscal rules (to constrain bad behavior) and fiscal support (to aid those in genuine need). The year 2021 showed that when push comes to shove, central support is crucial (no municipality collapsed financially because the central government and EU propped them up). But long-term, reliance solely on that isn't sustainable; local governments must maintain prudent practices. Also, Table 1 encapsulates variables for all municipalities; policymakers can use it as a reference to see broad trends. For example, if average rigid cost is, say, 55% (from Table 1 stats) and some are 80%, those outliers are where interventions could focus. The table's implication is that the entire "universe" is captured, so solutions can be municipality by municipality.

While the research provides a robust analysis of the municipal financial landscape in Portugal, there are several limitations that should be acknowledged:

- Single-Year Focus: The authors deliberately focused on the year 2021 to capture the COVID impact. This cross-sectional snapshot is powerful for our specific aims, but it does not capture trends over time. Financial distress is often a gradual process, and a one-year analysis might miss trajectory nuances. For example, a municipality might have a surplus in 2021 (good Factor 2) but had deficits for five years prior and might revert to deficit. Future research could extend this by performing a panel factor analysis.
- Exclusion of Qualitative Factors: Our analysis is quantitative. The authors did not directly include qualitative aspects like governance quality, leadership, political factors, or even quality of service provision. Some financial distress might be politically

driven (spending spree before elections). The authors did not incorporate an election cycle or political variables. Future research might integrate political economic variables

- Data Limitations: Despite using official data, there were some data issues, for example missing values for 2022 and 2023. Additionally, the authors did not separate the Azores and Madeira municipalities (they have slightly different finance regimes) – our data mostly covered mainland because the local finance law the authors reference is for mainland. Including autonomous regions might add complexity. For consistency, future studies could either explicitly compare mainland *versus* autonomous region municipality factors or confine to one regime.
- Factor Interpretability and Simplification: the authors chose varimax rotation and nice labels for factors, but real-world finances can be disordered. Some variables did not load "cleanly" perhaps (though the authors suppressed small loadings in Table 1 for clarity). There might be hybrid factors (e.g., a moderate correlation between debt and size – big cities can have big accounts just because scale). Future research could experiment with oblique factor analysis or other dimensional reduction like cluster analysis (grouping similar municipalities) to complement our factor approach. A clustering might directly identify groups like "high debt large cities" versus "low debt small towns," which could be another way to categorize distress profiles.
- Scope limited to Portugal: While our findings align with international literature at points, every country's local finance context is different. The fact that the authors found factors like what Spanish studies found suggests some generalizability, but caution is warranted. In countries where local taxation or autonomy is different, factor structures might vary. For example, in the U.S., municipal bankruptcy is a real thing, so factors might include pension liabilities, which in Portugal are centrally managed. Future research could apply a similar factor methodology to other jurisdictions to compare.
- COVID-Specific Effects: Our analysis implicitly attributes many 2021 patterns to COVID (which is logical), but there could be other coincident events, for example, 2021 was also an election year for municipalities. Election years often see higher spending as incumbents try to please voters, possibly contributing to deficits or capital projects. Some of Factor 2 deficits or Factor 3 projects might be partially explained by election cycle spending rather than COVID need alone. The authors did not explicitly control that.
- Outcome Validation: the authors identified risk factors but did not track outcomes like actual instances of financial distress. Ideally, one would validate the factor model by showing that municipalities scoring poorly on factors indeed had subsequent trouble (or vice versa). Since not much time has passed, we could use past data, for example, use factors in 2018 and see if they predicted who ended up under recovery by 2021, to validate the model's predictive power. The authors didn't do that due to our focus on the present, but it is a logical extension.
- Integration with AI models: Our introduction mentioned AI and an ambition for an AI-driven model. While our analysis is not a machine learning model per se, it creates a foundation. Future work might use these factor scores as inputs into predictive algorithms (like training a model to predict distress events using factor scores instead of raw ratios, which could be more stable). Zhao et al. [28] applied neural networks

to local government risk in China. Similar analysis could be applied for Portuguese municipalities using these factors to simplify the feature space. This would tie back to the original concept of advancing AI in public financial management by providing data-driven features to such models.

5 Final Considerations

The paper aims to contribute significantly to both academic research and political mayors in financial distress management. While advancing theoretical understanding through the integration of AI techniques with financial and behavioral models, it also seeks to address potential challenges that could arise during implementation. These challenges may include resistance from stakeholders unfamiliar with AI-driven tools, data quality issues that impact on predictive accuracy, and the need to ensure model scalability across diverse industries.

To mitigate these risks, the paper incorporates stakeholder discussion to establish rigorous data validation processes and adopt a modular design approach for the model. These measures enhance the robustness of the EFA but also ensure its practical applicability in real-world settings, providing organizations with a reliable solution for real-time crisis management and decision-making.

For municipal leaders, the findings have direct practical implications. First, the results underscore the importance of Prudential Financial Management: keeping debt at reasonable levels and avoiding operational deficits are not just accounting ideals but concrete practices that ensure long-term viability. Municipal leaders must treat debt and deficit metrics as key performance indicators and ensure compliance with such rules can greatly reduce distress incidence.

For policymakers (including the central government, Court of Auditors and other Municipal Authorities), the findings highlight the value of adopting AI-based analytical tools in detecting financial distress. An early warning system for municipal financial distress could be developed using the key indicators identified.

For citizens and society, the management of municipal financial distress is crucial. When a municipality falls into (economic and financial) crisis, the consequences can include cuts to public services, delayed infrastructure maintenance and layoffs of municipal employees, among others.

Also relevant are the ethical considerations. The research's algorithms have been designed to ensure transparency and accountability, avoiding biases in data analysis and decision-making of the municipality. Each municipality analysis is publicly available, but data has been stored securely, complying with GDPR [29] and other relevant regulations [30]. Access to data research prioritize inclusivity, ensuring affordability and usability for Municipalities and stakeholders (citizens, politics, communities and others). The feedback guides by the ethical and deontological principles of accountants and certified auditors. The planning of research has been developed with continuous improvement.

In conclusion, this paper contributes to a detailed and timely analysis of municipal financial distress. It confirms many tenets of public finance theory (the importance of debt and balanced budgets) and provides nuance (the role of investment and size) in the specific setting of Portuguese local governments during a global crisis. For future research, the authors aim to expand the longitudinally the database, integrating qualitative dimensions, and testing interventions that might emerge from our recommendations.

Acknowledgement. The authors aim to thank FCT and FCCN for the Research Project of Advance Computer nº 2024.18370.CPCA.A0 and the anonymous reviewers for comments and improvements of this research.

References

1. Zafra-Gómez, J., López-Hernández, A., Hernández-Bastida, A.: Developing an alert system for local governments in financial stress (II): empirical verification. Public Money Manage. **29**(4), 247–252 (2009)
2. Gavurová, B., Jenčová, S., Bačík, R., Miškufová, M., Letkovsky, S.: Artificial intelligence in predicting the bankruptcy of non-financial corporations. Oeconomia Copernicana. **13**, 1215–1251 (2022)
3. Kanász, R., Gnip, P., Zoričák, M., Drotár, P.: Bankruptcy prediction using ensemble of autoencoders optimized by genetic algorithm. PeerJ Comput. Sci. **8**(9), e1257 (2023)
4. Altman, E.I.: Financial ratios, discriminant analysis and the prediction of corporate bankruptcy. J. Financ. **23**(4), 589–609 (1968)
5. Ohlson, J.A.: Financial ratios and the probabilistic prediction of bankruptcy. J. Acc. Res. **18**(1), 109–131 (1980)
6. Barboza, F., Kimura, H., Altman, E.I.: Machine learning models and bankruptcy prediction. Expert Syst. Appl. **83**, 405–417 (2017)
7. Jones, S., Johnstone, D., Wilson, R.: Predicting corporate bankruptcy: an evaluation of alternative statistical frameworks. J. Bus. Financ. Acc. **44**(1–2), 3–34 (2017)
8. Groves, S.M., Godsey, W.M., Shulman, M.A.: Financial indicators for local government. Public Budg. Financ. **1**(2), 5–19 (1981)
9. Cabaleiro, R., Buch, E., Vaamonde, A.: Developing a method to assess the municipal financial health. Am. Rev. Public Adm. **43**(6), 729–751 (2013)
10. Turley, G., Robbins, G., McNena, S.: A framework to measure the financial performance of local governments. Local Gov. Stud. **41**(3), 401–420 (2015)
11. Zafra-Gómez, J., López-Hernández, A., Hernández-Bastida, A.: Developing an alert system for local governments in financial crisis. Public Money Manage. **29**(3), 175–181 (2009)
12. Jorge, S., Cerqueira, P., Furtado, S.: Municipal revenue over-budgeting: a dynamic analysis of its determinants. Local Gov. Stud. **49**(3), 644–675 (2022)
13. Jarmulska, B.: Random forest versus logit models: which offers better early warning of fiscal stress? J. Forecast. **41**(3), 455–490 (2022)
14. Chen, D.: Risk assessment of government debt based on machine learning algorithm. Complexity **2021**, 3686692 (2021)
15. Navarro-Galera, A., Rodríguez-Bolívar, M., Alcaide-Muñoz, L., López-Subires, M.: Measuring the financial sustainability and its influential factors in local governments. Appl. Econ. **48**(41), 3961–3975 (2016)
16. Shon, J., Junghack, K.: The impact of revenue diversification on municipal debts: comparing short-term and long-term debt levels. Local Gov. Stud. **45**(1), 241–261 (2019)
17. Directorate-General for Local Authorities (DGAL). Portuguese local government finance information system. Lisboa: DGAL (2025)
18. Hair, J.F., Black, W., Babin, B., Anderson, R.: Multivariate data analysis: a global perspective. Pearson Educ. (2023)

19. Araujo, J., Tejedo-Romero, F.: Local government transparency index: determinants of municipalities' rankings. Int. J. Public Sect. Manag. **29**(4), 327–347 (2016)
20. Cabasés, F., Pascual, P., Vallés, J.: The effectiveness of institutional borrowing restrictions: empirical evidence from Spanish municipalities. Public Choice **131**(3–4), 293–313 (2007)
21. Benito, B., Guillamón, M.-D., Bastida, F.: Non-fulfilment of debt limits in Spanish municipalities. Fisc. Stud. **36**(1), 75–98 (2015)
22. Lobo, F., Ramos, P., Lourenço, O.: Causes of financial distress of Portuguese municipalities: empirical evidence. Int. J. Monetary Econ. Financ. **4**(4), 390–409 (2011)
23. Santos, P.G., Albuquerque, F.: Does the Average Payment Period play a relevant role in explaining the Portuguese municipalities' financial distress? Economies **11**(7), 183 (2023)
24. Rodríguez Bolívar, M., Navarro-Galera, A., Alcaide Muñoz, L., López-Subirés, M.: Factors influencing local government financial sustainability: an empirical study. Lex Localis **12**(1), 31–54 (2014)
25. Rodríguez Bolívar, M., Navarro-Galera, A., Alcaide Muñoz, L., López-Subirés, M.: Risk factors and drivers of financial sustainability in local government: an empirical study. Local Gov. Stud. **42**(1), 29–51 (2016)
26. Auditor General of Canada: 2015 Fall Reports of the Auditor General of Canada. Office of Auditor General of Canada, Ottawa (2015)
27. OECD. OECD's Government at a Glance 2021 highlights. Paris: OECD (2021)
28. Zhao, Z, Dong, M., Bian, Q.: BP neural network algorithm based on big data for monitoring and early warning of China's public finance. Math. Probl. Eng. ID 2780122 (2022)
29. European Union (EU). Regulation (EU) 2016/679. General Data Protection Regulation. Official J. L119 (2016)
30. European Union (EU). Regulation (EU) 2024/1689. EU AI Act: first regulation on artificial intelligence. (2024)

Does Regional Heterogeneity Impact Default Probability? the Case of Portuguese Hotels

Inês Lisboa(✉) and Magali Costa

CARME—Centre of Applied Research in Management and Economics, School of Management and Technology, Polytechnic of Leiria, Campus 2-Morro Do Lena, Apartado 4163, Leiria 2411-901, Portugal
ines.lisboa@ipleiria.pt

Abstract. Companies should identify in advance determinants that impact their financial situation, to promote financial sustainability, and avoid situations of bankruptcy. The literature has shown that the probability of default is conditioned by several factors that go beyond companies' specific characteristics. This work aims to understand whether the region where the company is located and its economic development influence its probability of default. An unbalanced panel data of 1,205 hotels located in Portugal, over the period 2012–2022 is analyzed, using the Logit model. Results show that companies' location influences their financial constraints. Moreover, regional economic development, more precisely the regional gross domestic product growth, and the regional unemployment rate impact companies' default probability. Additionally, companies' liquidity, leverage, size, age, and asset structure are also determinants to explain the probability of default. These findings help to explain regional disparities and the concentration of some industries in a specific region.

Keywords: Default risk · hotels · regions · Logistic regression · Portugal

1 Introduction

Default risk has been an area that deserves to be addressed in the literature since 1960. It can be traced back to the works of Beaver (1966), Altman (1968), Ohlson (1980), and Zmijewski (1984). Over the past half century, the literature has attempted to understand the best model for predicting default risk and what determines it. Much of the empirical literature has focused on justifying the financial performance with firms' factors. Nonetheless, the literature has demonstrated that the context, namely the macroeconomic context, where the company operates influences the success of organizations (e.g., Kim & Gu, 2014; Fernández-Gaméz et al., 2016; Gémar et al., 2016; Cathcart et al., 2020; Abidin et al., 2021; García & Miguélez, 2021; Costa et al., 2022). However, the company is not only influenced by macroeconomic factors but also by regional differences. According to López-Bazo and Motellón (2018), the empirical analysis of the regional impact is an under-explored field. Where the company is located, its access to financial institutions, customers, and/or suppliers, its surroundings, and regional development can influence the way the company operates. To the best of our knowledge, the regional influence on default probability has not yet been addressed.

L. Almeida et al. (Eds.): ICAFI 2025, ISEM 80, pp. 270–283, 2026.
https://doi.org/10.1007/978-3-032-16268-7_15

Every country presents singularities, but within the same country, several differences exist. The theme of regional heterogeneity has been gaining prominence in recent years as it helps to explain why companies and persons are concentrated in some geographical spaces (Brenner & Niebuhr, 2021). This fact creates imperfect competition across companies, which can influence their financial situation (Ascani et al., 2021). The impact of regional heterogeneity has been addressed in other areas of corporate finance, such as capital structure (e.g. Palacín-Sánchez et al., 2013, Di Pietro et al. 2019), firms' innovation and productivity (Zhang et al. 2012), and performance (López-Bazo & Motellón, 2018). The literature presents empirical evidence for the impact of regional development on companies performance (López-Bazo & Motellón, 2018). As it conditions companies performance, it will also have an impact on their ability to meet financial obligations. Therefore, companies' probability of default can also be explained by regional economic development.

This study aims to expand the financial literature on default risk by analyzing the impact of regional factors on the default probability of companies located in the same country. This impact is special relevant for the tourism industry since the impact that tourism companies can have on local development depends on the facilities that surround a travel experience (Santos & Vieira, 2020).

The study focuses on hotel establishments with a turnover higher than € 500,000, to include only hotels with similar business models. This sector is of particular interest as 1) there are hotels/firms located over the seven Portuguese regions: Nort, Centre, Lisbon, Alentejo, Algarve (located in the continent), Azores and Madeira (islands–autonomous regions), which allows the analysis of regional influence; 2) the accommodation sector had an average mortality rate (12.4%) higher than the total average of Portuguese firms (11.8%) (Pordata, 2022); 3) the tourism sector has a high contribution to countries economic wealth and employment (Pordata, 2022); 4) the tourism industry is highly affect by the economic growth (García & Miguélez, 2021); 5) tourism companies depend on the surrounding (Santos & Vieira, 2020); and 6) guests prefer hotel establishments (INE, 2022), which justifies concentrating the analysis on this sub-sector.

Companies are classified as compliant or default using an ex-ante criterion (i.e., based on a set of financial ratios). This criterion allows for the detection of signs of financial difficulties in advance, helping managers to make timely decisions to promote the company's financial stability. To analyse the regional impact on the default probability, we included a variable that identifies the region where the company is located (following e.g., Gémar et al., 2019), the regional gross domestic product growth, the regional unemployment rate, and the regional firms' mortality rate (instead of these variables for the country, as for example, Cathcart et al, 2020; Costa et al., 2022). Additionally, following previous literature, we also include companies' specific variables (e.g., cash flow, liquidity, profitability, leverage, efficiency, size, and age) and specific characteristics of hotel establishments (e.g., growth opportunities, asset structure, and airport distance), and macroeconomic variables (e.g., inflation rate). Finally, the Logit model is estimated.

This work adds new knowledge to the literature by examining the impact of regional economic development on companies' probability of default, a theme that was not previously explored, as far as we know. Contributions are also added to practice. The government can understand that to minimize the greater discrepancy and heterogeneities among regions, it should create incentives and policies to attract companies to less developed regions. This will help to fix companies and persons, avoiding regional asymmetries.

The paper is structured as follows: after this introductory section, which presents the objective and relevance of the study, the next section presents the main literature review. The model, sample, and variables used are explained in Sect. 3, followed by the empirical results in Sect. 4. The paper concludes with the main conclusions and future research directions.

2 Literature Review

A default situation is defined as the company's difficulty in fulfilling its commitments (Beaver, 1966). It is a temporary situation, but if not solved, it can lead to bankruptcy (Tinoco & Wilson, 2013). Therefore, it is important to identify signs to prevent this situation and promote companies' financial sustainability.

Some works define that a company is in default based on its legal situation (ex-post criterion) (e.g., Altman, 1968; Kim & Gu, 2014; Fernández-Gámez et al., 2016). However, it depends on the laws of every country, and it can change over the years, not allowing for comparisons (Tinoco & Wilson, 2013). Other studies use a set of financial criteria to identify a company's financial difficulties (ex-ante criterion). This criterion allows for the early identification of financial problems, which, if corrected on time, can avoid situations of bankruptcy (e.g., Abinzano et al., 2020; Costa et al., 2022). Some studies use a mix of both criteria to have an overall picture of financial constraints (e.g., Pindado et al., 2008; Ashraf et al., 2019 and 2020).

The literature on default extends in several directions, on the one hand in an attempt to present the best prediction model and on the other in identifying determinants of default. Regarding the predictive model, previous studies on default risk focused essentially on the use of parametric model, such as multivariate discriminant analysis (MDA), logit, and probit (Beaver, 1966; Altman, 1968; Ohlson, 1980; Zmijewski, 1984). Later on, non-parametric models, such as Support Vector Machine (SVM) and Artificial Neural Network (ANN) were used to support the default risk analysis (Zieba et al., 2016; Jones et al., 2017; Du et al. 2020). Although, there is still no consensus about the model to predict default (Ashraf et al., 2020; Costa et al., 2022).

More recently, researchers focused on identifying the best determinants to explain default probability, exploring different models and contexts (Barboza et al., 2017). The most common determinants are related to the internal characteristics of companies, such as cash flow, liquidity, profitability, leverage, size, and age (e.g., Abinzano et al., 2020; Costa et al, 2022). Companies with more liquidity, cash flow, and profitability usually have more possibilities to cover their covenants, contributing to decreased situations of default, whether more indebted companies are more prone to fail their commitments, leading to situations of default (e.g., Fernandez-Gaméz et al., 2016; Gémar et al., 2016; Vivel Búa et al., 2018; Abidin et al., 2021). For specific samples, other variables can be added. For instance, tourism companies are highly affected by seasonality and high fixed cost structure (Wieprow & Gawlik, 2021). Therefore, companies' specific characteristics such as growth opportunities (e.g. Vivel-Búa et al., 2018), asset structure (Gu, 2002; Fernandez-Gaméz et al., 2016; Abidin et al., 2021; García & Miguélez, 2021), and the distance to the airport (Gémar et al., 2016 and 2019) have been also included in default probability of tourism companies. Establishments with more opportunities to grow and more near the airport have more chances to increase their business and reduce financial constraints.

As companies operate in a macroeconomic context, their factors also influence the success of the organizations (e.g., Kim & Gu, 2014; Fernandez-Gaméz et al., 2016; Gémar et al., 2016; Cathcart et al., 2020; Abidin et al., 2021; García & Miguélez, 2021; Costa et al., 2022). In addition to the internal determinants of company performance, factors specific to the location in which the company operates can influence both its innovative behaviour, productivity and performance, in particular the existence of a suitable labour force, a supportive industrial environment, a favourable institutional framework and the availability of local infrastructure (López-Bazo & Motellón, 2018).

The relationship between company development and the environment in which it operates is particularly evident in the tourism sector. Almeida et al. (2021) found that the number of tourists differs among the Spanish regions. Moreover, Gémar et al. (2016 and 2019) argue that the Spanish hotel's survival depends on its location. The authors suggest that companies that are located more distant from airports are more prone to present situations of financial constraints. Fernández-Gámez et al. (2016), and Vivel-Búa et al. (2018) also confirm that the region where the hotels are located impact default risk.

Tourism companies depend on the facilities that surround a travel experience. It is more convenient for tourists to stay where they have access to more opportunities to enrich their travel, and it also contributes to cost savings (Santos & Vieira, 2020). As these facilities increase, the number of tourists increases too, and with that tourism companies can benefit from a better financial situation. This fact can justify why there is a spatial agglomeration of tourism activities in some regions (Santos & Vieira, 2020). Andraz et al. (2015) argue that tourism companies are not only relevant at a macroeconomic level but also at a regional level. Tosun et al. (2003) add that tourism development can contribute to reducing regional asymmetries.

Some studies found that regional economic development impacts companies' opportunities and their access to financial sources (e.g., Palacín-Sánchez et al., 2013, Di Pietro et al. 2019). Companies located in different regions have singular investment opportunities (Di Pierto et al., 2019). These factors can explain why some regions have more companies and persons than others, justifying regional disparities (Mahadevan et al., 2017). According to Butzbach and Sarno (2019), Italy, Spain, and Portugal are the three regions with greater differences among regions concerning its economic development. Likewise, studying the regional development impact on default risk of Portuguese companies is relevant.

3 Methodology

3.1 Sample

Portugal is a European country with seven regions: North, Centre, Lisbon, Alentejo, Algarve (located in the continent), Azores, and Madeira (islands–autonomous regions). With more than 3,000 h of sunshine, different landscapes a short distance away, unique cultural heritage, and excellent cuisine, Portugal attracts visitors from all over the world (Visit Portugal). Portugal is one of the most visited destinations. It won the Europe's Leading Destination award in the years 2022–2023, 2017–2020, and the World's Leading Adventure Tourism Destination award in 2023 (World Travel Awards).

Tourism companies are located in all Portuguese regions, and tourists can benefit from art and culture, outdoor activities, gastronomy, golf, nature, sun and sea, nautical tourism, among other (Visit Portugal). However, there are significant disparities among the regions in the country, and spatial distribution of tourism flows (Santos & Vieira, 2020). Most companies are located in the Lisbon region, where the Portuguese capital is located, followed by the North region, where Oporto, the second-highest Portuguese city, is located (Pordata). These two regions have airports and ports, making it easier to receive tourists and for import and export. Regional development is also singular across regions (Pordata), which can justify the agglomeration of tourism activities in some regions. According to Almeida and Silva (2021), tourism is more prominent in the regions of Lisbon, Algarve, Azores, and Madeira.

An unbalanced panel data of 1,205 hotels located in Portugal, over the period 2012–2022 is analysed.

3.2 Variables

The dependent variable is a dummy variable that takes the value one if the firm is in a default situation, and zero otherwise. A set of financial ratios–ex-ante criterion-was used to classify companies as default or compliant. This criterion is less explored in the context of tourism sector. Although it allows us to detect signs of financial difficulties in advance helping managers to take timely decisions to promote the company's financial stability. Following Lisboa et al. (2021) and Costa et al. (2022), a company is in a compliant situation when capital ratio (total equity divided by total assets) is higher than 5%, earnings before interests, taxes, depreciations and amortizations (EBITDA) divided by financial expenses is greater than 1.3, and financial debt divided by EBITDA is higher than 0 and less than 10. The ratios are analyzed in the three previous years, each ratio should fulfill the criterion at least once in the three years and at least half of the ratios should be fulfilled in each year. The other companies are in a default situation.

As explanatory variables we include regional variables and control variables. To analyze the regional impact on the default probability, following Gémar et al. (2016 and 2019) we include a variable that identifies the region where the company is located. It is one if the company is located in the North region, 2 – Centre, 3 – Lisbon, 4 – Alentejo, 5 – Algarve, 6 – Madeira, 7 – Azores. Additionally, to understand the impact of regional heterogeneities, we included: the regional gross domestic product (GDP) growth, the regional unemployment rate, and the regional companies' mortality rate. These variables are usually included regarding the country analyzed (e.g., Cathcart et al, 2020; Costa et al., 2022). Although, as regions present singular economic development within the same country, we have added these regional variables instead of country variables.

Some control variables were also included, namely regarding companies' specific variables and other macroeconomic variables. In the group of companies' specific variables, following previous literature on default (e.g., Pindado et al., 2008; Ashraf et al., 2019; Abinzano et al., 2020; Abidin et al., 2021; García & Miguélez, 2021; Wieprow & Gawlik, 2021; Costa et al., 2022) we added cash flow (operational cash flow divided by total liabilities), liquidity (current assets divided by current liabilities), profitability (net income divided by turnover), leverage (total liabilities divided by total assets), efficiency (employees cost divided by turnover), size (natural logarithm of turnover), and age (natural logarithm of the firm's age). We also added specific characteristics of hotel establishments (e.g., Kim & Gu, 2014; Fernández-Gámez et al., 2016; Gémar et al., 2016 and 2019; Vivel-Búa et al., 2018), such as growth opportunities (annual growth of earnings before interest and taxes), asset structure (fixed assets divided by total assets), and airport distance (natural logarithm of airport distance). As a macroeconomic variable (e.g., Tinoco & Wilson, 2013; Costa et al. 2022), only the inflation rate was included as GDP growth was included as a regional variable.

3.3 Model

Following Pindado et al. (2008) and Costa et al. (2022), panel data is used as it allows for control unobservable heterogeneity and eliminates the bias of omitted variables. The model is estimated using the Logit model, which allows for the variation in time covariates and adjusts for unobservable heterogeneity and is estimated using robust standard errors.

$$P = \left(1 + \exp\left\{-\left[\alpha + \sum \beta_m \times \text{Regional}_{i,t} + \sum \gamma_n \times \text{Internal Characteristics}_{i,t} + \sum \delta_p \times \text{Hotel Characteristics}_{i,t} + \theta_l \times \text{Macroeconomic}_{i,t}\right]\right\}^{-1}\right) \tag{1}$$

4 Results

4.1 Sample Characterization

Figure 1 presents the percentage of companies in the sample in default and compliant by region.

The percentage of compliant companies is higher than the percentage of defaulters in all regions. Although some singularities emerge. Alentejo is the region with more cases of companies with financial difficulties, while Azores is the region with more cases of companies with a better financial situation. These results suggest that the region where the company is located can influence the probability of default.

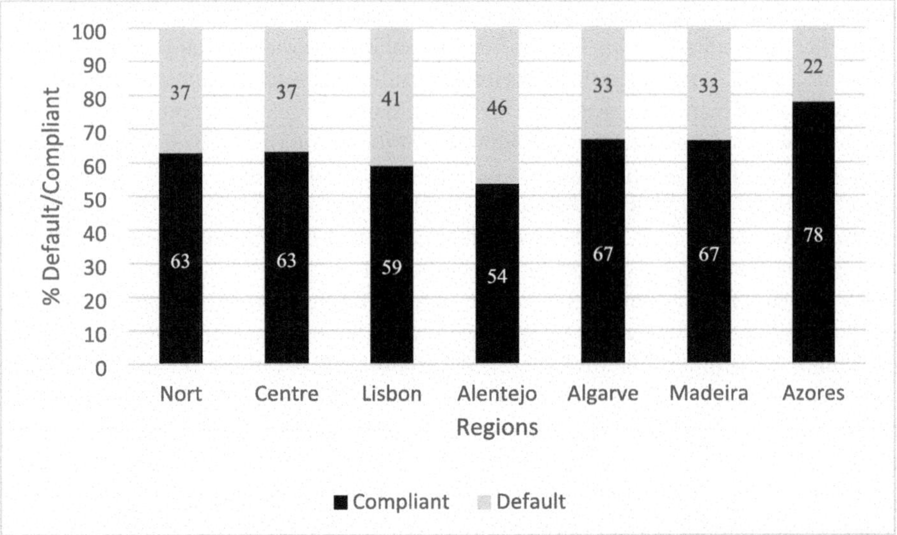

Fig. 1. Companies' characterization per region

4.2 Descriptive Statistics

Table 1 shows the descriptive statistics: mean, median, standard deviation, minimum, and maximum, for the total sample and for both groups: default and compliant. The Kolmogorov-Smirnov test was performed, and results show that the variables are not normally distributed. Therefore, to compare both groups, we performed Mann–Whitney nonparametric test (MW) to understand whether the medians for both groups are statistically equal.

Results suggest that Portuguese hotels are in all regions. In median, regions present economic growth, but also high levels of unemployment and a high mortality rate. Cash flows are around 0, suggesting few operational cash flows over total debt. Companies present more current assets than current liabilities, since liquidity is higher than 1. Companies are profitable and efficient. In median, 60.41% of total assets are financed through liabilities. The younger company in the sample is 3 years old, while the older is 127 years. Companies' growth opportunities decreased, which can be due to the pandemic situation caused by COVID-19. Most of the assets are fixed assets. The distance to the airport is low, but Portugal is a small country, which can explain this result. Finally, the inflation rate was around 0 in the period analyzed.

Table 1. Descriptive Statistics

Variable	Sample	Mean	Median	Standard Deviaton	Minimum	Maximum	MW test
Region	Total	**3.3546**	**3.0000**	**1.7110**	**1.0000**	**7.0000**	-
	Default	3.2349	3.0000	1.6104	1.0000	7.0000	0.0000
	Compliant	3.4251	3.0000	1.7639	1.0000	7.0000	
Regional GDP	Total	**0.0356**	**0.0429**	**0.0639**	**−0.1672**	**0.2129**	-
	Default	0.0320	0.0392	0.0612	−0.1672	0.2129	0.0000
	Compliant	0.0377	0.0447	0.0654	−0.1672	0.2129	
Regional Unempl	Total	**10.1970**	**8.3000**	**4.1296**	**5.3000**	**20.9000**	-
	Default	10.7518	8.6000	4.3027	5.3000	20.9000	0.0000
	Compliant	9.8705	7.9000	3.9887	5.3000	20.9000	
Regional Mortality rate	Total	**12.3834**	**12.1000**	**1.6927**	**9.7000**	**16.8000**	-
	Default	12.5292	12.3000	1.7192	9.7000	16.8000	0.0000
	Compliant	12.2977	12.1000	1.6711	9.7000	16.8000	
Cash Flow	Total	**0.0862**	**0.0000**	**0.4151**	**−9.9816**	**8.0357**	-
	Default	0.0278	0.0000	0.4100	−9.9816	8.0357	0.0000
	Compliant	0.1206	0.0000	0.4143	−6.3634	6.7550	
Liquidity	Total	**3.9326**	**1.1735**	**21.0906**	**0.0000**	**842.1527**	-
	Default	4.9435	0.9149	31.8625	0.0000	842.1527	0.0000
	Compliant	3.3394	1.3614	10.4498	0.0093	400.9425	
Profitability	Total	**−0.4019**	**0.0330**	**8.6576**	**−356.8349**	**256.6946**	-
	Default	−1.1122	-0.0432	13.3727	−356.8349	256.6946	0.0000
	Compliant	−0.0209	0.0640	4.3392	−143.5939	219.6550	
Leverage	Total	**0.6795**	**0.6041**	**1.0361**	**0.0000**	**54.7422**	-
	Default	1.0042	0.8604	1.6237	0.0000	54.7422	0.0000
	Compliant	0.4884	0.4931	0.2353	0.0046	1.0233	
Efficiency	Total	**0.6925**	**0.3074**	**11.7127**	**0.0000**	**774.3748**	-
	Default	1.1290	0.3226	19.2945	0.0000	774.3748	0.0000
	Compliant	0.4582	0.3029	3.2891	0.0000	193.9723	

(*continued*)

Table 1. *(continued)*

Variable	Sample	Mean	Median	Standard Deviaton	Minimum	Maximum	MW test
Size	Total	**15.2148**	**15.2684**	**1.5386**	**6.9102**	**19.8840**	-
	Default	15.2239	15.3573	1.7042	6.9102	19.8840	0.0032
	Compliant	15.2095	15.2187	1.4323	9.6156	19.8300	
Age	Total	**23.6046**	**20.0000**	**17.6592**	**3.0000**	**127.0000**	-
	Default	19.0318	14.0000	16.5167	3.0000	121.0000	0.0000
	Compliant	26.2961	23.0000	17.7572	3.0000	127.0000	
Growth Opportunities	Total	**0.2084**	**−0.1692**	**32.7261**	**-705.5685**	**1 310.7042**	-
	Default	0.2944	−0.2343	38.0163	-640.5928	1 290.2917	0.0188
	Compliant	0.1577	−0.1267	29.1732	-705.5685	1 310.7042	
Asset Structure	Total	**0.5609**	**0.6381**	**0.3326**	**0.0000**	**0.9990**	-
	Default	0.5522	0.6584	0.3680	0.0000	0.9986	0.2653
	Compliant	0.5660	0.6304	0.3099	0.0000	0.9990	
Airport distance	Total	**45.3996**	**22.0000**	**51.1874**	**0.0000**	**262.0000**	-
	Default	45.5543	16.0000	51.5660	2.0000	262.0000	0.5053
	Compliant	45.3085	22.0000	50.9669	0.0000	262.0000	
Inflation	Total	**0.0151**	**0.0060**	**0.0229**	**−0.0030**	**0.0780**	-
	Default	0.0146	0.0060	0.0224	−0.0030	0.0780	0.0019
	Compliant	0.0155	0.0060	0.0232	−0.0030	0.0780	

When comparing both groups, default and compliant firms present singular median values at a statistical level, except regarding asset structure and airport distance. Results suggest that the region where the company is located and its economic development influence the probability of default.

Default companies have less cash flow and liquidity, it means, less money to fulfill commitments. Moreover, those companies are less profitable, are more indebted, invest more in personal costs compared with turnover, and have fewer growth opportunities. Companies with more costs and less profits, and with less financial independence, present more risk and therefore have more probability of having financial difficulties (Kim & Gu, 2014; Abidin et al., 2021; García & Miguélez, 2021; Zhang & Xie, 2023; Costa et al., 2022).

Results also show that default companies are larger and younger. Older companies may have more difficulties in changing the way they operate, especially in downturn moments. Moreover, younger companies have fewer resources and experience, which can increase their probability of presenting financial difficulties (Zhang & Xie, 2023).

The correlation matrix and the VIF (Variance inflation factor) analysis were performed and results show that the variables are not highly correlated nor present higher VIF values (the greater is less than 5) which indicate that all variables can be used in the model.

4.3 Model Results

Table 2 reports the results of the model estimation.

Table 2. Model Results

Group of Variables	Variables	Results
	C	– 4.144 ***
Region	Region	– 0.047 ***
	Regional GDP	0.905 *
	Regional Unempl	0.062 ***
	Regional Mortality rate	– 0.003
Internal Characteristics	Cash Flow	– 0.166
	Liquidity	0.013 ***
	Profitability	– 0.011
	Leverage	3.911 ***
	Efficiency	– 0.001
	Size	0.103 ***
	Age	– 0.333 ***
Hotel characteristics	Growth Opport	– 0.001
	Asset Structure	– 0.274 ***
	Airport distance	– 0.005
Macroeconomic	Inflation	0.993
Sucess tax rate		79.4%

* 10% significance level; ** 5% significance level; *** 1% significance level.

Results suggest that the region where the hotel is located influences default probability. Tourism companies depend on the facilities that surround a travel experience, as tourists prefer to stay in regions with more opportunities to enrich their travel experience (Santos & Vieira, 2020). This explains why companies are mostly located in some geographical spaces, as larger cities (Brenner & Niebuhr, 2021). This result is similar to those obtain by Almeida et al. (2021), Gémar et al. (2016 and 2019), Fernández-Gámez et al. (2016), and Vivel-Búa et al. (2018).

Moreover, the region's economic growth and unemployment rate also influence default probability. In periods of high economic growth, hotels have greater probability of default. This result is contrary to the result found by Cathcart et al. (2020), but it can be explained since the period analyzed includes the COVID-19 impact. Despite the GDP growth seen in 2021 and 2022, according to the Bank of Portugal, the growth was essentially due to Gross Fixed Capital Formation (GFCF) with private consumption almost stagnant, which had a greater influence on GDP growth.

The higher regional unemployment rate increases the probability of default. More unemployment results in less development, which limits the financial situation of companies. We do not prove the contagious effect of bankruptcy, suggesting that an increase in firms' regional mortality rate does not mean an increase in hotels' financial difficulties.

Results also show that hotels with higher liquidity are less prone to present financial difficulties. Similar results were found by Costa et al. (2022) when analyzing Portuguese companies. This can be due to the fact that not all current assets are liquid, such as inventories, and therefore cannot be used to pay short-term debts.

Debt is also significant in explaining default probability. Gu (2002), and Kim and Gu (2014), suggest that for tourism companies, debt is one of the most important determinants of default. More indebted companies have more commitments to fulfill and more financial risk, which explains an increase in their financial difficulties.

Hotels with more fixed assets have a greater probability of default. To invest in fixed assets, companies usually need more financial resources, which causes more uncertainties. Moreover, companies with more fixed assets also present more fixed costs, making it more difficult to present profits.

Larger and younger companies are more prone to have financial problems. Larger companies have more difficulties in managing the business and monitoring employees, which increases their financial difficulties (Chancharat et al., 2010). Younger companies have less experience and resilience and less retained earnings, leading to more difficulties in paying their obligations (Altman, 1968).

Cash flow, profitability, efficiency, and growth opportunities do not impact the default probability of tourism companies at a significant statistical level. The distance to the airport also does not impact companies' financial problems, which can be explained as Portugal is a small country, and everything is near, with small distances to airports. The macroeconomic variable is also irrelevant in explaining default probability.

Finally, the model's success rate is 79.4%, which is smaller than those presented by the initial models (e.g., Altman, 1968; Ohlson, 1980), but similar to those that analyze Portuguese companies (e.g., Costa et al., 2022).

5 Conclusion

Companies' surroundings influence their business and financial situations. Although most works focus on macroeconomic impact, ignoring that within the same country several differences exist, which can explain the concentration of companies in a geographical space, and regional disparities (Ascani et al., 2021).

This work aims to understand whether the region where the company is located and its economic development influence its probability of default. For this an unbalanced panel data of 1,205 hotels located in Portugal, over the period 2012–2022, is analyzed.

Results show that the region where the hotel is located, and its economic development influence default probability. This is special relevant to tourism companies as tourists prefer to stay in places with more tourism experiences.

Results also show that hotels with higher liquidity and debt levels, more fixed assets, larger size, and younger, are more prone to have financial problems, i.e., more difficulties in fulfilling their commitments.

This work makes several contributions to the literature and to practice. First, it analyses the regional heterogeneity impact on default probability, a topic never addressed before, at least to our knowledge. The company location and its distance to the airport impact on default were previously analyzed by Almeida et al. (2021), Gémar et al. (2016 and 2019), Fernández-Gámez et al. (2016), and Vivel-Búa et al. (2018). Although regional development was not addressed before in this context. The companies' surroundings, their access to financial institutions, customers, and/or suppliers, can influence the way companies operate and their financial situation. Therefore, this study contributes to the debate on the theme of default risk.

Second, we explore a sector that is relevant to almost all countries and that make significant contributions to economic wealth and employment creation. Understanding how hotels can promote their sustainability is a relevant topic.

The results are also relevant to practice. It helps in the decision-making process of hotels' managers, as they can understand which determinants are relevant to avoid financial problems and difficulties. The main conclusions are also relevant to the Government to understand how it can help some regions to increase their development, minimizing regional disparities.

Funding. This research was supported by National Funds of the FCT-Portuguese Foundation for Science and Technology within the project [UID/04928/2025].

References

Abidin, Z.J., Abdullah, N.A.H., Khaw, K.L.-H.: Bankrupcy prediction: SMEs in the hospitality industry. Int. J. Bank. Financ. **16**(2), 51–80 (2021). https://doi.org/10.32890/ijbf2021.16.2.3

Abinzano, I., Corredor, P., Martinez, B.: Does family ownership always reduce default risk? Acc. Financ. 1–36 (2020). https://doi.org/10.1111/acfi.12725

Almeida, A., Golpe, A., Justo, R.: Regional tourist heterogeneity in Spain: a dynamic spatial analysis. J. Destin. Mark. Manag. **21**, 100643 (2021)

Altman, E.I.: Financial ratios, discriminant analysis and the prediction of corporate Bankruptcy. J. Financ. **XCV**(9), 561–572 (1968)

Andraz, J.M., Norte, N.M., Goncalves, H.S.: Effects of tourism on regional asymmetries: empirical evidence for Portugal. Tour. Manage. **50**, 257–267 (2015)

Ascani, A., Faggian, A., Montresor, S.: The geography of COVID-19 and the structure of local economies: the case of Italy. J. Reg. Sci. **61**(2), 407–441 (2021)

Ashraf, S., Félix, E.G.S., Serrasqueiro, Z.: Development and testing of an augmented distress prediction model: a comparative study on a developed and an emerging market. J. Multinatl. Financ. Manag. **57**, 100659 (2020). https://doi.org/10.1016/j.mulfin.2020.100659

Ashraf, S., G.S.Félix, E., Serrasqueiro, Z.: Do traditional financial distress prediction models predict the early warning signs of financial distress? J. Risk Financ. Manage. **12**(2), 55 (2019). https://doi.org/10.3390/jrfm12020055

Barboza, F., Kimura, H., Altman, E.: Machine learning models and bankruptcy prediction. Expert Syst. Appl. **83**, 405–417 (2017). https://doi.org/10.1016/j.eswa.2017.04.006

Beaver, W.H.: Financial ratios as predictors of failure. J. Acc. Res. **4**, 71–111 (1966). http://www.jstor.org/stable/2490171

Brenner, T., Niebuhr, A.: Policy options for lagging regions—effects, new approaches and emerging challenges: introduction to the special issue. Rev. Reg. Res. **41**(2), 125–130 (2021)
Butzbach, O., Sarno, D.: To what extent do regional effects influence firms' capital structure? The case of Southern Italian SMEs'. Int. J. Financ. Stud. **7**(1), 3 (2019). https://doi.org/10.3390/ijfs7010003
Cathcart, L., Dufour, A., Rossi, L., Varotto, S.: The differential impact of leverage on the default risk of small and large firms. J. Corp. Finan. **60**, 101541 (2020). https://doi.org/10.1016/j.jcorpfin.2019.101541
Chancharat, N., Tian, G., Davy, P., McCrae, M., Lodh, S.: Multiple states of financially distressed companies: tests using a competing-risks model. Australas. Acc. Bus. Financ. J. **4**(4), 27–44 (2010)
Costa, M., Lisboa, I., Gameiro, A.: Is the financial report quality important in the default prediction? SME Portuguese construction sector evidence. Risks **10**(5) (2022). https://doi.org/10.3390/risks10050098
Di Pietro, F., Bontempi, M.E., Palacín-Sánchez, M.J., Samaniego-Medina, R.: Capital structure across Italian regions: the role of financial and economic differences. Sustainability **11**(16), 4474 (2019). https://doi.org/10.3390/su11164474
Du, X., Li, W., Ruan, S., Li, L.: CUS-heterogeneous ensemble-based financial distress prediction for imbalanced dataset with ensemble feature selection. Appl. Soft Comput. J. 97 (2020). https://doi.org/10.1016/j.asoc.2020.106758
Fernández-Gámez, M.Á., Cisneros-Ruiz, A.J., Callejón-Gil, Á.: Applying a probabilistic neural network to hotel bankruptcy prediction. Tourism Manage. Stud. **12**(1), 40–52 (2016). https://doi.org/10.18089/tms.2016.12104
García, A.del.C., Miguélez, S.M.F.: Predictive potential of the global bankruptcy models in the tourism industry. Tourism Manage. Stud. **17**(4), 23–31 (2021). https://doi.org/10.18089/TMS.2021.170402
Gémar, G., Moniche, L., Morales, A.J.: Survival analysis of the Spanish hotel industry. Tour. Manage. **54**, 428–438 (2016). https://doi.org/10.1016/j.tourman.2015.12.012
Gémar, G., Soler, I.P., Guzman-Parra, V.F.: Predicting bankruptcy in resort hotels: a survival analysis. Int. J. Contemp. Hosp. Manag. **31**(4), 1546–1566 (2019)
Gu, Z.: Analyzing bankruptcy in the restaurant industry: a multiple discriminant model. Hospitality Manage. **21**, 25–42 (2002). https://doi.org/10.1016/S0278-4319(01)00013-5
INE. Estatísticas do Turismo. Instituto Nacional de Estatísticas (2022)
Jones, S., Johnstone, D., Wilson, R.: Predicting corporate Bankruptcy: an evaluation of alternative statistical frameworks. J. Bus. Financ. Acc. **44**(1–2), 3–34 (2017). https://doi.org/10.1111/jbfa.12218
Kim, H., Gu, Z.: A logistic regression analysis for predicting Bankruptcy in the hospitality industry. J. Hospitality Financ. Manage. **14**(1), 24 (2014). https://doi.org/10.1080/10913211.2006.10653812
Lisboa, I., Costa, M., Santos, F.: Analysis of family smes default risk: the portuguese case. Australas. Account. Bus. Finance J. **15**(4), 76–92 (2021). https://doi.org/10.14453/aabfj.v15i4.5
López-Bazo, E., Motellón, E.: Innovation, heterogeneous firms and the region: evidence from Spain. Reg. Stud. **52**(5), 673–687 (2018)
Mahadevan, R., Amir, H., Nugroho, A.: Regional impacts of tourism-led growth on poverty and income inequality: a dynamic general equilibrium analysis for Indonesia. Tour. Econ. **23**(3), 614–631 (2017)
Ohlson, J.A.: Financial ratios and the probabilistic prediction of Bankruptcy. J. Acc. Res. **18**(1), 109 (1980). https://doi.org/10.2307/2490395
Palacín-Sánchez, M.J., Ramírez-Herrera, L.M., Di Pietro, F.: Capital structure of SMEs in Spanish regions. Small Bus. Econ. **41**, 503–519 (2013). https://doi.org/10.1007/s11187-012-9439-7

Pindado, J., Rodrigues, L., Torre, C.de.la.: Estimating financial distress likelihood. J. Bus. Res. **61**, 995–1003 (2008). https://doi.org/10.1016/j.jbusres.2007.10.006
Pordata: Base de Dados de Portugal Contemporâneo. Lisboa: PORDATA, Fundação Francisco Manuel dos Santos (2022)
Santos, L.D., Vieira, A.C.: Tourism and regional development: a spatial econometric model for Portugal at municipal level. Port. Econ. J. **19**(3), 285–299 (2020)
Tinoco, M.H., Wilson, N.: Financial distress and bankruptcy prediction among listed companies using accounting, market and macroeconomic variables. Int. Rev. Financ. Anal. **30**, 394–419 (2013). https://doi.org/10.1016/j.irfa.2013.02.013
Tosun, C., Timothy, D.J., & &# xd6; zt&# xfc; rk, Y.X.K.: Tourism growth, national development and regional inequality in Turkey. J. Sustain. Tourism **11**(2–3), 133–161 (2003)
Visit Portugal. Homepage https://www.visitportugal.com/en/. Last accessed 23 Mar 2025
Vivel-Búa, M., Lado-Sestayo, R., Otero-González, L.: Risk determinants in the hotel sector: risk credit in MSMEs. Int. J. Hosp. Manag. **70**, 110–119 (2018). https://doi.org/10.1016/j.ijhm.2017.11.004
Wieprow, J., Gawlik, A.: The use of discriminant analysis to assess the risk of bankruptcy of enterprises in crisis conditions using the example of the tourism sector in Poland. Risks **9**(4) (2021). https://doi.org/10.3390/risks9040078
World Travel Awards. Homepage https://www.worldtravelawards.com/profile-28112-turismo-de-portugal. Last accessed 02 April 2025
Zhang, D., Xie, J.: Influence of tourism seasonality and financial ratios on hotels' exit risk. J. Hospitality Tourism Res. **47**(4), 714–733 (2023). https://doi.org/10.1177/1096348021101603
Zhang, R., Sun, K., Delgado, M.S., Kumbhakar, S.C.: Productivity in China's high technology industry: regional heterogeneity and R&D. Technol. Forecast. Soc. Chang. **79**(1), 127–141 (2012)
Zieba, M., Tomczak, S.K., Tomczak, J.M.: Ensemble boosted trees with synthetic features generation in application to bankruptcy prediction. Expert Syst. Appl. **58**, 93–101 (2016). https://doi.org/10.1016/j.eswa.2016.04.001
Zmijewski, M.E.: Methodological issues related to the estimation of financial distress prediction models. J. Acc. Res. 59–82 (1984)

Advantages and Disadvantages of Formal Accounting Harmonization Processes

Joana Ventura[1(✉)] and Graça Azevedo[2]

[1] Department of Economics, Management, Industrial Engineering and Tourism (DEGEIT), University of Aveiro, Aveiro, Portugal
joanaventura@ua.pt

[2] Higher Institute for Accountancy and Administration–University of Aveiro (ISCA-UA), Research Centre on Accounting and Taxation–CICF, Aveiro, Portugal

Abstract. This paper reviews the existing literature on formal accounting harmonization processes, particularly emphasizing the initial adoption of the International Financial Reporting Standards (IFRS) in European countries. It explores the advantages and disadvantages of accounting harmonization, the challenges faced during its implementation, and the benefits it provides.

A literature review was conducted to achieve the proposed objective, analysing studies on the effects of accounting harmonization, particularly the adoption of IFRS, across different countries and jurisdictions. The review identifies the advantages and disadvantages associated with implementing these standards.

We conclude that the main advantage of accounting harmonization, as highlighted in the literature, is improving the quality of financial reporting. This results in more market-relevant earnings and timely loss recognition, given that IFRS standards are principles-based. However, the process also presents several disadvantages, such as increased operational costs, the need for specific software to support the more complex accounting system, and costs associated with employee training and recruitment. On the other hand, the benefits often include reduced operational costs for multinational corporations, lower capital costs, and increased foreign direct investments. The outcomes of accounting harmonization processes are generally influenced by the adopting country's legal system and the internal incentives of the adopting entities.

Keyword: Accounting Harmonization Processes · IFRS adoption · Operating costs

1 Introduction

Accounting harmonization processes required by national regulators in jurisdictions with strong or weak enforcement capabilities have been deeply analyzed by academic researchers and practitioners, Despite the efforts made by local and international institutions in the 1970s, no proper accounting harmonization was achieved at the time, which hinders transparency and comparability in financial reporting. In 2002, to strengthen the common market, the EU forced the adoption of IFRS/IAS by all listed entities

L. Almeida et al. (Eds.): ICAFI 2025, ISEM 80, pp. 284–295, 2026.
https://doi.org/10.1007/978-3-032-16268-7_16

with securities traded on a European stock exchange with the aim of ensuring that the same accounting practices were used across jurisdictions to produce financial information, since financial reporting users rely on accounting information to make economic decisions.

Accounting harmonization, as defined by Nobes and Parker (2002), is the process of aligning accounting practices, reducing the number of alternatives, and promoting the comparability of financial reporting. Applying a specific set of accounting practices across different jurisdictions enhances the transparency, consistency, and comparability of financial reporting.

Adopting a common framework to prepare financial reporting that limits the possibilities of accounting practices and principles is the most efficient alternative to reduce the options available for recognizing similar accounting transactions (Van der Tas, 1988; Guerreiro, 2006). Harmonized accounting practices eliminate economic inefficiencies, as they provide investors with a common framework for analyzing financial reporting, thereby reducing the effort required to develop investment analysis. At the same time, external auditors can also develop their work with more focus.

Multinational companies also benefit from lower operating costs, as different local operating units implement the same accounting practices. Additionally, international trade becomes less challenging, as suppliers and clients can interpret financial reports without incurring extra costs.

However, the accounting harmonization process typically incurs non-negligible implementation costs due to the impact on companies' overhead costs and the loss of flexibility when adopting standardized practices that may not suit all companies, thereby limiting the ability to reflect their unique circumstances in their financial reporting.

Despite uniform accounting standards, the implementation process adopted in each country can generate inconsistencies due to the country's legal system, varying enforcement practices, and local interpretation of accounting standards performed by local regulators and preparers.

This essay examines the advantages and disadvantages of accounting harmonization processes, their challenges, and the benefits they offer.

2 Accounting Harmonization

According to Van der Tas (1988, 1992), the harmonization process can be formal, when there are specific accounting standards/legislation legally bidding that must be used by companies to prepare financial reporting, both for recognition, measurement and disclosure purposes, or; material, when corporations voluntarily prepare financial reporting using a similar accounting framework, namely techniques and/or practices, both at the recognition, measurement and disclosure levels to allow the comparability of their financial reporting with its peers.

Comparability of financial reporting is only possible when similar single events and/or transactions are grouped and/or recognized using the same accounting methodologies/techniques (Kriesment, 1997) by different preparers.

Typically, a formal harmonization process will lead to a material harmonization process. Thus, a formal harmonization process tends to be perceived as a bidding uniformization of accounting practices adopted by corporations, defining the suitable accounting policies, measurement criteria, and financial statement models (Guerreiro et al., 2008).

According to the International Accounting Standards Board (IASB), as of 2023, approximately 168 jurisdictions require IFRS for the preparation of financial reports by listed entities, with the adoption of IFRS in 2005 being a crucial step in the formal harmonization process worldwide.

The accounting harmonization process often involves aligning domestic regulations/accounting norms with international standards, which can be a challenging task due to differences in legal, economic, and cultural contexts across countries.

Countries' legal systems tend to influence the accounting system adopted in each country. Typically, in code-law countries, the accounting system is composed of strict accounting norms that are generally aligned with tax rules. In common-law countries, accounting systems are principles-based and are broadly accepted by accounting professionals. Developed domestic financial markets, which tend to have strong investor protection mechanisms, also promote accounting systems with more detailed disclosures to protect investors.

Mainly to foster common financial markets within a single market, jurisdictions decided to force IFRS adoption only for listed companies, leaving not only some industries out of scope (such as insurance) but also SMEs, since smaller entities tend to require standards of less complexity to record accounting transactions. Those adoption exceptions result in significant differences in the adoption process worldwide, as most corporations are excluded from the process (Camfferman & Zeff, 2018).

A diverse array of accounting practices, due to differing national regulations and frameworks, can lead to inconsistencies in financial reporting, making it challenging for stakeholders to interpret these documents. The lack of comparability can obscure the true financial position of listed companies, thereby increasing the risk of misinterpretation and inefficient resource allocation. Accounting harmonization aims to mitigate these issues, fostering greater transparency and consistency that benefits all stakeholders involved.

At the beginning of the harmonization process, stakeholders perceived IFRS as a high-quality accounting standard. Consequently, implementing IFRS is often associated with benefits in terms of improved accounting quality.

Under IFRS, accounting quality is generally denoted by how faithfully financial statements represent a company's underlying economic reality. High-quality accounting under IFRS is expected to offer transparent, relevant, reliable, and comparable information that supports stakeholders in making financial decisions.

The key determinants of accounting quality under IFRS are faithful representation, relevance, comparability, consistency, transparency, measurement accuracy, and improved disclosure.

When changing accounting systems to IFRS, Barth et al. (2008) demonstrate that listed companies tend to present less earnings management, more value relevance of earnings, and more timely loss recognition, which constitutes evidence of faithful representation of economic transactions while financial reporting is free from bias, misrepresentation, and error. Additionally, authors do not encounter evidence that IFRS safeguards that the financial reporting presented is relevant to users' needs, since IFRS contains a limited number of accounting methods that are less reflective of firms' performance and can be used by managers to manage earnings.

However, according to Ding et al. (2005), IFRS provides more measurement models and criteria, as well as more detailed disclosure requirements, thereby improving financial reporting quality. This is because the measurement bases provided by IFRS provide a more realistic assessment of companies' assets, liabilities, revenue, and expenses.

Even though fair value-based accounting models tend to provide more timely recognition of accounting gains and losses, providing more relevant numbers in the accounting statement of financial position, fair value gains and losses tend to be more volatile, which causes the income statement to be less informative for the decision-making process. Gomez-Ortega et al. (2022) provide evidence that fair-value measurement criteria for recognizing financial assets tend to incorporate prospective information, resulting in pro-cyclicality in financial statements, which negatively impacts the fair value of financial assets. However, the income statement becomes more stable without unexpected losses due to credit losses.

Consequently, financial reporting prepared under IFRS is more relevant for the decision-making process since it contains timely, forward-looking information that can influence stakeholders' decisions. Nonetheless, the IASB struggles to accommodate all financial users' needs, specifically those of debt holders and shareholders, since the latter tend to require more detailed accounting disclosures and more accurate recognition criteria than the former, as they have different alternatives to access private financial information (Florou et al., 2016).

As far as comparability is concerned, IFRS tends to facilitate comparability across different companies and periods, enabling stakeholders to evaluate financial performance and position in relation to their industry peers. Despite common standards being utilized to prepare financial reporting, Nobes (2013) provides evidence that international accounting differences continue to exist not only due to different formats of IFRS adopted in different jurisdictions, that may incorporate modifications to maintain previous accounting practices, but also incorrections in the standards translation to local languages, may also impose implementation differences across countries (Ball, 2006; 2016).

Unfortunately, there are some industries, such as the insurance and extractive industries, in which the comparability of financial reporting remains impossible since adopting IFRS is not yet possible or mandatory. Despite the importance of the extractive sector for several jurisdictions, no specific IFRS establishes the recognition, measurement, and disclosure criteria. According to Nobes (2013), further investigations are needed to understand the accounting practices employed in these industries.

Nobes and Stadler (2021) highlight that different accounting policies for booking exploitation and evaluation costs lead to comparability deficiencies in financial reporting, as there is no definitive accounting regulation for recording accounting transactions. However, authors state that preparers tend to use measurement models that are compliant with IFRS principles, despite the fact that disclosures are difficult for stakeholders to interpret, as there are no formal definitions prescribed in the IFRS for that industry.

Regarding the insurance industry, Grace and Basu (2022) highlight that the accounting standards applicable to insurance contracts are complex and, consequently, difficult to implement, as insurers can use different measurement models to record the same insurance contract liabilities. Therefore, due to the benefits of financial supervision, some regulators did not allow listed life insurance companies to prepare financial reports under that standard.

Nonetheless, IFRS promotes consistent application of accounting policies, providing transparency and consistency in financial reporting, which enhances trust in the information presented.

Additionally, IFRS requires companies to disclose detailed notes and explanations for complex transactions, improving information quality by providing context and clarity. Hodgeon and Wallace (2008) and Bischof (2009) examine how the adoption of IFRS 7 impacted the quality of bank disclosures in Europe and report an overall improvement in disclosure quality, particularly regarding credit risk, which gained prominence over market risk. Bischof (2009) alerts that despite IFRS 7 setting a framework for enhanced transparency, local regulators significantly influenced how IFRS are formally applied. Companies tend to show more standardized disclosure practices when regulators adopt interventionist approaches in their supervisory practices.

Barroso *et al.* (2015) demonstrate that the accounting culture present in each country affects accounting procedures and practices, which in turn impact the quality of financial reporting. Therefore, countries' institutional settings cannot be left aside when assessing the advantages and disadvantages of IFRS adoption.

In Europe, listed companies have been required to adopt IFRS standards since 2005 to promote single market transactions. This requirement allowed researchers a unique set to evaluate the impact of adopting IFRS standards, given that corporations from different countries were simultaneously developing efforts to prepare financial reports on the same accounting basis. Therefore, several researchers have developed pre- and post-IFRS adoption studies to properly assess the impact of IFRS standards' first-time adoption. However, most research studies are often focused on listed firms, typically larger and better prepared to implement those standards.

Lately, research studies have focused on single countries, usually developing countries, that are implementing IFRS standards for the first time (Albu & Albu, 2012; Silva et al., 2021).

In developing countries, the accounting harmonization process was implemented to enhance economic development and attract foreign direct investment. Investors are commonly drawn to emerging countries by high-yield investment opportunities and tend to favor reliable and transparent accounting frameworks that provide decision-makers with sound accounting information. Additionally, coercive pressures from local regulators, resulting from restrictions imposed by the IMF and the EU's adoption process, contribute to an institutional setting that favors the adoption of the IFRS (Albu & Albu, 2012).

Shortly after the mandatory adoption in Europe, Soderstrom and Sun (2007) analyzed the effect of adopting IFRS on accounting quality and argue that the financial reporting quality improvements following IFRS adoption tend to depend on three main elements:

the country's legal and political environment, corporations' financial reporting incentives and country-specific institutional factors, namely legal systems and enforcement mechanisms implemented by local regulators.

When IFRS standards are adopted for the first time, the accounting quality of financial reporting tends to improve, provided the company has an incentive to adopt those standards. Corporations with an incentive to adopt IFRS are those that deeply engage in implementing IFRS, thoroughly interpreting and applying it (Daske et al., 2013). Without inner incentives, accounting quality increments are not observed since standards are not properly implemented (Daske et al., 2013; Cao & Patel, 2020).

For Christensen et al. (2015), companies that are forced to adopt IFRS do not exhibit improvements in accounting quality, possibly because these companies may explore the IFRS principle-based approach and exploit opportunities to manage earnings. Additionally, entities without inner incentives to adopt IFRS standards tend to have close relationships with capital providers that do not require additional accounting information in their investment decision. Consequently, the latter companies do not have intrinsic reasons to comply with those standards and fail to do so.

The need to finance internal operations, mainly growth opportunities through external shareholders/debtholders, constitutes an internal incentive to adopt IFRS standards adequately. This is particularly the case for novel technological firms that want to participate in an initial public offering of shares and bonds to finance investment opportunities with external shareholders, and for that reason, tend to make more efforts to adequately implement IFRS to reduce information asymmetry for potential investors (Ball, 2006).

Jermakowicz (2004) recognizes that IFRS standards narrow the accounting practices available for booking similar economic transactions, thereby facilitating the investment decision assessments performed by bondholders and equity holders. Bassemir (2018) provides evidence that this also occurs in private firms when operations are financed through private equity funds.

Cao and Patel (2020) demonstrate that firms involved in international trade operations in a jurisdiction where IFRS standards are not mandatory tend to voluntarily adopt IFRS to foster international trade since customers and suppliers become more trusting in establishing business with them.

Regarding the institutional setting, common-law countries tend to have better accounting systems. They can provide significant shareholder protection, which positively influences the quality of financial reporting produced by corporations operating in those jurisdictions (La Porta et al., 1998), whether using IFRS or local standards. For those countries, the increment of accounting quality resulting exclusively from the adoption of IFRS is not clear. Brochet et al. (2013) document that in the case of a common law country with local GAAP based on principles similar to IFRS, the incremental improvement in accounting quality resulting from the implementation of IFRS is not significant.

Lourenço et al. (2018) highlight that the improvement of accounting quality also depends on strong enforcement mechanisms. Likewise, Van Tendeloo and Vanstraelen (2005) affirm that legal enforcement of accounting principles, typically higher in common law countries, also justifies the increase in financial reporting quality given the fact that the adoption of IFRS per se is not enough to assure the rise of accounting quality because some preparers may only communicate to stakeholders that implement IFRS without effectively modifying their reporting policies (Christensen et al., 2015; Daske et al., 2013; Daske et. al., 2008).

Christensen et al. (2015) corroborate that the benefits typically associated with IFRS adoption are mainly verified in countries that made significant modifications within the legal enforcement of financial reporting mechanisms before the IFRS mandatory adoption period and simultaneously had strong legal systems. Those public enforcement mechanisms were fully effective when companies began to prepare their financial reports in accordance with IFRS. However, private enforcement mechanisms, developed by auditing firms or private litigation, also contribute positively to enhancing accounting quality (Brown *et al.*, 2014).

Despite possible bias in the statistical inference performed (De George et al., 2015) and a lack of robust empirical support in some circumstances (Chua & Taylor, 2008), several academic studies attribute advantages to the adoption of IFRS standards. The benefits of the harmonization process are typically categorized into six broad areas.

According to Ball (2016), the adoption of IFRS promotes enhancements in corporate governance mechanisms, as the standards require the adoption of a more transparent reporting strategy. This allows for the close monitoring of companies' business activities by analysts/credit rating agencies, without the need for specific databases or accounting adjustments to make financial reporting comparable. Therefore, boards and management are held accountable to more precise and uniform financial reporting requirements.

The IFRS principle-based approach benefits corporations' decision-making process, as loss recognition is booked sooner, providing a clearer and more reliable view of the company's financial position and performance. This enables a more rigorous oversight process for stakeholders. Due to transparency, companies' Board of Directors will be more careful when making investment decisions, particularly in the case of unnecessary intangible assets (Ball, 2016).

The second category corresponds to the liquidity benefits, both in terms of frequency and trading volume increments, for entities with financial instruments traded in active financial markets (Cornell & Sirri, 1992), as investors can analyze the information prepared without requiring further adjustments. Therefore, by utilizing the same accounting standards, the economic position and performance of companies are properly reflected, which allows investors to evaluate them better and compare their investments (Daske et al., 2013).

Daske et al. (2008) and Christensen et al. (2015), mention that the liquidity benefits typically associated with IFRS adoption are mainly verified in EU countries. Still, those improvements have not been verified in all countries. Only the UK, Germany, the Netherlands, Norway, and Finland exhibit liquidity benefits around the IFRS adoption period. Those countries, prior to the IFRS mandatory adoption period, implemented significant modifications in the enforcement of financial reporting mechanisms and had robust legal systems simultaneously.

Additionally, after the mandatory adoption of IFRS, some authors also document more foreign direct investments (Gordon et al., 2013), an increase in equity investment of foreign investors (DeFond et al., 2010; Florou & Pope 2012), and an increase in mergers and acquisitions activity promoted by external investors (Francis et al., 2016; Louis & Urcan, 2014) that have easier access to local capital markets.

Following the efficient market hypothesis, IFRS promotes a decline in the cost of capital (Dhaliwal et al., 2019), as, from an investor's perspective, more transparent and detailed financial reporting with increased disclosures in annual reports will reduce existing information asymmetry. Consequently, those companies experience a reduction in the rate of return required by investors since the risk of investment is more precisely captured (Leuz & Verrecchia, 2000).

Additionally, with the adoption of IFRS, foreign equity ownership increases (DeFond et al., 2010; Florou & Pope, 2012) because investors can evaluate investment opportunities more clearly across borders, as the same accounting methodologies are implemented. Further foreign direct investments (Gordon et al., 2013) and increased cross-border M&A activity (Francis et al., 2016) are also expected to occur. Indeed, more financial integration due to a reduction in information frictions is perceived after IFRS mandatory adoption (Dhaliwal et al., 2019).

Finally, multinational companies benefit from harmonized standards because the same accounting practices are applied across all subsidiaries, reducing the cost of preparing financial reports. IFRS simplifies the process of preparing multiple sets of financial reports to meet specific national standards.

It is worth noting that the benefits associated with IFRS, as mentioned by several researchers, are only verifiable in studies focusing on publicly traded firms, where a streamlined enforcement environment is present to ensure full compliance with IFRS requirements. Additionally, strong results are verified in a reduced number of corporations that are in the process of being quoted, integrated in a listed group, or have cross-border activities.

Those benefits are frequently not verifiable for private firms (Bassemir, 2018). They may not be entirely captured by robust academic research (Nobes, 2013), Damant (2006) and Jermakowicz and Gornik-Tomaszewski (2006) encounter several disadvantages regarding IFRS adoption that may not outweigh the permanent increase in overhead costs attributable to IFRS adoption.

Fox et al. (2013) warn that training costs increase due to training provided to personnel directly involved in preparing financial reporting documents. Formal adoption processes usually require the creation of an internal IFRS implementation team, training IT staff, internal audit and management, taking tax advice from consultants since the introduction of new standards may turn accounting rules non-compliant with local tax rules, investments in software and accounting systems are also necessary to accommodate IFRS complexity.

Additionally, on-off costs associated with external communication to stakeholders, including press conferences and the preparation of specific financial documentation, are required to explain in detail the impact on financial reporting resulting from the adoption of the new standard. Adjusting financial reporting documents to accommodate novel disclosures is also significant and recurring.

Incurring supplementary costs for an external audit, since IFRS tends to be complex and challenging to implement. Audit works will take longer to develop and become more onerous, not only in the initial adoption year but also in subsequent years, increasing corporations' overall costs. This is particularly the case with recent accounting standards, namely IFRS 9 and IFRS 17, which require companies to improve the estimates used to produce accounting information and increase the complexity within financial reporting. Therefore, more pressure is placed on expert judgment and audit work, as further estimates and validation of external information are required (Gomez-Ortega et al., 2022).

For companies with outstanding debt instruments, the forced adoption of a new accounting system may trigger debt covenants linked to accounting numbers, such as net income, temporary earnings, and net worth, which may result in additional costs associated with renegotiating debt instruments (Li, 2010).

Moreover, in regulated industries such as insurance and banking, the initial adoption of a new standard may trigger additional supervisory intervention due to regulatory capital requirements. During economic recessions, fair-value measurement models may also lead to technical insolvencies in pure financial statement analysis (Onali & Ginesti, 2014; Beerbaum & Ahmad, 2015; Bischof & Daske, 2016), which can incur additional costs for regulated companies.

For certain standards, hiring specific IFRS-trained staff may also be necessary, which can be problematic due to the scarcity of personnel with those characteristics (Darenidou et al., 2006). Therefore, additional external consultancy services may also be essential to implement IFRS properly.

IFRS adoption is closely linked to future compliance costs, as companies must implement future IFRS developments to avoid penalties imposed by regulators/standard setters.

However, due to standard density, possible application inconsistencies may also be observed. Gomez-Ortega et al. (2022) report that the local regulator in Spain forced financial institutions to overprovision for credit losses, indicating that some Spanish financial institutions did not implement the former IAS 39 impairment recognition criteria.

Finally, singularities of specific accounting treatment that is developed to accommodate local accounting specificities may be jeopardized by IFRS since a global set of accounting standards is adopted by local companies.

Damant (2006) states that corporations must carefully analyze the expected costs and benefits derived from adopting IFRS standards when considering voluntary application, despite the comparability and transparency benefits that come from implementing IFRS. However, Nobes (2013) demonstrates that stakeholders are generally satisfied with the IFRS adoption process, primarily due to the improved comparability of financial reporting within EU operating firms, the more accurate and fair view provided under IFRS measurement requirements, increased transparency, and enhanced internal group accounting procedures.

3 Conclusions

In conclusion, some researchers consider the global harmonization process a success, as noted by Ball (2016), despite initial adoption efforts.

Overall, accounting harmonization offers significant advantages, including improved comparability of financial reporting, enhanced transparency for investors and financial analysts, and the facilitation of cross-border investments and global trade. Establishing common, high-quality accounting standards fosters trust and reduces the compliance costs for multinational corporations operating in diverse jurisdictions. Listed entities, with internal incentives to properly comply with accounting standards, also tend to exhibit a decline in the cost of capital after IFRS adoption.

However, harmonization is not without its challenges. Differences in cultural, legal, and economic environments can make standard adoption difficult, potentially leading to resistance from specific stakeholders. Furthermore, the one-size-fits-all approach may not always address the unique needs of individual countries or industries.

While accounting harmonization holds great potential to streamline financial reporting globally, its success depends on striking a balance between global uniformity and local flexibility. Ongoing dialogue, collaboration, and a willingness to adapt are essential to maximize its benefits while mitigating its drawbacks.

Indeed, successful accounting harmonization often requires that listed corporations in developing or developed countries adopt IFRS if their local GAAP is very dissimilar from IFRS, or if adoption of IFRS is required by a supranational entity. These entities tend to experience greater benefits from adoption, such as reduced costs for financial reporting preparation and auditing fees.

However, some of the analyzed studies lack control samples, which may bias the results since it is not possible to isolate economic events attributed to IFRS adoption. Additionally, most studies focus on only one jurisdiction, which may jeopardize the generalization of results (De George et al., 2015). Further research could examine the accounting practices followed by preparers of financial reporting when there is no specific standard applicable to that matter (Nobes, 2013) and investigate the impact of an accounting harmonization process on SME.

References

Nobes, C., Parker, R.: Comparative International Accounting, 7th edn. Prentice Hall, London (2002)

Van der Tas, L.: Measuring harmonization of financial reporting practices. Account. Bus. Res. **18**(70), 157–169 (1988)

Guerreiro, M.: Impacto da Adopção das International Financial Reporting Standards: Factores Explicativos do Nível de Informação Divulgada pelas Empresas Portuguesas Cotadas. Contabilidade e Gestão **3**, 7–32 (2006)

Van der Tas, L.: Evidence of EC financial reporting practice harmonization. Eur. Account. Rev. **1**, 69–104 (1992)

Kriesment, M.: An approach for measuring the degree of comparability of financial accounting information. Eur. Account. Rev. **6**(3), 465–485 (1997)

Guerreiro, M., Rodrigues, L., Craig, R.: The preparedness of companies to adopt international financial reporting standards: Portuguese evidence. Account. Forum **32**, 75–88 (2008)

Camfferman, K., Zeff, S.: The challenge of setting standards for a worldwide constituency: research implications from the IASB's early history. Eur. Account. Rev. **27**, 289–312 (2018)

Barth, E., Landsman, R., Lang, M.: International accounting standards and accounting quality. J. Account. Res. **46**(2), 467–498 (2008)

Ding, Y., Jeanjean, T., Stolowy, H.: Why Do National GAAP Differ from IAS? The role of culture. Int. J. Account. **40**, 325–350 (2005)

Gómez-Ortega, A., Gelashvili, V., Delgado, M., Menéndez, J.: Impact of the application of IFRS 9 on listed Spanish credit institutions implications from the regulatory, supervisory and auditing point of view. J. Risk Financ. **23**(4), 437–455 (2022)

Florou, A., Kosi, U., Pope, P.: Are international accounting standards more credit relevant than domestic standards? Account. Bus. Res. (2016)

Nobes, C.: The continued survival of international differences under IFRS. Account. Bus. Res. **43**(2), 83–111 (2013)

Ball, R.: International Financial Reporting Standards (IFRS): pros and cons for investors. Account. Bus. Res. **36**(1), 5–27 (2006)

Ball, R.: IFRS – 10 years later. Account. Bus. Res. **46**(5), 545–571 (2016)

Nobes, C., Stadler, C.: Towards a solution to the variety in accounting practices of extractive firms under IFRS. Aust. Account. Rev. **3**(4), 273–285 (2021)

Grace, M., Basu, S.: Insurance: In or Out of the "Too Difficult" Box? (2022). Available at SSRN: https://ssrn.com/abstract=4071072 or https://doi.org/10.2139/ssrn.4071072

Hodgeon, E., Wallace, P.: accounting for change: transparency in the midst of Turmoil (London: PwC International)

Bischof, J.: The effects of IFRS 7 adoption on bank disclosure in Europe. Account. Eur. **6**, 167–194 (2009)

Albu, N., Albu, C.: International financial reporting standards in an emerging economy: lessons from Romania. INTL: Econ. Financ. Issues (Topic) (2012)

Silva, A, Fontes, A., Martins, A.: Perceptions regarding the implementation of International Financial Reporting Standards in Portugal and Brazil Journal of International Accounting, Auditing and Taxation 44 (2021)

Soderstrom, N., Sun, K.: IFRS adoption and accounting quality: a review. Eur. Account. Rev. **16**(4), 675–702 (2007)

Daske, H., Hail, L., Leuz, C., Verdi, R.: Adopting a label: heterogeneity in the economic consequences around IAS/IFRS adoptions. J. Account. Res. **51**(3), 495–547 (2013)

Cao, J., Patel, C.: The role of the national institutional environment in IFRS convergence: a new approach. Account. Financ. **60**(4), 3367–3406 (2020)

Christensen, H., Lee, E., Walker, M., Zeng, C.: Incentives or standards: what determines accounting quality changes around IFRS adoption? Eur. Account. Rev. **24**(1), 31–61 (2015)

Jermakowicz, E.: Effects of adoption of international financial reporting standards in Belgium: the evidence from BEL-20 companies. Account. Eur. **1**, 51–70 (2004)

Bassemir, M.: Why do private firms adopt IFRS? Account. Bus. Res. **48**, 237–263 (2018)

La Porta, R., Lopez-de-Silanes, F., Shleifer, A., Vishny, R.: Law and finance. J. Polit. Econ. **106**(6), 1113–1155 (1998)

Brochet, F., Jagolinzer, A., Riedl, E.: Mandatory IFRS adoption and financial statement comparability. Contemp. Account. Res. **30**(4), 1373–1400 (2013)

Van Tendeloo, B., Vanstraelen, A.: Earnings management under German GAAP versus IFRS. Eur. Account. Rev. **14**(1), 155–180 (2005)

Lourenço, I., Sarquis, R., Branco, M., Magro, N.: International differences in accounting practices under IFRS and the influence of the US. Aust. Account. Rev. **28**(4), 468–481 (2018)

Daske, H., Hail, L., Leuz, C., Verdi, R.: Mandatory IFRS reporting around the world: early evidence on the economic consequences. J. Account. Res. **46**(5), 1085–1142 (2008)

De George, E., Li, X., Shivakumar, L.: A review of the IFRS-adoption literature. Fox School Bus. Res. 15–078 (2015)

Chua, W., Taylor, S.: The rise and rise of IFRS: an examination of IFRS diffusion. J. Account. Public Policy **27**(6), 462–473 (2008)

Cornell, B., Sirri, E.: The reaction of investors and stock prices to insider trading. J. Financ. **47**, 1031–1059 (1992)

Gordon, E., Greiner, A., Kohlbeck, M., Lin, S., Skaife, H.: Challenges and opportunities in cross-country accounting research. Account. Horiz. **27**(1), 141–154 (2013)
DeFond, M., Hu, X., Hung, M., Li, S.: The impact of mandatory IFRS adoption on foreign mutual fund ownership: the role of comparability. J. Account. Econ. **51**(3), 240–258 (2010)
Florou, A., Pope, P.: Mandatory IFRS adoption and institutional investment decisions. Account. Rev. **87**(6), 1993–2025 (2012)
Louis, H., Urcan, O.: The effect of IFRS on cross-border acquisitions. Working Pap. (2014)
Francis, J., Huang, S., khurana, I.: The role of similar accounting standards in cross-border mergers and acquisitions. Contemp. Account. Res. **33**(3), 1298–1330 (2016)
Dhaliwal, D., He, W., Li, Y., Pereira, R.: Accounting standards harmonization and financial integration. Contemp. Account. Res. **36**(4), 2437–2466 (2019)
Leuz, C., Verrecchia, R.: The economic consequences of increased disclosure. J. Account. Res. **38**, 91–124 (2000)
Damant, D.: Discussion of 'International Financial Reporting Standards (IFRS): pros and cons for investors. Account. Bus. Res. **1**, 29–30 (2006)
Jermakowicz, E., Gornik-Tomaszewski, S.: Implementing IFRS from the perspective of EU publicly traded companies. J. Int. Account. Audit. Tax. **15**(2), 170–196 (2006)
Fox, A., Hannah, G., Helliar, C., Veneziani, M.: The costs and benefits of IFRS implementation in the UK and Italy. J. Appl. Account. Res. **14**(1), 86–101 (2013)
Onali, E., Ginesti, G.: Pre-adoption market reaction to IFRS 9: a cross-country event-study. J. Account. Public Policy **33**(6), 628–637 (2014)
Beerbaum, D., Ahmad, S.: Credit risk according to IFRS 9: significant increase in credit risk and implications for financial institutions working paper (2015). Available at SSRN: https://ssrn.com/abstract=2654120
Bischof, J., Daske, H.: Interpreting the European Union's IFRS endorsement criteria: the case of IFRS 9. Account. Europe **13**(2), 129–168 (2016)
Darenidou, C., McLeay, S., Raonic, I.: Expected earnings growth and the cost of capital: an analysis of accounting regime change in the European financial market. Abacus **42**(3/4), 388–414 (2006)

Balancing Innovation and Sustainability: Financial Development, ICT, and Load Capacity Factor Dynamics

Louis David Junior Annor[1,3](✉), Margarita Robaina[1], and Elisabete Vieira[2]

[1] Department of Economics, Management, Industrial Engineering and Tourism, University of Aveiro, Aveiro, Portugal
lannor@ua.pt

[2] Higher Institute of Accounting and Administration of University of Aveiro, University of Aveiro, Aveiro, Portugal

[3] Department of Accounting, Banking and Finance, Ghana Communication Technology University, Accra, Ghana

Abstract. This paper explores the symbiotic relationship between ICT diffusion measured as (Mobile cellular subscriptions per 100 people and Individuals using the internet as a percentage of the population, and financial institutions' efficiency, access and depth on environmental quality (EQ). Focusing on 47 African economies, this study analyses data from 1990 to 2021 using a two-step system generalised method of moments (SGMM) estimation approach. The data were sourced from World Development Indicators, Global Footprints Network, and the International Monetary Fund. This study reveals that financial efficiency and access significantly improve the environment (it increases the load capacity factor (LCF)). However, financial depth hurts EQ in all three EQ proxies used. Ecological footprints are significantly reduced as financial efficiency and access increase. Furthermore, financial efficiency and depth are significant drivers of carbon emissions in the region. ICT diffusion (mobile cellular subscriptions per 100) harms EQ across all three proxies. Internet usage impedes EQ by reducing LCF. However, internet usage improves EQ by reducing ecological and carbon footprints. This study introduces LCF as a novel EQ proxy and explores ICT's moderating role in financial systems' environmental impact, offering new insights into Africa's sustainable development dynamics.

Keywords: ICT diffusion · Environmental Quality · Financial Efficiency · Sub-Saharan Africa · System GMM technique · Financial Access · Financial Depth

1 Introduction

Key drivers defining human livelihoods in recent times are financial development, climate change, ICT, and energy transition. The world is expected to be a better place by 2030, considering the comprehensive scope of the 17 United Nations Sustainable Development

L. Almeida et al. (Eds.): ICAFI 2025, ISEM 80, pp. 296–317, 2026.
https://doi.org/10.1007/978-3-032-16268-7_17

Goals (SDGs). Modernization and development can be costly due to their environmental impact, especially in Africa. [1] reveal that between 1995 and 2017, the SSA countries saw a notable 15.48% increase in CO_2 emissions. According to [2], SSA's carbon dioxide emissions are projected to rise by 50% by 2050.

The emissions of CO_2 and other greenhouse gases through the burning of fossil fuels exacerbate global warming, thereby threatening future sustainable goals. One formidable pillar which can augment the environmental sustainability agenda and foster sustainable growth is the proliferation of ICT [3, 4]. ICT growth is considered an enabler that increases financial sector activities and drives economic growth. [5] assert that ICT stimulates both economic growth and social development, though it harms the quality of the environment. Moreover, previous literature has identified urbanisation (URB), economic growth, industrialization, energy consumption, and air pollution as major environmental drivers. ICT encompasses the internet, mobile devices, and other communication channels, which are crucial for enhancing living standards. Smart technologies, such as touch displays, monitors, and tablets, are becoming increasingly in demand due to the diffusion of ICT. The eruption of ICT characterizes modern-era globalisation and fuels numerous socio-economic activities, including trade, foreign direct investment, productivity, FD, and international collaboration [5]. While ICT growth is considered a key driver of economic growth, it also has a significant impact on environmental sustainability [6, 7].

The path to zero emissions and sustainable development may be paved through radical innovation in ICT and technology, increasing the availability of renewable energy solutions for businesses and society in many nations [8, 9]. Moreover, due to ICT, financial institutions and markets can efficiently allocate assets, reduce costs, enhance credit and deposit facilities, facilitate transfers, and manage risks. Developed financial systems encourage firms and households to access funds to acquire modern and environmentally friendly technologies. Several studies have explored determinants of environmental sustainability, including ICT and EQ [10–12], energy transition, EQ and economic growth [11, 13], FD and EQ [14–16] the role of ICT in foreign direct investment and environmental quality nexus [17]. Other studies on African nations focus on economic growth and environmental sustainability [18, 19], with institutional quality serving as a moderator [18], as well as the nexus between financial inclusion and EQ [19].

While research on innovation and environmental sustainability remains scarce [20], the interactive effect of FD on ICT diffusion and EQ has not been studied in the African context, despite the phenomenon being relevant to the actualization of the carbon neutrality agenda. [21] applied the extended Stochastic impact by regression and found that ICT use – measured by mobile phones and internet penetrations significantly stimulates CO_2 emissions. Additionally, the mediation analysis demonstrates that ICT has a direct favourable impact on CO_2 emissions, a positive indirect impact on energy consumption and FD, and a negative indirect impact on trade openness. The overall impact is detrimental, indicating that ICT is degrading SSA's environment. According to [3], the use of ICT and fossil fuels increases carbon dioxide emissions, whereas the use of renewable energy and FD lowers emissions. The findings specifically indicate that a 1% increase in ICT activity will, over time, result in a 0.565% rise in CO_2 emissions in South Africa. The ICT-led emissions proposition essentially contends that the development, manufacture, and, in many cases, disposal of ICT infrastructure involve emissions that harm the environment. [22], focusing on Middle East and Northern African countries, concluded that ICT diffusion is a critical determinant of economic activity, job creation, and environmental sustainability. The study finds that the diffusion of ICT and renewable energy have a positive and significant impact on GDP. In assessing the impact of internet use, FD, economic growth, and trade openness on carbon emissions [5] revealed that internet use lowered EQ in EU countries from 2001 to 2014. Thus, the diffusion of ICT via internet usage stimulates environmental risk and threatens sustainable development. Unlike other regions in the world, such as Europe, Asia, and North America, the African sub-region is considered to have a higher ICT penetration [23, 24]. Empirical evidence on the impact of ICT and FD on EQ has yielded inconsistent results [3, 25], partly due to the varying study periods, econometric techniques, and sampled countries. Existing studies have ignored the role of FD through which ICT diffusion impacts EQ, particularly in Africa.

First, like other developed nations, Africa has ratified several agreements on environmental sustainability and climate change. Consistent with global expectations towards curbing global warming, several environmental frameworks and agreements have been signed by African countries to achieve the agenda. It is essential to understand climate change predictors and mitigation strategies in a timely manner. Second, even though SSA produces less than 3% of the world's carbon emissions, given the region's ongoing institutional and economic reforms aimed at boosting economic growth, advancing industrialization and economic diversification, enhancing transportation infrastructure, and addressing the energy crisis, it is anticipated that emissions will rise there over the next several decades. It is further expected that the modernization, automation, and digitisation of the production process required to meet these objectives will encourage the adoption and spread of ICTs, giving rise to justifiable worries about their environmental effects [6].

Moreover, this study examines both the linear and non-linear effects of ICT and FD on efforts to ensure ecological stability in Africa, thereby providing relevant policy directives to mitigate environmental impacts in Africa. This study measures FD by decomposing financial institutions' efficiency, access, and depth, making varied regional policy implications. The conflicting results of FD on environments highlight the need to further investigate new data on the dynamics of financial institutions, which addresses questions about how the efficiency, access, and effectiveness of financial institutions interact with ICT diffusion to improve EQ. By introducing the constituents of financial institutions in the finance-environment argument, this study contributes to the macroeconomic literature on how financial access, efficiency and access interact with ICT diffusion towards environmental sustainability. Fifth, to the best of our knowledge, unlike previous studies [5, 11], this study premiers on assessing the effect of ICT diffusion and FD on three environmental proxies, namely, the LCF, ecological footprint, and CO_2 emissions, thereby aiding environmental policies and sustainability frameworks in Africa. The study first departs from the traditional method of determining the variables influencing ecological sustainability by utilising a reliable proxy, the LCF, as a gauge of EQ. Conducting a comprehensive environmental assessment is challenging because the ecological footprint and CO_2 emissions only consider humans' demand for natural resources. Environmental analysis should consider both human demands (the demand side) and biocapacity (the supply side). [26] suggest the LCF, which is computed as biocapacity/ecological footprint. Because environmental aspects consider supply and demand, LCF enables a much more thorough assessment [27]. Compared to other metrics like CO_2 emissions, greenhouse gas emissions, and ecological footprints, this indicator might produce more accurate results [28].

This study focuses on emerging economies, such as Sub-Saharan Africa, where ICT penetration has higher potential [23, 24], and economic development, which may impact its already deteriorating environmental biocapacity. Despite the link between financial sector activities and ICT diffusion [29, 30], the literature on Africa lacks rigorous empirical works on the simultaneous effect of financial institutions —access, efficiency, and depth —on ICT diffusion and EQ. Moreover, this study uniquely uses the current data ending in 2021, unlike previous studies, which have used relatively old data. This suggests that the results from this study reflect current conditions on the subject matter. The remainder of the document is organized as follows: Sect. 2 presents a review of previous research, Sect. 3 illustrates the methodology, Sect. 4 presents the study's findings, and Sect. 5 concludes the paper.

2 Literature Review

2.1 Financial Development and Environmental Quality

Researchers have redefined the EKC hypothesis to examine the linear and non-linear effects of the financial sector's development on EQ, considering the significant role the financial sector plays in economic development and its cascading impact on environmental issues [31, 32]. [31] modified the EKC as FDEKC by categorising African economies into levels of human development. According to the study, FD and EQ have an inverse U-shaped relationship, with FD initially increasing and EQ (as indicated by an increase in the ecological footprint) decreasing. Nevertheless, additional FD eventually results in gains in EQ. This shows that the advantages of FD begin to offset its detrimental effects on the environment at higher levels of economic development.

Ex-ante literature on the finance-environment nexus has produced two major outcomes: (i) FD improves EQ, thereby reducing environmental risks, and (ii) FD increases environmental risks by augmenting carbon emissions, greenhouse gas emissions, and ecological footprints, among others. [31] concluded that the effect of FD on EQ largely depends on the level of development among African economies and the EQ proxy used. Specifically, they found that in both the high-human development index and middle-human development index economies in SSA, there is a non-linear (inverted U-shaped) relationship between FD and EQ (defined by ecological footprint). Exploring the dynamics of FD by decomposing it into financial institutions and financial markets, and examining their effect on EQ, [33] found that in SSA, the development of both financial institutions and financial markets improves EQ (increases the load capacity factor) in Africa. However, CO_2 emissions are exacerbated when financial institutions and markets become more efficient. Furthermore, they found that the development of financial institutions and markets are the transmission channels through which green energy consumption improves EQ in the region. Nonetheless, FD raises EQ in low human development index economies over the long run as well as during the initial phase of FD. [34] found that FD is at the expense of environmental pollution in Tunisia. The study employed the autoregressive distributed lag bound testing and error correction method to investigate the long-run relationship between CO_2 emissions, real GDP, energy consumption, FD, trade openness, and URB for the period 1971–2012. [35] explored the socio-economic determinants of environmental performance using data from 116 developing economies. They found that the financial sector, GDP growth, and energy consumption exacerbate the deteriorating effect of CO_2 emissions, while institutional quality, political freedom, and ethnic fractionalization play a crucial role in mitigating the environmental risks associated with these economies. [36] found that additional financial system development will worsen EQ in Asia-Pacific after a certain point. The development of the financial system decreases environmental degradation up to a specific threshold level. [37] concluded that FD accelerates environmental degradation in 44 African economies.

[38] explored the complementarity of FD and CO_2 emissions while accounting for good governance and found that the effect of FD on environmental pollution depends on the adopted indicator. Specifically, the study reveals that the FD index reduces environmental risk (reduces pollution levels), while FD triggers pollution via domestic credit to the private sector for Brazil, Russia, India, China, South Africa (BRICS); Mexico, Indonesia, Nigeria, and Turkey (MINT), and the G7 economies. [39] found that FD and renewable energy penetration significantly improved EQ (reducing environmental risk via CO_2 emissions) in Pakistan. [40] found that FD, industrialization, trade openness, and renewable energy improve EQ (CO_2 emissions) of BRICS economies in the long run. Moreover, whereas FD has a positive impact on EQ in the short run in Brazil, China, and Russia, it has a negative impact on EQ in South Africa and India. Moreover, [41] found that both FD and economic growth have a significant and negative influence on environmental degradation. The study concludes that the authorities should prioritize foreign direct investments to boost EQ and development. In the same vein, [42] reported that FD, URB, natural resources, and human capital all contributed to reducing ecological footprints in Nigeria. This suggests that progress in the banking, non-banking, and stock markets enhances the adoption of energy-efficient practices that protect the environment. [43] found that overall, FD reduces EQ. Additionally, the multidimensional analysis reveals that the dimensions of FD—depth, access, and efficiency—reduce EQ. Similarly, [44], focusing on the six world's top Islamic finance economies, found that Islamic FD contributes materially to increased energy consumption and the deterioration of EQ.

[45] found that FD has a positive impact in the short run and a negative impact in the long run. [46] aimed to investigate FD's (a)symmetric impacts on EQ in the presence of economic development, energy consumption, URB, and foreign direct investment. According to their symmetric analysis, FD reduces non-carbon footprints while stimulating ecological and carbon footprints in the medium term. Long-term FD has a significant detrimental impact on the ecological footprint and carbon footprint. [47] investigated the relationship between FD and environmental deterioration and found a U-shaped association between the three quantile groups based on FD and CO_2 emissions. Globalisation, the moderator, brings up the changes in the shift at the turning point and notes that it flattens at the middle quantile before the U-shaped curve's maturity, while

flattening at the upper quantile once it has reached maturity. We therefore hypothesize that:

H_1: Financial development has a positive effect on environmental quality

2.2 ICT Diffusion and Environmental Quality

Regarding the varied effects of ICT diffusion on economic growth, the environmental implications of ICT diffusion have also been explored, yielding two major conclusions from the existing literature. First, ICT diffusion aids in preserving the environment through improved production processes of renewable energy, increased energy efficiency, and enhanced productivity, ultimately leading to the development of smart cities and transportation systems. Nonetheless, ICT diffusion has negative environmental ramifications. It is argued that high internet penetration and innovation lead to increased production, energy consumption, and improved financial sectors, which in turn stimulate environmental deterioration. The proliferation of ICT over the years has had enormous economic implications globally. It has become a catalyst driving innovation and change in various industries, thereby propelling economic growth and ensuring inclusivity. The ICT revolution has emerged as a major force and catalyst in the world economy's energy sector [48]. For example, the diffusion of ICT has encouraged the use of more ICT devices, thereby increasing energy usage and emissions of greenhouse gases [49]. Exploring the non-linear effect of ICT on the environment, [50] found that the initial stage of ICT diffusion accelerates CO_2 emissions; however, at a certain threshold, environmental degradation subsides as CO_2 emissions are reduced. According to [51], the global financial sector has been the greatest buyer of ICT goods and services since the mid-1990s, having spent over $197 billion on them in 2014. [21] found that ICT diffusion and FD accelerate economic growth. Intuitively, economic growth resulting from FD and ICT diffusion spurs energy usage through the construction and expansion of ICT infrastructure, which in turn harms the environment. [6] assessed the direct and indirect effects of ICT diffusion on CO_2 emissions using data on 21 economies in the SSA region. The study reveals that the use of ICTs contributes to environmental degradation by increasing CO_2 emissions.

[52] found that whereas an increase in ICT leads to a decline in CO_2 emissions, it has an insignificant influence on energy efficiency. However, [3] found that ICT and fossil fuel consumption aggravate CO_2 emissions. [53] asserts that as e-banking and e-commerce become more popular, online transactions become easier, which lowers the need for physical travel and greenhouse gas emissions. In an analysis of emerging economies, [54] present interaction terms and note that FD was a significant channel by which ICT contributed to CO_2 emissions. However, an insignificant relationship has been established between ICT and environmental degradation [55, 56]. According to [57], China's environmental sustainability is enhanced by eco-digitalization, green technology, green finance, and renewable energy, as these measures reduce CO_2 emissions and ecological footprint. [58], focusing on the top five carbon-emitting African

nations, discovered that, based on visible warning signals, structural changes, technological advancements, and renewable energy, environmental pollutants can be prevented from surging. We therefore hypothesize that:

H_2: ICT diffusion has a positive effect on environmental quality
H_3: ICT diffusion has a positive moderating effect on the finance-environment nexus

3 Research Methodology

3.1 Panel Estimation Technique

This study uses the GMM dynamic panel estimation technique to analyze the data. Specifically, this study adopted the two-step system GMM due to its robustness and efficiency in panel data analysis. Consistent with previous studies [33, 59, 60, 61], this study employs the system GMM instead of the difference GMM because it is an unbiased estimator compared to fixed effects and ordinary least squares estimations. Other strengths of this estimator include its ability to mitigate omitted variable bias, its efficiency in handling missing data, its robustness to weak instruments, and its effective control of endogeneity. Thus, given that system GMM addresses endogeneity for the lagged dependent variable, autocorrelation error, unresolved measurement errors, heterogeneity in the panel, and bias from misspecification (omitted variable bias), it will produce more accurate estimates [7]. When handling endogeneity problems with panel data, System GMM is a valuable tool. Endogeneity concerns can be addressed by employing instruments for endogenous variables and lag values of the dependent variable. Hypothesis testing becomes more reliable as a result of the system GMM estimators' consistency and asymptotically normal behaviour, which under some circumstances converges to the true parameter values and has a normal distribution. [62] assert that the two-step system estimator provides a smaller asymptotic variance, making it more efficient than the one-step estimator. Apart from these features, the system GMM estimator offers pertinent post-estimation tests like the Hansen test to confirm the validity of the instrument, the AR (2) for serial correlation test, and multiple instruments to confirm that there is no instrument proliferation. These post-estimation test findings are evaluated to verify the accuracy of the study's estimations. Moreover, the method accounts for the inherent variances of the cross-country panel model [63, 64]. Lastly, the cross-sections in this study are greater than the time dimensions; hence, the application of a two-step system GMM estimator is justified due to its superior feature in handling datasets with this characteristic [59]. GMM becomes more suitable when the panel set contains more cross-sections (N) than periods (T). Considering this study, the N (47) outnumbered the T (1990–2021) [3].

3.2 Data Sources and Measurement

This study uses data from forty-seven (47) countries in SSA for the period 1990 to 2021 and sourced data from the Global Footprint Network (GFN), the World Development Indicator (WDI), and the International Monetary Fund (IMF). In this study, EQ is denoted by three proxies: load capacity factor (LCF), measured as ecological biocapacity/ecological footprint; ecological footprint (EFP), measured as ecological footprint (gha per person); and CO_2 emissions, measured as CO_2 emissions (metric tons per capita). Financial institutions' dynamics are proxied by financial depth (FID), access (FIA), and efficiency (FIE), respectively, whereas ICT diffusion is proxied by mobile cellular subscriptions (per 100 people) (ICT2) and internet usage access (ICT3) measured as individuals using the internet (% of population). Moreover, the study controlled for URB measured as urban population (% of total population), renewable energy consumption (REN) measured as renewable energy consumption (% of total final energy consumption), education (EDU) measured as school enrollment, primary (% gross), and economic growth (GDP) measured as GDP growth (annual %).

3.3 Theoretical Foundation and Functional Models

In assessing the effect of ICT diffusion on EQ, much consideration has been given to the theoretical underpinnings advanced by [65]. The diffusion of innovation theory provides a framework for comprehending how societies adopt new technologies, such as ICT, and how this process might impact environmental quality. The more companies and individuals adopt ICT, the more impact it has on the environment because it leads to systemic improvements like reduced greenhouse gas emissions or improved EQ. Several studies have assessed the environmental ramifications of the proliferation of ICT usage [7, 66, 53]. Based on the theoretical elucidation and as indicated earlier, this study examines the interaction between ICT diffusion and the dynamics of financial institutions on EQ in the SSA region.

This study, therefore, hypothesized that EQ is a function of ICT diffusion, financial efficiency, financial access, financial depth, and other control variables (Ctrls), as specified in Eq. 1 below.

$$EQ_{k,it} = f(ICT, FIE, FID, FIA, Ctrls) \tag{1}$$

Equations 2–4 show a detailed function that verifies the surmised relationships in linear form. The subscripts i and t denote country and time dimensions, respectively. The intercept is given as α_i, and $\gamma_1 - \gamma_{10}$ denote the coefficient of the explanatory variables while μ_{it} represents the composite error term.

$$\begin{aligned} LCF_{it} &= \alpha_i + \gamma_1 LCF_{it-1} + \gamma_2 ICT2_{it} + \gamma_3 ICT3_{it} + \gamma_4 FIE_{it} + \gamma_5 FID_{it} \\ &+ \gamma_6 FIA_{it} + \gamma_7 URB_{it} + \gamma_8 EDU_{it} + \gamma_9 GDP_{it} + \gamma_{10} REN_{it} + \mu_{it} \end{aligned} \tag{2}$$

$$\begin{aligned} EFP_{it} &= \alpha_i + \gamma_1 EFP_{it-1} + \gamma_2 ICT2_{it} + \gamma_3 ICT3_{it} + \gamma_4 FIE_{it} + \gamma_5 FID_{it} \\ &+ \gamma_6 FIA_{it} + \gamma_7 URB_{it} + \gamma_8 EDU_{it} + \gamma_9 GDP_{it} + \gamma_{10} REN_{it} + \mu_{it} \end{aligned} \tag{3}$$

$$\begin{aligned} CO2_{it} &= \alpha_i + \gamma_1 CO2_{it-1} + \gamma_2 ICT2_{it} + \gamma_3 ICT3_{it} + \gamma_4 FIE_{it} + \gamma_5 FID_{it} \\ &+ \gamma_6 FIA_{it} + \gamma_7 URB_{it} + \gamma_8 EDU_{it} + \gamma_9 GDP_{it} + \gamma_{10} REN_{it} + \mu_{it} \end{aligned} \tag{4}$$

Furthermore, the moderating effect of financial institutions' dynamics, specifically FIE, FID, and FIA, on the ICT diffusion (ICT3) and EQ nexus in Africa is presented in Eqs. 5, 6, and 7.

$$LCF_{it} = \alpha_i + \gamma_1 LCF_{it-1} + \gamma_2 ICT2_{it} + \gamma_3 ICT3_{it} + \gamma_4 FIE_{it} + \gamma_5 FID_{it}$$

$$+ \gamma_6 FIA_{it} + \gamma_7 URB_{it} + \gamma_8 EDU_{it} + \gamma_9 GDP_{it} + \gamma_{10} REN_{it} + \gamma_{11}(FIE \times ICT3)_{it} + \gamma_{12}(FID \times ICT3)_{it} + \gamma_{13}(FIA \times ICT3)_{it} + \mu_{it} \quad (5)$$

$$EFP_{it} = \alpha_i + \gamma_1 EFP_{it-1} + \gamma_2 ICT2_{it} + \gamma_3 ICT3_{it} + \gamma_4 FIE_{it} + \gamma_5 FID_{it} + \gamma_6 FIA_{it} + \gamma_7 URB_{it} + \gamma_8 EDU_{it} + \gamma_9 GDP_{it} + \gamma_{10} REN_{it} + \gamma_{11}(FIE \times ICT3)_{it} + \gamma_{12}(FID \times ICT3)_{it} + \gamma_{13}(FIA \times ICT3)_{it} + \mu_{it} \quad (6)$$

$$CO2_{it} = \alpha_i + \gamma_1 CO2_{it-1} + \gamma_2 ICT2_{it} + \gamma_3 ICT3_{it} + \gamma_4 FIE_{it} + \gamma_5 FID_{it} + \gamma_6 FIA_{it} + \gamma_7 URB_{it} + \gamma_8 EDU_{it} + \gamma_9 GDP_{it} + \gamma_{10} REN_{it} + \gamma_{11}(FIE \times ICT3)_{it} + \gamma_{12}(FID \times ICT3)_{it} + \gamma_{13}(FIA \times ICT3)_{it} + \mu_{it} \quad (7)$$

From Eqs. (5), (6), and (7); γ_{11}, γ_{12}andγ_{13} represents the coefficients of the interaction variables, whilst the remaining variables follow the definitions in Eqs. (2), (3), and (4).

4 Discussion of Results

Table 1 presents the descriptive statistics of the variables used in this study, including their means, standard deviations, minimum, and maximum scores.

Table 1. Descriptive Statistics.

Variables	Obs	Mean	Std. Dev	Min	Max	Skew	Kurt
LCF	1504	1.88	2.93	.13	24.246	3.997	22.15
ICT2	1492	36.256	43.505	0	171.01	1.005	2.869
ICT3	1465	9.335	15.813	0	88.13	2.224	7.632
FIE	1504	.491	.158	0	.971	−.555	3.271
FID	1504	.103	.154	0	.878	3.305	14.348
FIA	1504	.086	.107	0	.72	2.269	8.483
REN	1476	62.602	29.386	.06	98.34	−.781	2.275
URB	1501	40.02	17.659	5.416	90.423	.399	2.61
EDU	1227	95.477	25.266	21.708	156.445	−.425	3.007
GDP	1482	3.969	7.543	−50.339	149.973	5.87	117.538

Source: Authors' work (2025)

Table 1 shows an average biocapacity and ecological footprint ratio of 1.88 (>1), indicating a quality environment in Africa at other things being equal [26, 33]. However, there is a greater disparity between the minimum and maximum scores, with a standard deviation of 2.93%. ICT2 averaged 36.256 with a standard deviation of 43.505%, indicating a significant variation from the mean. Internet use or penetration is lower in Africa compared to mobile cellular subscriptions considering the average scores. ICT3 recorded a variation of 15.813% from the mean score of 9.335%. Further, the results reveal that among the dynamics of financial institutions in Africa, FIE has a mean (highest) of 49.1%, followed by FID and FIA, which are 10.3% and 8.6% respectively. Renewable energy consumption (% of total final energy consumption) averaged 62.602% indicating a relatively high use of clean energy among African countries. However, almost 30% of countries deviate from the average scores. Approximately 40% of Africans reside in urban centres in their respective countries. With a minimum of 5.416 and a maximum of 90.423%, there is a variation of 17.659% from the mean score. EDU averaged 95.477 with a standard deviation of 25.266. This implies that, although primary school enrolment is high in Africa, over 25% of countries cannot achieve this mean score. Over the study period, the average economic growth is 3.969% with a standard deviation of 7.543%. While some economies contracted by up to -50.339%, others progressed by up to 149.973% over the same period.

Table 2 shows tests for multicollinearity among the variables used in this study. Specifically, correlational tests and the variance inflation factor (VIF) are used. Our results show that financial efficiency, access, and depth negatively correlate with LCF. Similarly, ICT diffusion proxies negatively correlated with LCF. This inverse relationship implies that ICT diffusion and financial institutions' dynamics shrink, whereas the EQ improves, respectively. Further, renewable energy, URB, education, and economic growth are found to have a positive relationship with LCF. This implies that as EQ improves, the variables also improve, and vice versa. The strength of relationships among variables reveals no presence of multicollinearity [67, 68]. Consistent with the assertion opined by [15, 69] and [61] that a correlation coefficient below the 0.85 threshold suggests no multicollinearity, this study concludes as such. Moreover, Table 2 presents the VIF results, indicating the absence of multicollinearity among the explanatory variables in this study. Studies such as [15, 61, 70] and [31] have concluded that if VIF does not exceed 10, the presence of no multicollinearity can be confirmed.

Table 2. Pairwise correlations

Variables	(1)	(2)	(3)	(4)	(5)	(6)	(7)	(8)	(9)	(10)	VIF
(1) LCF	1.000										
(2) ICT2	−0.119	1.000									3.61
(3) ICT3	−0.135	0.811	1.000								4.23
(4) FIE	-0.266	0.207	0.195	1.000							1.31
(5) FID	−0.094	0.308	0.310	0.296	1.000						1.93
(6) FIA	−0.165	0.554	0.656	0.266	0.589	1.000					3.24
(7) REN	0.194	−0.364	−0.427	−0.345	−0.449	-0.627	1.000				3.31
(8) URB	0.282	0.437	0.433	0.067	0.168	0.485	-0.633	1.000			3.09
(9) EDU	0.070	0.300	0.236	−0.035	0.262	0.301	−0.263	0.251	1.000		1.25
(10) GDP	0.024	−0.067	-0.087	0.119	−0.045	−0.034	0.023	−0.038	0.107	1.000	1.06

Source: Authors' work (2025)

The lagged variables are positive and have significant coefficients, indicating that the previous year's environmentally destructive or promoting activities have a substantial impact on the current year's EQ. This study found that ICT2 and ICT3 have a negative and significant effect on LCF at a 1% alpha level. The results of ICT2 on LCF are consistent with the effects of ICT2 on EFP and CO_2 emissions, respectively, as they yield significantly positive coefficients. Thus, in all three models, EQ deteriorates as ICT2 increases, all other things being equal. However, increases in ICT3 significantly reduce EFP and CO_2 emissions in the region. Thus, the more individuals use the internet and related services for their transactions, the less the effect of CO_2 and EFP dwindles, thereby improving the environment.

Our results in Table 3 show that improvement in FIA and FIE stimulates EQ positively (by increasing LCF and reducing EFP, respectively). Similarly, [38] found that the FD index reduces environmental risks in BRICS, MINT, and G7 economies. Nonetheless, FID increases environmental risks by increasing EFP and CO_2 emissions and significantly reducing LCF. These results are corroborated by earlier studies [34–36] that show that financial system development worsens EQ. Furthermore, [37] found that in Africa, environmental degradation increases as FD improves. Accounting for the multidimensionality of FD, [43] found that FD–depth, access, and efficiency reduced EQ in 15 EAGLES from 1984 to 2018. In Nigeria, [46] noted that the use of energy-efficient machines and equipment, as well as ecologically friendly industrial activities, increases as FD improves. Specifically, the study found that FD significantly reduces ecological footprints. The varied results from the three constituents of financial institutions on the three EQ proxies suggest that the effect of financial system dynamics on EQ depends on the specific proxies being used [30].

This study found that both ICT2 and ICT3 significantly reduce LCF at a 1% alpha level. This implies that intensive ICT diffusion (increased mobile penetration and internet usage) has a detrimental effect on EQ. Furthermore, ICT2 significantly intensifies carbon and ecological footprints in the region. Thus, across all three models, ICT2 is found to be a significant determinant of environmental deterioration in Africa. Electrical waste rises in tandem with the growth of mobile devices. Electronic gadgets often contain toxic components; therefore, improper recycling or disposal of these devices can pollute soil and water. However, empirical results show that at 5% and 1% alpha levels, respectively, ICT3 lead to a decline in the EFP and CO_2 emissions, all other things being equal. This implies that as African countries increase internet connectivity in the region, the need for incessant individual and business travel is drastically reduced, thereby reducing carbon emissions associated with vehicular transportation. Again, the move to online learning lessens the demand for printed materials and hard-copy textbooks, which helps to cut down on paper waste and deforestation. Our results align with the results of [50]. They found that CO_2 emissions increase at the initial stages of ICT diffusion, but at a certain threshold, it improves EQ significantly. Other studies [6, 3, 54] have shown that the diffusion of ICT significantly aggravates environmental deterioration.

Furthermore, the interactions between ICT3 and FIE across all models show that EQ significantly improves in Africa. Thus, internet usage increases FIE and improves the environment in Africa, thereby serving as a coordinator that positively drives environmental progress. Similarly, the interaction between ICT3 and FID increases LCF and

Table 3. Financial institutions dynamics, ICT diffusion and EQ nexus

	(1)	(2)	(3)
	LCF	EFP	CO_2
L.LCF	0.787***		
	(30.36)		
L.EFP		0.337***	
		(3.24)	
L.CO2			0.409***
			(31.94)
ICT2	−0.0363***	0.00422***	0.00223***
	(−7.90)	(2.85)	(6.51)
ICT3	−0.0187***	−0.00650**	−0.00424***
	(−3.31)	(−2.46)	(−8.02)
FIE	6.689***	−0.0655	0.352**
	(12.81)	(−0.16)	(2.15)
FID	−24.69***	9.248***	5.534***
	(−7.62)	(4.88)	(29.21)
FIA	32.73***	−4.163***	−0.601***
	(6.44)	(−3.34)	(−2.78)
REN	−0.0115	−0.0334***	−0.0459***
	(−1.55)	(−3.33)	(−16.92)
URB	0.0759***	−0.0512*	−0.0403***
	(3.44)	(−1.83)	(−7.44)
EDU	0.00397	−0.00555	−0.0134***
	(0.44)	(−1.48)	(−4.64)
GDP	−0.0439***	0.00662***	0.0214***
	(−7.99)	(3.21)	(14.44)
_cons	−3.897**	4.952***	5.259***
	(−2.36)	(3.79)	(9.45)
Obs	1138	1138	1133
Countries	47	47	47
F-Stats	2920.4	85.37	1427.9
p-value	5.81e-61	2.86e-26	7.80e-54
Instruments	33	33	32

(*continued*)

Table 3. (*continued*)

	(1)	(2)	(3)
ar1p	0.0406	0.0104	0.0631
ar2p	0.897	0.699	0.244
hansenp	0.536	0.394	0.0934

t statistics in parentheses * $p < 0.1$, ** $p < 0.05$, *** $p < 0.01$
Source: Authors' work (2025)

reduces CO_2 emissions. [71] found that internet use and fixed broadband have a positive effect on FD. The development of financial technologies and products is stimulated as ICT diffusion increases, which leads to economic development [52, 72]. However, the interaction between ICT3 and FID yields a positive coefficient, implying environmental deterioration. The interaction between ICT diffusion and FIA has a detrimental effect on EQ by reducing LCF and increasing CO_2 emissions. This study argues that Africa's fight for EQ can be achieved via efficient financial institutions churning out cutting-edge financial products and services to economic agents. Again, access to financial products and services should be streamlined to ensure that economic agents engage in environmentally friendly activities as access to and the diffusion of ICT increase in the region.

Interestingly, this study reveals that whereas renewable energy consumption reduces EQ (LCF declines), it leads to a decline in EFP and CO_2 emissions. The potential of renewable energy to provide power with minimal or no greenhouse gas emissions is one of its key benefits. Thus, power plants that generate electricity using solar, wind, hydroelectric, and geothermal energy do not release CO_2 emissions or other pollutants that worsen global warming. In SSA, our results are corroborated by similar studies [29, 73] that renewable energy reduces CO_2 emissions and EFP. Again, [32] found that CO_2 emissions are reduced as renewable energy consumption increases. This study found that URB improves EQ in Africa. Specifically, results in Table 3 show that whereas URB increases LCF, it decreases EFP and CO_2 emissions at 10% and 1% alpha levels respectively. This result is corroborated by [43], who state that URB improves the environment in India in the long run. Similarly, [74] found that ecological footprints significantly reduced URB increases in BRICS economies. Thus, URB increases energy efficiency, lowers energy usage, and reduces environmental degradation, reducing CO_2 emissions and EFP. Conversely, [75] found that URB increases CO_2 emissions and leads to environmental degradation.

EQ is adversely affected by GDP growth across all three models. Specifically, an improvement in economic growth reduces LCF and increases EFP and CO_2 emissions significantly at a 1% alpha level. This suggests that African countries are particularly concerned with exploring resources and advancing the production, construction, and building of industries, with less emphasis on ecologically friendly alternatives. This finding is consistent with the EKC hypothesis [76] which implies that African countries are focused on economic growth at the expense of the environment [77]. Thus, African economies have adopted the so-called "*grow-now, clean-up-later approach*", which

[78] considers leaving behind irreversible environmental harm. In SSA, [15] found that whereas economic development increases CO_2 emissions, it has a significant negative impact on ecological footprints. Consistent with previous findings [79, 80], this study also found that education contributes to the reduction of CO_2 emissions. Similarly, [81] found a negative relationship between education and CO_2 emissions. Moreover, an inverted U-shaped relationship between education and CO_2 emissions in Australia [82]. Thus, the initial stages of education tend to pay less attention to environmentally friendly solutions and alternatives; however, education improves EQ significantly in the long term.

Interestingly, the direct effects of ICT3 and FID produced negative coefficients, indicating environmental deterioration (as measured by the reduced LCF proxy); however, their interactive term shows a significant and positive effect on LCF, as presented in Table 4. This implies that ICT drives the expansion of financial institutions in the region and has a favourable net effect on the environment (i.e., increasing LCF and reducing CO_2 emissions). Conversely, the interactive term of FIE and ICT3 has no material effect on ecological footprints. This study further explores the interactive effect of FIA and ICT3 on EQ proxies (LCF, EFP, and CO_2 emissions). Although FIA has a direct and significant positive effect on LCF, the interactive term of ICT3 and FIA has a negative impact on EQ. Thus, internet use negates the environmental benefits of FIA in the region. Similarly, the CO_2 emissions proxy has a significant and positive coefficient, indicating that as ICT3 and FIA interact, more CO_2 emissions are generated by economic agents. Financial institutions' access and internet development reduce the quantum of environmental pressure from financial institutions building office spaces and economic agents travelling to access financial services among others, thereby reducing EFP.

Table 4: The moderating effect of ICT diffusion on financial access, efficiency and depth and EQ in SSA

	(1)	(2)	(3)
	LCF	EFP	CO_2
L.LCF	0.629^{***}		
	(14.24)		
L.EFP		0.342^{***}	
		(5.27)	
L.CO2			0.740^{***}
			(28.54)
ICT2	-0.0485^{***}	0.00339	0.00275^{***}
	(−6.90)	(1.65)	(5.07)
ICT3	-0.271^{***}	0.0130	-0.0241^{***}
	(−3.24)	(0.75)	(−3.42)
FIE	3.047^{***}	-2.051^{***}	2.335^{***}
	(6.08)	(−6.21)	(9.14)

(continued)

Table 4: (*continued*)

	(1)	(2)	(3)
FID	−35.93***	−0.463	4.283***
	(−7.64)	(−0.36)	(10.01)
FIA	59.77***	0.569	-4.513***
	(6.52)	(0.42)	(-8.09)
REN	−0.00229	−0.0156***	0.00449**
	(−0.37)	(−3.76)	(2.04)
URB	0.142***	0.00697	0.0298***
	(5.70)	(1.38)	(5.92)
EDU	−0.0103	−0.00448	−0.0159***
	(−1.25)	(−1.41)	(−8.94)
GDP	−0.0508***	0.00165	0.0172***
	(−6.18)	(0.60)	(14.99)
FIE*ICT3	0.517***	−0.0147	−0.0612***
	(4.01)	(−0.50)	(−6.32)
FID*ICT3	0.322***	0.0405*	−0.0295***
	(3.27)	(1.70)	(−9.32)
FIA*ICT3	−0.639**	−0.121***	0.232***
	(−2.26)	(−3.38)	(17.72)
_cons	−4.172**	3.108***	−1.024***
	(−2.53)	(5.43)	(−3.14)
Obs	1138	1138	1133
Countries	47	47	47
F-Stats	1951.7	355.5	11814.0
p-value	1.83e-58	1.49e-41	1.98e-76
Instruments	33	40	35
ar1p	0.643	0.00104	0.0812
ar2p	0.322	0.615	0.411
hansenp	0.631	0.454	0.712

t statistics in parentheses * $p < 0.1$, ** $p < 0.05$, *** $p < 0.01$
Source: Authors' work (2025)

5 Conclusion and Policy Implications

This study examined the potential impact of ICT diffusion, financial efficiency, depth, and access on the dynamics of EQ among SSA economies. The data used for the estimations were sourced from a sample of 47 economies in the SSA subregion from 1990 to 2021, as listed in the prior sections. Empirical estimates of the research objectives were carried out using the two-step system GMM technique.

The empirical findings reveal that EQ deteriorates (reduces LCF) as the number of mobile cellular subscriptions increases in the region. This finding is corroborated when EFP and CO_2 emissions were also used as a proxy for EQ. Thus, ecological and carbon footprints increase as ICT2 increases. Additionally, Internet use (ICT3) reduces EQ by significantly reducing LCF at a 1% alpha level. Nonetheless, this study found that ICT3 significantly reduces both EFP and CO_2 emissions in the region, all other things being equal. FIA stimulates EQ (across all three proxies) significantly. Thus, among the dynamics of financial institutions, FIA increases LCF but reduces EFP and CO_2 emissions, implying that increasing financial access to Africa is a key driver and determinant in improving EQ. This study finds that FID has a significant negative impact on EQ across all three proxies. The deepening of the African financial sector, including pension funds, insurance firms, banks and mutual funds, creates avenues for the destruction of ecological stability. This suggests that an improvement in the size of these financial institutions may potentially make funds available to economic agents, leading to the consumption of resources that ultimately harm the environment. Specifically, FID reduces LCF while significantly increasing both EFP and CO_2 emissions. Furthermore, FIE improves EQ (by increasing LCF). However, the efficiency of financial institutions significantly contributes to increased CO_2 emissions in the region. The interactive effects of financial institutions' dynamics and ICT diffusion on the EQ nexus were explored. This study found that ICT diffusion has a significant positive effect on EQ, as measured by financial efficiency, in all three models presented in Table 4. Specifically, internet use positively moderates the FIE and LCF nexus, while it negatively moderates the relationship between FIE, EFP, and CO_2 emissions. Furthermore, whereas ICT3 positively moderates the FID and LCF nexus, which implies an improvement in EQ, the moderating effect of ICT3 on the FIA and LCF nexus is negative. This study confirms that the complementary role of internet use in enhancing the financial capabilities of institutions in Africa is crucial in the effort to preserve the environment. The expansion of financial service providers such as banks, mutual funds, pension funds, and insurance companies are driven by ICT usage. Thus, ICT can enhance the effectiveness and transparency of financial services and markets, facilitating the tracking and promotion of environmentally friendly investments. This combination may result in FD being more ecologically friendly. The direct effect of FIA improves EQ by significantly increasing LCF and decreasing both EFP and CO_2 emissions. However, the moderating effect of ICT3 on FIA and EQ (LCF and CO_2 emission proxies) reveals a significant detrimental impact on the environment. Thus, Africa continues to move farther from achieving carbon neutrality due to the proliferation of ICT (internal access) and financial access.

This study makes several practical policy recommendations. First, policymakers should pay critical attention to ICT diffusion and FD due to their enormous contribution to EQ in Africa. Considering the significant effect of internet usage in reducing ecological and carbon footprints, SSA economies should increase committed investment in ICT capital in mobile and broadband infrastructure and internet penetration, as well as education in ICT skills since governments can more successfully enforce environmental restrictions by using ICT to monitor and report infractions. Moreover, public education about environmental issues and promoting sustainable behaviours, including reducing plastic use or water conservation, can be accomplished using online platforms, social media, and mobile apps. This study makes the following essential policy considerations for SSA economies:

Financial institutions should be supported through tax rebates in the acquisition of environmentally friendly equipment that also increases their efficiency. Governments should consider allocating sufficient financial resources to the acquisition and development of ICT infrastructure, expanding network coverage, and protecting the internet space from criminal activities that erode public confidence in its adoption and comprehensive use. Environmental education will stimulate public understanding of ways to safeguard the environment and promote digital literacy; therefore, it should be harnessed. Moreover, education will increase public trust in virtual banking transactions and increase financial efficiency. Since financial services and products are utilized by both individuals and firms that invest in various lucrative ventures and participate in economic activities that tend to exacerbate environmental deterioration, it is imperative that environmentally friendly financial sector policies are strictly implemented. For example, environmentally responsible firms can be incentivized by lowering interest rates to reduce their environmental impact, increase transparency about ESG matters, and align their business plans with sustainable development objectives through the use of financial instruments.

Despite the fascinating results of this study, there are a few limitations to consider. Variables used in this study were sourced from various databases, and therefore, probable inherent shortfalls in the variables' computations cannot be accounted for in this study. Considering the levels of economic development in various African countries, future studies can consider segregating SSA economies based on income or development groups in studying this phenomenon.

References

1. Edziah, B.K., Sun, H., Adom, P.K., Wang, F., Agyemang, A.O.: The role of exogenous technological factors and renewable energy in carbon dioxide emission reduction in Sub-Saharan Africa. Renew. Energy (2022)
2. Leimbach, M., Roming, N., Schultes, A., Schwerhoff, G.: Long-term development perspectives of Sub-Saharan Africa under climate policies. EcologicalEconomics (2018). https://doi.org/10.1016/j.ecolecon.2017.07.033
3. Atsu, F., Adams, S., Adjei, J.: ICT, energy consumption, financial development, and environmental degradation in South Africa. Heliyon (2021)
4. Goundar, S., Appana, S.: Mainstreaming development policies for climate change in Fiji: a policy gap analysis and the role of ICTs. In: Sustainable Development: Concepts, Methodologies, Tools, and Applications. IGI Global, pp. 402–432 (2018)
5. Park, Y., Meng, F., Baloch, M.A.: The effect of ICT, financial development, growth, and trade openness on CO2 emissions: an empirical analysis. Environ. Sci. Pollut. Res. **2018**(25), 30708–30719 (2018)
6. Avom, D.H., Nkengfack, H.K., Fotio, T.A.: ICT and EQ in Sub-Saharan Africa: effects and transmission channels. Technol. Forecast. Soc. Change **155**, 120028 (2020). https://doi.org/10.1016/j.techfore.2020.120028

7. Chhabra, M., Verma, A., Giri, A.K.: ICT Diffusion, trade openness and growth: empirical evidence from Asian Countries. Indian Econ. J. 1–14 (2024). https://doi.org/10.1177/00194662241263316
8. Belkhir, L., Elmeligi, A.: Assessing ICT global emissions footprint: trends to 2040 & recommendations. J. Clean. Prod. **177**, 448–463 (2018)
9. Borowski, P.F.: Digitization, digital twins, blockchain, and Industry 4.0 as elements of management process in enterprises in the energy sector. Energies **14**, 1885 (2021). https://doi.org/10.3390/en14071885
10. Asongu, S., De Moor, L.: Financial globalisation dynamic thresholds for financial development: evidence from Africa. Eur. J. Dev. Res. **29**(1), 192–212 (2017). https://doi.org/10.1057/ejdr.2016.10
11. Arshad, Z., Robaina, M., Botelho, A.: The role of ICT in energy consumption and environment: an empirical investigation of Asian economies with cluster analysis. Environ. Sci. Pollut. Res. Int. **27**(26), 32913–32932 (2020). https://doi.org/10.1007/s11356-020-09229-7
12. Dahmani, M., Mabrouki, M., Youssef, A.B.: The ICT, financial development, energy consumption and economic growth nexus in MENA countries: dynamic panel CS-ARDL evidence. Appl. Econ. **55**,10, 1114–1128 (2023) https://doi.org/10.1080/00036846.2022.2096861
13. Assi, A.F., Isiksal, A.Z., Tursoy, T.: Renewable energy consumption, financial development, environmental pollution, and innovations in the ASEAN+ 3 Group: evidence from (P-ARDL) model. Renew. Energy **165**(1), 689–700 (2021). https://doi.org/10.1016/j.renene.2020.11.052
14. Dada, J.T., Adeiza, A., Ismail, N.A., Arnaut, M.: Financial development–ecological footprint nexus in Malaysia: the role of institutions. Management of EQ. Int. J. **33**(4), 913–937 (2022)
15. Abaidoo, R., Agyapong, E.K.: Development, effective governance and environmental sustainability risk: emerging markets perspective. Environ. Sci. Pollut. Res. (2023). https://doi.org/10.1007/s11356-023-25458-y
16. Pata, U.K.: Renewable energy consumption, urbanization, financial development, income and CO2 Emissions in Turkey: testing EKC hypothesis with structural breaks. J. Clean. Prod. **187**, 770–779 (2018). https://doi.org/10.1016/j.jclepro.2018.03.236
17. Jakada, A.H., Mahmood, S., Ali, U.A., Aliyu, D.I.: The moderating role of ICT on the relationship between foreign direct investment and the quality of environment in selected African countries. Cogent Econ. Financ. **11**(1), 2197694 (2023). https://doi.org/10.1080/23322039.2023.2197694
18. Ntow-Gyamfi, M., Bokpin, G.A., Aboagye, A.Q.Q., Ackah, C.G.: Environmental sustainability and financial development in Africa; does institutional quality play any role? Dev. Stud. Res. (2020). https://doi.org/10.1080/21665095.2020.1798261
19. Musah, M., Owusu-Akomeah, M., Nyeadi, J.D., Alfred, M., Mensah, I.A.: Financial development and environmental sustainability in West Africa: evidences from heterogeneous and cross-sectionally correlated models. Environ. Sci. Pollut. Res. (2021). https://doi.org/10.1007/s11356-021-16512-8
20. Ibrahim, M., Vo, X.: Exploring the relationships among innovation, financial sector development and environmental pollution in selected industrialized countries. J. Environ. Manage. **284**, 112057 (2021). https://doi.org/10.1016/j.jenvman.2021.112057
21. Shobande, O.A., Asongu, S.A.: Searching for sustainable footprints: does ICT Increase CO_2 emissions? Environ. Model. Assess. **28**, 133–143 (2023). https://doi.org/10.1007/s10666-022-09859-w(2022)
22. Dahmani, M., Mabrouki, M., Youssef, A.B.: The ICT, financial development, energy consumption and economic growth nexus in MENA countries: dynamic panel CS-ARDL evidence. Appl. Econ. **55**(10), 1114–1128 (2023). https://doi.org/10.1080/00036846.2022.2096861

23. Asongu, S.A., Nwachukwu, J.C.: The role of governance in mobile phones for inclusive human development in sub-Saharan Africa. Technovation **55–56**, 1–13 (2016)
24. Tchamyou, S.V.: The role of knowledge economy in African business. J. Knowl. Econ. (2016). https://doi.org/10.1007/s13132-016-0417-1
25. Asongu, S.A.: CO2 emission thresholds for inclusive human development in sub-Saharan Africa. Environ. Sci. Pollut. Res. **25**(26), 26005–26019 (2018). https://doi.org/10.1007/S11356-018-2626-6
26. Siche, R., Pereira, L., Agostinho, F., Ortega, E.: Convergence of ecological footprint and emergy analysis as a sustainability indicator of countries: Peru as case study. Commun. Non-linear Sci. Numer. Simul. **15**, 3182–3192 (2010)
27. Pata, U.K.: Do renewable energy and health expenditures improve load capacity factor in the USA and Japan? A new approach to environmental issues. Eur. J. Health Econ. **22**(9), 1427–1439 (2021). https://doi.org/10.1007/s10198-021-01321-0
28. Alola, A.A., Ozkan, O., Usman, O.: Role of non-renewable energy efficiency and renewable energy in driving environmental sustainability in India: evidence from the load capacity factor hypothesis. Energies **16**, 2847 (2023). https://doi.org/10.3390/en16062847
29. Salahuddin, M., Gow, J.: The effect of internet usage, financial development and trade openness on economic growth in South Africa: a time series analysis. Telematics Inform. **33**(4), 1141–1154 (2016)
30. Ofori, K.I., Osei, D.B., Alagidede, I.P.: Inclusive growth in Sub-Saharan Africa: exploring the interaction between ICT diffusion, and financial development. Telecommun. Policy (2022)
31. Annor, L.D.J., Robaina, M., Vieira, E.: Financial development, inclusive growth, and environmental quality: emerging markets perspective. Environ. Dev. Sustain. (2023). https://doi.org/10.1007/s10668-023-04198-6
32. Acheampong, A.O., Amponsah, M., Boateng, E.: Does financial development mitigate carbon emissions? Evidence from heterogeneous financial economies. Energy Econ. (2020). https://doi.org/10.1016/j.eneco.2020.104768
33. Annor, L.D.J., Robaina, M., Vieira, E.: Climbing the green ladder in Sub-Saharan Africa: dynamics of financial development, green energy, and load capacity factor. Environ Syst Decis. **2024** (2024). https://doi.org/10.1007/s10669-023-09959-2
34. Farhani, S., Ozturk, I.: Causal relationship between CO2 Emissions, Real GDP, Energy Consumption, Financial Development, Trade Openness, and Urbanization in Tunisia. Environ. Sci. Pollut. Res. **22**, 15663–15676 (2015)
35. Madni, G.R., Anwar, M.A., Ahmad, N.: Socio-economic Determinants of Environmental Performance in Developing Countries. J. Knowl. Econ. **13**(2), 1157–1168 (2022)
36. Vo, D.H., Vo, A.T., Le, Q.T.: Does the Kuznets curve apply for financial development and environmental degradation in the Asia-Pacific region? Heliyon **7**(7), e06708 (2021). https://doi.org/10.1016/j.heliyon.2021.e06708
37. Nathaniel, S.P., Adeleye, N.: Environmental preservation amidst carbon emissions, energy consumption, and urbanization in selected African countries: Implication for sustainability. J. Cleaner Prod. **285** (2021)
38. Ofori, E.K., Onifade, S.T., Ali, E.B., Alola, A.A., Zhang, J.: Achieving carbon neutrality in post COP26 in BRICS, MINT, and G7 economies: the role of financial development and governance indicators. J. Cleaner Prod. **387** (2023)
39. Usman, M., Kousar, R., Makhdum, M.S.A., Yaseen, M.R., Nadeem, A.M.: Do financial development, economic growth, energy consumption, and trade openness contribute to increase carbon emission in Pakistan? An insight based on ARDL bound testing approach. Environ. Dev. Sustain. **25**(1), 444–473 (2023)
40. Zoaka, J.D., Ekwueme, D.C., Güngör, H., Alola, A.A.: Will financial development and clean energy utilization rejuvenate the environment in BRICS economies? Bus. Strateg. Environ. **31**(5), 2156–2170 (2022)

41. Qamri, M.G., Sheng, B., Adeel-Farooq, R., Alam, G.M.: The criticality of FDI in environmental degradation through financial development and economic growth: Implications for promoting the green sector. Resour. Policy **78**, 102765 (2022). https://doi.org/10.1016/j.resourpol.2022.102765
42. Dada, J.T., Adeiza, A., Noor, A.I., Marina, A.: Investigating the link between economic growth, financial development, urbanisation, natural resources, human capital, trade openness and ecological footprint: evidence from Nigeria. J. Bioecon. **24**(2), 153–179 (2022)
43. Khan, M.A., Khan, M.A., Ahmed, M., Khan, K.: Environmental consequences of financial development in emerging and growth-leading economies: a multidimensional assessment. Borsa Istanbul Rev. **22**(4), 668–677 (2022)
44. Abduh, M., Buys, W.A., Aziz, S.A.: Exploring the relationship between islamic financial development, energy consumption, and environmental quality. Int. J. Energy Econ. Policy (2022)
45. Mesagan, E.P., Olunkwa, C.N.: Heterogeneous analysis of energy consumption, financial development, and pollution in Africa: the relevance of regulatory quality. Utilities Policy (2022)
46. Dada, J.T., Ojeyinka, T.A., Al-Faryan, M.A.S.: Does financial development has (a) symmetric effect on environmental quality: insights from South Africa. J. Econ. Stud. (2022b)
47. Rani, T., Amjad, M.A., Asghar, N., Rehman. H.U.: Revisiting the environmental impact of financial development on economic growth and carbon emissions: evidence from South Asian economies. Clean Technol. Environ. Policy **24**, 2957–2965 (2022). https://doi.org/10.1007/s10098-022-02360-8
48. IEA: Energy Efficiency 2019–The authoritative tracker of global energy efficiency trends. In: The authoritative tracker of global energy efficiency trends (2019). https://www.iea.org/reports/energy-efficiency 2019?msclkid=edf20355b9c911ec9b4b422ede507de2
49. IEA Energy Technology Perspectives 2020 – Part of Energy Technology Perspectives (2020). https://www.iea.org/reports/energy-technology-perspectives-2020?msclkid=9b7c27aeb9c911ec9f4ac58dbc3d07d7
50. Faisal, F., Azizullah, T., & Pervaiz, R.: Does ICT lessen CO2 emissions for fast-emerging economies? An application of the heterogeneous panel estimations. Environ. Sci. Pollut. Res. **27**(10), 10778–10789 (2020). https://doi.org/10.1007/S11356-019-07582-W
51. Cheng, C.Y., Chien, M.S., Lee, C.C.: ICT diffusion, financial development, and economic growth: an international cross-country analysis. Econ. Model. **94**(February 2020), 662–671 (2021)
52. Verma, A., Chhabra, M., Giri, A.K.: ICT diffusion, energy consumption, institutional quality, and environmental sustainability in 20 emerging economies during 2005–2019. Int. J. Environ. Sci. Technol. ISSN 1735–1472 (2023)
53. Kouton, J.: Information Communication Technology development and energy demand in African countries. Energy **189**, 116192 (2019)
54. Tsaurai, K., Chimbo, B.: The impact of information and communication technology on carbon emissions in emerging markets. Int. J. Energy Econ. Pol. **9**(4), 320 (2019)
55. Amri, F., Zaied, Y.B., Lahouel, B.B.: ICT, total factor productivity, and carbon dioxide emissions in Tunisia. Technol. Forecast. Soc. Change **146**, 212–217 (2019)
56. Mbarek, M.B., Zghidi, N.: Dynamic links between ICT, transport energy, environmental degradation and growth: empirical evidence from Tunisia. Environ. Econ. **8**(3), 76–83 (2017)
57. Han, F., Ibrahim, R.L., Al-Mulali, U., et al.: Tracking the roadmaps to sustainability: what do the symmetric effects of eco-digitalisation, green technology, green finance, and renewable energy portend for China? Environ. Dev. Sustain. **26**, 13895–21391 (2024). https://doi.org/10.1007/s10668-023-04289-4

58. Shen, J., Ibrahim, R.L., Ajide, K.B., Al-Faryan, M.A.S.: Tracking environmental sustainability pathways in Africa: do natural resource dependence, renewable energy, and technological innovations amplify or reduce the pollution noises? Energy Environ. **35**(1), 88–112 (2024). https://doi.org/10.1177/0958305X221124221
59. Ofoeda, I., Agbloyor, E.K., Abor, J.Y.: How do anti-money laundering systems affect FDI flows across the globe? Cogent. Econ. Financ. **10**(1), 2058735 (2022)
60. Abaidoo, R., Agyapong, E.K.: Financial institutions dynamics, investments and development. J. Emerging Financ. (2023). https://doi.org/10.1177/09726527231160856
61. Agyapong, E.K., Annor, L.D.J., Ohemeng, W.: Corporate social responsibility and performance among rural banks in Ghana: the moderating role of governance structures. Int. J. Soc. Econ. (2023). https://doi.org/10.1108/IJSE-02-2023-0116
62. Hwang, J., Sun, Y.: Should we go one step further? An accurate comparison of one-step and two-step procedures in a generalised method of moments framework. J. Econometrics **207**(2), 381–405 (2018)
63. Ibrahim, R.L., Ajide, K.B.: Trade facilitation and environmental quality: empirical evidence from some selected African countries. Environ. Dev. Sustain. **24**(1282–1312), 2022a (2022). https://doi.org/10.1007/s10668-021-01497-8
64. Ibrahim, R.L., Ajide, K.B.: Trade facilitation, institutional quality, and sustainable environment: renewed evidence from Sub-Saharan African Countries. J. Afr. Bus. **23**(2), 281–303 (2022). https://doi.org/10.1080/15228916.2020.1826886
65. Rogers, E.M.: Diffusion of Innovations. Free Press, New York (1962)
66. Avom, D.H., Nkengfack, H.K., Fotio, T.A.: ICT and EQ in Sub-Saharan Africa: effects and transmission channels. Technol. Forecast. Soc. Chang. **155**, 120028 (2020). https://doi.org/10.1016/j.techfore.2020.120028
67. Annor, L.D.J., Robaina, M., Vieira, E., Mota, J.: Pursuing the Carbon Neutrality agenda: the role of financial development, foreign direct investment, and effective governance. Eur. J. Dev. Res. **37**(666–697), 2025a (2025). https://doi.org/10.1057/s41287-025-00696-7
68. Annor, L.D.J., Kyei-Mensah, T., Addae, J.A., Robaina, M., Vieira, E.: Fostering energy transition in Africa: financial systems, inclusive growth, corruption control and ICT as catalysts. Sustain. Environ. **11**(1) (2025b). https://doi.org/10.1080/27658511.2025.2518791
69. Elith, J., et al.: Novel methods improve prediction of species' distributions from occurrence data. Ecography **29**(2), 129–151 (2006)
70. Liao, D., Valliant, R.: Variance inflation factors in the analysis of complex survey data. Surv. Methodol. **38**(1), 53–62 (2012)
71. Alshubiri, F., Jamil, S.A., Elheddad, M.: The impact of ICT on financial development: empirical evidence from the Gulf Cooperation Council countries. Int. J. Eng. Bus. Manage. **11**, 1–14 (2019)
72. Marszk, A., Lechman, E. (Eds.).: The digitalization of financial markets: the socioeconomic impact of financial technologies (1st ed.) (2021). Routledge. https://doi.org/10.4324/9781003095354
73. Adams, S., Nsiah, C.: Reducing carbon dioxide emissions; does renewable energy matter? Sci. Total. Environ. **693**, 133288 (2019)
74. Ulucak, R., Khan, S.U.D.: Does information and communication technology affect CO2 mitigation under the pathway of sustainable development during the mode of globalization? Sustain. Dev. (2020)
75. Hu, Z., Khan, H.: The effect of institutions and urbanization on environmental quality: evidence from the Belt and Road Initiative countries using dynamic panel models. Environ. Sci. Pollut. Res. **30**, 65746–65761 (2023). https://doi.org/10.1007/s11356-023-27031-z
76. Grossman, G.M., Krueger, A.B.: Does the Environmental Kuznets Curve describe how individual countries behave. Q. J. Econ. **110**(2), 353–377 (1995)

77. Udeagha, M.C., Breitenbach, M.C.: Exploring the moderating role of financial development in environmental Kuznets curve for South Africa: fresh evidence from the novel dynamic ARDL simulations approach. Financ. Innov. (2023). https://doi.org/10.1186/s40854-022-00396-9
78. Kahuthu, A.: Economic growth and environmental degradation in a global context. Environ. Dev. Sustain. **8**, 55–68 (2006). https://doi.org/10.1007/s10668-005-0785-3
79. Wu, Y., Wan, J., Yu, W.: Impact of environmental education on environmental quality under the background of low-carbon economy. Front. Public Health **11**, 1128791 (2023). https://doi.org/10.3389/fpubh.2023.1128791
80. Zafar, M.W., Zaidi, Syed, A.H., Mansoor, S., Sinha, A., Qin, Q.: ICT and education as determinants of environmental quality: The role of financial development in selected Asian countries (2022). Online at https://mpra.ub.uni-muenchen.de/111920/ MPRA Paper No. 111920, posted 11 Feb 2022 11:35 UTC
81. Bano, S., Zhao, Y., Ahmad, A., Wang, S., Liu, Y.: Identifying the impacts of human capital on carbon emissions in Pakistan. J. Clean Prod. (2018). https://doi.org/10.1016/j.jclepro.2018.02.008
82. Balaguer, J., Cantavella, M.: The role of education in the Environmental Kuznets Curve. evidence from Australian data. Energy Econ. **70**, 289–96 (2018). https://doi.org/10.1016/j.eneco.2018.01.021

Determinants of the Type of Opinion Issued by the Auditor - Empirical Evidence in the Portuguese Capital Market

Kátia Lemos(✉), Sara Serra, and Maria Francisca Abreu

Research Centre On Accounting and Taxation (CICF), Polytechnic University of Cávado and Ave, Campus Do IPCA, 4750-810 Barcelos, Portugal
klemos@ipca.pt

Abstract. This paper aims to identify the factors that influence the type of opinion issued by the auditor in the Audit Report of companies listed on Euronext Lisbon. The data were collected from the Reports and Accounts and the respective Audit Report of the companies listed on Euronext Lisbon for the period from 2018 to 2022. After content analysis, the data were statistically processed through bivariate statistical analyses. It was found that, of all the variables analyzed, the audit fee variable and the type of audit firm variable influence the type of opinion issued by the auditor. That is, when the fees are higher, there is a greater probability of issuing an unmodified opinion. In addition, companies audited by Big Four firms are more likely to receive an unmodified opinion.

Keywords: Type of opinion · Auditor · Audit report · determining factors

1 Introduction

Financial auditing aims to attest to the reliability of financial statements and is therefore fundamental to the functioning of the capital market. In Portugal, all entities listed on the capital market are subject to a statutory audit, conducted annually, which results in an Audit Report.

In this context, the aim of this study is to identify the factors that influence the type of opinion issued in the Audit Report of companies listed on Euronext Lisbon. To this end, the determining factors considered are the type of audit firm (whether or not it is a Big Four), the gender of the auditor, the rotation of the audit firm, and the existence of the pandemic.

The data were collected from the Reports and Accounts and the respective Audit Report of the companies listed on Euronext Lisbon for the period from 2018 to 2022. After content analysis, the data were statistically processed through bivariate statistical analyses.

The choice of this topic is due to the relevance of the Audit Report for all stakeholders, making it pertinent to understand their determining factors. Furthermore, the fact that this is a longitudinal study allows us to understand the evolution of results over time.

L. Almeida et al. (Eds.): ICAFI 2025, ISEM 80, pp. 318–336, 2026.
https://doi.org/10.1007/978-3-032-16268-7_18

Following this introduction, this paper includes a literature review on the type of auditor opinion, followed by a review of studies and research hypotheses, as well as the presentation of the empirical study, with its methodology, data processing, sample, and discussion of results. The paper concludes with the drafting of conclusions, limitations, and suggestions for future research.

2 Type of Opinion Issued by the Auditor

Once the audit process is completed, the auditor issues an opinion on the financial statements of the audited company. Consequently, the auditor must express a judgment on whether the financial statements are prepared, in all material respects, in accordance with the applicable financial reporting framework. To form this opinion, the auditor must determine whether they have obtained reasonable assurance that the financial statements as a whole are free from material misstatement, whether due to fraud or error.

In this regard, auditors issue an opinion that may be either modified or unmodified. An unmodified opinion, as set out in ISA 700 - Forming an Opinion and Reporting on Financial Statements, means that the financial statements are prepared, in all material respects, in accordance with the applicable financial reporting framework and therefore free from misstatements [1]. The unmodified opinion can be divided into two types: an unqualified opinion or an unqualified opinion with an emphasis of matter paragraph, when there are no misstatements, but there is some important matter the auditor wishes to highlight.

A modified opinion is issued when the financial statements contain material misstatements or when the auditor is unable to obtain appropriate and sufficient evidence to support their opinion [2]. According to ISA 705 - Modifications to the Opinion in the Independent Auditor's Report, the modified opinion can be classified as a qualified opinion, an adverse opinion, or a disclaimer of opinion [3].

A qualified opinion should be issued by the auditor when there are material misstatements, individually or in the aggregate, that are not pervasive to the financial statements; or when sufficient and appropriate audit evidence cannot be obtained to form a basis for the opinion [3].

An adverse opinion should be issued by the auditor when, having obtained sufficient and appropriate audit evidence, they conclude that the misstatements, individually or in the aggregate, are both material and pervasive to the financial statements [3].

A disclaimer of opinion should be issued when the auditor is unable to obtain sufficient and appropriate audit evidence on which to base their opinion and concludes that the possible effects of undetected misstatements could be both material and pervasive to the financial statements [3].

In continuation, ISA 705 (revised) - Modifications to the Opinion in the Independent Auditor's Report, presents in paragraph A10 the types of modified opinions, as shown in Table 1.

However, when referring to a qualified opinion, it is common to distinguish between a qualified opinion due to disagreement and a qualified opinion due to scope limitation, as mentioned in [4]. A qualification due to disagreement arises when there is an irreconcilable disagreement between management and the auditor on certain matters. Examples

Table 1. Types of modified opinions

Natur of matter giving rise to the modification	Aditor's judgement about the pervasiveness of the effects or possible effects on the financial statemets	
	Material but not pervasive	Material and pervasive
Financial statements are materially misstated	Qualified opinion	Adverse opinion
Inability to obtain sufficient appropriate audit evidence	Qualified opinion	Disclaimer of opinion

Fonte: [3].

include overvaluation of inventories, or management failing to disclose in the notes to the financial statements aspects that are relevant to the entity's going concern status, among others.

A scope limitation occurs when the auditor is unable to obtain sufficient and appropriate audit evidence. This may be due to events beyond the audited entity's control, such as fires, or due to limitations imposed by management [4].

3 Review of Studies and Research Hypotheses

There have been several studies on this subject, both in the public and private sectors. Most studies seek to analyse the association between variables associated with the auditor and the Audit Report. These variables are: the gender of the auditor, the type audit firm associated with whether or not it is Big Four, the rotation of the audit firm and the auditor's fees. In line with these studies, the following research hypotheses were defined.

3.1 Type of Audit Firm (Big Four or not)

According to the study by [5], audit clients tend to be more interested in larger and more reputable audit firms than in smaller ones. To protect their reputation, audit firms aim to avoid issuing distorted reports, since an inaccurate report that fails to fairly and appropriately represent the audited company can damage the auditor's reputation. That is, larger audit firms tend to issue undistorted reports. Consequently, the study found a correlation between the type of opinion issued by the auditor and the type and size of the audit firm, showing that the issuance of an unmodified opinion is strongly correlated with the auditor being part of a Big Four firm.

Similarly, among other variables analyzed, [6] found that the type of audit firm influences the opinion issued by the auditor, concluding that when the audit firm is national - i.e., not part of the Big Four - there is a higher probability of issuing a modified opinion.

According to DeAngelo (as cited by [5]), stakeholders place greater credibility in the opinions issued by larger audit firms.

[7] also analyzed the effects of the Big Four on auditor opinion in listed companies, finding that 75% of these companies were audited by Big Four firms. The authors concluded that the Big Four issued fewer modified opinions. Supporting this result, [7] cite

the study by Francis and Yu (2009), which found that companies audited by Big Four firms are less likely to receive a modified opinion.

Thus, research hypothesis 1 was formulated as follows:

H1: The type of audit firm (Big Four or non-Big Four) influences the type of opinion issued by the auditor in the Audit Report.

3.2 Audit Fees

The study conducted by [5], previously discussed, refers to audit fees. Thus, when making a decision, clients tend to choose larger audit firms, as they believe their staff are more qualified and these firms are more credible to stakeholders. This leads to higher fees being charged by larger audit firms. Due to charging higher fees, auditors feel compelled to perform higher-quality work. In other words, higher audit quality leads to higher fees. In their study, the authors concluded that there is a significant positive correlation between the type of opinion issued by the auditor and the audit fees, meaning that an unmodified opinion is associated with higher fees.

[8] conducted a review of empirical studies, in which they analyzed various studies by other authors on the topic of audit fees. These studies present diverse views regarding fees. There is a tendency among auditors to offer fee discounts when dealing with a new client. Consequently, some studies associate fee discounts (which result in lower fees) with lower audit quality, that is, there may be a higher likelihood of reports being issued with an unmodified opinion.

[9] conducted an investigation in which they concluded that the hours spent and the effort exerted by auditors improve the quality of the audit provided. These two variables are positively related to the fees paid to the audit firm and, consequently, to the type of opinion issued in the Audit Report. In this sense, the following research hypothesis was formulated:

H2: Audit fees influence the type of opinion issued by the auditor in the Audit Report.

3.3 Audit Firm Rotation

The study by [10], conducted in Brazil, found that when an auditor issues a modified opinion, the management of the audited company often responds negatively, which may result in a change of audit firm. Additionally, when a rotation of the audit firm is anticipated and the auditor is aware that they will no longer audit the client, they tend to feel more at ease issuing a modified opinion. In line with this perspective, [11] examined the impact of audit firm rotation on the type of opinion issued and concluded that there is a positive relationship between the two variables in Brazilian companies.

Expanding on this topic, [12] determined that in the year of audit firm replacement, auditors are motivated to "clean up" the financial statements. That is, the change of audit firm often leads to increased auditor diligence, thereby enhancing audit quality. This study also demonstrated that audit quality improves both in the period immediately before and after the rotation, supporting the benefits of mandatory audit firm rotation.

In a study comparing Chinese companies subject to mandatory audit firm rotation with those not subject to such regulations, [13] concluded that auditors are more likely to issue modified opinions when rotation is mandatory.

Similarly, [14], in a study covering Taiwanese public companies between 2016 and 2018, found that the likelihood of auditors issuing modified opinions and raising different key audit matters increases when an audit firm change occurs. Such a change may affect audit quality, as newly appointed auditors tend to exercise more critical judgment during their initial engagements.

In contrast, [15] analyzed the non-compliance with mandatory audit firm rotation in China and concluded that when audit firm rotation does not occur, auditors are less likely to issue modified opinions. In such cases, audit reports tend to be less critical, and higher audit fees are often charged.

Based on the empirical findings discussed, the following research hypothesis was formulated:

H3: Audit firm rotation influences the type of opinion issued by the auditor in the Audit Report.

3.4 Auditor Gender

Another variable examined in various studies is the gender of the auditor. [16] conducted a study to assess the potential effects of auditor gender on the audit report and the corresponding opinion issued. The authors found that female auditors identify more potential misstatements than their male counterparts, although they assess these misstatements with less precision. Furthermore, they found that female auditors do not report more misstatements than male auditors that is, they do not issue more modified opinions indicating no correlation between auditor gender and the type of audit opinion issued.

In contrast, [17] concluded that female auditors are less likely to issue a modified opinion. Conversely, [18], in a study conducted in the public sector, found that female auditors are more likely to issue a report with a modified opinion (usually due to scope limitations) when compared to male auditors.

The study by [19], focused on nonprofit organizations, also revealed a significant correlation between female auditors and the likelihood of issuing modified audit opinions. These modified opinions were found to arise due to errors rather than uncertainties. A similar conclusion was reached by [20], whose research focused on the Finnish market. Their results showed that female auditors tend to be more persistent and more inclined to issue modified opinions. This tendency was particularly evident in audits conducted for the first time following a change in audit firm.

Based on this body of research, the following hypothesis was formulated:

H4: The auditor's gender influences the type of opinion issued by the auditor in the Audit Report.

3.5 Impact of the Covid-19 Pandemic

Another variable to be analyzed in this study is the impact of the COVID-19 pandemic on the audit report, as outlined by the Portuguese Order of Statutory Auditors (Ordem dos Revisores Oficiais de Contas - OROC). During the pandemic period, auditors were expected to face various potential challenges.

Due to restrictions on movement, auditors might have encountered limitations in accessing the entity's premises, which could have made it impossible to attend physical inventory counts. In addition, entities might have restricted access to original documents, and auditors may have been unable to perform alternative procedures [21].

One example includes the accounting policies adopted with regard to government subsidies applied for by the entity, where the appropriate response could be a qualified opinion due to the instability present at the time. The going concern assumption may also have been called into question for several entities, given the material uncertainty that existed, raising significant doubt about the entity's ability to continue operating in the following year [21].

In this context, the OROC issued reporting recommendations for such situations. These included qualified opinions due to scope limitations resulting from insufficient audit evidence such as not being able to attend inventory counts. There could also be qualified opinions due to disagreement, when the selection of accounting policies was deemed inappropriate or where there was a significant lack of disclosures, for instance [21].

[22] conducted a study evaluating the impact of the COVID-19 pandemic on the performance of publicly listed European companies in 2020. Their analysis concluded that, as expected, the audit reports issued during the pandemic year were of lower quality compared to those from the previous year (2019, a non-pandemic year). From this perspective, audit quality serves as a safeguard against earnings management that is, less earnings manipulation implies a clean (unmodified) audit opinion.

As [23] notes, companies most affected by the adverse effects of the pandemic were more likely to receive modified opinions, either due to insufficient and inappropriate audit evidence or the identification of material misstatements.

When the pandemic emerged, audits related to the 2019 financial year treated the outbreak as a non-adjusting subsequent event. This led entities to disclose potential impacts of the pandemic in the financial statements, even though these impacts remained uncertain. In such cases, auditors chose to include an emphasis of matter paragraph in the audit report, drawing stakeholders' attention to these disclosures. Based on this, [23] states it is reasonable to expect that reports especially those of the most affected entities would contain modified opinions.

Accordingly, the following research hypothesis was developed:

H5: The COVID-19 pandemic influenced the type of opinion issued by the auditor in the Audit Report.

4 Research Methodology

4.1 Methodology and Data Processing

In the present empirical study, a qualitative methodology will be employed, with data collected through documentary content analysis. For this purpose, the Annual Reports and Accounts, as well as the corresponding Audit Reports of companies listed on Euronext Lisbon, will be analyzed for the period 2018 to 2022. The choice of Annual Reports is justified by the fact that they are comprehensive yearly documents that contain more complete and accessible information. The selected period -2018 and 2019 (pre-pandemic years), and 2020 to 2022 (pandemic years) - was chosen to allow for comparison before and during the COVID-19 pandemic.

For data processing, a quantitative methodology will be applied, specifically bivariate statistical analyses. This bivariate analysis will examine the relationship between the dependent variable (Type of Opinion) and the independent variables (Type of Audit Firm, Audit Fees, Audit Firm Rotation, Auditor Gender, and Pandemic Year) in order to assess any existing correlations.

Due to the inclusion of both quantitative and dichotomous variables (i.e., variables with only two possible values, in this case, 0 and 1), two types of statistical tests were conducted: the Chi-Square Test and the T-Test for means.

In this study, the T-Test is used when one variable is dichotomous and the other is quantitative. The Chi-Square Test is applied when both the dependent variable (Type of Opinion) and the independent variables (Type of Audit Firm, Audit Firm Rotation, Auditor Gender, and Pandemic Year) are dichotomous.

4.2 Population and Sample

The total population consists of 52 companies listed on Euronext Lisbon, as of April 10, 2024 (the date on which the website was accessed). Initially, 4 companies were excluded for belonging to the sports sector and therefore having a reporting period different from that of the other companies. Subsequently, 14 more companies were excluded due to the unavailability of their Annual Reports and Accounts for all five accounting periods under study, or due to the lack of complete data for all the variables analyzed across the five-year period.

Thus, the final sample consists of 34 companies, corresponding to a total of 170 Audit Reports, with 34 Audit Reports analyzed for each year from 2018 to 2022.

5 Presentation and Discussion of Results

5.1 Descriptive Analysis

For this study, a total of N = 170 observations were obtained for the period from 2018 to 2022.

Starting with an analysis of the dependent variable (type of audit opinion) over the five years under review, we can observe that, during the study period (see Table 2), auditors predominantly issued unmodified opinions, with 155 reports, representing 91% of the sample. Over the five years, only 15 audit reports contained modified opinions, accounting for 9% of the sample.

Table 2. Analysis of the type of opinion

		Frequency	%
Type of opinion	Unmodified	155	91%
	Modified	15	9%
	Total	170	100%

When we analyze the information by year in Table 3, we can observe that, for the most part, reports with an unmodified opinion were issued, with a frequency of more than 30 observations per year. Regarding reports with a modified opinion, the frequency was between 2 and 4 reports per year, with 2020 and 2021 being the years with the most reports containing a modified opinion.

Table 3. Analysing the type of opinion per year

		2018	2019	2020	2021	2022
		N	N	N	N	N
Type of opinion	Unmodified	32	32	30	30	31
	Modifed	2	2	4	4	3
	Total	34	34	34	34	34

Regarding the independent variables (Type of firm, Fees, Firm rotation, Auditor's gender, and Pandemic year), starting with the analysis of the type of auditing firm (see Table 4), that is, whether it is a Big Four or not, we can conclude that out of the 170 reports analyzed, 126 are reports issued by Big Four firms, which represents 74% of the sample. The remaining 44 reports are issued by non-Big Four firms, which accounts for 26% of the reports.

Table 4. Analysis of the type of audit firm variable

		Frequency	%
Big Four	0 – No Big Four	44	26%
	1 – Big Four	126	74%
	Total	170	100%

Concerning the fees, the total fees (which include the fees for all companies within the group) and the fees for the analyzed company (the company's fees within the total fees) were extracted from the Reports and Accounts of the analyzed companies.

Starting with the total fees variable, as shown in Table 5, the minimum fees charged were €6,000. The maximum fees were €10,286,310, which results in an average of €765,399 in fees charged for the years under analysis.

On the other hand, the company fees variable has lower fee values. The minimum fees charged were €1,000, and the maximum was €4,399,400, which gives an average of €271,746 for the years under analysis.

The next independent variable to analyze is the rotation of the auditing firm. In Table 6, we can observe that there were only 18 instances of auditing firm rotation out of the 170 observations analyzed, which corresponds to 11% of the sample. For the most

Table 5. Analysis of the audit fees variable

		Values in €
Total fees	Minimum	6.000
	Maximum	10.286.310
	Mean	765.399
Company fees	Minimum	1.000
	maximum	4.399.400
	Mean	271.746

Table 6. Analysis of the audit firm rotation variable

		Frequência	%
Audit firm rotation	0 – No	152	89%
	1 – Yes	18	11%
	Total	170	100%

part, there was no auditing firm rotation, with 152 observations, representing 89% of the sample.

Next is the auditor's gender, and as we can see in Table 7, it is the male gender that stands out more. Out of the 170 reports analyzed, 148 were issued by male auditors, accounting for 87% of the sample. Only 22 of the reports were signed by female auditors, representing 13% of the sample.

Table 7. Analysis of the gender variable

		Frequency	%
Gender	0 – Male	148	87%
	1 – Female	22	13%
	Total	170	100%

Regarding the pandemic variable, most of the reports were issued during this period, with 102 observations, accounting for 60% of the sample, thus reflecting the 3 years of the pandemic. Only 68 reports were issued during the non-pandemic period, which represents 40% (see Table 8).

Table 8. Analysis of the pandemic variable

		Frequency	%
Pandemic year	0 – No	68	40%
	1 – Yes	102	60%
	Total	170	100%

5.2 Presentation of Bivariate Analysis Test Results

In this second part of the analysis, we will use the Statistical Package for Social Science (SPSS) data processing software.

As previously mentioned, the T-test was used to check the correlation between the Opinion Type (dependent variable) and the quantitative variables (total fees and the company's fees within the total fees). To understand this test, we must first consider Levene's test, which allows us to assess the homogeneity of variances between different groups. Once the significance of Levene's test is validated, we need to check if the T-test is significant.

To test the association between the Opinion Type and the dichotomous variables, the Chi-square and Fisher's tests were performed, complemented by the association measures V of Cramer and Phi Coefficient.

At this point, the results of the statistical tests for the Opinion Type will be presented to identify which variables influence the type of opinion issued by the auditor in the Audit Report.

We will begin by presenting the results of the association between the opinion type variable and the variables Firm Type, Auditing Firm Rotation, Auditor's Gender, and Pandemic. Since these are dichotomous variables, the tests applied will be the Chi-square and Fisher's tests, complemented by the association measures V of Cramer and Phi Coefficient.

Regarding the type of auditing firm, as shown in Table 9, out of 170 observations, 15 are modified opinions, of which 11 are from non-Big Four firms, while there are only 4 modified opinions from Big Four firms.

Table 9. Type of opinion and Type of audit firm

		Type of audit firm		Total
		No Big Four	Big Four	
Type of opinion	Unmodified	33	122	155
	Modified	11	4	15
	Total	44	126	170

Next, we proceed with the Chi-square statistical test, which confirms that there is an association between the type of opinion and the type of firm. The p-value was significant, meaning it was below 0.05, as we can observe in Table 10.

In the study by [5], the authors concluded that issuing an unmodified opinion is strongly correlated with the auditor belonging to a Big Four firm, as is observed in this study. These results were also corroborated by [7].

However, as we can see in Table 11, although there is an association between the variables, the strength of the association is not very significant, as it is close to 0. This is because the V of Cramer ranges between 0 and 1- the closer it is to 0, the weaker the association, and the closer it is to 1, the stronger the association. The association is negative, meaning the probability of obtaining a modified opinion is higher in non-Big Four firms, which further reinforces the results obtained.

In this sense, we validate hypothesis H1: The type of auditing firm (Big Four or non-Big Four) influences the type of opinion issued by the auditor in the Audit Report.

Table 10. Chi-Square Fisher (Type of opinion and Type of audit firm)

	Value	df	Asymptotic Significance (2-sided)	Exact Sig. (2 sided	Exact Sig (1-sided)
Pearson Chi-Square	19.310	1	<0,01		
Continuity	16,692	1	<0,01		
Correction Likelihood	16,511	1	<0,01		
Ficher's Exact Test				<0,01	<0,01
Linear by linear	19,196	1	<0,01		
N of valid	170				

Table 11. Measures of association (Type of opinion and Type of audit firm)

		Value	Approximate Significance
Nominal by Nominal	Phi	0,337	<0,01
	Chramer's V	0,337	<0,01
	N of Valid cases	170	

We now move on to the analysis of the results for the type of opinion and the variable firm rotation.

In Table 12, we can see that there were no observations when the auditing firm rotation variable was crossed with the modified opinion. The majority of observations occurred when there was no firm rotation, and the opinions were unmodified.

Table 12. Type of opinion and Audit firm rotation

		Audit firm rotation		Total
		No	Yes	
Type of opinion	Unmodified	137	18	155
	Modified	15	0	15
	Total	152	18	170

We then proceed with the Chi-square statistical test, which confirms that there is no association between the type of opinion and the firm rotation. The p-value was not significant, meaning it was greater than 0.05, as we can observe in Table 13.

Studies by [11–13], and [14] concluded that when there is firm rotation (whether mandatory or not), there is a tendency to issue a report with a modified opinion. According to our study, the results obtained contradict the findings of the mentioned authors. Thus, hypothesis H3: The rotation of the auditing firm influences the type of opinion issued by the auditor in the Audit Report, was not validated.

Table 13. Chi-Square and Fisher (Type of opinion and Audit firm rotation)

	Value	df	Asymptotic Significance (2-sided)	Exact Sig. (2 sided	Exact Sig (1-sided)
Pearson Chi-Square	1.948	1	0,163		
Continuity	0,915	1	0,339		
Correction Likelihood	3,525	1	0,060		
Ficher's Exact Test				0,373	0,173
Linear by linear	1,937	1	0,164		
N of valid	170				

Next, we analyze the type of opinion with the auditor's gender. Out of the 170 observations, 15 correspond to modified opinions. Of these 15 observations, only 2 were issued by female auditors, while 13 were issued by male auditors (as shown in Table 14).

Table 14. Type of opinion and Auditor gender

		Auditor Gender		Total
		Male	Female	
Type of opinion	Unmodified	135	20	155
	Modified	13	2	15
	Total	148	22	170

The result of the Chi-square statistical test, presented in Table 15, confirms that there is no association between the type of opinion and the auditor's gender. This result does not validate hypothesis H4: The gender of the auditor influences the type of opinion issued by the auditor in the Audit Report.

These results are consistent with [16], who found that there is no correlation between the auditor's gender and the type of opinion issued, contrary to the studies of [17–19], and [20].

Table 15. Chi-Square and Fisher (Type of opinion and Auditor gender)

	Value	df	Asymptotic Significance (2-sided)	Exact Sig. (2 sided	Exact Sig (1-sided)
Pearson Chi-Square	0.002	1	0,962		
Continuity	0,000	1	1,000		
Correction Likelihood	0,002	1	0,962		
Ficher's Exact Test				1,000	0,607
Linear by linear	0,002	1	0,962		
N of valid	170				

Now, we move on to the analysis of the type of opinion with the pandemic.

Table 16 shows that more reports with a modified opinion were issued during the pandemic years than in the non-pandemic years, with 11 observations in the pandemic years and only 4 in the non-pandemic years.

Table 16. Type of opinion and Pandemic year

		Pandemic year		Total
		No	Yes	
Type of opinion	Unmodified	64	91	155
	Modified	4	11	15
	Total	68	102	170

In turn, the Chi-square statistical test shows that there is no association between the type of opinion and the pandemic (see Table 17). In this sense, H6: The COVID-19 pandemic influenced the type of opinion issued by the auditor in the Audit Report, cannot be validated.

[23] stated that companies most affected by the undesirable effects of the pandemic are likely to have issued modified opinions, due to insufficient audit evidence that was true and appropriate, or the identification of material distortions. However, this was not supported by the results of this study.

Table 17. Chi-Square and Fisher (Type of opinion and Pandemic year)

	Value	df	Asymptotic Significance (2-sided)	Exact Sig. (2 sided	Exact Sig (1-sided)
Pearson Chi-Square	1,219	1	0,270		
Continuity	0,685	1	0,408		
Correction Likelihood	1,278	1	0,258		
Ficher's Exact Test				0,409	0,206
Linear by linear	1,211	1	0,271		
N of valid	170				

Now, we analyze the relationship between the type of opinion and the fees. Since the variable "type of opinion" is dichotomous and the "fees" variable is quantitative, the test to be applied is the T-test for means.

As we can see in Table 18, Levene's test was significant, indicating that the variances of the groups are not equal. Therefore, we must examine the results of the T-test for unequal variances, where the p-value was less than 0.01, making it significant. This means there is a correlation between the type of opinion and the fees.

Moving on to Table 19, which cross-tabulates the type of opinion with the fees, we observe that the average fees are higher when the opinion is unmodified. In other words, fees are higher for an unmodified opinion. Therefore, hypothesis H2: The auditing fees influence the type of opinion issued by the auditor in the Audit Report, was validated.

These results are consistent with the study by [5], which concluded that there is a significant positive correlation between the type of opinion the auditor issues and the audit fees. Hence, an unmodified opinion is associated with higher fees.

Table 18. Levene's Test and T test (Type of opinion vs.Audit fees)

		Levene's Test of Equality of Variances			T-test for Equality of Means						
						Significance				95% Confidence Interval of the Difference	
F	Sig	t	df	One sided p	Two-sided p	Mean Difference	St. Error Difference	Lower	Upper		
Audit fees	Equal variances assumed	5,452	0,021	-1,654	168	0,050	0,100	−683.211,097	412.979,8	−1.498.509,7	132.087,5
	Equal variances not assumed			−5,035	166,519	<0,001	<0,001	−683.211,097	135.696,1	−951.117,7	−415.304,5

Table 19. Group Statistics (Type of opinion vs Audit fees)

	Type of opinion	N	Mean	St. Deviation	Std Error Mean
Audit fees	Modified	15	151.418,80	173.825,99	44.881,68
	Unmodified	155	834.629,90	1.594.320,27	128.058,89

The results obtained earlier are applicable to the company's fees within the total fees, as we can observe in Tables 20 and 21. Levene's test was significant, with a p-value of 0.032, which is less than 0.05.

Therefore, we need to examine the results of the T-test for unequal variances, where the p-value was below 0.01, making it significant, revealing a correlation between the type of opinion and the fees. By cross-referencing the type of opinion with the company's fees, we can see that the average fees are higher when the opinion is unmodified. In other words, higher fees are associated with an unmodified opinion.

Thus, hypothesis H2: The auditing fees influence the type of opinion issued by the auditor in the Audit Report, was validated.

These results further reinforce the study conducted by [5].

Table 20. Levene's Test and T test (Type of opinion vs.Audit company fees)

		Levene's Test of Equality of Variances			T-test for Equality of Means						
				Significance			95% Confidence Interval of the Difference				
F	Sig	t	df	One sided p	Two-sided p	Mean Difference	St. Error Difference	Lower	Upper	F	Sig
Audit company fees	Equal variances assumed	4,703	0,032	−1,276	168	0,102	0,204	−257.211,1	201.608,31	−655.223,23	140.801,04
	Equal variances not assumed			−4,092	156,915	<0,01	<0,01	−257.211,1	62.857,85	−381.376,77	−133.054,43

Table 21. Group Statistics (Type of opinion vs.Audit company fees)

	Type of opinion	N	Mean	St. Deviation	Std Error Mean
Audit company Fees	Modified	15	44.133,33	24.190,63	6.245,99
	Unmodified	155	301.344,43	778.700,89	62.546,76

As we can see in Table 22, the factors determining the type of opinion are shown. Since H1 and H2 were validated, we can conclude that the type of auditing firm and the fees are factors that determine the type of opinion issued by the auditor.

Table 22. Summary of validated hypotheses (Type of Opinion)

	Type of Opinion
Type of audit firm	H1: The type of audit firm (Big Four or non-Big Four) influences the type of opinion issued by the auditor in the Audit Report **Validated**
Audit fees	H2: Audit fees influence the type of opinion issued by the auditor in the Audit Report **Validated**
Autid firm rotation	H3: Audit firm rotation influences the type of opinion issued by the auditor in the Audit Report **Not validated**
Auditor gender	H4: The auditor's gender influences the type of opinion issued by the auditor in the Audit Report **Not validated**
Pandemic year	H5: The COVID-19 pandemic influenced the type of opinion issued by the auditor in the Audit Report **Not validated**

6 Conclusions, Limitations and Sugestions for Future Research

The main objective of this study is to identify the factors that influence the type of opinion issued in the audit report of companies listed on Euronext Lisbon.

In this context, several studies have been conducted, both in the public and private sectors. Most of the studies aim to analyze the association between variables related to the auditor and the Audit Report. These variables include the auditor's gender, the type and size of the audit firm (specifically whether it is a Big Four or not), audit firm rotation, and the auditor's fees. Despite the existence of various studies, few address the topic longitudinally or consider the impact of the pandemic.

For the present study, we used content analysis of the Annual Reports and the corresponding Audit Reports from companies listed on Euronext Lisbon between 2018 and 2022. Through a bivariate analysis using parametric tests, it was found that, of all the variables analyzed, the fees and the type of firm have an influence on the type of opinion issued by the auditor. Specifically, when fees are higher, there is a greater likelihood of issuing an unmodified opinion. Additionally, companies audited by Big Four firms are more likely to receive an unmodified opinion. However, it is important to note that we are dealing with listed companies, which generally treat financial information rigorously due to the impact it has on stakeholders' decision-making.

The variables of audit firm rotation and auditor gender did not present significant results, leading to the conclusion that there is no association between these variables and the type of opinion issued.

The choice of two external variables - those that are not dependent on the auditor's characteristics or actions - was intended to analyze how economic and social factors could influence the Audit Report. However, the results were not significant, leading to the conclusion that there is no association between the pandemic variable and the type of opinion issued.

One limitation of this study is the small sample size, driven by the limited size of the Portuguese capital market. Another limitation is that companies do not disclose all the information on their websites. In this sense, a suggestion for future research is to expand the sample to include other stock exchanges, such as Euronext Paris or the London Stock Exchange. Another suggestion could be to include variables such as asset profitability and equity, in order to analyze how company performance might influence the Audit Report.

Acknowledgment. This work is financed by national funds through the FCT - Foundation for Science and Technology, I.P., under project UIDB/04043:CICF and UIDP/04043:CICF.

References

1. International Auditing and Assurance Standards Board. ISA 700 – Formar uma opinião e relatar sobre demonstrações financeiras, in Manual das normas de controlo de qualidade, auditoria, revisão, outros trabalhos de garantia de fiabilidade e serviços relacionados (Parte I, pp. 803–858), Nova Iorque: IAASB (2018a)

2. Pereira, F., Machado, N., Pinheiro, L., Dutra, S.: Auditoria contábil: Um estudo acerca da relação entre o relatório dos auditores e as características da empresa auditada. Revista Capital Científico Vol. 17, Nº. 2 (2019). Disponível em: revistas.unicentro.br/index.php/capitalcientifico/index
3. International Auditing and Assurance Standards Board. ISA 705 – Modificações à opinião no relatório do auditor independente, in Manual das normas de controlo de qualidade, auditoria, revisão, outros trabalhos de garantia de fiabilidade e serviços relacionados (Parte I, pp. 885–917), Nova Iorque: IAASB (2018b)
4. Costa, C.B.: Auditoria Financeira - Teoria & Prática (13ª ed.). Rei dos livros (2023)
5. Ruban, D., Hategan, C.: Impact factors on the audit opinion–Case study. Audit Financ. **168**(4), 668–680 (2022). https://doi.org/10.20869/AUDITF/2022/168/024
6. Costa, M.R., Serra, S., Gomes, P.: Fatores determinantes da opinião do auditor: estudo empírico das empresas portuguesas. Ordem dos Contabilistas Certificados. pp. 1–23 (2013). https://www.occ.pt/news/comcontabaudit/pdf/124.pdf
7. Marques, V., Louzada, L., Amaral, H., Souza, A.: O poder da reputação: evidências do efeito big four sobre a opinião do auditor. Revista Conteporânea de Contabilidade **15**(35), 03–31 (2018). https://doi.org/10.5007/2175-8069.2018v15n35p3
8. Liu, W., Huang, H.: Internal control opinion shopping: does initial audit fee discounting matter? Int. Bus. Financ. **69**(102248), 1–14 (2024). https://doi.org/10.1016/j.ribaf.2024.102248
9. Caramanis, C., Lennox, C.: Audit effort and earnings management. J. Account. Econ. **45**(1), 116–138 (2008). https://doi.org/10.1016/j.jacceco.2007.05.002
10. Dantas, J., Barreto, I., Carvalho, P.: Relatório com modificação de opinião: risco para o auditor? Revista Contemporânea de Contabilidade **14**(33), 140–157 (2017). https://doi.org/10.5007/2175-8069.2017v14n33p140
11. Cunha, P.R., Leite, M., Moras, V.R.: Efecto del cambio de la firma de auditoría en la gestión de resultados de las compañías abiertas brasileñas. Contabilidad Negocios **14**(28), 70–87 (2019). https://doi.org/10.18800/contabilidad.201902.004
12. Lenox, C.S., Wu, X., Zhang, T.: Does mandatory rotation of audit partners improve audit quality? Account. Rev. **89**(5), 1775–1803 (2014)
13. Firth, M., Rui, O.M., Wu, X.: How do various forms of auditor rotation affect audit quality? Evidence from China. Int. J. Account. **47**(1), 109–138 (2012)
14. Lin, H., Yen, A.: Auditor rotation, key audit matter disclosures, and financial reporting quality. Adv. Account. **57**(100594), 1–14 (2022). https://doi.org/10.1016/j.adiac.2022.100594
15. Zhang, X., Yan, H., Hub, F., Wangc, H., Li, X.: Effect of auditor rotation violation on audit opinions and audit fees: evidence from China. Res. Int. Bus. Financ. **62**, 101715 (2022). https://doi.org/10.1016/j.ribaf.2022.101715
16. Breesch, D., Branson, J.: The effects of auditor gender on audit quality. IUP J. Account. Res. Audit Pract. **8**(3–4), 78–107 (2009)
17. Hardies, K., Breesch, D., Branson, J.: Are female auditors still women? Analyzing the sex differences affecting audit quality. SSRN – Elsevier (2010). https://doi.org/10.2139/ssrn.1409964
18. Carrera, N., Mareque, M.: Does gender affect qualifying decisions? Evidence from public sector audits. J. Bus. Res. **154**, 113358 (2023). https://doi.org/10.1016/j.jbusres.2022.113358
19. Reheul, A.-M., Van Caneghem, T., Van den Bogaerd, M., Verbruggen, S.: Auditor gender, experience and reporting in nonprofit organizations. Manag. Audit. J. **32**(6), 550–577 (2017). https://doi.org/10.1108/MAJ-01-2016-1296
20. Karjalainen, J., Niskanen, M., Niskanen, J.: The effect of audit partner gender on modified audit opinions. J. Auditing **22**, 449–463 (2018). https://doi.org/10.1111/ijau.12130
21. Ordem dos Revisores Oficiais de Contas. Potenciais impactos da COVID-29 na auditoria 2020 (2020)

22. Lassoued, N., Khanchel, I.: Impact of COVID-19 pandemic on earnings management: an evidence from financial reporting in European firms. Glob. Bus. Rev. (2021). https://doi.org/10.1177/09721509211053491
23. Proença, P.: Os impactos da COVID-19 na Auditoria. Revista da Ordem dos Revisores Oficiais de Contas, Nº95, pp. 36–51 (2021)

Evaluating Operational Strategies in Car Dealerships: A Case Study of Franchised Dealership Strategy in Northeast Brazil

Valentino Yves Corentin Clerici[1], Viviane Oliveira Vilela de Vasconcelos[1], Ana Lucia F. de S. Vasconcelos[2], and Caio Arruda Souza Henriques[2](✉)

[1] Escola Americana do Recife, Recife,, Brazil
[2] Universidade Federal de Pernambuco, Recife, Brazil
{ana.svasconcelos,caio.henriques}@ufpe.br

Abstract. This study explores the effectiveness of operational strategies in enhancing the competitiveness of car dealerships in Recife, Brazil. Adopting an exploratory sequential mixed-methods approach, the research involved semi-structured interviews and structured surveys conducted with managers from four franchised dealerships. The study developed and applied quantitative performance indices—such as the Sales Efficiency Index (SEI), Cost Efficiency Index (CEI), and Overall Dealership Performance Score (ODPS)—to measure dealership performance across key operational areas. Findings reveal that in the context of Recife, basic operational consistency and digital adaptability play a more decisive role in dealership competitiveness than sustainability initiatives or premium customer experiences. These results challenge the universal applicability of global best practices and underscore the need for locally adapted strategies in developing markets.

Keywords: Operational Strategies · Car Dealerships · Mixed Methods · Performance Indices · Northeast Brazil

1 Introduction

The automotive sector makes up almost 4% of the world's gross domestic product (GDP), making it one of the most significant and preeminent global industries. Additionally, this sector accounted for nearly 10% of pre-pandemic world trade [1]. Analogously, in Brazil, the automobile sector is of critical relevance. As the nation evolved into a car-centric country during post-colonial developments, railroads were deemed obsolete and public transportation negligible. Therefore, car dealerships have become indispensable retail and service hubs, serving customers demanding transportation.

Recife, a coastal city in northeast Brazil, has a unique economic status as a developing regional capital with a middle-income consumer base, distinguishing itself from international or national automotive trends. Most of the literature on dealership strategies focuses solely on developed markets with high-income levels and advanced infrastructure. These studies highlight global best practices, including sustainability, digital

L. Almeida et al. (Eds.): ICAFI 2025, ISEM 80, pp. 337–349, 2026.
https://doi.org/10.1007/978-3-032-16268-7_19

marketing, and customer-centric strategies. Although regarded as effective globally in contributing to overall success amongst dealerships, these strategies have been tested in vastly different contexts than those in developing regions like Recife. While national-level studies have explored dealership performance in Brazil, these studies mainly consider economies in major metropolitan hubs like São Paulo and Rio de Janeiro. There is limited research on how specific operational strategies succeed at the regional level. This context raises an important research question: How do operational strategies and market adaptation contribute to the success and competitiveness of franchised car dealerships in Recife?

To address this question, the study adopted an exploratory sequential mixed-methods design, combining qualitative data from semi-structured interviews with managers of four franchised dealerships and quantitative data from structured dealership surveys. The research used dealership performance indices, such as the Sales Efficiency Index (SEI), Cost Efficiency Index (CEI), or Overall Dealership Performance Score (ODPS), to quantitatively assess and compare performance across operational dimensions.

This study providing a developing-world perspective on dealership strategy effectiveness, and delivering actionable insights for dealership managers seeking context-appropriate strategies in Recife and similar urban markets in emerging economies. Our aim is to bridge a significant research gap by investigating how operational strategies contribute to the competitiveness of franchised car dealerships in Recife. Specifically, we seek to: Evaluate the effectiveness of locally adapted versus globally promoted practices; Identify which performance metrics best reflect dealership success in this context; Provide actionable insights for managers operating in similar emerging markets.

To achieve these objectives, the study employs an exploratory sequential mixed-methods design, combining qualitative interviews with dealership managers and quantitative surveys to develop performance indices. The subsequent sections are structured as follows: Sect. 2 reviews literature on dealership strategies globally and in Brazil; Sect. 3 outlines the methodology; Sects. 4 and 5 present findings and discuss their implications; and Sect. 6 concludes with recommendations for practitioners and future research.

2 Literature Review

Car dealers in Brazil experience a distinct set of challenges, given the high levels of competition and fluctuating economic conditions, changing consumer tastes and preferences. In analyzing the factors contributing to success, it becomes important to consider various key aspects. Diógenes et al. analyzed car dealerships in Brazil and identified that some factors like innovation, effective management, and customer satisfaction are vital for performance. Their research underscores the importance of adapting to market dynamics and maintaining operational efficiency to stay profitable in a competitive environment [2]. On the other side of the business, Corswant and Fredriksson's study on sourcing trends analyzes B2B, emphasizing the growing complexity of manufacturer-supplier relationships and the need for faster product development and increased supplier involvement to stay competitive globally [3]. These studies demonstrate how internal management practices and external partnerships impact dealership performance. Moreover, Borchardt et al. found that dealerships providing high-quality after-sales service tend to have higher revenue and greater customer satisfaction, indicating that post-purchase interactions are crucial for customer loyalty and dealership reputation [4]. Ultimately, the keys to a successful dealership strategy in Brazil are effective adaptability, efficient management, and strong customer and supplier relationships. Then, while Borchardt et al. emphasize the importance of after-sales service in developed markets, this study investigates how this practice applies in Recife, where socioeconomic factors may influence its effectiveness.

In the dynamic automotive industry, car dealerships must implement effective operational strategies and adapt to market changes to achieve sustained success and maintain a competitive edge, particularly in specific regional markets like Recife [5]. Some adaptation methods have been observed to work exceptionally well in rapidly evolving sectors. A study on agile market adaptability strategies highlights "flexibility to change," particularly in market demand, as a crucial strategy [6]. Moreover, the same research indicated market adaptability with the "execution based on continuous feedback," which may include customer desire. It creates an automotive business model centered around the core product and consumers—eventually leading to a positive business case.

Technology is revolutionizing the automotive retail sector. Zilber explores General Motors Brazil's use of e-business strategies during the launch of the Celta, showing how online platforms can transform traditional sales methods [7]. Adopting digital marketing and sales channels has enabled companies to cut costs and reach a wider audience, changing how vehicles are marketed and perceived. Hase and Busch discuss how incorporating digital tools into dealership operations underscores the transformative influence of technology [8]. They suggest that organizing sales teams around customer needs and promoting ethical sales practices are essential for navigating the challenges of a digitally-focused market. The interaction between digital transformation and changing consumer behavior emphasizes the need for dealerships to remain adaptable and creative to maintain their competitive edge. Although Zilber demonstrates the success of e-business in major Brazilian cities, the reality in Recife—with its less developed digital infrastructure—may require adaptations.

Historical and cultural circumstances also greatly influence car dealerships' operational tactics. Ferro offers a historical examination of the expansion of the Brazilian automobile sector after the Real Plan of 1994, which promoted industrial growth and stabilized the economy [9]. Manufacturers and dealerships were able to implement lean production methods and build collaborative partnerships with the government and labor unions during this economic transformation, highlighting the value of outside alliances in advancing the sector. Furthermore, Musacchio uses automobility as a prism to examine Brazil's quest for modernity, placing the development of the Brazilian automotive sector within larger societal shifts [10]. This perspective on culture emphasizes how national identity and consumer desires influence the car market, offering a comprehensive view of the dynamics within dealerships. By combining historical, cultural, and operational knowledge, dealerships can create strategies that align with the needs of the market and the expectations of society.

Understanding the dynamics of sales environments and employees is essential to comprehending dealership success. The impact of retail work surroundings on salesperson orientations and job performance is examined by Boles et al. [11]. According to their results, improving staff performance and customer happiness requires both organizational support and the development of a customer-centric approach. Furthermore, Scott Morton et al. investigate the elements influencing price negotiations in the United States vehicle retailing industry, highlighting the importance of consumer information and search habits in reaching beneficial outcomes [12]. Despite concentrating on a different geographical area, this study has clear implications for Brazilian dealerships: giving sales staff the resources and instruction they need to handle challenging discussions can boost customer confidence and increase profitability. Dealerships can develop plans promoting long-term success by considering employee and customer dynamics.

Environmental sustainability increasingly influences operational strategies in the automotive industry. Firnkorn and Müller discuss how automakers are shifting towards sustainable business models, such as car-sharing services, to reduce their environmental footprint [13]. This trend reflects a broader shift in consumer values, with an increasing emphasis on eco-friendly practices. Dealerships must adapt to these changes by promoting sustainable vehicle options and aligning their operations with environmental standards. Borchardt et al. also emphasize the role of sustainability in after-sales services, noting that eco-friendly practices can enhance customer satisfaction and contribute to a positive brand image [14]. These findings suggest that integrating sustainability into dealership operations is not only ethically responsible but also economically advantageous. As the automotive industry continues evolving, dealerships that prioritize environmental consciousness and align their strategies with consumer expectations will be better positioned to thrive in a competitive market.

Essentially, long-term dealership success depends on sustainability, innovation, cultural sensitivity, and customer satisfaction. Zilber or Diógenes et al., and other research provide an in-depth framework for understanding the various elements affecting the success of dealerships. Dealerships can successfully navigate the challenges of a rapidly changing industry by embracing digital transformation, adapting to market dynamics, and building excellent relationships with customers and employees. In addition, integrating sustainability into operating plans will make dealerships resilient and relevant to the shifting values of consumers and concerns about the environment. This comprehensive dealership management approach will provide insightful information to the industry participants seeking to enhance their competitive posture and achieve long-term success.

The reviewed literature highlights the importance of strategies such as operational efficiency, digital adaptation, and sustainability, but it pays little attention to middle-income regional markets like Recife. This study aims to fill this gap by examining how these strategies are applied and adapted in this specific context.

3 Methodology

The methodology was adjusted to concentrate primarily on internal dealership insights, exclusively with manager interviews and dealership data collection surveys, maintaining a mixed methods approach. So, a Sequential Explanatory Design was envisioned to be utilized for the methodological design. This involved quantitative data collection (structured surveys distributed to dealerships) and qualitative exploration (semi-structured interviews with dealership managers). The quantitative data would help identify trends in operational performance across dealerships.

Then, the qualitative phase provided explanatory context, deepening the understanding of the strategic reasons behind quantitative findings. This design would have been well-recognized for effectively utilizing quantitative data to establish broad, generalizable results, while qualitative data offered richer insights to further interpret and validate those results. However, as this investigation entered the outreach phase, it became clear that managers were reluctant to distribute sensitive and confidential dealership data without personally meeting the researcher. Accordingly, data collection was inverted, and an exploratory sequential design was implemented; interviews with dealership managers were conducted, and then surveys with quantitative questions were sent. The first phase of the conducted research—the in-depth, semi-structured online interviews with dealership managers, gathering qualitative data including but not being limited to market strategies, day-to-day operations, and risk management—offered an in-depth understanding of the dealership and how it is led.

To gather data, dependent and independent variables were identified and categorized. The dependent variables included consumer purchase decisions, defined by the number and frequency of sales; customer satisfaction, measured indirectly through dealership returning customer percentages and survey responses regarding perceived satisfaction levels; dealership trustworthiness, assessed qualitatively through manager interviews; and overall dealership performance. One of our limitations was that as customer interviews became impracticable, customer satisfaction was measured solely by the percentage of returning customers. The study's independent variables were simple yet crucial to determining strategic operational prosperity. They included sales tactics, incorporating pricing strategies, promotions, customer interactions, digital marketing initiatives, and dealership types, either franchised dealerships or independent operations. However, later, franchised dealerships became the primary focus of the study, and we study four of them.

The study focused on dealership personnel, specifically targeting dealership managers within Recife's automotive market. Initially, a purposive sampling strategy was aimed to be implemented to ensure representation across different dealership types (franchised and independent), yet the study eventually focused entirely on franchised stores. The purposive strategy was administered to represent different operational models (mass-market and luxury segments). The study aimed to recruit participants through professional networks, industry contacts, and direct outreach to dealerships, ensuring diverse representation to minimize potential bias.

Some softwares were employed during the creation and implementation of the study. So, Google Forms for surveys; Google Meet for interviews; "Voice Memos" Apple app recorded each interview; Interview recordings were uploaded to the "Transkriptor".

Finally, for clarity and methodological consistency, the study defined operational costs exclusively as overhead expenses. Product acquisition expenses (cost of goods sold [COGS]) were excluded. This choice separated dealership overhead and fixed operational efficiency from varying inventory costs, thus showcasing the direct effect of strategy alone on dealership competitiveness.

4 Findings

To better protect the identity and confidentiality of managers and car dealerships participating in the study, they have been renamed. They will be referred to as "Interviewee A," "Interviewee B," "Interviewee C," and "Interviewee D," following the order of interview conducted from first to last. Likewise, the dealerships they manage have also been renamed "Dealership A," "Dealership B," "Dealership C," and "Dealership D." Interviewee A manages DA (Dealership A), Interviewee B manages DB (Dealership B), Interviewee C manages DC (Dealership C), and finally Interviewee D manages DD (Dealership D).

To accurately compare and score each dealership's competitiveness and performance quantitatively, a series of equations were derived. There was the Sales Efficiency Index (SEI) to show how well the dealership utilizes physical area, salespeople, and vehicle stock to generate sales; Cost Efficiency Index (CEI), to measure how efficiently each dealership converts spending into revenue; Marketing and Relationship Efficiency Index (MREI), to assess how effective the dealership is at marketing and customer retention, relative to its marketing spend; and Overall Dealership Performance Score (ODPS), to find a composite dealership performance score by combining all indices into a single one. CEI and SEI are versions of classic efficiency ratios (output/input) widely used in economics and finance (e.g., ROI). MREI mirrors a marketing ROI but was expanded to include customer retention. Finally, ODPS is a composite score (similar to index scoring in business or finance) used to combine multiple performance measures into one metric.

The following are the expanded equations, followed by the explanations of variables and how to interpret each result.

$$SEI = \frac{TotalCarsSoldperMonth(new + used)}{\alpha A + \beta SE + \gamma S} \tag{1}$$

A, SE, and S are dealership surface area, number of sales employees, and vehicles in stock, respectively. α, β and γ are weights assigned based on relative importance. β was weighted most heavily, followed by γ and α. A high SEI represents more sales with relatively fewer resources, indicating high efficiency. A low SEI means the dealership requires more resources to achieve the same number of sales, indicating lower efficiency.

$$CEI = \frac{MonthlyRevenue}{MonthlyCosts} \tag{2}$$

A higher **CEI** means the dealership generates more revenue than its total monthly cost (efficient and profitable). A lower **CEI** means the dealership spends more than it earns (inefficient, operating at a loss).

$$MREI = \frac{ConversionRate + DigitalChannelSales\% + ReturningCustomer\%}{MarketingSpendas\%ofTotalMonthlyCosts} \tag{3}$$

A higher **MREI** means the dealership is highly effective at marketing and customer retention relative to their marketing spend. A lower **MREI** means the dealership's marketing strategy and customer relationships are less effective relative to their marketing spend.

$$ODPS = w_1 * SEI_{norm} + w_2 * CEI_{norm} + w_3 * MREI_{norm} \tag{4}$$

w_1, w_2, w_3 are weights assigned equal values (1/3). The indices were normalized with min-max normalization.

It is challenging to determine if greater revenue is the product of strategies employed by managers (mentioned in interviews) or simply greater investment from each dealership. Therefore, to affirmatively confirm that increased revenue was a direct result of tactics adopted, isolating the impact of revenue from raw investment alone, another equation was designed: Revenue Due to Strategy (RDS). The following is the expanded equation, followed by the explanations of variables and how to interpret the results.

$$RDS = R_{actual} - R_{expected}$$

where

$$R_{expected} = \delta(SurfaceArea) + \in (SalesEmployees) + \zeta(vehiclestock) + \eta(MonthlyTotalCost)$$

"*R*" represents revenue. Positive RDS: Actual revenue exceeds what you would typically expect based on investments alone, clearly indicating that strategic decisions (pricing, customer service, marketing techniques) are positively influencing sales. Zero or Negative RDS: Actual revenue is equal to or below the expected revenue based on investments alone, suggesting strategies are not particularly effective or that investment alone explains positive performance.

Table 1 outlines general reference ranges for each index's high, moderate, or low performance.

Table 1. *Interpretation thresholds for each dealership performance index.*

Index	Low Performance	Moderate Performance	High Performance	Font size and style
SEI (Sales Efficiency Index)	<0.0200	0.0200–0.0400	>0.0400	Values >0.04 show highly efficient use of space, staff, and inventory
CEI (Cost Efficiency Index)	<15.0	15.0–20.0	>20.0	>1.0 always means profit; >20.0 reflects exceptional cost efficiency
MREI (Marketing & Relationship Efficiency Index)	<13.0	13.0–17.0	>17.0	Includes customer return, digital use, and conversion rate relative to spend
ODPS (Overall Dealership Performance Score)	<0.3	0.3–0.7	>0.7	A higher ODPS means the dealership performs exceptionally well across multiple areas. Lower ODPS means dealerships are underperforming

Dealerships answered surveys through ranges rather than precise amounts, limiting precision. Exact values were needed for calculations (SEI, CEI, MREI, ODPS, and RDS). Hence, midpoints of reported ranges were employed to reasonably represent financial trends. This approach, however, has limitations as it can mislead conclusions, cause potential misinterpretation, and lose granular detail, especially since response ranges were often significant. For example, the average variation in dealership area measurements was 28.75%, with the highest variation being 40.00%.

The table below shows values for each dealership's score in formulas, rounded to the nearest ten thousandths when necessary:

Table 2. Performance Scores for Dealerships A–D Across Key Indices

	DA	DB	DC	DD
SEI	0.0305	0.0387	0.0138	0.0621
CEI	19.5	18.6923	17.1429	24.3571
MREI	19	16.1538	11.5	16.4103
ODPS	0.6021	0.5001	0	1.0964
RDS	3.9488	1.3039	-6.2957	2.2724
Overall Quantitative Performance	Notable (Marketing effectiveness, customer relationships)	Stable, not excessively competitive	Lowest Overall	Top Overall Performer (Strong cost efficiency, strategic digital marketing)
Key Qualitative Strategies	Personalized promotions and employee training	Aggressive digital marketing and financial flexibility	Premium experience and sustainability, but without local market adaptation	Price-based promotions and proprietary financing methods

Table 2 showcases how, although DA dominates the MREI scores, DD dominates the SEI, CEI, and ODPS. On the other hand, DC ranks last in all categories. The quantitative analyses demonstrate that DD is the top overall performer, primarily due to strong cost efficiency and strategic digital marketing. DD's total monthly revenue greatly outweighs its operational expenses, explaining the high cost efficiency. DA also demonstrated notable performance, particularly in marketing effectiveness and customer relationships. Conversely, DC scored the lowest overall despite its premium market positioning.

Qualitatively, notable strategies emerged throughout the interviews. In particular, as Interviewee A said, "When it comes to discounts, rewards, and bonuses, there's a lot of variability. It's really about communication—determining what customers value most in the purchase, whether they prioritize a higher trade-in value for their used car, a lower price for the new car, a better financing rate, additional gifts, or accessories. Ultimately, the discounts are primarily applied in-store because customers typically don't purchase vehicles online". Essentially, DA emphasized customized in-store promotions. Additionally, Interviewee A gives crucial insight into the importance of employee training in car dealership success as it helps increase customer satisfaction and conversion rate. Interview A adds that although the franchisor standardizes its training, her team makes sure to increment its unique touch.

DB implemented aggressive digital marketing and focused heavily on financing flexibility, ensuring a stable yet not excessively competitive position against other dealerships in Recife. Following the franchised luxury brand's image, DC prioritized premium experiences, sustainability initiatives, and minimal direct discounting, creating a bubble of exclusivity. Accordingly, Interviewee C highlights, "Our strategy for managing fluctuations in demand has been to invest heavily in the physical dealership experience, exclusive events, personalized test drives, and highly targeted communication toward a high-income clientele". While stating the dealership has also "made significant investments in training related to company's electric and hybrid vehicles and have partnered with local companies to provide sustainable solutions integrated with our vehicles." In contrast, DD leveraged frequent price-based promotions and integrated proprietary financing methods. This allowed DD to vary prices based on demand, sell regardless of the time period, and outperform the competition's stagnant prices. All dealerships recognized the importance of employee training, but implementation varied, with dealerships A and C notably emphasizing tailored, brand-specific training, reflecting higher scores in customer retention metrics.

5 Discussion

When analyzing results, DD's success emerges in various areas. This success can be attributed to multiple factors, one of them being globally functioning and accepted operations. For example, Interviewee D discussed their store's focus on adaptability and efficiency, which are supported by Diógenes et al.'s study, showcasing how these themes are key to industry success. Additionally, the dealership's operations closely relate to Zilber's digital marketing and financing techniques, supporting DD's high revenue. Furthermore, DA's high scores on the indexes can be explained by Borchardt et al.'s conclusions that customer retention and after-sales care directly impact revenue. As interviewee A mentions, after-sales is critical to their dealership's sales process, aligning with Borchardt et al.'s findings and corroborating DA's positive results. Thus, some globally accepted practices can relate to dealership performance.

However, previous research's findings were not always applicable to Recife's market. Though Firnkorn and Müller's research emphasizes sustainability and exclusivity, these strategies proved purposeless. Interviewee C highlighted the brand's push for sustainable, premium, and exclusive experiences, yet revenue and index scores remained low. Although DC applied "best practice" models, they failed in this context. Furthermore, collaborative partnerships, employee well-being prioritization, and creative business strategies are either absent or never mentioned in interviews. Business strategies are generally monotone and similar between different dealerships, showcasing the lack of implementation of proven successful strategies.

These divergences between accepted sales approaches and actually implemented methods are likely due to Brazil and Recife's specific market and customer preferences. These discrepancies may be linked to culture, economic disparities, ethics, ethnicity, religion, education level, and personal preferences. For example, DC's poor overall productivity may not be directly linked to ineffective tactics but rather to Recife's lower income levels and, therefore, a smaller luxury market. Recife is still a developing city, with a GDP per capita of less than eight thousand dollars [13]. Hence, Recife's citizens are less prone to spend money on luxury cars DC sells when basic needs like food and utilities are a constant concern. Overall, automobiles have a high price elasticity of demand because of their nonessential nature, and in Recife's market, that elasticity is intensified. Additionally, the cross-price elasticity of demand in cars is high, meaning if a price increases for luxury cars, then the demand for more accessible cars will increase. Overall, Recife's luxury goods are in the niche, so dealerships like DC are greatly affected.

Another underperforming strategy evident when analyzing dealership efficiency was the sustainable practices employed by DC. Unfortunately, environmental concerns are not yet a purchase driver in Brazil. Economic priorities, lack of education and awareness, absence of government policies, perception of nature as abundant, and agribusiness and deforestation normalization cause sustainable strategies to lack traction with current consumers, suggesting that even validated strategies may fail without contextual alignment. This is highlighted by the lack of green spaces in Recife [14] and the fact that although environmental goals are outlined, they lack clear deadlines and funding, making proposals ineffective in practice [15]. Accordingly, it is imperative to recognize that strategies must be tailored to local socio-economic conditions. This will create stores shaped around their clients while establishing shared foundations for sales representatives to collaborate effectively. Recognizing Recife as a developing urban center is crucial—successful strategies differ from those in other more developed European or American contexts. To combat this, dealerships should adopt "glocal" strategic thinking; employees will have global insight with local adaptation.

Ultimately, the key to a successful business in the retail automotive industry is boringness. Although it may seem counterintuitive, creating a boring dealership allows for a competitive advantage in a massive market. A boring dealership must not be equated with a store that does not excite clients, create urgency, or have an unmotivated staff. Instead, it relates to a store that successfully operates and manages its fundamentals. These basics are as simple as updating car prices, checking inventory, finishing paperwork, ensuring websites work, and fixing features such as inoperative phone numbers or incorrect car model pictures. Before plunging into extravagant and distinct strategies, management staff must first set their foundations, concentrating on everyday operations that allow for greatness to come from boring, everyday tactics. As highlighted by DD's simplistic tactics, doing the basics brilliantly sets a dealership apart, as many times, management staff jumps to innovative strategies to fix problems caused by simple issues. The greatest threat to success is not failure but boredom. If a dealership gets bored of daily tasks, it is prone to failing. Thinking outside the box should continue staying outside the box; these operations should be made infrequently; if you employ outside-the-box strategies constantly, the outside of the box becomes the inside.

While these insights offer valuable strategic guidance, it is important to acknowledge the limitations of this study. Firstly, only four dealerships were used for data collection. Additionally, all of the stores are franchised. These constraints greatly limit generalizability. Internal validity remains relatively strong as insights are based on real data and interviews. However, the same is not applied to external validity. Generalizing these results to all dealerships in Recife, let alone in Brazil, is challenging given the differences in dealership size, brand, and age. The small number of dealerships also restricts the ability to find various strategies, performance patterns, or management styles. With more study participants, more creative findings could have been observed. Finally, certain conclusions may be specific to dealerships rather than universally applicable. It is imperative that the study's findings are interpreted as exploratory, not definitive, and that, consequently, broader conclusions must be drawn cautiously.

Furthermore, customer perspective is a limitation of this study as customer interviews could provide external validations from buyers. Without customer feedback on dealerships, it is harder to understand what methods clients view most positively and how they affect client acquisition. Essentially, the study may be subject to being biased toward the internal viewpoint of management. Not capturing real-world clientele opinion limits generalizability across consumer segments, indicating conclusions about customer satisfaction (e.g., from returning customer % or interviewee comments) should be treated as indirect inferences, not first-hand accounts.

Finally, with car acquisition costs excluded, the CEI might overstate profitability, especially for high-volume dealerships, distorting comparisons between dealerships with different ticket prices and car sales. Not including COGS may limit construct validity as dealerships with tighter margins, but more sales may appear more efficient than they truly are. It also creates a divide between the study and similar studies that include COGS. The decision is justifiable for isolating operational strategy, but readers should know that CEI and RDS do not reflect total financial performance. Despite the foregoing, it is imperative to clarify that the presented assumptions and correlations do not constitute causal evidence. Consequently, longitudinal investigations would be requisite to definitively establish causality. While this partially circumscribes the scope of our conclusions, it does not detract from the study's intrinsic relevance. Although it slightly narrows the scope of your conclusion, it does not affect the study's relevance.

For future research, it is essential to expand the sample size to include independent dealerships and increase geographic, social, educational, and economic diversity across Brazil, increasing generalizability. To properly analyze and study quantitative data, further research should seek more precise financial metrics instead of ranges if access allows. Consumer-side research should be incorporated to triangulate findings and prove internal perceptions. Researching the effects of strategies over long periods to track the evolution of strategies could be a great benefit of longitudinal research. Eventually, exploring how emerging sustainability awareness may grow and begin influencing dealership performance in the coming years would be advantageous.

6 Conclusion

With the study's findings, it should seem clear that specific practices in various areas of dealerships can affect revenue, customer return rates, and other metrics indicating excellence if they are specific, contextually aligned strategies. Still, it cannot be asserted with certainty that the relationship between techniques employed and strategic advantage is causation over correlation. The key takeaway is that "boring," well-executed fundamentals and adaptable, customer-oriented practices outperformed trendier techniques. Our study's primary contribution underscores a critical insight: global strategic frameworks do not universally translate to regional markets, as exemplified by the automotive sector in Recife. Consequently, effective managerial approaches must be meticulously tailored to local economic, cultural, and consumer dynamics. We assert that the success of franchised dealerships in such contexts hinges on their capacity to integrate fundamental operational discipline with regional awareness. This finding challenges conventional wisdom regarding standardized global strategies and highlights the localized adaptation.

Initially, the study intended to adopt a mixed-methods approach, combining quantitative dealership surveys with qualitative semi-structured interviews involving store managers and customers. This could provide comprehensive insights from both managerial and consumer perspectives. However, customer interviews proved infeasible due to practical constraints, including confidentiality concerns —limiting access to customer information— resource and time limitations, and logistical barriers. Furthermore, due to limitations such as small sample size, lack of COGS, and missing consumer voice, the study has restricted external validity and data granularity. However, the value of the core insights is maintained regardless. Thus, it is crucial to emphasize that the findings are exploratory and not generalizable.

Through this investigative structure, there are some significant observations for both the operators and the involved parties: i) Local dealership managers should critically evaluate whether their strategic choices resonate with local market realities rather than mindlessly replicating national or global models; ii) Policymakers may benefit from promoting regionally adaptive business support, prioritizing context-specific solutions over top-down standardization; iii) Automotive brands are encouraged to reconsider the one-size-fits-all deployment of operational strategies across franchises, especially in socioeconomically diverse markets like Recife.

To further corroborate and expand on this study's findings while gaining a deeper understanding of the automotive sector, further research is essential, given the exploratory nature of this study. Specifically, research including larger, more diverse samples of dealerships would enhance external validity and capture a broader spectrum of strategic practices. Additionally, incorporating consumer-side perspectives would allow for more robust conclusions about customer satisfaction, loyalty, and perceived value. Longitudinal studies could also track the evolving effectiveness of operational strategies over time to substantiate correlation. Finally, future research might integrate COGS and profitability metrics more fully, bridging the gap between operational efficiency and financial performance. These expansions would not only reinforce the insights of this study but also contribute meaningfully to the global conversation around strategy localization in emerging markets.

Taken together, success in Recife's car market is not built on reinvention—it is built on doing the fundamentals brilliantly in a place that demands local vision over global mimicry. As interviewee B says, "You don't see an ad for a soggy pizza or a warm beer, right?" asserting that the basics should be a dealer's primary worry. Adaptation must precisely match each dealership's local context. Ultimately, success is not about being the loudest—it is about being the clearest in a market that's already full of noise.

References

1. Dimitrakopoulos, G., Tsakanikas, A., Panagiotopoulos, E.: A path of structural transformation for the automotive and insurance industries toward autonomous vehicles. Auton. Veh. 69–83 (2021). https://doi.org/10.1016/b978-0-323-90137-6.00003-5

2. Diógenes, J.R.F., Queiroz, F.C.B.P., Queiroz, J.V., Furukava, M., Lima, N.C., Souza, G.H.S.: Cultura da qualidade nas concessionárias automotivas brasileiras. Gestão Produção **26**, e2046 (2019). https://doi.org/10.1590/0104-530x2046-19
3. Von Corswant, F., Fredriksson, P.: Sourcing trends in the car industry: a survey of car manufacturers' and suppliers' strategies and relations. Int. J. Oper. Prod. Manag. **22**(7), 741–758 (2002). https://doi.org/10.1108/01443570210433526
4. Borchardt, M., Souza, M., Pereira, G.M., Viegas, C.V.: Achieving better revenue and customers' satisfaction with after-sales services: how do the best branded car dealerships get it? Int. J. Qual. Reliab. Manage. **35**(9), 1686–1708 (2018). https://doi.org/10.1108/ijqrm-01-2017-0016
5. Almohri, H., Chinnam, R. B., Colosimo, M.: Data-Driven Analytics for Benchmarking and Optimizing Retail Store Performance (2018). arXiv preprint arXiv:1806.05563
6. Aaltio, I., Eskelinen, M. T.: Proceedings of The 11th European Conference on Innovation and Entrepreneurship, pp. 15–16 (2016)
7. Zilber, S. N.: E-business and the company strategy: The case of the Celta at GM Brazil. In: Challenges in the Management of New Technologies, pp. 163–180 (2007). https://doi.org/10.1142/9789812770318_0012
8. Hase, S., Busch, C.: The quintessence of sales. Springer, Cham, Springer Nature Switzerland AG (2018)
9. Ferro, J.R.: Current developments of the Brazilian automotive industry (1995). https://dspace.mit.edu/bitstream/handle/1721.1/1636/ferro-1.pdf
10. Wolfe, J.: Autos and progress: The Brazilian search for modernity. Oxford University Press (2010). https://doi.org/10.1111/j.1468-0289.2010.00597_27.x
11. Boles, J.S., Babin, B.J., Brashear, T.G., Brooks, C.: An examination of the relationships between retail work environments, salesperson selling orientation-customer orientation and job performance. J. Mark. Theory Pract. **9**(3), 1–13 (2001). https://doi.org/10.1080/10696679.2001.11501893
12. Scott Morton, F., Silva-Risso, J., Zettelmeyer, F.: What matters in a price negotiation: evidence from the US auto retailing industry. Quant. Mark. Econ. **9**, 365–402 (2011)
13. Secretaria de Desenvolvimento Econômico – SDEC: IBGE Aponta Recife Com O Maior PIB per Capita Do Nordeste. Prefeitura Do Recife (2023). https://www2.recife.pe.gov.br/noticias/18/12/2023/ibge-aponta-recife-com-o-maior-pib-capita-do-nordeste
14. Melo, I.S., de Carvalho, R.M.C.M., Sobral, M.D.C.M., Lyra, M.R.C.C., Silva, H.P.D.: Adaptation to the impacts of climate change from the perspective of the City of Recife master plan. Revista Brasileira de Estudos Urbanos e Regionais **23**, e202140 (2022). https://doi.org/10.22296/2317-1529.rbeur.202140pt
15. Turmena, L., Lusieux, A.: Transformative Urban Coalitions (TUC) city profile: Recife, Brazil. Transformative Urban Coalitions (2022). https://www.idos-research.de/externe-publikationen/article/transformative-urban-coalitions-tuc-city-profile-recife-brazil/

Factors Associated with Mortality in Patients with Oral Cavity and Oropharyngeal Cancer: Indirect Evidence of Financial Toxicity

Danielle Karla Vieira e Silva(✉), Wenner Gláucio Lopes Lucena, and Yuri Wanderley Cavalcanti

Federal University of Paraíba, Cidade Universitaria, Castelo Branco, João Pessoa/Paraíba, Brazil
{danielle.vieira,wenner.lucena}@academico.ufpb.br,
yuri@ccs.ufpb.br

Abstract. Objective: To analyze the factors associated with mortality in patients with oral cavity and oropharyngeal cancer, considering death as an indirect outcome of financial to-xicity. **Methods:** This was an analytical cross-sectional study using data from the Hospital Cancer Registry (RHC) for the period from 2019 to 2023. A total of 11,189 adult patients diagnosed with oral cavity (C00–C06) or oropharyngeal (C09–C10) cancer were included. The independent variables were: education level, alcoholism, clinical staging, type of healthcare (public or private), year of diagnosis, tumor location, and region of treatment. The association between these variables and the outcome (death) was evaluated using binary logistic regression, with a significance level of 5%. **Results:** Advanced clinical staging (OR = 3.53; 95% CI: 3.04–4.10), low education level (OR = 1.94; 95% CI: 1.25–3.01), severe alcoholism (OR = 1.28; 95% CI: 1.15–1.44), and treatment in the Northern region of the country (reference) were factors associated with higher mortality risk. Patients diagnosed in 2022 had a higher likelihood of death (OR = 1.26; 95% CI: 1.11–1.43), while those diagnosed in 2023 had a lower risk (OR = 0.75; 95% CI: 0.62–0.90). **Conclusion:** Low education level, severe alcoholism, advanced staging at diagnosis, and regional disparities were associated with increased mortality risk, reflecting the indirect impact of financial toxicity on cancer patient survival.

Keywords: Financial Toxicity · Oral Neoplasms · Mortality · Social Determinants of Health · Socioeconomic Factors

1 Introduction

Head and neck cancer, particularly those located in the oral cavity and oropharynx, represents a serious public health problem, with high global incidence and mortality rates, especially in low- and middle-income countries [1]. In Brazil, the pattern of illness and death from these tumors is strongly associated with social, economic, and territorial inequalities, which contribute to delayed diagnosis and adverse clinical outcomes [2, 3].

L. Almeida et al. (Eds.): ICAFI 2025, ISEM 80, pp. 350–364, 2026.
https://doi.org/10.1007/978-3-032-16268-7_20

Factors such as low educational attainment, poverty, and limited access to healthcare are identified as central determinants of diagnostic delay and worsened prognosis [4–6]. Patients with lower educational levels, for instance, tend to show lower adherence to preventive practices and delayed care-seeking behavior, in addition to facing greater difficulties in understanding therapeutic guidelines [2, 7].

In this context, the concept of financial toxicity stands out. It encompasses the direct and indirect economic burdens faced by cancer patients during treatment, including costs related to transportation, food, income loss, job abandonment, and therapy interruption [8, 9]. These factors compromise not only treatment continuity but also patients' quality of life and survival [10, 11]. National studies support this perspective, showing that patients treated under the Brazilian Unified Health System (SUS) experience high levels of financial toxicity, even with publicly funded treatment, especially among those with lower income and continuous medication use [12]. Furthermore, a negative correlation was identified between financial toxicity and quality of life ($\rho = -0.41$; $p < 0.001$), highlighting the multidimensional impact of economic burden [13]. These effects were exacerbated during the COVID-19 pandemic, as reported in an integrative review that emphasized the worsening of economic barriers to accessing care [14].

Studies indicate that the risk of death may be intensified by the interaction between clinical characteristics (such as advanced staging at diagnosis), behavioral factors (such as alcohol use), and structural elements, including the geographic region of treatment, reflecting the unequal distribution of healthcare services across Brazilian territory [15–17].

Given the absence of direct data on the patients' economic status in the available records, the present study adopts death during the first course of treatment as a proxy outcome to investigate potential inequalities in access, continuity, and effectiveness of oncological care, which may be associated with financial toxicity. Although it does not replace standardized measures such as the COST questionnaire, early mortality is understood to indirectly reflect unmeasured social and economic barriers.

The literature also highlights that the population most affected by these inequalities includes Black patients, residents of rural or peripheral areas, and individuals who rely exclusively on the public health system, all of whom face compounded difficulties that result in poorer clinical outcomes [10, 18].

Given this scenario, understanding the factors associated with mortality in patients with oral cavity and oropharyngeal cancer requires a broader analysis that goes beyond clinical data and incorporates the social determinants of health and structural inequalities that affect cancer care. Thus, the present study aims to analyze the factors associated with mortality in these patients in Brazil, considering death as an indirect expression of financial toxicity.

2 Theoretical Framework

2.1 Epidemiology of Oral Cavity and Oropharyngeal Cancer

Oral cavity and oropharyngeal cancers are part of the group of malignant head and neck neoplasms and represent a significant public health issue, especially in low- and middle-income countries. Global estimates indicate that in 2020, there were more than 377,000 new cases of oral cavity cancer and approximately 98,000 cases of oropharyngeal cancer, with a combined mortality exceeding 230,000 deaths annually [19]. In Brazil, these neoplasms are among the most incident in men, particularly in the North and Northeast regions, reflecting the unequal distribution of risk factors and healthcare services [20].

These cancers are frequently diagnosed at advanced stages, which contributes to worse prognoses and higher mortality. Staging, in this context, refers to the classification that determines the extent of the disease at the time of diagnosis, considering the size of the primary tumor, involvement of regional lymph nodes, and presence of distant metastases (TNM system). Early-stage tumors (I and II) usually offer better treatment prospects and prognosis, whereas advanced stages (III and IV) are associated with greater morbidity, more aggressive treatment needs, and lower survival rates [21].

Late presentation is linked to the absence of specific symptoms in the early stages, lack of awareness of warning signs, and limited access to specialized services, especially in areas with low oncological care coverage [22]. Moreover, there is evidence that the five-year survival rate for oral cavity cancer may be below 50% when diagnosis occurs at a late stage [23].

The main risk factors include chronic use of tobacco and alcohol, as well as infection with human papillomavirus (HPV), especially in oropharyngeal cases [24]. These factors are strongly associated with socially adverse conditions, such as low educational attainment and high-risk occupational environments, which reinforce the socially determined nature of these neoplasms. Studies indicate that the prevalence of such risk behaviors is higher in populations with lower socioeconomic status, contributing to disparities in disease burden [25].

Beyond clinical and etiological characteristics, head and neck cancers present a regionally unequal distribution in Brazil. The North and Northeast regions report higher mortality rates despite lower recorded incidence, pointing to difficulties in early diagnosis and continuity of care [26]. The scarcity of specialized services, long travel distances to referral centers, and lack of integration across levels of care contribute to the worsening of clinical conditions [9].

Therefore, understanding the epidemiology of oral cavity and oropharyngeal cancer requires consideration not only of biomedical factors but also of the social and regional determinants that shape its distribution, outcomes, and access to the oncological care pathway.

2.2 Social Inequalities and Access to Oncological Care

Access to diagnosis and treatment of head and neck cancer in Brazil is deeply influenced by social, economic, and territorial inequalities. Patients living in regions with lower density of specialized oncological services, such as the North and Northeast, face geographic and structural barriers that directly impact the continuity and effectiveness of care [11, 15]. In these regions, long travel distances to referral centers, extended waiting times for tests and procedures, and disorganized referral and regulation networks are common, all contributing to delayed diagnoses and increased mortality [3, 11].

The literature shows that the most vulnerable populations such as individuals with low educational attainment, informal workers, Black populations, and residents of rural or peripheral areas have less access to prevention, early diagnosis, and adequate treatment [1, 7]. These conditions represent both economic and symbolic barriers, affecting the bond with healthcare services and the capacity to navigate care pathways. Low schooling levels, for example, are associated with reduced comprehension of clinical guidance and difficulty recognizing early signs of disease, which may lead to delayed care-seeking behavior [11].

Moreover, although the Brazilian health system is universal, it exhibits significant disparities between the public (SUS) and private sectors. Patients who rely exclusively on SUS are more exposed to infrastructure failures, shortages of healthcare professionals, and fragmented oncological care, especially in less developed regions [5, 6]. In contrast, individuals with access to private health insurance are more likely to undergo exams with shorter waiting times and initiate treatment more promptly, which positively affects clinical outcomes [25].

The findings of the present study reinforce this scenario of inequality, showing that the proportion of deaths was significantly higher among patients treated exclusively by SUS compared to those who accessed private services or paid out of pocket. This difference can be interpreted as a manifestation of structural financial toxicity experienced by individuals without supplemental coverage, particularly when combined with treatment delays and advanced stage at diagnosis.

International studies support that disparities in access to early diagnosis and appropriate treatment are among the main determinants of survival differences between population groups [4, 18]. In Brazil, this inequality is even more pronounced due to the combination of structural deficiencies, inadequate public health funding, and the absence of specific policies targeting oncology care in vulnerable populations [5, 27].

Therefore, social and regional inequalities are central elements in determining oncological care for head and neck cancer. They influence the entire care trajectory from the initial recognition of symptoms to treatment completion and contribute to the perpetuation of health inequities that translate into avoidable deaths and a significant burden of individual and collective suffering.

2.3 Clinical Staging and Prognosis

Clinical staging is a fundamental step in the management of head and neck cancer, as it determines the extent of disease and guides treatment choices, prognosis, and follow-up strategies. The most widely used system is the TNM, proposed by the American Joint Committee on Cancer (AJCC), which assesses the size of the primary tumor (T), the involvement of regional lymph nodes (N), and the presence of distant metastases (M), classifying cases into stages ranging from I to IV [21].

Studies have shown that advanced staging at diagnosis (stages III and IV) is strongly associated with increased mortality and reduced chances of therapeutic success [5, 8]. In the present study, advanced staging was the main clinical factor associated with death, reinforcing the notion that late disease identification represents a critical barrier to controlling oral and oropharyngeal cancer.

Patients diagnosed at early stages have a higher likelihood of receiving curative treatment with lower morbidity, such as surgery alone or radiotherapy. In contrast, those in advanced stages often require combined therapies radical surgery, chemotherapy, and radiotherapy which impose greater functional, psychological, and economic burdens on the patient [24].

Late diagnosis is closely related to factors such as low risk perception, neglect of initial symptoms, lack of screening, shortage of specialized services, and long waiting times for diagnostic tests [22]. These barriers predominantly affect vulnerable populations, contributing to inequalities in access to care and amplifying the effects of financial toxicity, particularly among patients without private health insurance or living in regions with limited healthcare infrastructure [9, 20].

Thus, clinical staging should not be interpreted solely as a biomedical variable, but also as an indicator of the social and structural inequities that hinder early diagnosis and compromise timely treatment, ultimately influencing patient survival outcomes.

2.4 Financial Toxicity in Cancer Care

Financial toxicity refers to the economic hardship caused by the direct and indirect costs of cancer treatment, negatively affecting patients' quality of life, treatment adherence, and clinical outcomes [19, 28]. This concept, originally coined to describe the reality of countries with insurance-based healthcare systems, has been widely adopted in universal healthcare contexts such as Brazil, to describe situations where, even with free treatment, patients still bear significant costs related to transportation, food, income loss, and other illness-related expenses. Studies conducted in Brazil corroborate this perspective, showing that cancer patients treated within the Brazilian Unified Health System (SUS), especially those with lower income, experience high levels of financial toxicity, with a direct impact on treatment adherence and quality of life [12, 13]. Moreover, the COVID-19 pandemic worsened this scenario by increasing inequalities and hindering the continuity of cancer care [14].

Financial toxicity acts as a social mediator of illness, disproportionately affecting individuals in situations of socioeconomic vulnerability. Among the most common impacts are indebtedness, interruption of work or studies, strain on family activities, and psychological distress [26, 28]. Recent studies show that patients experiencing higher levels of financial toxicity report more depressive symptoms, higher rates of treatment discontinuation, and even increased mortality, even in technologically advanced healthcare environments [20, 29].

Recognizing financial toxicity as a clinical dimension not merely a socioeconomic one is essential for the development of fairer and more effective public policies. Strategies for early screening of financial vulnerability, psychosocial support, transportation assistance, and institutional accommodations should be incorporated into care protocols, particularly within public healthcare systems like Brazil's SUS, where the promise of universality must confront invisible barriers that hinder equitable care [13, 19, 30].

Patients with head and neck cancer are among the most affected by financial toxicity due to the mutilating nature of treatments, the frequency of severe side effects, and the need for prolonged functional rehabilitation [26]. Communication, swallowing, and facial appearance difficulties resulting from treatment significantly impact the productive and social lives of these patients, leading to work leave, social exclusion, and family overload factors that amplify financial toxicity even in contexts where clinical treatment is provided.

Thus, when discussing financial toxicity in head and neck cancer, the issue goes beyond costs it is also a matter of social justice. It means acknowledging that material deprivation, when left unaddressed, can be as decisive for outcomes as the cancer stage itself. Incorporating this perspective broadens the understanding of illness and reinforces the urgency of providing comprehensive, coordinated, and socially sensitive cancer care in Brazil.

3 Materials and Methods

This is an analytical cross-sectional study based on secondary data from the Brazilian Hospital Cancer Registry (Registro Hospitalar de Câncer–RHC), covering the period from 2019 to 2023. The study included adult patients ($\geq$18 years) diagnosed with oral cavity cancer (C00–C06) or oropharyngeal cancer (C09–C10), with complete records regarding outcome (death) and the following independent variables: education level, clinical staging, alcoholism, year of diagnosis, treatment orientation, and treatment region. Patients under the age of 18 or with incomplete records were excluded from the analysis.

The main independent variables included: geographic region, clinical stage, and type of funding (SUS or private). Control variables included age group, sex, and educational level, aiming to adjust for potential confounding factors in the association with the outcome (death).

The final sample consisted of 11,189 patients. The outcome variable was death (yes/no). Independent variables included education level, alcoholism, treatment orientation (public/private/self-funded), clinical stage at diagnosis, year of diagnosis, tumor location, and treatment region.

Data were tabulated and organized using Microsoft Excel 2019, and statistical analyses were conducted using the Jamovi software (version 2.6). Descriptive analyses were performed using measures of central tendency (median) and dispersion (25th and 75th percentiles). The Kolmogorov Smirnov test was applied to assess the normality of continuous variables, which showed non-normal distributions, thus justifying the use of the non-parametric Mann Whitney test to compare tumor locations.

The association between independent variables and the outcome was evaluated using binary logistic regression. Results were expressed as odds ratios (ORs) with 95% confidence intervals. Model fit was assessed using deviance statistics, the Akaike Information Criterion (AIC), and Cox & Snell's R^2. Multicollinearity was tested using Variance Inflation Factor (VIF) and Tolerance values. A significance level of 5% ($p < 0.05$) was adopted.

In accordance with Resolution nº 466, dated December 12, 2012, of the Brazilian National Health Council (CNS), the study was exempt from review by a Research Ethics Committee, as it used only publicly available and anonymized data.

4 Results

The final sample consisted of 11,189 adult patients diagnosed with oral cavity cancer (C00–C06) or oropharyngeal cancer (C09–C10) between 2019 and 2023, with 8,025 cases (71.7%) of oral cavity cancer and 3,164 cases (28.3%) of oropharyngeal cancer.

As shown in Table 1, most patients with oral cavity and oropharyngeal cancer survived during the follow-up period. However, a higher absolute number of deaths was observed among patients with oral cavity cancer (2,103 cases) compared to those with oropharyngeal cancer (940 cases). This difference can be partially attributed to the higher proportion of oral cavity cancer cases in the sample. Nevertheless, when considering the mortality rate within each group, it is noteworthy that the mortality rate was also higher among patients with oral cavity cancer (18.8%) compared to those with oropharyngeal cancer (8.4%).

These findings indicate that oral cavity cancer, in addition to being more frequent in the sample, had a worse prognosis. This observation justifies the comparative analysis between the groups to investigate the factors associated with mortality.

Table 1. Frequency of death among patients by tumor location.

Tumor Location	Survivors (N)	% of Total	Deceased (N)	% of Total
Oral Cavity (C00–C06)	5,922	52.90%	2,103	18.80%
Oropharynx (C09–C10)	2,224	19.90%	940	8.40%

The normality of continuous variables was assessed using the Kolmogorov–Smirnov test, which indicated a violation of the normal distribution assumption. Therefore, the non-parametric Mann Whitney U test was applied to compare tumor site groups.

A statistically significant difference was found between patients with oral cavity cancer and those with oropharyngeal cancer regarding mortality occurrence ($U = 1.23 \times 10^7$; $p < 0.001$), suggesting that the distribution of deaths differs according to tumor location. The test indicates that patients with oral cavity cancer and those with oropharyngeal cancer have distinct mortality risks, which may reflect clinical, biological, and possibly socioeconomic differences between these groups.

Subsequently, a multivariate analysis was performed to identify factors associated with mortality. The binomial logistic regression model yielded a deviance of 12,587, an AIC of 12,625, and a Cox & Snell R^2 of 0.0388, indicating satisfactory model fit.

Patients with lower education levels had a higher risk of death compared to those with higher education. The "education" variable was categorized into six ascending levels: level 1 referred to patients with no formal education; level 2, incomplete primary education; level 3, complete primary education; level 4, incomplete secondary education; level 5, complete secondary education; and level 6, complete higher education, which served as the reference category in the analysis. Compared to patients with higher education, those classified at levels 1, 2, and 3 showed statistically significant associations with increased mortality, with odds ratios of 1.944 (95% CI: 1.255–3.011; $p = 0.003$), 1.591 (95% CI: 1.041–2.432; $p = 0.032$), and 1.550 (95% CI: 1.007–2.386; $p = 0.046$), respectively. Levels 4 and 5 showed a trend toward statistical significance, with odds ratios of 1.449 (95% CI: 0.938–2.237; $p = 0.094$) and 1.790 (95% CI: 0.964–3.326; $p = 0.065$), suggesting that even among patients with secondary education, mortality may still be influenced by social barriers and access to healthcare. These findings reinforce the link between low educational attainment and poor clinical outcomes, highlighting the impact of social determinants of health on cancer survival.

Clinical staging was the main factor associated with mortality: patients diagnosed at stages III or IV had a risk of death more than three times higher compared to those diagnosed at stages I or II (OR = 3.530; 95% CI: 3.042–4.097; $p < 0.001$).

Among behavioral factors, severe alcoholism (category 3) was associated with a higher risk of death compared to category 1 (OR = 1.284; 95% CI: 1.147–1.439; $p < 0.001$).

Moderate alcoholism (category 2), however, did not show a statistically significant association (OR = 1.090; 95% CI: 0.967–1.229; $p = 0.159$). Regarding the year of diagnosis, patients diagnosed in 2022 had a higher risk of death compared to those diagnosed in 2019 (OR = 1.261; 95% CI: 1.101–1.433; $p < 0.001$). Diagnoses in 2020 and 2021 showed no statistically significant differences. In 2023, a reduction in the risk of death was observed (OR = 0.745; 95% CI: 0.620–0.895; $p = 0.002$).

With regard to treatment region, all regions presented a lower risk of death when compared to the North region: Central-West (OR = 0.526; 95% CI: 0.338–0.818; p = 0.004), Northeast (OR = 0.520; 95% CI: 0.412–0.658; p < 0.001), Southeast (OR = 0.554; 95% CI: 0.442–0.695; p < 0.001), and South (OR = 0.661; 95% CI: 0.522–0.838; p < 0.001).

The variables *treatment orientation* (public SUS vs. others) (OR = 0.976; 95% CI: 0.849–1.122; p = 0.729) and *tumor site* (oropharynx vs. oral cavity) (OR = 1.092; 95% CI: 0.994–1.199; p = 0.065) did not show statistically significant associations with the outcome (Table 2).

Table 2. Logistic Regression for Death

Variables	p	OR	95% CI (Lower- Upper)
EDUCATION *(reference = Level 6)*			
1 – 6	0.003	1.944	1.255–3.011
2 – 6	0.032	1.591	1.041–2.432
3 – 6	0.046	1.550	1.007–2.386
4 – 6	0.094	1.449	0.938–2.237
5 – 6	0.065	1.790	0.964–3.326
ALCOHOLISM *(reference = Category 1)*			
2 – 1	0.159	1.090	0.966–1.229
3 – 1	<0.001	1.284	1.146–1.439
TREATMENT ORIENTATION			
Other vs. SUS	0.729	0.976	0.849–1.122
CLINICAL STAGING			
III-IV vs. I-II	<0.001	3.530	3.042–4.097
YEAR OF DIAGNOSIS *(reference = 2019)*			
2020 – 2019	0.454	1.049	0.925–1.190
2021 – 2019	0.234	1.080	0.951–1.226
2022 – 2019	<0.001	1.261	1.101–1.433
2023 – 2019	0.002	0.745	0.620–0.895
TUMOR LOCATION *(reference = C00–C06)*			
C09-C10 vs. C00-C06	0.065	1.092	0.994–1.199
REGION *(reference = North)*			
Central-West vs. North	0.004	0.526	0.337–0.818
Northeast vs. North	<0.001	0.520	0.411–0.658
Southeast vs. North	<0.001	0.554	0.442–0.695
South vs. North	<0.001	0.661	0.521–0.838

In Fig. 1 presents the interaction between tumor location (oral cavity and oropharynx) and treatment region. It is observed that, regardless of tumor site, patients treated in the Northern region have the highest adjusted probabilities of death, while the lowest are concentrated in the Central-West and Southeast regions. The Southern region shows intermediate values, ranking above the Central-West and Southeast, but below the North and, in some instances, the Northeast. This pattern may be related to both the distribution of cases and the installed capacity of oncology services in each region. The observed differences in probability highlight persistent regional disparities in the structure of the healthcare network, likely influenced by wait times, distance to specialized centers, and resource availability factors that are central to understanding the financial toxicity experienced by patients.

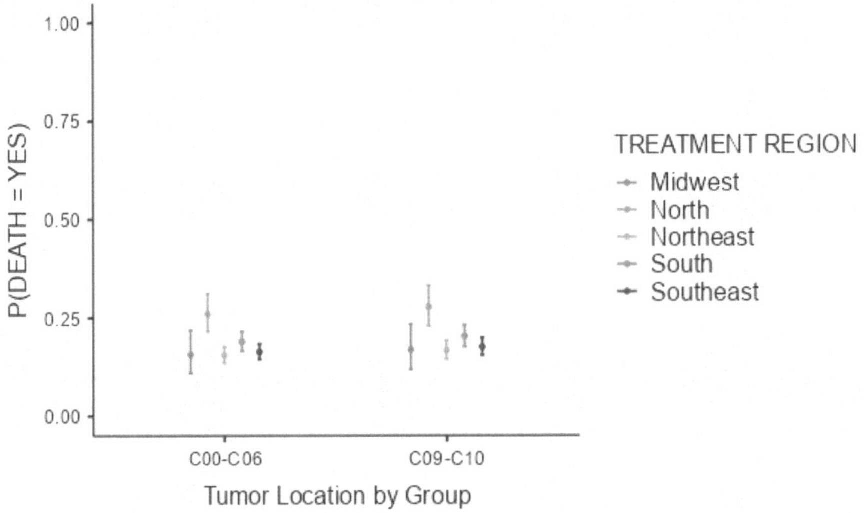

Fig. 1. Estimated marginal means for death occurrence according to tumor location (C00–C06 and C09–C10) and treatment region.

In Fig. 2, an interaction is observed between clinical staging (I/II vs. III/IV) and treatment region. A general trend indicates a significantly increased risk of death in more advanced stages (III/IV), particularly among patients treated in the North region. This pattern supports the hypothesis that patients from regions with lower healthcare infrastructure face greater challenges in achieving early diagnosis and, by the time they enter the care pathway, are already at advanced stages resulting in a direct impact on mortality.

These visualizations help to understand how the interaction between clinical and territorial aspects contributes to mortality outcomes, reinforcing the need for public policies that consider regional disparities as a central factor in mitigating financial toxicity in cancer care.

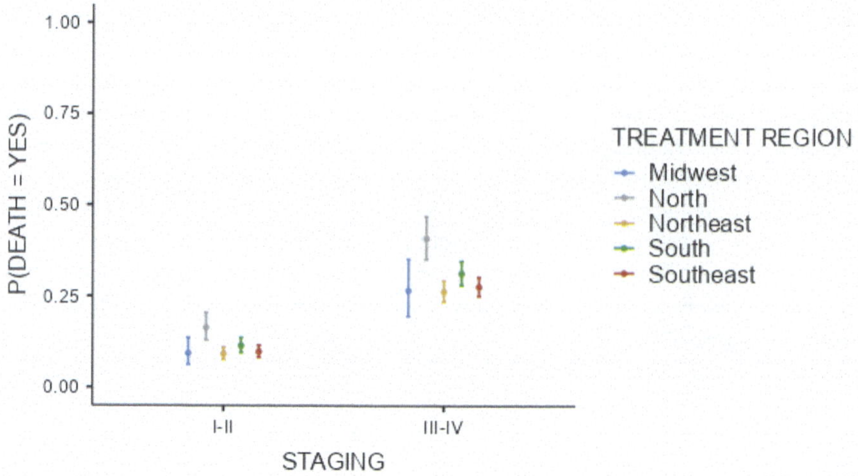

Fig. 2. Estimated marginal means for the probability of death according to clinical staging and treatment region.

5 Discussion

The use of mortality as an indirect outcome of financial toxicity, although not conventional, is justified in contexts where data on individual costs are limited. In middle-income countries like Brazil, the early interruption of cancer care due to financial, geographic, or structural barriers can lead to adverse outcomes, including early death. Thus, this study adopts mortality during the first course of treatment as an extreme marker of failure in access to or continuity of care, potentially influenced by socioeconomic factors.

The results of this study demonstrate that sociodemographic, behavioral, clinical, and regional factors significantly influence mortality among patients with oral cavity and oropharyngeal cancer in Brazil. Recognizing these variables as social determinants of health allows death to be understood not only as a clinical outcome but also as a reflection of the inequalities that permeate access to, continuity of, and quality in oncological care [4, 6].

Low educational attainment was identified as one of the main factors associated with increased mortality risk, which aligns with previous studies showing the relationship between education level and access to health services, treatment adherence, and comprehension of clinical guidance [5, 7]. Individuals with lower education levels tend to seek care at more advanced stages of the disease, show lower engagement with prevention strategies, and face greater barriers to understanding the therapeutic process [2, 11].

Severe alcoholism, identified as another predictor of mortality, reflects not only the biological impact on disease progression but also its association with social vulnerability, lower family support, and cumulative occupational risks [10, 23]. Moreover, alcoholism may serve as a marker of social exclusion, hindering patients' navigation through the health system and their adherence to treatment [17].

Advanced staging at diagnosis remained the primary clinical factor associated with mortality, reinforcing previous findings that late diagnosis is one of the main barriers to controlling oral cancer [5, 8]. Diagnostic delays are not only linked to low risk perception but also to the scarcity of specialized services, long waiting times for diagnostic tests, and financial barriers that particularly affect vulnerable populations [6, 8].

The association between treatment region and mortality, especially in the North and Northeast regions, aligns with studies demonstrating the unequal distribution of healthcare resources across Brazil. These regions have a lower density of high-complexity oncology services and require greater travel to specialized centers, creating logistical and financial barriers for patients [15, 17, 25]. This context directly contributes to financial toxicity, understood here as a structural phenomenon that combines poverty, geographic barriers, and lack of institutional support [8, 9].

Financial toxicity, therefore, emerges in this study as a transversal axis that amplifies the negative effects of social determinants on clinical outcomes. More than an individual economic burden, it reflects a structural overload that disproportionately penalizes patients with lower economic, social, and educational capital [4, 18]. This notion is reinforced by studies indicating that income loss, transportation costs, and the need to interrupt work are common among patients undergoing cancer treatment, potentially leading to care discontinuation [9, 10].

Although the logistic regression model identified statistically significant associations between clinical, social, and regional variables and the outcome of death, its explanatory power (Cox & Snell $R^2 = 0.0388$) was relatively low. This finding may reflect the absence of key variables needed to capture the complexity of social determinants of cancer mortality, such as household income, time between diagnosis and treatment initiation, presence of comorbidities, social support, hospital complexity level, and treatment adherence. In this regard, recent studies underscore the importance of analyzing structural inequalities in cancer mortality, showing that cancer care in Brazil remains strongly marked by regional disparities. Cancer treatment under the Brazilian Unified Health System (SUS) presents significant territorial inequalities, with more limited access in the North and Northeast regions, reflecting the unequal distribution of services, equipment, and specialized professionals [31]. Data from the Ministry of Health indicate that, although federal investment in specialized treatments has exceeded R$ 74 billion, important challenges remain regarding the equitable allocation of these resources, addressing unmet demand, and ensuring comprehensive care. These issues add to the economic barriers faced by patients, exacerbating the effects of financial toxicity on clinical outcomes and reinforcing the urgency of public policies focused on equity [32].

Additionally, the increased mortality among patients diagnosed in 2022 may be related to the direct and indirect effects of the COVID-19 pandemic, which resulted in reduced access to health services, diagnostic delays, and disruption of care networks [8, 27]. The pandemic intensified pre-existing inequalities, exacerbating financial toxicity in various contexts, especially among patients who rely exclusively on the public health system.

Taken together, these findings highlight the need for public policies that consider financial toxicity as a critical factor in cancer care planning. Strategies such as decentralizing specialized services, strengthening primary care, expanding financial support for patients, and promoting health education in vulnerable populations may mitigate the effects of inequality and contribute to reducing mortality from oral cavity and oropharyngeal cancer [1, 6, 10].

This study presents some limitations that should be considered when interpreting the results. First, it relied on a secondary data source (Hospital Cancer Registry – RHC), which may contain inconsistencies in data entry, particularly in regions with limited hospital infrastructure. The absence of variables directly related to the patients' socioeconomic status, such as income, education, and occupation, required the use of proxy variables to infer aspects of financial toxicity, which may limit the precision of the inferences.

Furthermore, the cross-sectional design of the analysis does not allow for the establishment of direct causal relationships between the investigated factors and the outcome of death during the first course of treatment. Other potentially relevant factors, such as the presence of comorbidities, time between diagnosis and treatment initiation, psychosocial support, and quality of care provided, were not available in the dataset and therefore could not be analyzed.

It is also worth noting that the use of mortality as an indirect measure of financial toxicity, although innovative, presents important limitations. This outcome can be influenced by multiple clinical and contextual factors not captured in the data. In addition to the aforementioned limitations regarding the use of secondary data, it is important to consider that the model's low R^2 indicates that relevant variables for understanding mortality may not be available in the dataset. Information such as income, waiting time to begin treatment, hospital type (reference center or not), availability of psychosocial support, and level of treatment adherence would be important to refine the analysis.

As future perspectives, primary studies integrating clinical, economic, and psychosocial variables are recommended to deepen the understanding of the mechanisms linking social vulnerability to early mortality. Qualitative and longitudinal investigations could expand the analysis of the impacts of financial toxicity on the cancer care journey, contributing to the development of more equitable public policies within the SUS context, with special attention to regional inequalities and the comprehensiveness of care.

6 Conclusion

This study analyzed factors associated with mortality among patients with oral cavity and oropharyngeal cancer in Brazil, highlighting the impact of financial toxicity as an indirect determinant of oncological mortality. Low educational attainment, severe alcoholism, advanced staging at diagnosis, and regional inequalities particularly treatment in the North were identified as independent predictors of poor prognosis.

These findings reinforce that financial toxicity goes beyond the direct costs of treatment, encompassing socioeconomic barriers, structural limitations, and disparities in access that hinder early diagnosis, continuity of care, and therapeutic effectiveness. The premature mortality observed in this study therefore reflects the intersection of social vulnerability and the fragility of support systems for cancer patients.

Although the use of secondary data limits access to more detailed clinical variables, the large sample size and robust multivariate analysis provide strength to the findings.

These results underscore the urgent need for public policies aimed at mitigating financial toxicity, whether through strengthening the oncology care network, promoting financial support strategies for patients, or reducing regional health inequities.

Initiatives focused on expanding early diagnosis, health education, and social support are essential to improving survival in potentially treatable cancers such as those of the oral cavity and oropharynx. Overcoming the economic and structural barriers identified in this study is a necessary condition to ensure equity and effectiveness in cancer care across Brazil.

References

1. Zhang, S.Z., Xie, L., Shang, Z.J.: Burden of oral cancer on the 10 most populous countries from 1990 to 2019. Int. J. Environ. Res. Public Health **19**(2), 875 (2022)
2. Silva, D.S., et al.: Deaths from head and neck cancer according to schooling in Rio de Janeiro, Brazil, 2010–2018. Cad. Saude Publica. **41**(2), e00140023 (2025)
3. Viana, L.P., et al.: Trend of the Burden of Larynx Cancer in Brazil, 1990 to 2019. Rev. Soc. Bras. Med. Trop. **55**(suppl 1), e0269 (2022)
4. Freire, A.R., et al.: Socioeconomic indicators and economic investments influence oral cancer mortality in Latin America. BMC Public Health **21**, 377 (2021)
5. Groome, P.A., et al.: A population-based study of factors associated with early versus late stage oral cavity cancer diagnoses. Oral Oncol. **47**(7), 642–647 (2011)
6. Lima, F.L.T.D., O'Dwyer, G.: Atenção ao câncer bucal na região de saúde Metropolitana I do Rio de Janeiro Brasil Ciênc. Saúde Coletiva **28**(3), 875 (2023)
7. Akinkugbe, A.A., et al.: Lifestyle risk factor-related disparities in oral cancer examination in the U.S. BMC Public Health **20**, 153 (2020)
8. Matos, L.L., et al.: The hidden curve behind COVID-19 outbreak: the impact of delay in treatment initiation in cancer patients. Cancer Causes Control **32**(5), 459–471 (2020)
9. Maia, F., et al.: Acesso desigual ao tratamento do câncer e toxicidade financeira no Brasil. Arch. Oncol. Immunol. **1**(1), 1–10 (2024)
10. McDermott, J.D., et al.: Elderly Black non-Hispanic patients with head and neck squamous cell cancer have the worst survival outcomes. J. Natl. Compr. Canc. Netw. **19**(1), 57–67 (2020)
11. Lara-Carrillo, E., et al.: Effect of human development index on oral cavity cancer mortality in Mexican states. BMJ Open **11**(6), e042376 (2021)
12. Nogueira, L.A.S., et al.: Assessment of financial toxicity (FACIT-COST) in cancer patients in southern Brazil. Cogitare Enferm. **27**, e89763 (2022)
13. Nogueira, L.A.S., et al.: Financial toxicity and health-related quality of life among cancer patients: a correlational study. Aquichan **24**(1), e2411 (2024)
14. Nogueira, L.A.S., et al.: Impact of financial toxicity in adults with cancer during the COVID-19 pandemic: an integrative review. Rev. Bras. Enferm. **77**, e20230318 (2023)
15. Raymundo, M.L., et al.: Trend of hospitalized cases of oral cancer in Brazil and its relationship with oral health coverage. Med. Oral Patol. Oral Cir. Bucal **26**(2), e78–e83 (2021)
16. Naik, A.N., et al.: Postoperative radiation therapy in oral cavity verrucous carcinoma. Laryngoscope **132**(10), 1953–1961 (2022)
17. Khanna, D., et al.: Rural–urban disparity in cancer burden and care: findings from an Indian cancer registry. BMC Cancer **24**(1), 1–10 (2022)
18. Bindra, G.S., et al.: Assessing social vulnerabilities of salivary gland cancer care, prognosis, and treatment in the United States. Head Neck **46**(9), 2152–2166 (2024)
19. Smith, G.L., Lopez-Olivo, M.A., Lin, Y.-L., et al.: Navigating financial toxicity in patients with cancer: a multidisciplinary team approach. CA Cancer J. Clin. **72**(5), 407–424 (2022)

20. Long, A., McQueen, A., Jones, R.M., et al.: Long-term impact of financial distress and mental health on treatment adherence and survival in patients with head and neck cancer. Oral Oncol. **144**, 106417 (2023)
21. Amin, M.B., Edge, S.B., Greene, F.L., et al.: AJCC Cancer Staging Manual, 8th edn. Springer, New York (2017)
22. Burgers, J.A., van Dongen, J.A.C., et al.: Financial distress and health outcomes in head and neck cancer patients: a multicenter study. Support. Care Cancer **30**, 8457–8467 (2022)
23. Zafar, S.Y., Abernethy, A.P.: Financial toxicity, part I: a new name for a growing problem. Oncology (Williston Park) **27**(2), 80–81 (2013)
24. Shah, J.P., Gil, Z.: Current concepts in management of oral cancer surgery. Oral Oncol. **45**(4–5), 394–401 (2009)
25. Curioso, P.A.B., et al.: Comparative analysis of clinicopathological characteristics in young and elderly patients with oral squamous cell carcinoma. Braz. Oral Res. **38**, e138 (2024)
26. Ehsan, M., et al.: The role of geographic factors and health care access in head and neck cancer treatment disparities. Otolaryngol. Head Neck Surg. **169**(1), 1–10 (2023)
27. Araújo, E.C.F.D., et al.: Cost of implementation and maintenance of dental care in a hospital setting. J. Serv. Sci. Manag. **17**(4), 377–385 (2024)
28. Pisu, M., et al.: Economic hardship of minority and nonminority cancer survivors one year after diagnosis: another long-term effect of cancer. Cancer **126**(14), 3192–3200 (2020)
29. Mazor, K.M., et al.: The long-term impact of financial distress and mental health on cancer survivors. Cancer Med. **9**(2), 594–605 (2020)
30. de Souza, J.A., et al.: Measuring financial toxicity as a clinically relevant patient-reported outcome: the validation of the COST measure. Cancer **123**(3), 476–484 (2017)
31. Cancer treatment in the Brazilian Unified Health System (SUS) shows regional disparities. Oncomonitor (2024). Available at: https://oncomonitor.org.br/relatorio-sus-2024 (Last accessed: 16 Jun 2025)
32. Brazil. Ministry of Health. Federal funding for specialized treatments reaches BRL 74 billion. Brasília: Ministry of Health (2024) Available at: https://www.gov.br/saude/pt-br/assuntos/noticias/financiamento-federal-para-tratamentos-especializados (Last accessed: 16 Jun 2025).

From Reports to Resilience: Evaluating the Role of Regional Councils in Flood Protection Through Disclosure, Performance, and Engagement

Radiah Othman[1(✉)] and Rashid Ameer[2]

[1] School of Accountancy, Economics, and Finance, Massey University, Palmerston North 4410, New Zealand
r.othman@massey.ac.nz

[2] IPU New Zealand, Palmerston North 4410, New Zealand

Abstract. This study examines how regional councils in New Zealand incorporate financial transparency, performance evaluation, and community engagement into the governance of flood protection and control schemes. Through a qualitative analysis of annual reports and related documentation, the research assesses the extent to which councils adhere to regulatory obligations, implement key performance indicators, and engage with stakeholders, including Māori, local communities, and government agencies. The findings indicate a broad commitment to compliance, yet reveal considerable variation in the clarity, depth, and consistency of reporting practices across different regions.

Keywords: Accounting · Flood · Local Government · New Zealand · Regional Authorities · Climate Change

1 Introduction

Research on climate change adaptation has progressively shifted from a focus on developing countries to include developed nations such as New Zealand [41, 45, 58, 59]. While accounting scholars have made commendable efforts to address environmental concerns, much of the research remains confined within traditional accounting and reporting frameworks. Criticisms persist that organisations often fail to apply robust due diligence processes to identify, mitigate, and disclose social and environmental impacts [13, 14, 36, 63]. Previous studies have also highlighted how institutional fragmentation, and administrative traditions can hinder flood resilience efforts [41, 54, 69]. For instance, [69] found that governance deficits in England, New Zealand, and the Netherlands limited the adoption of adaptive approaches to flood risk. Moreover, public authorities are often held accountable for disseminating inaccurate flood risk information and for inadequate evacuation planning [64].

In the Anthropocene, the divide between social and natural sciences is increasingly blurred, necessitating integrated and collaborative responses [1]. Flooding represents a

L. Almeida et al. (Eds.): ICAFI 2025, ISEM 80, pp. 365–384, 2026.
https://doi.org/10.1007/978-3-032-16268-7_21

grand challenge, a wicked problem and a classic example of the tragedy of the commons [7, 24, 31, 33]. Anticipation and preparedness are essential in mitigating the impacts of extreme weather events, particularly as climate change intensifies flood risks globally [20, 34]. It is therefore critical to challenge the status quo and reframe flood events as opportunities for transformative intervention. As a profession grounded in public interest, accounting has a vital role to play in enhancing accountability and supporting sustainability efforts beyond the corporate sector [1, 39]. Scholars have called for greater examination of how accounting is applied in the public sector, particularly by regional authorities, and how effectively these responsibilities are carried out [8, 52]. This study responds to that call by investigating how New Zealand regional councils integrate financial disclosure, performance measurement, and community engagement in flood protection governance, with a focus on the interplay between regulatory compliance, operational transparency, and the social dimensions of flood risk management. Evaluating the effectiveness of these practices is essential for building future resilience [56].

Framed by an interpretive perspective on the role of accounting, this paper problematises flood prevention and control works within the context of regional authorities' accountability. Although New Zealand regional councils are required to disclose a funding impact statement related to flood prevention and control works (FPCW), the nature and extent of these disclosures have not been thoroughly researched. This study addresses that gap by exploring how and what information should be disclosed for councils to effectively discharge their accountability in mitigating flood risk. The research question guiding this inquiry is: "How do New Zealand regional councils integrate financial disclosure, performance measurement, and community engagement in the governance of flood protection and control works schemes?"[1].

Noting the limited support in the accounting literature regarding accounting and accountability in flood risk mitigation, this paper aims to achieve the following objectives:

- To evaluate the extent and consistency of financial disclosure practices related to flood protection and control works across New Zealand regional councils, with a focus on funding sources, expenditure reporting, and budget comparisons.
- To analyse the use and variation of key performance indicators (KPIs) in measuring the effectiveness, efficiency, and long-term goals of flood protection schemes, including alignment with national standards and sustainability goals.
- To explore the role of community engagement and inter-agency collaboration in shaping flood risk governance, including how councils involve Māori, local communities, and other stakeholders in planning, decision-making, and implementation.
- To assess the balance between technical accounting practices and social accountability in flood protection reporting, identifying how councils integrate both dimensions in their annual reporting and strategic planning.
- To identify best practices and gaps in current flood protection governance frameworks that could inform more transparent, inclusive, and effective flood risk management strategies.

[1] Regional councils and authorities are used interchangeably.

This study is exploratory and descriptive in nature; however, it offers several key contributions. First, it provides a comprehensive analysis of how financial reporting, performance measurement, and community engagement intersect in the governance of flood protection schemes. It contributes to academic literature by bridging technical accounting practices with social and environmental governance. Second, by comparing the approaches of regional councils, the study identifies best practices and inconsistencies in disclosure, performance indicators, and community involvement. These insights can inform future policy development and support greater standardisation and transparency in local government reporting. By showcasing exemplary practices, the study aims to facilitate shared learning and address the gap in practice-relevant knowledge within the local government context.

Third, the research highlights how accounting functions not only as a technical tool but also as a social practice that fosters accountability, trust, and collaboration between councils, communities, and iwi. This deepens our understanding of how the complex and destructive nature of flooding can open dialogue among actors in managing environmental risk. Evidence suggests that accounting for flood damage extends beyond technical reporting to include social considerations, supporting better coordination during flood events and demonstrating the socialising effect of accounting [6, 39]. Finally, the findings offer actionable recommendations for policymakers, auditors, and council officials to strengthen flood risk governance. This includes improving reporting frameworks, enhancing stakeholder engagement, and aligning local practices with national and global sustainability goals. From a public accountability perspective, the study underscores the importance of collaboration, coordination, and innovation in enabling local governments to respond effectively to climate-related risks. The insights and lessons learned may also inform international knowledge-sharing efforts, including those led by the UNDP, and contribute to global discussions on adaptive flood risk management. As [3] aptly stated, it is valuable "to borrow from the wisdom and practice of others" in the public interest.

The next section provides the literature review, followed by a brief description of the local government context in Sect. 3. Section 4 outlines the research method, and the main findings are presented in Sect. 5. The final section offers discussion and conclusion in Sect. 6.

2 Literature Review

Accountability is a fundamental pillar of a democratic political system [9]. At its core, democracy relies on collaboration and cooperation among individuals and groups with diverse perspectives and interests, making accountability essential to the functioning of a healthy democratic society. In the context of climate change adaptation, intergovernmental collaboration and coordination are crucial for effective responses [38]. Climate risk is inherently complex, ambiguous, and scientifically uncertain [60]. Climate change research transcends disciplinary, institutional, and organisational boundaries [16, 19, 25, 37, 70]. Although accounting research has addressed areas such as carbon and greenhouse gas emissions [48], performance measurement [21], climate change costing [2], and carbon cost structures [15], there remains a notable gap in the literature on flood risk mitigation. Preventive measures have long been central to Indigenous philosophies, offering pathways for meaningful and emancipatory change [4]. The Indigenous concept of the Seven Generations teaches that decisions made today will have lasting impacts, positive or negative, on the next seven generations [40]. Aligned closely with this principle is the aim of sustainable development to meet the needs of the present without compromising the ability of future generations to meet their own needs [72].

While sustainability accounting and reporting initiatives have gained traction, much of the accounting literature remains focused on organisational-level benefits, often with limited impact on broader social and environmental systems [5, 27, 28, 30, 35]. Furthermore, the literature continues to advocate for a broader understanding of organisational accountability in the context of sustainability [6, 10, 13, 21, 49].

Previous studies on flood prevention and risk management have predominantly been conducted by scholars in engineering rather than accounting. These studies have explored the increasing severity and frequency of flood events, attributing them to factors such as population growth and infrastructure development on floodplains [50], as well as the limitations of current mitigation practices [71]. The impacts of flooding similarly demand an interdisciplinary approach, moving beyond the traditional domain of engineering to include accounting's potential contribution to sustainable development. Integrating accounting with sustainability opens new ways of thinking and allows discipline to evolve beyond its conventional boundaries. [49] have called on accounting researchers to engage with and support other disciplines.

Flood events exemplify complexity, uncertainty, and unpredictability [24], with the potential to affect both current and future ecosystems and communities [37]. A shared challenge lies in understanding how to identify, mitigate, and disclose the social and environmental impacts of such events [14, 63]. In the New Zealand context, there is a recognised fragmentation and inconsistency in the collection of climate impact and adaptation data, with no centralised, easily accessible repository available for councils [42]. Yet, the provision of information is a critical component of the accountability mechanisms expected of public agencies [29, 32].

To hold authorities accountable for disaster risk management such as in the case of flooding, there must be clear systems in place to ensure accurate and timely information is made available to local communities. The relationship between information provision and accountability is symbiotic; both depend on each other to function effectively and, when integrated, can enhance transparency and trust in government. [44] advocated for regional authorities to share information with district and city councils to support development planning in flood-prone areas. Effective coordination across different levels of government is vital for establishing an integrated approach to flood risk management, particularly when considering the efficient use of public resources and balancing costs and benefits [53].

Citizen engagement has transformed accountability from a traditional top-down model into a more horizontal, participatory structure [51]. International research shows that communities often hold public authorities primarily responsible for preventing and mitigating flood damage [66] and expect to be accurately informed, holding authorities accountable when information is incorrect [64]. Within the accounting discipline, there is a growing call to move beyond the technical aspects of accounting (i.e., how it is done) to consider its social functions and normative responsibilities, what accounting does and what it ought to do [17, 18]. In the context of flood risk governance, [64] refer to moral responsibility as the obligation to avoid harm, to assist within one's capacity, and to take responsibility for flood risk based on one's role, such as being a member of a community (see also [57]). [39] demonstrated the socialising effect of accounting through flood recovery accountability processes, showing how accounting can extend beyond its technical domain. As recent findings from different geographical contexts are still emerging, this study contributes to the literature by offering insights into flood risk mitigation practices from a New Zealand perspective. With the exception of [39], accounting scholars have shown limited engagement in flood risk accounting and reporting.

[39] emphasise the role of accounting in supporting the monitoring of disaster-specific funds, ensuring transparency and accountability throughout the recovery process. Transparency in how authorities manage funding allocated to flood prevention is therefore a critical component of any accountability framework. Despite this, there is a noticeable gap in the literature regarding how accounting and reporting practices contribute to sustaining accountability in flood risk mitigation. To our knowledge, no study has yet comprehensively examined this intersection. This research seeks to address that gap by exploring the role of accounting in supporting transparent and accountable flood risk governance within New Zealand's regional councils.

3 Local Government Context in New Zealand

According to the Local Government New Zealand website, all local authorities contributed 3.4% (NZ$9.7 billion) of the total expenditure of the economy (GDP, year ending March 2018). There are 78 local authorities in New Zealand, comprising 11 regional councils and 67 territorial authorities (including unitary authorities, city, and district councils). All regional councils are included in this study. Collectively, their populations accounted for 63.97% of New Zealand's total population of 5.09 million in 2020. Table 1 shows that the majority of the population is concentrated in the Canterbury, Wellington, Waikato, and Bay of Plenty regions, while the West Coast has the smallest population, with fewer than 50,000 people.

The role of regional authorities differs from that of other local councils within the region. District and city councils are primarily responsible for services such as water supply, refuse collection, sewerage, road maintenance, land use and subdivisions, and community services. In contrast, regional councils are tasked with delivering FPCW to reduce the risks associated with river flooding and erosion, protect human life, and enhance the region's productivity. Figure 1 illustrates the councils with environmental responsibilities.

A 'regional influence' approach is common in New Zealand, offering a regional perspective rather than treating the country as a single entity. This helps to avoid boundary issues associated with fixed regions [12, 65]. At the regional level, authorities are responsible for managing the environmental effects of human activities on freshwater, land, air, and coastal waters through regional rules, policies, and the issuing of consents. Most importantly, regional authorities are responsible for issuing flood warnings [26, 61], based on the principle that decision-making should rest with those directly affected and most familiar with the local context [23].

Table 1. Regional Council's Population

	Regional Council	No. of population (June 2020)
1	Canterbury	628,600
2	Wellington	527,800
3	Waikato	496,700
4	Bay of Plenty	477,700
5	Manawatu-Whanganui	258,200
6	Otago	236,200
7	Northland	194,600
8	Hawke's Bay	178,600
9	Taranaki	122,700
10	Southland	102,600
11	West Coast	32,600
	Total	3,256,300

Fig. 1. Councils with responsibilities for the environment. Source: Wilson (2023).[2]

[2] **This item is licensed under a Creative Commons Attribution-NonCommercial 3.0 New Zealand Licence;** https://teara.govt.nz/en/map/2740/regional-council-boundaries**.**

4 Research Method

This study adopts a qualitative document analysis approach, which is well-suited for exploring the content, context, and meaning embedded in publicly available documents, particularly annual reports, long-term plans, and funding impact statements.

The primary data sources include publicly available annual reports and related planning documents from all regional councils in New Zealand, covering the period from 2010 to 2022. The year 2010 was selected as the starting point, as it marks the implementation of a requirement under the Local Government Act 2002 for all flood protection and control activities to be reported as a separate group of activities. Supplementary materials, such as government regulations, performance reporting standards, and relevant legislation (e.g., the Local Government Act 2002), were also reviewed to provide contextual grounding.

The analysis involved manually coding and categorising the content following the research objectives. Particular attention was given to identifying best practices, inconsistencies, and innovative approaches across councils in relation to financial disclosure, performance measurement, and community engagement.

5 Findings

The following section reports how the regional councils account for and report their achievements in flood protection and control works, as legislated, as part of their accountability to local communities.

5.1 Disclosure of Flood Protection and Control Works (FPCW)

The Local Government (Financial Reporting) Regulations 2011, which applied to local authorities' financial statements in their annual plans, long-term plans, and annual reports, were revoked and replaced by the Local Government (Financial Reporting and Prudence) Regulations 2014 on 1 May 2014 1. Under the updated regulations, local authorities are required to use accrual accounting in accordance with generally accepted accounting practice (GAAP). Additionally, the total sources of funding must match the total application of funding, ensuring transparency and accountability in financial reporting.

The format for presenting funding impact statements in annual reports is prescribed in Form 5 of the 2014 regulations and must align with Clause 30, Schedule 10 of the Local Government Act 2002. To assess compliance with these requirements, the most recent annual reports of regional councils were reviewed. As shown in Table 2, most councils adhered to the prescribed format, though minor variations in content were observed. Notably, some councils did not provide detailed disclosures for flood protection and control works. For instance, Taranaki Regional Council presented a single combined statement for civil defence emergency management and river control schemes under the heading "Flood Management and General River Control." Similarly, the West Coast Regional Council used the term "Hydrology and Flood Warning Services" instead of the standard terminology.

Table 2. Regional Authorities' Funding Impact Statement Items reported for *Flood Protection and Control Work* activity.

Source of Operating Funding	BOP	Cant.	HB	MW	NLand	Otago	Wai	Well	West	SLand	Tara.
General rates, uniform annual general charges, rates penalties	●	●	●	●	●	●	●	●	●	●	●
Targeted rates	●	●	●	●	●	●	●	●	●	●	●
Direct charges											●
Subsidies and grants for operating purposes	●	●	●	●	●	●	●	●	●	●	
Fees and charges	●	●	●	●	●	●	●	●	●	●	
Internal charges and overheads recovered	●	●		●	●	●				●	
Interest and dividends from investments			●				●	●			
Local authorities fuel tax, fines infringement fees and other receipts	●	●	●	●	●	●		●	●	●	
Government grants							●				●
Investment income							●		●		●
Other income							●				
Transfer to and from reserve							●				●
Application of Operating Funding											
Payments to staff and suppliers	●	●	●	●	●	●		●	●	●	
Finance costs	●	●	●	●	●	●		●	●	●	
Internal charges and overheads applied	●	●		●	●	●		●		●	
Other operating funding applications	●	●	●	●	●	●			●	●	
(Operating) expenditure							●				●
Sources of capital funding											
Subsidies and grants for capital expenditure	●	●	●	●	●	●			●	●	
Development and financial contributions	●	●	●	●		●			●	●	
Increase (decrease) in debt	●	●	●	●	●	●		●	●	●	

(*continued*)

Table 2. (*continued*)

Gross proceeds from sale of assets	●	●	●	●	●	●		●	●	●	●
Lump sum contributions	●	●	●	●	●	●			●	●	
Other dedicated capital funding	●	●	●	●	●	●			●	●	
Applications of capital funding											
Capital expenditure. - to meet additional demand - to improve levels of service - to replace existing assets	●	●	●	●	●	●		●	●	●	●
Increase (decrease) in reserves	●	●	●	●	●	●		●	●	●	
Other operating funding applications	●										
Increase (decrease) of investments	●	●	●	●	●	●		●	●	●	
Other distinct features:											
Comparison with Long-term Plan	●	●			●	●			●		
Comparison with Annual Plan	●	●	●	●			●				
Comparison with budget								●	●	●	●
Explanation of budget variances from Long-term plan	●				●						
Reconciliation of operating funding to Statement of Comprehensive Revenue and Expense		●									

BOP: Bay of Plenty; Cant: Canterbury; HB: Hawkes Bay; MW: Manawatu-Whanganui; NLand: Northland; Otag: Otago; SLand: Southland; Wai: Waikato; Well: Wellington; West: West Coast; Tara: Taranaki.

As anticipated, all regional councils reported similar sources of income and demonstrated a high degree of consistency in disclosing their sources of capital funding, as well as the application of both operating and capital funding. However, only Waikato and Taranaki provided a detailed breakdown of their operating funding sources, specifying categories such as government grants, investment income, and transfers to and from reserves. Furthermore, only Waikato, West Coast, and Taranaki explicitly reported their investment income. In terms of the application of operating funding, most councils followed a consistent reporting format. However, Waikato and Taranaki deviated by presenting only the total operating expenditure, without itemizing payments to staff, suppliers, finance costs, or other components. These two councils also omitted disclosures on the sources of capital funding.

Table 2 also highlights inconsistencies in how councils compared actual results with budgeted figures from their annual and long-term plans. Only Bay of Plenty and Canterbury included comparisons with both plans, while the remaining councils referenced just one. Additionally, only four councils: Wellington, West Coast, Southland, and Taranaki included funding impact information benchmarked against their budgets. Notably, Bay of Plenty and Northland went a step further by explaining variances between actual results and long-term plans. Canterbury was the only council to reconcile its funding impact statement for flood protection and control works with its Statement of Comprehensive Revenue and Expense. Among all councils, Bay of Plenty stood out for providing the most comprehensive and transparent financial information, as illustrated below (Fig. 2).

Bay of Plenty Regional Council: Funding Impact Statement for the year ended 30 June 2022 for Flood Protection and Control

	LTP 2020/21 $000	LTP 2021/22 $000	Actual 2021/22 $000
Sources of operating funding			
General rates, uniform annual general charges, rates penalties	2,021	2,988	2,988
Targeted rates	12,434	11,025	10,961
Subsidies and grants for operating purposes	0	0	0
Fees and charges	12	12	(13)
Internal charges and overheads recovered	1,410	1,681	2,770
Local authorities fuel tax, fines infringement fees and other receipts	3,380	5,030	5,725
Total operating funding (A)	**19,257**	**20,737**	**22,431**
Applications of operating funding			
Payments to staff and suppliers	7,876	8,321	9,160
Finance costs	3,567	1,691	1,547
Internal charges and overheads applied	3,499	6,206	7,719
Other operating funding applications	0	0	0
Total applications of operating funding (B)	**14,942**	**16,219**	**18,427**
Surplus (deficit) of operating funding (A-B)	**4,314**	**4,518**	**4,004**
Sources of capital funding			
Subsidies and grants for capital expenditure	0	5,020	2,025
Development and financial contributions	0	0	0
Increase (decrease) in debt	14,284	12,988	8,564
Gross proceeds from sale of assets	0	0	0
Lump sum contributions	0	0	0
Other dedicated capital funding	5,701	1,848	1,052
Total sources of capital funding (C)	**19,985**	**19,856**	**11,641**

	LTP 2020/21 $000	LTP 2021/22 $000	Actual 2021/22 $000
Applications of capital funding			
Capital expenditure			
- to meet additional demand	0	4,198	0
- to improve levels of service	8,053	5,682	6,472
- to replace existing assets	11,932	3,108	2,092
Increase (decrease) in reserves	4,314	11,386	7,081
Other operating funding applications	0	0	0
Increase (decrease) of investments	0	0	0
Total applications of capital funding (D)	**24,299**	**24,374**	**15,645**
Surplus (deficit) of capital funding (C-D)	**(4,314)**	**(4,518)**	**(4,004)**
Funding balance (A-B) + (C-D)	**0**	**0**	**0**
Note 1: This financial statement excludes:			
Depreciation and amortisation	1,465	1,456	2,388
Asset revenue gain	0	0	(253)
Vested asset revenue	0	0	(8)
Note 2: This financial statement includes:			
Internal interest	3,567	1,691	1,547

Te Mahere ā-Tau Annual Report 2021/22 | 55

Fig. 2. Bay of Plenty's Funding Impact Statement for Flood Protection and Control. Source: Bay of Plenty Annual Report 2021/22 (p. 55)

5.2 Key Performance Indicators for Flood Protection and Control Work

Under Sect. 261B of the Local Government Act 2002, the Secretary for Local Government is required to establish a standard set of performance measures for key local government activities, including flood protection and control works. These measures are intended to ensure consistency, transparency, and accountability in how councils report on their service delivery. The term "major flood protection and control works" refers to projects that meet at least two of the following criteria: Annual operating expenditure exceeding $250,000; Annual capital expenditure exceeding $1 million; Asset replacement value exceeding $10 million; Directly benefiting a population of at least 5,000 people.

In 2022, the Taranaki Regional Council reported that none of its flood protection and control schemes met the thresholds to be classified as "major." As a result, it did not apply the standard performance measures prescribed under the Act. Among the councils reviewed, only the Bay of Plenty Regional Council explicitly aligned its flood protection and control activities with Sustainable Development Goal 11: "Make cities and human settlements inclusive, safe, resilient, and sustainable." This alignment is illustrated in Fig. 3 below.

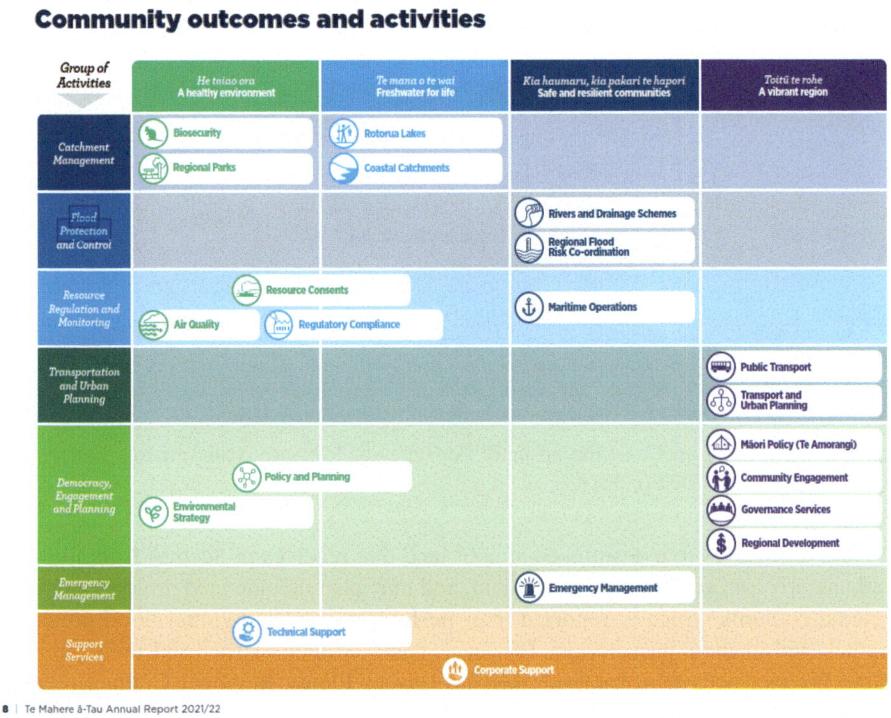

Fig. 3. Bay of Plenty's Community Outcomes and Activities. Source: Bay of Plenty Annual Report 2021/22 (p.8)

All regional councils reported their performance against key indicators in various formats, with most repeating similar measures annually. Generally, the flood warning manual served as the primary benchmark. Other reference points included the Asset Management Plan (e.g., Manawatū-Whanganui) and agreed service standards (e.g., Otago).

Among the councils, Northland provided the most comprehensive performance reporting, comparing current results with those from the previous year. Examples of its indicators included: Issuing accurate and timely flood warnings; Updating flood level monitoring at priority rivers within one month of significant flood events; Recording zero failures of flood protection systems for schemes below specified design levels; Acknowledging nationally issued warnings within 30 min.

Taranaki offered detailed explanations and justifications for both achievements and shortfalls, along with the target years for each measure. The West Coast Regional Council was particularly specific in reporting its flood risk mitigation performance, referencing benchmarks such as the flood warning manual, update frequency, and year-on-year comparisons. Non-conformances were clearly explained, as illustrated in Fig. 4 below.

Hydrology and Flood warning performance measures

Level of Service
Maintain a level of staff preparedness so that Council can respond to significant events in a timely manner

Measure	Baseline	Outcome 2022	Outcome 2021
Staff response to high flow events.	Deliver flood warning alerts as required in accordance with the Flood Warning Manual - ***100%***	Not achieved*	Achieved
Availability of information about high flow events.	River level data available on the Council website for the 6 key rivers on the West Coast (updated 12-hourly, or 3-hourly during flood events) ***>90%***	Achieved	Achieved

Non-conformances

- *An early flood alert on the Hokitika River went unheard by a staff member during this period. Later alerts were picked up and a procedure has been implemented to ensure this does not re-occur. Due to a technical issue with electronic monitoring systems, the Arnold River alerted but was not received within the required timeframe. This was noted and restored within the same day but breached the 60 minutes timeframe.*

Fig. 4. Example of West Coast Reporting of Performance Measure Achievement. Source: West Coast Annual Report 2022 (p.33)

A scheme refers to a designated area of land that receives protection from flooding, riverbank erosion, and channel movement, and may also include land drainage services. Regional councils typically reported their performance by individual flood protection schemes within their jurisdiction. However, some councils structured their reporting around broader activities, for example, the Bay of Plenty Regional Council reported according to its Rivers and Drainage Scheme.

Several councils focused on the effectiveness and efficiency of flood risk management. Examples include: The percentage of flood warnings issued at pre-determined levels in accordance with the flood warning manual (e.g., Bay of Plenty, West Coast); Ensuring the flood warning network was operational 100% of the time during flood events (e.g., Southland); The number of requests for advice and information responded to within a specified timeframe, such as five working days (e.g., Bay of Plenty and Taranaki); The West Coast Regional Council also ensured that real-time river level data was accessible via the Council's website and information line, with updates every 12 h under normal conditions and at least every 3 h during flood events.

Some councils also reported on expenditure performance, indicating whether actual spending was under, over, or on budget (e.g., Canterbury). However, several performance measures were relatively generic and lacked specificity. Examples include Broad goals such as reducing flood risk (e.g., Canterbury, Otago); The percentage of coastal erosion areas identified and mapped (e.g., Canterbury); Providing the community with timely flood event information and situation reports, and responding effectively to flood events (e.g., Waikato); The number of requests for flood hazard advice received (e.g., Wellington); Maintaining flood protection schemes to mitigate flood impacts (e.g., Wellington); Whether a procedural review was conducted after each significant flood event (e.g., Canterbury).

Some councils adopted a longer-term, strategic approach to performance measurement. For instance, Hawke's Bay set a goal to manage flood risk in a way that adapts to foreseeable climate change impacts through to 2100, with a target date of 2030. Similarly, Taranaki established a planning horizon extending to 2065.

5.3 Socialising Effect of Flood Accounting

Approximately 64% of New Zealand's total stopbank length is located within five key regions: Canterbury, Waikato, Southland, Manawatū-Whanganui, and Bay of Plenty (Crawford-Flett et al., 2021). The presence of extensive flood protection infrastructure within these regions highlights the importance of collaborative efforts, both within and across regional boundaries to effectively manage and mitigate flood risk.

A review of regional council's annual reports reveals that collaboration and coordination occur across three key dimensions: Between local authorities; Between central government (and its agencies) and regional authorities; Between regional authorities and local communities. Several examples illustrate these collaborative efforts: In 2022, Canterbury advocated for stronger national leadership in flood management and reported working with other regional and unitary councils to seek $150 million per annum in central government co-investment to strengthen flood protection schemes (Annual Report 2021/22).

Councils also partnered with various government agencies—including the Department of Conservation, iwi, district and city councils, the New Zealand Transport Agency, the Historic Places Trust, and Civil Defence Emergency Management—to support activities such as river channel maintenance, flood risk management, technical advisory services, and the classification of residential structures in high-risk floodway corridors. These collaborations were noted in the annual reports of Bay of Plenty (2013/14), Canterbury (2019/20), Hawke's Bay (2021/22), Northland (2013/14), Otago (2014/15), and Wellington (2013/14). A unique example comes from Waikato, which collaborated with energy provider Mighty River Power to enhance communication and understanding of flood roles and operations during high-flow events. This partnership aimed to improve preparedness for potential large-scale flooding along the Waikato River (Annual Report 2013/14, p. 22).

Regional councils demonstrated transparency in how and under what circumstances they engaged in dialogue and held forums with local communities and stakeholders. For example, Manawatū-Whanganui reported consulting with ratepayers to identify a preferred water pumping solution to improve drainage from the Taonui Basin (Annual Report 2009/10). The council also acknowledged the strong support of the local community and the significant contribution of Palmerston North City Council in completing the 100-year flood protection standard for Ashhurst township (Annual Report 2011/12). Councils also showed accountability to Māori communities through close engagement and consultation. Māori representatives were appointed to council committees to contribute to a wide range of issues, including flood protection scheme upgrades, flood risk management strategies, and broader catchment-related matters. This was evident in the annual reports of Bay of Plenty (2015/16), Northland (2014/15), Otago (2020/21), Taranaki (2013/14, 2016/17), and Wellington (2021/22).

Engagement extended to the farming sector, community groups, and individuals, with councils providing education on flood hazards, advice for building sites and land use planning, support for emergency management planning, and initiatives to enhance infrastructure resilience. Community participation was also encouraged through activities such as tree planting events and the refinement of flood hazard mapping. These efforts were documented in the annual reports of Canterbury (2009/10), Northland (2016/17), Otago (2013/14), Southland (2014/15), and Waikato (2014/15). Only Canterbury and Waikato explicitly disclosed the social dimension of their budgeting processes. For instance, Waikato detailed the number of submissions received, the percentage of ratepayer representation, the outcomes of those submissions, and how they influenced the final budget approved in the 2014/15 Annual Plan. This suggests that, in some cases, the social engagement aspect of flood risk mitigation was prioritised over purely technical or financial considerations.

6 Discussion and Conclusion

This study explores how New Zealand's regional councils integrate financial disclosure, performance measurement, and community engagement in the governance of flood protection and control schemes. It highlights the interplay between regulatory compliance, operational transparency, and the social dimensions of flood risk management.

Among the councils, Otago stood out as a well-rounded performer. It demonstrated strong collaboration and coordination with central government, ministries, other government agencies, district and local councils, and local communities. Notably, Otago was the only council to explicitly declare a significant commitment to enhancing knowledge and understanding of the likely impacts of climate change in the region, aligning its efforts with the National Climate Change Risk Assessment. It produced a comprehensive public report detailing projected changes in climate variables under various scenarios and actively shared this information with a wide range of organisations. Otago also added a dedicated climate change section to its website to support public awareness and engagement.

Regional councils generally maintained transparency in financial disclosures, which is particularly important when managing disaster-specific funds [39]. A review of their annual reports indicated broad compliance with the Flood Protection and Control Works (FPCW) reporting requirements, although the presentation of information varied. Most councils reported similar income sources and were consistent in disclosing the application of capital and operating funding. However, only a few provided detailed breakdowns of operating funding sources and investment income. Furthermore, budget comparisons were inconsistently reported, with only four councils comparing actual results against their budgets.

All regional councils reported on performance indicators, though the formats and benchmarks varied, suggesting that past adverse climate events may have influenced their approaches [47]. The focus of performance reporting also differed: some councils reported by individual schemes, while others grouped activities by functional areas such as rivers and drainage. The range of measures included both technical indicators, such as flood warning accuracy and system uptime, and broader strategic goals, like climate change adaptation. This diversity reflects the need to consider both physical and non-physical factors from multiple perspectives [46, 62, 69].

In terms of the socialising effect of flood accounting, annual report narratives revealed that regional councils engaged in coordination at three levels: Between local authorities; Between regional and central government agencies; With local communities. Such collaboration is essential, as local-level action alone is insufficient for effective climate change adaptation [43]. Councils demonstrated accountability to Māori communities through consultation and representation on committees addressing flood protection and catchment management. Broader community engagement included outreach to farmers, community groups, and individuals, aiming to raise awareness of flood hazards, support land use and emergency planning, and enhance infrastructure resilience. This socialising aspect of flood governance reflects a "sense of social solidarity" [55], where collaboration plays a central role in preventing and mitigating flood risks. The findings also highlight how accountability has extended beyond the technical dimensions of accounting to embrace community-centred approaches, fostering better coordination and a stronger sense of security for local populations [6, 11, 39].

In conclusion, the study reveals a general commitment to regulatory compliance, but also notable inconsistencies in the depth and clarity of financial reporting and performance indicators. While some councils align closely with national standards and sustainability goals, others lack transparency in budget comparisons and funding disclosures. These variations underscore the need for more standardised reporting frameworks to ensure comparability and accountability across regions. Beyond technical compliance, the research underscores the critical role of social engagement in flood risk governance. It contributes to the growing body of literature on public sector accounting and environmental governance, offering a more holistic understanding, one that balances technical rigour with inclusive, community-focused governance in the face of escalating environmental challenges.

This study is subject to several limitations. Firstly, it relies primarily on publicly available annual reports, which may not fully capture the internal decision-making processes, informal practices, or unpublished community engagement efforts of regional councils. The quality and depth of reporting varied significantly across councils, potentially affecting the comparability of findings. Secondly, the research adopts a qualitative, document-based approach, which limits the ability to assess the actual effectiveness or outcomes of the reported practices. Additionally, the study focuses on a specific timeframe, which may not reflect recent developments in policy, practice, or climate-related challenges.

Future research could benefit from a mixed-methods approach, incorporating interviews with council staff, community stakeholders, and iwi representatives to gain deeper insights into the motivations, challenges, and outcomes of flood risk governance. Longitudinal studies could also track changes in reporting practices and flood resilience over time. This study opens pathways for further investigation into the effectiveness of different engagement models; the long-term impact of performance reporting on flood resilience; and the role of indigenous knowledge in shaping sustainable flood management strategies. Comparative studies across different countries or regions could also help identify the best global practices and inform improvements in regional governance. As noted by the United Nations Development Programme [67, 68], effective information sharing within and between countries, along with systematic reviews of data quality, reporting frequency, and dissemination channels, is critical for improving Climate Information and Early Warning Systems.

Policy and Practice Implications. For policymakers, the findings highlight the need for more consistent and standardised reporting frameworks across regional councils. Clearer guidelines on financial disclosures, performance indicators, and community engagement practices would enhance transparency, comparability, and accountability, especially in the context of climate change, where public trust and coordinated action are essential. For regional councils and practitioners, the research underscores the importance of integrating technical and social dimensions in flood risk management. Councils that actively engage with communities, iwi, and other stakeholders not only improve the quality of decision-making but also foster resilience and shared responsibility. The study suggests that councils should move beyond compliance-driven reporting to embrace more participatory and inclusive governance models. As [22, p. 11] aptly stated:

> "...the one certainty is this: if there is a non-zero probability of a damaging flood event, it will happen."

References

1. Alawattage, C., et al.: Opening accounting: a manifest. Account. Forum **45**(3), 227–246 (2021). https://doi.org/10.1080/01559982.2021.1952685
2. Atkins, J., Atkins, B.C., Thomson, I., Maroun, W.: "Good" news from nowhere: imagining utopian sustainable accounting. Account. Auditing Accountability J. **28**(5), 651–670 (2015). https://doi.org/10.1108/AAAJ-09-2013-1485
3. Bardach, E.: Presidential address – the extrapolation problem: how can we learn from the experience of others? J. Policy Anal. Manage. **23**(2), 205–220 (2004)
4. Barrett, M., Watene, K., McNicholas, P.: Legal personality in Aotearoa New Zealand: an example of integrated thinking on sustainable development. Account. Auditing Accountability J. **33**(7), 1705–1730 (2020). https://doi.org/10.1108/AAAJ-01-2019-3819
5. Bebbington, J., et al.: Accounting and accountability in the Anthropocene. Account. Auditing Accountability J. **33**(1), 152–177 (2020)
6. Bebbington, J., Brown, J., Frame, B., Thomson, I.: Theorizing engagement: the potential of a critical dialogic approach. Account. Auditing Accountability J. **20**(3), 356–381 (2007)
7. Bebbington, J., Unerman, J.: Achieving the United Nations sustainable development goals: an enabling role for accounting research. Account. Auditing Accountability J. **31**(1), 2–24 (2018)
8. Bovens, M.: Analysing and assessing accountability: a conceptual framework. Eur. Law J. **13**(4), 447–468 (2007). https://doi.org/10.1111/j.1468-0386.2007.00378.x
9. Brandsma, G.J., Schillemans, T.: The accountability cube: measuring accountability. J. Public Adm. Res. Theory **23**(4), 953–975 (2013). https://doi.org/10.1093/jopart/mus034

10. Brown, J., Dillard, J., Hopper, T.: Accounting, accountants and accountability regimes in pluralistic societies: Taking multiple perspectives seriously. Account. Auditing Accountability J. **28**(5), 626–650 (2015)
11. Burby, R.J.: Hurricane Katrina and the paradoxes of government disaster policy: bringing about wise governmental decisions for hazardous areas. Ann. Am. Acad. Pol. Soc. Sci. **604**(1), 171–191 (2006)
12. Burn, D.H.: Evaluation of regional flood frequency analysis with a region of influence approach. Water Resour. Res. **26**, 2257–2265 (1990)
13. Busco, C., Giovannoni, E., Granà, F., Izzo, F.: Making sustainability meaningful: aspirations, discourses and reporting practices. Account. Auditing Accountability J. **31**(8), 2218–2246 (2018)
14. Busco, C., Granà, F., Achilli, G.: Managing and measuring social impact through integrated thinking and reporting: the case of a European university. In: De Villiers, C., Hsiao, P.C.K., Maroun, W. (eds.) The Routledge Handbook of Integrated Reporting, pp. 251–268. Routledge, London (2020)
15. Cadez, S., Guilding, C.: Examining distinct carbon cost structures and climate change abatement strategies in CO_2 polluting firms. Account. Auditing Accountability J. **30**(5), 1041–1064 (2017). https://doi.org/10.1108/AAAJ-03-2015-2009
16. Campbel, N., McHugh, G., Ennis, P.J.: Climate change is not a problem: speculative realism at the end of organization. Organ. Stud. **40**(5), 725–744 (2019)
17. Carnegie, G., Parker, L. Tsahuridu, E.: It's 2020: what is accounting today? Australian Account. Rev. **31**(1), 65–73 (2021). https://onlinelibrary.wiley.com/doi/epdf/https://doi.org/10.1111/auar.12325
18. Carnegie, G.D., Gomes, D., McBride, K.: COVID-19 and accounting as multidimensional technical, social and moral practice: a framework for future research. Meditari Accountancy Res. (ahead-of-print) (2023). https://doi.org/10.1108/MEDAR-10-2022-1826
19. Chakhovic, T., Virtanen, T.: Accountability for sustainability – An institutional entrepreneur as the representative of future stakeholders. Crit. Perspect. Account. **9**, 102399 (2021)
20. Chang, L.C., Chang, F.J., Yang, S.N., Tsai, F.H., Chang, T.H., Herricks, E.E.: Self-organizing maps of typhoon tracks allow for flood forecasts up to two days in advance. Nat. Commun. **11**(1), 1–13 (2020)
21. Cooper, S., Pearce, G.: Climate change performance measurement, control and accountability in English local authority areas. Account. Auditing Accountability J. **24**(8), 1097–1118 (2011)
22. Day, T.: Managing flood risk: the case for change, Report prepared for the Flood Risk Management Governance Group, The Centre for Advance Engineering, University of Canterbury, Christchurch (2005)
23. Fenemor, A., Robb, C.: Groundwater Management in New Zealand. In: Rosen, M.R., White, P.A. (eds.) Groundwaters in New Zealand, pp. 273–289. New Zealand Hydrological Society Inc., Wellington (2001)
24. Ferraro, F., Etzion, D., Gehman, J.: Tackling grand challenges pragmatically: robust action revisited. Organ. Stud. **36**, 363–390 (2015)
25. Frey-Heger, C., Barret, M.: Possibilities and limits of social accountability: the consequences of visibility as recognition and exposure in refugee crises. Acc. Organ. Soc. **89**, 101197 (2021)
26. Golnaraghi, M.: An overview: Building a global knowledge base of lessons learned from good practices in multi-hazard early warning systems. In: M. Golnaraghi (Ed.), Institutional Partnerships in Multi-Hazard Early Warning Systems: A Compilation of Seven National Good Practices and Guiding Principles, Springer, pp. 1–8 (2012). https://doi.org/10.1007/978-3-642-25373-7_1
27. Gray, R.: Social, environmental and sustainability reporting and organisational value creation? Whose value? Whose creation? Account. Auditing Accountability J. **19**(6), 793–819 (2006)

28. Gray, R.: Is accounting for sustainability actually accounting for sustainability…and how would we know? An exploration of narratives of organisations and the planet. Account. Organ. Soc. **35**(1), 47–62 (2010)
29. Greiling, D., Spraul, K.: Accountability and the challenges of information disclosure. Public Adm. Quart. **34**(3), 338–377 (2010). http://www.jstor.org/stable/41288352
30. Guthrie, J., Ball, A., Farneti, F.: Advancing sustainable management of public and not for profit organizations. Public Manag. Rev. **12**(4), 449–459 (2010)
31. Guthrie, J., Dumay, J.: Wicked problems involve social justice, social change, climate change and the social economy. J. Behav. Econ. Soc. Syst. **3**(1), 11–16 (2021)
32. Han, Y.: The impact of accountability deficit on agency performance: performance-accountability regime. Public Manag. Rev. **22**(6), 927–948 (2020). https://doi.org/10.1080/14719037.2019.1679237
33. Hardin, G.: The tragedy of the commons: the population problem has no technical solution; it requires a fundamental extension in morality. Science **162**, 1243–1248 (1968)
34. Hirabayashi, Y., et al.: Global flood risk under climate change. Nat. Clim. Chang. **3**, 816–821 (2013). https://doi.org/10.1038/nclimate1911
35. Hsiao, P.C.K., De Villiers, C., Horner, C., Oosthuizen, H.: A review and synthesis of contemporary sustainability accounting research and the development of a research agenda. Account. Financ. **62**(4), 4453–4483 (2022)
36. Korca, B., Ericka, C., Federica, F.: From voluntary to mandatory non-financial disclosure following Directive 2014/95/EU: an Italian case study. Account. Eur. **18**(3), 353–377 (2021)
37. Kosmala, K., McKernan, J.F.: From care of the self to care for the other: neglected aspects of Foucault's late work. Account. Auditing Accountability J. **24**(3), 377–402 (2011)
38. Kraft, M.E.: Institutions and environmental change: principal findings, applications, and research frontiers. In: O. R. Young, L.A. King, H. Schroeder (eds.), Perspectives on Politics. Cambridge, MA, MIT Press, pp. 152–153 (2011). https://doi.org/10.1017/S153759271000366X
39. Lai, A., Leoni, G., Stacchezzini, R.: The socializing effects of accounting in flood recovery. Crit. Perspect. Account. **25**(7), 579–603 (2014)
40. Lavallee, L.F., Poole, J.M.: Beyond recovery: Colonization, health, and healing for Indigenous people in Canada. Int. J. Ment. Heal. Addict. **8**(2), 271–281 (2010). https://doi.org/10.1007/s11469-009-9239-8
41. Lawrence, J., Sullivan, F., Lash, A., Ide, G., Cameron, C., McGlinchey, L.: Adapting to changing climate risk by local government in New Zealand: institutional practice barriers and enablers. Local Environ. **20**(3), 298–320 (2015). https://doi.org/10.1080/13549839.2013.839643
42. Lawrence, J., Manning, M.: Developing adaptive risk management for our changing climate', NZCCRI 2012 Report: 2012-1, The New Zealand Climate Change Research Institute, Victoria University of Wellington, pp. 1–54 (2012)
43. McDonald, J.: The role of law in adapting to climate change. Wiley Interdisc. Rev.: Clim. Change **2**(2), 283–295 (2011)
44. McSweeney, J.: Overview of flood management legislation, Report prepared for the Ministry Environ. 1–37 (2006), https://riversgroup.org.nz/wp-content/uploads/2018/06/2.1-Overview-of-flood-management-legislation-in-NZ.pdf (Accessed 20 March 2024)
45. Measham, T., Brooke, C., Gorddard, R., Withycombe, G., Morrison, C.: Adapting to climate change through local municipal planning: barriers and challenges, [Open access]. Mitig. Adapt. Strat. Glob. Change **16**(8), 889–909 (2011)
46. Messner, M.: The limits of accountability. Account. Organ. Soc. **34**(8), 918–938 (2009)
47. Milly, P., et al.: Climate change: stationarity is dead: whither water management? Science **319**, 573–574 (2008)

48. Milne, M.J., Grubnic, S.: Climate change accounting research: keeping it interesting and different. Account. Auditing Accountability J. **24**(8), 948–977 (2011)
49. Milne, M., Guthrie, J., Parker, L.: Into the light and engagement: two decades of interdisciplinary perspectives on accounting, auditing and accountability research. Account. Auditing Accountability J. **21**(2), 117–128 (2008). https://doi.org/10.1108/09513570810854374
50. Ministry of the Environment (MfE): Meeting the challenges of future flooding in New Zealand' Ministry for the Environment and the flood risk management and river control review steering group, Wellington, New Zealand, pp. 1–48 (2008). https://environment.govt.nz/assets/Publications/Files/meeting-challenges-of-future-flooding-in-nz.pdf
51. Mizrahi, S., Minchuk, Y.: Accountability and performance management: citizens' willingness to monitor public officials. Public Manag. Rev. **21**(3), 334–353 (2019). https://doi.org/10.1080/14719037.2018.1473478
52. Mulgan, R.: Accountability': an ever-expanding concept? Public Adm. **78**(3), 555–573 (2000)
53. Organisation for Economic Co-operation and Development (OECD): Disaster Risk Financing: A global survey of practices and challenges, OECD Publishing, Paris, pp. 1–135 (2015). https://doi.org/10.1787/9789264234246-en
54. Organisation for Economic Co-operation and Development (OECD): Financial Management of Flood Risk, OECD Publishing, Paris, pp. 1–133 (2016). https://doi.org/10.1787/9789264257689-en
55. Panozzo, F.: Accountability and identity: accounting and the democratic organization. In: Munro, R., Mouritsen, J. (eds.) Accountability: power, ethos and the technologies of managing, pp. 182–195. Thompson International Business Press, Oxford (1996)
56. Paulik, R., Wild, A., Zorn, C., Wotherspoon, L: Residential building flood damage: Insights on processes and implications for risk assessments. J. Flood Risk Manage. **15**(4), 1-15 (2022). https://doi.org/10.1111/jfr3.12832
57. Pellizzoni, L.: Responsibility and environmental governance. Environ. Polit. **13**(3), 541–565 (2004). https://doi.org/10.1080/0964401042000229034
58. Preston, B.L., Westaway, R.M., Yuen, E.J.: Climate adaptation planning in practice: an evaluation of adaptation plans from three developed nations. Mitig. Adapt. Strat. Glob. Change **16**(4), 407–438 (2011)
59. Reisinger, A., Wratt, D., Allan, S.: The role of local government in adapting to climate change: lessons from New Zealand. In: Ford, J.D., Berrang-Ford, L. (eds.) Climate change adaptation in developed nations, pp. 303–319. Springer, Toronto, Canada (2011)
60. Renn, O., Klinke, A., van Asselt, M.: Coping with complexity, uncertainty and ambiguity in risk governance: a synthesis. Ambio **40**(2), 231–246 (2011)
61. Rogers, D.P., Tsirkunov, V.V.: Weather and climate resilience: effective preparedness through national meteorological and hydrological services. Environment and Sustainable Development, World Bank Publications, pp. 1–141(2013). https://documents.worldbank.org/en/publication/documents-reports/documentdetail/308581468322487484/weather-and-climate-resilience-effective-preparedness-through-national-meteorological-and-hydrological-services
62. Scorzini, A.R., Frank, E.: Flood damage curves: new insights from the 2010 flood in Veneto Italy. J. Flood Risk Manage. **10**, 1753–2318 (2017)
63. Serafeim, G.: Social-impact efforts that create real value. Harv. Bus. Rev. **98**, 37–48 (2020)
64. Snel, K.A., Priest, S.J., Hartmann, T., Witte, P.A., Geertman, S.C.: Do the resilient things. Residents' perspectives on responsibilities for flood risk adaptation in England. J. Flood Risk Manage. 1**4**(3), 1–14 (2021). https://doi.org/10.1111/jfr3.12727
65. Svensson, C., Jones, D.A.: Review of rainfall frequency estimation methods. J. Flood Risk Manage. **3**, 296–313 (2010). https://doi.org/10.1111/j.1753-318X.2010.01079.x
66. Terpstra, T., Gutteling, J.M.: Households' perceived responsibilities in flood risk management in the Netherlands. Int. J. Water Resour. Dev. **24**(4), 555–565 (2008). https://doi.org/10.1080/07900620801923385

67. United Nation International Strategy for Disaster Reduction (UNISDR): The human cost of weather-related disasters: 1995–2015, Centre for Research on the Epidemiology of Disasters (CRED), The United Nations Office for Disaster Risk Reduction, pp. 1–27 (2015). https://www.unisdr.org/files/46796_cop21w. (Accessed 20 Feb 2023)
68. United Nation Development Program (UNDP): UNDP Issues Brief Climate Information and Early Warning Systems (2020). https://www.undp.org/publications/undp-issues-brief-climate-information-and-early-warning-systems (Accessed 20 Feb 2024)
69. van Buuren, A., Lawrence, J., Potter, K., Warner, J.F.: Introducing adaptive flood risk management in England New Zealand, and the Netherlands: the impact of administrative traditions. Rev. Policy Res. **35**(6), 907–929 (2018)
70. Vollmer, H.: Public value and the planet: accounting in ecological reconstitution. Account. Auditing Accountability J. **34**(7), 1527–1554 (2021)
71. Wenger, C.: Better use and management of levees: reducing flood risk in a changing climate. Environ. Rev. **23**, 240–255 (2015)
72. World Commission on Environment and Development (WCED): Our Common Future, Oxford, New York (1987) Journal **2**(5), 99–110 (2016). https://www.are.admin.ch/are/en/home/media/publications/sustainable-development/brundtland-report.html

Impacts and Future Perspectives of Remote Auditing: A Critical Literature Review

Rita Sousa[1](✉) and Carla Carvalho[2]

[1] Higher Institute of Accounting and Administration, University of Aveiro, Aveiro, Portugal
ritarsousa@ua.pt

[2] GOVCOPP - Research Unit On Governance, Competitiveness and Public Policies, Higher Institute of Accounting and Administration, University of Aveiro, Aveiro, Portugal
carla.carvalho@ua.pt

Abstract. The COVID-19 pandemic has significantly accelerated the adoption of remote auditing, a practice enabled by digital technologies and characterized by the absence of auditors' physical presence at client sites. This study presents a comprehensive and critical literature review of remote auditing, structured around five key dimensions: (1) the implementation of new technologies, (2) changes in audit procedures and professional profiles, (3) implications for audit quality, (4) impacts on auditors' work-life balance, and (5) associated opportunities, risks, and challenges. The review synthesizes empirical findings and assesses the methodological approaches, sample characteristics, and geographical distribution of existing studies. The study identifies gaps for each dimension and proposes cues for future research. Findings indicate that while ICT tools facilitate remote audits, disparities in their use result in uneven adoption across firms. Remote auditing has reshaped auditor roles, emphasizing autonomy and digital skills, yet challenges persist in replicating in-person procedures. The impact on audit quality remains inconclusive, often based on subjective perceptions. Remote work improves flexibility for some auditors but raises concerns about stress and social isolation for others. The transition presents opportunities for a hybrid audit model, though it introduces ethical, supervisory, and cybersecurity risks. The study highlights the need for regulatory frameworks and calls for future research using mixed methods and cross-contextual comparisons to assess the effectiveness and long-term effects of remote auditing. Practical insights are provided for audit firms, educators, and policymakers to support sustainable remote audit practices.

Keywords: Remote Auditing · Remote Work · ICT · Work-Life Balance · Auditors · Literature Review

1 Introduction

The COVID-19 pandemic has led to profound societal transformations, significantly altering personal routines and professional practices. One of the most notable outcomes of this disruption has been the accelerated adoption of digital technologies within organizational contexts (Lontsikh et al., 2021; Rahman & Ziru, 2023). Remote work, once

L. Almeida et al. (Eds.): ICAFI 2025, ISEM 80, pp. 385–400, 2026.
https://doi.org/10.1007/978-3-032-16268-7_22

a contingency, has become a normalized way of working (Li et al., 2023). Among the professions most affected by this shift is auditing, where remote auditing practices are rapidly replacing traditional, in-person audits (Castka & Searcy, 2023; Sian, 2022; Subhan & Suyanto, 2023).

Remote auditing, the execution of audit procedures without the auditor's physical presence at the client's premises, supported by information and communication technologies (ICT), has emerged in response to pandemic-induced restrictions (Eulerich et al., 2022; Lorentzon et al., 2023). Auditors now face the dual challenge of adapting to this evolving landscape while preparing for the broader digital transformation imposed by Industry 4.0 (Rahman & Ziru, 2023; Ribeiro & Oliveira, 2024). As noted by Handoko and Kurniawan (2022), the pandemic has not only accelerated the use of remote audits but also presented an opportunity to rethink how audits are conducted.

Despite the increase in studies, the literature on remote auditing remains essentially descriptive, offering limited critical reflection on its broader implications. This study contributes to filling this gap identified in the literature by conducting a comprehensive and critical literature review on remote auditing. In order to provide a more transparent and more objective analysis of the topics investigated, this literature review was organized into five fundamental topics:

1. Analyze the implementation and implications of emerging technologies within remote audit practices.
2. Examine changes in the auditing profession, including the required skills and the profile of the auditor.
3. Assess the impact of remote auditing on audit quality.
4. Explore the implications for work-life balance, including the necessary adjustments and their impact on daily routines.
5. Identify the opportunities, risks, and challenges associated with remote auditing.

Additionally, the review systematically analyzes the empirical studies across three dimensions: (i) the key findings in each thematic area, (ii) methodological approaches adopted, and (iii) sample size, composition, and geographic distribution. For each dimension, the study identifies gaps and proposes directions for future research.

The study contributes to both theory and practice by providing a comprehensive overview of the current state of remote auditing research, identifying existing gaps in the literature, and offering insights for future research and professional development.

This paper is organized into nine sections. After this Introduction, Sect. 2 defines the concept of remote auditing. Sections 3 to 7 explore the five fundamental topics of research: technological implementation, professional transformation, audit quality, personal implications for auditors, and future opportunities and risks. Section 8 presents a critical analysis of the empirical studies, highlighting methodological and sample patterns, identifying gaps, and proposing directions for future research. Finally, Sect. 9 presents the main conclusions and contributions.

2 Definition of Remote Auditing

COVID-19 changed everyone's lives, and the auditing profession was no exception. Remote auditing is now a reality for many auditors, who have had to adapt to this new way of performing their tasks, with audits—or parts of them—being conducted remotely. Although remote auditing is not a new process, prior to the pandemic, 98% of audits were conducted in person (Aivazi, 2022).

Teeter et al. (2010) define remote auditing as the process by which auditors combine information and communication technologies (ICT) with analytical procedures to electronically gather audit evidence, interact with the audited entity, and ensure the reliability of data and internal controls, regardless of the auditor's physical location. A remote audit involves using electronic methods such as videoconferencing, telephone calls, web platforms, and emails to collect audit evidence. Remote auditing can thus be defined as a more technologically enhanced form of auditing, where auditors utilize technologies to obtain evidence and analyze audit data (Castka et al., 2020). It differs from conventional audits mainly because in-person interactions are unnecessary (Eulerich et al., 2022), changing how visits, interviews, and other audit procedures are conducted (Teeter et al., 2010).

Eulerich et al. (2022) define remote auditing as the use of ICT to execute audit procedures at locations other than the audited entity's premises. In this context, work is considered remote when employees of an entity spend most of their time working independently, managing their workspace and equipment, and communicating with their supervisors from locations such as home (Nickson & Siddons, 2012). Remote workers are expected to possess more specialized skills than on-site workers, particularly communication skills on virtual platforms and managing tasks without supervision. Lorentzon et al. (2023) summarize the concept by defining remote auditing as the process by which auditors perform audit procedures from a location outside the audited entity's premises, utilizing digital tools for support.

3 Implementation of New Technologies

The concepts of digitization, Industry 4.0, and artificial intelligence have gained popularity, and their application in auditing is increasingly drawing attention (Rahman & Ziru, 2023). Before the COVID-19 pandemic, audits were conducted using more traditional procedures, which typically began with a review of documents followed by visits to the audited entities. The aim was to collect evidence through visual inspections, interviews, and other procedures, ultimately supporting the preparation of the audit report (Castka & Searcy, 2023). With the onset of the pandemic, auditors were compelled to adopt new methodologies and procedures to minimize physical contact with their clients and comply with existing health restrictions, thereby shifting to conducting audits remotely. Although videoconferencing via digital platforms, such as Zoom and Microsoft Teams, is essential in reshaping ICT, their usage reveals that merely replicating traditional audit practices online is insufficient. Many expected that the introduction of remote auditing would reduce or eliminate travel costs; however, this may not be the case, as the primary goal of remote auditing is not cost reduction, particularly in the short term. Nonetheless, the pandemic provided an opportunity to transition from traditional audits to remote auditing (Castka & Searcy, 2023).

Remote auditing is not new for large firms, such as the Big Four, which operate in multiple locations worldwide. These firms have used ICT to audit processes for over a decade (Castka & Searcy, 2023). However, the restrictions imposed by the pandemic forced many small and medium-sized audit firms to adopt the emerging technologies used by larger firms. This implementation of ICT will drive further investment in even more advanced technologies, such as Big Data and Artificial Intelligence (AI), enhancing the review of data in volume and speed (Castka & Searcy, 2023).

With the increasing complexity of current information systems, the need to adapt auditing to these computerized environments has emerged. Computer-Assisted Audit Tools (CAATs) have been developed to help automate the audit process and extract data from companies' Enterprise Resource Planning (ERP) systems (Ribeiro & Oliveira, 2024). Auditors use CAATs to retrieve data and analyze transactions to detect potential anomalies or evaluate automated controls, such as verifying who performed a specific control, since modern systems may lack documentary evidence. These tools are also helpful in selecting samples and leveraging computing speed to process large volumes of data. However, they are primarily data extraction tools for analysis conducted through queries performed after transactions have occurred and the data have been processed in the system (Byrnes et al., 2014).

As fully paper-based traditional audits have become nearly impossible due to the electronic storage of documents, new data analysis tools (Data Analytics) have emerged. These tools allow auditors to query and test a higher volume of transactions, reconciling accounts and/or transactions (CPAB Exchange, 2019). Furthermore, they produce visual analyses that help auditors understand data more clearly and efficiently through the use of trend charts and the identification of outliers (Financial Reporting Council, 2017; Salijeni et al., 2021).

It is a fact that digitization is increasingly present in our daily lives, with a growing number of business processes occurring online. ICT offers many new opportunities, including conducting financial audits (Lontsikh et al., 2021). According to the literature (Al-Aroud & Falah, 2020; Cooper et al., 2019; Huang & Vasarhelyi, 2019; Issa et al., 2016), technologies are expected to have a positive impact on auditors' everyday work. With remote audits, professionals can focus on higher-responsibility tasks and assign the remaining ones to robotic processes, improving efficiency and effectiveness. Auditors believe that new technological tools allow them to conduct high-quality audits that are accurate, complete, timely, and compliant with audit standards, governmental regulations, and professional expectations (Sharma et al., 2022). However, the main limitations in implementing remote audits relate to auditors' insufficient technological knowledge and the lack of client familiarity with technology-supported audit processes. Furthermore, permanent updating and training in ICT are necessary, considering its rapid and constant evolution. Nevertheless, these obstacles are offset by cost and time savings, as well as improved audit efficiency and quality (Sharma et al., 2022).

4 Changes in the Profession and Audit Procedures

According to the International Auditing and Assurance Standards Board (IAASB), COVID-19 triggered uncertainties and unpredictability that created new risks or intensified existing ones, requiring auditors to redesign how audit procedures were executed to ensure the sufficiency and appropriateness of audit evidence. During the pandemic, remote work—and consequently remote audits—required the adoption of new technologies and communication methods with clients and among audit teams, due to the difficulty or impossibility of conducting in-person audits. Furthermore, COVID-19 created an economically uncertain environment with higher risks and required ongoing assessment (Morris et al., 2023).

Audit work is an inferential practice that gathers and analyzes evidence to conclude financial statements (Sian, 2022). At the same time, professional regulations require independence, relevance, reliability, and sufficiency in evidence collection, subject to materiality assessments and the risk of material misstatement. Audit evidence is considered more reliable if obtained from external sources, documented, or collected directly and personally by the auditor. However, due to the pandemic's restrictions, evidence collection became more challenging, forcing auditors to adopt a higher degree of professional skepticism and rely on oral evidence and detailed analysis of relevant documentation (Sian, 2022).

The choice of communication method can have significant implications for conducting a remote audit. In their study, Carlisle and Hamilton (2021) reveal that auditors expressed concerns about potentially relying too heavily on email communications, which may lead to insufficient follow-up questions or weakened client relationships. Moreover, when auditors request information by email, clients tend to respond more favorably to them (i.e., with bias) than to in-person requests. However, email communications can also enhance audit work, as they are textual and lack social and personal cues, allowing auditors to assess a client's explanation more objectively than in face-to-face interactions (Carlisle & Hamilton, 2021).

Remote audits can save time and money because not all procedures need to be performed in person at the audited company, thereby reducing travel and accommodation costs and allowing for more efficient time management. Additionally, remote audits are more flexible (Serag & Daoud, 2021), as auditors can work independently, unlike traditional auditing, where auditors are highly dependent on the client's availability (Lontsikh et al., 2021). In remote audits, auditors benefit from data transfer via online platforms, enabling them to conduct interviews, observe processes, and analyze documents from anywhere in the world without the need for travel (Teeter et al., 2010). However, they also present challenges, such as network issues, requiring auditors to reconnect and resolve technical problems. In digital platform conversations, the connection signal may be weak, participants' voices may be unclear, and the video may be delayed (Mugabe et al., 2022).

Thus, combining in-person and remote audits emerges as an approach that offers various benefits, including cost reduction, improved time management, and the use of technology for efficient data analysis. Indeed, auditing at the audited entity's premises may not be fully replaceable due to its psychological impact and ability to assess certain documents and processes physically. Therefore, companies must carefully evaluate which audits can be conducted remotely, taking into account risk assessment, organizational culture, the client's industry sector, and country-specific aspects. A balanced approach is needed to leverage the benefits of remote auditing while ensuring the auditor's on-site presence for an effective audit (Jarva & Zeitler, 2023).

In their study, Eulerich et al. (2022) highlight that remote audit limitations include the inability to replace in-person observations, build client relationships, and lack personal interaction, all of which are important for fraud detection. It also restricts the performance of audit teams, as in-person checks are essential to identify nonconformities that can only be verified in person (Barretto et al., 2022). All these factors, and others, can trigger stress and frustrate auditors (Subhan & Suyanto, 2023).

Although conducting audit procedures remotely has become a new reality in the auditing profession, no regulations currently exist to guide remote audits. Moreover, research has not systematically studied how the shift from in-person to remote audits affects audit success, despite many companies now allowing their auditors to work remotely indefinitely (Li et al., 2023).

5 Impact on Audit Quality

Auditors' competence in managing flexibility leads to high-quality and efficient remote auditing (Li et al., 2023). Furthermore, a physical work environment that is favorable for focus during remote auditing tasks is positively associated with audit efficiency and auditors' job satisfaction. Thus, remote work can enhance audit quality and efficiency when audit firms adequately support their employees (Eulerich et al., 2022). However, the lack of guidance, training, and financial resources from the audit firm poses challenges to the remote execution of audit procedures and to maintaining their high quality and efficiency (Li et al., 2023). The approach of remote auditing—whether fully remote or hybrid—also directly affects its efficiency, but not the audit quality or the auditors' job satisfaction (Lorentzon et al., 2023).

Auditors believe that remote audits are as effective and efficient as in-person audits (Li et al., 2023) and consider that stakeholders place equal trust in the outcomes of both types of audits (Eulerich et al., 2022). Remote auditing can enhance audit quality through its association with new ICTs (Lorentzon et al., 2023), as there appears to be a positive relationship between the clients' level of digitalization and audit quality (Rahman & Ziru, 2023). Therefore, as a company integrates more digital aspects into its business operations, its audit quality tends to improve. Nevertheless, some studies (e.g., Gong et al., 2022) suggest that remote auditing may lead to lower audit quality.

Efforts to increase auditors' sense of responsibility are beneficial to audit quality. Responsibility leads to greater auditor effort and performance, increases conservatism, reduces judgment variability, and contributes to more complete audits (Sorensen & Ortegren, 2021). Auditors' professional skepticism also significantly impacts the quality of remote audits (Handoko & Sardjono, 2022). The higher the level of skepticism, the more likely the auditor is to seek in-depth information from management regarding its decisions. Therefore, to maintain audit quality, the auditor must adopt high professional skepticism when performing remote audit procedures (Handoko & Sardjono, 2022). Higher levels of social presence also lead to increased auditor skepticism (Sorensen &

Ortegren, 2021). Face-to-face communication fosters a higher sense of social presence due to the visual cues it enables. In contrast, written communication channels, which do not allow for visual or auditory signals, such as email, contribute to lower levels of social presence.

Auditors' competence also affects the quality of the remote audits they perform. For example, during the pandemic period, auditors were required to be more able to analyze limited evidence (Handoko & Sardjono, 2022). Auditors faced difficulties transitioning to remote auditing, and their ability to identify, correct, and disclose matters that could lead to material financial misstatements was compromised. However, in some cases, delays in accounting records allowed auditors to improve audit quality (Morris et al., 2023). The pandemic has significantly increased auditors' awareness of risks and prudence in audit reports, resulting in a higher likelihood of modified audit opinions in listed companies in China (Huang, 2023). The possibility of correcting significant accounting errors has increased considerably during the two years of the pandemic, mainly due to the implementation of remote audit procedures (Huang, 2023).

On the other hand, audit risk does not appear to affect the quality of remote auditing. Although the expectation during the pandemic was that audit risk would increase, this does not seem to have occurred, as it may have been mitigated by the adoption of ICTs (Handoko & Sardjono, 2022). Thus, compared to auditors in more digitally developed countries, auditors in countries with lower ICT development may experience reduced audit efficiency and quality (Rahman & Ziru, 2023).

6 Implications for Auditors' Personal and Family Life

The pandemic required individuals to make adjustments and face restrictions in their personal and family lives, mainly due to the need to work remotely. This remote work context, especially when working from home, presented new challenges for auditors, requiring them to be creative in obtaining sufficient and appropriate audit evidence to support their opinion (Handoko et al., 2023) while maintaining audit quality (Handoko et al., 2023). According to evidence gathered by Lorentzon et al. (2023), auditors believe that remote auditing has improved work-life balance by offering greater schedule flexibility, increased autonomy in task execution, and a more efficient use of time.

Currently, audit firms face a significant shortage of auditors due in part to difficulties in recruiting new professionals, high turnover in the sector, the deterioration that the profession causes, and auditors' growing desire for a better work-life balance, increasingly prioritizing their personal and family life over their professional responsibilities (Stuart et al., 2024). Thus, as audit firms attempt to balance client obligations with remote and hybrid work models, they must understand and address the impact of remote auditing on auditors' work-life balance (Lorentzon et al., 2023).

Achieving a work-life balance remains crucial for auditors, as it can have a direct impact on the quality of audits. Auditors with low work-life balance experience higher stress and burnout symptoms, greater difficulty concentrating, reduced energy, and a decline in overall job satisfaction (Lorentzon et al., 2023). Achieving a healthy work-life balance requires auditors to effectively manage their time, establish clear limits, and make thoughtful decisions about allocating attention across the different dimensions of their lives. In their study, Lorentzon et al. (2023) found that this balance improves when auditors work remotely, particularly from home, since it allows them to shift focus more easily and conveniently between work and family.

Several researchers have identified the requirement to be physically "present" at the workplace as contributing to closing off career progression opportunities, for example, for working mothers with children in their care. However, post-pandemic studies indicate that women, in general, are now more interested in working from home, as remote work offers greater autonomy and control over their work-life balance (Sian, 2022).

The cumulative effect of stress caused by conflict, uncertainty, and remote work overload can harm auditors' performance, compromising the excellence and efficiency of their work (Subhan & Suyanto, 2023). However, during the pandemic, sharing and managing information among auditors from small audit firms proved to be fundamental and strategic in improving their performance. In remote auditing and through the frequent use of digital platforms, knowledge sharing among auditors is considered easier and faster (Subhan & Suyanto, 2023). Thus, auditor engagement in knowledge-sharing activities, such as participation in social media groups and community support, can enhance job satisfaction and reduce stress. Knowledge sharing increases auditors' positive feelings when facing workplace disputes and reduces negative emotions when confronted with uncertainty and work overload (Subhan & Suyanto, 2023). Nonetheless, while remote communication (e.g., via email) may cause less anxiety for auditors, it may also be less effective in many cases—for instance, in urgent matters and in building client relationships (Carlisle & Hamilton, 2021). Therefore, it is essential to acknowledge that on-site presence during audits cannot be entirely replaced, particularly due to its psychological impact (Jarva & Zeitler, 2023). ICTs only partially substitute for physical testing and human interaction and remain limited when auditing non-digitized processes.

7 Opportunities, Risks, and Future Challenges of Remote Auditing

Considering the changes caused by the pandemic and their implications for the workforce, Barreto et al. (2022) concluded that some changes imposed by remote work have yielded significant benefits for all stakeholders. From an organizational standpoint, cost reductions associated with optimizing physical spaces, infrastructure, and resources are highlighted as a key advantage of remote auditing. However, from a workforce management perspective, organizations must seek alternatives to reduce or mitigate the side effects of remote auditing, such as emotional distancing from the team and the organizational culture, as well as other impacts associated with decreased productivity, which can even affect audit efficiency and quality (Barreto et al., 2022). As the number of people working remotely increases and, in some cases, becomes the new normal, audit firms must adapt and reassess their strategies to save money and mitigate risks (Farcane et al., 2023).

Future auditors must develop ICT and data analytics skills because remote auditing represents the future (Kokina & Davenport, 2017). Companies must invest in training and professional development programs to enable auditors to acquire the necessary skills to work with emerging ICT. Auditing is evolving rapidly, and continuous training is crucial to ensure that professionals are prepared to face challenges and capitalize on opportunities arising from technological innovations and the rise of remote auditing (Barreto et al., 2022). According to Vitali and Giuliani (2024), the evolution and adaptation of classical auditors' skills to new technologies may lead to a new concept of auditor - the "ICT auditor" - capable of working with modern technological platforms and tools. Implementing new technologies is neither easy, inexpensive, nor universally accessible (Issa et al., 2016). Therefore, large audit firms have a competitive advantage in operationalizing remote audits. With more human and financial resources, they have a greater capacity to implement emerging technologies and digital programs and applications, such as artificial intelligence.

Remote auditing tends to be more efficient as work becomes more task-oriented, travel is reduced, data analysis is facilitated, and collaboration among audit team members is enhanced (Lorentzon et al., 2023). However, although technological advancements allow auditors to intensify online communication, reduced physical presence may limit the learning and development of junior auditors. A lower frequency of face-to-face meetings reduces the opportunity for junior auditors to learn through observation and their ability to ask questions, which may result in less support from senior auditors to juniors. According to Lorentzon et al. (2023), this is a significant challenge that audit firms must address and overcome, as learning from seniors is a crucial part of developing and training junior auditors. Conversely, Sian (2022) collected evidence that remote auditing promotes greater communication with clients and among audit teams, with increased contact between senior and junior auditors to support and motivate the latter. While some aspects of communication, such as visual or emotional cues, may be lost in online meetings, they are compensated for by the convenience and autonomy of working from home (Sian, 2022).

Although Mugabe et al. (2022) have recognized many benefits of remote auditing, they also noted that unreliable network connections may produce significant risks. One critical risk and a major challenge for firms involves the security of remote auditing and related ethical concerns (Lorentzon et al., 2023). Using digital platforms for communication and document sharing requires auditors to protect clients' sensitive information from unauthorized access, ensuring data confidentiality. In their study, Jarva and Zeitler (2023) identify strategic risks for those conducting audits remotely, particularly the potential loss of audit legitimacy if they do not adapt to stakeholders' expectations.

8 Critical Analysis of Empirical Studies

8.1 Topics Addressed and Future Research

Implementation of New Technologies. Most studies highlight the importance of technology as a facilitator of remote auditing. Communication and document-sharing platforms have been widely adopted. Auditors' technological competence is a critical success factor in remote auditing, as it is associated with their perceived effectiveness (Baatwah et al., 2023; Handoko & Kurniawan, 2022). However, studies such as those by Lorentzon et al. (2023) and Morris et al. (2023) point to limitations related to technological dependence and difficulties in applying specific procedures remotely. While the literature agrees on the facilitating and crucial role of technology in remote auditing, it remains scarce regarding the costs and inequalities associated with the intensive use of digital platforms. Most studies take a descriptive approach to technology, lacking a more critical analysis of the power that technology has within teams, auditor-auditee relationships, and stakeholder interactions.

Changes in the Profession and Audit Procedures. Empirical evidence shows that remote auditing has led to profound changes in traditional audit procedures, team management, and supervision (Carlisle & Hamilton, 2021; Castka et al., 2020). The literature emphasizes auditors' ability to adapt to remote auditing, yet few studies examine how these changes affect auditor training, performance evaluation, and career progression. It is also suggested that studies be conducted to understand how remote auditing might reinforce systemic inequalities within the profession, given differences in digital literacy, firm size, and other factors.

Impact on Audit Quality. The literature presents conflicting views regarding the impact of remote auditing on audit quality. Some studies conclude that quality is maintained with technological support and appropriate skills (Handoko et al., 2023; Sharma et al., 2022), while others warn of weaknesses in collecting physical evidence (Mugabe et al., 2022). Moreover, most study conclusions are based on perceptions, primarily those of auditors, rather than robust quantitative empirical data. Therefore, there is a need for studies that adopt a more objective approach to analyzing remote audit quality, using comparative quality indicators between remote and in-person audits within the same audit firm and across firms.

Implications for Auditors' Personal and Family Life. The literature indicates that remote auditing has mixed effects on the personal and family lives of auditors. On the one hand, difficulties in separating work from personal life and feelings of isolation have been reported (Carlisle & Hamilton, 2021). On the other hand, some studies highlight improved work-life balance and increased autonomy (Handoko & Thomas, 2022; Lorentzon et al., 2023). These opposing results justify further investigation into the impact of remote auditing on auditors' work-life balance. Furthermore, these studies must consider different contexts and dimensions, such as the professionals' socioeconomic backgrounds, gender diversity, maternity, parenting, and career progression.

Opportunities, Risks, and Future Challenges of Remote Auditing. There is an optimistic consensus on the potential of remote auditing as a hybrid and economically efficient model. Studies such as those by Castka et al. (2022) and Vitali and Giuliani (2024) indicate a transformation of the profession driven by remote auditing, highlighting the need for greater adaptation, ICT training, and regulation (Sian, 2022; Subhan & Suyanto, 2023). The studies also identify technical risks, including insufficient evidence collection, as well as ethical and security risks related to digital platforms. Thus, the vision for the future of remote auditing is optimistic but insufficiently critical. There is a lack of empirical studies addressing who benefits most from remote auditing, why, what inequalities are being exacerbated, and how to respond to new ethical, security, and regulatory challenges. Most studies are descriptive and do not propose control mechanisms or new normative and regulatory frameworks.

8.2 Research Methodologies

An analysis of the methodologies employed in empirical studies on remote auditing reveals a predominance of qualitative and mixed-method approaches. Online questionnaires are the most frequently used data collection technique, particularly in the context of auditors' perceptions of the impact, effectiveness, and challenges of remote auditing (e.g., Baatwah et al., 2023; Carlisle & Hamilton, 2021; Lorentzon et al., 2023; Subhan & Suyanto, 2023). The data analysis follows a descriptive approach, although some studies have employed statistical techniques to test hypotheses (Handoko & Sardjono, 2022; Handoko & Kurniawan, 2022).

Individual interviews are another commonly used methodology (e.g., Castka et al., 2022; Eulerich et al., 2022; Jarva & Zeitler, 2023; Sharma et al., 2022), providing deeper and more contextualized narratives, although they lack generalizability. Interviews are beneficial for understanding subjective aspects such as adaptation to change, professionals' well-being, or their perceptions of remote auditing risks. Finally, some studies have adopted content analysis of documents or organizational websites (Castka et al., 2020).

The reviewed studies reveal a predominance of qualitative data collection approaches and descriptive data analysis. Online questionnaires are the most common method; however, they do not facilitate in-depth analysis and discussion of the issues under study. Interviews, also used in some studies, provide interpretive richness but lack generalizability. Therefore, it would be beneficial for future research to combine questionnaires with interviews to explore the same issue, enabling the integration of auditors' perceptions with other sources of information. It is also important to conduct case studies, which would allow for a more contextualized and in-depth analysis of the impact and efficiency of remote auditing. Furthermore, it is necessary to conduct more robust analyses, specifically examining correlation and causal relationships between variables, which will enable hypotheses to be tested regarding the determining factors and consequences of remote auditing. For example, it would be interesting to critically explore how remote auditing might differentially impact auditors based on their gender, seniority, socioeconomic background, or firm type.

8.3 Sample Size, Composition, and Location

The samples used in the empirical studies on remote auditing vary in size and composition. In terms of size, some studies relied on larger samples (Baatwah et al., 2023; Subhan & Suyanto, 2023), while others used smaller, strategically focused samples, such as internal auditors (Eulerich et al., 2022), professionals from international firms (Castka et al., 2022), or members of audit teams working on sustainability matters (Castka et al., 2020). Regarding sample composition, most studies included external auditors as participants. However, some focused on internal auditors (e.g., Eulerich et al., 2022) or combined professionals from different functions related to the audit and control process. In many cases, however, insufficient details about inclusion criteria and participant characteristics are provided, which limits the interpretation of each study's findings.

The geographical distribution of the samples spans several countries, reflecting broad global coverage, with the following countries highlighted by continent:

- America: United States (Carlisle & Hamilton, 2021; Li et al., 2023; Sian, 2022);

- Europe: Germany (Eulerich et al., 2022; Jarva & Zeitler, 2023);
- Asia: Indonesia (various studies by Handoko et al., between 2022 and 2023); Malaysia (Rahman & Ziru, 2023); Saudi Arabia (Baatwah et al., 2023); and China (Castka et al., 2022);
- Africa: Egypt (Serag & Daoud, 2021) and Uganda (Mugabe et al., 2022);
- Oceania: New Zealand (Castka et al., 2022).

It is worth noting that some of these studies cover samples from more than one country, as is the case with Castka et al. (2022), which includes individuals from both New Zealand and China.

In conclusion, the reviewed studies reflect a vast geographical diversity, with samples from countries across all continents. This dispersion reinforces the perception of remote auditing as a global phenomenon. However, few studies contextualize their findings based on the unique characteristics of each country, including its institutional, regulatory, technological, social, and cultural contexts. This lack of critical contextual analysis compromises the in-depth understanding of the national factors influencing remote auditing and affects the comparative analysis of results across countries. For future research, it is recommended that sample jurisdictions be contextualized to understand better the determining factors and practical implications of remote auditing in each country. Another suggestion is to conduct systematic comparisons between jurisdictions, as this represents a significant gap identified in the literature.

9 Conclusions

This study provides a comprehensive and critical review of remote auditing literature, highlighting its impacts and future perspectives across five thematic dimensions: the implementation of new technologies, changes in audit procedures and professional roles, implications for audit quality, effects on auditors' work-life balance, and emerging opportunities and risks.

The review revealed that remote auditing is fundamentally enabled by the integration of ICT, including CAATs and data analytics platforms. These technologies facilitate evidence-gathering, communication, and analytical procedures in the absence of physical presence. However, the effectiveness of technological integration varies significantly across audit firms due to disparities in digital literacy and infrastructure. While large firms are often better positioned to leverage advanced digital tools, smaller entities may have difficulties with adoption, raising concerns regarding equity and accessibility within the audit profession.

The transition to remote auditing has reshaped audit procedures and the professional profile of auditors. Greater autonomy, digital proficiency, and asynchronous communication tools have become essential. Despite demonstrated adaptability among auditors, the inability to replicate specific in-person audit procedures remains a challenge. Moreover, the change has implications for performance evaluation, supervision, and knowledge transfer, particularly affecting junior auditors who rely on face-to-face mentorship and in-person learning experiences.

Evidence on the impact of remote auditing on audit quality remains inconclusive. While some studies report that audit quality is maintained—or even enhanced—when supported by appropriate technologies and professional skepticism, others identify insufficiencies in physical evidence collection that may compromise audit quality. Much of the existing research is based on auditors' perceptions rather than objective performance data, which underscores the need for future studies to employ more rigorous empirical methods.

Remote auditing has had ambivalent effects on auditors' well-being. It has improved work-life balance for some by offering flexibility and autonomy. For others, however, it has exacerbated stress, reduced work-life boundaries, and led to social isolation. These effects may be influenced by factors such as gender, caregiving responsibilities, and socioeconomic status, but remain underexplored in the literature, suggesting a need for further research in this area.

Remote auditing presents a strategic opportunity for the profession to transition toward a hybrid model that combines technological efficiency with in-person procedures. Nonetheless, this evolution introduces several risks, including reduced team spirit and the sharing of knowledge, cybersecurity issues, and ethical concerns regarding data confidentiality and audit integrity. Addressing these challenges will require robust regulatory frameworks, effective digital ethics, and comprehensive security training to maintain audit quality and stakeholder trust.

This study contributes to theory and practice by providing a comprehensive and critical overview of the current state of remote auditing research. Also, identify gaps in the literature and propose directions for future research. It highlights the need to integrate technological, organizational, and human dimensions to understand the holistic impacts of remote auditing. Following this, it provides valuable insights for audit firms, professionals, and educators on successfully adapting to and implementing remote auditing practices. Ultimately, it underscores the importance of establishing regulatory frameworks and ethical guidelines in remote auditing.

Future studies should adopt mixed-method and longitudinal approaches to follow the evolution from traditional to remote auditing and to investigate the effectiveness and efficiency of remote audits. Furthermore, comparative studies across countries and firm sizes can help assess the influence of regulatory, technological, and organizational environments. It is essential to examine how technological infrastructure, cultural attitudes, and legal frameworks impact remote auditing in a particular region. To enhance the analytical depth of these studies, applying existing theoretical frameworks (e.g., socio-technical systems theory and institutional theory) is recommended. Additionally, research should focus on developing frameworks for remote auditing ethics, cybersecurity, and performance measurement, as well as exploring its long-term impact on auditor life, learning, and inclusion.

Acknowledgments. This work was supported by the UID Research Unit on Governance, Competitiveness and Public Policies (GOVCOPP), funded by national funds through FCT - Fundação para a Ciência e a Tecnologia.

References

Aivazi, R.: The Pros and Cons of Remote Auditing (2022). https://www.quality.org/knowledge/pros-and-cons-remote-auditing

Al-Aroud, Falah, S.: The impact of artificial intelligence technologies on audit evidence. Acad. Account. Financ. Stud. J. **24**(2), 1–11 (2020)

Baatwah, S.R., Al-Ansi, A.A., Almoataz, E.S., Salleh, Z.: Self-efficacy, remote audit proficiency, effort, and performance in the COVID-19 crisis: an auditor's perspective. Manag. Audit. J. **38**(6), 832–862 (2023). https://doi.org/10.1108/MAJ-05-2022-3570

Barretto, C.R., Drumond, G.M., Méxas, M.P.: Remote audit in the times of COVID-19: a successful process safety initiative. Braz. J. Oper. Prod. Manag. **19**(3), 1–17 (2022). https://doi.org/10.14488/BJOPM.2021.048

Byrnes, P., Criste, T., Stewart, T., Vasarhelyi, M.A.: Reimagining Auditing in a Wired World. AICPA, New York (2014)

Carlisle, M., Hamilton, L.: The role of communication mode in auditor-client interactions: insights from staff auditors. Curr. Issues Audit. **15**(1), A19–A37 (2021). https://doi.org/10.2308/CIIA-2020-007

Castka, P., Searcy, C., Fischer, S.: Technology-enhanced auditing in voluntary sustainability standards: the impact of COVID-19. Sustainability (Switzerland) **12**(11), 4740 (2020). https://doi.org/10.3390/su12114740

Castka, P., Zhao, X., Bremer, P., Wood, L.C., Mirosa, M.: Supplier audits during COVID-19: a process perspective on their transformation and implications for the future. Int. J. Logist. Manag. **33**(4), 1294–1314 (2022). https://doi.org/10.1108/IJLM-05-2021-0302

Castka, P., Searcy, C.: Audits and COVID-19: a paradigm shift in the making. Bus. Horiz. **66**(1), 1–5 (2023). https://doi.org/10.1016/j.bushor.2021.11.003

Cooper, L.A., Holderness, D.K., Sorensen, T.L., Wood, D.A.: Robotic process automation in public accounting. Account. Horiz. **33**(4), 15–35 (2019). https://doi.org/10.2308/acch-52466

CPAB Exchange: Enhancing Audit Quality Through Data Analytics (2019). https://cpab-ccrc.ca/docs/default-source/thought-leadershippublications/data_analytics_tlp_en_20190401.pdf?sfvrsn=ac6110c3_28

Eulerich, M., Wagener, M., Wood, D.A.: Evidence on internal audit quality from transitioning to remote audits because of COVID-19. J. Inf. Syst. **36**(3), 219–234 (2022). https://doi.org/10.2308/ISYS-2021-021

Farcane, N., et al.: Auditor´s perceptions on work adaptability in remote audit: A COVID-19 perspective. Economic Research-Ekonomska Istrazivanja **36**(1), 422–459 (2023). https://doi.org/10.1080/1331677X.2022.2077789

Financial Reporting Council: Audit Quality Thematic Review: the Use of Data Analytics in the Audit of Financial Statements (2017). https://www.frc.org.uk/getattachment/4fd19a18-1beb-4959-8737-ae2dca80af67/AQTR_AuditData-Analytics-Jan-2017.pdf

Gong, S., Ho, N., Jin, J.Y., Kanagaretnam, K.: Audit quality and COVID-19 restrictions. Manag. Audit. J. **37**(8), 1017–1037 (2022). https://doi.org/10.1108/MAJ-11-2021-3383

Handoko, B.L., Kurniawan, E.: Analysis of impact of competency, professionalism, and audit deadline on remote audit effectiveness. In: 6th International Conference on E-Business and Internet, pp. 90–95, Singapore (2022). https://doi.org/10.1145/3572647.3572661

Handoko, B.L., Thomas, G.N.: How mental health program moderate remote audit, computer literacy and CAATs impact on auditor work performance during Pandemic Covid-19. In: 6th International Conference on E-Commerce, E-Business and E-Government (ICEEG) (2022). https://doi.org/10.1145/3537693.3537754

Handoko, B.L., Lindawati, A.S.L., Muljo H.H.: Financial auditor efficiency and effectiveness in remote audit transition due to the pandemic of Covid-19. In: International Conference on Informatics, Multimedia, Cyber and Information System (ICIMCIS) (2023). https://doi.org/10.1109/ICIMCIS56303.2022.10017564

Handoko, B.L., Sardjono, C.T.: The impact of auditor's competency, audit risk, and professional skepticism to audit quality in fraud detection using remote audit. In: 8th International Conference on Industrial and Business Engineering, pp. 160–166 (2022). https://doi.org/10.1145/3568834.3568845

Huang, F., Vasarhelyi, M.A.: Applying robotic process automation (RPA) in auditing: a framework. Int. J. Account. Inf. Syst. **35**, 1–11 (2019). https://doi.org/10.1016/j.accinf.2019.100433

Huang, R.: Impact of the COVID-19 pandemic on audit quality: evidence from China. Appl. Econ. Lett. **32**(1), 67–72 (2023). https://doi.org/10.1080/13504851.2023.2257023

Issa, H., Sun, T., Vasarhelyi, M.A.: Research ideas for artificial intelligence in auditing: The formalization of audit and workforce supplementation. J. Emerg. Technol. Account. **13**(2), 1–20 (2016). https://doi.org/10.2308/jeta-10511

Jarva, H., Zeitler, T.: Implications of the COVID-19 pandemic on internal auditing: a field study. J. Appl. Acc. Res. **25**(2), 355–370 (2023). https://doi.org/10.1108/JAAR-12-2021-0333

Kokina, J., Davenport, T.H.: The emergence of artificial intelligence: how automation is changing auditing. J. Emerg. Technol. Account. **14**(1), 115–122 (2017). https://doi.org/10.2308/jeta-51730

Li, Y., Goel, S., Williams, K.: Impact of remote audit on audit quality, audit efficiency, and auditors' job satisfaction. Int. J. Audit. **27**(2–3), 130–149 (2023). https://doi.org/10.1111/ijau.12306

Lontsikh, P.A., Safonova, O.M., Koksharov, A.V., Lontsikh, N.P., Golovina, E.Y.: Remote audit improvement methods in system-oriented information and security analysis. In: International Conference on Quality Management, Transport and Information Security, Information Technologies, pp. 178–182. Yaroslavl, Russia (2021) https://doi.org/10.1109/ITQMIS53292.2021.9642900

Lorentzon, J.I., Fotoh, L.E., Mugwira, T.: Remote auditing and its impacts on auditors' work and work-life balance: auditors' perceptions and implications. Account. Res. J. **37**(1), 1–18 (2023). https://doi.org/10.1108/ARJ-06-2023-0158

Morris, L., Hoitash, R., Hoitash, U.: The effectiveness and efficiency of auditors' remote work during COVID-19. Audit. J. Pract. Theory **42**(4), 223–245 (2023). https://doi.org/10.2308/AJPT-2022-019

Mugabe, T., Ndlovu, J., van Zijl, W.: COVID-19 and electronic audit evidence gathering in South Africa. South. Afr. J. Account. Audit. Res. **24**(1), 97–115 (2022). https://doi.org/10.54483/sajaar.2022.24.1.7

Nickson, D., Siddons, S.: Remote Working: Linking People and Organizations. Routledge, London (2012)

Rahman, M.J., Ziru, A.: Clients' digitalization, audit firms' digital expertise, and audit quality: Evidence from China. Int. J. Account. Inf. Manag. **31**(2), 221–246 (2023). https://doi.org/10.1108/IJAIM-08-2022-0170

Ribeiro, M.C., Oliveira, J.: Auditoria contínua: um novo paradigma - parte I [continuous auditing: a new paradigm – part I]. Revista da Ordem dos Revisores Oficiais de Contas **104**, 40–47 (2024)

Salijeni, G., Samsonova-Taddei, A., Turley, S.: Understanding how big data technologies reconfigure the nature and organization of financial statement audits: a sociomaterial analysis. Eur. Account. Rev. **30**(3), 531–555 (2021). https://doi.org/10.1080/09638180.2021.1882320

Serag, A.A.E., Daoud, M.M.: Remote auditing: an alternative approach to face the internal audit challenges during the COVID-19 pandemic. Account. Thought J. **25**(2), 228–259 (2021). https://doi.org/10.21608/atasu.2021.181073

Sharma, N., Sharma, G., Joshi, M., Sharma, S.: Lessons from leveraging technology in auditing during COVID-19: an emerging economy perspective. Manag. Audit. J. **37**(7), 869–885 (2022). https://doi.org/10.1108/MAJ-07-2021-3267

Sian, S.: Remote audit: the challenges of re-creating the audit room during the Covid 19 pandemic. Account. Forum **48**(3), 506–535 (2022). https://doi.org/10.1080/01559982.2022.2151963

Sorensen, K., Ortegren, M.: The next best thing: Social presence and accountability's impact on auditor professional skepticism. J. Corp. Account. Financ. **32**(2), 39–51 (2021). https://doi.org/10.1002/jcaf.22484

Subhan, M., Suyanto, S.: Mediating role of work stressors between auditor knowledge-sharing activities and job satisfaction in Indonesian small audit firms. Probl. Perspect. Manag. **21**(3), 300–315 (2023). https://doi.org/10.21511/ppm.21(3).2023.24

Stuart, I.C., Olsen, C., Der, B.A.: Auditing during COVID-19: evidence from Ghana and Norway. J. Int. Account. Res. **23**(1), 199–221 (2024). https://doi.org/10.2308/JIAR-2022-037

Teeter, R.A., Alles, M.G., Vasarhelyi, M.A.: The remote audit. J. Emerg. Technol. Account. **7**(1), 73–88 (2010). https://doi.org/10.2308/jeta.2010.7.1.73

Vitali, S., Giuliani, M.: Emerging digital technologies and auditing firms: opportunities and challenges. Int. J. Account. Inf. Syst. **53**, 1–13 (2024). https://doi.org/10.1016/j.accinf.2024.100676

Innovation Networks in European Regions: Insights for Innovation Policies

Pedro Miguel Domingos Duarte de Oliveira[1,2,3],
Maria Manuela Santos Natário[4,5,6,7](✉), João Pedro Almeida Couto[8,9],
and Elisabeth Teixeira Pereira[10]

[1] Instituto Politécnico de Santarém, Santarém, Portugal
pedro.oliveira@esg.ipsantarem.pt
[2] DINÂMIA'CET-ISCTE– Instituto Universitário de Lisboa, Lisboa, Portugal
[3] CIEQV, Santarém, Portugal
[4] Instituto Politécnico da Guarda, Guarda, Portugal
m.natario@ipg.pt
[5] GOVCOPP, Universidade de Aveiro, Aveiro, Portugal
[6] CICF-IPCA, Instituto Politécnico do Cávado e do Ave, Barcelos, Portugal
[7] CITUR, Guarda, Portugal
[8] Universidade dos Açores, Centre of Applied Economics Studies of the Atlantic – University of the Azores, Ponta Delgada, Portugal
joao.pa.couto@uac.pt
[9] Advance/CSG, ISEG - Lisbon University, Lisbon, Portugal
[10] Universidade de Aveiro, GOVCOPP- Universidade de Aveiro, Aveiro, Portugal
melisa@ua.pt

Abstract. The aim of this study is to analyse innovation networks in European regions using the 2023 Regional Innovation Scoreboard (RIS). To study how European NUTS II regions ranked in terms of innovation networks, cluster analysis and the mean comparison test were applied to find out which factors are associated with the best network dynamics. Subsequently, multiple linear regression was used to realise that collaboration networks influence the regions' capacity for innovation. The results show that three clusters of regions with different network dynamics were identified: cluster 1, with 53 NUTS II, has strong network dynamics, cluster 2, with 88 NUTS II, has medium network dynamics and cluster 3, with 98 NUTS II, has weak network dynamics. The most dynamic clusters in collaboration/cooperation networks are associated with stronger performance in innovation outputs (Innovation Index, product and process innovators, business process innovators, PCT patent applications, trademark applications, and design applications), human-resources framework conditions (population with tertiary education, lifelong learning, digital skills, and IT specialists), and innovation impacts (employment impacts and sales impacts). It is also concluded that collaborative networks influence the innovation capacity of regions, particularly that Innovative SMEs collaborating with others and Public-private co-publications have a very positive effect on the innovation capacity of regions and that International scientific co-publications have no influence on this capacity.

Keywords: Innovation Networks · European Regions · Innovation · Innovation Capacity

L. Almeida et al. (Eds.): ICAFI 2025, ISEM 80, pp. 401–422, 2026.
https://doi.org/10.1007/978-3-032-16268-7_23

1 Introduction

Innovation networks have become essential for regional development and the economic sustainability of European cities, enabling knowledge flow and resource sharing that strengthens the innovation capacity of companies and regions alike [1–3]. These networks play a crucial role in creating innovation ecosystems and reducing uncertainties and costs related to innovation processes [4, 5]. Recent studies have highlighted specific drivers, particularly for small and medium-sized enterprises (SMEs), that vary across EU regions, emphasizing the importance of local dynamics in shaping innovation outcomes. [6] examine how tailored policies can strengthen innovation networks, supporting sustainable growth across diverse contexts, while [7] underscore the significance of collaborative networks for SME innovation in varied regional settings.

Additionally, recent research has explored how innovation networks support reindustrialization efforts, particularly in regions facing structural challenges. [8] illustrate how collaborative networks can foster regional growth by linking technological advancements to spatial development, reinforcing the role of these networks in sustaining innovation in resource-limited areas.

Despite the established benefits of innovation networks, significant gaps persist in understanding how these networks can be effectively integrated into diverse regional contexts. This gap is especially relevant in the interplay between innovation networks and regional public policies, given the disparities between highly innovative and less-developed regions in Europe. These disparities present challenges to cohesive and equitable development, highlighting the need for public policies that support collaborative networks in an adaptive and inclusive manner [9, 10]. [11] further emphasize the resilience of cluster policies during economic crises, reinforcing the importance of adaptive public policies in sustaining innovation networks, particularly in regions with limited resources.

This study aims to analyze innovation networks across European regions, using data from the *Regional Innovation Scoreboard 2023* [12]. Through a quantitative approach, the study identifies factors associated with robust network dynamics and examines how cooperation among regional actors enhances innovation capacity. This empirical analysis provides a strong basis for policy recommendations to improve regional innovation ecosystems.

Therefore, beyond its academic contribution, this study offers practical insights for developing public policies to promote cohesive and sustainable growth across European regions [13–15].

The chapter is organised as follows. After this introduction, Sect. 2 discusses the conceptual and empirical contributions on co-operation networks and innovation, with particular attention to regional disparities and the role of public policy. Section 3 presents the data, variables, and empirical strategy adopted in the study. Section 4 reports and discusses the results of the analysis of innovation networks across European NUTS II regions, drawing on the 2023 Regional Innovation Scoreboard. Finally, Sect. 5 concludes by summarising the main findings and outlining policy implications and avenues for future research.

2 Co-operation Networks and Innovation

The study of innovation networks is particularly associated with the approaches to: innovative milieu (developed in the 1980s by [16–18], among others); innovation systems introduced by [4] and [19]; innovation clusters [20, 21]; the triple helix model [22, 23]; and open innovation [24, 25].

From the perspective of the innovative environments of [17] and [26], innovation networks between the different actors in the environment energise and strengthen the environments, making them innovative [27]. Innovation networks enrich the environment and contribute to the growth of the environment's creative capacity, transforming it into an innovative environment. In turn, the innovative environment is a favourable context for the dynamisation of innovation networks. Innovation networks help generate a dynamic, localised process of collective learning.

The systemic approach to innovation emphasises the role of the interaction networks of different agents in producing innovation [4, 6, 28, 29]. Networks of institutions participate in the creation of innovation and facilitate the transfer of knowledge. [6] analyse the effects of knowledge networks on local performance and highlight collaborative networks in cities as a source of innovation. [30] link global innovation networks to the creation of regional innovation systems.

In innovation clusters, the focus is on interaction, technological dynamics and innovation [20]. In this sense, for [20] clusters are an instrument for analysing the configuration and dynamics of the networks of actors involved in the innovation process. The geographical concentration of organisations - such as suppliers, service providers, higher education institutions, and associations - contributes to the development of competitive advantages by facilitating the sharing of specialised knowledge and resources among cluster members, as argued by [21] and [31]. In fact, the study of innovation clusters is associated with innovation networks and interaction between the different players to create competitive advantages. Within the cluster, knowledge networks emerge that facilitate learning and knowledge transfer [32].

With regard to the triple helix model, the interaction networks between the 3 helices: government, industry and academia are fundamental to fueling the innovation process [23], facilitating the transfer of knowledge to industry. The interactions between the propellers, with an emphasis on strengthening the role of academia, contribute to the transition from an industrial society to a knowledge-based society [33].

The open innovation model also reinforces the role of internal and external interaction networks in the development of the innovation process [25]. Open innovation is a paradigm that assumes that companies can and should utilise external and internal ideas, and internal and external routes to market. The business model uses external and internal ideas to create value, while defining internal mechanisms to claim a share of that value [24]. The open innovation model facilitates the creation of external networks that help develop innovation dynamics.

At regional level, regional and external interaction networks between companies and organisations are important for boosting regional innovation, as they facilitate access to and use of financial support for innovation [34]. In this context, [6] emphasise the importance of public policies aimed at dealing with the uncertainty inherent in the process of opening up and supporting the cost of creating networks.

[7] also reinforce the need to develop policies that foster collaboration to maximise SME innovation in the variety of European regional contexts. The authors study the innovation drives of SMEs in European regions and emphasise the importance of collaboration as a central innovation activity to complement research and development and that it has a greater effect than R&D in SMEs.

Networks of interaction and cooperation between regions facilitate access to external knowledge and combat the isolation of these regions [35] and contribute to the development of innovation dynamics and the competitiveness of regions. Networks lead to the co-creation of innovative projects and processes with very interesting results [36] for both the parties involved and the regions.

According to [37] there is a general consensus that co-operation for innovation between different players in a region promotes growth in that region. Collaborative networks are part of the Network paradigm, which emphasises that regional innovation networks stimulate innovation and growth both from the perspective of companies and from a regional/territorial perspective.

3 Methodology

The aim of this study is to analyse innovation networks in European regions using the 2023 Regional Innovation Scoreboard (RIS).

After cleaning the 2023 Regional Innovation Scoreboard (RIS) database, removing countries, the European Union average and regions (NUTS II) that had no data or only zero values, 239 NUTS II regions remained (Appendix I). The RIS 2023 variables are shown in Table 1.

Table 1. RIS 2023 variables

Variables
Population with tertiary education
Lifelong learning
International scientific co-publications
Most-cited publications
Digital skills
R&D expenditures public sector
R&D expenditures business sector
Non-R&D innovation expenditures
Innovation expenditures per person employed
Information Technologies specialists
Product process innovators
Business process innovators

(continued)

Table 1. (*continued*)

Variables
Innovative SMEs collaborating with others
Public-private co-publications
Patent Cooperation Treaty (PCT) patent applications
Trademark applications
Design applications
Employment knowledge-intensive activities
Employment in innovative SMEs
Sales of new-to-market and new-to-firm innovations
Air emissions by fine particulates
Innovation index (2023)

Source: RIS database 2023 [12] (https://research-and-innovation.ec.europa.eu/statistics/performance-indicators/regional-innovation-scoreboard_en)

To study innovation networks in European regions, cluster analysis was applied (in SPSS) to group the regions in terms of innovation network performance and considering 3 variables: International scientific co-publications, Innovative SMEs collaborating with others and Public-private co-publications. Subsequently, an attempt was made to analyse the characteristics of these groupings and which factors were associated with the performance of innovation networks. To this end, the innovation outputs (namely Innovation Index 2023, Product process innovators Business process innovators PCT patent applications Trademark applications Design applications) from the framework conditions were considered: human resources (Population with tertiary education, Lifelong learning, Digital skills, IT specialists) and impacts on employment, sales, R&D and environmental sustainability (Employment knowledge-intensive activities, Employment in innovative SMEs, Sales of new-to-market and new-to-firm innovations, Most-cited publications, Air emissions by fine particulates). In this case, the Scheffe test was used to identify the characteristics of each cluster.

According to [38] "cluster analysis is an exploratory technique that detects homogeneous groups in the data, based on the differences presented by a set of variables" (p.575) and makes it possible to interpret the characteristics of each cluster (group) through associations between its elements.

Cluster analysis was carried out using a hierarchical approach followed by k-means partitioning, based on the variables international scientific co-publications, collaborations between innovative SMEs, and public-private co-publications. Following on from the analysis and in order to identify the characteristics of each group, to understand which factors differentiate them, the differences in means were analysed using the Scheffé test. The aim is to understand which factors are associated with the different network dynamics and the performance of co-operation networks.

The dendogram (Appendix II) and the elbow method were applied to identify the optimum number of clusters, while the F-statistic was used to validate the differences between clusters. The inter-group and intra-group variance was compared to ensure that the clusters reflect significant differences in the performance of regional innovation networks. In addition, the Scheffé test was applied for multiple comparisons, ensuring that the differences observed between the clusters were robust and statistically significant.

The Scheffé test was chosen for its robustness in multiple comparisons between groups, making it possible to guarantee that the differences between the clusters are reliable even in the presence of variability in cluster sizes. This test was applied to validate that the clusters identified differ significantly in terms of collaborations and co-publications, providing a solid basis for the study's conclusions.

Following on from the study, the linear regression method was also used to analyse the relationship between the regional innovation capacity (using PCT patent applications) and innovation networks (using Public-private co-publications, Innovative SMEs collaborating with others, International scientific co-publications).

Multiple linear regression was used to explore how different independent variables contribute to regional innovation capacity using patents. In order to avoid multicollinearity, the regression model was adjusted by excluding variables that were strongly correlated with each other.

Multiple linear regression was used to explore regional innovation capacity, represented by the number of patents registered as the dependent variable. This indicator was chosen because it directly reflects a region's innovation performance.

The variables were selected on the basis of literature reviews and the recommendations of the reviewers, with the aim of ensuring that each variable included had a clear and independent role in the model.

The variables selected are all directly relevant to measuring regional innovation capacity, without introducing redundancies:

- *International Scientific Co-publications*: this indicator was selected as it represents the capacity for cross-border cooperation and the level of engagement of regions in cutting-edge research networks. Studies such as [9] highlight that international scientific collaboration is associated with greater knowledge diffusion and strengthening the regional innovation base.
- *Innovative SME Collaborations*: Small and medium enterprises are considered fundamental drivers of innovation, particularly when they collaborate among themselves. Collaborations between SMEs allow the sharing of resources, knowledge, and risks, facilitating incremental innovation and adaptation to market changes. [39] argue that the presence of structured collaborations among SMEs is one of the most critical factors for regional competitiveness, especially in moderately innovative regions.
- *Public-Private Co-publications*: Interaction between the public and private sectors is essential to create environments favorable to innovation. This indicator measures the extent of collaboration between academic, public institutions, and companies, which is crucial for technology transfer and knowledge dissemination, as suggested by [40]. Regions with high levels of public-private co-publications are generally better prepared to face complex economic challenges, as they combine diverse capabilities and knowledge to solve practical problems.

4 Analysis of Innovation Networks in European NUTS II

To study how the European NUTS II regions were positioned in terms of innovation networks, cluster analysis was applied.

The application of cluster analysis (in SPSS) to study innovation networks in European regions revealed 3 homogeneous groups with regard to the 3 variables considered: International scientific co-publications, Innovative SMEs collaborating with others and Public-private co-publications (Table 2). As can be seen in Table 2, these variables behave differently in the 3 groups, as indicated by the F-Statistic tests, with significance levels of 0.000 and a p-error of less than 0.001. The level of significance is associated with a low probability of analysis rejection.

Table 2. ANOVA analysis for the identified clusters

ANOVA	Cluster		Error		F	Sig
	Sum of Square	Df	Mean Square	df		
International scientific co-publications	7.987	2	0.015	236	522.962	0.000
Innovative SMEs collaborating with others	3.399	2	0.034	236	100.972	0.000
Public-private co-publications	5.596	2	0.012	236	454.201	0.000

The identification of homogeneous groups of regions, considering the selected variables, can be seen in Table 3 and in Appendix III.

Table 3. Average of each variable in each of the clusters created and number of regions in each cluster

	Cluster 1	Cluster 2	Cluster 3
	N = 53	N = 88	N = 98
International scientific co-publications	0.835	0.376	0.154
Innovative SMEs collaborating with others	0.663	0.629	0.300
Public-private co-publications	0.891	0.588	0.326

Cluster 1, with 53 European regions (NUTS II), is the one with the best results in terms of co-operation networks, in the 3 variables selected, and is therefore also the one with strong network dynamics. Cluster 2, with 88 European regions (NUTS II), shows average behaviour in terms of collaborative networks, and has been called medium network dynamics. Cluster 3, with 98 European regions (NUTS II), shows the worst performance in terms of networks in the 3 selected variables and has been labelled weak network dynamics. Annex II shows the location in the clusters of the different European NUTS II.

Following on from the analysis and in order to identify the characteristics of each group, to understand which factors differentiate them, the differences in means were analysed using the Scheffe test. The aim is to understand which factors are associated with the different network dynamics and the performance of co-operation networks.

The factors studied were innovation activities in terms of innovation Outputs (innovation Index, Product process innovators, Business process innovators, PCT patent applications, Trademark applications and Design applications); framework conditions in Human resources (Population with tertiary education, Lifelong learning, Digital skills and Information Technologies specialists and impacts in terms (Employment impacts and Sales impacts).

Table 4 shows the difference in averages between the 3 clusters in terms of innovation outputs.

Table 4. Mean differences between groups: Outputs of Innovation

ANOVA						
	Sum of Square	Df	Mean Square	F	Sig	Obs
Innovation Index 2023	144902	2	72451	250.579	0.000	1 > 2 > 3
Product process innovators	4.401	2	2.201	55.594	0.000	1,2 > 3
Business process innovators	5.817	2	2.909	63.057	0.000	1,2 > 3
PCT patent applications	6.537	2	3.268	79.585	0.000	1 > 2 > 3
Trademark applications	1.795	2	0.897	21.143	0.000	1 > 2 > 3
Design applications	0.700	2	0.350	6.549	0.002	1,2 > 3

It can thus be seen that the two clusters with the best network dynamics are also those with the best averages in innovation activities in terms of innovation Outputs (innovation Index, Product process innovators, Business process innovators, PCT patent applications, Trademark applications and Design applications).

Table 5 summarises the results for the framework conditions in Human resources.

Table 5. Mean differences between groups: framework conditions in Human resources

ANOVA						
	Sum of Square	Df	Mean Square	F	Sig	Obs
Population with tertiary education	3.547	2	1.773	38.879	0.000	1 > 2 > 3
Lifelong learning	5.258	2	2.629	47.474	0.000	1 > 2 > 3
Digital skills	2.122	2	1.061	19.768	0.000	1 > 2 > 3
IT specialists	7.207	2	3.604	61.796	0.000	1 > 2 > 3

Table 5 shows that the cluster with the best performance in terms of network dynamics is the one with the best averages in the framework conditions in Human resources (Population with tertiary education, Lifelong learning, Digital skills and IT specialists).

Table 6 shows the results for innovation impacts.

Table 6. Mean differences between groups: impacts

ANOVA						
	Sum of Square	Df	Mean Square	F	Sig	Obs
Employment knowledge-intensive activities	3.558	2	1.779	36.856	0.000	1 > 2 > 3
Employment in innovative SMEs	5.317	2	2.659	86.890	0.000	1, 2 > 3
Sales of new-to-market and new-to-firm innovations	0.785	2	0.393	9.052	0.000	2 > 1 > 3
Most-cited publications	3.569	2	1.784	55.122	0.000	1 > 2 > 3
Air emissions by fine particulates	0.650	2	0.325	5.986	0.003	1 > 2,3

Analysing Table 6, it can be seen that in terms of impacts (Employment impacts and Sales impacts) the most dynamic clusters in networks are the ones that perform best in these dimensions.

Calculating the multiple linear regression using the Dependent Variable: regional innovation capacity (using PCT patent applications) and the variables innovation networks (Public-private co-publications, Innovative SMEs collaborating with others, International scientific co-publications), we obtain Tables 7, 8 and 9.

Table 7 Model Summary shows whether the model is adjusted. The correlation coefficient R measures the quality of the prediction of the Dependent Variable: regional innovation capacity. A value of R = 0.736 indicates a good level of consistency, so it is clear that there is a linear relationship between the variables under study. The R square value shows that the independent variables explain 54.2 per cent of the variation in the Dependent Variable: regional innovation capacity.

Table 7. Model Summary

Model	R	R Square	Adjusted R Square	Std. Error of the Estimate
1	.736[a]	0.542	0.536	0.177843

[a] Predictors: (Constant), International scientific co-publications, Innovative SMEs collaborating with others, Public-private co-publications

Table 8 is the analysis of variance table and shows that the independent variables (Public-private co-publications, Innovative SMEs collaborating with others, International scientific co-publications) significantly predict the dependent variable (regional innovation capacity) because p is less than 0.05. The regression model is therefore considered to adjust well to the data.

Table 8. ANOVA

ANOVA[a]						
Model		Sum of Squares	df	Mean Square	F	Sig
1	Regression	8.796	3	2.932	92.705	<,001[b]
	Residual	7.433	235	0.032		
	Total	16.229	238			

[a] Dependent Variable: PCT patent applications
[b] Predictors: (Constant), International scientific co-publications, Innovative SMEs collaborating with others, Public-private co-publications

The test performed by ANOVA is: H0: $b1 = 0$ vs. H1: $b1 \neq 0$ Since p-value $= 0 \Rightarrow$ for any significance level rejects H0.

Then $b1 \neq 0 \Rightarrow$ the linear regression has significance for any significance level.

Table 9 shows the regression coefficients.

Table 9. Regression coefficients

Coefficients[a]						
Model		Unstandardized Coefficients		Standardized Coefficients		
		B	Std. Error	Beta	t	Sig
1	(Constant)	– 0.081	0.037			0.031
	Public-private co-publications	1.283	0.126	1.196	10.185	0.000
	Innovative SMEs collaborating with others	0.101	0.054	0.096	1.854	0.065
	International scientific co-publications	– 0.545	0.104	– 0.599	– 5.256	0.000

[a] Dependent Variable: PCT patent applications

The non-standardised coefficients indicate how much the dependent variable varies with an independent variable, when the other independent variables are held constant. The non-standardised coefficient for Public-private co-publications is equal to 1.283, which means that for every increase of 1 (or 100) in Public-private co-publications there is an increase of 1.283 in regional innovation capacity. The t-test tests whether the coefficients are equal to zero. If p is less than 0.05, the coefficients are significantly different from zero.

The coefficients for Public-private co-publications and the variable International scientific co-publications are significantly different from zero, i.e. they show differences with statistical value.

A multiple linear regression was then fitted to predict the regional innovation capacity as a function of innovation networks (Public-private co-publications, Innovative SMEs collaborating with others, International scientific co-publications) and was obtained:

$$\begin{aligned} \textit{Regional Innovation Capacity} = &- 0,081 - 0,545\ \textit{International scientific co-publications} \\ &+ 1,283\ \textit{Public-private co publications} \end{aligned}$$

The variable Innovative SMEs collaborating with others is not statistically significant as it has a p-value of 0.065, which is greater than the 0.05 significance level.

The results of the multiple linear regression showed that public-private co-publications have a statistically significant impact on regional innovation capacity.

5 Conclusions

Innovation networks are fundamental both in promoting innovation in regions and in co-creation processes, and they foster the regions' capacity for innovation and therefore their competitiveness.

This study sought to understand how European regions (NUTS II) position themselves in terms of innovation networks and their influence on regional innovation dynamics.

The study concluded that 3 clusters of regions with different network dynamics were identified: cluster 1, with 53 NUTS II, has a strong network dynamic, cluster 2, with 88 NUTS II, has a medium network dynamic and cluster 3, with 98 NUTS II, has a weak network dynamic.

The most dynamic clusters in collaboration/cooperation networks are associated with innovation Outputs (innovation Index, Product process innovators, Business process innovators, PCT patent applications, Trademark applications and Design applications), framework conditions in Human resources (Population with tertiary education, Lifelong learning, Digital skills and IT specialists) and impacts in terms (Employment impacts and Sales impacts) and are the best performers in these dimensions.

It is also concluded that collaborative networks influence the innovation capacity of regions, particularly Public-private co-publications have a positive effect on the innovation capacity of regions and International scientific co-publications have a negative effect on the innovation capacity of regions and that Innovative SMEs collaborating with others have no influence on this capacity.

These results indicate that public policies should encouraging partnerships between the public and private sectors, especially in areas with low innovation performance. It is recommended that public investment in R&D also be geared towards facilitating these collaborations, thus ensuring a wider positive impact on regional innovation.

This study contributes to consolidating the importance of innovation networks in promoting the innovation capacity of regions and can serve as a tool for defining public policies, particularly those of the European Union, capable of boosting the dynamisation of networks and thus stimulating innovation in European regions.

The results reinforce the conclusions of [37] regarding sectoral networks and the relationship between networks and regional development with a focus on the conventional producer-supplier linkages, and also those of [30] regarding the role of global innovation networks to the creation of regional innovation systems. They also go in the direction of the effects of knowledge networks on the performance of regions in terms of innovation capacity achieved by [6].

The study's limitations include the fact that it only uses secondary data and the 2023 Regional Innovation Scoreboard (RIS) database, and that it is a static analysis. Despite these limitations, the results are consistent with those achieved by [30]. It is therefore suggested that future research include other elements to help explain the constitution of the different types of networks, in order to identify which ones, have the greatest impact on the regions' capacity for innovation.

As avenues for future research, we also suggest longitudinal studies to address the limitations of the static RIS 2023 dataset and delve deeper into the types of innovation networks that have the greatest impact on regional innovation capacity.

Acknowledgments. We gratefully acknowledge financial support from FCT- Foundation for Science and Technology (Portugal) through research grants - Advance/CSG, ISEG – Lisbon School of Economics and Management, - Centre of Applied Economics Studies of the Atlantic, School of Business and Economics of the University of the Azores, + Governance, Competitiveness and

Public Policies and Research Center on Accounting and Taxation (CICF) and Centre for Tourism Research, Development and Innovation (CITUR).

Appendix

Appendix I- European Regions NUTS II

EU27_2020	EU27	DEB2	Trier
BE	Belgium	DEB3	Rheinhessen-Pfalz
BE1	Région de Bruxelles-Capitale / Brussels Hoofdstedelijk Gewest	DEC	Saarland
BE2	Vlaams Gewest	DED2	Dresden
BE3	Région Wallonne	DED4	Chemnitz
BG	Bulgaria	DED5	Leipzig
BG31	Severozapaden	DEE	Sachsen-Anhalt
BG32	Severen tsentralen	DEF	Schleswig-Holstein
BG33	Severoiztochen	DEG	Thüringen
BG34	Yugoiztochen	EE	Estonia
BG41	Yugozapaden	IE	Ireland
BG42	Yuzhen tsentralen	IE04	Northern and Western
CZ	Czech Republic	IE05	Southern
CZ01	Praha	IE06	Eastern and Midland
CZ02	Strední Cechy	EL	Greece
CZ03	Jihozápad	EL3	Attiki
CZ04	Severozápad	EL41	Voreio Aigaio
CZ05	Severovýchod	EL42	Notio Aigaio
CZ06	Jihovýchod	EL43	Kriti
CZ07	Strední Morava	EL51	Anatoliki Makedonia, Thraki
CZ08	Moravskoslezsko	EL52	Kentriki Makedonia
DK	Denmark	EL53	Dytiki Makedonia
DK01	Hovedstaden	EL54	Ipeiros
DK02	Sjælland	EL61	Thessalia
DK03	Syddanmark	EL62	Ionia Nisia
DK04	Midtjylland	EL63	Dytiki Ellada
DK05	Nordjylland	EL64	Sterea Ellada
DE	Germany	EL65	Peloponnisos

(*continued*)

(*continued*)

EU27_2020	EU27	DEB2	Trier
DE11	Stuttgart	ES	Spain
DE12	Karlsruhe	ES11	Galicia
DE13	Freiburg	ES12	Principado de Asturias
DE14	Tübingen	ES13	Cantabria
DE21	Oberbayern	ES21	País Vasco
DE22	Niederbayern	ES22	Comunidad Foral de Navarra
DE23	Oberpfalz	ES23	La Rioja
DE24	Oberfranken	ES24	Aragón
DE25	Mittelfranken	ES3	Comunidad de Madrid
DE26	Unterfranken	ES41	Castilla y León
DE27	Schwaben	ES42	Castilla-la Mancha
DE3	Berlin	ES43	Extremadura
DE4	Brandenburg	ES51	Cataluña
DE5	Bremen	ES52	Comunidad Valenciana
DE6	Hamburg	ES53	Illes Balears
DE71	Darmstadt	ES61	Andalucía
DE72	Gießen	ES62	Región de Murcia
DE73	Kassel	ES63	Ciudad Autónoma de Ceuta
DE8	Mecklenburg-Vorpommern	ES64	Ciudad Autónoma de Melilla
DE91	Braunschweig	ES7	Canarias
DE92	Hannover	FR	France
DE93	Lüneburg	FR1	Île de France
DE94	Weser-Ems	FRB	Centre - Val de Loire
DEA1	Düsseldorf	FRC	Bourgogne - Franche-Comté
DEA2	Köln	FRD	Normandie
DEA3	Münster	FRE	Hauts-de-France
DEA4	Detmold	FRF	Grand Est
DEA5	Arnsberg	FRG	Pays de la Loire
DEB1	Koblenz	FRH	Bretagne
FRI	Nouvelle-Aquitaine	NL33	Zuid-Holland
FRJ	Occitanie	NL34	Zeeland
FRK	Auvergne - Rhône-Alpes	NL41	Noord-Brabant
FRL	Provence-Alpes-Côte d'Azur	NL42	Limburg
FRM	Corse	AT	Austria

(*continued*)

(*continued*)

EU27_2020	EU27	DEB2	Trier
FRY	Régions ultrapériphériques françaises	AT1	Ostösterreich
HR	Croatia	AT2	Südösterreich
HR02	Panonska Hrvatska	AT3	Westösterreich
HR03	Jadranska Hrvatska	PL	Poland
HR05	Grad Zagreb	PL21	Malopolskie
HR06	Sjeverna Hrvatska	PL22	Slaskie
IT	Italy	PL41	Wielkopolskie
ITC1	Piemonte	PL42	Zachodniopomorskie
ITC2	Valle d'Aosta/Vallée d'Aoste	PL43	Lubuskie
ITC3	Liguria	PL51	Dolnoslaskie
ITC4	Lombardia	PL52	Opolskie
ITH1	Provincia Autonoma Bolzano/Bozen	PL61	Kujawsko-Pomorskie
ITH2	Provincia Autonoma Trento	PL62	Warminsko-Mazurskie
ITH3	Veneto	PL63	Pomorskie
ITH4	Friuli-Venezia Giulia	PL71	Lódzkie
ITH5	Emilia-Romagna	PL72	Swietokrzyskie
ITI1	Toscana	PL81	Lubelskie
ITI2	Umbria	PL82	Podkarpackie
ITI3	Marche	PL84	Podlaskie
ITI4	Lazio	PL91	Warszawski stoleczny
ITF1	Abruzzo	PL92	Mazowiecki regionalny
ITF2	Molise	PT	Portugal
ITF3	Campania	PT11	Norte
ITF4	Puglia	PT15	Algarve
ITF5	Basilicata	PT16	Centro
ITF6	Calabria	PT17	Lisboa
ITG1	Sicilia	PT18	Alentejo
ITG2	Sardegna	PT2	Região Autónoma dos Açores
CY	Cyprus	PT3	Região Autónoma da Madeira
LV	Latvia	RO	Romania
LT	Lithuania	RO11	Nord-Vest
LT01	Sostinės regionas	RO12	Centru

(*continued*)

(*continued*)

EU27_2020	EU27	DEB2	Trier
LT02	Vidurio ir vakarų Lietuvos regionas	RO21	Nord-Est
LU	Luxembourg	RO22	Sud-Est
HU	Hungary	RO31	Sud - Muntenia
HU11	Budapest	RO32	Bucuresti - Ilfov
HU12	Pest	RO41	Sud-Vest Oltenia
HU21	Közép-Dunántúl	RO42	Vest
HU22	Nyugat-Dunántúl	SI	Slovenia
HU23	Dél-Dunántúl	SI03	Vzhodna Slovenija
HU31	Észak-Magyarország	SI04	Zahodna Slovenija
HU32	Észak-Alföld	SK	Slovakia
HU33	Dél-Alföld	SK01	Bratislavský kraj
MT	Malta	SK02	Západné Slovensko
NL	Netherlands	SK03	Stredné Slovensko
NL11	Groningen	SK04	Východné Slovensko
NL12	Friesland	FI	Finland
NL13	Drenthe	FI1B	Helsinki-Uusimaa
NL21	Overijssel	FI1C	Etelä-Suomi
NL22	Gelderland	FI19	Länsi-Suomi
NL23	Flevoland	FI1D	Pohjois- ja Itä-Suomi
NL31	Utrecht	FI2	Åland
NL32	Noord-Holland	SE	Sweden
SE11	Stockholm		
SE12	Östra Mellansverige		
SE21	Småland med öarna		
SE22	Sydsverige		
SE23	Västsverige		
SE31	Norra Mellansverige		
SE32	Mellersta Norrland		
SE33	Övre Norrland		
NO	Norway		
NO02	Innlandet		
NO06	Trøndelag		
NO07	Nord-Norge		

(*continued*)

(continued)

EU27_2020	EU27	DEB2	Trier
NO08	Oslo og Viken		
NO09	Agder og Sør-Østlandet		
NO0A	Vestlandet		
NO0B	Jan Mayen and Svalbard		
CH	Switzerland		
CH01	Région lémanique		
CH02	Espace Mittelland		
CH03	Nordwestschweiz		
CH04	Zürich		
CH05	Ostschweiz		
CH06	Zentralschweiz		
CH07	Ticino		
RS	Serbia		
RS11	Belgrade		
RS12	Vojvodina		
RS21	Šumadija and Western Serbia		
RS22	Southern and Eastern Serbia		
UK	United Kingdom		
UKC	North East		
UKD	North West		
UKE	Yorkshire and The Humber		
UKF	East Midlands		
UKG	West Midlands		
UKH	East of England		
UKI	London		
UKJ	South East		
UKK	South West		
UKL	Wales		
UKM	Scotland		
UKN	Northern Ireland		

Appendix II- Dendogram

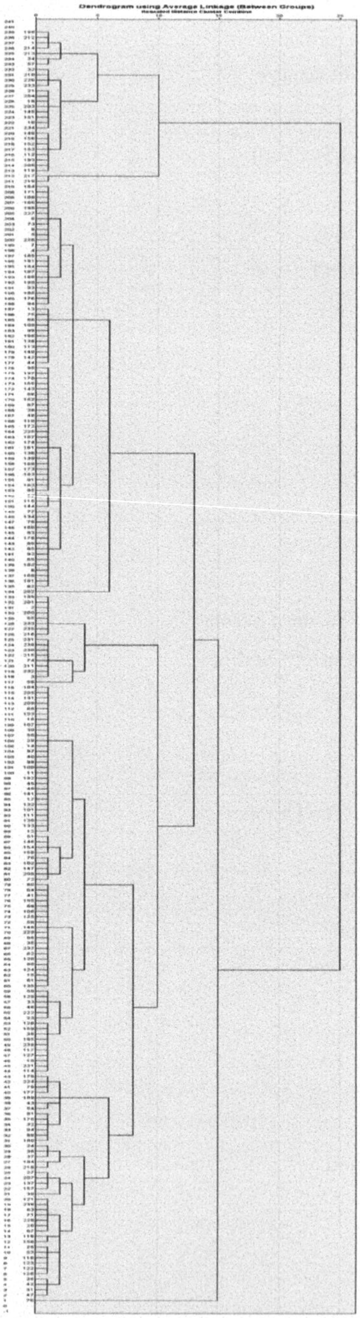

Appendix III- NUTS II by Cluster

Cluster 1	Cluster 2	Cluster 3	Cluster 1	Cluster 2	Cluster 3
AT1	AT3	BG31		ITC3	ITC2
AT2	BE2	BG32		ITC4	ITF4
BE1	BE3	BG33		ITF1	ITF6
CH01	CH05	BG34		ITF2	ITG1
CH02	CH06	BG41		ITF3	ITG2
CH03	CZ03	BG42		ITF5	LT02
CH04	CZ06	CZ02		ITH1	NL12
CH07	CZ07	CZ04		ITH3	NL34
CZ01	DE11	CZ05		ITH5	PL21
DE12	DE13	CZ08		ITI1	PL22
DE14	DE23	DE22		ITI2	PL41
DE21	DE24	DE73		ITI3	PL42
DE25	DE26	DE93		ITI4	PL43
DE3	DE27	DE94		LT01	PL51
DE5	DE4	DEA3		NL13	PL52
DE6	DE71	DEB2		NL21	PL61
DE72	DE8	EL41		NL23	PL62
DE91	DE92	EL42		NL41	PL63
DEA2	DEA1	EL62		NO02	PL71
DED2	DEA5	EL64		NO09	PL72
DED5	DEB1	ES11		PL91	PL81
DK01	DEB3	ES12		PT16	PL82
DK04	DEC	ES13		PT17	PL84
DK05	DED4	ES23		RS11	PL92
FI1B	DEE	ES24		SE21	PT11
FI1C	DEF	ES41		SE31	PT15
FI1D	DEG	ES42		SE32	PT18
HR05	DK02	ES43		UKC	PT2
HU11	DK03	ES52		UKD	PT3
IE06	EL3	ES53		UKE	RO11
ITH2	EL43	ES61		UKF	RO12
ITH4	EL51	ES62		UKG	RO21
NL11	EL52	ES63		UKK	RO22

(*continued*)

(*continued*)

Cluster 1	Cluster 2	Cluster 3	Cluster 1	Cluster 2	Cluster 3
NL22	EL53	ES64		UKL	RO31
NL31	EL54	ES7		UKN	RO32
NL32	EL61	FI2			RO41
NL33	EL63	FRB			RO42
NL42	EL65	FRC			RS12
NO06	ES21	FRD			RS21
NO07	ES22	FRE			RS22
NO08	ES3	FRF			SI03
NO0A	ES51	FRM			SK02
SE11	FI19	FRY			SK03
SE12	FR1	HR02			SK04
SE22	FRG	HR03	**53**	**88**	**98**
SE23	FRH	HR06			
SE33	FRI	HU12			
SI04	FRJ	HU21			
SK01	FRK	HU22			
UKH	FRL	HU23			
UKI	IE04	HU31			
UKJ	IE05	HU32			
UKM	ITC1	HU33			

References

1. Cooke, P.: Regional innovation systems, clusters, and the knowledge economy. Ind. Corp. Chang. **10**(4), 945–974 (2001)
2. Asheim, B., Gertler, M. S.: The geography of innovation: Regional innovation systems. In: Fagerberg, F., Mowery, D., Nelson, R. (eds.). The Oxford handbook of innovation (pp. 291–317). Oxford University Press (2005)
3. Moulaert, F., Sekia, F.: Territorial innovation models: A critical survey. Reg. Stud. **37**(3), 289–302 (2003)
4. Lundvall, B.A.: National systems of innovation: towards a theory of innovation and interactive learning. Pinter, London (1992)
5. Cooke, P.: Regional innovation systems: Origin of the species. International Journal of Technological Learning, Innovation and Development **1**(3), 393–409 (2008)
6. Bianchi, C., Galaso, P., Palomeque, S.: The trade-offs of brokerage in inter-city innovation networks. Reg. Stud. **57**(2), 225–238 (2023)
7. Hervás-Oliver, J., Parrilli, M.D., Rodríguez-Pose, A., Sempere-Ripoll, F.: The drivers of SME innovation in the regions of the EU. Res. Policy **50**(9), 104316 (2021)

8. Popescu, C., Covrig, M., Persu, M.: Encounters of technology and space in the context of reindustrialization in Romania. Applied Geography, 157, Article 103361 (2024)
9. Rodríguez-Pose, A.: Do institutions matter for regional development? Reg. Stud. **47**(7), 1034–1047 (2013)
10. Crescenzi, R., Rodríguez-Pose, A., Storper, M.: The territorial dynamics of innovation: A Europe-United States comparative analysis. Journal of Economic Geography **7**(6), 673–709 (2007)
11. Kiese, M., Ketels, C., Fornahl, D.: The resilience of clusters and cluster policies. Competitiveness Review: An International Business Journal **34**(3), 489–495 (2024)
12. European Commission: Regional innovation scoreboard 2023 – Databases Regional innovation indexes (2023)
13. Asheim, B.T., Boschma, R., Cooke, P.: Constructing regional advantage: Platform policies based on related variety and differentiated knowledge bases. Reg. Stud. **45**(7), 893–904 (2011)
14. Morgan, K.: The learning region: Institutions, innovation and regional renewal. Reg. Stud. **41**(S1), S147–S159 (2007)
15. Wolf, A.: Framework conditions for net-zero industry clusters in Europe. Intereconomics **59**(5), 300–306 (2024)
16. Aydalot, P.: Milieux innovateurs en Europe. Gremi (Groupe de recherche européen sur les milieux innovateurs), Paris (1986)
17. Maillat, D.: Territorial dynamic, innovative milieus and regional policy. Entrep. Reg. Dev. **7**(2), 157–165 (1995)
18. Camagni, R.: Innovation networks: spatial perspectives. Belhaven-Pinter, Bristol (1991)
19. Edquist, C.: Systems of innovation: Technologies. Institutions and Organizations. Pinter, London (1997)
20. Preissl, B., Solimene, L.: Dynamics of Clusters and Innovation: Beyond Systems and Networks. Physica-Verlag, Heidelberg (2003)
21. Engel, J. S., del-Palacio, I.: Global networks of clusters of innovation: Accelerating the innovation process. Business Horizons, 52(5), 493–503 (2009)
22. Leydesdorff, L., Etzkowitz, H.: The triple helix as a model for innovation studies. Science and Public Policy **25**(3), 195–203 (1998)
23. Leydesdorff, L.: The knowledge-based economy and the triple helix model. Ann. Rev. Inf. Sci. Technol. **44**, 367–417 (2010)
24. Chesbrough, H.: Open innovation: The new imperative for creating and profiting from technology. Harvard Business Press (2003)
25. Chesbrough, H., Bogers, M.: Explicating open innovation: Clarifying an emerging paradigm for understanding innovation, pp. 3–28. New Frontiers in Open Innovation. Oxford University Press, Oxford (2014)
26. Camagni, R., Maillat, D., Matteaccioli, A., Perrin, J.: Le paradigme du milieu innovateur dans l'économie spatiale contemporaine. Revue D'Economie Regionale Et Urbaine **3**, 425–428 (1999)
27. Fisher, M., Suarez-Villa, L., Steiner, M.: Innovation, networks and localities. Springer Science & Business Media (1999)
28. Freeman, C.: Networks of innovators: A synthesis of research issues. Res. Policy **20**(5), 499–514 (1991)
29. Lundvall, B. Å.: The COVID-19 Crisis, National Innovation Systems, and World Development. Science, Technology and Society, 09717218231178201 (2023)
30. Chaminade, C., Plechero, M.: Do regions make a difference? Regional innovation systems and global innovation networks in the ICT industry. Eur. Plan. Stud. **23**(2), 215–237 (2015)
31. Porter, M.: The competitive advantage of nations. Macmillan, London (1990)
32. Fioravanti, V.L.S., Stocker, F., Macau, F.: Knowledge transfer in technological innovation clusters. Innovation & Management Review **20**(1), 43–59 (2023)

33. Cai, Y., Etzkowitz, H.: Theorizing the Triple Helix model: Past, present, and future. Triple Helix **7**(2–3), 189–226 (2020)
34. Cooke, P.: Regional innovation systems: general findings and some new evidence from biotechnology clusters. J. Technol. Transf. **27**(1), 133–145 (2002)
35. Balland, P.A., Boschma, R.: Complementary interregional linkages and Smart Specialisation: An empirical study on European regions. Reg. Stud. **55**(6), 1059–1070 (2021)
36. Abbasiharofteh, M., Kogler, D.F., Lengyel, B.: Atypical combinations of technologies in regional co-inventor networks. Res. Policy **52**(10), 104886 (2023)
37. Sternberg, R.: Innovation networks and regional development - evidence from the European Regional Innovation Survey (ERIS): theoretical concepts, methodological approach, empirical basis and introduction to the theme issue. Eur. Plan. Stud. **8**(4), 389–407 (2000)
38. Pestana, M. H., Gageiro, J.: Análise de Dados para Ciências Sociais. A Complementaridade do SPSS. Edições Silabo, Lisboa (2014)
39. Tödtling, F., Trippl, M.: One size fits all?: Towards a differentiated regional innovation policy approach. Res. Policy **34**(8), 1203–1219 (2005)
40. Boschma, R.A.: Does geographical proximity favour innovation? Économie et institutions **6–7**, 111–128 (2005)

Investigating Technological Solutions in Internal Auditing: A Systematic Review of the Literature

Alexandre Rodrigues[1(✉)], Liliane Cristina Segura[1], and Rute Abreu[2]

[1] Universidade Presbiteriana Mackenzie, São Paulo, Brazil
{alexandre.rodrigues,lilianecristina.segura}@mackenzie.br
[2] Instituto Politécnico da Guarda, CICF-IPCA, CISeD-IPV, CITUR, Guarda, Portugal
ra@ipg.pt

Abstract. This research performs a systematic review of the literature on the integration of technological solutions in internal auditing, focusing on the application of tools such as data analysis (Analytics), Big Data, Robotic Process Automation (RPA) and other emerging technologies. The aim of the study is to investigate whether existing research presents practical technological solutions that can be applied or adapted in areas of internal audit in different organizations. The literature review covered research published between 2020 and 2024, with an emphasis on the accelerated digitization of audits due to the COVID-19 pandemic. The research initially identified 172 studies, of which 44 were selected after an in-depth analysis. The research concludes that, despite the increasing use of technological tools in internal audits, a large part of the publications focuses on theoretical approaches. Only 27% of the articles analyzed presented technological solutions that can be effectively replicated or adapted for internal auditing. In addition, most research focuses on the audit examination phase, indicating that technology has been more applied at this stage of the audit process. Finally, it is highlighted that studies tend to ignore the practical barriers to access to data by internal auditors, treating data analysis as a naturally accessible process, even though data constitutes the main raw material of auditing.

Keywords: Systematic review · Internal Audit · Technology · Technological Solutions · Digitalization of Auditing

1 Introduction

In an increasingly digitized corporate environment, the adoption of technological solutions by internal audit areas has become not only a trend, but a strategic necessity. This movement seeks to meet the growing demands for efficiency, agility and autonomy in conducting audit work. Understanding how technology integrates into audit processes is essential to assess its applicability, challenges, and potential.

Before properly investigating the recent use of technology by internal auditing based on a systematic review of the literature, it is initially appropriate, for a better understanding of technological applicability, to understand the auditing process in organizations.

L. Almeida et al. (Eds.): ICAFI 2025, ISEM 80, pp. 423–445, 2026.
https://doi.org/10.1007/978-3-032-16268-7_24

According to the Global Internal Audit Standards (IIA, 2024), a recently updated standard focused on the practice of internal auditing, the audit process encompasses four phases for conducting a given audit project: (1) Planning the work effectively, (2) Conducting the audit work, (3) Communicating the findings, and (4) Monitoring the action plans.

Based on the activities developed in these stages, internal auditors, based on information collected, analyses and evaluations carried out in the production of evidence, (a) provide evaluation services and carry out the findings, (b) determine the causes, consequences and materiality of the findings, (c) develop recommendations and assist management in determining action plans and, finally, (d) draw conclusions.

From a technological perspective, it is observed that all four phases of the audit process are, in practice, capable of being supported by technological solutions. Regardless of the stage in which the intensive use of technology is concentrated, its adoption presupposes the existence of a minimally structured process of development, implementation and evaluation of technological artifacts.

Although technological advancement is part of the daily life of organizations, including internal auditing, its uncontrolled use can generate more problems than solutions, especially at a time when companies are under permanent cyber-attacks.

Considering this challenge, the audit activity, whether on its own initiative or motivated by the needs of the organization, goes through an accelerated process of incorporation of technologies. However, this movement often remains restricted to the corporate environment, without wide dissemination, limiting the sharing of practices and solutions.

In view of this gap, the academic environment, as a proliferator of content and public knowledge, can play a relevant role in providing models that contribute to the development of technological solutions in internal auditing.

Despite the advancement in academic production in internal auditing, there is no clarity regarding the growth of studies that can be effectively replicated by professionals in the area, especially in technological issues. In this context, the following research question is formulated: do academic studies on internal auditing and technology offer practical, applicable and replicable technological solutions for use by internal audit executives, or are they mostly restricted to perception analysis?

In view of this question, it is proposed to adopt a systematic review of the literature as a means of understanding the relationship between information technology and internal audit functions and practices in organizations, including the adoption, or not, of methodologies for implementing technologies, such as Design Science Research (DSR).

In view of the situation exposed, the objective of the research is to investigate whether researchers, when dealing with the relationship between information technology and internal auditing, objectively present practical technological solutions that could be developed, replicated or adapted in internal audit areas of other organizations, to the detriment of studies that would focus on the perception of technology use by internal auditors.

In an increasingly digitalized corporate landscape, strengthening the autonomy and independence of internal audit is essential, aligning with the Global Internal Audit Standards (IIA, 2024), which emphasize obtaining accurate and reliable information directly from the auditors themselves.

In addition, the advancement of more accessible technologies, such as Python and its free libraries, generates a favorable environment for internal audit to develop its own solutions, without depending on the IT areas of organizations.

Thus, the main contribution of this article is to offer, both internal audit executives and academics, an in-depth understanding of the relationship between internal audit and technology, generating insights that enable practical technological solutions applicable to the areas of internal audit.

2 Literature Review

From the identification of the objective of this study, according to the introduction, this topic seeks to understand the researchers' approach to the integration of technological tools, such as Data Analytics, Big Data, Robotic Process Automation (RPA) and Artificial Intelligence, in internal audit practices. This understanding is fundamental to understanding the technological approach practiced for internal audit functions in organizations.

The relationship between internal audit and technology basically occurs through two approaches, sometimes using technology as an object of analysis by internal auditors, as occurs, for example, in cybersecurity audits, sometimes adopting technology as a work tool to achieve a certain objective foreseen in the work.

As an example of the first approach, Slapnicar et al. (2022) developed a study with the objective of evaluating the effectiveness of internal audit in cybersecurity work. Through the development of a cybersecurity audit index, composed of the planning, execution and report variables, and hypothesis testing, the auditors concluded that the planning and execution phases are strongly correlated with each other, but do not correlate strongly with the report, the product generated in the audit.

Another conclusion of Slapnicar et al. (2022), validating the hypothesis presented, is that the formulated index is positively associated with the maturity of the organization's cyber risk management process, however, contrary to the initial forecast, the index is not related to the probability of a cyber-attack, demonstrating a certain fragility of the audit as it does not significantly influence the adoption of preventive actions by the organization.

As for the second approach of linking technology to internal auditing, the object of this study, the use of technological solutions in audit work is not a recent phenomenon. At least twenty years ago, data analysis tools (Analytics) began to gain space in the internal audit environment (Andrade et al., 2023).

More recently, with the popularization of Python and R languages and free access to their respective libraries, Robotic Process Automation (RPA) and Artificial Intelligence solutions are beginning to emerge in the audit environment (Wassie & Lakatos, 2024).

Situating technology in relation to the phases of the audit process, Zhang et al. (2020) consider planning to be the most important phase of the audit work, playing a prominent role in the effectiveness of the work.

By assessing all existing risks, the auditor will be able to classify and prioritize them when conducting the audit itself. It is also in this phase that human and financial resources will be raised and, as a product, a detailed schedule will be prepared describing the activities necessary to start and complete the audit work.

As for the second phase, "conducting the audit work", in order to fully achieve the initial objectives of the work, the internal auditors must execute the work program detailed in the previous phase – planning.

Álvarez-Foronda et al. (2023) emphasize that, in the execution phase, auditors carry out examinations, capture evidence, analyze data and information to obtain the first conclusions.

In the relationship between auditing processes and information technology, the IIA itself, when detailing Standard 14.1, which deals with the collection of information for analysis and evaluation, highlights the importance of working with accurate information (IIA, 2024, p. 105):

"To conduct analysis and evaluation, internal auditors must collect information that is:

- Relevant – that are consistent with the objectives of the work, are within the scope of the work, and contribute to the development of the results of the work;
- Reliable – factual and current. Internal auditors use professional skepticism to assess whether information is reliable. Reliability is enhanced when the information is:
- Obtained directly from an internal auditor or from an independent source.
- Corroborated.
- Collected from a system with effective governance, risk management and control processes".

In addition to being mandatory for auditors to assess whether the information is relevant and reliable, the IIA (2024), with the update of the standard in 2024, textually provides that reliability is reinforced when the information is obtained directly by an internal auditor.

Without ignoring other sources of data collection, including with the help of professionals external to internal audit, the IIA takes an important step towards encouraging the auditor himself to have greater autonomy and independence in accessing data, however the institute does not detail, nor does it suggest, as the auditor, from an operational point of view, it should obtain and analyze the data independently.

As for the "communication of conclusions" phase, without specifying in detail how to communicate, the IIA (2024) emphasizes that internal auditors must necessarily issue a final communication when the work is completed, informing the results achieved to management. In practice, these communications can occur through structured meetings and/or through the issuance of conclusive reports.

Lenz & O'Regan (2024) point out that, as it is a key step, internal audit communication, in order to be considered effective, must be timely, frequent, and oriented with a focus on problem solving.

At the end of the audit cycle, in relation to the follow-up of action plans, the IIA (2024) reminds that auditors must constantly monitor the achievements of the suggested actions and, likewise, periodically communicate to management the status of the actions (overdue, in progress, completed), since, in essence, the objective of the auditor, and of the Administration itself, is that the actions are actually implemented.

Regarding the planning of the work, the use of a computerized project management system contributes to the management of audit projects, mapping the processes to be audited, measuring the associated costs, dimensioning the execution deadlines and detailing the work script to be developed by the auditors (Asare & Wright, 2019).

Regarding the conduct of internal audit work, Moura et al. (2023) point out that test automation, via RPA, is a trend in the evaluation of internal controls that are periodically and repeatedly tested.

The use of a data repository, via Data Lake, can provide greater autonomy to auditors in the evaluation processes via Analytics, in addition to enabling the application of Machine Learning techniques in the interpretation and grouping of data.

Regarding the phases of communication of results and monitoring of action plans, Johnson & Lee (2019) and Smith & Brown (2020) address the possibility of developing interactive dashboards demonstrating, in real time, the findings found and the pending issues of the audited areas in the execution of action plans.

Silva & Costa (2017) highlight the importance of internal audit management systems, which, in addition to functioning as a repository of projects, allow detailing the audits carried out over the years, and can also trigger communications and store the plans, the evidence found by the auditors and the evidence of execution of action plans by the organizational units.

Finally, as a methodological approach, DSR stands out in the integration between academic studies and professional practices, with emphasis on areas with intensive use of Information Technology, which would be the case of internal auditing itself.

According to Peffers et al. (2007), in the DSR methodology, technology is created to solve problems through an approach determined by principles, practices and procedures.

Based on four case studies, the authors considered the application of the DSR through the development of six steps, (1) identification and motivation of the problem, (2) definition of the objectives for a solution, (3) design and development, (4) demonstration, (5) evaluation, and (6) communication.

In summary, the literature review shows that, although there are advances in the discussion about the adoption of technologies in internal auditing, there are significant limitations regarding the offer of robust, permanent and effectively applicable technological solutions to the areas of internal auditing. This panorama reinforces the relevance of the research question detailed in the introduction and supports the need for the present investigation, which aims to understand the extent to which the current literature contributes to the construction of technological artifacts capable of strengthening autonomy, efficiency and independence of internal audit in the face of the challenges of an increasingly digitized corporate environment.

3 Method

According to Tranfield et al. (2003), there would be a reasonable consensus among researchers as to the methodological characteristics to be followed when adopting the systematic review of literature. In general terms, the authors reinforce the need to conduct the review through a protocol segregated into stages.

In the specific case of this study, based on the authors cited, a protocol was adopted that covers the following phases (stages):

Phase 1 – Review planning – Due to the accentuated importance of technology nowadays, it is necessary to explore how research deals with the technological issue in internal audit routines, with the main objective of understanding whether researchers investigate technological solutions developed and customized for internal auditors in organizations.

Phase 2 – Reviewing the research – The search for studies was carried out in the Web of Science database, a portal with a recognized reputation in the academic and professional environment. Because it addresses technological issues, in which there are constant changes, the analysis covered studies published in the last five years, between the years 2020 and 2024. In addition, in 2020, the global coronavirus pandemic occurred, in which organizations, including the internal audit areas, had an accelerated process of digitalization (Jarva & Zeitler, 2024).

Phase 3 – Identification of the research – Based on the survey of the Web of Science portal database, the filters were applied considering the relationship between technology and internal audit. To this end, from the fields "abstract", "keywords" and "title", the logical link between "internal audit" and "technology" was applied. As a result, 172 studies were initially identified (Table 1):

Table 1. Search result "technology and internal auditor"

Document Type	n	%
Article	153	89%
Other	19	11%
Total	172	100%

Source: data analysis (research)

Phase 4 – Selection of studies – Focusing on the articles, the 153 studies were analyzed for their applicability to this study. From the detailed analysis of the articles, which included the verification of the abstracts and, when inconclusive, there was a need to analyze the general content of the article, studies were excluded that, although dealing with technology, presented other aspects: (1) technology as an object of evaluation (audit) by the internal audit itself (example: audit of wi-fi networks), (2) technology for independent auditing without any application, Although adapted, for internal audit, (3) cyber security audit, (4) systems engineering audit, (5) secondary citation of technological issues for internal audit. From this analysis, 51 articles remained.

Subsequently, in a more in-depth evaluation of the 51 documents, there was a need to discard seven more studies, as three were duplicate, one dealt with tests of controls in the IT environment, and three with an eminent focus on independent auditing, with no possibility of application in internal auditing, including mentioning the applicable accounting standards. Therefore, for the final evaluation, 44 studies remained (Table 2):

Seeking to understand how internal audit relates to technological issues, in addition to responding to the stated objective, research was carried out focusing on technologies

Table 2. Results of the evaluation of the studies

Content analysis	N
Initial selection	172
Remnants (partial)	51
Remnants (final)	44

Source: data analysis (research)

supporting the audit work, whose results were segregated and presented in four groups: Analytics, Big Data, RPA and IT General.

Table 3. Searches segmented by technology

Technology	n	%
Analytics	22	50,0%
Big Data	4	9,1%
RPA	5	11,4%
IT General	13	29,5%
Total	44	100%

Source: data analysis (research)

From the grouping carried out, it is observed that the theme "data analysis" (Analytics) is the most recurrent, representing exactly half of the studies analyzed. It is inferred, therefore, that, when it comes to technology, data analysis is more present in the daily lives of internal auditors than other technologies, many of which are considered more recent from a technological point of view, such as Big Data and RPA.

Concluding the evaluation in the context of this study, according to the Tranfield protocol, Denyer & Smart (2003), phases (1) Evaluation of the quality of the study and (2) Data synthesis was executed, presented in conjunction with the next topic.

4 Results and Discussion

4.1 Analytics

To interpret the studies, the technologies listed above - Table 3 - were grouped as follows, considering the title, the year of publication, and the authors. In this first table - Table 4 - the articles that dealt with data analysis techniques (Analytics) linked to audit functions are listed.

Subsequently, seeking to understand the different approaches of the authors in relation to data analysis (*Analytics*) and internal audit, the contents of the articles that presented practical solutions, which could be adopted and replicated by the internal audit

Table 4. Two22 grouped articles on Analytics technology

Title	Author(s)	Year
An AI-based automated continuous compliance awareness framework (CoCAF) for procurement auditing	Wang, Zipperle, Becherer, Gottwalt & Zhang	2020
Correlates of the use of data analysis by the internal audit function: global evidence	Rakipi, Santis & D'Onza	2020
Implementation of Continuous Audit at Hering Company in Brazil	Codesso, de Freitas, Wang, de Carvalho & da Silva	2020
Internal audit in the digital age: opportunities, risks and challenges	Luís, Drogalas, Karagiorgos, & Tsikalakis	2020
Analysis and application of audit data based on the correlation analysis algorithm	Chen & Talha	2021
Assessing the Purpose of GAS Use: An Insight into Internal Audit Functions in Australia	Smidt, Steenkamp, Ahmi, van der Nest & Lube	2021
Use of data analysis tools to increase efficiency in the portfolio of internal audit projects	Andrade, Penha & da Silva	2021
Audit data analysis, machine learning, and thorough population testing	Huang, Won, Vasarhelyi & Yan	2022
Benford's Law for High-Volume Database Integrity Testing: An Internal Audit Case Study in a State-Owned Enterprise	Morales, Porporato, & Epelbaum	2022
Determinants of the adoption of computer-aided audit tools and techniques among internal audit units in Ghana	Ah, Onumah & Duho	2022
Discussion on the construction of computerization of corporate internal audit	Li	2022
Factors that play a role in improving the efficiency of continuous audits using computer-aided auditing techniques in companies, including healthcare facilities	Sagar & Ramanathan	2022
The impacts of the use of data analytics and the performance of consulting activities on the perceived quality of internal audit	Betti, DeSimone & Gray	2022
Visual text analysis in auditing: data analysis for release testing	Lee, Zhang, Liu & Vasarhelyi	2022
Data analysis implementation model as a tool for improving internal audit processes	Álvarez-Foronda, De-Pablos-Heredero, Rodríguez-Sánchez	2023

(*continued*)

Table 4. (*continued*)

Title	Author(s)	Year
Factors Determining Internal Auditors' Behavioral Intent to Use Computer-Aided Auditing Techniques: An Extension of the UTAUT Model and an Empirical Study	Almagrashi, Mujalli, Khan & Átia	2023
Governance of artificial intelligence applications in an enterprise audit through a fuzzy multiple-rule merger decision-making model	Hu, Chen, Hsu & Tzeng	2023
Navigating the Digital Landscape: Unraveling Technological, Organisational and Environmental Factors Affecting Digital Audit Readiness in the Malaysian Public Sector	Noor, Abd Razak, Jusoh & Hasan	2023
The impact of audit data analysis on audit quality and audit review continuity in Thailand	Ditkaew & Suttipun	2023
Assessing the impact of digitalization on the internal audit function	Karimallah & Drissi	2024
Enhancing Continuous Auditing with Large Language Models: AI-Assisted Real-Time	Li, Freitas, Lee & Vasarhelyi	2024
Examining the Critical Factors of Adopting Computer-Aided Auditing Tools and Techniques in the Post-COVID-19 Period: Internal Auditors' Perspective	Al-Okaily, Alqudah, Al-Qudah & Alkhwaldi	2024

Source: author

areas, were evaluated in more detail. The articles that presented these characteristics are analyzed below.

Relating Project *Portfolio Management* (PPM), which consists of a project management approach in organizations, to the implementation of continuous auditing in a leading company in the segment of paper and hygiene products, Andrade et al. (2021) sought to explore *Analytics* as a means of increasing the area's productivity, including with the aim of avoiding delays in the completion of audits. Due to the characteristics of the practical case – continuous audit – the authors classified this study as Action Research.

By measuring the results achieved in different audited processes, the authors objectively concluded that the massive adoption of the *Analytics platform* generated a time gain of 44% in relation to the traditional process, in addition to an 80% increase in audit types and coverage in 100% of the records examined (Andrade et al., 2021).

Seeking greater precision in the data analysis process, Li et al. (2024), based on the *Design Science Research* (DSR) methodology, sought to define a model aimed at performing auditing by crossing data from two databases of completely different origins and formats. Operationally, the model was developed from the creation of *Robotic Process Automation* (RPA), through the Python programming language, and making use of the LLM (*Large Language Model*), an artificial intelligence model based on the analysis of a large number of texts.

Unlike traditional data crossings, in which there is a need for fields with common content, including character extension, Li et al. (2024) were able to cross similar texts (not the same) between two databases, one of which is typical running text (information from the official gazette, newspaper that discriminates acts of the public administration), thus making it possible to identify possible inconsistencies to be investigated by the auditors. In addition to the time savings, the results showed 96% accuracy in the comparison, proving to be a technique of high applicability.

With the advancement of technologies related to auditing activities, Huang et al. (2022) presented an approach aimed at applying data analysis and machine learning in full-population tests, to the detriment of the classic model of using sampling in audit examinations.

By minimizing the use of sampling, Huang et al. (2022) prioritize, as a form of work, the "examination by exception", leaving in the background the traditional "transaction examination", in which the elements of the sample are selected, most of the time, in a random manner and, after being analyzed, inference is made from the results analyzed in the sample to the population that served as the basis for the audit.

In the conceptual model demonstrated, the technologies would be applied at different times, starting with the analysis of data from the application of filters by the auditors themselves, such as, for example, filtering purchase orders made without adequate approval (outside the limits of managerial authority).

Then, using the database with the results generated by the execution of filters, machine algorithms would be applied to identify suspicious transactions. One approach suggested by Huang et al. (2022) would be "*clustering*", which would allow the consolidation of transactions into groupings, differentiating groups with discrepant values, which would be the final transactions to be examined by the auditors.

Despite viewing it as a solution to sampling, Huang et al. (2022) consider, however, some associated difficulties that could be encountered:

- Possible increase in costs associated with data extraction, especially when there is a large volume of data involved.
- Existence of non-standardized and inconsistent data, especially when the scope of the audit covers several systems.
- Increased risk to information security when auditors themselves, without the appropriate knowledge, directly access the systems to obtain audit data.

As an evolutionary form of data analysis, Lee et al. (2022) developed the method called *Text Visual Analysis* (TVA), which consists mainly of applying data analysis to text fields in databases to identify existing patterns or anomalies.

Through the use of tools such as Orange and Tableau and the Python programming language, Lee et al. (2022) sought to identify, in accounting entries, certain patterns and apparent anomalies in textual information present in the "description" and "journal description" text fields of each entry.

The TVA starts with the import of data and foresees, in all, the completion of five steps: (1) text import; (2) pre-processing of text; (3) creation of new text attributes; (4) creation of visualizations; and (5) analyses.

With access to the databases of accounting entries, this action aims to standardize apparently different texts that mean, in practice, the same nature – such as, for example, the descriptions "Conference Meal" and "Conference – Meals". Near the end of the process, at the time of the "analysis" stage, the objective is to create *clusters* that make it possible to identify, according to the frequencies presented, patterns and apparent exceptions.

Finally, without detailing the audit "findings" in the application of the solution and without discarding traditional tests, Lee et al. (2022) considers that the article will contribute to the integration of the textual analysis of the data into the examinations usually performed by auditors.

Codesso et al. (2020) prepared a study focused on the implementation of continuous auditing in a Brazilian organization, with the purpose of making the use of technology frequent in the audit department. The study, focusing on practical issues, demonstrated the use of an *Analytics* tool, in this case the IDEA software, aimed at identifying any inconsistencies between internal invoice information against a set of tax data extracted directly from the government's tax and finance department.

As a result of the project, the authors concluded that internal auditors were able to review all electronic invoices in a timely manner, thus improving the quality of the information and, especially, the company's internal controls.

Mentioning the manual process of auditing purchases in Australian companies, Wang et al. (2020) proposed an automated artificial intelligence-based approach to conduct compliance audits in procurement processes.

The method, titled CoCAF, is defined as an automated method for comparing compliance audits of procurement records with procurement policies and requirements.

From the purchasing database, a series of information about the purchase is compared, such as purchase date, quantity, description, unit price, total price and exchange rate, with the supporting documentation (purchase order) converted into pdf. Later, with the extraction of the text through Optical Character Recognition (OCR) techniques, they are carried out through text mining and machine learning techniques.

Based on the results of the study, the authors conclude that the adoption of the technological solution allows auditing, in real time, 500 purchases in just five minutes with a level of accuracy in the exams of 95.6%, resulting in a high level of efficiency and quality in the procurement audit process.

Already Morales et al. (2022) proposed a study with the objective of evaluating the technical feasibility of using Benford's law to validate the integrity of large volumes of data.

Used to evaluate the behavior of a set of numbers, Benford's law assumes that certain digits appear more frequently than others in a given set of data.

To identify possible anomalies, the authors applied Benford's law, through the Audit Command Language (ACL) software, to several tables that make up the Oracle database of an integrated management system of an Argentine state-owned company.

Among the results, the authors concluded that the law is valid when there are large volumes of data in numerous tables, revealing that, for modules with less than 350 tables, there is low compliance with Benford's law.

Much authors researched, including Rakipi et al. (2021), Ditkaew & Suttipun (2023) and Álvarez-Foronda et al. (2023), agree that, by using *Analytics* tools, internal auditors improve the effectiveness and efficiency of internal audit areas in organizations.

Another point highlighted, related to the scope of audit examinations, is the gains that *Analytics* provides to internal audit, since auditors can evaluate the entire population of a given process, with no need to restrict tests on a sample basis (Ditkaew & Suttipun, 2023; Huang et al.; 2022).

The studies analyzed, regarding technological complexity, addressed basic issues, such as the performance of analyses through market software, to advanced forms, such as the combination of *Analytics* and Artificial Intelligence, especially *clustering analysis* (Nana, 2022; Huang et al.; 2022).

Regarding the objective proposed for this study, it is found that, of the 22 articles evaluated, only seven studies (32%) objectively presented technological solutions that could be developed and applied in areas of internal auditing.

It should be emphasized, however, that the technological solutions presented were not limited to demonstrating simple program codes executed in traditional Analytics tools. Huang et al. (2022), for example, presented auditing solutions that, by applying *Analytics* techniques, used *Machine Learning* as a means of identifying anomalies from the creation of *clustering*.

On the other hand, Lee et al. (2022) and Li et al. (2024) used RPA techniques to import and prepare data files to be processed using *Analytics techniques.*

In addition, from a methodological point of view, only Li et al. (2024) related the implementation of the technological solution to the adoption of the DSR.

4.2 Big Data

Given the comprehensive nature of Big Data, it is better to elucidate the concepts before relating it to audit work.

With the world increasingly connected, Big Data, as a business tool, has gained more and more prominence in the daily lives of organizations. Treating and analyzing large volumes of data has become, in the view of Nadikattu (2020), an extremely valued activity, including as one's own career options to be pursued.

Citing the strategic potential and renewal of business models, Wiener et al. (2019) conducted a study, through a critical review of the literature, which, based on types, dimensions, and implementation of Big Data business models, proposes to evaluate the state of the literature and identify any deficiencies.

Regarding the types, the authors observed that the focus fell on data users, not covering the external environment (suppliers, for example). Regarding size, the analyses focused on value creation, to the detriment of value capture. Regarding implementation, the review by the authors pointed out that the focus of the studies examined is focused on internal challenges, in addition to a superficial approach to the implementation process itself.

With a certain similarity in relation to the previous study, Fiorini et al. (2018) developed a study with the objective of improving the literature, evaluating the impact of Big Data on business growth, relating several theories that could be applied to understand the consequence on business from the application of this new technological concept.

Studying Big Data practices in auditing, et al. (2022) cite ten areas, under the scope of internal audit and external audit, that could benefit from the use of Big Data analytics: (i) accounts payable and accounts receivable, (ii) duplicate detection, (iii) sampling, (iv) data import/extraction and analysis of large data sets, (v) continuous auditing and monitoring, (vi) fraud detection and forensic auditing, (vii) card spend analysis, (viii) payrolls and time sheets, (ix) joins and comparisons, and (x) inventory audits.

Without prevailing any stage of the audit, the authors also mention that the three phases of the internal audit – planning, execution (fieldwork) and preparation of reports – could benefit from the use of Big Data analysis.

Party of the classical definition, Shabani et al. (2022) expanded on the concepts of Big Data such as:

A dataset that can be characterized by six V's: Velocity, Volume, Value, Variety, Veracity, and Variability. As the name suggests, the size of Big Data should be "big", which is measured as Volume. Velocity refers to the increasing speed at which new data is being generated and the corresponding requirement for that data to be ingested and analyzed in near real-time. Value measures the value of Big Data, meaning that insights generated from Big Data should lead to quantifiable improvements. Variety refers to the enormous diversity of data types. The reliability of the data attributes it to Veracity. The variability suggests that the way data is captured varies from time to time and from place to place. In addition, variability suggests that the interpretation of data may change depending on the context (Shabani et al., 2022, p. 3).

Regarding the challenges of incorporating Big Data analysis into auditing (Shabani et al., 2022) cites nine obstacles that must be overcome for the project to be considered effectively implemented, including Data Access, Data Compatibility, Data Relevance, Data Integrity and Veracity, Data Management, Data Confidentiality and Privacy, Regulations, and Training.

From a practical point of view, the authors mention difficulties associated with organizations that operate autonomously in the form of a business unit, in which data is stored in different systems, in addition to the auditors themselves often not having full access to the system's transactions. Regarding compatibility, they mention obstacles associated with the different formats in which data files are made available, and there is a need to normalize them to enable accurate comparisons.

In relation to "integrity and veracity", the difficulty of operationalization is associated with the data extraction process, in which companies may not have the most appropriate tools, in addition to the fact that it is plausible that those responsible for extraction, in situations of low autonomy of internal auditors, may manipulate the data before making it available to audit professionals.

Conceptualized technology, including its application in the audit environment, the articles that dealt with Big Data and internal auditing are listed below (Table 5).

Table 5. Four grouped articles on Big Data

Title	Author(s)	Year
A study of Big Data *Analytics* in Internal Audit	Shabani, Munir, & Mohanty	2022
Challenges of Big Data Transformation in the Inter-Business Audit Process in Malaysia	Sivalingam, Arumugam & Uzza	2022
Integration and development of enterprise internal audit and Big Data based on data mining technology	Nan	2022
Optimization of internal control of budget operations in public institutions based on the random forest algorithm	Jin	2024

Source: author

Seeking to understand the various forms of approaches of the authors in relation to Big Data and internal auditing, the contents of the articles that presented practical solutions were evaluated below.

Regarding the use of Big Data, Nan (2022) proposes the construction of a Big Data audit system that necessarily goes through the following layers: infrastructure, data, data analysis, and application.

Once the infrastructure is established, Nan (2022) demonstrates the optimization of data mining by improving the information grouping algorithm (*clustering*). Despite detailing the statistical results of the accuracy achieved in four data sets, the author does not demonstrate, with business data, how the *clustering* process enabled the internal auditor to find apparent anomalies in certain processes, such as billing, payments, human resources, among others.

In the context of internal auditing, data mining emerges as an efficient, convenient, and viable technique that would allow internal auditors to search for hidden information in large databases (Nan, 2022). Such a situation could generate audit "findings" without the internal auditor previously formulating a hypothesis of error or fraud.

As a result of the experiments carried out, Nan (2022) concludes, among other aspects, that the improvement of the algorithm, by generating greater *clustering capacity*, enables greater promotion of internal auditing in the organization.

From an operational point of view, Nan (2022) mentions that the data processing aimed at machine learning was carried out from an unstructured repository existing in the organization's own internal audit.

With a practical and functional approach, Jin (2024) focused the study on the creation of a Big Data solution aimed at the evaluation, by internal auditors, of the internal control of budgetary operations of public institutions.

Big Data, prepared from three phases (transformation, cleaning and verification of data), aimed to perform audits based on the analysis of internal control standards for administrative institutions. With the application of algorithms developed by the Python platform and supported by SQL, the auditors found anomalous situations in large databases. In the asset disposal audit, as an example, two "dubious" operations were highlighted, through *clustering*, in a population of 1419 elements.

The author concludes that the Big Data tool effectively helps internal auditors to find control failures, reducing the workload and increasing the effectiveness of audits.

From the analysis of the four articles, it is observed that the presented practical solutions that could be adopted by the audit areas.

Differentiating Big Data from *Analytics*, the articles with practical approaches initially dealt with the phases that precede the performance of data analysis, describing the processes of obtaining and storing data in their own repositories.

4.3 RPA (Robotic Process Automation)

Continuing the interpretation of the studies, in Table 6, the following are listed the articles that dealt with *Robotic Process Automation (RPA)* techniques related to audit functions.

Table 6. Five articles grouped on Robotic Process Automation technology

Title	Author(s)	Year
Participation in process automation in auditing: a framework and a demonstration	Zhang, Thomas & Vasarhelyi	2020
A framework for using robotic process automation for auditing tasks	Eulerich, Pawlowski, Waddoups & Wood	2022
The impact of implementing robotic process automation (RPA) on the internal audit function: a study from Malaysia	Xin, Asokan, & Balasingam	2022
Challenges of using RPA in auditing: a socio-technical systems approach	Dahabiyeh & Mowafi	2023
Doublethink in Government Accounting: Developing an RPA to Identify Inconsistencies in Financial Reporting	Schmall, Cardoso, Buchbinder, Vasarhelyi	2024

Source: author

Subsequently, seeking to understand the different approaches of the authors in relation to RPA and internal auditing, the contents of the articles with a practical bias were evaluated in more detail.

Exploring Assisted Automation in the Planning Phase of an Audit (Zhang & Vasarhelyi, 2020) propose a framework, entitled Attended Process Automation (APA), aimed at implementing assisted automation in audits, reinforcing the role of auditors in situations that deserve judgment.

The authors also comment that existing studies focus mostly on autonomous automation, with little intervention by the auditor, noting that many tasks, because they require judgment from auditors, are not easily automatable.

Assuming that technology is essentially created to solve problems (Zhang et al., 2020) adopts DSR as a methodology. In this specific case, the adoption of assisted automation would solve the problem associated with a type of audit that takes place several times a year – evaluation of the application of government resources – and, consequently, the work would be carried out with greater speed and precision.

In the application developed, from DSR, the technology must be used to solve organizational problems, which, in this case, would be the limitation of autonomous automation, since most audit activities are not structured and could not be automated. The adoption of assisted automation would solve this problem and allow the auditor to interact with the robot ("bot"), making it less "manual work" and more focused on "analytical thinking", thus achieving human-machine synergy.

For application purposes, the planning phase of the audit was chosen, considered by the authors to be the main stage of the process, which involves, for example, risk assessment, a task that, because it involves judgment, is not subject to automation. Still, at this stage there are many repetitive activities, such as documentation and file reference.

The study, classified as interventionist, concludes that the audit firm's decision to use a prototype validated the proposed framework, concluding that other audit teams could easily adapt the practice to the framework (Zhang et al., 2020).

Schmall et al. (2024) developed a *framework* to download financial reports from *websites*, extract the data, organize them, and check for possibles consistencies. Through program code and the use of utilities such as AWS Textract, auditors were able to (1) download files, (2) prepare files (organize them into folders and name them), (3) extract the data through OCR in pdf format, (4) organize the data in database format, (5) compare the data between reports, and (6) report the inconsistencies identified between the databases, these being the "findings" of the audit.

The sequence with the six steps presented was only possible to be executed after the creation of an RPA through the Python language.

From the point of view of the result, as a product of the audit work, the authors conclude that there are reports prepared with many inconsistencies, demonstrating weaknesses in the information presented. In relation to the form, it is concluded that the developed tool is effective.

Regarding the objective proposed for this study, it is observed that, of the five articles analyzed, two (40%) presented solutions that could be adapted to the areas of internal audit, with apparent productivity gains in the work (Zhang et al., 2020; Schmall et al., 2024).

Despite the apparent low frequency of articles that treated RPA as an exponent technology in internal auditing, several studies have dealt with the use of this technology to automate part of the data analysis, functioning as an accessory technology for conducting audit examinations (Lee et al., 2022; Li et al., 2024; Wassie & Lakatos, 2024).

Zhang et al. (2020) presented the RPA solution as assisted automation that would allow the audit area to improve the planning of work through interaction with the solution. Schmall et al. (2024), focusing on the testing phase, demonstrated a solution that made it possible to extract documents in unstructured format and transform them into a database that could be analyzed.

Regarding the methodological aspect, Zhang's article stands out (Thomas & Vasarhelyi, 2020), who emphasized the use of DSR in all phases of the project.

4.4 Technologies in General

Concluding the interpretation of the studies collected, in Table 7, the articles that dealt with technologies as heterogeneous foci related to internal auditing are listed below.

Table 7. Threee13 grouped articles on technologies in general

Title	Author(s)	Year
An examination of the factors associated with investing in internal audit technology	Garven & Scarlata	2020
State-of-the-art artificial intelligence in the context of internal audit	Couceiro, Pedrosa & Marini	2020
Using PowerApps to design file management for internal auditing of quality standards in the control group (QGM)	Hari, Welya & Refdi	2020
Critical auditor expertise for blockchain-based business environment	Tušek & Jezovita	2021
Sustainable technologies for the audit transition to a circular economy	Imoniana, Silva, Reginato, Slomski & Slomski	2021
Understanding the Function of Internal Audit in a Digitized Business Environment	Betti & Sarens	2021
Internal audit and the use of new technologies in higher education	Marques & Morais	2022
The Impact of Audit Technology on Task Outcome Auditing: Evidence for Technology-Based Auditing Techniques	Eulerich, Masli, Pickerd & Wood	2022
Digitalization of auditing and its consequences on the audit expectations gap: a critical perspective	Fotoh & Lorentzon	2023
A Research Study on the Adoption of Blockchain Technology by Iraqi Auditors	Majeed &Taha	2024
Artificial intelligence and the future of the internal audit function	Wassie & Lakatos	2024
Factors associated with the intention to use information technology in auditing in Iraq	Allami, Almaqtari, Al-Hattami & Sapra	2024
The Important Role of Information Technology and Internal Audit in Risk Management: Evidence from Greece	Drogalas, Pazarskis, Lazos & Golidopoulos	2024

Source: author

In general, studies on this topic focused on the perception of internal auditors, internal audit leaders, and audit committee members regarding the application of technologies in internal audit functions in organizations, especially extremely generic issues focusing on digitalization, artificial intelligence, and even blockchain.

Thus, and even due to the heterogeneous characteristic of this topic, it was to be expected that there would be a low rate of articles that dealt with applicable technological solutions, given also the scope of the previous topics that dealt with *Analytics*, often combined with *Machine Learning*, in addition to Big Data and RPA.

Still, thinking about a solution for managing work papers (Hari & Refdi 2021) designed an application, through Microsoft's PowerApps, to manage evidence for internal auditing in a mobile environment.

The solution developed for the university itself aimed to solve the problem of lack of standardization in the collection of evidence for the internal auditor, since, until the development of the application, the documentation was still available on paper and involved a series of collaborators, from professors to administrative employees.

According to Hari & Refdi (2021), the choice of PowerApps was due to the simplicity of use that the tool enables in the development of basic applications. In addition to being easy to use, Microsoft's solution is like Microsoft Office, including the logic used and the use of Excel formulas.

After carrying out the pre-operation test and the entry into production of the solution, Hari & Refdi (2021) considered that PowerApp successfully created the platform for file management for the internal audit, excepting, however, the absence of access restriction control, since the data editing function was enabled for all users (actors).

It should be noted that, contrary to the practical solutions demonstrated in the previous topics, Hari & Refdi (2021) developed a technological solution to organize the audit process, to the detriment of solutions that improved audit examinations.

Despite the restrictions inherent to this topic, of a more generic nature, only one article (7.7%), out of a total of thirteen, presented a technological solution that could be replicated in internal audit departments.

5 Final Considerations

Circulating between theoretical and practical issues, this study demonstrates, from different angles, how the internal auditor uses and visualizes technology as a work tool.

Through a systematic review of the literature, this study aimed to investigate whether researchers, when dealing with the relationship between information technology and internal auditing, objectively present practical technological solutions that could be developed, replicated or adapted in internal audit areas of other organizations, to the detriment of studies that would focus, for example, on the perception of technology use by internal auditors.

Based on the analysis of 44 studies, it was possible to verify that, even though there are different technology approaches to auditing, with different degrees of complexity, only 12 studies (27% of the total) addressed, in a practical way, the development and adoption of solutions that could be replicated, or at least subject to evaluation of adoption by the internal audit areas.

Analyzing by grouping, it can be seen that, except for Big Data (2 practical and 2 theoretical), in all segments there was a predominance of theoretical approaches - *Analytics* (7 practical and 15 theoretical), RPA (2 practical and 3 theoretical) and Technologies in General (1 practical and 12 theoretical).

Evaluating the present stages of the audit process, which begins with planning and ends with the monitoring of action plans, it is observed the prevalence of the application of technologies in the audit examination phase – 31 articles, including those of a theoretical nature, which represent 70% of the total, dealt with this phase of the audit.

Despite the benefits associated with using DSR as an approach to creating and evaluating technological solutions that solve practical problems, only two practical articles (17%) have broadly demonstrated the use of this important approach.

Still focusing on the studies that presented practical solutions, it is observed the prevalence of analysis techniques (*Analytics*), by internal auditors, in large volumes of data with the objective of identifying anomalies, many of which are combined with *Machine Learning* techniques, which enable, without previously labeling, the grouping of similar data into sets (*clusters*).

Regarding the theoretical approach, the authors sought to analyze the impact of technology around internal auditing itself or in the competencies of the internal auditor, without, however, in most cases, presenting the technologies involved in a very clear way.

Finally, from the aspect of access to data, studies, in general, treat the relationship between internal audit and data analysis as something that does not present natural barriers. They often emphasize the analysis process itself but fail to describe the practical difficulties that internal auditors have in autonomously obtaining a series of databases that will support the audit itself, neglecting that, essentially, data is the internal auditor's raw material. Consequently, no study directly addressed the use by the Internal Audit of its own data repository (structured or unstructured), such as the optimized use of a Data Lake aimed at the use of internal auditors.

Although comprehensive, this study has some limitations. First, the restriction to the databases used may have inhibited any relevant studies present in other academic sources. Another point is the time interval, which may not reflect emerging technological solutions or trends that have not yet been disclosed.

As significant contributions, by conducting a systematic review of the literature, the study provides an overview, structured and analytical view of how technology has been approached in internal audit research. In addition, the study demonstrates which technologies were most used, bringing a scenario that can help researchers and professionals identify gaps and define priorities for the development of technological artifacts in internal audits.

Based on the findings and the limitations pointed out, opportunities for future research arise. It is recommended to carry out empirical studies with solution-oriented methodologies, such as Design Science Research (DSR), to develop, test and validate technological artifacts that can be effectively applied in internal auditing. In addition, it is very important to deepen the investigation into the practical barriers that affect the adoption of technologies.

References

Allami, K.K.J., Almaqtari, F.A., Al-Hattami, H.M., Sapra, R.: Factors associated with the intention to use information technology in audit in Iraq. Inf. Discov. Deliv. **52**(2), 197–212 (2024). https://doi.org/10.1108/IDD-12-2022-0128

Almagrashi, A., Mujalli, A., Khan, T., Attia, O.: Factors determining internal auditors' behavioral intention to use computer-assisted auditing techniques: an extension of the UTAUT model and an empirical study. Future Bus. J. **9**(1), 74 (2023). https://doi.org/10.1186/s43093-023-00231-2

Al-Okaily, M., Alqudah, H.M., Al-Qudah, A.A., Alkhwaldi, A.F.: Examining the critical factors of computer-assisted audit tools and techniques adoption in the post-COVID-19 period: internal auditors perspective. VINE J. Inf. Knowl. Manag. Syst. **54**(5), 1062–1091 (2024). https://doi.org/10.1108/VJIKMS-12-2021-0311

Álvarez-Foronda, R., De-Pablos-Heredero, C., Rodríguez-Sánchez, J.-L.: Implementation model of data analytics as a tool for improving internal audit processes. Front. Psychol. **14**, 1140972 (2023). https://doi.org/10.3389/fpsyg.2023.1140972

Andrade, A., Penha, R., Ferreira da Silva, L.: Use of data analytics tools for increased efficiency in the internal audit project portfolio. J. Innov. Sustain. RISUS **12**(3), 138–149 (2021). https://doi.org/10.23925/2179-3565.2021v12i3p138-149

Andrade, A.C.R., Penha, R., Silva, L.F., Bizarrias, F.S., Kniess, C.T.: The adoption of technological tools in continuous audit projects. Revista de Administração da UFSM **16**(4), e5 (2023). https://doi.org/10.5902/1983465971587

Asare, S.K., Wright, A.M.: Auditors' risk assessments and sensitivity to the precision of audit evidence. J. Account. Res. **57**(3), 671–699 (2019)

Awuah, B., Onumah, J.M., Duho, K.C.T.: Determinants of adoption of computer-assisted audit tools and techniques among internal audit units in Ghana. Electron. J. Inf. Syst. Dev. Ctries. **88**(2), e12203 (2022). https://doi.org/10.1002/isd2.12203

Betti, N., Sarens, G.: Understanding the internal audit function in a digitalised business environment. J. Account. Organ. Chang. **17**(2), 197–216 (2021). https://doi.org/10.1108/JAOC-11-2019-0114

Chen, J., Talha, M.: Audit data analysis and application based on correlation analysis algorithm. Comput. Math. Methods Med. **2021**, 2059432 (2021). https://doi.org/10.1155/2021/2059432

Codesso, M., De Freitas, M.M., Wang, X., De Carvalho, A., Da Silva Filho, A.A.: Continuous audit implementation at Cia. Hering in Brazil. J. Emerg. Technol. Account. **17**(2), 103–118 (2020). https://doi.org/10.2308/JETA-2020-006

Couceiro, B., Pedrosa, I., Marini, A.: State of the art of artificial intelligence in internal audit context. In: 2020 15th Iberian Conference on Information Systems and Technologies (CISTI), pp. 1–7. IEEE, Seville (2020). https://doi.org/10.23919/CISTI49556.2020.9140863

Dahabiyeh, L., Mowafi, O.: Challenges of using RPA in auditing: a socio-technical systems approach. Int. Syst. Account. Financ. Manag. **30**(2), 76–86 (2023)

Ditkaew, K., Suttipun, M.: The impact of audit data analytics on audit quality and audit review continuity in Thailand. Asian J. Account. Res. **8**(3), 269–278 (2023). https://doi.org/10.1108/AJAR-04-2022-0114

Drogalas, G., Pazarskis, M., Lazos, G., Golidopoulos, K.: The important role of information technology and internal auditing in risk management: evidence from Greece. J. Oper. Risk (2024). https://doi.org/10.21314/JOP.2023.012

Eulerich, M., Masli, A., Pickerd, J., Wood, D.A.: The impact of audit technology on audit task outcomes: evidence for technology-based audit techniques. Contemp. Account. Res. **40**(2), 981–1012 (2023). https://doi.org/10.1111/1911-3846.12847

Eulerich, M., Pawlowski, J., Waddoups, N.J., Wood, D.A.: A framework for using robotic process automation for audit tasks. Contemp. Account. Res. **39**(1), 691–720 (2022). https://doi.org/10.1111/1911-3846.12723

Fiorini, P.C., Seles, B.M.R.P., Jabbour, C.J.C., Mariano, E.B., Jabbour, A.B.L.S.: Management theory and big data literature: from a review to a research agenda. Int. J. Inf. Manag. **43**, 112–129 (2018). https://doi.org/10.1016/j.ijinfomgt.2018.07.003

Fotoh, L.E., Lorentzon, J.I.: Audit digitalization and its consequences on the audit expectation gap: a critical perspective. Account. Horiz. **37**(1), 43–69 (2023). https://doi.org/10.2308/HORIZONS-2021-027

Garven, S., Scarlata, A.: An examination of factors associated with investment in internal auditing technology. Manag. Audit. J. **35**(7), 955–978 (2020). https://doi.org/10.1108/MAJ-06-2019-2321

Hari, P.D., Refdi, C.W.: The use of PowerApps to design archive management for internal quality standard audit in Quality Control Group (GKM) of Department of Agricultural Product Technology, Faculty of Agricultural Technology, Andalas University. In: Proceedings of the 5th International Conference on Engineering, Technology and Applied Science (ICETAS) (2021). https://doi.org/10.2991/assehr.k.210202.073

Hu, K.-H., Chen, F.-H., Hsu, M.-F., Tzeng, G.-H.: Governance of artificial intelligence applications in a business audit via a fusion fuzzy multiple rule-based decision-making model. Financ. Innov. **9**(1), 117 (2023). https://doi.org/10.1186/s40854-022-00436-4

Huang, F., No, W.G., Vasarhelyi, M.A., Yan, Z.: Audit data analytics, machine learning, and full population testing. J. Financ. Data Sci. **8**, 138–144 (2022). https://doi.org/10.1016/j.jfds.2022.05.002

Imoniana, J.O., Silva, W.L., Reginato, L., Slomski, V., Slomski, V.G.: Sustainable technologies for the transition of auditing towards a circular economy. Sustainability **13**(1), 218 (2020). https://doi.org/10.3390/su13010218

Institute of Internal Auditors (IIA): The IIA's Three Lines Model. IIA, Altamonte Springs (2020)

Institute of Internal Auditors (IIA): Normas Globais de Auditoria Interna. IIA, Altamonte Springs (2024)

Jicang, W.: Challenges and responses to internal auditing of public institution in the new situation. Account. Corp. Manag. **5**(8), 63 (2023). https://doi.org/10.23977/acccm.2023.050810

Jin, Y.: Optimization of internal control for budget operations in public institutions based on random forest algorithm. Scalable Computing: Practice and Experience **25**(3), 1863–1871 (2024). https://doi.org/10.12694/scpe.v25i3.2775

Johnson, K., Lee, C.: Real-time dashboards in internal audit: enhancing communication and monitoring. Int. Audit J. **25**(1), 103–120 (2019)

Karimallah, K., Drissi, H.: Assessing the impact of digitalization on internal auditing function. Int. J. Adv. Comput. Sci. Appl. **15**(6), 864 (2024)

Kaya, I., Akbulut, D.H., Ozoner, K.: Big data analytics in internal audit. PressAcademia Procedia **7**, 260–262 (2018). https://doi.org/10.17261/Pressacademia.2018.893

Lee, H., Zhang, L., Liu, Q., Vasarhelyi, M.: Text visual analysis in auditing: data analytics for journal entries testing. Int. J. Account. Inf. Syst. **46**, 100571 (2022). https://doi.org/10.1016/j.accinf.2022.100571

Lenz, R., O'Regan, D.J.: The global internal audit standards – old wine in new bottles? EDPACS **69**(3), 1–28 (2024). https://doi.org/10.1080/07366981.2024.2322835

Li, H., De Freitas, M.M., Lee, H., Vasarhelyi, M.: Enhancing continuous auditing with large language models: AI-assisted real-time accounting information cross-verification. SSRN Electron. J. (2024). https://doi.org/10.2139/ssrn.4692960

Li, H., Vasarhelyi, M.A.: Applying large language models in accounting: a comparative analysis of different methodologies and off-the-shelf examples. J. Emerg. Technol. Account. **21**(2), 133–152 (2024). https://doi.org/10.2308/JETA-2023-065

Li, S.: Discussion on the construction of enterprise internal audit informatization. J. Adv. Transp. **2022**, 4700817 (2022). https://doi.org/10.1155/2022/4700817

Lois, P., Drogalas, G., Karagiorgos, A., Tsikalakis, K.: Internal audits in the digital era: opportunities, risks and challenges. EuroMed J. Bus. **15**(2), 205–217 (2020). https://doi.org/10.1108/EMJB-07-2019-0097

Majeed, R.H., Taha, A.A.D.: A survey study of Iraqi auditors' adoption of blockchain technology. Asian Rev. Account. **32**(3), 521–546 (2024). https://doi.org/10.1108/ARA-01-2023-0015

Marques, D., Morais, G.: Auditoria interna e o uso de novas tecnologias no ensino superior. In: Conferência Ibérica sobre Sistemas e Tecnologias de Informação (CISTI), pp. 1–8. Madrid (2022)

Morales, H.R., Porporato, M., Epelbaum, N.: Benford's law for integrity tests of high-volume databases: a case study of internal audit in a state-owned enterprise. J. Econ. Financ. Admin. Sci. **27**(53), 154–174 (2022). https://doi.org/10.1108/JEFAS-07-2021-0113

Moura, D.S., Martins, G.A., Pimentel, R.M.: Automação e tendências em auditoria interna: o uso de RPA na avaliação de controles internos. Revista Brasileira de Auditoria Interna **19**(2), 45–60 (2023)

Nadikattu, R.R.: Research on data science, data analytics and big data. Int. J. Eng. Sci. Technol. **9**(5), 99–105 (2020)

Nan, N.: Integration and development of enterprise internal audit and big data based on data mining technology. Comput. Intell. Neurosci. **2022**, 8138046 (2022). https://doi.org/10.1155/2022/8138046

Peffers, K., Tuunanen, T., Rothenberger, M.A., Chatterjee, S.: A design science research methodology for information systems research. J. Manag. Inf. Syst. **24**(3), 45–77 (2007)

Rakipi, R., De Santis, F., D'Onza, G.: Correlates of the internal audit function's use of data analytics in the big data era: global evidence. J. Int. Account. Audit. Tax. (2023)

Saka, H., Krahel, J.P., Vasarhelyi, M.A.: The impact of emerging technologies on internal auditing. J. Inf. Syst. **34**(3), 129–152 (2020). https://doi.org/10.2308/isys-18-053

Schmitz, J., Leoni, G.: Accounting and auditing at the time of blockchain technology: a research agenda. Aust. Account. Rev. **29**(2), 331–342 (2019)

Seles, B.M.R.P., Jabbour, C.J.C., Fiorini, P.C., et al.: Business opportunities and challenges in adopting circular economy: a theory-driven analysis and review of the empirical literature. J. Clean. Prod. **263**, 121865 (2020)

Sidorov, A., Skachkova, T., Kulik, S., et al.: Internal audit of credit risk assessment and monitoring in the digital transformation era. Sustainability **15**, 8495 (2023)

Silva, A.P., Costa, F.R.: Gestão de auditoria interna: repositórios e sistemas integrados para auditorias. Gest. Controle **12**(3), 120–140 (2017)

Silvola, H., et al.: The role of internal audit in assurance provision and management control. Eur. Account. Rev. (2024)

Slapnicar, S., Loncar, M., Cadez, S.: Enterprise risk management: bridging risk and performance in management control systems. Br. Account. Rev. **54**(5), 101104 (2022)

Steinbart, P.J., Raschke, R.L., Gal, G., Dilla, W.N.: The influence of a good relationship between the internal audit and information security functions on information security outcomes. Acc. Organ. Soc. **71**, 15–29 (2018)

Supriadi, P.D., Resmi, S.: The role of the internal audit in fraud prevention. J. Econ. Sustain. Dev. **14**(2), 8–16 (2023). https://doi.org/10.7176/JESD/14-2-02

Susanti, A., Hardiningsih, P., Udin, U., Rachmawati, S., Haryono, S.: Internal audit function role in corporate governance: the impact of using artificial intelligence and data analytics on fraud prevention. Sustainability **14**(13), 7980 (2022). https://doi.org/10.3390/su14137980

Tajudin, T., Awaluddin, M., Erawati, T.: The effect of big data analytics, competence and internal auditor independence on the quality of audit results in the Indonesian banking sector. Int. J. Res. Bus. Soc. Sci. **12**(2), 415–423 (2023)
Tao, Y., Zhao, X., Zhang, W.: Application of intelligent audit based on blockchain and edge computing technology in the enterprise accounting information system. Wirel. Commun. Mob. Comput. **2022**, 6538782 (2022)
Taufik, T.: The role of internal audit in fraud prevention: a case study at PT. Angkasa Pura II. J. Account. Financ. Audit. Stud. **8**(4), 168–182 (2022)
Tuo, J., Hu, W., Zhu, Q., Liao, Z.: The role of internal audit functions in the digital transformation process: evidence from China. J. Int. Account. Res. (2024)
Tzur-David, S., Gilad, T., Zahavi, Y.: A framework for data-driven audits: impact of big data on the audit profession. Int. J. Account. Inf. Syst. **43**, 100556 (2021). https://doi.org/10.1016/j.accinf.2021.100556
Tzur-David, S., Zahavi, Y.: The impact of technology on the audit process. J. Emerg. Technol. Account. **18**(2), 99–122 (2021). https://doi.org/10.2308/JETA-19-019
Uyar, A., Gungormus, A.H., Kuzey, C.: Digital transformation in business organizations: the role of internal audit in the digital transformation process. Sustainability **15**(1), 236 (2023). https://doi.org/10.3390/su15010236
Vasarhelyi, M.A., Warren, J.D., Jr., Teeter, R.A.: Audit automation: past, present, and future. J. Inf. Syst. **29**(1), 5–20 (2015)
Wassie, F.A., Lakatos, L.P.: Artificial intelligence and the future of the internal audit function. Humanit. Soc. Sci. Commun. **11**(1), 386 (2024). https://doi.org/10.1057/s41599-024-02905-w
Weber, R.: The rhetoric of positivism versus interpretivism: a personal view. MIS Q. **28**(1), iii–xii (2004)
Wiener, M., Remus, U., Heumann, J., Mähring, M.: The effective use of control configurations in IT project risk management: a resource-based view. MIS Q. **40**(3), 741–774 (2016)
Williams, L., Wood, D.A.: The effect of IT controls on financial reporting quality. J. Inf. Syst. **37**(1), 73–92 (2023)
Xu, W.: Enhancing internal audit through big data analytics: a case study of a large Chinese manufacturing company. J. Account. Org. Change (2024)

Large Language Models in Accounting: An Exploratory Case Study

Miguel Resende(✉)

Higher Institute of Accounting and Administration, University of Aveiro, 3810-193 Aveiro, Portugal
miguelresende@ua.pt

Abstract. This study investigates the use of Large Language Models (LLMs) in accounting by focusing on their potential to automate financial analysis tasks. The primary goal is to demonstrate how LLMs can streamline accounting processes by reducing the time and complexity involved in analyzing financial data and generating accessible financial reports. This research follows the Design Science Research (DSR) methodology proposed by [1] and involves the development of a Generative Pre-trained Transformer (GPT) designed specifically for performing vertical analysis of the Profit and Loss statement (P&L). The chatbot's performance was benchmarked against traditional methods to assess its efficiency and interpretative improvements. The results show a significant time reduction in the vertical analysis, from 12 min to just 47 s, a 93% decrease. In addition, GPT enhances the accessibility of financial insights for users without advanced accounting expertise, such as managers and small business owners. Limitations include dependency on structured data formats, token and time constraints in large documents, and challenges in user adoption. Despite these challenges, this study illustrates the practical viability of LLMs in accounting contexts and shows the benefits of advanced prompting strategies, including Chain-of-Thought (CoT) reasoning, in improving analysis reliability. This study suggests that LLM-based tools can be extended to other accounting areas, such as horizontal and industry-level comparisons.

Keywords: Large Language Models · Accounting · Artificial Intelligence · Automation · Chain-of-Thought

1 Introduction

Technological advancements have significantly transformed the accounting paradigm, particularly with the advent of Artificial Intelligence (AI) and Large Language Models (LLMs). These technologies facilitate the automation of routine tasks and redefine the role of accountants, thus necessitating the acquisition of new analytical and strategic competencies [2]. The integration of LLMs in accounting has markedly enhanced operational efficiency, reshaping professionals' interpretation of financial information and requiring greater adaptation to emerging technological demands [3]. LLMs possess the capability to automate processes such as bank reconciliation, data analysis, forecasting, and preparation of corresponding reports [4]. However, their implementation poses

L. Almeida et al. (Eds.): ICAFI 2025, ISEM 80, pp. 446–460, 2026.
https://doi.org/10.1007/978-3-032-16268-7_25

certain challenges, including algorithmic bias (commonly referred to as hallucinations) and resistance from professionals [5]. While financial analysts and auditors tend to adopt these tools more swiftly, accountants and tax consultants exhibit greater resistance owing to a lack of training and concerns regarding job displacement [6].

The increasing application of AI in accounting has created ethical and operational challenges. Some studies emphasize the responsible use of LLMs, aiming to mitigate risks such as model hallucinations and lack of transparency in processes and data usage [5]. Continuous professional training for accountants is essential to ensure oversight and regulatory compliance [6]. Digital transformation driven by LLMs represents not only technological advancement but also a paradigm shift in the accountant's role, making it more focused on data analysis and strategic consulting [7].

This study examines the application of LLMs to accounting. Utilizing the Design Science Research (DSR) methodology of [1], a Generative Pre-trained Transformer (GPT) was developed to perform a vertical analysis of the Profit and Loss statement (P&L). This implementation reduced the analysis time from 12 min to 47 s (-93%), delivering faster and more reliable results. Additionally, LLMs enable the expansion of financial analysis into other areas, paving the way for the full automation of accounting information analysis. This study adopts the Chain-of-Thought (CoT) prompting strategy instead of Tree-of-Thought (ToT), given the linear and structured nature of vertical analysis. According to [8], CoT enhances the logical consistency of responses by simulating step-by-step human reasoning, which is ideal for interpreting financial items relative to the total revenue. This methodological choice improved clarity, processing speed, and accuracy in interpreting P&L.

This study contributes to the literature by demonstrating the feasibility of LLMs in accounting and by providing practical guidance for their implementation. These results underscore the importance of professional adaptation and appropriate regulations to fully leverage the benefits of this technology in the accounting field.

This study is structured into five sections. Following this introduction, the second chapter presents a literature review. The third and fourth chapters detail the methodology and results, respectively, along with their discussion. The fifth chapter presents the study's main conclusions, limitations, and suggestions for future research.

2 Literature Review

The adoption of LLMs in accounting redefines accountants' roles by automating repetitive tasks and demanding greater analytical and strategic capabilities [2]. Artificial intelligence not only enhances operational efficiency, but also transforms the interpretation and use of financial information, requiring advanced technological skills and critical thinking [3]. LLMs have demonstrated significant potential in bank reconciliations, financial reporting, and strategic and predictive analyses [4]. However, their implementation faces challenges, such as lack of regulation, algorithmic bias, and resistance from professionals [5]. Beyond automation, LLMs assist in interpreting accounting standards and formulating strategic insights, acting as a support for professional judgment rather than replacing it [9]. Nevertheless, a critical analysis of their outputs is essential because of the risk of algorithmic bias, which may compromise the reliability of collected data, analyses, and subsequent decision-making [10]. During auditing, LLMs enhance the continuous analysis of financial data, facilitating fraud detection and anomaly identification [11]. However, the opacity of their decision-making processes raises concerns, highlighting the importance of frameworks that incorporate human validation and security layers [3, 5, 12]. Continuous training of accountants is crucial to ensure that they understand the potential and limitations of this technology [5, 12].

The application of LLMs in accounting has been analyzed with a focus on prompt engineering techniques aimed at optimizing the performance of these models in complex tasks. Three primary approaches stand out: one-shot prompting, CoT, and ToT, each of which different implications for the effectiveness of LLMs.

One-shot prompting involves presenting a single example before requesting a response. [13] demonstrated that LLMs, such as GPT-3, can learn new tasks without requiring parameter reconfiguration. In accounting, this technique is used to automate the classification processes, allowing a model to record transactions based on a single example. [14] further highlighted its application in financial statement analysis, reducing manual review time and delivering faster results. CoT prompting is more effective for tasks that require structured reasoning. [15] demonstrated that this technique breaks down problems into smaller steps, thereby improving the accuracy of the results.

ToT prompting extends CoT by structuring reasoning into a tree format, thus enabling the exploration of multiple hypotheses before reaching a final decision. [16] finds that this technique is particularly useful in forensic accounting and financial auditing, where evaluating diverse scenarios enhances the detection of irregularities. [17] indicated that ToT improves regulatory compliance accuracy by allowing models to adjust their responses based on new information.

CoT prompting guides the model through sequential reasoning steps, before presenting the final output. [8, 18] emphasizes that CoT is more appropriate for structured problems such as financial statement analysis, where each item must be interpreted in relation to a total figure. As shown in Table 1, unlike ToT, which is designed for complex decision paths and multiple hypotheses, CoT aligns better with the logic required in the P&L vertical analysis.

Table 1. Comparison Between CoT and ToT Techniques [18].

Criteria	Chain-of-Thought (CoT)	Tree-of-Thought (ToT)
Reasoning structure	Linear, step-by-step	Exploratory, branching logic
Best suited for	Tasks with clear logical progression	Ambiguous tasks with multiple valid paths
Interpretability	High (easy to follow each reasoning step)	Moderate (can be complex to track all branches)
Computational efficiency	More efficient (fewer paths to explore)	Higher resource demand due to tree expansion
Ideal use case in accounting	Financial statement analysis (e.g. vertical P&L)	Forensic analysis, fraud detection
Prompt engineering complexity	Low to moderate	High

Thus, the adoption of chatbots in accounting has been growing, driven by advancements in AI and LLMs. [19] emphasized that prompt engineering influences the AI model response accuracy, impacting risk management and operational efficiency. Although chatbots have reduced the time required for financial operations, challenges, such as algorithmic bias remain. [20] analyzed the application of AI in automated financial statement analysis and concluded that model customization and human supervision are essential for ensuring reliable analyses. The interpretation of highly specific accounting data remains a challenge requiring continuous chatbot adjustments to prevent errors and inaccurate conclusions. [21] specifically examines the impact of chatbots on accounting, highlighting their usefulness in automating repetitive tasks such as bank reconciliation and invoice processing. However, concerns about data security and professional resistance have hindered their widespread adoption. [22] demonstrated that chatbots can automate tasks such as tax calculations, bank reconciliations, and financial report generation, facilitating tax compliance and improving financial management. [23] studied a chatbot application for bank reconciliations and accounts payable/receivable management, revealing a positive user perception, with users finding the tool intuitive and effective in reducing their manual effort. The authors estimated that automation could save up to 1,500 working hours per year. [7] highlighted that chatbots can revolutionize accounts receivable and payable reconciliation and financial reporting, allowing accountants to assume more strategic roles. However, professional resistance, data privacy, and regulatory constraints remain as obstacles to their adoption. [24] analyzed the use of ChatGPT in forensic accounting and demonstrated that AI can synthesize large volumes of information and detect suspicious financial patterns. However, limitations such as errors in financial calculations and difficulties in interpreting accounting standards reinforce the need for human oversight to prevent decision-making failures. [25] warns of the challenges faced by small and medium-sized enterprises in adopting new technologies, emphasizing the role of accountants in this transition and the necessity of continuous training for these professionals.

3 Methodology

This study adopts the DSR methodology, following the framework outlined by [1], to develop a GPT capable of performing vertical analysis of P&L statements using data extracted from a Customer Relationship Management (CRM) platform. The purpose is to reduce processing time and enhance the practical applicability of financial analysis tasks (see Fig. 1).

Empresa demonstração - Compra e venda de mercadorias - produtos informáticos		NIF: 999999999	
Demonstração dos resultados por naturezas em 31 de dezembro de 2024			(em euros)
Rendimentos e Gastos	**Notas**	**2024**	**2023**
Vendas e serviços prestados		€ 1 667 563,42	€ 1 883 449,65
Custo das mercadorias vendidas e das matérias consumidas		-€ 1 133 197,84	-€ 1 301 561,37
Fornecimentos e serviços externos		-€ 194 813,35	-€ 172 647,82
Gastos com o pessoal		-€ 190 758,20	-€ 188 908,78
Imparidade de dívidas a receber (perdas / reversões)		-€ 24 780,68	-€ 76 843,41
Outros rendimentos		€ 17 866,29	€ 16 704,00
Outros gastos		-€ 20 254,70	-€ 4 180,66
Resultado antes de depreciações, gastos de financiamentos e impostos		€ 121 624,94	€ 156 011,61
Gastos / reversões de depreciação e de amortização		-€ 21 760,15	-€ 29 082,25
Resultado operacional (antes de gastos de financiamentos e impostos)		€ 99 864,79	€ 126 929,36
Juros e rendimentos similares obtidos		€ -	€ 9,50
Juros e gastos similares suportados		-€ 44 490,35	-€ 55 681,79
Resultado antes de impostos		€ 55 374,44	€ 71 257,07
Imposto sobre o rendimento do período		-€ 13 289,87	-€ 16 802,93
Resultado líquido do período		€ 42 084,57	€ 54 454,14

Fig. 1. P&L statement structure used as input.

The DSR approach, traditionally applied in software engineering, was adapted to the accounting context to design a technological artifact that automates the vertical analysis of an income statement (see Fig. 2). Given that the processes of data extraction, calculation execution, and analytical interpretation are sequential and well-defined, the solution was implemented using a "one GPT, one task" strategy. For more complex, multi-step tasks requiring adaptive reasoning or planning, the use of AI agents is more suitable [26].

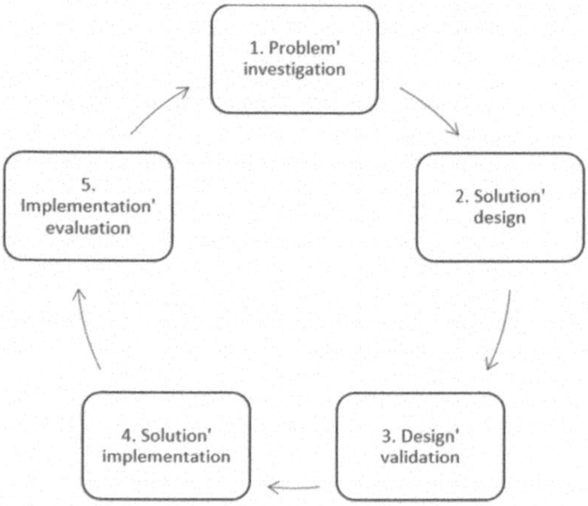

Fig. 2. DSR methodology framework (Zaiden et al., 2016, pp 3762).

3.1 Application of the DSR Methodology

As shown in Fig. 2, the application of the DSR methodology followed five phases:

Problem identification (step 1): The analysis of CRM logs showed that performing a manual vertical analysis of P&L required approximately 12 min per case. This pointed to an opportunity for automation to reduce the time and effort expended on routine calculations and free the accountant to focus on value-added tasks.

Design of the solution (step 2): The artifact developed was a GPT-based chatbot, created through a sequence of structured prompts that guided the model to:

- Understand and apply the principles of vertical analysis.
- Extract relevant parameters from a standard income statement.
- Execute calculations to express each line item as a percentage of total revenue.
- Formulate an executive summary.

The CoT prompting technique was applied to ensure that the model processed each item in a logical, step-by-step fashion, consistent with accounting reasoning. This is particularly useful for tasks that involve structured numerical comparisons [8].

Validation of the design (step 3): Validation focused on ensuring the accuracy of calculations and interpretative consistency of results. The outputs were compared to manually computed examples to verify that the chatbot applied the correct formulas and accounting logic. This process confirms the capability of the model to perform the task reliably.

Implementation of the solution (step 4): The chatbot was implemented using the XML prompt infrastructure provided by GPT, without reliance on a preloaded knowledge base. The configuration settings included the use of code interpretation (Python) and web browsing capabilities. The entire process—from prompt engineering to technical deployment, is illustrated in Figs. 3, 4, 5, 6, 7, 8, 9, 10, and 11 on the sequential prompting strategy.

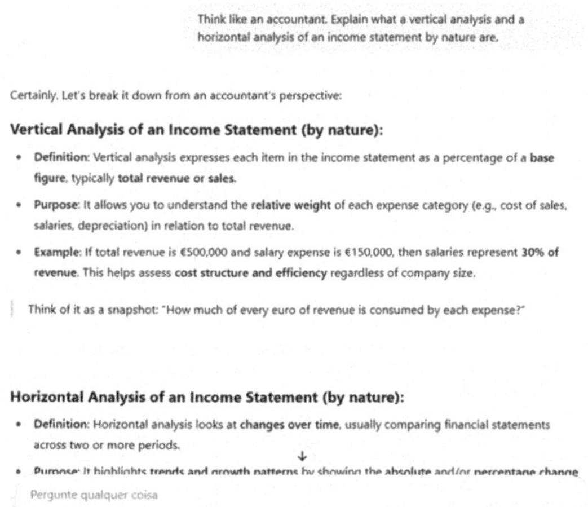

Fig. 3. Prompt 1 – Conceptual definition of vertical and horizontal analysis.

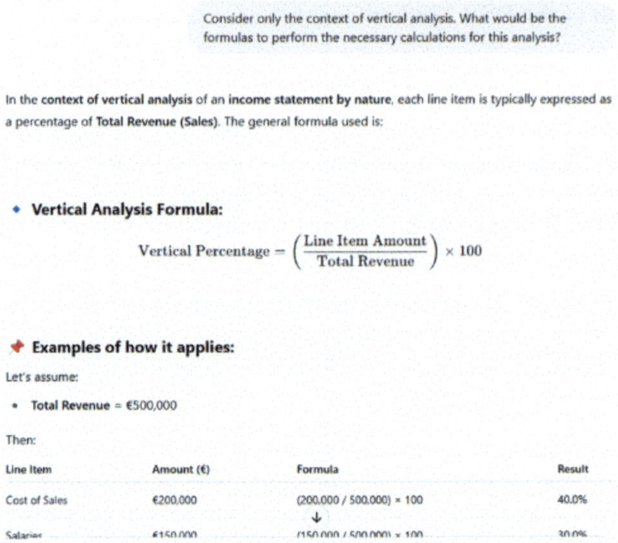

Fig. 4. Prompt 2 – Formulas for vertical analysis.

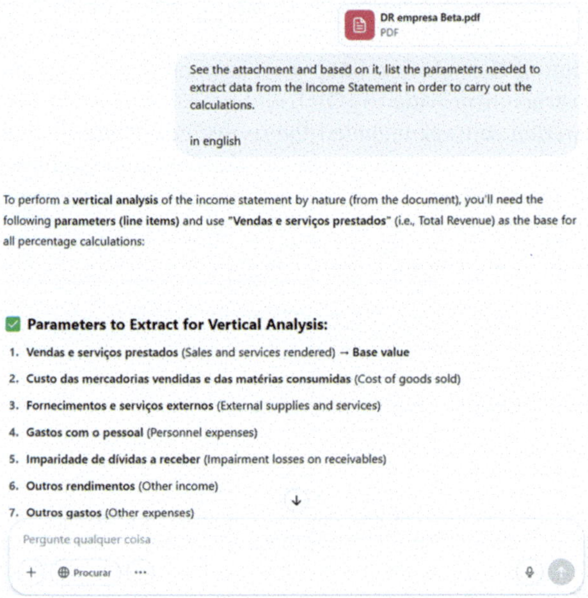

Fig. 5. Prompt 3 – Parameters required for data extraction from the income statement.

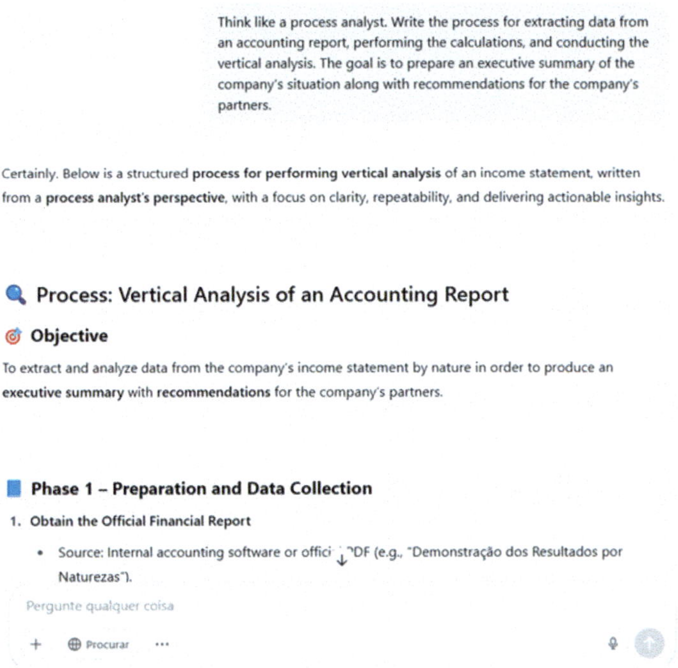

Think like a process analyst. Write the process for extracting data from an accounting report, performing the calculations, and conducting the vertical analysis. The goal is to prepare an executive summary of the company's situation along with recommendations for the company's partners.

Certainly. Below is a structured **process for performing vertical analysis** of an income statement, written from a **process analyst's perspective**, with a focus on clarity, repeatability, and delivering actionable insights.

🔍 **Process: Vertical Analysis of an Accounting Report**

🎯 **Objective**

To extract and analyze data from the company's income statement by nature in order to produce an **executive summary** with **recommendations** for the company's partners.

📘 **Phase 1 – Preparation and Data Collection**

1. **Obtain the Official Financial Report**
   - Source: Internal accounting software or offici ↓ DF (e.g., "Demonstração dos Resultados por Naturezas").

Pergunte qualquer coisa

+ Procurar ···

Fig. 6. Prompt 4 – Process for analysis and summary generation.

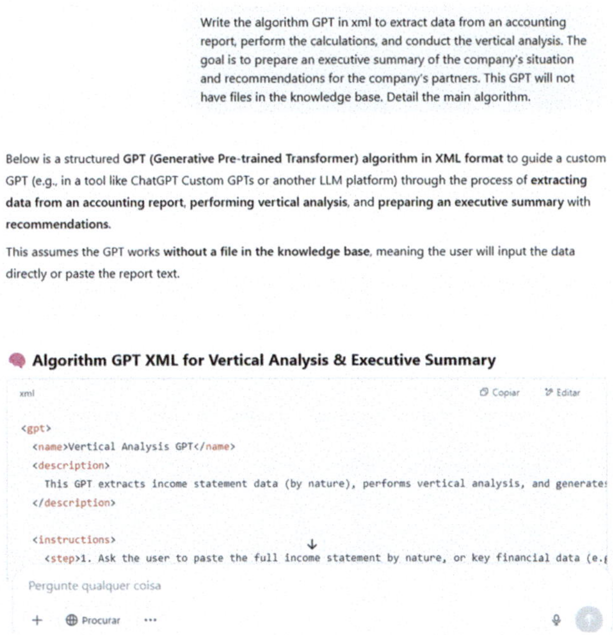

Write the algorithm GPT in xml to extract data from an accounting report, perform the calculations, and conduct the vertical analysis. The goal is to prepare an executive summary of the company's situation and recommendations for the company's partners. This GPT will not have files in the knowledge base. Detail the main algorithm.

Below is a structured **GPT (Generative Pre-trained Transformer) algorithm in XML format** to guide a custom GPT (e.g., in a tool like ChatGPT Custom GPTs or another LLM platform) through the process of **extracting data from an accounting report**, **performing vertical analysis**, and **preparing an executive summary** with **recommendations**.

This assumes the GPT works **without a file in the knowledge base**, meaning the user will input the data directly or paste the report text.

🧠 **Algorithm GPT XML for Vertical Analysis & Executive Summary**

xml Copiar Editar

```
<gpt>
  <name>Vertical Analysis GPT</name>
  <description>
    This GPT extracts income statement data (by nature), performs vertical analysis, and generate
  </description>

  <instructions>
    <step>1. Ask the user to paste the full income statement by nature, or key financial data (e.
```

Pergunte qualquer coisa

+ Procurar ···

Fig. 7. Prompt 5 – Algorithm written in XML for GPT execution.

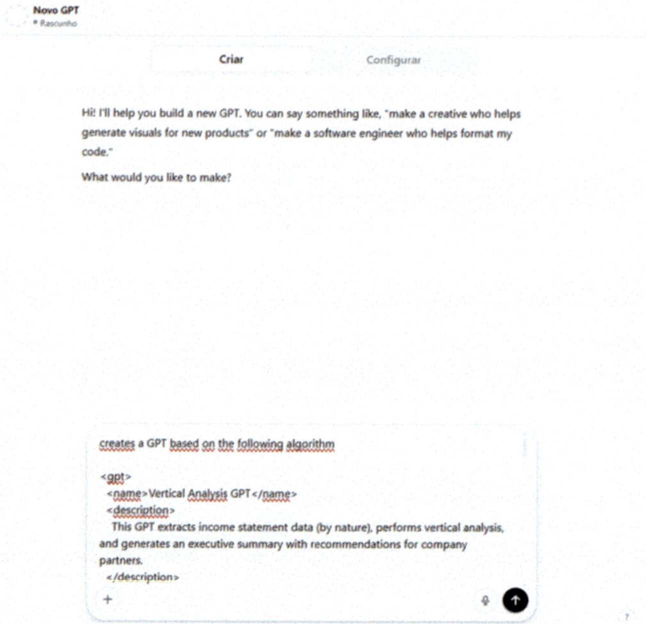

Fig. 8. Create GPT.

Fig. 9. GPT configuration – Prompt loading.

Fig. 10. GPT configuration – Final algorithm integration.

Fig. 11. GPT capabilities – Web access, screen reading, and code interpreter activation.

Evaluation of the implementation (step 5): A test was conducted using a sample income statement (Fig. 12). The processing time was reduced from 12 min to 47 s, confirming the efficiency of the artifact. A human-in-the-loop approach was retained to allow accountants to review the outputs before presenting them to the managers and entrepreneurs.

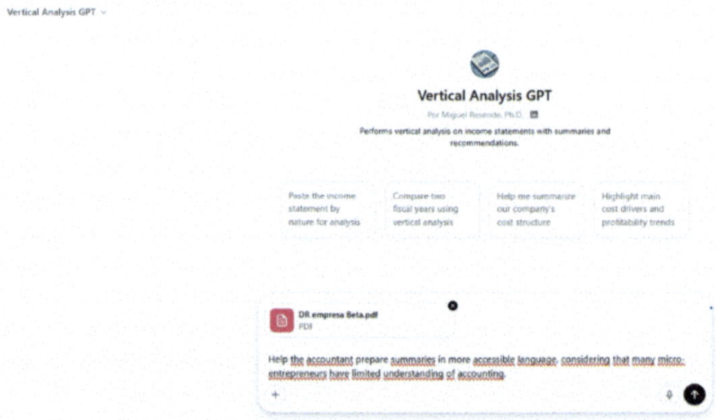

Fig. 12. P&L statement structure used as input.

3.2 Sequential Prompting Strategy

The design logic behind the chatbot followed a three-stage sequential prompting strategy, as illustrated in Figs. 3, 4, 5, 6, 7, 8, 9, 10, 11, 12, and 13.

Stage 1 - Conceptual foundation: Prompts were developed to guide the model in defining vertical and horizontal analyses, identifying relevant formulas, and extracting the necessary parameters (Figs. 3, 4, and 5).

Stage 2 - Process structuring: The model was prompted to act as a process analyst, outlining the full workflow from data extraction to the final summary (Figs. 6 and 7).

Stage 3 - Technical automation: Detailed prompts were used to construct the XML algorithm, configure the GPT environment, and integrate the tool into accounting workflows. The chatbot was tested with real data and evaluated for output clarity and accessibility (Figs. 8, 9, 10, 11, 12 and 13).

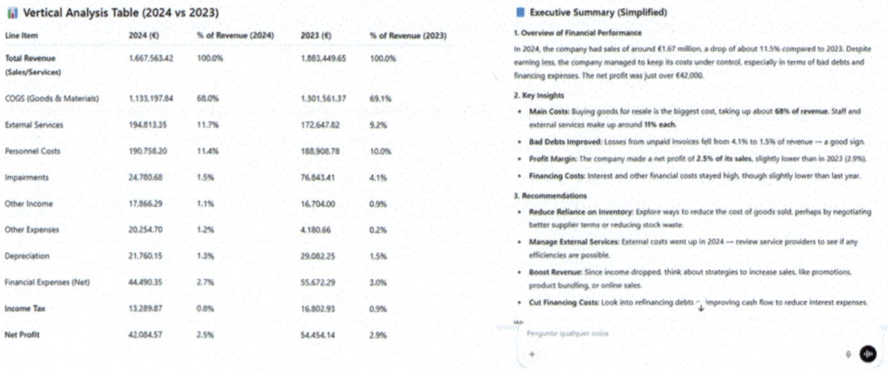

Vertical Analysis Table (2024 vs 2023)

Line Item	2024 (€)	% of Revenue (2024)	2023 (€)	% of Revenue (2023)
Total Revenue (Sales/Services)	1,667,563.42	100.0%	1,883,449.65	100.0%
COGS (Goods & Materials)	1,133,197.84	68.0%	1,301,561.37	69.1%
External Services	194,813.35	11.7%	172,647.82	9.2%
Personnel Costs	190,758.20	11.4%	188,908.78	10.0%
Impairments	24,780.68	1.5%	76,843.41	4.1%
Other Income	17,866.29	1.1%	16,704.00	0.9%
Other Expenses	20,254.70	1.2%	4,180.66	0.2%
Depreciation	21,760.15	1.3%	29,082.25	1.5%
Financial Expenses (Net)	44,490.35	2.7%	55,672.29	3.0%
Income Tax	13,289.87	0.8%	16,802.93	0.9%
Net Profit	42,084.57	2.5%	54,454.14	2.9%

Executive Summary (Simplified)

1. Overview of Financial Performance

In 2024, the company had sales of around €1.67 million, a drop of about 11.5% compared to 2023. Despite earning less, the company managed to keep its costs under control, especially in terms of bad debts and financing expenses. The net profit was just over €42,000.

2. Key Insights

- **Main Costs:** Buying goods for resale is the biggest cost, taking up about **68% of revenue.** Staff and external services make up around **11% each.**
- **Bad Debts Improved:** Losses from unpaid invoices fell from 4.1% to 1.5% of revenue — a good sign.
- **Profit Margin:** The company made a net profit of **2.5% of its sales**, slightly lower than in 2023 (2.9%).
- **Financing Costs:** Interest and other financial costs stayed high, though slightly lower than last year.

3. Recommendations

- **Reduce Reliance on Inventory:** Explore ways to reduce the cost of goods sold, perhaps by negotiating better supplier terms or reducing stock waste.
- **Manage External Services:** External costs went up in 2024 — review service providers to see if any efficiencies are possible.
- **Boost Revenue:** Since income dropped, think about strategies to increase sales, like promotions, product bundling, or online sales.
- **Cut Financing Costs:** Look into refinancing debts or improving cash flow to reduce interest expenses.

Pergunte qualquer coisa

Fig. 13. Output: Vertical analysis and summary in accessible language.

4 Results and Discussion

The results indicate an increase in financial analysis efficiency through the use of the chatbot developed in this study. Compared with the manual method, which required an average of 12 min to perform the vertical analysis of the P&L, the chatbot reduced this time to 47 s, representing a 93% reduction in the time required for data processing and interpretation. This optimization reinforces the applicability of LLMs in accounting, validating previous research on their potential for automating financial processes [2, 4, 8, 13, 18, 21, 22].

Beyond time efficiency, a qualitative improvement in the results was also observed, as the chatbot not only performed the necessary calculations but also presented the analyses in an accessible manner (Fig. 13). This linguistic adaptation facilitates the understanding of managers and micro-entrepreneurs without financial training, making accounting a practical decision support tool [3]. By generating a clear and structured executive summary, the chatbot enables accountants to validate and provide timely and comprehensible information to managers, offering a transparent view of the company's economic and financial situation without the need for specialized software.

There is potential to expand the scope of this study by incorporating horizontal and comparative analyses with industry benchmarks and competitors. To achieve this, the method developed in this study can be applied to create specialized chatbots for different types of analyses, enabling broader automation. This approach would further enhance process efficiency and provide a more comprehensive perspective of a company's economic and financial position, aligning with best practices in data-driven financial management [4, 5, 9, 22, 24].

Despite these advancements, several challenges remain, such as the need for linguistic adaptation to avoid Brazilian Portuguese terminology when using the Portuguese language, ensuring the reliability of analyses, the 518.000-character (128.000-token) conversation limit, and one minute per task where the task can be completed for timeout, which may restrict the analysis of extensive documents. The CoT technique, as proposed by [15] and supported by [8], enhances the accuracy of financial interpretation by structuring the reasoning path step by step. This approach proved more appropriate than ToT in contexts where each financial variable is linked in a linear dependency chain. However, it is essential to mitigate the risks related to algorithmic bias and model overconfidence, which have already been identified as challenges in accounting for LLMs applications [5, 10, 12, 19]. As this was a small task, where the processes of data extraction, calculation execution, and analytical interpretation are sequential and well-defined, the solution was implemented using a "one GPT, one task" strategy. For larger and more complex multi-step tasks requiring adaptive reasoning or planning, the use of AI agents is more suitable [26].

5 Conclusion

This study aimed to analyze the application of LLMs in accounting and explore their potential in the professional practice of accountants. Using the DSR method proposed by [1], a GPT for the vertical analysis of P&L was developed by comparing traditional manual analysis with an automated approach. The findings demonstrate that the use of LLMs in accounting has the potential to significantly transform the profession by increasing the efficiency and automation of repetitive tasks. The results confirm the existing literature, highlighting that LLMs not only reduce the time required for financial data analysis but also improve information accessibility, enabling managers and business owners to make more informed and timely decisions.

The implementation of a GPT-based chatbot reduced the time required for the vertical analysis of the P&L from 12 min to 47 s (– 93%), demonstrating substantial efficiency gains. In addition to time optimization, the chatbot generated analyses in a more accessible language, streamlined the accountant's intervention, and facilitated the comprehension of financial information by managers and entrepreneurs. The decision to use CoT over ToT proved effective, given the structured nature of the vertical analysis. CoT increases coherence and reduces computational load in tasks that do not require exploring alternative decision paths. This approach paves the way for the development of new specialized chatbots for various accounting functions, potentially leading to an integrated system for generating more comprehensive analyses.

Despite these advantages, the present study had several limitations. The need for continuous model training and dependence on high-quality data are challenges that may affect the effectiveness of LLMs, particularly in environments in which financial information is not well-structured. Moreover, the 512.000-character (128.000-token) limit per conversation and 1 min per task may restrict data analysis to more extensive accounting documents. Additionally, the adaptation of accounting professionals to digital transformation requires investment in training and skill development, making it essential to promote initiatives that encourage technological upskilling in this sector.

Future research should analyze the impact of LLMs in other areas, such as auditing and taxation, to assess how auditors can leverage LLMs in financial data analysis and how tax professionals can utilize them. Furthermore, a study of more complex tasks involving the development of integrated AI agents can be conducted to investigate the feasibility of fully automating financial analysis processes.

References

1. Zaiden, F., Bax, M., Parreiras, F.: Design science research: application in a research and development project. In: 13th International Conference on Information Systems & Technology Management - Contecsi – 2016 (2016). https://doi.org/10.5748/9788599693124-13CONTECSI/PS-4163
2. Kroon, N., Alves, M.C., Martins, I.: The Impacts of emerging technologies on accountants' role and skills: connecting to open innovation—a systematic literature review. J. Open Innov. Technol. Market Complex. **7**(3), 163 (2021). https://doi.org/10.3390/joitmc7030163
3. Anica-Popa, I.-F., Vrîncianu, M., Anica-Popa, L.-E., Cismasu, I.-D., Tudor, C.-G.: Framework for integrating generative AI in developing competencies for accounting and audit professionals. Electronics **13**, 2621 (2024). https://doi.org/10.3390/electronics13132621
4. Li, H., Vasarhelyi, A.: Applying large language models in accounting: a comparative analysis of different methodologies and off-the-shelf examples. J. Emerg. Technol. Account. **21**(2), 133–152 (2024). https://doi.org/10.2308/JETA-2023-065
5. Mwachikoka, T.: Effects of artificial intelligence on financial reporting accuracy. World J. Adv. Res. Rev. **23**(03), 1751–1767 (2024). https://doi.org/10.30574/wjarr.2024.23.3.2791

6. Toumeh, A.A.: Assessing the potential integration of large language models in accounting practices: evidence from an emerging economy. Future Bus. J. **10**, 82 (2024). https://doi.org/10.1186/s43093-024-00368-8
7. Odonkor, B., Kaggwa, S., Uwaoma, P.U., Hassan, A.O., Farayola, O.A.: The impact of AI on accounting practices: a review - exploring how artificial intelligence is transforming traditional accounting methods and financial reporting. World J. Adv. Res. Rev. **21**(1), 172–188 (2024). https://doi.org/10.30574/wjarr.2024.21.1.2721
8. Duarte, R.D: Aprenda a técnica Thought Chain Prompting na análise de balanço e DRE com ChatGPT. Roberto Dias Duarte (2024). Retrieved April 30, 2025, from https://www.robertodiasduarte.com.br/aprenda-a-tecnica-thought-chain-prompting-na-analise-de-balanco-e-dre-com-chatgpt/
9. Huang, X.: Analysis of the impact of artificial intelligence on enterprise financial accounting work. Trans. Econ. Bus. Manag. Res. **7**, 44–48 (2024). https://doi.org/10.62051/jhedh119
10. Yoo, M.: How Much Should We Trust Large Language Model-Based Measures for Accounting and Finance Research? University of Pennsylvania, SSRN Electronic Journal (2024). https://doi.org/10.2139/ssrn.4983334
11. Antwi, B., Adelakun, B., Eziefule, A.: Transforming financial reporting with AI: enhancing accuracy and timeliness. Int. J. Adv. Econ. **6**, 205–223 (2024). https://doi.org/10.51594/ijae.v6i6.1229
12. Barna, L.E.L., Hurducaci, C.C.: The impact of using artificial intelligence and ERP systems in the work of accounting professionals and auditors. The annals of the University of Oradea. Econ. Sci. TOM XXXIII **33**(1), 246–258 (2024). https://doi.org/10.47535/1991AUOES33(1)028
13. Brown, T.B., et al.: Language Models are Few-Shot Learners (2020). arXiv preprint arXiv:2005.14165. https://doi.org/10.48550/arXiv.2005.14165
14. Fung, S., Wong, M., Tan, C.W.: Chain-of-thoughts prompting with language models for accurate math problem-solving. In: 2023 IEEE MIT Undergraduate Research Technology Conference (URTC), Cambridge, MA, USA, 2023, pp. 1–5 (2023). https://doi.org/10.1109/URTC60662.2023.10534945
15. Wei, J., Wang, X., Schuurmans, D., Bosma, M., Ichter, B., Xia, F., Chi, E. H., Le, Q.V., Zhou, D.: Chain-of-Thought Prompting Elicits Reasoning in Large Language Models (2023). arXiv preprint arXiv:2201.11903. https://doi.org/10.48550/arXiv.2201.11903
16. Long, J.: Large Language Model Guided Tree-of-Thought (2023). arXiv preprint arXiv:2305.08291. https://doi.org/10.48550/arXiv.2305.08291
17. Chen, B., Zhang, Z., Langrené, N., Zhu, S.: Unleashing the Potential of Prompt Engineering in Large Language Models: A Comprehensive Review (2023). arXiv preprint arXiv:2310.14735. https://doi.org/10.48550/arXiv.2310.14735
18. Duarte, R.D.: Guia para aplicação das técnicas Chain of Thought e Tree of Thoughts em prompts. Roberto Dias Duarte (2024). Retrieved April 30, 2025, from https://www.robertodiasduarte.com.br/guia-para-aplicacao-das-tecnicas-chain-of-thought-e-tree-of-thoughts-em-prompts
19. Dhar, A., Datta, A., Das, S.: Analysis on enhancing financial decision-making through prompt engineering. In: Proceedings of the 7th International Conference on Electronics, Materials Engineering & Nano-Technology (IEMENTech). IEEE (2023). https://doi.org/10.1109/IEMENTech60402.2023.10423447
20. Noels, S., Merlevede, A., Fecheyr, A., Vanhalst, M., Meerlaen, N., Viaene, S., De Bie, T.: Automated Financial Analysis Using GPT-4. In: De Francisci Morales, G., et al. (eds.) ECML PKDD 2023, LNAI 14175, pp. 345–349. Springer (2023)
21. Resende, M.: O impacto dos chatbots na Contabilidade: Aplicações, benefícios e desafios, 10th GECAMB/P-CSEAR & XXIV Grudis Conference and Doctoral Colloquium (2025)

22. Zadorozhnyi, Z.-M., Muravskyi, V., Muravskyi, V., Shevchuk, A., Pochynok, N., Majda, M.: Application of Chatbots with Artificial Intelligence in Accounting. In: 13th International Conference on Advanced Computer Information Technologies (ACIT), pp. 196–200 (2023). https://doi.org/10.1109/ACIT58437.2023.10275395
23. Bavaresco, R.S., et al.: Machine learning-based automation of accounting services: an exploratory case study. Int. J. Account. Inf. Syst. **49**, 100618 (2023). https://doi.org/10.1016/j.accinf.2023.100618
24. Street, D.A., Wilck, J.: "Let's Have a Chat": Applying ChatGPT and Other Large Language Models to the Practice of Forensic Accounting. SSRN Electronic Journal (2023). https://ssrn.com/abstract=4351817
25. Resende, M.: Impacts of accounting digitization on the profession of accountants and productivity: a case study. Eur. J. Appl. Bus. Manag. **10**, 1 (2024). https://doi.org/10.58869/EJABM10(1)/06
26. Fujitsu: AI Agents and the Transformation of the Financial Industry (2025). Retrieved April 30, 2025, from https://global.fujitsu/en-global/insight/tl-aiagents-financial-industry-20250418

Legal, Accounting, and Tax Structuring of e-sports in Brazil

Leni Regina Segura[1(✉)], Henrique Formigoni[1], Liliane Cristina Segura[1], and Rute Abreu[2]

[1] Universidade Presbiteriana Mackenzie, São Paulo, Brasil
leni@opbo.com.br, {henrique.formigoni, liliane.segura}@mackenzie.br
[2] Instituto Politécnico da Guarda, Guarda, Portugal
ra@ipg.pt

Abstract. The exponential growth of e-Sports in Brazil and worldwide has positioned this digital phenomenon as a central element of the creative economy. Despite its economic relevance and professional reach, the Brazilian legal and institutional framework lacks specific regulation, generating uncertainties for the accounting, legal, and tax treatment of entities operating in the sector. This theoretical essay aims to analyze the main normative and structural challenges faced by e-Sports organizations in Brazil, highlighting gaps in their legal recognition, limitations in applying accounting standards (IFRS 15 and IAS 38), and the complexity of tax obligations under current legislation and the upcoming tax reform (EC 132/2023). Through comparative analysis with foreign regulatory experiences—including South Korea, France, the United States, and Latin American countries—the study identifies models that combine legal security with economic incentives. The research proposes a comprehensive agenda for regulatory development in Brazil, suggesting guidelines for legislation, technical guidance for accounting and taxation, and a checklist for internal professionalization of clubs and digital sports associations. It concludes that effective structuring of the legal–accounting–tax framework is essential to ensure transparency, investor confidence, and sustainable development of the Brazilian e-Sports ecosystem.

Keywords: e-Sports · legal framework · IFRS · taxation · governance · Brazil · tax reform · intangible assets

1 Introduction

The consolidation of e-Sports as an economic sector in Brazil faces structural obstacles that undermine its institutionalization, transparency, and sustainability. Even though the market has shown significant dynamism, as noted in previous sections, many clubs and organizations operate on the fringes of the legal system, without formal registration, without proper accounting records, and without adequate contracts with athletes, sponsors, and platforms. This reality not only limits access to investments, but also

L. Almeida et al. (Eds.): ICAFI 2025, ISEM 80, pp. 461–477, 2026.
https://doi.org/10.1007/978-3-032-16268-7_26

heightens labor, tax, and reputational risks—characteristics common to sectors in a pre-regulatory stage. Brazil's e-Sports and gaming industry also show strong growth. Data from Abragames indicate that Brazil is already among the main global hubs of this market (Brazil Games Export Program and Abragames, 2022). In 2021, Brazil's gaming industry earned approximately US$2.7 billion in revenue and had more than 103 million players—equivalent to about 74.5% of the Brazilian population (Sebrae, 2023). A study by the same organization shows that the sector generated around US$2.3 billion in 2021, a figure expected to continue growing in coming years. PricewaterhouseCoopers (2024) projects that e-Sports revenue in Brazil will double from US$9.4 million in 2022 to US$20 million in 2027. This economic dynamism is reflected in the popularization of electronic competitions: local events have drawn thousands of fans—for example, a national League of Legends tournament that gathered around 10,000 people in 2015 (Fenati, 2025)—and have sparked the interest of sponsors and media platforms. The production chain expanded rapidly: the number of game studios in Brazil jumped from 133 in 2014 to 1,042 in 2024, an increase of 683% (Brazil Games Export Program and Abragames, 2022).

However, this promising scenario contrasts with the absence of specific e-Sports regulation in Brazil. The national sports legislation does not yet explicitly include electronic games as an official sport, which creates legal uncertainty and undefined rights and duties for athletes, teams, and organizers. This regulatory gap brings about significant accounting and tax challenges: there are no clear rules on the taxation of prize money and sponsorships, nor on the application of ISS or ICMS taxes to the operations of digital platforms and online events. Valério et al. (2023) note that the high revenues of these competitions led the Brazilian government to seek ways to tax them, but current provisions still do not establish specific rates or tax procedures. In the accounting sphere, the lack of dedicated standards makes it difficult to properly recognize intangible assets—such as broadcasting rights, team brands, and athletes' image rights—and undermines the transparency of financial results for the organizations involved. As Coutinho and Rodrigues (2023) point out, any future regulation must ensure economic freedom and avoid unjustified barriers to the sector's development.

Considering this panorama, the present study is justified by the academic and social relevance of the topic. By gathering up-to-date market data and specialized literature, the article aims to highlight the strategic role of e-Sports in Brazil's creative economy and to discuss the main accounting, tax, and legal hurdles faced in the country. As Fenati (2025) and Valério et al. (2023) emphasize, the current prominence of e-Sports demands a careful analysis of its legal and economic challenges. The approach of this work is based on the knowledge gap regarding the normative treatment of this modality in Brazil and the urgent need to provide input for the formulation of public policies and accounting practices appropriate to a rapidly expanding sector.

2 Theoretical Foundations

2.1 Legal Classification of e-Sports

The absence of specific legislation for electronic sports in Brazil raises a series of controversies about their legal classification. The applicable constitutional framework is article 217 of the Constitution of the Federative Republic of Brazil of 1988, which recognizes sport as an activity of relevant social interest, granting autonomy to sports entities regarding their organization and functioning (Brazil, 1988). This provision, although originally intended for traditional modalities like football, volleyball, basketball, tennis, etc., does not impose any restriction on the nature of the sporting activity, opening space for the inclusion of e-Sports as long as they meet the requirements of regular practice, competitiveness, and economic purpose.

Law No. 9,615/1998—known as the Lei Pelé (Pelé Law)—regulates the organization of sports entities and the hiring of professional athletes, establishing rights, duties, and possible legal forms for sports organizations in Brazil (Brazil, 1998). Article 26 of that law guarantees freedom to practice professional sports, regardless of the modality. Therefore, there is no express prohibition on including digital modalities among the sports activities governed by this legislation. In practice, Brazilian law adopts a broad notion of sport, which would allow the incorporation of e-Sports, provided they are institutionalized according to legal principles.

This interpretation is already supported by administrative actions of public authorities. The Ministry of Sport, for example, has approved projects related to e-Sports under the Sports Incentive Law (Law No. 11,438/2006), and the Ministry of Labor authorized the issuance of visas for foreign players to work professionally in Brazil as e-athletes, using the category of "professional athlete" (Silva, 2023). Such precedents reinforce the understanding that e-Sports can be recognized as a sports modality from a legal-administrative perspective.

However, the absence of a specific national law addressing e-Sports still generates legal uncertainty for those involved. Several Bills have been presented in recent years with the goal of giving electronic sports their own legal status. Bill No. 70/2022, for example, proposes the formal recognition of e-Sports as a sports modality, granting players athlete status and regulating the rights and duties of organizations that promote competitions (Câmara dos Deputados, 2022). Bill No. 383/2017 similarly provided for institutional recognition and included guidelines for the formation of specific federations and leagues. The progress of these legislative proposals demonstrates a growing effort by lawmakers to adapt the legal system to the technological and sociocultural reality of electronic sports (Valério et al., 2023).

From a business law perspective, the structuring of e-Sports entities can take different forms allowed by the legal system. According to article 44 of the Brazilian Civil Code, non-profit associations or business companies (such as limited or joint-stock companies) are valid legal forms for institutional organization (Brazil, 2002). The Pelé Law itself provides that clubs and sports entities can be organized under any of these forms, depending on the organization's objectives (Brazil, 1998, art. 27, §9º). This means that e-Sports clubs may adopt an associative model (more common in amateur leagues or grassroots organizations) or a corporate model (more suitable for professional clubs with profit motives and large commercial contracts), without incompatibility with sports legislation.

Nonetheless, the regulatory vacuum regarding the legal definition of e-Sports implies practical difficulties in applying the Consolidation of Labor Laws (CLT) and the Pelé Law to electronic athletes. The absence of a formal employment contract with players, or even the lack of any recognizable legal bond, leaves these athletes outside the labor and social security protections, a situation akin to precarious employment (Silva, 2023). This problem is aggravated in the case of underage players, for whom the legal system requires special protection and strict rules for participation in work activities (Brazil, 1990).

Moreover, comparative legal analysis shows that many countries have advanced in specific e-Sports regulation. South Korea, for example, implemented sector-specific legislation through the "e-Sports Promotion Act," which sets guidelines for player protection, competition regulation, and financial transparency of organizations (Korean Ministry of Culture, Sports and Tourism, 2009). France also recognized e-Sports as an official sport in 2016, guaranteeing professional players a series of social and contractual rights (Loirat, 2020). These international experiences provide valuable examples that can inspire clearer and more sector-adapted Brazilian legislation.

In summary, the legal framework for e-Sports in Brazil is still characterized by normative ambiguity. Although there is constitutional and legal leeway for recognizing the modality, the absence of specific regulation compromises the legal certainty of relationships between clubs, athletes, sponsors, and event organizers. The progress of legislative projects and the mobilization of public and private entities are fundamental measures to ensure the proper institutionalization of e-Sports in the national legal system.

2.2 Accounting Structure for e-Sports

The growth of e-Sports as a significant economic sector demands the application of accounting practices capable of faithfully reflecting the financial reality of the entities involved—clubs, teams, event organizers, and platforms. However, the absence of accounting standards specific to the sector poses considerable challenges, especially regarding revenue recognition, valuation of intangible assets, and structuring of the accounting information system. The accounting of e-Sports entities must, therefore, be analyzed in light of the harmonization between Brazilian and international standards, such as those issued by the Accounting Pronouncements Committee (CPC) and the International Financial Reporting Standards (IFRS) Foundation.

From a revenue recognition standpoint, IFRS 15—Revenue from Contracts with Customers, adopted in Brazil as CPC 47, is applied. This standard establishes that revenue must be recognized based on a five-step model: (i) identify the contract; (ii) identify the performance obligations; (iii) determine the transaction price; (iv) allocate the price to the performance obligations; and (v) recognize revenue when the obligation is satisfied (IFRS Foundation, 2014). In e-Sports, this model applies to various revenue streams, such as sponsorships, broadcasting rights, ticket sales for in-person and online events, licensed merchandise sales, prize money, and licensing contracts.

Contractual particularities make this application more complex. For example, sponsorship contracts often combine different performance obligations: displaying logos on uniforms, activations on social media, athletes' appearances at events, and mentions in live broadcasts. Proper allocation of the transaction price among these obligations and the timing of revenue recognition requires qualified professional judgment (Grant Thornton Brasil, 2020). Similarly, broadcasting rights or contracts with streaming platforms involve the transfer of intangible rights that can generate recurring revenues, often tied to performance in tournaments—which makes revenue recognition dependent on performance criteria.

Regarding the measurement and recognition of intangible assets, IAS 38—Intangible Assets, corresponding to CPC 04 in Brazil, is applied. According to this standard, an intangible asset is an identifiable non-monetary resource without physical substance that is controlled by the entity and from which future economic benefits are expected to flow (IFRS Foundation, 2004). Intangible assets relevant in the e-Sports context include team brands and logos, software licenses, intellectual property for games developed, broadcasting rights, proprietary technologies, contracted image rights, and image rights related to the avatars used in such games. Nonetheless, rights related to the development of athletes who train their skills in gaming houses sponsored by the teams' parent companies should also be considered.

However, not all assets used by e-Sports entities can be recognized in accounting. An athlete's image rights, for example, belong to the athlete by virtue of article 5, item X of the Federal Constitution, and can only be licensed. Even if a club pays for the right to use an image, that payment is classified as a contractual expense, not as an asset, except when it is tied to an identifiable and separable contract with a measurable value and exclusive control by the entity (CFC, 2024). Thus, recognizing such rights as assets depends on the existence of objective measurement and control criteria.

Regarding expenditures on athlete development and training—a common practice in youth teams and "gaming houses"—the Brazilian standard ITG 2003 (R2), issued by the Federal Accounting Council (CFC), recently appears to have eliminated the possibility of capitalizing these costs, aligning with the understanding of IAS 38. By this interpretation, athlete development expenses are no longer considered intangible assets and must be recorded directly as expenses of the period (CFC, 2024). This applies to e-Sports clubs that invest in talent development programs or in their own academies, preventing the recognition of these expenditures as assets on the books. We do not agree with this interpretation.

In contrast, e-Sports entities that develop technological products—such as tactical analysis tools, proprietary applications, or performance management systems—can record development costs as intangible assets, provided the requirements of IAS 38 are met, such as technical feasibility, intent to complete, and the ability to generate future economic benefits (IFRS Foundation, 2004). In these cases, distinguishing between the research phase (expense) and the development phase (potential asset) is essential for proper accounting treatment.

Another relevant aspect is the valuation of a club's brand as an intangible asset. Although brands generate significant value for e-Sports organizations—especially when they have a large digital reach and appeal among young fans—the capitalization of a brand as an asset is only allowed in business combination transactions (acquisition of control) or when it is developed internally under strict recognition criteria. In most cases, it is an asset not recognized on the balance sheet, though still relevant for the economic and financial analysis of the entity.

Finally, there is a noticeable lack of accounting standards specific to e-Sports entities in Brazil. ITG 2003 (R2), aimed at sports entities, was developed with a focus on football clubs and still does not adequately cover digital operations, licensing contracts, and international revenues common in the electronic gaming ecosystem. Adapting these standards—or issuing technical guidance geared toward e-Sports—is an increasingly necessary step as the economic complexity of the sector grows.

In summary, the application of accounting principles to e-Sports requires interpretative effort and adaptation of existing standards. The recognition of revenues and assets must follow technical criteria defined by IFRS and CPC, but current normative gaps hinder the standardization of practices and compromise the comparability of financial statements across different entities in the sector. The professionalization of accounting in e-Sports organizations is, therefore, a necessary condition for their sustainability and transparency.

2.3 Tax Regime and Regulatory Uncertainties

Entities operating in the e-Sports sector in Brazil—whether clubs, event organizations or platforms—are subject to the same tax framework applicable to service-providing legal entities. However, the lack of specific regulation for e-Sports creates significant uncertainties regarding the proper classification of activities in the tax system, exposing sector actors to risks of penalties and disproportionate tax regimes.

The first aspect to consider is the applicable tax regime. E-Sports organizations can choose among the Lucro Real (Actual Profit), Lucro Presumido (Presumed Profit), or Simples Nacional regimes, provided they meet legal requirements. Under the Presumed Profit regime, which is often adopted by small and medium-sized companies in the sector, operational revenues—such as sponsorships, prize winnings, image rights, and product sales—are taxed based on fixed presumptive rates (8% for corporate income tax (IRPJ) and 12% for the social contribution on profits (CSLL) in the case of services), to which PIS (0.65%) and COFINS (3%) are added (Brazil, 1996; Receita Federal do Brasil, 2012). The total effective rate can range between 7.33% and 16.33%, depending on the nature of the activity and the composition of the revenue.

However, difficulty in correctly classifying the income received—especially amounts from sponsorships or contracts with foreign platforms—may lead the Federal Revenue Service to reclassify them as extraordinary or non-operational income, raising the presumptive base and, consequently, the total tax burden (Contábeis, 2023). For instance, a sponsorship received by a team that does not have event promotion or advertising services listed in its corporate purpose could be deemed atypical revenue. In such cases, the tax authority may apply a presumptive rate of 32% for IRPJ and CSLL, increasing the total burden to about 11,33 a 19,33% (Valério et al., 2023). This scenario illustrates the fiscal risk arising from the lack of legal and accounting standardization in the sector's organizational structures.

Prize winnings are another point of concern. According to Cosit Divergence Ruling No. 09/2012, prizes paid to individuals in sports competitions are subject to withholding income tax (IRRF), applying the progressive rate table (rates between 7.5% and 27.5%)

(Receita Federal do Brasil, 2012). If the prize is awarded to a legal entity, the amount becomes part of the company's operational revenue and is not withheld at source, but must be declared and taxed under the chosen tax regime. This distinction has significant practical implications, especially because many e-Sports players are contracted as self-employed or via legal entities, without a formal employment relationship. The contractual arrangement affects not only the taxation of prize money, but also the calculation base for social security (INSS) and the FGTS, depending on the regime adopted.

Furthermore, in the case of prize money paid to players residing in Brazil by foreign entities—such as international tournaments with payouts in dollars or cryptocurrency—the Carnê-Leão system is applied, according to Federal Revenue Instruction No. 1,500/2014. The tax must be paid monthly by the recipient in Brazil, based on the official currency conversion rate on the date of receipt. In the case of cryptocurrencies, the taxpayer must use the market exchange rate and report the transaction in the tax return schedule for income received from abroad. The absence of withholding at source and the informality of contracts increase the complexity of tax calculations and the risk of omission or tax penalties (Oscorp, 2020).

Municipal taxation also presents challenges. The Tax on Services of Any Kind (ISSQN), governed by Complementary Law No. 116/2003, applies to activities such as event organization, promotion of entertainment events, and commercialization of digital content. Although the law does not explicitly mention e-Sports, several city governments have started to levy ISS on tournaments, ticket sales, registration fees, and licensing agreements (Valério et al., 2023). The controversy lies in the tax's territoriality criterion, which defines the service recipient's domicile as the place of incidence. In online events—where the service recipient could be in any municipality in Brazil or abroad—applying this rule becomes complex and subject to multiple interpretations. Initiatives like Bill PLP 202/2019 attempted to explicitly include e-Sports in the list of taxable services, but there is still no law in force clarifying the application of ISS in this context.

Regarding ICMS (Tax on the Circulation of Goods and Services), which is levied on the circulation of goods and digital goods, there are doubts about its applicability to the sale of games, licenses, and digital assets by platforms based outside the country. The Supreme Federal Court (STF), in ruling ADI 5958, upheld the constitutionality of charging ICMS on software sold via download, consolidating the position that digital goods are taxable in Brazil (STF, 2021). This has implications for Brazilian companies that resell game licenses or offer virtual products such as "skins," season passes, and character upgrades.

Streaming platforms, often used by players and teams as a source of monetization, also face tax risks. Revenues from subscriptions, donations, and bits are treated by the tax authority as service income, being subject to income tax (IR), CSLL, PIS, and COFINS, and in some cases ISS (Contábeis, 2023). The tax treatment of these revenues depends on the formal structure of the recipient (individual, micro-entrepreneur (MEI), or a company under the Simples regime) and the origin of the payments (domestic or international). The lack of specific regulation makes it difficult to standardize the treatment and leaves room for disputes and audits.

Additionally, there is the risk of an employment relationship being legally characterized between players and clubs, which would entail the payment of payroll taxes and contributions (INSS, FGTS, 13th salary, paid vacation, etc.) if the elements of subordination, regularity, exclusivity, and compensation are present (Silva, 2023). The informality of contracts—especially with young and underage athletes—increases the exposure of organizations to actions by the Federal Revenue Service, the Ministry of **Some aspects to be considered to Brazilian legislation about e-Sports in Brazil.**

Regarding to Labor, and the Labor Prosecutor's Office, it is important, in summary, to observe that the tax regime applicable to e-Sports in Brazil is fragmented and uncertain. The absence of specific legislation and the diversity of business models in the sector make proper tax classification difficult at both the federal and municipal levels. The existing gaps increase the risk of tax assessments and raise the compliance costs for organizations, jeopardizing the sector's sustainability. As Valério et al. (2023) conclude, "the absence of clear rules exposes the e-Sports sector to varied interpretations by tax authorities, heightening litigiousness and legal uncertainty".

The consolidation of e-Sports as an economic sector in Brazil encounters structural obstacles that compromise its institutionalization, transparency, and sustainability. Even though the market has demonstrated expressive dynamism, as observed in earlier sections, many clubs and entities operate outside the legal system, without formal registration, without bookkeeping, and without proper contracts with athletes, sponsors, and platforms. This reality not only limits access to investments, but also amplifies labor, fiscal, and reputational risks—characteristics common to sectors in a pre-regulatory stage.

Also, another aspect to be considered is the informality and Precarious Organizational Structure. A study by Ananias (2022) reveals that a significant portion of Brazilian e-Sports teams do not have an active CNPJ (business registration number), nor any registration with commercial boards or formally written bylaws. Frequently, these organizations are structured informally, based on verbal agreements among friends or small groups of investors who manage social media and register athletes in tournaments without adopting any legally recognized form. Informality is particularly common in amateur teams or those just starting out, but it also occurs in professional teams that achieve national notoriety without developing internal governance.

Such a makeshift structure prevents access to public incentives (such as the Sports Incentive Law), hinders the securing of sponsorships from private companies, and compromises fiscal compliance, especially with regard to issuing tax invoices, fulfilling ancillary obligations, and paying taxes. As Silva (2023) highlights, this organizational precariousness reflects not only the youth of the sector, but also the lack of regulatory and tax incentives to encourage formalization.

Specially the contractual fragility and labor risks are other points to be considered. Informality also affects the relationship between organizations and players. Many athletes are recruited without signing a formal written contract—a situation that results in the violation of fundamental labor rights, such as annual paid leave, 13th salary, FGTS, and social security coverage. Although the Pelé Law (Law No. 9,615/1998) establishes a special legal regime for professional athletes, this legal framework is still applied only precariously to e-Sports, since the category was not explicitly contemplated in the legislation (Brazil, 1998).

In many cases, players live in "gaming houses," train full-time, and follow direct instructions from managers and coaches, clearly meeting the elements of an employment relationship: subordination, regularity, personal exclusivity, and remuneration (Silva, 2023). The lack of formalization creates hidden liabilities that can lead to labor lawsuits, inspections by the Ministry of Labor, or fines from the Federal Revenue Service. The situation is even more delicate in the case of minors who participate in official championships, often without judicial authorization or adequate legal oversight—in breach of the Child and Adolescent Statute (Brazil, 1990).

The main aspect to be observed is the difficulties in applying accounting and tax standards in e-sports companies. The difficulties faced by e-Sports organizations are also reflected in the application of accounting and tax rules. The diversity of revenue streams—such as sponsorships, prize winnings, subscriptions, contracts with digital platforms, and product sales—requires a level of accounting sophistication that is often not adopted by these entities. The absence of proper accounting records prevents the recognition of intangible assets, such as brands and licenses, as well as the correct calculation of revenues according to IFRS 15 and CPC 47 (IFRS Foundation, 2014).

From a tax perspective, as seen in the previous section, the difficulty of fitting activities into the correct CNAE (National Classification of Economic Activities) codes, the absence of formal contracts, and the receipt of funds from abroad without the due collection of IRRF (withholding income tax) or reporting via Carnê-Leão are common practices that aggravate audit risks. As Valério et al. (2023) conclude, the regulatory vacuum and fragile structure of many teams result in an environment of legal uncertainty, which discourages the entry of investors and the professionalization of the sector.

It is important to observe the impacts of the tax reform (Constitutional Amendment 132/2023). The enactment of Constitutional Amendment No. 132/2023, which institutes a Tax Reform on consumption, represents an important milestone for all economic sectors—including e-Sports. The reform proposes the replacement of five current taxes (PIS, COFINS, IPI, ICMS, and ISS) with two new taxes: the Contribution on Goods and Services (CBS), under federal jurisdiction, and the Tax on Goods and Services (IBS), under state and municipal jurisdiction (Brazil, 2023). These taxes will adopt a Value-Added Tax (VAT) system, which brings specific advantages and risks to the sector.

For e-Sports entities, the new model could bring greater simplicity and rationality to taxation, especially if a clear national legislation on digital services is established. However, several challenges emerge:

- Broad scope of incidence: CBS and IBS will apply across the entire consumption chain, including event organization services, image rights, and digital licensing. This may increase the tax burden on organizations that operate on multiple fronts (events, media, products), depending on the final rate to be set.
- Uncertainty about the location of consumption: In the case of online services like broadcasts and virtual tournaments, there remains doubt about the spatial criterion for taxation (the location of the customer, the provider, or the server). The complementary legislation still under debate will need to address this issue, at the risk of maintaining the uncertainty that exists with the ISS tax (Fenati, 2025).

- End of credit accumulation for informal entities: Many teams operate outside the tax credit system. With the full adoption of the Tax Reform, organizations without regular accounting will not be able to recover CBS/IBS credits on inputs, sponsorships, or operating expenses, which increases the effective cost and incentivizes informality.
- Risk of interpretative disputes: The absence of sector-specific legislation for e-Sports may lead to disputes regarding the classification of operations in the new tax system (service, licensing, or product), revisiting the problems currently observed with ICMS and ISS.

2.4 International Reference: The European VAT Model

The European Value-Added Tax (VAT) model, adopted by the European Union, can serve as a benchmark for constructing a more effective tax regime for e-Sports. In EU countries, VAT is levied on all stages of the production and distribution chain of goods and services, allowing for input tax credits and ensuring greater fiscal neutrality. Directive 2006/112/EC establishes specific rules for digital services, including streaming, electronic games, and online licensing, with an emphasis on the place of consumption as the criterion for tax jurisdiction (European Union, 2006).

Some countries, such as France and Germany, have developed specific interpretative guides for the taxation of e-Sports and digital content, addressing the definition of the place of consumption, the classification of tournaments as cultural events, and rules for credit offsets (Loirat, 2020). This pragmatic approach has favored the formalization of the sector and the alignment of tax and accounting rules. Brazil could benefit from these experiences when regulating the effects of CA 132/2023, especially with regard to the technology, media, and digital entertainment sector.

2.5 International Comparisons and Regulatory References

The absence of specific e-Sports legislation in Brazil contrasts with the regulatory evolution observed in several other countries, which have adopted distinct models of legal recognition, sector-specific accounting, and tax treatment. A comparative analysis of these systems can offer valuable input for the formulation of Brazilian public policies, helping to avoid ad-hoc regulations and ensuring legal and accounting security for entities involved in the e-Sports value chain.

South Korea: State Regulation and Athlete Protection: South Korea is widely recognized as a pioneer in the institutionalization of e-Sports. In 2000, the government created the Korea e-Sports Association (KeSPA), an entity with semi-governmental status responsible for organizing championships, accrediting athletes, and promoting ethics in competitions. In 2009, the e-Sports Industry Promotion Act was promulgated, a legal framework that regulates the sector, establishes guidelines for contractual transparency, and requires the registration of professional players' employment contracts (Korean Ministry of Culture, Sports and Tourism, 2009).

In addition, KeSPA requires minimum clauses in contracts—such as limits on training hours, medical leave, and anti-doping controls—and maintains a league certification system. The coordinated action among the state, federations, and private companies has given South Korea a solid governance structure and international prestige. From an accounting perspective, e-Sports organizations are treated as cultural enterprises, with tax incentives and specific financial reporting standards (Loirat, 2020).

In the case of France: Legal Recognition and Protected Contracts. France was the first European country to officially recognize e-Sports as a category distinct from betting or games of chance. Law No. 2016-1321, known as the "Loi pour une République Numérique" (Law for a Digital Republic), included e-Sports as a legitimate activity, defining criteria for its professional practice (France, 2016). The law requires standardized contracts between players and teams, with social security protection, work hour controls, and legal safeguards. It also differentiates between amateur and professional competitions, allowing for more appropriate tax and accounting treatment.

Since 2018, the French government has been granting tax incentives to companies in the sector, based on technological innovation and job creation criteria. Club accounting follows IFRS standards, with adaptations for image rights contracts, software licenses, and game-related intellectual property. This regulatory model balances social protection and economic promotion, with strong state involvement in public–private partnerships (Loirat, 2020).

Regarding to United States: Contractual Freedom and Tax Federalism. In the United States, e-Sports are organized in a decentralized manner, with self-regulation largely by the major leagues (such as the Overwatch League or Call of Duty League) and private contracts between organizations and athletes. There is no specific federal legislation on e-Sports, but various states have adopted rules on visas, intellectual property, and revenue recognition (Scholz, 2019). The sector is treated as a branch of the entertainment industry, based on classic contractual principles.

From a tax perspective, the federal system results in different regimes across states, especially regarding the sales tax, which applies to digital goods and services. Accounting follows US GAAP (Generally Accepted Accounting Principles), with disclosure requirements for revenues from sponsorships, merchandising, and prize winnings. The focus is on transparency and private regulation of minimum compliance standards.

Observing Canada: Associative Governance and Innovation Incentives. Canada has adopted a mixed approach. Although it does not yet have national legislation on e-Sports, several provinces recognize the practice as part of the creative economy. The Canadian Sport Policy (2012) allows unconventional disciplines to be included in sport and culture development programs. Canadian e-Sports organizations often take the form of non-profit associations, especially in youth categories (Stewart, 2021).

The Canadian government also maintains technology innovation incentive programs applicable to the gaming and e-Sports sector, with tax deductions for companies that develop digital solutions or train athletes. Accounting is based on IFRS standards, and organizations operating for profit must report event revenues, sponsorships, and product sales according to Canada Revenue Agency (CRA) rules.

When it comes to Latin America we may observe that there are emerging experiences in Chile and Argentina. In Latin America, countries are at different stages of institutionalizing e-Sports. Chile, through the Ministry of Sport, officially recognized e-Sports as a sport in 2019, allowing organizations to enroll in public incentive programs and access infrastructure (Gobierno de Chile, 2019). The country also began discussions on creating a national federation, linked to the Chilean Olympic Committee, with a view to regulating the sector's contracts and accounting practices.

Argentina, in turn, published Resolution 202/2021 of the National Communications Agency (ENACOM) in 2021, which included e-Sports in public policies on digital inclusion and youth. Some provinces, such as Buenos Aires, have launched professional training programs and tournament certification initiatives. In the tax domain, there is still no specific rule, but there are efforts to adapt taxes on services and software to the digital games and sports sector (CIPPEC, 2022).

3 Method

This essay adopts a theoretical and analytical approach, grounded in the examination of existing academic literature, legal frameworks, and international accounting standards. The study does not involve empirical data collection; instead, it is based on a critical review of normative documents, including the Brazilian Constitution, the Pelé Law (Law No. 9,615/1998), IFRS standards (IFRS 15 and IAS 38), and technical guidance from the Federal Accounting Council (CFC).

In addition, the analysis incorporates a comparative perspective, drawing on regulatory experiences from jurisdictions such as South Korea, France, the United States, and selected Latin American countries. These cases were selected for their relevance in structuring e-Sports as a regulated activity within diverse legal and economic systems.

The objective is to identify legal and institutional gaps in the Brazilian context, assess limitations in the application of accounting principles to e-Sports entities, and evaluate tax-related uncertainties under current and forthcoming legislation, including Constitutional Amendment No. 132/2023.

This methodological approach is appropriate given the normative and exploratory character of the research, which aims to inform future regulatory developments and provide technical guidance for the formalisation and professionalisation of the e-Sports ecosystem in Brazil.

4 Recommendations and Proposed Studies for the Legal, Accounting, and Tax Structuring of e-Sports in Brazil

Based on the analyses developed in this essay, several gaps are identified that require further academic and technical investigation. The following proposes lines of research and applied projects that can support the consolidation of an appropriate legal, accounting, and tax framework for the Brazilian e-Sports ecosystem.

Legal Studies

- Comparative study of e-Sports legal frameworks: Analyze the regulatory models adopted by South Korea, France, the USA, Canada, and Latin American countries, identifying best practices and challenges in incorporating e-Sports into the legal system.
- Critical analysis of pending Bills in the National Congress (PL 70/2022, PL 383/2017, etc.): Evaluate their legal viability and potential impact on the formalization of clubs, contracts, and athletes' rights.

- Study on employment contracts and e-Sports players' rights: Examine the application of the Consolidation of Labor Laws (CLT) and the Pelé Law (Law No. 9,615/1998), considering the particularities of digital practice, hybrid work schedules, and international competition by athletes.
- Proposal for a federative governance model for e-Sports in Brazil: Develop guidelines for creating a national coordinating entity, inspired by the models of KeSPA (South Korea) and the Direction Générale du Numérique (France).

Accounting Studies

- Analysis of the applicability of IFRS 15 (CPC 47) to sponsorship and streaming contracts in e-Sports: Conduct real case studies and develop a revenue recognition model for contracts with multiple performance obligations.
- Evaluation of intangible asset recognition criteria in e-Sports clubs: Apply CPC 04 (IAS 38) to brands, logos, broadcasting rights, and proprietary software.
- Feasibility study for issuing a specific accounting standard (ITG) for e-Sports entities: Provide theoretical and technical groundwork for drafting a standard similar to ITG 2003 (R2), adapted to the digital realities of the sector.
- Research on bookkeeping and transparency practices in Brazilian e-Sports entities: Conduct an empirical survey on the level of accounting maturity of organizations and their compliance with current standards.

Tax Studies.

- Mapping of the effective tax burden in typical e-Sports operations: Identify the taxes levied on prize winnings, sponsorships, licensing, merchandising, and digital services.
- Study of the impacts of the Tax Reform (CA 132/2023) on e-Sports: Run simulations based on the new CBS/IBS model and provide a comparative analysis with international VAT systems, especially that of the European Union.
- Proposal of a simplified tax regime for micro and small e-Sports entities: Develop an alternative model inspired by the Simples Nacional, focusing on encouraging formalization and tax fairness.
- Analysis of the tax treatment of international prize winnings and income via foreign platforms: Focus on IRRF, Carnê-Leão, and related obligations, with special attention to receipts in foreign currency and cryptocurrency.

5 Discussion and Final Remarks

This theoretical essay examined the growing importance of e-Sports in the national and international context, with special attention to the legal, accounting, and tax challenges faced by the entities operating in this sector. Noting that e-Sports move billions of dollars in revenue and mobilize audiences comparable to those of traditional sports, it is paradoxical that specific regulation is absent in the Brazilian legal framework. This gap compromises not only legal certainty and the protection of athletes, but also proper accounting compliance and appropriate taxation of the sector's operations.

On the legal front, it was observed that although the Federal Constitution and the Pelé Law provide interpretative leeway to recognize e-Sports as a sports modality, the lack of specific legislation creates uncertainty regarding the nature of contracts, the rights of players, and the legal form of organizations. From an accounting perspective, the application of international standards IFRS 15 (CPC 47) and IAS 38 (CPC 04) to e-Sports clubs and platforms imposes technical challenges, especially in recognizing multi-faceted revenues and non-standard intangible assets. In the tax sphere, the lack of a clear normative framework results in divergent interpretations by tax authorities, unwarranted assessments, and high tax burdens that discourage formalization.

Furthermore, the Brazilian Tax Reform (CA 132/2023) promises to significantly transform the consumption tax regime, which could represent both an opportunity for simplification and a risk of increased costs for entities that operate informally or with low compliance capacity. The Value-Added Tax model, if well structured, could benefit the sector—provided that specific rules are established for digital services and for the proper classification of e-Sports.

International analyses demonstrated that several countries already recognize e-Sports as a legitimate activity, with their own regulatory framework, tax incentives, and contractual protection rules. South Korea, France, Canada and, more recently, Latin American countries like Chile and Argentina, offer valuable experiences that can be adapted to the Brazilian context. Building a national regulatory framework requires, therefore, a multidisciplinary approach coordinated among the Legislative branch, regulatory agencies, professional councils, private entities, and the academic community.

Accordingly, the following agenda is proposed:

- Legal recognition of e-Sports as an autonomous sports modality, through a specific federal law defining the legal contours of the activity, the rights of athletes, and the organizational form of e-Sports entities;
- Issuance of accounting norms applicable to the sector, with technical guidelines published by the CFC and aligned with international standards, adapted to the digital and global reality of e-Sports;
- Specific tax regulation for prize winnings, sponsorships, and digital content licensing, to be implemented via complementary legislation to CA 132/2023 and normative instructions from the Federal Revenue and local tax authorities;
- Creation of public programs for development and training for clubs, athletes, and industry professionals, promoting digital inclusion, access to credit, educational partnerships, and the incubation of creative startups;
- Promotion of academic and technical research on e-Sports governance, accounting, law, and taxation, integrating universities, research institutes, and the productive sector;
- Establishment of a federative e-Sports governance body with equal representation from government, the private sector, and civil society, capable of formulating guidelines, regulating competitions, and promoting transparency in the sector.

In summary, the sustainable development of e-Sports in Brazil depends on overcoming the current legal, accounting, and tax asymmetries. By structuring a regulatory framework that respects the specificities of the sector and promotes its integration into the formal economy, the country can transform a cultural phenomenon into a strategic pillar of the contemporary digital economy.

References

Albino, J.C.A., Carrieri, A.P., Figueiredo, D.S., Bastos, F.L.R.S.: Tornando-se clube-empresa: uma análise das práticas estratégicas e de branding do Sport Club Internacional. Anais do IV Encontro de Estudos em Estratégia, Recife, 21–23 junho (2009)

Andon, P., Free, C.: Accounting and the business of sport: Past, present and future. Accounting, Auditing & Accountability Journal **32**(7) (2019). https://doi.org/10.1108/AAAJ-08-2018-3627

Araujo, O.N., Dantas da Silva, F.J.: A contabilidade aplicada em clubes de futebol, com ênfase em ativos intangíveis. Periódicos UESB (2022). http://periodicos2.uesb.br

Asiri, M., Ah-Hadi, A., Taylor, G., Duong, L.: Is corporate tax avoidance associated with investment efficiency? North American J. Econ. Fin. **52** (2020). https://doi.org/10.1016/j.najef.2020.101167

Barrainkua, I., Espinosa-Pike, M.: Antecedentes de conflitos profissionais organizacionais enfrentados por contadores. Revista Brasileira de Gestão de Negócios – FECAP **22**(1), 5–23 (2020)

Bascón-Seda, A., Ramírez-Macías, G.: ¿Son los e-sports un deporte? Término "deporte" en jaque. Movimento 26 (2020). https://doi.org/10.22456/1982-8918.104932

Beusch, P., Frisk, J.E., Rosén, M., Dilla, W.: Management control for sustainability: Towards integrated systems. Management Accounting Research 54 (2022)

Brunner, K.M.: Accounting issues for publicly traded gaming companies. Gaming Res. Rev. J. **1**(1), 43–53 (1994)

Cani, J.P., Meneghetti, T.: Transformação das associações desportivas em sociedades econômicas – os clubes empresas. Revista Eletrônica de Iniciação Científica – UNIVALI **5**(1), 436–460 (2014)

Carvalho, M.J., Moura, J.C., Oliveira, N.: Empresas municipais de desporto. Revista Porto Ciências e Desporto **9**(2), 33–45 (2021)

CFC – Conselho Federal de Contabilidade: ITG 2003 (R2) – Entidade Desportiva é publicada no DOU (2024). https://cfc.org.br/noticias/itg-2003-r2-entidade-desportiva-e-publicada-no-dou

CIPPEC: Videojuegos en Argentina: Estado del arte y desafíos regulatorios. Buenos Aires (2022)

Constancio, D., Santos Neto, J.P., Grecco, M.C.P.: Contabilização de bitcoins à luz das IFRS e aspectos tributários. V Encontro dos Mestrados Profissionais em Administração, São Paulo (2018)

Contábeis: Receita de patrocínio pode ter alíquota maior no Lucro Presumido (2023). https://www.contabeis.com.br

Cristofol, F.J., Martínez-Ruiz, A., Román-Navas, I., Cristofol Rodríguez, C.: Evolución de las estrategias de patrocinio en los esports en España: 2013–2021. Ámbitos. Revista Internacional de Comunicación **48**, 188–204 (2020)

Cruz Junior, A.T., Carvalho, M.M., Laurindo, F.J.B.: Estratégia e estrutura: Em busca do alinhamento organizacional em um clube social esportivo. Revista Gestão e Produção **12**(3), 429–441 (2005)

Damásio Educacional: Sociedade Anônima do Futebol (SAF) [E-book] (2021)

Egelman, S., Peer, E., Harbach, M., Malkin, N., Mathur, A., Frik, A.: Nudge me right: Personalizing online security nudges to people's decision-making. Comput. Hum. Behav. **109**, 106347 (2020)

European Union: Directive 2006/112/EC on the common system of value added tax. Official Journal of the European Union (2006)

Fenati, R.: Esportes eletrônicos no Brasil: panorama econômico e desafios jurídicos. Atlas, São Paulo (2025)

Franco, P.L.: E-sports: Do entretenimento ao plano de carreira [Trabalho de Conclusão de Curso, UniCEUB] (2017)

Gobierno de Chile: Reconocimiento de los e-Sports como práctica deportiva. Ministerio del Deporte (2019)

Grant Thornton Brasil: IFRS 15: cinco considerações sobre o reconhecimento de receitas (2020). https://www.grantthornton.com.br

IFRS Foundation: IFRS 15 – Revenue from Contracts with Customers. IFRS Foundation, Londres (2014)

IFRS Foundation: IAS 38 – Intangible Assets. Londres: IFRS Foundation (2004)

Isler, O., Rojas, A., Dulleck, U.: Easy to shove, difficult to show: Effect of educative and default nudges on financial self-management. J. Behav. Exp. Financ. **34**, 100639 (2022)

Juniper Research: eSports & Games Streaming: Emerging Opportunities & Market Forecasts 2021–2025 (2021). https://www.juniperresearch.com

Loirat, A.: Les politiques publiques de l'e-sport: vers un encadrement juridique européen. Revue Internationale de Droit Comparé **72**(1), 123–143 (2020)

Löfgren, Å., Nordblom, K.: A theoretical framework of decision making explaining the mechanisms of nudging. J. Econ. Behav. Organ. **172**, 248–263 (2020)

Mazzioni, S., Klann, R.C.: Aspects of the quality of accounting in international context. Revista Brasileira de Gestão de Negócios – FECAP **19**(63), 262–279 (2017)

Melo Neto, F.P.: Administração e marketing de clubes esportivos: a base para a criação do clube-empresa. Sprint, Rio de Janeiro (1998)

Meyer, P.E.: Human assets accounting in the professional sports industry. Auditing and Reporting, Boston University (2018)

Morais, H.C.B., Macedo, M.A.: Relationship between earnings management and abnormal book-tax differences in Brazil. Revista de Contabilidade e Finanças **31**(82), 97–112 (2020)

Nishi, Y.: O mercado de Esportes Eletrônicos: a eficiência alocativa e o caso do League of Legends [Monografia, UFOP] (2018)

Oscorp: A tributação das premiações em torneios de e-Sports (2020). https://www.oscorp.com.br

PwC: Global Entertainment & Media Outlook 2023–2027 (2024). https://www.pwc.com/outlook

Raab, M., Bar-Eli, M., Plessner, H., Araújo, D.: The past, present and future of research on judgment and decision making in sport. Psychol. Sport Exerc. **42**, 25–32 (2019)

Raschka, I.M., Wallner, R.J.G., Costa, K.B.: Contabilidade esportiva: um estudo sobre a evidenciação das demonstrações contábeis dos clubes paulistas. Anais do Congresso FIPECAFI de Contabilidade Esportiva (2010). https://congressousp.fipecafi.org

Receita Federal do Brasil: Solução de Divergência COSIT nº 09, de 2012 (2012). https://normas.receita.fazenda.gov.br

Santoro Neto, G.: A Sociedade Anônima do Futebol (SAF) e a Monetização da Paixão [Monografia, Faculdade de Direito de Curitiba] (2021)

Sampaio, M.S.: Clubes de Futebol e o Desenvolvimento do Esporte Eletrônico no Brasil [Monografia, PUC-Rio] (2019)

Scholz, T.M.: eSports is Business: Management in the World of Competitive Gaming. Springer International Publishing (2019)

Sebastian, J.P.: Triggering the emergence of digital ecosystems: the role of mobile and video games in emerging economies. Digital Policy, Regulation and Governance **20**(5), 449–478 (2018)

Sebrae: Mercado de jogos digitais no Brasil (2023). https://sebrae.com.br/sites/PortalSebrae/estudosmercado

Silva, B.O.: E-Sports: a ausência de regulamentação específica. Centro Universitário Uninorte (2023)

Simon, J.P.: Making sense of a complex world: online gaming. Digital Policy, Regulation and Governance **20**(5), 449–478 (2018)

Souza, A.L.R., Alvarez, G., Andrade, J.C.S.: Mercado regulado de carbono no Brasil: um ensaio sobre divergências contábil e tributária dos créditos de carbono. Revista OES (2013). www.revistaoes.ufba.br

Souza, F.: Dilemas e Perspectivas do Clube Empresa: o Caso do Figueirense Futebol Clube [Monografia, UFSC] (2019)
Steenkamp, M.: Capturing the value of football players in financial reporting [Magister Commercii, University of Johannesburg] (2020)
Stewart, K.: The state of Canadian e-Sports. Canadian Journal of Media Studies **3**(1), 55–72 (2021)
Valério, M.A.G., Braga, G., Gallo, D.: A indústria dos jogos eletrônicos: a tributação de e-games e e-sports no Brasil. Revista Tributária e de Finanças Públicas **159**, 150–159 (2023)

Predictive Analysis and Decision Making: A New Perspective in the Management of Municipal Services

Bernardo Fernando[1(✉)], Augusta Ferreira[1,2], and Carlos Santos[1,2]

[1] ISCA-UA, Universidade de Aveiro, Aveiro, Portugal
{bernardofernando75,augusta.ferreira,carlos.santos}@ua.pt
[2] CICF, Instituto Politécnico Do Cávado E Do Ave, Barcelos, Portugal

Abstract. This article analyzes the potential of predictive analysis as a tool to support the management of municipal services, using the Municipal Services of Abrantes as a case study. This study uses the application of three forecasting models (ARIMA, ETS and the Power BI forecast functionality) applied to five economic-financial indicators (general liquidity, reduced liquidity, financial autonomy, solvency and return on assets). The analysis is based on a time series of data collected between 2011 and 2023, focusing the predictive analysis on the years 2024 and 2025 (data for 2024 is not yet available and the year 2025 is still in progress). The performance of the models was evaluated using three error metrics: RMSE; MAE and MAPE, allowing to compare the accuracy of the models used. The results obtained demonstrate that predictive models offer a high degree of reliability and usefulness for financial planning, particularly in the context of municipal services. The predictions obtained point to a trend of stability in the analyzed indicators, with controlled levels of uncertainty, thus validating the application of these models in supporting decision-making. It is concluded that predictive analysis constitutes a significant added value for the modernization of the management of the entities under analysis, promoting more informed and rigorous decision-making.

Keywords: Predictive Models · Advanced Analytics · Municipal Services · Economic-Financial Indicators · Decision Making

1 Introduction

The increasing complexity of the challenges faced by public services, such as increased demand for essential services, budgetary constraints and the need for efficiency in resource allocation, requires more robust, data-driven and future-oriented management approaches. In this context, predictive analysis is seen as a strategic tool, as it allows us to anticipate future behaviors based on historical data, contributing to more rigorous, assertive and well-founded decision-making.

This article aims to analyze the potential of predictive analysis in the management of municipal services (services offered by city halls with the aim of meeting the population's needs, managed in a business-like manner, with autonomous organization).

L. Almeida et al. (Eds.): ICAFI 2025, ISEM 80, pp. 478–490, 2026.
https://doi.org/10.1007/978-3-032-16268-7_27

The Municipal Services of Abrantes were used as a case study (the others can be analyzed in a similar way). To this end, predictive time series models were applied, the ARIMA model, ETS and the Power BI forecasting functionality, applied to a set of economic-financial indicators considered relevant for assessing the sustainability and financial efficiency of these entities. The aim is also to assess the reliability of future projections generated by each of the models, identifying those that perform best, with a view to supporting the management of municipal services.

The expected results of this study consist of providing a perspective of the behavior of the analyzed indicators, with the reduction of possible errors, and thus facilitating the planning of institutions with a view to obtaining the best results.

In addition to this introduction, this work is organized into four more sections. Section 2 presents the literature review. Section 3 explains the methodology used. In Sect. 4, the results obtained by applying the selected predictive models are analyzed to analyze the behavior of each of the economic-financial indicators and, finally, in Sect. 5, the final considerations are presented.

2 Literature Review

Predictive analytics is an analytical process that uses historical data and uses statistical techniques and machine learning algorithms to anticipate future events and thus allow more rigorous decision-making [1, 2]. This type of approach allows the identification of patterns and relationships between different variables, identifying behaviors that allow responding to future challenges with a high level of precision. In a very complex situation with frequent uncertainties in public management, the ability to predict behaviors and conceive possible scenarios is considered increasingly relevant.

For the process of modernizing and updating public services, in a broad sense, and municipalized services, in a strict sense, to be successful, decision-making processes must be agile and duly supported, considering several factors: demographic growth; climate change; economic recessions and increased demand for social services considered essential to local communities. Therefore, predictive models can be relevant instruments, allowing the management bodies of public entities to adopt more proactive and assertive approaches, minimizing risks and optimizing the decision-making process [3–5].

Analysis based on predictive models has been consolidating themselves as relevant and considering in the decision-making process in public administration, thus providing managers with the ability to anticipate certain scenarios by identifying patterns using a large volume of historical data. The use of these models not only provides rational and efficient management of public resources, but also ensures better allocation of resources, providing higher quality and more transparent services.

Given the variety of forecasting models available, the choice of the most appropriate one will depend on the specific characteristics of the data used. Therefore, to ensure that the analysis results are reliable, it is important to be aware of some properties of time series (data), such as seasonality, trend and randomness. Furthermore, the choice of forecasting method should take into account the available resources [6].

The behavior of time series can be satisfactorily described by different forecasting models. In certain situations, choosing the most suitable model is relatively simple, especially when one of them clearly stands out from the others. However, there are cases in which this distinction is not obvious, making it necessary to resort to criteria that can facilitate this decision. These criteria are, most of the time, based on the analysis of the residuals of the adjusted model or on the evaluation of the forecast errors, thus ensuring a more informed and rigorous choice [6–9].

Cunha (2022), suggests that in time series analysis, it is common for more than one model to meet the established assumptions and prove suitable for predicting a given set of data. When this occurs, the choice tends to fall on the model with the fewest parameters, prioritizing simplicity without compromising predictive capacity. However, this decision is not always evident, becoming one of the central challenges of statistical modeling. Therefore, identifying the model that best fits each series requires a balance between technical rigor and practical criteria.

Some studies suggest the use, among others, of the following metrics to evaluate the accuracy (precision) of the model used [6–9]: mean absolute error (MAE); root mean square error (RMSE); mean absolute percentage error (MAPE).

MAE is a statistical metric used to quantify forecast accuracy in time series, representing the average of the absolute differences between observed values and predicted values. This metric provides a direct measure of the average deviation between predicted values and actual values, being expressed in the same unit as the variables analyzed. The lower the MAE, the greater the accuracy of the predictive model [11].

The RMSE corresponds to the mean of the squares of the differences between the predicted values and the observed values. By squaring the differences, MSE penalizes larger deviations more significantly, making it particularly sensitive to outliers. Here too, smaller values indicate greater accuracy of the model [11].

MAPE calculates the average of the absolute differences between predicted and actual values, expressed as a percentage of the actual values. This is a metric that facilitates the interpretation of the error at different scales, but it can be sensitive to very low observed values, making the analysis less reliable under these conditions. Values below 10% are considered excellent, between 10% and 20% good, and above 50% inadequate [7].

Power BI is a business intelligence platform that facilitates the processing, analysis and visualization of data through interactive dashboards that enhance decision-making [12]. On this platform, the native functionality for making predictions is available exclusively in the "line chart" visual element. This resource uses specific statistical techniques for time series analysis, projecting future values based on historical data [13].

In addition to native forecasting functionality, the Power BI platform allows integration with the R and Python languages, widely used by statisticians and data analysts. Integration with these languages significantly expands the tool's analytical capabilities, enabling richer analyses and the generation of valuable information. The Power BI platform automatically recognizes the installation of the R language, making its use quite accessible even for users with little experience or programming skills [14].

The use of economic and financial indicators is important for identifying the economic and financial health of entities, helping to understand whether public resources are allocated rationally and whether there is efficient financial management [15–17].

The relevance of these indicators is also evident in their ability, on the one hand, to facilitate comparative analysis between different entities and, on the other hand, to allow monitoring of an entity's evolution over the years, constituting essential instruments for informed decision-making and continuous improvement [16, 18].

The evaluation and analysis of the performance of a given entity, such as municipal services, can, consequently, be conducted largely using a set of indicators in order to obtain a clear view of its economic and financial health. Economic and financial indicators allow for the monitoring and evaluation of results in an objective and simplified manner, helping managers to make decisions that enable a better allocation of resources [16–19].

According to [20] general liquidity, reduced liquidity, solvency and financial autonomy are important indicators that allow measuring the existing balance between resources and short-term obligations, thus assessing the entity's ability to meet its short-term commitments. Return on assets is referred to as [17] with an important indicator given that shows the capacity to remunerate the investment in resources. The calculation formula for each indicator is shown in Table 1.

General liquidity measures the balance between resources and short-term obligations. Reduced liquidity allows us to assess an entity's ability to meet its short-term commitments. Solvency assesses an entity's financial capacity to honor its long-term obligations. Financial autonomy allows us to assess the entity's independence in relation to third-party capital. Return on assets allows you to assess financial sustainability [17, 20, 21].

Table 1. Economic and financial performance indicators

Indicador	Fórmula
General Liquidity	$\frac{\text{Current Assets}}{\text{Current Liabilities}}$
Reduced Liquidity	$\frac{\text{Current Assets} - \text{Inventories}}{\text{Current Liabilities}}$
Solvency	$\frac{\text{Equity}}{\textit{Liabilities}}$
Financial Autonomy	$\frac{\text{Equity}}{\text{Total Assets}}$
Return on Assets	$\frac{\text{Net Profit}}{\text{Total Assets}}$

Within the scope of this study, this set of indicators offers a good basis for evaluating and comparing the performance of municipal services, thus contributing to more detailed diagnoses and implementation of continuous improvement plans.

3 Methodology

This study adopts an exploratory approach that, according to [22, p. 27] It is developed "with the aim of providing a general, approximate view of a given fact". In the context of this study, this approach aims to understand how predictive analysis can be used to optimize decision-making and improve the management of municipal services. Furthermore, the present study assumes a quantitative aspect, justified by the use of statistical models that allow for a more direct and objective assessment of the benefits of using predictive models. Such models are particularly crucial for identifying patterns and trends, thus contributing to more assertive decision-making.

As they are widely used, we will take into account, in comparison with the native functionalities of the Power BI business intelligence platform, the following predictive models: exponential smoothing and autoregressive integrated moving average:

- **Exponential Smoothing (ETS)**. This method integrates adaptive forecasting methods capable of monitoring changes in level, slope and/or seasonal pattern and, therefore, is considered one of the most efficient when applied to series whose components, trend and seasonality, vary over time. In these methods, predictions are nothing more than a weighted combination of previous observations, where more recent observations receive a greater "weight" than older observations. That is, the more recent the observation, the greater its contribution to the forecast [8, 23]. The most commonly used exponential smoothing methods are three: (i) simple exponential smoothing algorithm, suitable for locally constant series; (ii) Holt's algorithm, which is suitable for series that present a trend; and (iii) Holt-Winters algorithm, suitable for series composed of seasonality and trend [24].
- **Autoregressive Integrated Moving Average (ARIMA)**. This model is a statistical approach widely used in time series analysis to model and predict temporal patterns [8, 25], being composed of three main components: AutoRegressive component (AR), Integrated Component (I) and Moving Averages component (MM). For the effective implementation of the ARIMA model, it is crucial to understand how the parameters "p", "d" and "q" work, since the adjustment of the model is influenced by the choice of values for these parameters [8, 26], in which:

  – the parameter "p" corresponds to the order of the AR (AutoRegressive) component;
  – the parameter "d" corresponds to the order of the component I (Integrated, number of differences required to make the time series stationary), and;
  – the "q" parameter corresponds to the order of the MM (Moving Averages) component.

The data used in the analysis of the indicators considered in this study constitute time series, observations made sequentially over time and were provided by the Center for Research in Accounting and Taxation (CICF), and include the Balance Sheet and Income Statement, corresponding to the years 2011 to 2023 of eighteen (18) municipalized services with valid data in all years of the referred time series.

For data processing and analysis, the Excel spreadsheet and the Power BI business intelligence platform were used, both Microsoft software. Additionally, Rstudio software, an integrated development environment for R, was used to build predictive models and visualize results. The selection of these tools is justified by their widespread use and their great capacity to connect with multiple sources of information and the possibility of providing multiple visualization elements that facilitate the construction of dashboards. With these tools, predictive models were built, tested and compared using the following native forecasting algorithms from the Power BI platform and made available by the ETS and ARIMA models.

Although the study involves data relating to eighteen (18) municipal services, for the reasons mentioned above, for the purposes of validation and detailed analysis, it was decided to present the analysis of the indicators for a single entity, namely the **Municipal Services of Abrantes**. For other entities the analysis will be similar.

To make predictions using native Power BI resources, the **Forecast** functionality integrated into the tool was used. This feature allows you to automatically generate forecasts based on time series. The configuration used defined the year as the forecast unit, with the projected horizon being two years (2024 and 2025). The seasonality parameter was kept in automatic mode, allowing Power BI to identify and adjust seasonal patterns dynamically, according to historical data. Furthermore, a 95% confidence interval was defined, a value widely used in statistics as it represents an adequate balance between precision and reliability of projections. This defined interval indicates that there is a 95% probability that the future real values will be within the interval predicted by the tool. Furthermore, the choice of this interval reflects a common practice in quantitative studies, as it guarantees a high level of confidence without making the intervals excessively wide, which would compromise the practical usefulness of the forecasts.

To evaluate the performance of the ARIMA, ETS and Power BI models, error metrics were calculated based on historical and predicted data for each of the indicators mentioned above. For the ETS and ARIMA models, the summary() command provided by the R language was used. Regarding the forecast functionality, the errors were calculated based on the actual and predicted values for two years, 2022 and 2023, and 2024 and 2025, respectively, see Table 2.

Table 2. Error metrics associated with Power BI forecasting functionality

Indicator	Year	Real	Predicted	RMSE	MAE	MAPE
General Liquidity	2022	1,41	1,37	0,29	0,23	12,74
	2023	1,81	1,40			
Reduced Liquidity	2022	1,25	1,27	0,23	0,18	10,86
	2023	1,64	1,31			
Solvency	2022	7,91	6,55	1,80	1,76	20,95
	2023	8,70	6,55			
Financial Autonomy	2022	0,87	0,85	0,03	0,03	3,40
	2023	0,89	0,85			
Return on Assets	2022	-3,90	0,00	3,15	3,03	100,00
	2023	2,15	0,00			

Based on the error metrics summarized in Table 3 for the indicator, overall liquidity, the ETS and ARIMA models present identical performances in the 3 metrics considered. The model based on the native features of the Power BI platform, in turn, presents better performance in any of the metrics. Thus, for the indicator, **general liquidity** the model is the one that should be considered.

Regarding the indicator, **reduced liquidity**, the model based on the native functionalities of the Power BI platform is the one that presents the lowest error values in the 3 metrics used and should, for this reason, be the model to be considered.

For the **solvency** indicator, the ARIMA model presents a lower error in the RMSE metric. However, the ETS model presents lower error values in the other 2 metrics MAE and MAPE. Therefore, to evaluate the future behavior of this indicator, the model with the best performance is the ETS model.

In the case of the indicator, **financial autonomy**, the model based on the native functionalities of the Power BI platform is also the one that presents the lowest error values in the 3 metrics considered. Therefore, for this indicator, the use of this model should be considered.

Finally, regarding the **return on assets** indicator, the ETS model is the one that records the lowest error values in the 3 metrics considered, which makes it an eligible model to evaluate the behavior of this indicator.

Table 3. Accuracy metrics of the 3 models considered in this study

Indicator	Model	RMSE	MAE	MAPE
General Liquidity	ETS	0,31	0,26	22,14
	ARIMA	0,31	0,26	22,14
	Power BI	**0,29**	**0,23**	**12,74**
Reduced Liquidity	ETS	0,29	0,24	20,56
	ARIMA	0,29	0,25	23,58
	Power BI	**0,23**	**0,18**	**10,86**
Solvency	ETS	1,22	**0,91**	**15,05**
	ARIMA	**1,15**	0,95	17,17
	Power BI	1,80	1,76	20,95
Financial Autonomy	ETS	0,05	0,03	3,87
	ARIMA	0,04	**0,03**	3,91
	Power BI	**0,03**	**0,03**	**3,40**
Return on Assets	**ETS**	**0,02**	**0,01**	**87,03**
	ARIMA	**0,02**	**0,01**	100
	Power BI	3,15	3,03	100

4 Presentation and Analysis of Results

The results obtained with the Power BI and ETS platform models are presented and discussed below, as previously mentioned, because they are the ones that present the lowest errors for the indicators: general liquidity; reduced liquidity and financial autonomy and solvency and return on assets respectively. Thus, using these models, the future behavior (2024 and 2025) of the indicators described in Table 1 within the scope of municipalized services in Abrantes was evaluated.

These models were applied to the time series corresponding to each indicator in the period between 2011 and 2023 using the algorithm inherent to the ETS model provided by the R language and the algorithms used in the forecast functionality provided by the Power BI platform. These algorithms automatically select the most suitable configuration based on the time series used, adjusting the error, trend and seasonality components in an optimized way. This approach allows flexible modeling, capable of capturing different dynamics of the time series, without the need for manual specification by the user.

Forecasts for **general liquidity** indicate stability in the indicator, with estimated values of 1.57 for 2024 and 1.61 for 2025, see Fig. 1.

The shaded area represents the 95% confidence interval, reflecting the degree of uncertainty associated with the estimates generated. The absence of a sharp broadening indicates relative stability in the projections, suggesting that the model maintains an acceptable level of confidence for the estimated values, within limits of variation considered controlled.

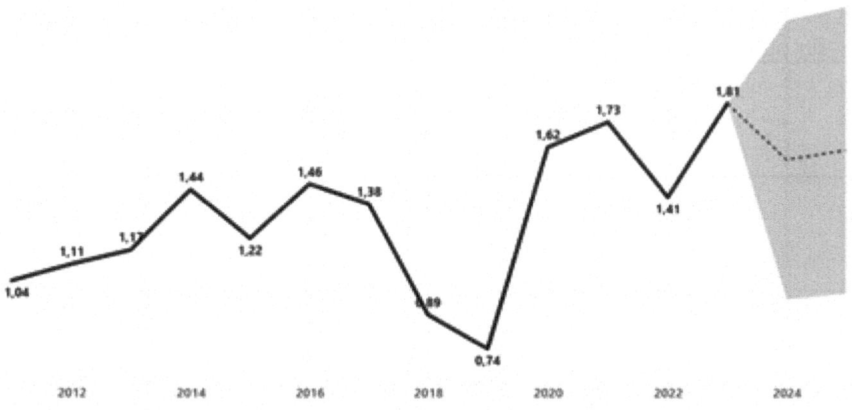

Fig. 1. General liquidity forecast

Figure 2 presents the results obtained for **reduced liquidity**. Forecasts for 2024 and 2025 suggest a slight recovery after the fall recorded compared to the 2023 peak (1.64), with values of 1.44 expected for 2024 and 1.49 for 2025. Despite this decrease, the indicator remains consistently above threshold 1, which reinforces the positive assessment of the solid short-term solvency capacity.

This expected stability translates into a positive signal in the context of financial management, suggesting that, even in more demanding scenarios, the entity should maintain a balanced profile in terms of fulfilling its immediate obligations. There is a slight widening of the confidence band, especially in 2025, which is consistent with the expected behavior in time series models, where uncertainty tends to increase as the time horizon extends.

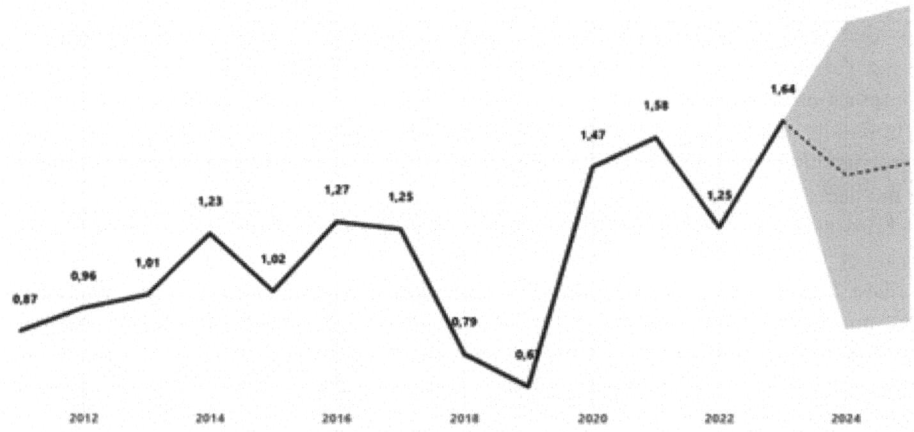

Fig. 2. Reduced liquidity forecast

In Fig. 3 we can see that the predicted values for the **solvency** indicator are equally constant for both years (8.70). However, it is worth noting that the shaded area shows a widening in 2025, compared to 2024, indicating greater uncertainty in the forecast and, therefore, that there may be greater future fluctuation than suggested.

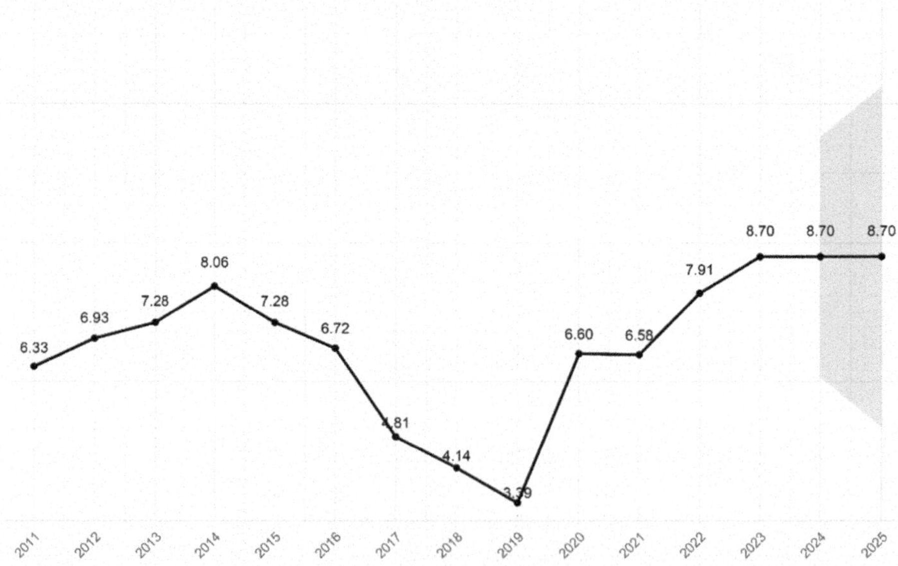

Fig. 3. Solvency forecast

Figure 4 shows the forecast for **financial autonomy**. The observed behavior is consistent with that of the indicators analyzed previously, that is, stability (0.88), signaling the consolidation of the trajectory observed in previous years. This result suggests a continued strengthening of the entity's financial structure, reflecting a low dependence on third-party capital and a high degree of autonomy in the management of its resources.

Regarding the shaded area, a more pronounced widening is observed in 2025, indicating greater uncertainty associated with the estimate for that year compared to 2024, suggesting, as previously mentioned, greater uncertainty in the value predicted for 2025.

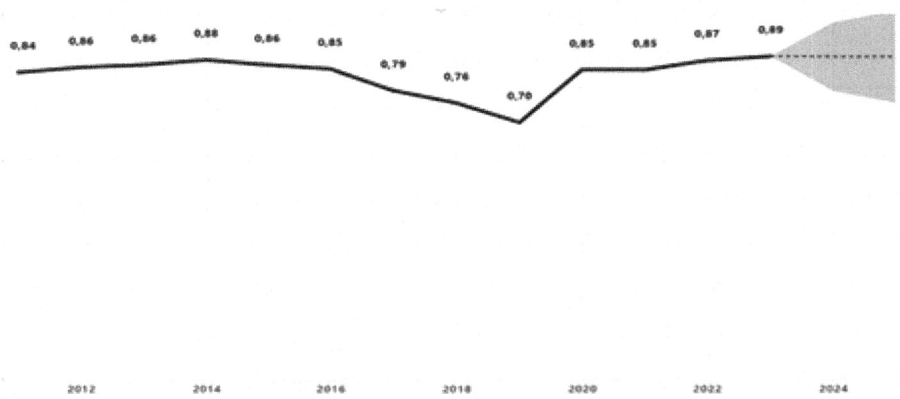

Fig. 4. Financial autonomy forecast

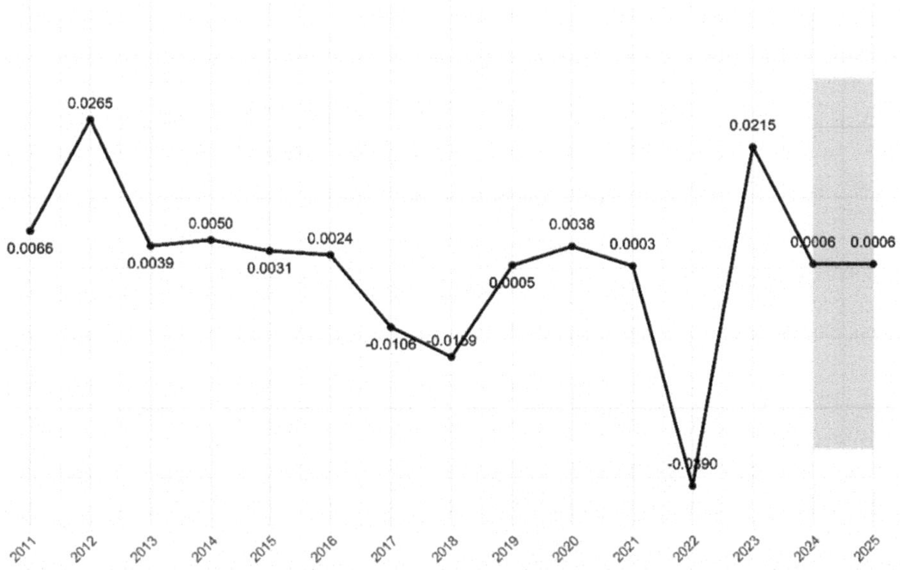

Fig. 5. Return on assets forecast

Finally, regarding the return on assets, as can be seen in Fig. 5, the forecast values are also constant (0.0006). The fact that the shaded band does not show a widening means that the forecast uncertainty has not increased over time.

5 Conclusion

The results obtained show that the application of predictive models, in the context of municipal public management, is not only viable, but also recommended. Comparative analysis using different approaches revealed different performances depending on the indicators analyzed.

The ETS model proved to be effective in predicting **solvency** and **return on assets** indicators, while the native Power BI functionality demonstrated greater accuracy in **general liquidity**, **reduced liquidity** and **financial autonomy** indicators, simultaneously constituting a practical and accessible solution for institutional contexts with more limited technical resources.

The forecasts for the years 2024 and 2025 point to a trend of stability in the main indicators studied, accompanied by controlled levels of uncertainty, which reinforces the usefulness of these models as instruments to support decision-making. From a theoretical point of view, this research contributes to the reflection on the methodological adequacy in the use of predictive techniques applied to local public management, demonstrating that the choice of models must be sensitive to the characteristics of the time series analyzed.

In practical terms, the results support the relevance of integrating advanced statistical algorithms into accessible platforms, such as Power BI, enhancing analytical capacity and the quantity of forecasts generated.

In this sense, predictive analysis constitutes a concrete added value for municipal services and should be systematically integrated into planning and management processes. The adoption of these methodologies represents a significant contribution to the modernization and professionalization of local public administration, with direct impacts on financial sustainability and the improvement of services provided to the population.

As future work, it is proposed to test other prediction models and develop methodologies that allow the use of other artificial intelligence algorithms in predictive analysis.

Acknowledgements. This work was financed by the CICF of the Instituto Politécnico do Cávado e do Ave by national funds through FCT – Fundação para a Ciência e a Tecnologia, I.P., within the scope of multi-annual financing UIDB/04043:CICF e UIDP/04043:CICF.

References

1. Monteiro, M.: A análise preditiva sob o aspecto da regulação," Instituto Superior de Gestão (2018)
2. Freitas, C., Barddal, J.: Análise Preditiva e Decisões Judiciais: controvérsia ou realidade? Revista Democracia Digital e Governo Eletrônico **1**(18) (2019)

3. Oliveira, Â.: A Relevância da Informação Contablística no Processo de Tomada de Decisão do Setor Público - Estudo de Caso da Marinha. Universidade de Aveiro (2017)
4. Pinho, S.: A Contabilidade de Gestão nas Autarquias Locais: O Caso do Município de Espinho. Instituto Politécnico do Porto (2014)
5. Queiroz, F., Hékis, H., Andrade, D., Queiroz, J., Macêdo, D.: Previsão e Séries Temporais para Tomada de Decisão Empresarial em uma Indústria Moveleira da Região de Criciúma–SC. Revista Catarinense da Ciência Contábil **11**(32), 26–42 (2024). https://doi.org/10.16930/2237-7662/rccc.v11n32p26-42
6. Cunha, T.: Métodos de Previsão: aplicação prática no volume de negócios de uma empresa de consultoria. Universidade do Porto (2022)
7. Camelo, H.D.N., Lucio, P.S., Leal Junior, J.B.V., De Carvalho, P.C.M.: Métodos de previsão de séries temporais e modelagem híbrida ambos aplicados em médias mensais de velocidade do vento para regiões do nordeste do Brasil. Revista Brasileira de Meteorologia **32**(4), 565–574 (2017). https://doi.org/10.1590/0102-7786324005
8. Lima, S.: Métodos de previsão de séries temporais - uma aplicação a dados do segmento do Retalho. Universidade do Minho, Dissertação de Mestrado (2018)
9. Marques, A.: Desafios na Previsão de Séries Temporais Financeiras: o Caso da Taxa de Câmbio EUR/USD. Universidade de Ciências / Instituto Universitário de Lisboa, Dissertação de Mestrado (2017)
10. da Cunha, T.M.T.: Métodos de Previsão: aplicação prática no volume de negócios de uma empresa de consultoria. Universidade do Porto (2022)
11. Lima, J.: Análise e Previsão de Séries Temporais - uma abordagem com modelos de espaço de estados. Universidade do Minho, Dissertação de Mestrado (2023)
12. Palma-Ruiz, J.M., Torres-Toukoumidis, A., González-Moreno, S.E., Valles-Baca, H.G.: An overview of the gaming industry across nations: using analytics with power BI to forecast and identify key influencers. Heliyon **8**(2) (2022). https://doi.org/10.1016/j.heliyon.2022.e08959
13. Pinheiro, S.: Potencialidades do Power BI Desktop na Análise Preditiva. Universidade Católica Portuguesa (2020)
14. Pinheiro, S.: Potencialidades do Power BI Desktop na Análise Preditiva. Universidade Católica Portuguesa, Porto, Dissertação de Mestrado (2020)
15. Marques, A.: Um contributo para a caraterização da sua posição financeira. Dissertação de Mestrado, Instituto Poliutécnico de Porto (2012)
16. Rezende, S.: Soraia Ortet II Agradecimentos. Dissertação de Mestrado, Instituito Politécnico de Setúbal (2019)
17. Teixeira, A.B., Mata, C., Pardal, P.N., Teixeira, N.: Avaliação e Divulgação de Indicadores de Desempenho dos Municípios Portugueses: O Caso do Distrito de Setúbal. Revista Universo Contábil, 147–168 (2013). https://doi.org/10.4270/ruc.2013109
18. Rocha, C.: Os Indicadores de Gestão na Prestação de Contas dos Municípios Portugueses. Universidade do Minho, Dissertação de Mestrado (2015)
19. Baião, L.: Transparência e Desempenho Económico-Financeiro e Orçamental nos Municípios em Portugal. Universidade de Coimbra, Coimbra, Dissertação de Mestrado (2020)
20. Fernandes, C., Peguinho, C., Vieira, E., Neiva, J., Edição, A.: Análise Financeira Teoria e Prática Aplicação no âmbito do SNC (2022). www.silabo.pt
21. Correia, A.: Análise do Desempenho na Administração Pública com recurso a Rácios Financeiros: O caso do Município de Matosinhos. Dissertação de Mestrado, Instituto Politécnico do Porto (2015)
22. Gil, A.: Métodos e Técnicas de Pesquisa Social 6a (2008)
23. Hyndman, R.J., Athanasopoulos, G.: Forecasting: Principles and Practice, 3th ed. (2021). Accessed: 10 Mar. 2025. https://otexts.com/fpp3/
24. Bayer, F., Souza, A.: Wavelets e Modelos Tradicionais de Previsão: um Estudo Comparativo. Revista Brasileira de Biometria **28**(10) (2010)

25. Possato: Ajuste do ARIMA para análise e previsões de séries temporais. RPubs by RStudio
26. Possato: RPubs - Ajuste do modelo ARIMA (ideal) - Tutorial. Accessed: 12 Dec. 2024. https://rpubs.com/Possato/ARIMA_ajuste_tutorial

Promoting Job Retention: Strengthening Tax Incentives for Corporate Social Spending

Ana Arromba Dinis[1(✉)] and Sara Luís Dias[2,3]

[1] Research Centre On Accounting and Taxation (CICF), School of Management, Polytechnic University of Cávado and Ave, Barcelos, Portugal
adinis@ipca.pt

[2] School of Management, Polytechnic University of Cávado and Ave, Barcelos, Portugal
sldias@ipca.pt

[3] JUSGOV, University of Minho, Braga, Portugal

Abstract. Several authors have demonstrated a negative relationship between unemployment and economic growth, suggesting that companies should implement strategies to retain their workers within the organization. This strategy involves organizations' concern for providing jobs that reduce employee turnover and companies' commitment to offering attractive job opportunities. However, the role of the Government is also essential in enabling employment by regulating the protection of workers while offering support to companies to hire. The use of taxation is a powerful public policy tool. In this way, Governments should increasingly seek to integrate social aspects into their tax policies to encourage companies to use different forms of remuneration tailored to the specific needs of their employees, thereby improving their corporate social responsibility (CSR) and job creation. In this context, the Social Utility Regime (hereafter Social Tax Incentive) appears in Portugal as a tax incentive for companies to establish social, educational, and cultural support infrastructures, among others, for the benefit of the company's employees, their families, and pensioners. Therefore, this study examines the impact of tax incentives on corporate social spending and job retention in Portugal. Specifically, it examines the Social Utility Regime, a fiscal policy tool enabling companies to invest in social, educational, and cultural support for their employees. Using data from 2011 to 2021 provided by the Portuguese Tax Authority and the SABI database, the study analyses the impact of these incentives on employment and wage-related indicators. Bivariate and regression analyses were performed to test four hypotheses concerning employment levels, wage growth, research, and development (R&D) employment, and gender hiring trends. The results suggest a significant positive correlation between social tax incentives and increased R&D employment, wage growth, and female recruitment. However, awareness of and uptake of the incentive remain low, indicating a need for greater Government promotion.

Keyword: Tax Incentives; Job Retention; Employment

L. Almeida et al. (Eds.): ICAFI 2025, ISEM 80, pp. 491–504, 2026.
https://doi.org/10.1007/978-3-032-16268-7_28

1 Introduction

The employment rate has been shown to have a significant impact on the economic development of a nation. So, it is recommended that states should increasingly seek to incorporate social aspects in their taxation policies because this would encourage companies to offer different forms of remuneration geared to the specific needs of their employees. Employment plays a critical role in national economic development, with multiple studies demonstrating the negative correlation between unemployment and growth. As a result, Governments are increasingly called upon to embed social considerations into their tax policies to stimulate corporate responsibility and workforce stability. This research explores how such integration—particularly through Portugal's Social Utility Regime—can influence corporate behaviour related to job retention.

The motivation for this study stems from the limited empirical evidence on how Portuguese companies perceive and use tax incentives to retain employees. Existing research tends to overlook corporate perspectives on social tax policies, particularly in the context of Portugal. Thus, this study contributes to both national and international literature on tax policy and employment management. So, the present study is relevant for several reasons. Within the national context, the present study aims to make a further contribution to the body of knowledge surrounding Portuguese tax policy. In the context of international relations, the present study aspires to contribute to the extant literature on tax policy, with a particular focus on its impact on corporate employment management decisions. Moreover, international literature on job retention is scarce in the views of companies. To the best of our knowledge, no other empirical study has hitherto examined how access to social tax incentives can be explained by the ability of Portuguese firms to retain jobs, which lends further emphasis to the present research. This study aims to address the following research problem: How do social tax incentives affect corporate employment decisions in Portugal? The research objectives of this work are, to highlight the importance of employment to economic growth, to explain how political strategies promote employment protection, to debate the role of taxation in job retention, to introduce tax incentives for social utility expenses in the Portuguese context; The fifth and last one, to study the Portuguese companies that have accessed tax incentives for social utility expenses and their impact on employee retention. The research questions that have been formulated are as follows: Do social tax incentives increase the level of employment? Do these incentives promote wage appreciation? Do they support R&D hiring? Do they influence gender hiring patterns?

The main contribution of this study is to provide empirical insights into the effectiveness of tax incentives in promoting social responsibility and labour market stability. The text is structured into five main sections: it begins with an introduction that frames the topic of corporate social spending and the Portuguese government's role in promoting it through fiscal incentives. This is followed by a results analysis section that explores the influence of R&D employees, female employees, and male employees on social spending outcomes, highlighting the significance and strength of each variable. A critical discussion is also included regarding the government's limited effectiveness in disseminating information about the tax benefit, suggesting that insufficient communication and outreach efforts may have contributed to the low level of corporate participation.

Finally, the conclusions section synthesizes these findings and presents the limitations and perspectives for future research.

2 Literature Review

2.1 Empirical Studies on Tax Incentives and Employment

About the literature review, we note that several authors argue that the level of employment affects the economic development of any country, and there is a negative relationship between unemployment and economic growth. As a result, companies should implement strategies to retain workers in organizations [1–3]. In this regard, the role of the Government is essential in enabling employment, by ruling employees´ protection and at the same time offering companies support for hiring [4]. Therefore, the use of taxation is a powerful instrument of public policy. This may even positively influence the allocation of human resources to Research and Development (R&D), fundamental for businesses' competitiveness [5]. Indeed, tax incentives are a valuable mechanism for States to foster employment [6, 7], and workplace-based tax incentives can be effective in stimulating local employment [8]. As a result, [9], reducing the corporate income tax rate can increase overall employment by affecting employee wages. Also, [6] states that a profit generated by a large corporate tax incentive explains strong wage gains for workers. At the same time [10] argues that tax incentives and R&D investments in technology-based SMEs positively affect their employment absorption. By that, [11] also claim that all taxes and transfers significantly reduce gender income inequality, but cannot compensate for high gender earnings gaps. Following the previous literature review, it is also possible to conclude that incentives also shape gender dynamics in hiring. Targeted fiscal incentives can encourage inclusive hiring, reducing gender disparities. [12] examined the impact of South Africa's Employment Tax Incentive (ETI) using administrative data from companies and employees between 2014 and 2016. The authors concluded that the tax incentive had a modest yet statistically significant effect on job preservation, particularly within small businesses. They estimated that around 35,000 jobs were preserved thanks to the ETI. The impact was greater in companies with fewer than ten employees, suggesting that small businesses are more sensitive to this type of public policy. The research of [13] evaluated the South African Employment Tax Incentive (ETI), focusing specifically on the employability of young people. While no significant average effects were found on the total number of young people employed, positive and statistically significant effects were observed in companies with up to two hundred employees. This suggests that tax incentives are more effective in certain organisational contexts, such as small and medium-sized enterprises (SMEs), and highlights the importance of targeted policies. In their study, [14] analysed the impact of an Italian tax reform that offered tax incentives for productivity-based pay. The authors demonstrated that the tax incentive increased labour productivity and average wages in beneficiary companies. They also found that the positive effects were more significant in family businesses located in northern Italy. This suggests that the local institutional and economic environment can influence the success of such policies.

These empirical insights lead to the following hypotheses: (i) Hypothesis One: Tax policy can increase the level of employment; (ii) Hypothesis Two: Tax incentives can promote wage appreciation; (iii) Hypothesis Three: Tax policies that encourage job retention can increase the number of R&D employees; (iv) Hypothesis Four: Companies that benefit from social tax incentives are more likely to employ more women than men.

2.2 Legislative Context: The Portuguese Social Utility Regime

The Portuguese Corporate Tax Code, under Article 43, provides tax deductions for socially beneficial expenditures such as childcare, education, and healthcare for employees. So, as part of a broader policy mix, the Social Utility Regime appears in Portugal as a tax incentive to companies for the establishment of social, educational, and cultural support infrastructures, among others, for the benefit of company employees, their families, and pensioners. In the Portuguese context, in article 43 of the Corporate Tax Code, the tax legislator listed a set of deductions from taxable profit for social nature expenses incurred by companies in payments to their employees (in service or already retired) and, in some cases, also to their family members, associated with the household's costs with nurseries, kindergartens, canteens, libraries and schools, as well as other socially useful activities recognized as such by the Directorate-General for Taxation. Underlying this are extra-fiscal concerns related to improving the social security of workers and their families. These costs are related to the promotion of social practices by companies, which, on the one hand, promote social assistance for families and, on the other, indirectly benefit companies by contributing to their image and the well-being of workers and, consequently, increasing their productivity. These payments must be general (i.e., they must be granted uniformly to all workers) and may not take the form of income from dependent work. It must be registered in the company's bookkeeping, regardless of its treatment in terms of social security or any other tax. These measures aim to enhance job quality and corporate social responsibility while improving productivity. Up to 15% of total personnel costs are deductible for benefits like insurance and pension contributions, provided they apply to workers. So, this measure considers the deduction, in article 43(2) of the IRC Code, up to a limit of 15% of the personnel costs accounted for as remuneration, wages or salaries for the tax period, of costs related to the payment of premiums for sick-ness insurance, personal accidents, pension funds and supplementary social security schemes, provided that they are enjoyed by all workers, or only by certain professional classes, in the latter case if the existence of such benefits is the result of a collective labour regulation instrument and their granting complies with objective and identical criteria for all beneficiaries. Regarding the amount of expenses considered for the calculation of the limit provided for in this article, the jurisprudence established that the relevant staff costs are those which, in accounting terms, must be registered as remuneration, wages or salaries, and there is no other criterion in the law which allows only costs which are subject to social security deductions to be considered (Administrative Supremo Court decision of 29/11/2017, process no 013/14). After the adoption of this interpretation, the Tax Administration updated its administrative orientations in this way (e.g., administrative information no 2020 000298, sanctioned by Order no. 227/2020 - XXII, of the Secretary of State for Tax Affairs). This provision allows the accounting recognition of certain expenses, which are not accepted for tax purposes under Article 23 of the CIRC. It can be deductible in the determination of tax results, justified by the social function performed by the company. Therefore, these expenses are accepted in

IRC but excluded from taxation on IRS (personal income taxation). Article 43 of the CIRC has a flexible framework, rather than a closed one as, at the beginning of paragraph 1, the expression "including" is used (i.e. it mentions, in an exemplary way, as deductible expenses: "depreciations or amortizations and rents of real estate, related to the optional maintenance of crèches, nurseries, kindergartens, canteens, libraries and schools, as well as other achievements of social utility as such recognized by the Directorate General of Taxes, made for the benefit of the staff or retirees of the company and their families", but nothing prevents other expenses from being included here, as long as they assume a general character (in the terms defined above)and do not constitute income from dependent work. As mentioned in this article, "other socially useful activities recognized as such by the Directorate-General for Taxes and carried out for the benefit of the company's staff or pensioners and their families are allowed." However, this does not include the social contributions referred to in paragraph 2 of this article (such as, for example, contributions to health and personal accident insurance contracts, as well as life insurance contracts, contributions to pension funds and the like or any supplementary social security schemes, among others), to which the 15% limit provided for here is applicable. As Portuguese courts have held, only proven and actual costs are tax deductible. Proof of accounted costs can be provided by any type of evidence, "and it doesn't have to be supported by a typical expense document" (in this sense, decision of the Central Administrative Court South of 14/01/2020, case no. 282/03.6BTLRS). In summary, judicial interpretations and updates, such as Order 227/2020, have clarified the deductible scope. However, legal complexity may still discourage firm participation.

2.3 Hypotheses and Research Questions

Based on the literature reviewed and the legislative context of the Portuguese Social Utility Regime, four hypotheses were developed to guide this research. Each hypothesis was designed to test a specific potential outcome of the social tax incentive policy and is linked to a corresponding research question aimed at empirically validating its premise (Table 1).

Table 1. Hypotheses and Research Questions

Hypothesis	Corresponding Research Question
H1: Tax policy can increase the level of employment	**RQ1**: Do social tax incentives increase the level of employment?
H2: Tax incentives can promote wage appreciation	**RQ2**: Do tax incentives for social utility expenses promote wage appreciation?
H3: Tax policies encouraging job retention can increase R&D employment	**RQ3**: Do companies hiring R&D benefit from social tax incentives?
H4: Companies benefiting from tax incentives are more likely to employ more women	**RQ4:** Do companies using these incentives hire more women or men?

The first hypothesis (**H1**) posits that tax policy can increase the level of employment. This hypothesis is examined through Research Question 1 (**RQ1**): Do social tax incentives increase the level of employment? This seeks to determine whether companies that benefit from the regime tend to have higher overall employment figures. The second hypothesis (**H2**) suggests that tax incentives can promote wage appreciation. It is addressed by Research Question 2 (**RQ2**): Do tax incentives for social utility expenses promote wage appreciation? This question investigates whether the financial advantages provided by the policy are associated with higher employee compensation. The third hypothesis (**H3**) assumes that tax policies designed to encourage job retention can increase employment in research and development (R&D). Research Question 3 (**RQ3**) asks: Do companies hiring for R&D benefit from social tax incentives? This question is essential for evaluating the policy's potential impact on innovation-driven employment. Finally, the fourth hypothesis (**H4**) claims that companies benefiting from social tax incentives are more likely to employ more women. This is evaluated through Research Question 4 (**RQ4**): Do companies using these incentives hire more women or men? The goal here is to assess whether the regime contributes to gender inclusion in the workplace. These hypotheses and research questions form the analytical foundation of the study, which is addressed in the subsequent results and discussion sections through quantitative analysis of firm-level data.

3 Research Methodology

This section presents the research methodology, which includes research hypotheses, data collection, sample characterization, and results. This study adopts a quantitative research design using secondary data sourced from the Portuguese Tax Authority and the SABI database (Iberian Balance Sheet Analysis System), which has allowed the authors to create a database of all the Portuguese companies that have taken advantage of the tax incentives for corporate social spending. An empirical study was developed on the Portuguese companies that took advantage of social tax incentives in the period 2011 to 2021. Using the Portuguese Tax Authority database, information was gathered on Portuguese companies that have taken advantage of tax incentives for corporate social spending. The authors used the software Statistical Package for the Social Sciences for Windows (SPSS).

3.1 Sample Description

During the study period, 2,114 Portuguese firms accessed the social tax incentive, totalling €12,573,110.65, as shown in Table 2.

Table 2 provides a quantitative overview of the use of the corporate tax benefit for social spending in Portugal between 2011 and 2021. Over these 11 years, a total of 2,114 companies accessed the benefit. When contextualized, these figures reveal the limited scale of corporate participation in the fiscal incentive scheme. Despite spanning more than a decade, the number of beneficiary companies is low, especially when considering the total number of active businesses in Portugal. On average, fewer than two hundred

Table 2. Data sources

period	corporate beneficiaries	Total (€)
2011–2021	2,114	12,573,110.65

companies per year made use of the incentive—a fraction of the national corporate landscape. The total financial value of just over €12.5 million across 11 years also suggests that the fiscal instrument has not yet achieved its intended scale or influence. The small number of firms actively claiming the tax benefit suggests that many socially engaged or innovation-oriented firms may not be leveraging available fiscal tools, either due to a lack of awareness, complexity in accessing the benefit, or perceived misalignment with their operational goals.

3.2 Research Questions

As far as research methodology is concerned, and to achieve the objectives initially defined, four research questions were outlined, to which an answer was sought: (i) Do social tax incentives increase the level of employment? (ii) Does the tax incentive for social utility expenses promote wage appreciation? (iii) Do companies hiring resources allocated to research and development have access to the tax incentive for social utility expenses? (iv) Do Companies that access the tax incentive for social utility expenses hire more women or men?. A bivariate analysis was applied. The objective was to study the relationship between the social tax incentives gained on average by Portuguese companies in the period under analysis and the social balance indicators.

3.3 Correlation Results

The correlations of the variables and their constituent items were assessed using Pearson correlation (see Table 3).

Table 3. Bivariate analysis

		r&d employeesmean 2011_2021	female employeesmean 2011_2021	male employees mean 2011_2021	employee costs mean 2011_2021	part-time employees mean 2011_2021	profit per employeemean 2011_2021
soctxb_mean 2011_2021	**R**	**0.522****	**0.072****	**0.102****	**0.128****	**0.360****	**0.075****
	p	< 0.001	0.001	< 0.001	< 0.001	< 0.001	0.001

Notes: ** Pearson correlations are significant at the 0.01 level (2-tailed). * Pearson correlations are significant at the 0.05 level (2-tailed)

The results reveal that there is a significant positive correlation between social tax incentives gained on average by Portuguese companies in the period under analysis and the social balance indicators. The highest significant and positive correlation relates to the average number of people employed in Research and Development. These results show the strength that social tax incentives have in encouraging businesses to hire human resources for Research and Development. Another positive correlation that stands out has to do with the average number of part-time employees. These results show the revolution of part-time work and the social aspect; more people are thinking about their family and social life. The results obtained in this study are consistent with the theoretical expectations outlined in the literature review. So, the results of this study are in line with recent empirical research on the effects of tax incentives on employment and wages. For instance, evidence from South Africa's Employment Tax Incentive (ETI) shows that such policies can lead to measurable, though moderate, increases in employment, particularly among small firms. According to [12], ETI helped preserve approximately 35,000 jobs between 2014 and 2016 and had the most impact on firms with fewer than ten employees (Bhorat et al., 2020). Likewise, [13] found that although the average effects of the incentive on youth employment were limited, firms with fewer than two hundred employees showed significant employment growth, suggesting that tax benefits can be especially impactful in specific business segments. Regarding wage effects, [14] showed that a tax break on incentive pay in Italy had a positive impact on both productivity and average wages, confirming the relevance of fiscal policies in shaping remuneration strategies. These findings support the positive relationships identified in this study between tax incentives and employment levels, wage appreciation, R&D hiring, and gender inclusion. They reinforce the broader conclusion that well-structured tax policies can contribute meaningfully to corporate social responsibility and labour market outcomes. These consistencies indicate that the Portuguese Social Utility Regime, despite being underutilized, aligns with global findings on the effectiveness of tax incentives as instruments of public employment policy and corporate social responsibility.

3.4 Regression Analysis

The authors also employed a model in which the variable to be explained (social tax incentives) is related to explanatory variables (social balance indicators). A linear regression model identified key employment predictors of social tax incentive uptake (Table 4).

The model set-up confirms that the average number of persons employed in R&D and the average number of female employees are important determinants of companies' eligibility for the social tax benefit under study. Specifically, the regression model presented shows the relationship between the average number of Research and Development (R&D) employees, female employees, and male employees from 2011 to 2021, and a dependent variable referred to as "soctxb_mean_2011_2021." The unstandardized coefficients indicate that all these independent variables have a positive effect on the dependent variable. Specifically, for each additional unit increase in the average number of R&D employees, the dependent variable is expected to increase by approximately 39,855.92 units, holding the other variables constant. Similarly, for each unit increase in the average number of female employees, the dependent variable is expected to rise by about 26,103.28 units, while for male employees, the increase is estimated at 6,503.91 units. All these coefficients are statistically significant, with p-values less than

Table 4. Regression model

Model	Unstandardized Coefficients			
	Coeff	Std. Error	t	Sig
R&D employees_mean_2011_2021	39,855.920	8,003.876	4.980	< 0.001
female employees_mean_2011_2021	26,103.285	4,890.746	5.337	< 0.001
male employees_mean_2011_2021	6503,908	2,071.226	3.140	0.002
R^2	0.671			
Adjusted R^2	0.451			

Notes: a. Dependent Variable: **soctxb_mean_2011_2021 (tax benefits granted on average between 2011 and 2021)**

0.01, which indicates that the observed relationships are unlikely to be due to chance and that the variables truly have a meaningful impact on the dependent variable. Additionally, the t-values associated with each coefficient are high, further reinforcing the reliability of these results. Regarding the model's overall quality, the coefficient of determination, R2, is 0.671. This means that approximately 67.1% of the variance observed in the dependent variable can be explained by the independent variables included in the model. However, the adjusted R2 value is 0.451, which suggests that when accounting for the number of variables and the sample size, the model's explanatory power drops to around 45.1%. This difference implies that although the model is explanatory, there is room for improvement, and some variables may be contributing less than expected to explain the variance in the dependent variable. Overall, the model indicates that the average number of R&D employees has the strongest impact on the dependent variable, followed by female employees with an intermediate impact, and finally male employees, whose impact is positive but smaller. These results may suggest that R&D teams and female participation play a particularly key role in the phenomenon captured by the variable "soctxb_mean_2011_2021."

3.5 Results and Research Hypotheses

The existence of a significant positive correlation between social tax incentives gained by Portuguese companies in the period under analysis and several key performance indicators, including the average number of full-time and part-time employees, the average number of employees paid and unpaid employees, the average number of employees, the average employee's costs, the average total assets per employee and the average equity per employee, it is consistent with the literature review. The analysis confirms that each research question is directly addressed by the empirical findings, allowing for the validation or rejection of the corresponding hypotheses. Regarding **RQ1**, the positive and significant correlation between the average tax incentives and overall employment indicators supports **H1**, indicating that tax policy can indeed stimulate employment levels. In response to **RQ2**, the observed correlation between tax incentives and employee wage-related variables confirms **H2**, suggesting that these fiscal benefits contribute to wage appreciation. **RQ3** is strongly supported by the highest correlation found in the study—between tax incentives and the number of R&D employees—thus confirming **H3** and highlighting the role of fiscal policy in fostering innovation-based employment. Lastly, the data reveal a statistically significant, albeit moderate, relationship between social tax incentives and female employment, affirming **H4** in the context of **RQ4**. The results highlight a **disconnect between policy design and implementation**. While the Government has created a mechanism to reward corporate social spending, the low level of participation indicates that **structural and communication barriers remain**. This supports the conclusion that **Portugal's Government still has substantial work to do** to activate the full potential of the private sector in contributing to social development through fiscal incentives.

4 The Portuguese Government's Approach to Strengthening Corporate Social Spending: A Brief Critical Analysis of the Results

The results reinforce prior research on the effectiveness of tax incentives as a tool for employment and social policy. The positive correlation between R&D employment and the Social Utility Regime supports the hypothesis that tax incentives can stimulate innovation-driven job creation. Similarly, the statistically significant relationship between female employment and social tax benefits confirms previous findings that tax policy can promote gender inclusion. Although the correlation is smaller, it is robust enough to affirm research hypotheses. The findings also suggest that firms with higher employee costs and more part-time roles are more likely to benefit from these incentives. This dual economic and social appeal supports corporate investment in employee well-being and workforce flexibility. Notwithstanding this fact, and in the case of Portugal, the number of companies that have availed themselves of social tax incentives remains disproportionately low in comparison to the number of Portuguese companies. Despite their benefits, the limited uptake of the Social Utility Regime may reflect awareness gaps, legal ambiguity, or administrative hurdles. These barriers align with broader literature indicating that for tax policies to be effective, they must be well-promoted and communicated. From our point of view, the Portuguese Government must establish measures to disseminate these tax incentives, so that companies may become more involved, especially because these benefits improve job retention. Considering the results presented, it is essential to reflect on how they relate to the initial research framework. This study was guided by four research questions, each connected to a specific hypothesis, aiming to assess the effectiveness of Portugal's social tax incentives in promoting employment and corporate social responsibility. The statistical analyses carried out—correlation tests and regression models—offered solid empirical evidence that enables the validation of each hypothesis. By evaluating the relationships between tax benefits and key employment indicators such as R&D staff, wage levels, gender composition, and overall workforce characteristics, the study was able to respond directly to each research question. The regression analysis demonstrates that the average number of R&D employees, female employees, and male employees each has a significant positive effect on the dependent variable "soctxb_mean_2011_2021." Among these, R&D employees exert the strongest influence, suggesting that investments in research and development staff are most strongly associated with increases in the outcome measured by the dependent variable. Female employees also have a substantial positive impact, greater than that of male employees, indicating that gender composition within the workforce plays a significant role in driving the dependent variable. The smaller yet positive coefficient for male employees suggests they contribute as well, but to a lesser extent. The model explains a meaningful portion of the variability in the dependent variable, with about 67% of the variance accounted for by the predictors. However, the adjusted R2 indicates that when accounting for model complexity, about 45% of the variation is explained, highlighting that other factors not included in this model may also be important. Overall, these findings emphasize the value of R&D personnel and female workforce participation in influencing the studied outcome. Organizations aiming to improve this outcome should consider strengthening their R&D teams and supporting gender diversity as key strategic priorities. This coherence between the empirical results and the theoretical propositions reinforces the robustness of the study and provides a comprehensive understanding of how fiscal mechanisms can influence organizational behaviour and employment strategies. Collectively, these findings align the research questions with the proposed hypotheses and provide empirical grounding for the study's core arguments. Given the strong positive impact of R&D employees on the outcome variable, organizations should prioritize investments in research and development teams. This could include hiring more skilled R&D staff, providing ongoing training and development, and fostering an innovative work environment that encourages creativity and experimentation. The substantial influence of female employees suggests that promoting gender diversity within the workforce can also be beneficial. Companies should implement policies and practices that support the recruitment, retention, and advancement of women. This might involve flexible work arrangements, mentorship programs, and initiatives to reduce gender bias and create an inclusive culture. Although the impact of male employees is positive but smaller, ensuring balanced workforce composition and collaboration between all employees, regardless of gender, can help maximize organizational performance. The findings from the regression analysis align closely with the Portuguese Government's recent initiatives aimed at boosting corporate social spending, particularly through investments in human capital and inclusive employment practices. The strong positive impact of R&D employees on the dependent variable suggests that supporting innovation and research activities within companies is vital for enhancing corporate social outcomes. This supports Government policies that encourage firms to allocate resources toward research and development, recognizing that such investments not only drive economic growth but also contribute to broader social benefits. Additionally, the significant role of female employees in positively influencing the outcome highlights the importance of gender diversity within the workforce. The Portuguese Government has emphasized the promotion of equal opportunities and gender inclusion as central pillars of its social agenda. By encouraging companies to strengthen the participation of women in the labour market and decision-making positions, the Government reinforces a corporate culture that values diversity, which is shown by the data to have tangible positive effects on social outcomes. The smaller, yet positive effect of male employees further indicates that a balanced and diverse workforce contributes to better corporate social performance. The Government's approach to fostering inclusive employment policies, flexible work environments, and anti-discrimination laws helps create conditions where all employees can contribute effectively, thus maximizing social impact. Finally, the model's explanatory power suggests that while human capital and gender composition are key factors, other elements also influence corporate social spending outcomes. This underlines the importance of comprehensive Government strategies that combine financial incentives, regulatory frameworks, and social policies to holistically strengthen corporate social responsibility in Portugal. However, despite the alignment between these results and governmental goals, there remains a considerable gap between policy intentions and corporate behaviour. In particular, the uptake of the tax benefit for corporate social investment—introduced to encourage firms to contribute more actively to societal well-being—has been strikingly low. This indicates that the existing fiscal incentives are either insufficiently attractive, inadequately communicated, or not aligned with firms' strategic priorities. The low participation rate suggests that the Portuguese Government still has a long road ahead in strengthening the mechanisms that translate policy into action. It may be necessary to redesign or expand fiscal incentives, provide more targeted support to firms (especially SMEs), and launch awareness campaigns to clarify the strategic and financial benefits of engaging in corporate social spending. Furthermore, integrating social objectives more explicitly into Environmental, Social, and Governance (ESG) frameworks and procurement standards could help normalize and mainstream corporate investment in social value creation.

5 Conclusions, Limitations, and Future Research Perspectives

This study provides empirical evidence that social tax incentives in Portugal positively impact employment, particularly in the hiring of research and development (R&D) staff and the inclusion of women in the workforce. The Social Utility Regime is confirmed as a viable policy tool for promoting job quality and corporate social responsibility. The authors of the study make the case for the existence of a significant positive correlation between social tax incentives gained by Portuguese companies in the period under analysis and several key performance indicators. The results obtained imply that human capital, especially in terms of R&D capabilities and gender diversity, is a critical driver of the outcome measured by the dependent variable. For policymakers and business leaders, this underscores the importance of fostering environments that value both technical expertise and diverse talent pools. By focusing on these areas, organizations may improve their competitive advantage, innovation capacity, and overall performance in their respective industries. The moderate adjusted R^2 also suggests that other factors not captured by this model could influence the outcome, so a comprehensive approach that considers additional variables such as technology adoption, organizational culture, or market conditions might provide further insights. While the results validate the direction of current public policy, they also reveal structural barriers that must be addressed. For the Portuguese Government to fully unlock the social potential of its corporate sector, a more robust, responsive, and visible policy framework is required—one that not only promotes best practices but also makes participation accessible and worthwhile for a broader spectrum of companies.

This study has some limitations. This study only looked at tax benefits granted until 2021, which is the period that could be analysed so far. The analysis is limited to data up to 2021, excluding more recent developments and potential post-pandemic effects. The absence of a control group restricts the ability to establish stronger causal relationships. Furthermore, legal and administrative complexity may have discouraged broader participation among companies. Meanwhile, the Portuguese Tax Authority has already provided data up to 2023, which should be the subject of future studies. The preliminary analysis shows a decrease in tax benefits from 2021 (COVID effect?) to 2022 and a slight increase from 2022 to 2023 (recovery period). In each year, they amount to €2,052,343.26 and €2,607,097.96, respectively. The number of companies continuing to benefit from social tax incentives remains low in 2022 and 2023 (around 1,500 companies). There are still some legal disputes between taxpayers and the tax authorities, which may also be the reason for taxpayers' mistrust of the benefit. This should also be investigated, especially among companies. So, future research should aim to include updated data, particularly from 2022 and 2023, to assess the impact of the COVID-19 pandemic on the use of social tax incentives. Introducing a control group would also strengthen causal inference. Additionally, it would be valuable to examine companies' awareness and trust in the regime and to explore the long-term effects of these incentives on business productivity.

References

1. Wijaya, A., Tasenţe, T., Darma, D.C., Kasuma, J.: Labor force and economic growth based on demographic pressures, happiness, and human development: Empirical from Romania. J. Eastern European and Central Asian Res. **8**(1), 40–50 (2021)
2. Christiano, L.J., Trabandt, M., Walentin, K.: Involuntary unemployment and the business cycle. Rev. Econ. Dyn. **39**, 26–54 (2021)
3. Park, C., McQuaid, R., Lee, J., Kim, S., Lee, I.: The impact of job retention on continuous growth of engineering and informational technology SMEs in South Korea. Sustainability (Switzerland) **11**(18) (2019)
4. Lucas, D.S., Boudreaux, C.J.: National regulation, state-level policy, and local job creation in the United States: a multilevel perspective. Res. Policy **49**(4), 103952 (2020)
5. Teirlinck, P.: Enhancing R&D employment slack during environmental turbulence: Triggers and firm performance consequences for R&D-active SMEs. Technovation **118**(August), 102622 (2022)
6. Carbonnier, C., Malgouyres, C., Py, L., Urvoy, C.: Who benefits from tax incentives? The heterogeneous wage incidence of a tax credit. J. Pub. Econ. 206 (2022)
7. Gurmu, S., Sjoquist, D.L., Wheeler, L.: The effectiveness of job creation tax credits. Reg. Sci. Urban Econ. **90**(August), 103721 (2021)
8. Ku, H., Schönberg, U., Schreiner, R.C.: Do place-based tax incentives create jobs? J. Pub. Econ. **191**(104105) (2020)
9. Zuo, S., Wu, B., Feng, J.: Does government reduction of the corporate income tax rate increase employment? Evidence from China. Int. Rev. Econ. Fin. **83**(August 2022), 365–372 (2023)
10. Mao, Y.: Tax incentives, R&D investment, and employment absorption: evidence from chinese technology-based SMEs. SAGE Open **14**(2), 1–14 (2024)

11. Avram, S., Popova, D.: Do taxes and transfers reduce gender income inequality? Evidence from eight European welfare states. Soc. Sci. Res. **102**(September 2021), 102644 (2022)
12. Bhorat, H., Hill, R., Khan, S., Lilenstein, K., Stanwix, B.: The Employment Tax Incentive Scheme in South Africa: An Impact Assessment, Working Papers 202007. University of Cape Town, Development Policy Research Unit (2020)
13. Ebrahim, A., Leibbrandt, M., Ranchhod, V.: The effects of the Employment Tax Incentive on South African employment, UNU-WIDER (2017)
14. Damiani, M., Pompei, F., Ricci, A.: Tax breaks for incentive pay, productivity and wages: evidence from a reform in Italy. Br. J. Ind. Relat. **61**, 188–213 (2022)

Relationship Between Tax Benefits and European Funds in Portuguese Companies: An Exploratory Study

Sandra Nunes[1](✉) and Carla Carvalho[2]

[1] Higher Institute of Accounting and Administration, University of Aveiro, Aveiro, Portugal
sign@ua.pt

[2] GOVCOPP - Research Unit On Governance, Competitiveness and Public Policies, Higher Institute of Accounting and Administration, University of Aveiro, Aveiro, Portugal
carla.carvalho@ua.pt

Abstract. This exploratory study examines the potential overlap between projects funded through European funding instruments—specifically, the Recovery and Resilience Plan (RRP) and Portugal 2020—and those used to justify granting Investment Tax Benefits (ITB). It also aims to identify and characterize companies that simultaneously benefited from both ITB and European funds between 2020 and 2023. The purpose is to assess whether public support mechanisms (tax incentives and European funding) are being allocated to the same companies, possibly for the same projects, and to detect patterns in the characteristics of these beneficiaries. Results indicate that approximately 30% of companies receiving ITB also implemented at least one project financed by European funds during the same period. Moreover, 42% of the total ITB amount granted was allocated to companies that also received European funding, suggesting a significant overlap in public financing. Most of these companies are geographically concentrated in the North and Center regions of Portugal, with a strong presence in the manufacturing sector, particularly in terms of project volume and funding amount. These findings contribute to the academic and policy debate on public investment support by providing empirical evidence on the interaction between fiscal and financial incentives. The findings offer evidence to support the eventual revision of eligibility and oversight criteria for these instruments, aiming to enhance the equity and efficiency of public resource allocation.

Keywords: Tax Benefits · European Funds · Recovery and Resilience Plan · Portugal 2020

1 Introduction

Between 2020 and 2023, marked by significant instability, the Portuguese State reinforced its commitment to fiscal instruments to stimulate private investment and support economic recovery. The aim was to create more favorable conditions for productive investment, promote business modernization, and preserve employment and the competitiveness of the national economy. Tax benefits such as the *Sistema de Incentivos*

L. Almeida et al. (Eds.): ICAFI 2025, ISEM 80, pp. 505–526, 2026.
https://doi.org/10.1007/978-3-032-16268-7_29

Fiscais à Investigação e Desenvolvimento Empresarial (SIFIDE, in english System of Tax Incentives for Business R&D), the *Regime Fiscal de Apoio ao Investimento* (RFAI, in english Tax Regime for Investment Support), and the *Benefício Contratual ao Investimento Produtivo* (BCIP, in english Contractual Tax Benefits for Productive Investment), which were in force during this period, were heavily promoted. The *Crédito Fiscal Extraordinário ao Investimento* (CFEI II, in english, Extraordinary Tax Credit for Investment) was also created specifically to support projects developed during the COVID-19 pandemic. These fiscal instruments thus played a significant role in the economic strategy, being more widely used by companies seeking to maintain or expand their operations in an adverse context.

In parallel, Portugal benefited from significant European funding through programs such as Portugal 2020 and the Recovery and Resilience Plan (RRP) to promote sustainable growth, digital and climate transitions, and territorial cohesion. The combination of these support instruments – tax benefits and European funds – raises questions about their coordination and complementarity, especially when both target the same investment projects, which may be part of corporate tax planning strategies.

Tax planning is a management tool designed to mitigate the fiscal impact of business decisions and minimize tax liabilities through strategic legal measures that reduce fiscal costs (Silva, 2008). The objective is to select business activities or make decisions resulting in lower or no taxation. This concept has gained growing relevance in companies as managers are increasingly concerned with adopting strategies and leveraging incentives and benefits to avoid or reduce tax payments (Oliveira, 2021).

This exploratory study primarily aims to analyze whether the projects subsidized by instruments such as the RRP and Portugal 2020 were the same ones used as the basis for granting Investment Tax Benefits (ITB). Simultaneously, the study seeks to identify and characterize the companies that benefited from ITB and received funding from European funds. The analysis period for tax benefits was from 2020 to 2023. However, since deduction policies allow investments made as early as 2008 to generate tax benefits in 2020 (as shown in Table 2), the analysis period for European funds was extended from 2008 to 2023. Portugal 2020 officially started in 2014, and our analysis covers projects from that year onward until 2023. Regarding the RRP, since its launch in 2021, our analysis of RRP-funded projects spans the period from 2021 to 2023.

This analysis enables the identification of overlaps in public support granted to the same company, potentially for the same project, and the determination of patterns and trends in the characteristics of companies that utilize both types of support (tax benefits and European funds). The relevance of this study lies in the importance of evaluating the coherence and effectiveness of the public investment incentive system, ensuring that national or European funds are applied fairly, strategically, and responsibly. Moreover, the study stands out for its innovative nature, as it analyzes, for the first time and systematically, the existence of overlap between ITB and European funds granted to the same company for the potential financing of the same investment project.

For the empirical study, data were cross-referenced between the "*Portal Mais Transparência*" and the list of tax benefit recipients from the Tax and Customs Authority. The goal was to identify companies that, between 2020 and 2023, simultaneously received ITB (specifically SIFIDE II, RFAI, BCIP, or CFEI II) and funding from European funds (RRP or Portugal 2020). The analysis of the RRP focused on components C05, C11, and C16, which were selected for their strong links to business productive investment, innovation, industrial decarbonization, and digitalization. Regarding Portugal 2020, all support measures were considered, given its broad scope in supporting investment. This study contributes to the theoretical and practical understanding of public investment support policies by analyzing the dynamics of overlap between instruments that have been explored to a limited extent until now.

The work is structured into five sections. Following this introduction, Section Two provides the theoretical framework for European funds and tax benefits. Section Three is dedicated to defining the research objectives and questions, as well as describing the methodology used in the empirical study. Section Four presents the exploratory analysis of the main results obtained, and in the final section, the conclusions, contributions, and suggestions for future research are outlined.

2 Theoretical Framework

2.1 Portugal 2020

Portugal 2020 is the partnership agreement established between Portugal and the European Commission, which operationalized the application of the European structural and investment funds from 2014 to 2020, with implementation extended until 2023. This strategic framework aligns with the objectives of the Europe 2020 Strategy, which is focused on intelligent, sustainable, and inclusive growth, aiming to promote the competitiveness of the European economy, employment, and social and territorial cohesion.

Portugal 2020 mobilized around 26.8 billion euros in European Union (EU) funding, distributed across various thematic and regional operational programs. The support provided targeted areas such as business innovation and competitiveness, internationalization, human capital development, social inclusion, energy efficiency, and environmental sustainability (Portugal 2021, 2025).

Regarding the business sector, Portugal 2020 played a central role in stimulating investment, particularly through incentive systems covering productive innovation, research and development (R&D), internationalization, qualified entrepreneurship, and infrastructure upgrading. These supports were especially relevant in the context of economic recovery, strengthening companies' investment capacity and their ability to adapt to the challenges of digitalization and the green transition.

In the context of this study, Portugal 2020 is particularly relevant, as it represents one of the primary sources of European public funding available to Portuguese companies during the analysis period. Among the various operational programs under Portugal 2020, notable examples include *Compete 2020*, the Operational Program for Competitiveness and Internationalization (POCI), and the regional programs (such as *Norte 2020* and *Centro 2020*), which mobilized a significant portion of the funds to support productive business investments, with a strong focus on innovation, qualification, internationalization, technological modernization, and energy efficiency.

Support from Portugal 2020 was accessible to individuals or legal entities from the public, cooperative, social, or private sectors, with or without profit-making purposes, provided they met the established eligibility criteria. These included being up to date with tax and social security obligations, having no outstanding unpaid debts, and complying with the specific conditions outlined in the calls for applications (Portugal, 2021, 2025). These supports were primarily granted as non-repayable or partially repayable financial incentives aimed at stimulating investment in tangible and intangible assets that contribute to increasing productivity and competitiveness of the national economy (IAPMEI, 2025).

2.2 Recovery and Resilience Plan (RRP)

The RRP is a national program designed to boost the country along a path of recovery, sustained economic growth, and convergence with Europe over the coming decade, with sustainability and the fulfillment of the United Nations Sustainable Development Goals (SDGs) as its main pillars (Recuperar Portugal, 2024).

Portugal's financial situation has not been the same since the 2007–2008 financial crisis and the ensuing decade of economic crisis (Vlachos, 2022). Among the 14 countries that, up to 2022, had not recovered their Gross Fixed Capital Formation (GFCF) levels, Vlachos (2022) identifies Portugal as one with the lowest indicators compared to pre-crisis levels. This suggests that Portugal was one of the countries most affected by that crisis and one of the most vulnerable when the COVID-19 pandemic crisis struck. Declared in 2020, the pandemic had immeasurable impacts on national, European, and global economies. To mitigate these impacts, restore the stability of European countries, and foster a greener, more digital, and more resilient economy (Wolff and Ladi, 2020), the EU created the Recovery and Resilience Facility to fund reforms and investments in EU Member States from the onset of the pandemic until December 31, 2026 (European Commission, 2025b).

According to the European Commission (2021), this mechanism combines the EU's long-term budget for 2021–2027 and the NextGenerationEU (NGEU) initiative. It is a temporary instrument to support Member States' recovery by funding investments and reforms that accelerate the green and digital transitions, support education, and promote greater gender, territorial, and generational equality (Di Pirro et al., 2022). To access NGEU funds, each Member State was required to prepare a national recovery and resilience plan for 2021–2026, in accordance with the criteria established in Article 18 of Regulation (EU) 2021/241 of 18 February (European Commission, 2021). This EU regulation defines six key pillars that all national RRPs must focus on: green transition (EU, 2025a), digital transformation, intelligent, sustainable, inclusive growth, social and territorial cohesion, health and economic, social, and institutional resilience, and policies for the next generation.

In this context, Portugal's economic recovery from the COVID-19 crisis is financed through the Portuguese RRP, which prioritizes investments in infrastructure and innovation (Vlachos, 2022). The national RRP is organized into 21 components, comprising 44 reforms and 117 investments, totaling 22.2 billion euros. These components and their respective reforms and investments are grouped into three structural dimensions:

resilience (15.1 billion euros), climate transition (4.39 billion euros), and digital transition (2.72 billion euros). Climate and digital transitions represent 32% of the total funding program, while the remaining resources are allocated to the resilience dimension, which encompasses social vulnerabilities and economic and territorial resilience (Recuperar Portugal, 2024).

The access criteria for RRP funding are diverse but generally relate to the financial health of the beneficiary organization, the targets defined in the program, the feasibility and sustainability of the proposed projects, and the capacity to execute them within the established deadlines. The support granted under the RRP mainly takes the form of non-repayable financial incentives, i.e., grants that do not require repayment, provided that beneficiaries meet the objectives and milestones defined in their respective funding contracts (Recuperar Portugal, 2024).

In Portugal, one of the primary focuses of the RRP is promoting economic growth and fostering the development of technology and innovation. Through these mechanisms, innovative projects developed by the business sector that drive national economic growth are supported. This program has gained increasing importance in developing and expanding the Portuguese business fabric, enabling significant investments aimed at, among other things, decarbonization, innovation, and industrialization. What remains uncertain, however, is the extent to which the RRP may contribute to increasing companies' use of tax benefits. Therefore, this empirical study will focus on investments within the scope of these three areas.

2.3 Tax Benefits

Portuguese tax policy encompasses various tax benefit schemes aimed at encouraging business investment, particularly in strategic areas such as innovation, industrialization, sustainability, and territorial cohesion (Tax and Customs Authority, 2023). In the context of the present study, it is important to understand the main fiscal mechanisms available to Portuguese companies during the 2020–2023 period. In this regard, the following four schemes are particularly noteworthy.

Tax Incentive System for Business R&D II (SIFIDE II). Regulated under articles 35 to 42 of the Investment Tax Code (CFI, Decree-Law No. 162/2014), SIFIDE II allows companies to deduct from Corporate Income Tax (IRC) a percentage of expenses incurred with R&D activities to foster technological innovation in the private sector. The deduction comprises:

- A base rate of 32.5% of the eligible expenditure incurred in the tax period and
- An incremental rate of 50% of the increase in these expenses, compared to the average of the previous two years (limited to 1.5 million euros).

Expenses related to technical staff, the acquisition of fixed assets allocated to R&D, participation in scientific institutions, and hiring external entities for technological development are eligible. The regime's management is the responsibility of the National Innovation Agency (2023). This incentive has gained relevance, especially in the manufacturing industry (Ruano, 2025), and can coexist with other public support as long as the cumulation limits defined in European regulations are respected.

Investment Support Tax Regime (RFAI). The RFAI is provided for in articles 22 to 26 of the CFI and applies to investments in tangible and relevant intangible fixed assets carried out by companies in productive sectors (such as industry, tourism, transport, and energy). Key benefits include:

- Deduction from the IRC of 25% to 40% of the relevant investments (depending on the location of the investment) and
- Exemption from Municipal Property Tax (IMI), Municipal Property Transfer Tax (IMT), and Stamp Duty (IS) on real estate acquired within the project's scope.

The regime is subject to constraints, namely the maintenance of assets for a minimum period (three years for small and medium-sized companies and five years for large companies) and the non-reduction of jobs. The RFAI can be combined with other incentives, provided that it does not exceed the State Aid ceilings set by the European Commission.

Contractual Benefits for Productive Investment (BCIP). The BCIP is regulated in Articles 2 to 21 of the CFI and is intended for investment projects with a value of equal to or greater than 3 million euros, which have a relevant economic interest for the country. Its concession depends on the conclusion of a contract between the company and the State after a binding opinion from the Agency for Investment and Foreign Trade of Portugal (AICEP) or a competent entity. Support includes:

- Deduction from the IRC liability (up to 10% of the relevant investment);
- Exemption or reduction of IMI, IMT, and Stamp Duty; and
- Possibility of additional financial support, if compatible with the European State aid scheme.

These benefits aim to attract significant national or foreign investments with a relevant impact on job creation, innovation, and territorial cohesion. The contractual nature allows for a case-by-case analysis adapted to the country's strategic objectives.

Extraordinary Investment Tax Credit II (CFEI II). Created by Law no. 27-A/2020, of 24 July, in the context of the 2020 Supplementary Budget, CFEI II was exceptional and transitory. It was aimed at investments made between July 1, 2020, and June 30, 2021, and allowed the deduction from the IRC liability of 20% of the amount invested in assets allocated to exploration (excluding land, vehicles, and non-productive assets), with a maximum limit of 70% of the IRC liability.

CFEI II is a temporary tax benefit introduced in Portugal to stimulate productive investment, applicable to both national and foreign companies, provided they are legally established in the country and make investments in tangible or intangible fixed assets allocated to economic activity within the national territory. The company must maintain the jobs for a minimum period of 3 years and be in a regular tax and contributory situation.

3 Objectives and Methodology

This exploratory study examines whether the projects funded through instruments, such as the RRP and Portugal 2020, were the same as those used to justify granting Investment Tax Benefits (ITB), which constitutes the central research question. Additionally, the study seeks to identify and characterize the companies that benefited from ITB while simultaneously receiving funding from European sources between 2020 and 2023.

To this end, the following sub-objectives are established:

- To analyze the proportion of tax benefits that may be, directly or indirectly, associated with projects financed with European public resources;
- Identify the profile of companies that use the RRP and/or Portugal 2020 and that, at the same time, benefit from ITB;
- Characterize the geographical distribution and sectors of activity of companies that have benefited from both tax benefits and European funds.

The research questions that are intended to be answered are:

- Is there an overlap between ITB and funding allocated by European funds (RRP and Portugal 2020)?
- What proportion of the tax benefits granted can be directly or indirectly associated with projects already financed with European public resources?
- What is the sectoral and geographical distribution of companies that have benefited from both tax benefits and European funds?

A search was conducted in the "*Portal Mais Transparência*"[1] database and in the list of tax beneficiaries made available by the Tax and Customs Authority (AT[2]) to answer these research questions. By cross-referencing these data, companies that simultaneously obtained any of the ITB (specifically, SIFIDE II, RFAI, BCIP, or CFEI II, as they are the most related to productive investment and, consequently, to European funds) between 2020 and 2023 and projects financed by European funds (RRP and/or Portugal 2020) from 2014 to 2023 (Portugal 2020) and from 2021 to 2023 (RRP) were selected.

Regarding the RRP, the analysis focused exclusively on investments made under the following components: C05 – Capitalization and Business Innovation, C11 – Decarbonization of Industry, and C16 – Companies 4.0. The choice of these three components, which together represent approximately 30% of the total allocation of the RRP, is justified by their close articulation with the objectives of productive business investment, in line with the tax benefits analyzed, namely those that promote innovation, modernization, and sustainable economic development. Indeed, component C05 aims to strengthen company capitalization and foster innovation through instruments that directly support investment and sustained growth, thereby promoting structural conditions for increased competitiveness. Component C11, in turn, supports the energy transition and resource efficiency, encouraging productive investments in the industry with a focus on decarbonization, process modernization, and the adoption of more sustainable technologies. Finally, component C16 is oriented toward digitalizing the business, supporting the

[1] https://transparencia.gov.pt/pt/.

[2] Portaldasfinancas.gov.pt.

adoption of digital technologies and productive transformation processes aligned with Industry 4.0. Regarding Portugal 2020, no filtering was carried out on this support, as all support measures, in their immense diversity, are directly or indirectly related to productive investment.

The companies' names were obtained by cross-referencing data in Excel from the two mentioned databases, using the formula "VLOOKUP" so that the Legal Person Identification Numbers (NIPC) were found in both databases. This work began by analyzing the companies that obtained ITB between 2020 and 2023, integrating the values, and removing duplicates. With this list, the data provided on the "*Portal Mais Transparência*" regarding the final beneficiaries of the RRP and Portugal 2020 were cross-referenced. Therefore, the study's sample is the companies that simultaneously obtained tax savings through ITB between 2020 and 2023 and investments financed by European funds between 2014 and 2023.

In a master Excel sheet, the annual amounts of tax benefits and European funds received by each beneficiary company were organized and consolidated. For ITB and Portugal 2020, the data were aggregated by company and year. For the RRP, the organization was made by year and by program component. Subsequently, this sheet was exhaustively analyzed under several dimensions, namely the location and sector of activity of the companies with the most significant number of projects, the highest value of projects, and the most significant benefit obtained, among others.

The study is exploratory because it addresses a topic that is still under research, as no previous systematic studies have analyzed the possible overlap of different types of public support for the same company. In addition, the exploratory analysis identifies patterns, trends, and relationships between the fiscal support mechanisms and European funds, providing an initial view of the articulation and potential duplications in allocating incentives.

Tables 1 and 2 show the periods during which the companies' investments took place, considering the years of tax benefit deduction and the years of European fund receipt. This information was essential for identifying and collecting data on projects carried out in the relevant years, that is, since 2008. However, it was concluded that there are only projects funded by Portugal 2020 since 2014.

Table 1. Investment periods of tax benefits and European funds.

Tax benefits/European funds	Investment period
BFCI	In the year of deduction or up to 10 years before. The deduction can be extended for another 10 years (maximum term of the contract)
SIFIDE	In the year of deduction or up to 12 years before
RFAI	In the year of deduction or up to 10 years before
CFEII	Between July 1, 2020 and June 30, 2021
RRP	Between 1 year before receipt of the advance and 2 years after approval of the application

(continued)

Table 1. (*continued*)

Tax benefits/European funds	Investment period
Portugal 2020	Between 1 year before receipt of the advance and 2 years after approval of the application

Source: Prepared by the authors

Table 2. Investment periods of the ITB and European funds associated with each deduction/notice year.

Tax benefits/European funds	Year of deduction (ITB) / Notice (European funds)			
	2020	2021	2022	2023
BFCI	2010 - 2030	2011 - 2031	2012 - 2032	2013 - 2033
SIFIDE	2008 - 2032	2009 – 2033	2010 - 2034	2011 - 2035
RFAI	2010 - 2030	2011 – 2031	2012 - 2032	2013 - 2033
CFEII	2020 - 2021	2020 – 2021	2020 - 2021	2020 - 2021
RRP	2020 - 2022	2021 – 2023	2022 - 2024	2023 - 2025
Portugal 2020	2020 - 2022	2021 – 2023	2022 - 2024	2023 - 2025

Source: Prepared by the authors

4 Analysis and Discussion of Results

4.1 Obtaining ITB and European Funds: General Analysis

The first cross-referencing of data between the companies benefiting from ITB and projects financed by European funds allowed us to draw some relevant conclusions. Figure 1 illustrates the distribution of ITB beneficiary companies that have developed investment projects versus those that have not. Figure 2 shows the distribution of the support received in companies that developed at least one project.

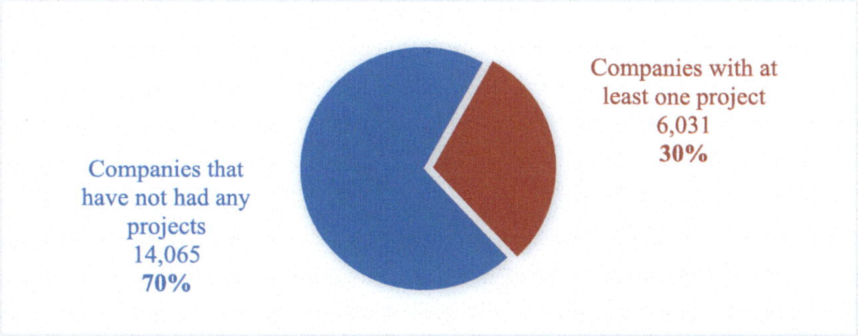

Fig. 1. ITB beneficiary companies between 2020 and 2023.

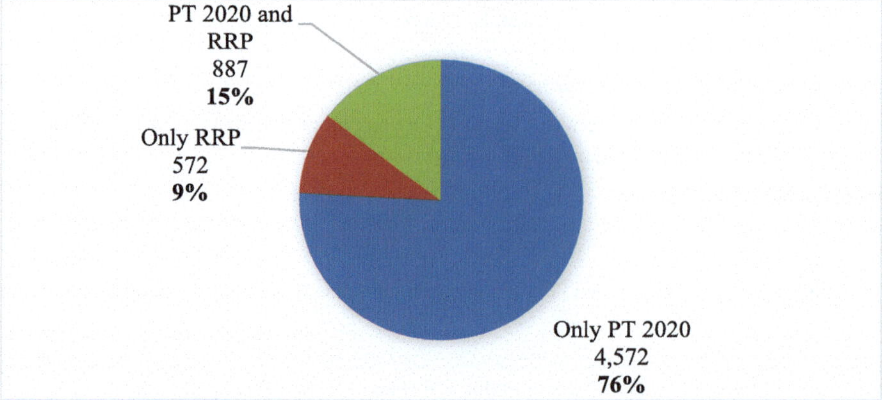

Fig. 2. Companies benefiting from ITB and European funds between 2020 and 2023.

As shown in Fig. 1, between 2020 and 2023, 20,096 distinct companies benefited from at least one ITB. Of these, 6,031 companies combined, during that period, at least one ITB with the realization of at least one project financed by European funds between 2014 and 2023, corresponding to 30% of the companies. It should be noted that, among the 20,096 companies, some obtained more than one type of ITB in the same year or different years. From the analysis of Fig. 2, it can be seen that among those 6,031 companies that benefited from at least one ITB, 887 (15%) simultaneously had at least one project financed by the RRP and another by Portugal 2020; 572 (9%) companies had only projects financed by the RRP; and 4,572 (76%) companies had only projects funded by Portugal 2020.

4.2 Evolution of ITB

This section analyzes the evolution of the absolute and relative weight of each ITB in relation to the total, highlighting the most commonly used schemes and the proportion of ITB directed to companies that, at the same time, benefited from projects financed by European funds.

Table 3. Evolution of ITB

ITB	2020					2021					2022					2023					Total				
	With projects			Total		With projects			Total		With projects			Total		With projects			Total		With projects			Total	
	Value	%		Value	%	Value	%		Value	%	Value	%		Value	%	Value	%		Value	%	Value	%		Value	%
BFCI	6,685	60	3	11,224	2	10,356	57	3	18,127	2		0	0	13,205	2	4,233	38	1	11,067	1	21,274	40	2	53,622	2
SIFIDE	109,204	33	45	326,530	52	157,533	39	45	407,611	51	189,946	36	54	521,240	61	240,226	37	60	656,896	71	696,909	36	52	1,912,276	60
RFAI	81,020	57	33	142,254	23	115,816	58	33	199,423	25	139,094	57	39	245,723	29	145,860	65	36	225,595	24	481,789	59	36	812,995	25
CFEII	46,384	32	19	144,491	23	66,785	37	19	178,248	22	24,214	33	7	72,504	9	13,079	35	3	37,462	4	150,463	35	11	432,705	13
Total	**243,294**	**39**	**100**	**624,499**	**100**	**350,489**	**44**	**100**	**803,408**	**100**	**353,254**	**41**	**100**	**852,672**	**100**	**403,399**	**43**	**100**	**931,019**	**100**	**1,350,435**	**42**	**100**	**3,211,598**	**100**

(Amounts in thousands of euros)

Table 3 reveals that, between 2020 and 2023, the absolute values of all ITB increased; however, in general, the relevance of ITB attributed to companies with projects financed by European funds decreased, except in the case of SIFIDE. The percentage shown in the third column represents the weight of the ITB value obtained by companies that also obtained projects financed by European Funds in the total ITB of that year. In the second column, the total representation of each ITB is presented. In total, 3,211.6 million euros were granted in tax benefits, of which 1,350.4 million (42%) were intended for companies that simultaneously carried out projects supported by European funds. This relationship suggests a significant alignment between tax incentives and European direct investment support. It can be concluded that 42% of the ITB awards are linked to projects partially financed by the State.

The RFAI is the most relevant tax benefit for companies with financed projects. Of the 813 million euros allocated under this scheme, 59% (equivalent to 481.8 million euros) were obtained by companies that developed projects supported by European funds. This result demonstrates a clear orientation of the RFAI to assist companies with structured investment activities. It is also important to highlight the consistency of this relationship over time, with annual percentages consistently above 57%.

4.3 Evolution of European Funds

This section analyzes the evolution of the value of European funds obtained by companies and those that simultaneously benefited from at least one ITB.

RRP. Table 4 presents the financing of business projects under the RRP from 2021 to 2023. It is possible to analyze the evolution of the absolute and relative value of the funding allocated under the RRP, distributed among its three components, highlighting the most frequently used components and the proportion of support directed to companies that, at the same time, benefited from at least one ITB.

Table 4. Evolution of financing through the RRP.

RRP	2021			2022			2023			Total		
	With ITB		Total value	With ITB		Total value	With ITB		Total value	With ITB		Total value
	Value	%		Value	%		Value	%		Value	%	
C05		-	49,492	492,915	24	2,023,125	621	0,22	280,403	493,535	21	2,353, 019
C11	400	4	9,317	371,412	47	798,393				371,412	46	807,710
C16				48,067	15	322,145	62,323	28	221,213	110,390	20	543,358
Total	400	1	58,809	912,394	29	3,143,663	62,944	13	501,616	975,338	26	3,704, 087

(Amounts in thousands of euros).

In total, the volume of investment supported by the RRP amounted to approximately 3,704 million euros, of which around 975 million (26%) correspond to projects carried out by companies that benefited from tax incentives between 2020 and 2023. This information indicates that, although most projects were carried out without a direct link to tax benefits, more than a quarter of the investment supported by European funds can result from an integrated public support strategy that combines tax benefits with direct project financing.

The annual evolution shows different dynamics. In 2021, the articulation between financing via RRP and tax benefits was practically non-existent, with only 400 thousand euros allocated to projects with ITB, representing 0.68% of the total value of the RRP for the year. However, in 2022, a significant change occurred: the value of the RRP allocated to companies that benefited from ITB reached 912.4 million euros, representing 29% of the total annual ITB. This suggests a substantial improvement in coordination between the different investment support instruments. In 2023, despite a relative reduction (13%), the financing volume to companies with ITB remained significant, totaling 62,9 million euros.

The data reveal a positive trend in articulating tax benefits and financing through the RRP, especially in 2022. The consolidation of 26% of the financing value by the RRP through companies that simultaneously obtained ITB over the total period indicates a significant presence of integrated public support. However, there is room to reinforce this convergence in the future. The analysis by component shows that C05 presents the highest absolute values in all years. Conversely, C11 is the highest relative financing weight for companies benefiting from ITB.

PT 2020. In this topic, similarly to the previous one, the evolution of the absolute and relative value of the funding allocated under Portugal 2020 between 2014 and 2023 is analyzed, highlighting the proportion of support directed to companies that simultaneously benefited from at least one ITB.

Table 5. Evolution of financing through Portugal 2020.

	2014	2015	2016	2017	2018	2019	2020	2021	2022	2023	Total
Values with ITB	211	9,576	97,423	451,529	443,652	335,019	329,198	428,940	255,510	567,964	2,919,021
Total	287	17,896	498,488	1,397,526	1,321,597	1,727,649	1,570,721	3,362,793	2,156,528	3,774,076	15,827,561
% (Values with ITB / Total)	73	54	20	32	34	19	21	13	12	15	18

(Amounts in thousands of euros)

As shown in Table 5, between 2014 and 2023, the total amount of support for projects under Portugal 2020 was 15,828 million euros. This value has been assigned to all types of entities, not just companies. Of this total, 2,919 million euros were allocated to projects by companies that also obtained ITB between 2020 and 2023. This amount represents 18% of the total support for projects through Portugal 2020, from 2014 to 2023, highlighting the relevance of tax benefits in the context of projects financed by this program.

4.4 Companies with ITB and European Funds, by Region

Nomenclature of Territorial Units for Statistics Purposes (NUTS) I. This subsection presents the territorial distribution of the number of companies the number of projects financed by European funds (RRP and Portugal 2020) and their values at the NUTS I level allowing a geographical analysis of the European support granted to companies in Portugal ().

The analysis of Table 6 reveals a concentrated allocation of European funds on the Continent (PT1), representing 95% of the companies, 93% of the projects, and 98.4% of the total project value. At first glance, this distribution may seem uneven, but it is important to contextualize it in light of the geographical and demographic characteristics of the different regions. The Continent is significantly larger in population, area, and industry than the Autonomous Regions of Madeira (PT3) and the Azores (PT2). Thus, it naturally holds most of the allocated resources, reflecting its greater capacity to generate and execute projects. Despite their smaller size, the autonomous regions have a proportionally reasonable participation. Madeira represents 3% of companies, 5% of projects, and 1,3% of the total value of financing, while the Azores represent 2% of companies, 2% of projects, and 0,3% of the value. This distribution indicates that, although the values attributed to the islands are lower, they are consistent with their smaller economic scale. In the case of Madeira, the percentage of projects is higher than that of companies, indicating a high level of application submissions by entities in this region.

Table 6. Distribution of companies with ITB and European funds by NUTS I.

NUTS	No. Companies	%	No. Projects	%	Value projects (thousands of euros)	%	Average projects value (thousands of euros)
PT1 - Continent	5,754	95	12,201	93	3,830,726	98.35	314
PT2 - Autonomous Region of Azores	126	2	283	2	12,372	0.32	44
PT3 - Autonomous Region of Madeira	151	3	625	5	51,660	1.33	83
Total	6,031	100	13,109	100	3,894,759	100	297

Regions have significant differences when analyzing the average funding per project. On the Continent, the average value per project is around 314 thousand euros, while in Madeira, it is approximately 83 thousand euros, and in the Azores, approximately 44 thousand euros. This disparity reflects the smaller scale and complexity of the projects developed on the islands, which is expected given their socio-economic realities.

In short, although the Continent concentrates most of the financing from European funds, this centralization is justified by the region's size and productive capacity. Although smaller, the autonomous regions' participation is proportional to their scale and demonstrates adequate integration into the European funding system.

NUTS II. This subsection, similarly to the previous one, presents the territorial distribution of the number of companies, the number of projects financed by European funds (RRP and Portugal 2020), and their values at the NUTS II level, allowing a more specific geographical analysis of the European support granted to companies in Portugal (Table 7).

The North, Centre, and Greater Lisbon regions concentrate the highest number of companies and projects funded by European funds, as well as the total amount of funding received. The North and Greater Lisbon stand out due to major urban and industrial centers, such as Porto and Lisbon, which attract significant investments and business initiatives. Although characterized by a lower population and business density, the Center region has also been recording an increase in the number of projects funded by European funds, particularly in innovation, technology, and sustainability. These regions represent the country's leading economic growth and value-creation hubs.

Table 7. Distribution of companies with ITB and European funds by NUTS II.

NUTS II	No. Companies	%	No. Projects	%	Value projects (thousands of euros)	%	Average projects value (thousands of euros)
PT11 - North	2,781	46.1	6,159	47.0	1,679,869	43.1	273
PT19 - Center	1,377	22.8	3,010	23.0	1,111,834	28.5	369
PT1A - Greater Lisbon	661	11.1	1,246	9.4	522,407	13.4	419
PT1D - West and Tejo Valley	459	7.6	959	7.2	317,197	8.1	331
PT15 - Algarve	208	3.4	372	3.0	51,498	1.3	138
PT1C - Alentejo	171	2.8	302	2.2	113,708	2.9	377
PT30 - Autonomous Region of Madeira	151	2.5	625	4.8	51,660	1.4	83
PT20 - Autonomous Region of Azores	126	2.1	283	2.2	12,372	0.4	44

(continued)

Table 7. (*continued*)

NUTS II	No. Companies	%	No. Projects	%	Value projects (thousands of euros)	%	Average projects value (thousands of euros)
PT1B - Setúbal Peninsula	97	1.6	153	1.2	34,214	0.9	224
Total	6,031	100.0	13, 109	100.0	3,894,758	100.0	297

4.5 Companies with ITB and European Funds, by Sector of Activity

This section analyzes how European funds are distributed across the various economic sectors of the beneficiary companies. The aim is to identify the sectors with the most significant number of companies, projects, and funding amounts, highlighting investment trends and priority areas within the scope of European funds. This sectoral framework is essential to understanding the alignment between public support and the productive dynamics of the economy.

From the analysis of Table 8, it can be concluded that the main sectors of activity of companies that benefited simultaneously from tax incentives and projects funded by European funds were manufacturing, wholesale trade, and the accommodation and food services sectors. Among these, manufacturing stands out, comprising the most significant number of companies and funded projects. It reflects its strategic importance within the national productive industry, as well as the strong emphasis on modernization, innovation, and internationalization through public incentives. Manufacturing remains the most significant sector in terms of total funding allocated, reflecting the scale of investments associated with productive modernization and innovation. However, two other distinct sectors also show a high relative weight: consulting and financial activities. Consulting stands out primarily due to the presence of specialized engineering and innovation firms, which lead high-tech and strategic projects, often in partnership with the industrial sector. On the other hand, the relevance of financial activities primarily stems from the subscription to Venture Capital Funds, which aim to finance projects in priority areas such as Agrobusiness, Smart Mobility, and Information Technologies, reflecting a strong link between private financing and productive investment supported by public funds.

Table 8. Distribution of companies with ITB and European funds by sector of activity.

Sector of Activity	No. Companies	No. Projects	Projects value (thousand euros)	Average projects value (thousand euros)
Manufacturing	2,705	6,698	2,675,831	400
Wholesale and retail trade; motor vehicle and motorcycle repair	1,022	1,729	141,717	82
Accommodation, restaurants and similar	847	1,511	131,715	87
Consulting, scientific, technical and similar activities	358	899	239,066	266
Information and communication activities	339	758	188,845	249
Construction	163	267	28,096	105
Administrative and support services activities	146	281	50,467	180
Human health and social support activities	114	169	17,110	101
Artistic, entertainment, sports and recreational activities	101	183	18,590	102
Transport and warehousing	75	113	46,723	414
Water collection, treatment and distribution; sanitation, waste management and depollution	50	137	82,546	603
Education	46	266	59,941	225
Real estate activities	33	52	6,217	120

(*continued*)

Table 8. (*continued*)

Sector of Activity	No. Companies	No. Projects	Projects value (thousand euros)	Average projects value (thousand euros)
Other service activities	18	29	1,327	46
Financial and insurance activities	12	14	205,871	14,705
Electricity, gas, steam, hot and cold water, and cold air	2	3	695	232
Total	6,031	13,109	3,894,758	297

The sector in which each company obtains the most funding is, on average, financial activities due to the high amounts associated with the subscription of these Venture Capital Funds. These financial instruments, often intended to finance innovative and high-potential projects, involve significant capital contributions from the underwriting entities, substantially increasing the average financing value per company in this sector. This ranking is followed by the water collection, treatment, distribution, and education sectors, with equally high average financing values per company. In the first case, these values are due to the capital-intensive nature of the infrastructure and water management systems. In the second, the growing investment in projects for technological modernization, digitalization, and requalification of educational infrastructures is often supported by European funds in conjunction with tax incentives.

For the above reasons, the projects that add the most significant financing value are those associated with the financial activities sector, followed by the water collection, treatment, and distribution sector. This is followed by the transport and storage sector, which focuses on advanced logistics, sustainable mobility, and transport infrastructure, often associated with large-scale projects with a substantial economic and territorial impact.

4.6 Top Companies with 10 or More Funded Projects

This analysis focuses exclusively on companies that have benefited from at least one ITB and that simultaneously had 10 or more projects funded, with the intention of identifying patterns of action and investment strategies among agents with a more intensive profile in attracting public support. This criterion enabled us to identify a subset of companies that are particularly active and experienced in formulating, submitting, and executing applications for European funds. These companies tend to possess greater technical capacity, a consolidated organizational structure, and a long-term strategic vision. By focusing on these companies, the analysis offers a clearer perspective on their behavior in the combined use of tax benefits and European funding. It enables us to evaluate the intensity of the support received and the potential for alignment between various public policy instruments aimed at promoting productive investment.

Table 9. The TOP key indicators of companies with 10 or more projects.

	Top companies with 10 or more projects	% of the total sample	Total sample
No. of companies	57	0.95	6,031
Total number of projects funded	730	5.57	13,109
Amount of funding obtained (thousands of euros)	289,063	7.42	3,894,759
Value of ITB (thousands of euros)	37,340	2.77	1,350,435

As stated in Table 9, 57 companies secured 10 or more projects funded under the RRP and/or Portugal 2020, distinguishing themselves as particularly active agents in mobilizing public funds. These companies were responsible for developing 730 projects, representing over 7% of the total funding from companies that benefited from ITB. This figure highlights the concentration of public resources and execution capacity in a relatively small number of entities.

Distribution of the TOP Companies by Region.

NUTS II. This subsection presents the territorial distribution, at the NUTS II level, of the TOP companies and the number of projects they promote, enabling a geographical analysis of the support provided to the leading project promoters.

Table 10. Distribution of companies in the TOP by NUTS II.

Region	No. Companies	%	No. Projects	%
PT11 - North	30	1.08	375	6.09
PT19 - Center	14	1.02	174	5.78
PT1D - West and Tejo Valley	7	1.53	87	9.07
PT1A - Greater Lisbon	3	0.45	47	3.77
PT15 - Algarve	1	0.48	26	12.50
PT1B - Setúbal Peninsula	1	1.03	11	7.19
PT30 - Autonomous Region of Madeira	1	0.66	10	1.60
Total	57	6.25	730	46.00

The only change observed in this ranking (Table 10), compared to the distribution of companies by NUTS II already made (Table 7), occurs in the third position, where the region of the West and Tejo Valley appears instead of the previous presence of Greater Lisbon. This change indicates that in the analysis of this TOP, the West and Tejo Valley surpass Greater Lisbon in terms of the average number of projects per company. The percentages represent the weight of companies and projects in the TOP in relation to the total number of companies and projects in each region.

NUTS III. In this subsection, the territorial distribution is presented similarly to the previous one, but now at the NUTS III level, along with the TOP companies and the number of projects they promoted. The analysis at the NUTS III level is essential because it enables the identification of inequalities and local dynamics in the distribution of investment support, providing a more precise basis for public policies tailored to the specificities of each sub-region.

Table 11. Distribution of companies in the TOP by NUTS III.

NUTS III	No. of companies in the TOP	%	No. of funded projects	%
PT11A - Metropolitan Area of Porto	15	1.1	179	5.9
PT112 - Cávado	7	1.7	88	8.9
PT191 - Aveiro Region	6	1.4	72	7.0
Total	28	4.2	339	21.8

As shown in Table 11, the Metropolitan Area of Porto stands out from the others, both in the number of projects and companies that benefit from European funds. This leadership can be attributed to the high business density, sectoral diversity, and the region's capacity to develop and execute investment projects, positioning Porto as one of the country's key economic engines in attracting public support.

Distribution of the TOP Companies by Sector of Activity. As in the previous subsection, this one analyses the distribution of support granted to TOP companies according to their respective sectors of activity.

Table 12. Distribution of companies in the TOP by sector of activity.

Sector of activity	No. of companies in the TOP	%	No. of projects funded	%	Amount of funding (thousands euros)	%	Mean value per company (thousands euros)	Mean value per project (thousands euros)
Manufacturing	30	1	347	5	164,355	57	5,479	473,646
Consulting, scientific, technical and similar activities	9	3	126	14	44,254	15	4,917	351,225

(*continued*)

Table 12. (*continued*)

Sector of activity	No. of companies in the TOP	%	No. of projects funded	%	Amount of funding (thousands euros)	%	Mean value per company (thousands euros)	Mean value per project (thousands euros)
Education	10	22	154	58	37,796	13	3,780	245,428

Table 12 shows that the manufacturing and consulting sectors hold an exceptionally significant share in funding amounts received from European funds, accounting for 72% of the total amount allocated to the companies included in the analyzed TOP group. These figures reflect the scale of investments made in the manufacturing sector and the growing importance of consulting, especially in engineering, innovation, and support for implementing complex projects using public funding.

5 Conclusions

Between 2020 and 2023, the Portuguese State strengthened its use of fiscal instruments to stimulate private investment and support economic recovery during significant instability. The aim was to create more favorable conditions for productive investment, promote business modernization, and preserve employment and the competitiveness of the national economy. The ITB played a crucial role in the country's economic strategy, while Portugal received European funding to support its restructuring. The joint use of these instruments raises questions about their overlap, mainly when both focus on the same investment projects.

This exploratory study examines whether the projects funded through instruments, such as the RRP and Portugal 2020, were the same as those used to justify granting ITB. Additionally, the study seeks to identify and characterize the companies that benefited from ITB while simultaneously receiving funding from European sources between 2020 and 2023. To this end, this study began by cross-referencing data between the "*Portal Mais Transparência*" and the list of tax beneficiaries provided by the Tax and Customs Authority. The objective was to identify companies that simultaneously benefited from ITB (specifically SIFIDE II, RFAI, BCIP, and/or CFEI II) and European funds (RRP and Portugal 2020). The analysis of the RRP focused on components C05, C11, and C16, given their strong link to business productive investment, innovation, industrial decarbonization, and digitalization. Regarding Portugal 2020, all measures were considered, given their scope in terms of investment support.

The results of this study allow us to answer the three research questions formulated. Firstly, concerning the overlap between ITB and funding allocated by European funds, data indicate that around 30% of companies that received ITB between 2020 and 2023 also benefited from at least one project financed by European support (RRP or Portugal 2020) between 2014 and 2023. This suggests an accumulation of public support by these companies. Secondly, in response to the second research question, it is found that 42% of the total value of the ITB awarded in the period corresponds to companies that have also benefited from European funds, which may indicate that a significant part of the ITB is associated with projects that are already partially financed by public resources. Finally, regarding the geographical and sectoral distribution of these companies, there is an intense concentration in the North and Center of the country, with a clear predominance in the manufacturing sector, particularly in terms of the overall value of projects and the number of projects financed. This concentration reflects the national productive structure, where these regions and this sector play a central role, suggesting the need for public policies that promote greater territorial and sectoral diversification of the support granted.

This study brings important contributions by highlighting how different instruments of public support for investment are articulated, revealing overlaps that have been little analyzed so far. Thus, it deepens the reflection and provides data on the efficiency, equity, and complementarity of public policies, contributing to a fairer and more effective management of public resources, with an impact on how tax incentives and European funds are accessed and utilized. The preliminary results of this study indicate a doubling of public support for the same financing project, reinforcing the need for a more in-depth analysis of the articulation and effectiveness of public policies to encourage investment.

For the future, it is intended to deepen the analysis of overlapping support by exploring the mechanisms of business decision-making in the choice and combination of available instruments through correlation and regression studies. On the other hand, it will be essential to conduct follow-up studies to evaluate the actual impact of this overlap on the efficiency of investments and the productivity of beneficiary companies. Finally, it will also be relevant to investigate how these practices affect equity in access to support, comparing different sectors of activity and regions.

Acknowledgments. This work was supported by the UID Research Unit on Governance, Competitiveness and Public Policies (GOVCOPP), funded by national funds through FCT - Fundação para a Ciência e a Tecnologia.

References

1. Decree-Law No. 162/2014, October 31st: Approves the Investment Tax Code, https://diariodarepublica.pt/dr/detalhe/decreto-lei/162-2014-58683382
2. Di Pirro, E., et al.: The embeddedness of nature-based solutions in the Recovery and Resilience Plans as multifunctional approaches to foster the climate transition: The cases of Italy and Portugal. Land **11**(8), 1254 (2022)
3. Commission, E.: The EU's 2021–2027 Long-Term Budget and NextGenerationEU. Facts and Figures. Publications Office of the European Union, Luxembourg (2021)
4. European Commission: European Green Deal (2025a). https://commission.europa.eu/strategy-and-policy/priorities-2019-2024/european-green-deal_en. Last accessed 03 April 2025
5. European Commission: Recovery and Resilience Facility (2025b). https://commission.europa.eu/business-economy-euro/economic-recovery/recovery-and-resilience-facility_en. Last accessed 03 April 2025
6. IAPMEI: Portugal 2020: System of Incentives for Companies. Institute for the Support of Small and Medium Enterprises and Innovation. https://www.iapmei.pt/. Last accessed 03 May 2025

7. Law No. 27-A/2020, July 24th: Approves the Supplementary Budget for 2020. https://diariodarepublica.pt/dr/detalhe/lei/27-a-2020-138762310
8. National Innovation Agency. https://ani.pt/sifide-sistema-de-incentivos-fiscais-a-ied-empresarial. Last accessed 10 April 2025
9. Oliveira, C.S.N.: Tax management and autonomous taxation in Portugal: Empirical evidence in Portuguese companies [Master's Thesis, Higher Institute of Accounting and Administration of Coimbra] (2021). http://hdl.handle.net/10400.26/36482
10. Portugal 2020: Annual Implementation Report 2020. Agency for Development and Cohesion (2021). https://www.portugal2020.pt
11. Portugal 2020 website: https://portugal2020.pt/perguntas-frequentes/. Last accessed 15 April 2025
12. Recuperar Portugal: RRP (2024). https://recuperarportugal.gov.pt/RRP/
13. Ruano, F.: Analysis of the tax benefits of IRC taxpayers: 2020–2023 (Occasional Publication No. 01/2025). Portuguese Public Finance Council (2025). https://www.cfp.pt/uploads/publicacoes_ficheiros/po-01-2025.pdf
14. Silva, A.F.: O direito dos contribuintes ao planeamento fiscal. Revista TOC **104**, 42–45 (2008)
15. Tax and Customs Authority: Tax Guide to Investment in Portugal (2023). https://info.portaldasfinancas.gov.pt/pt/apoio_contribuinte/Folhetos_informativos/Documents/Folheto_Investimento_em_Portugal.pdf
16. Vlachos, V.: The effect of institutions on R&D investment: the case of four Mediterranean euro area states. J. Appl. Acc. Res. **23**(4), 788–805 (2022)
17. Wolff, S., Ladi, S.: European union responses to the COVID-19 Pandemic: adaptability in times of permanent emergency. J. Eur. Integr. **42**, 1025–1040 (2020)

Strategic Management Accounting and Corporate Sustainability: Enabling Green Transition

Maria C Tavares[1,2](✉), Catarina Fernandes[1], Diana Costa[1], and José Vale[3]

[1] Higher Institute of Accounting and Administration of the University of Aveiro (ISCA-UA), Aveiro 3810-902, Portugal
mariatavares@ua.pt, isca5506@live.ua.pt, Diana.Costa@yazaki-europe.com

[2] Research Unit On Governance, Competitiveness and Public Policies (GOVCOPP), Aveiro 3810-193, Portugal

[3] CEOS.PP, ISCAP, Polytechnic of Porto, Matosinhos 4465-004, Portugal
josevale@iscap.ipp.pt

Abstract. This article explores the role of strategic management accounting in creating sustainable value in an era marked by the green transition and the increasing complexity of environmental challenges. Based on an exploratory literature review, integrating both classical and recent authors, the study assesses the role of management accounting in supporting strategic management oriented toward corporate sustainability and sustainable development. Normative instruments such as the Climate Framework Law (CFL) and IFRS 2 are examined, alongside exemplary corporate practices, including those adopted by Yazaki Corporation. The findings indicate that strategically oriented management accounting not only contributes to resource optimization and sustainable cost management but also plays a crucial role in implementing corporate strategies aligned with the Sustainable Development Goals (SDGs), while fostering greater awareness of the relevance of management accounting within the context of the green transition. The study concludes with reflections on the challenges and opportunities facing the accounting profession in promoting more responsible and resilient business models.

Keywords: Climate Law · Corporate Sustainability · Green Transition · IFRS 2 · Management Accounting · Sustainable Value Creation · SDG

1 Introduction

The growing social and market demands concerning social, environmental, and sustainability responsibilities have led organizations to reconfigure their business models in pursuit of sustainable value creation. In this context, management accounting plays a key strategic role by providing relevant and reliable information that supports decision-making aligned with sustainability principles [1, 2]. Companies are increasingly seeking to address the challenges of sustainable development through practices such as corporate governance, corporate social responsibility, and environmental management. In fact,

L. Almeida et al. (Eds.): ICAFI 2025, ISEM 80, pp. 527–539, 2026.
https://doi.org/10.1007/978-3-032-16268-7_30

several studies have been exploring the relationship between these practices and corporate sustainability [3–6]. Kuhndt [7] emphasizes that sustainability within organizations requires managers to develop new competencies to lead transformation processes and engage with multiple stakeholders. Complementarily, Gassenferth et al. [8] stress the importance of an institutional approach that considers the interactions among various internal and external actors.

When integrated into a sustainable business strategy, management accounting contributes not only to cost control and reduction but also to enhancing organizational competitiveness [9]. Its practical application in industrial contexts allows for the alignment of economic performance with environmental and social goals, which is particularly relevant in today's regulatory landscape, shaped by the implementation of the Climate Framework Law (CFL) and the adoption of international standards such as IFRS 2 [10, 11]. For instance, Pumiviset and Suttipun [12] examined the impact of strategic management accounting on the sustainable performance of green manufacturing firms in Thailand, concluding that strategic management accounting can be embedded in business strategies to promote sustainability, particularly in green manufacturing environments. Hence, this study aims to enhance understanding of management accounting as a tool that supports strategic management oriented toward corporate sustainability and sustainable development, reinforcing the role of accounting in the creation of sustainable value. In doing so, its practical application in industrial contexts is stressed, as well as its relationship with regulatory instruments such as the Climate Change Act and IFRS 2 [10].

To achieve this goal, an exploratory literature review is undertaken, with a focus on publications from recent decades. It addresses management accounting, sustainability, and organizational competitiveness, with special emphasis on the industrial sector. The present research offers relevant contributions to the advancement of the sustainable management accounting literature and to business practice by supporting decision-making aimed at minimizing environmental impacts and promoting more balanced and responsible development. Practical insights for managers and accountants are provided, namely showing how management accounting can support sustainable strategic decisions—particularly through strategic cost management, environmental impact measurement, and Environmental, Social, and Governance (ESG) integration across the value chain. As Nartey and Poll [13] point out, the modern accountant must be prepared to respond to societal challenges by incorporating contemporary practices that promote sustainability and the creation of shared value.

2 Methodology

To understand the role of management accounting as a tool that supports strategic management oriented towards corporate sustainability and sustainable development, this study adopted a qualitative and exploratory approach. This methodology is justified by the topicality and complexity of the subject, allowing for an in-depth understanding of the interactions between accounting, sustainability, and organizational performance, particularly within the context of the green transition. The research design is guided by the methodological contributions of Gil [14] and Saunders et al. [15], who emphasize the usefulness of exploratory research in fields that are still undergoing theoretical consolidation. More specifically, an integrative literature review was conducted, aimed at identifying, selecting, and analyzing scientific studies and normative documents published between 2015 and 2025. The inclusion criteria were: (i) thematic alignment with the topics of strategic management accounting, corporate sustainability, and sustainable value creation; (ii) academic or regulatory relevance; and (iii) publication in indexed scientific sources (such as Scopus, Web of Science, ScienceDirect, Springer, and Wiley) or official normative frameworks (such as the IFRS Foundation and national legislation).

The literature review was complemented by the analysis of a practical case in the Portuguese context—the Yazaki Corporation— based on public data and corporate sustainability reports, aiming to illustrate the real-world application of management accounting practices and ESG criteria in a leading industrial company. The selection of research documents and the interpretation of their contributions were based on scientific rigor and recency, highlighting authors such as Schaltegger et al. [2], Lukka and Vinnari [16], and Gomes et al. [17] whose recent work has helped consolidate the link between strategic management practices and corporate sustainability. This cross-analysis of sources enabled the construction of a robust and up-to-date analytical framework with the potential to contribute to both the theoretical and practical advancement of accounting in the era of the green transition.

3 Sustainability-Oriented Management Accounting: Theoretical Perspective

Historically, the main objective of accounting was to provide information on companies' financial and patrimonial position, focusing on overall results and accountability to external stakeholders [18, 19]. Financial accounting, grounded in strict legal and fiscal standards, offers historical and aggregated data which, although useful, have increasingly proven insufficient to support decision making, especially in the context of growing operational complexity and competitiveness.

The need for more detailed, relevant, and timely information to support planning, control, and decision-making led to the development of management accounting. This field evolved beyond mere cost analysis, assuming a strategic role by providing data across all stages of the production process and supporting the evaluation of organizational performance [20]. Strategic management accounting is expected to interact with key dimensions of organizational change—strategy, structure, and restructuring—in promoting more sustainable and resilient business models [21].

In recent years, management accounting has taken on an increasingly central role in integrating sustainability into corporate strategy. Authors such as Ghosh et al. [22] and Schaltegger et al. [2] demonstrate that by incorporating environmental and social metrics into management control systems, organizations can align economic performance with sustainability goals, enhancing operational efficiency and fostering sustainable value creation. Furthermore, Lukka and Vinnari [16] stress the impact of growing regulatory pressure—including standards such as IFRS 2—on the transformation of management information systems, reinforcing their relevance as instruments for guiding sustainable strategy.

In this regard, management accounting is no longer a mere technical tool for cost calculation. It is increasingly recognized as an essential instrument for defining, monitoring, and evaluating business strategies oriented toward sustainability, particularly in the industrial sector, where environmental impacts and regulatory challenges are more pronounced. On a different stance, strategic management accounting can be considered as a key tool within the context of integrated reporting, enabling more effective communication of sustainable value creation to all stakeholders. By incorporating customer-centric and business model perspectives, strategic management accounting can play a pivotal role in advancing corporate sustainability in the era of the green transition [23].

3.1 Sustainability in Business

Sustainability has emerged as a central concept in contemporary management, although its definition remains fluid and multidimensional. According to the Brundtland Report, published in 1987 under the auspices of the World Commission on Environment and Development, sustainable development is defined as that which "meets the needs of the present without compromising the ability of future generations to meet their own needs" This integrative approach presupposes a balanced coexistence of three fundamental dimensions: economic, social, and environmental [24].

With the popularization of the concept, particularly following the publication of the report Our Common Future, sustainability became a core value for business, influencing how companies structure their strategies and operational models [25, 26]. Although the concepts of sustainability and sustainable development exhibit subtle distinctions, they are often used interchangeably within the corporate context [27]. One of the pivotal milestones in the evolution of sustainability in business was introduced by Elkington [28–30] through the concept of the Triple Bottom Line (TBL), which proposes measuring corporate performance across three pillars: people, planet, and profit. This framework reinforces the necessity of integrating social and environmental responsibilities into organizations' economic strategies, promoting the creation of long-term, shared value [31]. The adoption of the United Nations' 2030 Agenda in 2015 brought the SDGs to the forefront of public and corporate policy. These 17 goals represent a global call to action involving governments, businesses, and civil society in addressing structural challenges such as poverty eradication, environmental protection, and inclusive economic growth [32, 33]. Integrating the SDGs into corporate practice enables companies not only to contribute to a more equitable and resilient development model but also to strengthen their legitimacy among stakeholders [34].

Within this framework, industry—both in the context of Industry 4.0 and the emerging paradigm of Industry 5.0—plays a decisive role in advancing corporate sustainability. Sustainable production entails designing products with environmentally conscious life cycles that minimize negative impacts and maximize socioeconomic value [35–37]. This perspective introduces new management challenges, requiring greater integration between technological innovation, environmental responsibility, and social justice.

Increasing regulatory pressure, combined with growing public awareness and climate change, has driven companies to reassess their processes toward more sustainable pathways [38, 39]. Recent studies confirm that corporate sustainability is not merely an ethical imperative but also a strategic opportunity to enhance competitiveness and organizational resilience [40, 41]. As such, the incorporation of sustainable practices into corporate culture and management systems is now recognized as a promising pathway to ensuring long-term sustainability. It is therefore crucial that businesses align their economic goals with social responsibility and environmental stewardship [37, 42–44].

3.2 Strategic Levers of Industrial Sustainability

The transition to a sustainable economy requires not only technological and operational changes but also a robust regulatory framework to guide economic agents. In this context, the publication of Law No. 98/2021–the CFL, in force since February 1, 2022, represents a significant legal milestone in consolidating Portugal's commitment to climate action [45]. The CFL formally acknowledges the climate emergency, enshrines the right to climate balance, and establishes ambitious greenhouse gas emission reduction targets: 55% by 2030, between 65% and 75% by 2040, and at least 90% by 2050 [46]. The CFL imposes cross-cutting obligations on companies, requiring them to incorporate assessments of their carbon footprint and climate impacts in their management reports, thereby reflecting the three pillars of sustainability: economic, environmental, and social. Beyond mandatory non-financial reporting—particularly relevant for large companies and carbon-intensive sectors—the law introduces the concept of a carbon budget, which sets quantitative emission limits aligned with national and international climate mitigation goals [46].

For Portuguese companies, at the level of corporate governance, these obligations are supported by the Portuguese Companies Code (Código das Sociedades Comerciais – CSC), particularly in the duties of loyalty and care (Art. 64), which bind directors and managers to provide adequate and transparent accountability regarding environmental risks and impacts (Arts. 65 and 72 of the CSC) [47]. Corporate management thus transcends its traditional economic-financial role to include strategic and legal environmental responsibilities, with failure to comply potentially resulting in civil liability and reputational damage. Even in cases where legal requirements are not yet mandatory, many companies are proactively adopting climate reporting practices to meet more demanding ESG criteria across value chains—particularly as suppliers to large corporate groups or to ensure access to sustainable financing [48, 49]. Within this framework, management accounting plays a decisive role in operationalizing climate strategy. By enabling the quantification, control, and analysis of costs and impacts throughout the value chain, management accounting provides critical information for sustainable decision-making and for formulating environmentally conscious cost strategies [50]. This integration is essential not only for legal compliance but also to meet market expectations and stakeholders' demands.

Cost structure rationalization and optimization based on sustainability criteria allow firms to achieve sustainable competitive advantages, securing long-term profitability without compromising the environment or social cohesion [13, 51]. This approach aligns with the principles set out in IFRS 2 – Share-based Payment, by emphasizing the need for appropriate recognition and measurement of transactions involving not only financial value but also human and environmental capital [11]. IFRS 2, by introducing greater rigor and transparency in value measurement, serves as a relevant reference for designing incentive mechanisms and accountability within a sustainable business model [10, 11]. Sustainable management accounting, when integrated with the robust legal framework of the CFL and international standards such as IFRS 2, thus contributes to strengthening a new corporate culture—one that is more resilient, transparent, and oriented toward a shared and sustainable future.

Sustainability in Industry: The Case of Yazaki Corporation
The CFL – Law No. 98/2021, in force since February 2022, marks a decisive step in Portugal's environmental policy by establishing national targets for CO_2 emissions reduction and binding all sectors of society to the climate transition. This legislation is embedded in a broader context of corporate responsibility, aligned with the SDGs and international standards such as IFRS 2 – a sustainability reporting standard issued by the International Sustainability Standards Board (ISSB). IFRS 2 sets disclosure requirements related to sustainability-linked risks and opportunities, promoting greater transparency and comparability of economic, environmental, and social impacts in financial statements [10, 11].

Since 2022, several companies have begun reporting non-financial information regarding the climate impact of their operations. This reporting has enabled the creation of carbon budgets and enhanced the accuracy in assessing sustainable performance. Even companies not yet legally required to do so are experiencing pressure from value chains and investors to adopt sustainable practices and disclose their ESG performance. In this scenario, management accounting plays a crucial role in cost measurement and in supporting the formulation of sustainable strategies, particularly in industrial contexts. Efficient resource use, underpinned by sustainability-oriented management methods, enables the generation of economic value in line with environmental and social goals [13, 50, 52]. Strategic cost management thus becomes a critical decision-making tool for achieving sustainable competitive advantages [51]. In this study, Yazaki Corporation, a Japanese multinational and leading player in the automotive sector, is used to illustrate how management accounting can drive organizational sustainability by integrating legal standards and effective business practices in the creation of economic, social, and environmental values. With a strong environmental sustainability commitment, Yazaki promotes technological innovation and quality through practices aligned with ESG principles. Its corporate philosophy is anchored in the mottos "A Corporation in Step with the World" and "A Corporation Needed by Society", reinforcing its mission of environmental and social responsibility. With over 75 years of environmental commitment, Yazaki develops electrical systems and components for hybrid and electric vehicles, promotes recycling, and adheres to international standards for quality, environment, health, and safety, such as ISO 14001, ISO 9001, and OHSAS 18001. The company's quality management system exemplifies a strategic management accounting practice, ensuring efficient processes, sustainable materials, and continuous improvement methodologies. Yazaki was awarded the "Family Business Award 2007," highlighting its integrated approach to corporate governance and intergenerational values [53, 54].

The CFL acknowledges the climate emergency and assigns responsibilities to all sectors of society, including businesses, to incorporate environmental considerations into their operations and management reports [46]. In this context, companies, such as Yazaki, are encouraged to adopt sustainability practices that encompass economic, environmental, and social dimensions. Thus, management accounting plays a critical role by providing detailed information on the costs and environmental impacts of its activities, thereby enabling strategic decision-making aligned with the SDGs [50, 52].

Another practical example of Yazaki Corporation's sustainable industrial strategies refers to the implementation of various initiatives to reduce its environmental impact and promote sustainability, specifically [53, 54]:

- Greenhouse Gas Emissions Reduction: In 2022, Yazaki Europe achieved a 16% reduction in GHG emissions compared to the 2017 baseline, despite a 15% year-on-year increase due to post-pandemic recovery. Direct (Scope 1) and indirect (Scope 2) emissions totaled approximately 34,147 tons of CO_2 equivalent, accounting for 5% of the company's total carbon footprint.
- Water Resource Management: In 2022, the company consumed 402,420 m^3 of water, with 84% sourced from municipal supplies and 16% from private sources. Although this represented an 11% increase compared to the previous year, it marked an 8% reduction relative to 2017. Measures included employee awareness programs, installation of efficient technologies, and regular maintenance to optimize water usage.
- Energy Consumption: Total energy consumption at Yazaki Europe in 2022 was 425,103 GJ, representing a 20% reduction compared to 2017 and a 3% decrease compared to 2021. Electricity accounted for 79% of total energy use.
- Environmental Certifications: All Yazaki manufacturing facilities maintain Environmental Management System (EMS) certifications in accordance with ISO 14001, showing the company's ongoing commitment to improving environmental performance.

These initiatives reflect the integration of sustainable practices into Yazaki's corporate strategy, aligning with the principles established by the CFL and serving as a model for other companies pursuing sustainability.

3.3 Strategic Management Accounting

Management accounting provides essential information for decision-making at the strategic, tactical, and operational levels, supporting managers in the continuous improvement of economic performance and value creation [55]. Strategic management accounting stands out as a branch that integrates cost, market, and competitive analysis to support decisions with long-term impact. This approach is fundamental in a context where competitiveness strongly depends on effective product development and organizational innovation.

Since the foundational contributions of Bromwich [56] and Drury [57], strategic management accounting has evolved beyond mere data collection, incorporating the organizational context, external changes, and long-term sustainable objectives. Recent literature has further developed this domain, emphasizing that strategic decision-making relies not only on accounting information but also on the ability to interpret and apply such information in an integrated manner [58, 59]. The complex and contingent nature of organizations makes it difficult to apply universal theoretical models for strategic decision-making [60], requiring a holistic understanding of cost structures, value chains, and a company's environmental and social performance [50, 61]. In this regard, strategic management accounting also serves as a sustainability management tool. According to Golov [62], this approach should support organizational diagnostics, strategic planning, and control processes, integrating efficiency objectives with social and environmental responsibility goals [63].

Strategic management accounting has a direct and significant impact on the economic, environmental, and social performance of companies. It can serve as a strategic tool for firms in emerging economies to address environmental and social challenges, promote sustainable practices, and enhance competitiveness [12]. Consequently, strategic cost management, as outlined by Duçi [64], enables companies to identify opportunities for competitive differentiation and resource optimization, reinforcing the role of accounting in sustainable value creation [65]. This strategic alignment is particularly relevant for industrial organizations, whose responsibility in transitioning toward sustainable models is critical [66]. As such, strategic management accounting becomes a key instrument for responding to the demands of the new green and circular economy.

Recent studies (see [17, 67]) have shown that integrating sustainability into business strategy enhances organizational performance, strengthens corporate reputation, and increases long-term resilience. However, the full adoption of this approach still faces resistance, often due to a lack of sustainability-oriented organizational culture [68–70]. Therefore, for strategic management accounting to effectively contribute to corporate sustainability, it is essential to promote its integration into planning and control systems, valuing not only financial outcomes but also the social and environmental impacts generated by economic activity [37, 42, 43, 71].

4 Final Considerations

This article aimed to analyze the role of management accounting in creating sustainable value, focusing on its practical application in industrial contexts and relating it to regulatory instruments such as the Climate Change Act and IFRS 2. The methodology adopted was exploratory and qualitative. It was based on a review of current literature, combining classic authors with more recent contributions, and included the study of practical cases such as that of Yazaki Corporation, which illustrates the use of management accounting practices to monitor environmental and strategic objectives. The choice of this approach has made it possible to identify and integrate different theoretical and empirical perspectives that support the relevance of accounting as a tool for the transition to a more sustainable economy.

The results of the study confirm the growing importance of strategic management accounting as an intermediary between financial objectives and sustainability objectives of organizations. Management tools such as environmental activity-based costing, ESG performance dashboards, and integrated strategic reporting not only improve organizational performance but also increase transparency and accountability to stakeholders (see [72–74]). However, the transition to sustainable management accounting practices is not without challenges. Organizations often encounter internal resistance due to entrenched routines, lack of awareness, or uncertainty about the long-term benefits of integrating ESG dimensions. Additionally, the absence of standardized frameworks and the complexity of measuring non-financial indicators can create dilemmas for practitioners attempting to balance comparability with context-specific relevance.

There are also sector-specific barriers: in industrial contexts, where short-term financial metrics dominate, sustainability investments are sometimes perceived as costs rather than strategic opportunities. Moreover, the integration of sustainability into accounting may conflict with existing incentive systems or organizational cultures that prioritize financial outcomes. To overcome these barriers, companies at the early stages of sustainability integration may benefit from a phased approach. This includes initiating pilot projects with environmental costing, developing internal ESG training programs, and gradually aligning performance indicators with sustainability objectives. Establishing multidisciplinary teams that include accounting, operations, and sustainability experts can also foster dialogue and innovation.

Therefore, the practical impact of sustainable management accounting depends not only on technical tools but also on the organization's capacity to adapt its governance structures, invest in human capital, and embrace a broader accountability culture. These considerations are critical to enabling the full transformative potential of accounting in fostering a sustainable economy. Moreover, recent studies show (see [50, 52]) that management accounting can play an active role in integrating the SDGs into business decisions through practices that make visible the environmental and social impacts of production activities. This evolution implies a redefinition of the skills and profile of the accounting professional, who must be able to interpret financial and non-financial data critically and strategically [71]. The training of accounting professionals to integrate ESG criteria into planning, control, and reporting practices is now deemed crucial.

This study contributes to the scientific and practical advancement of accounting for sustainability. As a contribution to the scientific community, this article highlights the need for the further development of interdisciplinary approaches in the field of accounting that combine the knowledge of strategic management, sustainability, and international standards. The study contributes to the consolidation of a new accounting paradigm geared towards organizational transformation and the pursuit of a green and responsible economy by highlighting the importance of strategic management accounting in creating sustainable value.

References

1. Burritt, R.L., Herzig, C., Tadeo, B.D.: Environmental management accounting: the significance of contingent variables for adoption. J. Clean. Prod. **234**, 1175–1188 (2019)
2. Schaltegger, S., Christ, K.L., Wenzig, J., Burritt, R.L.: Corporate sustainability management accounting and multi-level links for sustainability–A systematic review. Int. J. Manag. Rev. **24**(4), 480–500 (2022)
3. Epstein, M.J., Roy, M.J.: Sustainability in action: identifying and measuring the key performance drivers. Long Range Plan. **34**(5), 585–604 (2001)
4. Figge, F., Hahn, T., Schaltegger, S., Wagner, M.: The sustainability balanced scorecard–linking sustainability management to business strategy. Bus. Strateg. Environ. **11**(5), 269–284 (2002)
5. Johnson, M.P., Schaltegger, S.: Two decades of sustainability management tools for SMEs: how far have we come? J. Small Bus. Manage. **54**(2), 481–505 (2016)

6. Govindan, K., Seuring, S., Zhu, Q., Azevedo, S.G.: Accelerating the transition towards sustainability dynamics into supply chain relationship management and governance structures. J. Clean. Prod. **112**, 1813–1823 (2016)
7. Kuhndt, M.: Sustainable business development. In: J.-D. Seiler-Hausmann (ed.), Eco-Efficiency and Beyond: Towards the Sustainable Enterprise, pp. 64–73 (2004)
8. Gassenferth, W., Conceição, C.M., Machado, M.A.S., Pereira, S., Krause, W.: Gestão de negócios e sustentabilidade: Textos selecionados. Brasport (2015)
9. Laine, M., Tregidga, H., Unerman, J.: Sustainability accounting and accountability in the public, private and third sectors: current practice and future research. Sustain. Account. Manage. Policy J. **11**(4), 715–744 (2020)
10. IFRS Foundation: IFRS S2 Climate-related Disclosures, (2023a). Last accessed 25 Apr 2025
11. IFRS Foundation: IFRS 2—Share-based Payment. International Financial Reporting Standards (2023b). Retrieved from https://www.ifrs.org. Last accessed 25 Apr 2025
12. Pumiviset, W., Suttipun, M.: Sustainability and strategic management accounting: evidence of green manufacturing in Thailand. Cogent Bus. Manage. **11**(1), 2302794 (2024)
13. Nartey, S.N., van der Poll, H.M.: Innovative management accounting practices for sustainability of manufacturing small and medium enterprises. Environ. Dev. Sustain. **23**(12), 18008–18039 (2021)
14. Gil, A.C.: Métodos e técnicas de pesquisa social (7ª ed.). Atlas (2017)
15. Saunders, M., Lewis, P., Thornhill, A.: Research methods for business students (8th ed.). Pearson Education (2019)
16. Lukka, K., Vinnari, E.: On the interplay of accounting and sustainability regulation: a field-level perspective. Crit. Perspect. Account. **83**, 102347 (2022)
17. Gomes, J., Leite, J., Carmo, C.: Integrating sustainability and certification requirements in management control systems: insights from the construction sector in Portugal (2025)
18. Pereira, J., Franco, J.: Contabilidade Financeira e Contabilidade de Gestão: As duas faces da contabilidade empresarial. Texto Editora (1994)
19. Drury, C.: Introdução à contabilidade de gestão (5.ª ed.). Monitor (2000)
20. Caiado, A.L.M.: Contabilidade de gestão: Informação relevante para a tomada de decisão. Escolar Editora (2015)
21. Alsharari, N.M.: The interplay of strategic management accounting, business strategy and organizational change: as influenced by a configurational theory. J. Account. Organ. Chang. **20**(1), 153–176 (2024)
22. Ghosh, B., Herzig, C., Mangena, M.: Controlling for sustainability strategies: findings from research and directions for the future. J. Manag. Control. **30**(1), 5–24 (2019)
23. Roslender, R., Hart, S., Nielsen, C.: Strategic management accounting revisited: building on insights from the business model field. J. Account. Organ. Chang. **20**(1), 1–20 (2024)
24. United Nations: Our Common Future (1987). http://www.un-documents.net/our-common-future.pdf. Last accessed 23 Apr 2025
25. Aras, G., Crowther, D.: Governance and sustainability: an investigation into the relationship between corporate governance and corporate sustainability. Manag. Decis. **46**(3), 433–448 (2008)
26. Christofi, A., Christofi, P., Sisaye, S.: Corporate sustainability: historical development and reporting practices. Manag. Res. Rev. **35**(2), 157–172 (2012)
27. Gasparatos, A., El-Haram, M., Horner, M.: The sustainability assessment of indicators: an evaluation using household electricity consumption. Ecol. Econ. **68**(8–9), 2106–2115 (2009)
28. Elkington, J.: Cannibals with forks: the triple bottom line of 21st century business. Capstone (1997)
29. Elkington, J.: Partnerships from cannibals with forks: the triple bottom line of 21st-century business. Environ. Qual. Manage. **8**(1), 37–51 (1998)

30. Elkington, J.: Enter the triple bottom line. In: Henriques, A., Richardson, J. (eds.), The Triple Bottom Line: Does It All Add Up? Earthscan (2004)
31. Gao, S.S., Zhang, J.J.: Stakeholder engagement, social auditing and corporate sustainability. Bus. Process. Manag. J. **12**(6), 722–740 (2006)
32. United Nations: Transforming our world: The 2030 Agenda for Sustainable Development. United Nations (2015). https://sdgs.un.org/2030agenda. Last accessed 23 Apr 2025
33. United Nations: Our Common Agenda and the Sustainable Development Goals (SDGs) (2024). https://www.un.org/en/common-agenda/sustainable-development-goals. Last accessed 23 Apr 2025
34. Galeazzo, A., Miandar, T., Carraro, M.: SDGs in corporate responsibility reporting: a longitudinal investigation of institutional determinants and financial performance. J. Manage. Governance **28**(1), 113–136 (2024)
35. Martín-Gómez, A.M., Agote-Garrido, A., Lama-Ruiz, J.R.: A framework for sustainable manufacturing: integrating industry 4.0 technologies with Industry 5.0 values. Sustainability **16**(4), 1364 (2024)
36. Xu, X., Lu, Y., Vogel-Heuser, B., Wang, L. Toward sustainability and resilience with Industry 4.0 and Industry 5.0. Front. Manuf. Technol. **2**, 951643 (2022)
37. Lenort, R., Wicher, P., Zapletal, F.: On influencing factors for sustainable development goal prioritisation in the automotive industry. J. Clean. Prod. **387**, 135718 (2023)
38. Baumgartner, R.J.: Managing corporate sustainability and CSR: a conceptual framework combining values, strategies and instruments contributing to sustainable development. Corp. Soc. Responsib. Environ. Manag. **21**(5), 258–271 (2014)
39. Moon, J.: The contribution of corporate social responsibility to sustainable development. Sustain. Dev. **15**(5), 296–306 (2007)
40. Baumgartner, R.J., Rauter, R.: Strategic perspectives of corporate sustainability management to develop a sustainable organization. J. Clean. Prod. **140**(1), 81–92 (2017)
41. Margherita, A., Braccini, A.M.: Digital transformation continuous change capability: a multiple case study. J. Enterp. Inf. Manag. **33**(6), 1289–1310 (2020)
42. Stead, J.G., Stead, W.E.: Eco-enterprise strategy: standing for sustainability. J. Bus. Ethics **24**(4), 313–329 (2000)
43. Cioca, L.I., Ivascu, L., Turi, A., Artene, A., Găman, G.A.: Sustainable development model for the automotive industry. Sustainability **11**(22), 6447 (2019)
44. Agwu, U.J., Bessant, J.: Sustainable business models: a systematic review of approaches and challenges in manufacturing. Revista de Administração Contemporânea **25**, e200202 (2021)
45. Lei n.º 98/2021, de 31 de dezembro. Lei de Bases do Clima. Diário da República n.º 253/2021, 1.ª série. Entrada em vigor: 1 de fevereiro de 2022
46. European Union: Regulation (EU) 2021/1119 of the European Parliament and of the Council of 30 June 2021 establishing the framework for achieving climate neutrality and amending Regulations (EC) No 401/2009 and (EU) 2018/1999 ("European Climate Law"). Official J. Eur. Union L **243**, 1–17 (2021). https://eur-lex.europa.eu/legal-content/EN/TXT/?uri=CELEX%3A32021R1119. Last accessed 30 Nov 2024
47. Serens, M.C.N.: Código das Sociedades Comerciais (comentado e anotado). Editora Almedina (2025)
48. Commission, E.: EU Taxonomy for Sustainable Activities: Technical Screening Criteria. European Commission, Brussels (2023)
49. Farhan, K.: ESG integration in supply chains: Strategic shifts and reporting practices. Sustainability **16**(1), 205 (2024)
50. Appannan, J.S., Mohd Said, R., Ong, T.S., Senik, R.: Promoting sustainable development through strategies, environmental management accounting and environmental performance. Bus. Strateg. Environ. **32**(4), 1914–1930 (2023)

51. Anderson, S.W.: Managing costs and cost structure throughout the value chain: research on strategic cost management. Handbooks Manage. Account. Res. **2**, 481–506 (2006)
52. Fuzi, N.M., et al.: Sustainability management accounting and organizational performance: the mediating role of environmental management system. Sustainability **14**(21), 14290 (2022)
53. Yazaki Corporation: Sustainability Report e About us–Yazaki Europe (2022). https://www.yazaki-europe.com,
54. Yazaki Europe: Sustainability Report 2023 (2023). https://www.yazaki-europe.com/about-us/sustainability. Last accessed 08 Oct 2024
55. Varaniute, V., Giriūnas, L., Šiškienė, L.: Strategic management accounting in product development: the Lithuanian context. Entrepreneurial Bus. Econ. Rev. **10**(1), 65–79 (2022)
56. Bromwich, M.: The case for strategic management accounting: the role of accounting information for strategy in competitive markets. Acc. Organ. Soc. **15**(1/2), 27–46 (1990)
57. Drury, C.: Management and Cost Accounting. International Thomson Business Press, London (1997)
58. Krutova, A., Tarasova, T., Nesterenko, O., Bliznyuk, O., Nosach, N.: Strategic management accounting as an information basis of effective management of enterprise activities. Acad. Account. Financ. Stud. J. **24**(2) (2020)
59. Shmatkovska, T., et al.: Strategic management of the enterprise using the system of strategic management accounting in conditions of sustainable development. AD ALTA. Vol. **12**(2), XXIX. 123–128 (2022)
60. Junqueira, E., Dutra, E.V., Zanquetto Filho, H., Gonzaga, R.P.: The effect of strategic choices and management control systems on organizational performance. Revista Contabilidade & Finanças **27**(72), 334–348 (2016)
61. Major, M., Vieira, R.: Accounting and control in Portuguese public hospitals: from prescription to action. J. Account. Organ. Chang. **14**(1), 2–21 (2018)
62. Golov, S.: Strategic management accounting: theory and practice. Financial Times Press (2006)
63. AECA: La contabilidad de gestión en la empresa. Asociación Española de Contabilidad y Administración de Empresas (2003)
64. Duçi, A.: Strategic cost management and business competitiveness. Albanian J. Econ. Bus. **14**(1), 31–42 (2021)
65. Hansen, D., Mowen, M., Guan, L.: Cost management: accounting and control. Cengage Learn. (2007)
66. Stock, T., Obenaus, M., Kunz, S., Kohl, H.: Industry 4.0 as enabler for a sustainable development: a qualitative assessment of its ecological and social potential. Sustainability **10**(12), 4515 (2018)
67. Rahi, A.F., Johansson, J., Fagerström, A., Blomkvist, M.: Sustainability reporting and management control system: a structured literature review. J. Risk Financ. Manage. **15**(12), 562 (2022)
68. Kiron, D., Kruschwitz, N., Haanaes, K., Reeves, M., Goh, E.: Sustainability nears a tipping point. MIT Sloan Manage. Rev. (2012)
69. Kiron, D., Kruschwitz, N., Reeves, M., Goh, E.: The innovation bottom line. MIT Sloan Manage. Rev. (2013)
70. McKinsey, Company: Sustainability's Strategic Worth: McKinsey Global Survey Results (2014)
71. Hasan, S.A.S., Waghule, S.N., Al Koliby, I.S., Al-Bukhrani, M.A., Al Haifi, M.M., Hasan, M.B.: Innovating for sustainability: the role of environmental management accounting in driving environmental performance. Discover Sustain. **5**(1), 183 (2024)
72. Ortiz-Cea, V., Dote-Pardo, J., Geldres-Weiss, V.V., Peña-Acuña, V.: The role of activity-based costing in reducing environmental impact: a systematic literature review. Sustainability **17**(3), 1275 (2025)

73. Sun, Y.: The real effect of innovation in environmental, social, and governance (ESG) disclosures on ESG performance: an integrated reporting perspective. J. Clean. Prod. **460**, 142592 (2024)
74. Huynh, Q.L., Nguyen, V.K.: The role of environmental management accounting in sustainability. Sustainability **16**(17), 7440 (2024)

Sustainability in SMEs: A Descriptive Analysis of Practices and Indicators Based on Academic Literature

Pierre da Silva Marinho[1](✉), Ana Lúcia F. de S. Vasconcelos[2], Liliane Cristina Segura[1], and Eduardo da Silva Flores[3]

[1] Universidade Presbiteriana Mackenzie, São Paulo, Brasil
pierre.marinho10@gmail.com, liliane.segura@mackenzie.br
[2] Universidade Federal de Pernambuco (UFPE), Recife, Brasil
ana.svasconcelos@ufpe.br
[3] Universidade de São Paulo, São Paulo, Brasil
eduardoflores@usp.br

Abstract. The article focuses on understanding how small and medium-sized enterprises (SMEs) can adopt sustainable practices and measure their performance. This involvement of SMEs in sustainability efforts is essential for achieving broader environmental and social goals. This study highlights a gap in existing literature, which predominantly focuses on large corporations. It points out that while large companies have integrated sustainability into their operations, SMEs often lack the resources and knowledge to do the same. The main objective of the study is to identify sustainability practices and indicators from academic literature that can be adapted for SMEs. The authors aim to provide a systematic review that bridges the gap between theory and practice, helping SMEs managers implement feasible sustainability initiatives. The authors conducted a systematic literature review to gather insights on sustainability in SMEs. They focused on recent studies while also considering seminal works to provide a comprehensive overview of the topic. Many SMEs do not perceive immediate benefits from sustainability efforts, which can hinder their motivation to adopt such practices. The findings suggest that SMEs can start their sustainability journey with simplified indicators and gradually expand their efforts. The article emphasizes the importance of education and training to empower SME managers to understand sustainability as a vital part of effective business management. The article concludes that integrating sustainability practices into SMEs is not only possible but also necessary for long-term success. By adopting tailored sustainability indicators, SMEs can improve their performance and contribute to a more sustainable economy.

Keyword: Sustainability · descriptive analysis · SME

1 Introduction

Small and medium-sized enterprises (SMEs) represent the vast majority of businesses globally and play a crucial role in the economy. According to a report by the International Finance Corporation [1], SMEs account for approximately 9 out of every 10 companies

L. Almeida et al. (Eds.): ICAFI 2025, ISEM 80, pp. 540–560, 2026.
https://doi.org/10.1007/978-3-032-16268-7_31

worldwide, contributing around 40% of GDP in developing countries and generating two-thirds of global employment. Given this significant representation, it becomes clear that any meaningful progress in global corporate sustainability depends on the active participation of SMEs [2]. However, while large corporations have already incorporated sustainability indicators and re-ports into their routines, this is not yet the reality for most SMEs. Academic publications on sustainability reporting and indicators are predominantly focused on large enterprises, indicating a gap in understanding how SMEs can adopt sustainable practices and measure their performance.

Nevertheless, despite evidence suggesting that many sustainable practices developed by large corporations could, with appropriate adaptations, be applied to smaller businesses, SMEs often face limitations in financial, human, and knowledge re-sources to implement complex sustainability systems [3, 4]. Moreover, there are constraints or a lack of perceived immediate benefits among small business owners when it comes to adopting such practices [5]. These factors contribute to making the implementation of sustainability in SMEs a challenge, despite its strategic importance for a more sustainable economy.

Against this backdrop, this article aims to identify, in a descriptive manner, the sustainability practices and associated performance indicators documented in academic literature as relevant and applicable to SMEs. In other words, it seeks to answer the following research question: What sustainability practices and indicators have been identified in academic literature, and how can they be adopted by SMEs to enhance their organisational performance? To address this objective, the article conducts a systematic and descriptive review of recent academic literature on corporate sustainability, with special attention to the context of small and medium-sized enterprises. By synthesizing these findings in a comprehensible manner, the study aims to help bridge the gap between theory and practice, offering support for SME managers to implement feasible sustainability initiatives aligned with performance improvement.

The article is structured as follows: the Theoretical Framework presents a conceptual overview of corporate sustainability, including its historical development, main reporting frameworks, and the relationship between environmental, social, and governance (ESG) practices and corporate performance, with emphasis on the particularities of SMEs. The Methodology describes the research procedures, especially the systematic literature review carried out. Next, in the Descriptive Analysis of Results, the findings from the literature are discussed: trends, empirical evidence, and examples of relevant indicators and practices for SMEs. Finally, the Conclusion summarises the key insights, practical implications, and suggestions for future research.

2 Theoretical Framework

The concept of corporate sustainability has emerged and consolidated over decades, incorporating economic, environmental, and social considerations. An early milestone was the development of the Socially Responsible Investment (SRI) movement, whose roots trace back to ethical and religious values centuries ago. SRI practices have existed for at least two millennia; the Hebrew Bible already emphasized ethical considerations in financial investments. In the twentieth century, particularly from the 1960s and 1970s, the debate around corporate social responsibility gained momentum. In 1968, Garrett Hardin published "The Tragedy of the Commons", warning that the unrestrained exploitation of shared resources would lead to environmental degradation unless there was a shift in collective consciousness [6]. A few years later, Milton Friedman [7] summarized the traditional view by stating that "the social responsibility of business is to increase its profits", clearly expressing the shareholder theory approach and relegating social and environmental concerns to a secondary position. This strictly financial view was later challenged by alternative approaches.

During the 1980s, the idea of stakeholders gained prominence. In 1984, R. Edward Freeman published "Strategic Management: A Stakeholder Approach", emphasizing that managers should consider the interests of all stakeholders (customers, employees, the community, suppliers, etc.) and not just shareholders, to ensure long-term success for the company [8]. Freeman's perspective broadened the understanding of corporate responsibility, aligning it with societal expectations–a precursor to the modern concept of sustainability. Also in the 1980s, environmental concerns gained global prominence with the United Nations' 1987 report "Our Common Future" (also known as the Brundtland Report), which defined sustainable development as development that meets the needs of the present without compromising the ability of future generations to meet their own needs [7].

In the 1990s, the understanding that economic performance, social responsibility, and environmental management are interconnected became widely accepted. In 1997, John Elkington coined the expression Triple Bottom Line, referring to the three pillars: economic prosperity, environmental quality, and social equity. Elkington argued that companies wishing to thrive in the twenty-first century would need to balance these three dimensions in their strategies [9]. This concept significantly influenced how sustainability came to be measured, encouraging organizations to report not only financial results but also environmental and social indicators. Also in 1997, reinforcing the environmental agenda, the Kyoto Protocol was signed under the United Nations framework–the first global agreement aimed at reducing greenhouse gas emissions [10].

In the twenty-first century, the ESG (Environmental, Social, and Governance) agenda gained prominence in finance and corporate management. The acronym "ESG" formally appeared in 2004 in a report titled "Who Cares Wins", an initiative led by Kofi Annan at the UN that brought together financial institutions to propose the integration of environmental, social, and governance factors into investment analysis [11]. In the following years, several key developments strengthened the global sustainability agenda: The COP21 in 2015, which led to the Paris Agreement (a commitment to limit global warming to 2 °C) [12]; and also in 2015, the creation of the United Nations Sustainable Development Goals (SDGs)–17 goals and 169 targets to be achieved by 2030, covering issues such as poverty eradication, clean energy, responsible consumption, climate action, among others [13]. These goals fostered concrete business initiatives.

At the same time, corporate governance and sustainable management became increasingly intertwined. Companies began voluntarily adopting sustainability reports to demonstrate their commitment to ESG criteria. Europe, for example, moved from voluntarism to mandatory requirements: the Directive 2014/95/EU required large companies to disclose non-financial information from 2018 onwards, introducing the concept of double materiality, which considers both the impact of sustainability issues on the company (outside-in) and the company's impact on society and the environment (inside-out) [14]. This reflects a growing expectation for transparency regarding ESG aspects.

Sustainability Reporting Structures and Key Indicators.
As the ESG agenda progressed, standardizing the way sustainability is measured and reported became essential to ensure comparability and credibility. Various sustainability reporting structures and standards have been developed globally and adopted by organizations:

Global Reporting Initiative (GRI). Launched in 1997, GRI is one of the leading sustainability reporting frameworks. It provides guidelines and indicators for disclosing economic, environmental, and social impacts. Originating from an initiative in response to an environmental disaster (the Exxon Valdez oil spill in 1989), GRI has evolved into a globally recognized benchmark. GRI standards are periodically updated to reflect best practices and are applicable to organizations of all sizes, although the level of detail may be challenging for many SMEs [3].

International Integrated Reporting Framework (IIRF). Developed by the International Integrated Reporting Council (IIRC) in 2013, integrated reporting aims to combine financial and non-financial information into a concise report showing how the organization creates value over time. It introduced the concept of the six capitals (financial, manufactured, intellectual, human, social, and natural) to guide thinking on resources and value. While initially focused on large corporations and investors, pilot experiences show that some SMEs can also benefit from this approach to communicate value holistically [15].

Sustainability Accounting Standards Board (SASB) and International Sustainability Standards Board (ISSB). SASB developed ESG indicators tailored to specific industries. In 2021, during COP26, the IFRS Foundation announced the creation of the ISSB, integrating SASB and other initiatives to create global sustainability disclosure standards. The ISSB's first standards were published in 2023 (IFRS S1 and S2, covering general sustainability disclosures and climate-related disclosures, respectively) [16, 17] and are already being adapted to national contexts, such as in Brazil through the Brazilian Sustainability Pronouncements Committee (CBPS) [18, 19]. However, experts note that in the short term, these standards are unlikely to be adopted by most SMEs due to resource and capacity constraints in this segment.

European Sustainability Reporting Standards (ESRS). Developed by EFRAG, the ESRS emphasize double materiality and impose specific obligations on European companies. They are broader in scope and target a wide range of stakeholders, while ISSB standards focus on financial materiality. Unlisted European SMEs have proportionated guidelines and extended timelines for compliance, but even so, adapting to these standards remains a work in progress [20].

Carbon Disclosure Project (CDP). Founded in 2000, the CDP is a non-profit organization that collects standardized data on climate change, water use, and deforestation through voluntary questionnaires sent to companies and cities. The CDP has aligned with other initiatives, incorporating GRI guidelines and the UN Global Compact, becoming part of a broader reporting framework [21].

Greenhouse Gas Protocol (GHG Protocol). Created in 2001 by the World Resources Institute (WRI) and the World Business Council for Sustainable Development (WBCSD), this protocol established a globally accepted methodology for quantifying greenhouse gas emissions. Although technical, it is worth mentioning as it forms the basis for objective environmental indicators that even larger SMEs or those within international supply chains may be pressured to report [22].

These structures serve as a conceptual framework for sustainability reporting. However, their implementation in SMEs is not always straightforward. Many guidelines were originally designed for large corporations with dedicated sustainability departments. Studies show that SMEs struggle to follow complex frameworks: Permatasari and Kosasih [3], for instance, report that current guidelines contain "many indicators that are difficult to interpret and disclose" for most SMEs. Hence, there have been calls for simplification.

Sustainability in Small and Medium-Sized Enterprises.
Although the findings presented above are encouraging, it is crucial to note that most traditional studies focus on publicly listed or large corporations. SMEs have distinct characteristics: smaller operational scale, informal managerial structures, limited financial resources, and lower public visibility. These differences mean that the ESG–performance dynamic in SMEs may not be identical to what is observed in large firms [4]. However, in recent years, academic interest in understanding sustainability within the SME context has grown, and some patterns have begun to emerge.

Firstly, although they are generally not legally required to do so, many SMEs are voluntarily adopting sustainable practices and even publishing non-financial reports. Ortiz-Martínez and Marín-Hernández [23] analysed all European SMEs that voluntarily published sustainability reports following the GRI between 2016 and 2019 and confirmed that the trend of non-financial disclosure is not limited to large corporations.

According to Journeault et al. [24], investing in more sustainable technology or initiatives may be profitable. However, SMEs often lack the necessary working capital and funds to go ahead with such projects. Limited human and financial resources and time constraints are significant barriers to the implementation of sustainability initiatives within SMEs. As a practical example, SMEs employees often multitasking across various functions, make it challenging for this kind of enterprises to focus on strategic planning and sustainability initiatives.

Another fact is that we have complexity of guidelines as previously mentioned, frameworks like GRI, IIRC, and others may appear overly complex for these companies. Often, SMEs do not know where to start due to the multitude of possible indicators. Permatasari and Kosasih [3] found this to be the case in Indonesia: guidelines with too many indicators made it difficult for SMEs to produce reports. However, when a simplified and adapted list was provided, over 50% of the surveyed SMEs reported being able to implement it with ease and expressed willingness to continue using the approach in subsequent years.

They also have a perception of low return. Afolabi et al. [5] interviewed SME owners in the UK and found that many did not perceive tangible benefits from adopting sustainability reporting, particularly among micro and small enterprises. The voluntary nature of regulations in this area and the lack of clear rewards made such activities seem "nice to have", but not essential compared to daily business demands. In particular, there was scepticism about the "net zero" carbon emissions goal, which was viewed as a government imposition disconnected from the realities of these businesses.

It is also true that SMEs also have lower external pressure considering that, unlike large companies, which are under the spotlight of investors, the media, and NGOs, SMEs operate with less public scrutiny. This can reduce the pressure to improve ESG performance. However, this scenario is changing over the years, banks, insurers, and corporate clients are becoming indirect pressure agents. Thus, while SMEs may primarily focus on immediate financial survival (securing credit, retaining clients), these same factors can become drivers for sustainable practices as market conditions evolve.

Also, it is important to mention that the SMEs present a cultural and knowledge-related component. Some SME managers are simply unaware of environmental management accounting techniques or sustainability tools suited to smaller businesses. Education and training initiatives can empower them to understand that sustainability is not merely a cost or philanthropy, but an integral part of modern effective management [25, 26]. Krawczyk [26] argues that accounting can no longer be limited to financial data alone – it must encompass sustainability information to meet the demands of both internal and external users. In summary, a mindset shift is needed in some cases, including overcoming the initial scepticism that ESG topics may generate among certain SME owners [3].

On the other hand, several recent studies highlight pathways and advantages for SMEs that embrace sustainability. Edeigba and Arasanmi [27], in their examination of triple bottom line practices in New Zealand SMEs, found that when small business managers genuinely implemented sustainable policies (covering economic, social, and environmental aspects), they were able to "consider themselves fully responsible" across all three dimensions and also better measure their financial performance alongside environmental impacts. Importantly, these authors found that greater awareness of sustainability regulations correlates with stronger engagement in sustainable practices – suggesting that if governments and institutions clearly communicate expectations (even if not mandatory), this can voluntarily encourage SMEs to act.

Additionally, there is a strong argument for efficiency and innovation: Jones et al. [28] emphasize that an SME, by implementing a sustainable performance management system (with indicators such as energy consumption per unit produced, waste generation, employee satisfaction, etc.), can identify inefficiencies and improvement opportunities that would otherwise go unnoticed. This broader management approach can lead, for example, to reduced input and waste costs, improved employee productivity (through engagement in responsible practices), and even access to new green market niches. Shields and Shelleman [29] have even proposed the use of sustainability dashboards for SMEs, featuring a concise set of balanced indicators to make tracking easier for managers and to facilitate intuitive communication of results. This tool would be a sustainability-focused analogue of the balanced scorecard, tailored to the SME context [29].

3 Method

To achieve the proposed objective, this study adopted a systematic and descriptive literature review approach. According to Lakatos and Marconi [30], descriptive research seeks to outline "what is" – by recording, analysing, and interpreting current phenomena without interfering in them – and is appropriate for understanding a phenomenon in the present. In this case, the phenomenon in focus is the incorporation of sustainability practices by SMEs as portrayed in academic studies. This is also an exploratory study, as it investigates a relatively under-researched topic (sustainability in SMEs) with the aim of gathering knowledge that may later support more precise hypotheses or applied empirical studies.

The methodological strategy followed the guidelines of systematic literature review in the field of management [31]. Initially, the database and search criteria were defined. The Web of Science (WoS) was chosen as the main source of bibliographic data, as it is considered one of the most reliable and comprehensive databases for academic surveys [32–34]. The research focused on recent years to capture the state of the art, while not excluding classic and seminal works cited in the latest literature.

The keywords used included terms in both Portuguese and English related to sustainability and SMEs, such as: "sustentabilidade", "sustainability", "ESG", "pequenas e médias empresas (PME)", "small and medium enterprises (SME)", "indicadores", "relatórios", "performance", "contabilidade gerencial", among others. Boolean combinations were used to broaden the reach. For example, searches such as ("PME" OR "small and medium business") AND "sustainab*" were used to identify studies on sustainability involving SMEs, and also "ESG" AND "performance" AND "indica*" to find works addressing ESG indicators and performance. The search primarily covered the period from 2019 to 2024, targeting the most recent and relevant literature – but without ignoring earlier important references that appeared in the reviewed studies. The initial selection of publications was based on their citation scores.

Moreover, inclusion and exclusion criteria were applied during the selection process: included were (national and international) articles that dealt with sustainability indicators, practices, or reporting, particularly those linked to business performance or the SME context. Articles that were overly similar to others (in terms of results or citation base) or clearly out of scope were excluded to avoid redundancy and maintain focus. After reading titles and abstracts for an initial screening, a preliminary set of relevant studies was compiled.

Subsequently, the exploratory reading and "snowball" procedure was adopted: key references cited in the selected articles but not found in the initial search (for example, because they were from earlier years or not indexed in WoS) were tracked and included in the analysis. This ensured the inclusion of seminal works and others considered essential by authors in the field. This process added various classic sources - such as Jensen and Meckling [35], Freeman [8], Elkington [9], among others - as well as some recent studies outside the initial time frame but conceptually relevant.

As a result, a final portfolio of 229 academic publications was obtained, covering the period from 1976 to 2024. Among these, 145 were identified as central or key publications – those interconnected through citation relationships, indicating that they form the core of the academic discussion on the topic. It should be noted that these figures derive from bibliometric treatment (citation mapping) conducted during the review, but in this article the data are explored only descriptively, without applying clustering techniques or network analyses. That is, we use the information from the studies to extract trends, examples, and evidence, rather than performing advanced quantitative citation analyses.

Finally, the results of the review were organized thematically to be presented in the next section. Where useful, charts and tables were created to summarize the findings and support the descriptive analysis. All care was taken to comply with academic research standards, including proper referencing of all sources (in APA style) and critical analysis of the information collected.

4 Discussion of the Results

This section presents an integrated summary of the main findings extracted from the literature on sustainability and performance in SMEs. First, we discuss a general overview of the selected publications—including the temporal evolution and thematic focus. Then, we examine the empirical evidence and recommendations identified in the studies, with particular emphasis on the SME context. The findings are structured around key themes, aiming to answer the proposed research question.

General Overview of the Publications.
The systematic review confirmed that academic interest in the relationship between ESG (sustainability) and business performance has grown exponentially in recent years. While in the early 2000s this topic was explored in relatively few studies, the second half of the 2010s and the early 2020s witnessed a surge in research. One of the search queries in this review identified 437 publications that relate ESG practices to business performance, of which 314 (approximately 72%) were published in 2024 alone. This illustrates just how current and dynamic the subject has become.

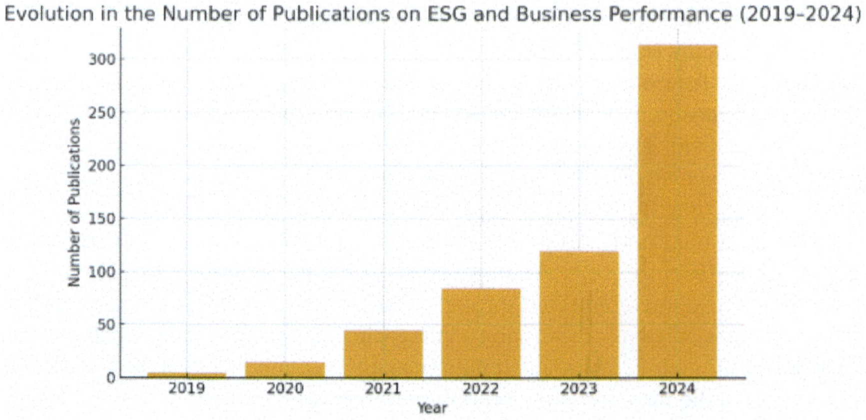

Fig. 1. Evolution in the number of publications on ESG and business performance in the Web of Science (2019–2024). The data shows a significant increase in 2024, reflecting growing attention to the topic in contemporary scientific literature.

As illustrated in Fig. 1, until 2019, the annual volume of articles was modest. However, from 2020 onwards, a notable increase occurred, culminating in 2024 with hundreds of papers published in a single year. Several factors may explain this growth:

(i). the materialization of global agendas (post-2015 SDGs, Paris Agreement) driving applied research.
(ii). the availability of more standardized ESG data (e.g., ratings and reports), which has enabled large-scale quantitative studies.
(iii). pressure from investors and society raising the practical relevance of the topic, translating into academic interest; and
(iv). the COVID-19 pandemic (2020–2021), which likely triggered reflections on resilience and corporate sustainability, leading to studies on the performance of sustainable companies during crises (e.g., Broadstock et al. [36]).

Another observable trend in the literature is the presence of consolidated review and meta-analysis papers, as well as studies focused on specific contexts or emerging issues. For example, Friede et al. [37] aggregated evidence from 2,000 studies; Gillan et al. [38] reviewed ESG/CSR research in corporate finance; and Khan [39] applied bibliometrics and meta-analysis to map ESG disclosure and performance research. These synthesis studies help to structure the field, highlighting convergences (e.g., a general consensus on the positive ESG–performance relationship) and gaps (e.g., the need for research in certain regions or business types).

As it is possible to see in Table 1, these are the most important studies presented in our research.

Table 1. Significant studies with an emphasis on SMEs

Study (Year)	Context / Ap-proach	Main Conclusions Relevant to SMEs
Permatasari & Kosasih (2022)	SMEs in Indonesia; analysis of sustainability reporting	Existing guidelines with numerous indicators are difficult to apply to SMEs. Adoption of a simplified list of indicators proved viable: > 50% of SMEs found the simplified guidance easy to implement and said they would use it again. Authors recommend developing tailored guides for SMEs to improve adoption
Ortiz-Martínez & Marín-H. (2022)	European SMEs; analysis of GRI reports (2016–2019)	A growing number of SMEs voluntarily disclose non-financial information. The trend of sustainability reporting is not limited to large companies. Transparency via reports (following GRI) is essential to communicate ethical practices, even for SMEs. Investment in ESG data disclosure and assurance is recommended
Javed et al. (2022)	Systematic review of barriers to Environmental Management Accounting (EMA) in Chinese SMEs	Identified key barriers: lack of knowledge, implementation costs, and absence of regulation. Strict regulation and flexible financing (green credit lines, incentives) stimulate the adoption of environmental practices and EMA in SMEs. Public policies promoting support and training are recommended

(*continued*)

Table 1. (*continued*)

Study (Year)	Context / Ap-proach	Main Conclusions Relevant to SMEs
Edeigba & Arasanmi (2022)	Quantitative research on New Zealand SMEs; triple bottom line practices	When SMEs effectively implement sustainability policies, they achieve positive economic, social, and environmental outcomes simultaneously (triple bottom line). Regulatory awareness has a direct effect: companies more aware of rules are more engaged in sustainable practices. Moderate regulatory pressure can encourage improvement
Afolabi et al. (2023)	Interviews with SME owners in the UK (perceptions of green economy and net zero)	Highlighted scepticism among micro and small firms about the benefits of sustainability reporting. Voluntary nature of rules and lack of perceived benefits lead to low priority. External stakeholder pressure (banks, customers) is emerging as an effective driver. Engagement by financiers and buyers is recommended to encourage ESG adoption
Jones et al. (2023)	Review and conceptual analysis (EU SMEs); strategic management accounting for sustainability	Found that SMEs tend to show lower environmental compliance and limited knowledge of impacts, but this presents an opportunity: implementing sustainable performance management can significantly boost efficiency and reduce costs (e.g., less waste, lower energy consumption). Proposes using management accounting tools to enhance SME sustainability

(*continued*)

Table 1. (*continued*)

Study (Year)	Context / Ap-proach	Main Conclusions Relevant to SMEs
Porciúncula & Andreoli (2023)	Study in Paraná (Brazil); analysis of local SME reports and indicators	Some SMEs already use simplified sustainability reports. Authors identified selected indicators that can serve a dual purpose: for simplified reporting and internal self-assessment. This helps SMEs reflect and improve their practices even if they don't publish formal reports
Lisi et al. (2023)	Systematic review of non-financial disclosure in global SMEs	Multiple factors influence the implementation of non-financial reporting in SMEs (e.g., company size, sector pressure, organisational culture), but they require deeper empirical investigation. Suggests future research agenda to study challenges and develop tailored solutions for SMEs (e.g., simplified frameworks, targeted training)
Shields & Shelleman (2020)	Conceptual proposal of sustainability dashboards for SMEs	Practical tool to help SMEs measure and manage sustainability in an integrated way. Recommends dashboards with balanced key indicators (environmental, social, economic) that are easy to update and understand, fostering the integration of sustainability into day-to-day management and communication with local stakeholders and investors

Table 1 above summarizes the contributions of some of the most relevant studies identified. Based on these and other sources from the review, we next provide an integrated discussion of the key insights regarding the challenges and opportunities for implementing sustainable practices in SMEs, as well as examples of feasible indicators and metrics for this segment.

Challenges and Opportunities for Sustainability in SMEs.
The literature consistently emphasizes that while SMEs face real challenges in adopting the ESG agenda, these challenges are not insurmountable. Frequently cited obstacles include limited resources, lack of expertise, absence of tailored standards, weak market incentives, and cultural resistance. However, alongside each challenge, many studies highlight corresponding strategies or opportunities. Below are some challenge-opportunity pairings identified in the literature:

Complexity vs. Simplification.
Challenge: Existing frameworks and technical jargon are often complex, leaving SMEs unsure of what to report.

Opportunity: Develop simplified guides with a basic set of adapted ESG indicators - such as those tested by Permatasari and Kosasih [3]. A reduced list focused on essential aspects allows SMEs to begin their sustainability journey in a pragmatic way. Suggested indicators by various authors include, for example: energy (kWh) and water (m3) consumption, waste generation (kg or number of pickups), estimated CO_2 emissions (tonnes), employee training hours, staff turnover (%), number of community initiatives, or customer ESG-related feedback. Shields and Shelleman [29] stress the importance of selecting indicators that align with the SME's strategic goals and that can be updated periodically. Once basic metrics are in place, companies can gradually incorporate more sophisticated ones.

Costs vs. Financial Benefits.
Challenge: Implementing sustainable practices implies monetary and personnel costs, which are difficult to justify without clear returns.

Opportunity: Highlight and internalize the economic benefits of ESG practices. Empirical studies show that many sustainability initiatives lead to savings: reduced consumption of inputs and energy (impacting costs directly), process improvement, less rework and waste, etc. For example, responsible practices can open access to green financing (some banks offer reduced rates to companies with strong ESG performance or green certifications) and prevent future penalties in supply chains (e.g., large customers demanding ESG compliance).

Scepticism vs. Stakeholder Pressure.
Challenge: "Why should I, as a small business, do this if no one demands it?" - a sceptical view found by Afolabi et al. [5].

Opportunity: Pressure may not come from regulators in the short term, but key stakeholders, investors, major clients, and consumers—are becoming more influential. SMEs that take the lead and adopt ESG practices or reports may use them as marketing differentiators and relationship-building tools. For instance, communicating simple environmental commitments (e.g., partially offsetting emissions, using sustainable materials) can attract sustainability-conscious clients. And, as predicted by Afolabi et al. [5], if banks and credit lines begin demanding ESG information for loans or rate negotiations, SMEs with structured data will gain a competitive edge. Thus, the narrative is shifting from "no one demands this" to "my business partners are beginning to value it."

Lack of Knowledge vs. Capacity Building and Collaboration.
Challenge: SME managers often lack training or access to corporate sustainability knowledge.

Opportunity: Invest in capacity building (training, affordable consultancy) and sectoral collaboration. SMEs in the same sector or region can form working groups, share best practices, and even pool consultancy costs. Collective initiatives, such as business associations providing report templates or sector-specific benchmarks, are mentioned by Das et al. [4] as key mechanisms to promote sustainability in SMEs. Partnerships with universities and research centres (e.g., via outreach projects) may also offer tailored support at low cost.

Lack of Specific Standards vs. Incentive Policies.
Challenge: Most standards and frameworks are designed for large corporations; SMEs are left without adequate regulation.

Opportunity: Develop public policies that offer incentives and guidance. Some countries are already creating voluntary SME-specific guidelines (e.g., ISO 26000 on social responsibility, or local sustainable business labels). Governments can offer tax incentives or preferential treatment in public procurement for sustainable SMEs, increasing the payoff for investing in ESG. While mandatory regulations for SMEs may be excessive, incentives and support are widely seen as balanced alternatives [40].

In summary, the literature clearly states that SMEs not only face sustainability challenges, but also stand to gain significantly by overcoming them. The gradual and strategic adoption of ESG practices can strengthen these firms' resilience and competitiveness. Moreover, the studies reviewed offer both anecdotal evidence (successful cases) and structured recommendations to guide this process.

Practical Sustainability Indicators for SMEs.
Various sustainability indicators and metrics are cited in the literature as suitable for small and medium-sized enterprises—often simplified versions of those used by large corporations, focusing on aspects directly controllable by SMEs. Here we highlight some key indicators, organized according to the ESG pillars, that have been considered useful or are already being applied by SMEs:

Environmental Indicators

- Electricity and fuel consumption (kWh, liters)
- Water consumption (m3)
- Waste volume generated per month (kg) and recycling rate
- Greenhouse gas emissions (CO_2 equivalent), often estimated from energy/fuel usage—some SMEs use free online calculators offered by government programmers

- Use of sustainable raw materials (% of recycled or certified inputs in total)

These indicators are reported in case studies and simplified reports analyzed by Porciúncula and Andreoli [41] and Ortiz-Martínez and Marín-H. [23]. Even if SMEs don't publish formal reports, tracking internal data—for instance, monthly kWh per unit produced—can inform decisions (e.g., investing in energy-efficient machinery or solar panels).

Social Indicators

- Employee turnover rate (annual %, staff leaving)
- Absenteeism rate or number of sick days lost
- Training hours per employee per year
- Employee satisfaction (through simple internal surveys)
- Workforce diversity (e.g., % of women in supervisory roles)
- Number of workplace accidents or safety incidents

In SMEs, many of these indicators may be tracked informally, but systematizing them helps identify human resource issues. Jones et al. [28] note that improving social indicators (like reducing turnover and boosting engagement) leads to productivity gains that enhance financial performance. An external social metric may also include community actions—e.g., number of local initiatives supported (donations, events, volunteering), which helps build local reputation.

Governance Indicators.
Although SMEs rarely have formal boards, governance indicators can include:

- Existence of a code of ethics or conduct (yes/no)
- Grievance or whistleblowing procedures (yes/no)
- Legal compliance level (qualitative indicator or number of identified and corrected non-compliances)
- Stakeholder engagement (e.g., regular employee meetings to discuss performance and strategy—yes/no; or customer satisfaction surveys—yes/no, with average score)

Some studies (e.g., Krawczyk [26]) recommend that SMEs implement governance practices gradually, starting with basic internal policies and compliance, before progressing to more complex structures. Corporate governance is crucial for achieving strategic goals, although it requires a long-term process to establish a good corporate culture in SMEs. Implementing governance practices can also improve SME's reputation and legitimacy. Good governance practices can facilitate SMEs access to capital by improving stakeholder relations and enhancing the company's reputation, making it more attractive to investors and capital lenders. Besides, effective governance practices may help in identifying and mitigating risks, which is a core tangible benefit for SMEs.

It is important to stress that indicator selection should be strategic and relevant to the company's reality. An industrial SME will likely prioritize environmental indicators (e.g., hazardous waste, energy use), while a service SME might focus more on social ones (e.g., employee wellbeing, client satisfaction). The literature does not impose a universal list but offers examples of good practices.

Shields and Shelleman [29] advocate for a balanced dashboard containing 8 to 12 indicators in total, spanning all three pillars, to allow managers to monitor performance at a glance. This avoids information overload and focuses attention where it matters. The dashboard can also include targets for each indicator, encouraging continuous improvement—e.g., reduce energy consumption by 5% next year; reach at least 80% employee satisfaction; ensure zero serious workplace accidents, etc.

Another point emphasized by authors like Lisi et al. [42] is comparability and transparency: whenever possible, SMEs should compare their indicators over time (to track progress) and, where industry data is available, benchmark against sectoral averages or similar firms. This helps assess whether their sustainable performance is below or above average, motivating action or serving as a market differentiator if ahead of peers.

In terms of reported outcomes, SMEs that adopted even simple indicators noted qualitative benefits such as: increased internal awareness (employees start suggesting waste reduction ideas when they see the numbers); improved organizational climate; enhanced image among partners (e.g., a small supplier that began reporting environmental indicators received praise and preferential treatment from a larger client)—as noted by Afolabi et al. [5] and Porciúncula & Andreoli [41].

Table 2 below shows a simple example of dashboard:

Table 2. Example of Financial and Sustainability Dashboard

Financial and Sustainability Dashboard (in KUSD or otherwise specified)	May - 2025	Target
Cash Position	**45,3**	**50,0**
Bank loans	30,0	-
Trade Receivables	35,0	20,0
Gross Revenues	**120,0**	**150,0**
Orders Received	200,0	180,0
Backlog	80,0	100,0
Female Employees (n°)	10	15
Male Employees (n°)	45	40
Safety incidents (n°)	**3**	**-**
Employees sick days lost	15	-
Employees turnover rate (annual %)	10%	5%
Waste materials to recycling (kg)	50	60
Electricity (kWh)	**350**	**250**
Fuel (L)	800	700
Water (L)	**500**	**550**
GHG (CO_2 equivalent)	250	200
Customer complaints (n°)	**5**	**1**

Table 2 includes feasible indicators and metrics that could be applied to SMEs. In this example - with fictitious data - there is a comparison between the current month (actual "May - 2025") and the respective targets ("Target").

Finally, it is worth noting that technology is making it easier for small companies to monitor indicators—there are affordable apps and software for tracking sustainability metrics, and even free tools (spreadsheet templates, online calculators) provided by NGOs and government bodies. Digitalization facilitates data collection (e.g., sensors measuring machine consumption) and automated reporting, reducing manual effort.

5 Final Remarks

This article presented a comprehensive descriptive analysis of academic literature on corporate sustainability, with a particular emphasis on the context of small and medium-sized enterprises (SMEs). We synthesized the main concepts, evidence, and recommendations relevant to the implementation of sustainability practices and indicators in SMEs.

The findings confirm that there is a growing body of knowledge aimed at understanding and promoting sustainability in SMEs, although still small compared to the extensive literature focused on large corporations. We identified that:

There is a historical gap in accounting and management research regarding SMEs and sustainability—a gap that has only recently begun to be filled, given the recognition that global goals (such as the SDGs or the Paris Agreement) cannot be achieved without involving SMEs. This study contributes by consolidating and giving visibility to this emerging knowledge.

The main theories linking sustainability and performance (e.g., stakeholder theory, resource-based view, triple bottom line) are applicable to SMEs, with appropriate scale adaptations. That is, the logic that well-implemented ESG practices can generate competitive advantages and performance improvements holds true even for smaller firms—including benefits such as operational efficiency and access to capital.

The barriers faced by SMEs are significant (limited resources, complexity, skepticism, weak external pressure), yet practical solutions have been mapped: adoption of simplified frameworks, training and awareness-raising, use of technology to facilitate monitoring, partnerships, and external support (financial and technical), among others. Public policies and market initiatives—whether "pulling" (incentives, stakeholder pressure) or "pushing" (proportional regulations, voluntary standards)—can act as catalysts.

There is a set of sustainability practices and indicators considered viable and useful for SMEs. This set ranges from tracking resource consumption and employee satisfaction to implementing simplified sustainability reports and internal dashboards. Several documented cases show that, by incorporating such indicators into daily management, SMEs achieved concrete improvements in both ESG aspects and financial performance, supporting the argument that sustainability is synonymous with good management.

Practical Implications.

For SME managers and policymakers, this article offers a compiled reference of what the recent literature suggests. For small business owners or managers, the findings indicate that beginning (or improving) the sustainability journey is worthwhile, starting with basic actions and evolving gradually. The evidence contradicts the fear that investing in ESG is a waste of resources—on the contrary, it tends to strengthen the business in the medium and long term. Additionally, with likely growing demands from clients and investors for socio-environmental responsibility, being ahead of the curve can result in competitive advantage.

From an academic perspective, we reinforce Lisi et al.'s [42] recommendation that more empirical research focused on SMEs is needed. Future studies could explore specific sectors (e.g., industrial vs service SMEs), regional differences (developed vs emerging countries), and develop refined theoretical frameworks tailored to SME sustainability.

SMEs can also benefit from interdisciplinary research paths. Studies on behavioral and cultural aspects of sustainability could investigate how the internal behaviour, culture, and reasoning of SMEs influence their sustainability practices. Besides, new studies may explore the economic motivations behind SMEs' sustainability-oriented behaviour, such as cost reductions through energy efficiency and competitive advantages from aligning with customer preferences. This can help to understand how economic incentives drive environmental actions. The social and ethical dimensions of sustainability could also be explored. The role of social drivers, such as ethical convictions of owners and managers, and their engagement in local communities, in shaping SMEs' sustainability actions.

Longitudinal studies following SMEs as they implement ESG practices would be especially useful to assess causal performance impacts. Furthermore, exploring the role of sustainable supply chains—that is, how large companies can "pull" their SME suppliers towards sustainability—is another promising research avenue.

Limitations and Final Remarks.
This study's main limitation is that it is based primarily on published literature, potentially overlooking undocumented practical initiatives or tacit business knowledge. Nonetheless, by compiling a wide range of international and national sources, we believe it provides a robust and up-to-date overview of the state of the art.

In conclusion, sustainability in SMEs is not only possible but necessary and potentially advantageous. As mentioned, SMEs account for nearly 90% of companies worldwide, so a truly sustainable development will not be achieved without the active participation of this segment [2]. We hope this work serves as both an incentive and a guide for SME managers, as well as for researchers and policymakers, to continue advancing in the journey to integrate sustainability practices into SMEs—generating value not only for the businesses themselves but for society and the planet.

References

1. International Finance Corporation – IFC. Sustainable MSME Finance Reference Guide. Washington (2023)
2. Saleem Q, International Finance Corporation (IFC): SME finance: opportunities and challenges. World Bank Group, Washington (2023)

3. Permatasari, P., Kosasih, E.: Analysis of Small and Medium Enterprises (SMEs) sustainability reporting to develop sustainability reporting guideline. Int. J. Entrepreneurship Sustain. Stud. **2**, 1–24 (2022). https://doi.org/10.31098/ijeass.v2i1.822
4. Das, M., Rangarajan, K., Dutta, G.: Corporate sustainability in small and medium-sized enterprises: a literature analysis and road ahead. J. Indian Bus. Res. **12**, 271–300 (2020). https://doi.org/10.1108/JIBR-09-2017-0166
5. Afolabi, H., Ram, R., Hussainey, K., et al.: Exploration of small and medium entities' actions on sustainability practices and their implications for a greener economy. J. Appl. Acc. Res. **24**, 655–681 (2023). https://doi.org/10.1108/JAAR-09-2022-0252
6. Hardin, G.: The tragedy of the commons. Science (1979) **162**, 1243–1248 (1968)
7. Friedman, M.: The social responsibility of business is to increase its profits. The Times, The New York Times Archives pp. 17–17 (1970)
8. Freeman, R.E.: Strategic Management: A Stakeholder Approach. Pitman, Boston (1984)
9. John, E.: Cannibals with forks: the triple bottom line of 21st century business. Capstone, Oxford (1997)
10. Brasil, Ministério da Ciência e Tecnologia, Ministério das Relações Exteriores. Protocolo de Quioto Editado e traduzido pelo Ministério da Ciência e Tecnologia com o apoio do Ministério das Relações Exteriores da República Federativa do Brasil. Brasília (1998)
11. United Nations (UN). Global Compact. Who Cares Wins: Connecting Financial Markets to a Changing World. New York (2004)
12. United Nations Framework Convention on Climate Change (UNFCCC). Paris Agreement. Bonn (2015)
13. United Nations General Assembly. Resolution adopted by the General Assembly on 25 September 2015: Transforming our world: the 2030 Agenda for Sustainable Development. New York (2015)
14. Commission, E.: Directive of the European Parliament and of the Council: amending Directives 2006/43/EC, 2013/34/EU, (EU) 2022/2464 and (EU) 2024/1760 as regards certain corporate sustainability reporting and due diligence requirements. European Commission, Brussels (2025)
15. Del Baldo, M.: The implementation of integrating reporting <IR> in SMEs: insights from a pioneering experience in Italy. Meditari Accountancy Res. **25**, 505–532 (2017). https://doi.org/10.1108/MEDAR-11-2016-0094
16. International Sustainability Standards Board (ISSB). IFRS S1 - General Requirements for Disclosure of Sustainability-related Financial Information. London (2023)
17. International Sustainability Standards Board (ISSB). IFRS S2 - Climate-related Disclosures. London (2023)
18. Comitê Brasileiro de Pronunciamentos de Sustentabilidade Pronunciamento Técnico (CBPS). CBPS 01 Requisitos Gerais para Divulgação de Informações Financeiras Relacionadas à Sustentabilidade. São Paulo (2024)
19. Comitê Brasileiro de Pronunciamentos de Sustentabilidade Pronunciamento Técnico (CBPS). CBPS 02: Divulgações Relacionadas ao Clima - Correlação à Norma Internacional de Sustentabilidade-IFRS S2. São Paulo (2024)
20. EFRAG European Sustainability Reporting Standards (ESRS). https://www.efrag.org/en/sustainability-reporting/about-sustainability-reporting. Accessed 11 Aug 2025
21. Carbon Disclosure Project (CDP) CDP Home. https://www.cdp.net/en. Accessed 11 Aug 2025
22. World Resources Institute (WRI), World Business Council for Sustainable Development (WBCSD). The GHG Protocol Corporate Accounting and Reporting Standard. Washington (2004)

23. Ortiz-Martínez, E., Marín-Hernández, S.: European SMEs and non-financial information on sustainability. Int. J. Sust. Dev. World **29**, 112–124 (2022). https://doi.org/10.1080/13504509.2021.1929548
24. Journeault, M., Perron, A., Vallières, L.: The collaborative roles of stakeholders in supporting the adoption of sustainability in SMEs. J. Environ. Manage. **287**, 112349 (2021). https://doi.org/10.1016/j.jenvman.2021.112349
25. Tinoco, J.E.P.: Balanço Social: uma abordagem da transparência e da responsabilidade pública das organizações, 1st ed. Editora Atlas, São Paulo (2001)
26. Krawczyk, P.: Non-financial reporting—Standardization options for SME sector. J. Risk Financ. Manage. **14**, 417 (2021). https://doi.org/10.3390/jrfm14090417
27. Edeigba, J., Arasanmi, C.: An empirical analysis of SMES' triple bottom line practices. J. Account. Organ. Chang. **18**, 238–259 (2022). https://doi.org/10.1108/JAOC-12-2020-0206
28. Jones, B., Rasha, A., Dyczkowska, J., Dyczkowski, T.: Sustainable performance management in the eu sme sector. A review and analysis of concepts and methods of strategic management accounting. Zeszyty Teoretyczne Rachunkowosci **47**, 191–215 (2023). https://doi.org/10.5604/01.3001.0054.0890
29. Shields, J.F., Shelleman, J.M.: SME sustainability dashboards: an aid to manage and report performance. J. Small Bus. Strateg. **30**, 106–114 (2020)
30. Lakatos, E.M., Marconi, M.deA.: Técnicas de pesquisa, 9th ed. Editora Atlas, São Paulo (2021)
31. Tranfield, D., Denyer, D., Smart, P.: Towards a methodology for developing evidence-informed management knowledge by means of systematic review. Br. J. Manag. **14**, 207–222 (2003). https://doi.org/10.1111/1467-8551.00375
32. Paul, J., Criado, A.R.: The art of writing literature review: what do we know and what do we need to know? Int. Bus. Rev. (2020). https://doi.org/10.1016/j.ibusrev.2020.101717
33. Pranckutė, R.: Web of Science (WoS) and Scopus: the titans of bibliographic information in today's academic world. Publications (2021). https://doi.org/10.3390/publications9010012
34. Zhu, J., Liu, W.: A tale of two databases: the use of Web of Science and Scopus in academic papers. Scientometrics **123**, 321–335 (2020). https://doi.org/10.1007/s11192-020-03387-8
35. Jensen, M.C., Meckling, W.H.: Theory of the firm: managerial behavior, agency costs and ownership structure. J financ econ **3**, 305–360 (1976). https://doi.org/10.1016/0304-405X(76)90026-X
36. Broadstock, D.C., Chan, K., Cheng, L.T.W., Wang, X.: The role of ESG performance during times of financial crisis: evidence from COVID-19 in China. Financ. Res. Lett. (2021). https://doi.org/10.1016/j.frl.2020.101716
37. Friede, G., Busch, T., Bassen, A.: ESG and financial performance: aggregated evidence from more than 2000 empirical studies. J. Sustain. Financ. Investment **5**, 210–233 (2015). https://doi.org/10.1080/20430795.2015.1118917
38. Gillan, S.L., Koch, A., Starks, L.T.: Firms and social responsibility: a review of ESG and CSR research in corporate finance. J. Corp. Finan. (2021). https://doi.org/10.1016/j.jcorpfin.2021.101889
39. Khan, M.A.: ESG disclosure and Firm performance: a bibliometric and meta analysis. Res. Int. Bus. Financ. (2022). https://doi.org/10.1016/j.ribaf.2022.101668
40. Javed, F., Yusheng, K., Iqbal, N., et al.: A Systematic review of barriers in adoption of environmental management accounting in Chinese SMEs for sustainable performance. Front. Public Health (2022). https://doi.org/10.3389/fpubh.2022.832711

41. da Porciúncula, S.A., Andreoli, C.V.A.: Proposal for a simplified sustainability report for small and medium-sized enterprises. Revista Brasileira de Ciências Ambientais **58**, 67–80 (2023). https://doi.org/10.5327/z2176-94781513
42. Lisi, S., Mignacca, B., Grimaldi, M.: Non-financial reporting and SMEs: a systematic review, research agenda, and novel conceptualization. J. Manag. Organ. **30**, 600–622 (2023). https://doi.org/10.1017/jmo.2023.43

The Ethical Implications of Artificial Intelligence in Accounting and Auditing: A Systematic Literature Review

Paula Gomes dos Santos[1,2] and Fábio Albuquerque[1,2](✉)

[1] ISCAL/Instituto Politécnico de Lisboa, Lisboa, Portugal
{pasantos,fhalbuquerque}@iscal.ipl.pt
[2] CICF, School of Management, IPCA, Barcelos, Portugal

Abstract. The increasing use of artificial intelligence (AI) technologies in accounting and auditing has been accompanied by growing concerns regarding their ethical risks. This paper aims to review research on the ethical implications of AI use in accounting and auditing. The characteristics of the papers gathered, the research methods followed, as well as the purposes, main conclusions, and suggestions for future avenues, are within the scope of this analysis. For this purpose, it uses a systematic literature review, using the Scopus database to collect published papers. The findings provide valuable insights into the ethical implications of AI in accounting and auditing, offering useful perspectives for practitioners, academics, researchers and society in general. They also indicate a vast field for researchers in several areas, which can guide future investigations. The ethical challenges of AI in accounting and auditing should remain a priority as the profession must balance technological advancements with a strong commitment to human ethical competence to maintain its integrity and societal responsibility.

Keywords: Accounting · Artificial Intelligence · Auditing · Ethical Implications · Literature Review

1 Introduction

The accounting and auditing profession has long struggled with ethical dilemmas, as its legitimacy and trustworthiness are crucial to financial transparency and societal confidence [1, 2]. High-profile corporate misconduct cases have further underscored the need for strong ethical standards [3–6], prompting professional bodies such as the International Federation of Accountants (IFAC) through its International Ethics Standards Board for Accountants (IESBA) [7], the Financial Reporting Council (FRC) [8] or The Institute of Internal Auditor (The IIA) [9], to publish ethics principles, intending to promote transparency and integrity through an ethical culture in the profession.

Several similarities between those frameworks can be found. For instance, the Ethical Standard of the FRC [8] promotes integrity, objectivity and independence, and the Global Internal Audit Standards of The IIA [9] establishes, as principles, integrity, objectivity, competency, due professional care, and confidentiality, while IESBA [7] code of ethics

L. Almeida et al. (Eds.): ICAFI 2025, ISEM 80, pp. 561–572, 2026.
https://doi.org/10.1007/978-3-032-16268-7_32

emphasizes principles such as integrity, objectivity, competence, confidentiality, and professional behaviour. However, these principles are not always sufficient in guiding auditors and accountants through complex moral dilemmas [10, 11].

Amidst these existing ethical challenges, the adoption of Artificial Intelligence (AI) in several industries has raised new ethical implications [e.g., 12–14]. For instance, in the healthcare field that has been incorporating AI, these concerns have been especially stressed since, although these technologies may help, they are accompanied by growing concerns regarding their ethical implications [14].

As happens with the healthcare industry, accounting and auditing are also knowledge-intensive fields [15], requiring professional judgment and critical thinking [16], and are equally integrating AI technologies to improve efficiency [17–22]. Furthermore, AI is raising the same concerns regarding the ethical risks of its use and the necessity of preserving human critical thinking [e.g., 6, 23–30].

Therefore, the integration of AI into the accounting and auditing profession has been capturing broader attention beyond ethicists, namely by academia, professionals, policymakers, and regulators [31].

For example, academia has been evaluating the ability of AI-driven tools to address accounting proficiency exams, with mixed findings regarding their ethical reasoning capabilities. Some studies indicate that AI performs well in ethical assessments [32–34], while others suggest that AI struggles with professional judgment and technical complexity [35]. This inconsistency raises concerns about AI's ability to support ethical decision-making in the field.

Besides, the literature has highlighted that a significant limitation of AI models is their dependence on data quality, which can lead to biased outcomes when training data lacks diversity or representativeness. As a result, these models struggle to achieve deep contextual understanding, which is necessary to replace human analytical thinking [6]. Additionally, AI lacks the capacity for self-reflection, a key component of ethical reasoning [31].

From the regulators' side, the European Commission (EC) has addressed this ethical concern by issuing guidelines to promote trustworthy AI [36]. Thus, a trustworthy AI should be lawful, ethical, and robust, and comply with ethical principles, such as respecting human autonomy, preventing harm, being fair, and being explicable [36]. EC also offered guidance for adopting an ethically focused approach in AI-based solutions, defining six general ethical principles that any AI system must preserve and protect: respect for human agency; privacy, personal data protection and data governance; fairness; individual, social, and environmental well-being; transparency; accountability and oversight [37].

Hence, the continuous advancement of AI technologies, particularly since 2020, has raised new challenges and captured researchers' attention in the accounting and auditing area [38]. Thus, the question of whether and how AI can be ethical remains [14]. In this context, this paper aims to systematically review the research on ethics implications concerning AI implementation in the accounting and auditing area, by collecting papers from the Scopus database. The following research questions (RQ) were defined to perform this review:

RQ1. What are the overall characteristics of the papers selected (publication year and source, authors' country and affiliation, and number of citations)?
RQ2. What are the research methods used by researchers in the studies analysed?
RQ3. What are the main purposes and conclusions of the studies analysed?
RQ4. What are the main suggestions for future investigations proposed by researchers in the studies analysed?

In addition to this introduction, this paper is structured into three sections. The next provides the materials and methods. The third presents the findings. Finally, the last section discusses those findings and outlines the main conclusions.

2 Material and Methods

This research uses a systematic literature review. All available information on 19 March 2025 was considered. No initial or final years were defined as a period. To select the appropriate papers for analysis, this research globally follows the Preferred Reporting Items for Systematic Reviews and Meta-Analyses (PRISMA) framework as proposed by Page et al. [39]. The search syntax initially contains the terms related to ethics in the papers' titles, namely the words "ethic", "ethics", and "ethical". Then, the terms related to AI matters, such as "artificial intelligence", and its acronym "AI" were also included. Finally, the terms related to accounting and auditing within the papers' abstract were also proposed, including the likely associated terms with them, namely the words "accountant", "audit", and "auditor" and their plurals. Further conditions were included in the search syntax to increase the comparability, namely the language (English), the type of publication (article) and the stage of publication (final).

The search initially delivered eleven papers. Notwithstanding, three papers were excluded to increase the comparability of the findings, considering they exclusively focused on human resources [40, 41] and healthcare fields [14], not covering accounting and auditing areas. Figure 1 summarizes this process, including the complete search syntax with the Boolean operators used.

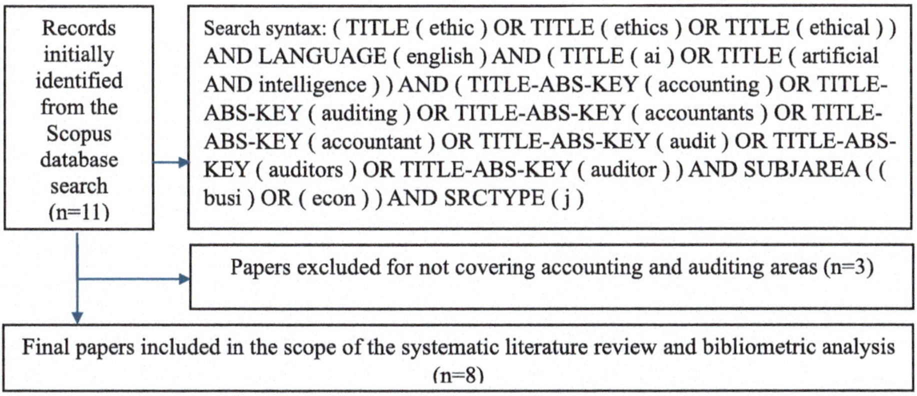

Fig. 1. Search criteria for collecting papers.

The next section provides the findings from the analysis of the papers gathered.

3 Results

Regarding **RQ1**, Table 1 presents the overall characteristics of the selected papers in terms of the publication year, number of citations, and source.

Table 1. Publication year, number of citations, and source.

Paper	Citations	Journal (publisher)	Rank; CiteScore; Percentile (quartile)
[42]	245	Journal of Business Ethics (Springer)	#6; 12.8; 99th
[43]	60	Auditing and Accountability Journal (Emerald)	#6; 10.3; 96th
[44]	40	International Journal of Accounting Information Systems (Elsevier)	#6; 12.8; 99th
[45]	19	Journal of Business Economics and Management (Vilnius Gediminas Technical University)	#31; 5.8; 83rd
[46]	18	Journal of Sustainable Finance & Investment (Taylor & Francis)	#4; 10.6; 98th
[47]	1	Business and Professional Ethics Journal (Philosophy Documentation Center)	#221; 1.0; 72nd
[48]	10	Information and Management (Elsevier)	#5; 17.9; 96th
[49]	0	Cogent Business and Management (Taylor & Francis)	#49; 4.4; 74th
Total	**393**		

Source: Own elaboration

Table 1 shows that the first paper that addressed this topic was published in 2020 [42], with half of the papers in 2023 [44–47], one in 2022 [43], and the remaining two in 2024 [48, 49], the last year in which publications were found.

Table 1 also indicates that five out of eight journals that had published those papers are ranked amongst the tenth most cited journals in their categories. Moreover, those five papers gathered 95% of the papers' total citations, with the first publication found [42] being the most cited with 62% of the total citations. Only one paper [49] has no citations until now.

Still concerning **RQ1**, Table 2 presents the countries and affiliations of the selected papers' authors.

Table 2 indicates the prevalence of the United States (with three papers), followed by Finland (with two papers). Globally, the United States and European countries have

Table 2. Authors' countries and affiliations.

Paper	Country	Affiliation
[42]	United States	Rutgers University
[43]	Finland; Austria	Hanken School of Economics; KPMG Audit
[44]	China	Hefei University of Technology; Southwestern University of Finance and Economics; Anhui Cnbisoft Software Technology Co
[45]	Romania; Spain	Babeş-Bolyai University; 1 Decembrie 1918 University; Institut of Multidisciplinary Research for Science and Technology; University of Murcia
[46]	United States	Pace University
[47]	United States	University of Baltimore
[48]	Finland	Turun yliopisto; Turun Kauppakorkeakoulu
[49]	Lebanon	Rafik Hariri University; Holy Spirit University of Kaslik

Source: Own elaboration

published most of the papers on this topic (six out of eight). The authors' affiliations show that several papers bring together authors from various institutions, including from different countries [43, 45]. As expected, most authors are affiliated with higher education and research institutions. However, it should be highlighted that two authors come from outside of academia, namely from KPMG Audit and Anhui Cnbisoft Software Technology.

Regarding **RQ2**, Table 3 presents the research methods used in developing the selected papers.

Table 3. Papers' research methods.

Paper	Research methods
[42]	Qualitative research based on two established ethical frameworks
[43]	Qualitative research based on a semi-systematic literature review
[44]	Qualitative research based on interviews
[45]	Qualitative research based on interviews
[46]	Qualitative research (essay)
[47]	Qualitative (case study) and quantitative research (based on a survey)
[48]	Qualitative research based on systematic literature review
[49]	Quantitative research using financial datasets

Source: Own elaboration

Table 3 shows that papers mostly used qualitative methods, namely based on literature review and interviews. The paper by Diab and El Hajj [49] is the only one that follows a quantitative method, and there is one that uses a mixed method [47].

Concerning **RQ3**, Table 4 presents the main purposes of the studies analysed.

Table 4. Papers' main purposes.

Paper	Main purpose
[42]	Forecast the ethical implications of the use of AI in auditing
[43]	Identify ethical challenges of using artificial intelligence (AI)-based accounting systems for decision-making
[44]	Explore the ethical impact of AI in managerial accounting at both pre-and post-adoption stages
[45]	Address the ethics of using artificial intelligence in accounting firms
[46]	Propose a model for ethical evaluation of AI-intensive companies, for investment decision-making
[47]	Examine the impact of AI on professional responsibility and ethical decision-making in small Certified Public Accountant (CPA) firms
[48]	Understand and consolidate the current literature on ethics-based AI auditing
[49]	Examine the role of organizational ethics in AI adoption and optimization

Source: Own elaboration

Table 4 shows that the papers globally address three main purposes. The first one, with six out of eight papers, discusses the ethical principles and challenges associated with the use of AI technology in accounting and auditing [42–45, 47, 48]. Secondly, Brusseau [46] develops a model for the ethical evaluation of AI-intensive companies, for investment guidance. Lastly, Diab and El Hajj [49] explore the role of organizational ethics in AI adoption and optimization, namely regarding financial decision-making models.

Concerning the main conclusions obtained by researchers in the studies analysed (**RQ3**), the papers that addressed the ethical principles and challenges associated with the use of AI technology in accounting and auditing (first purpose found) defend that AI's current and potential applications hold great promise for the profession. They highlight, however, that the ethical concerns about the technology may constrain its expected benefits [42]. Therefore, humans must retain control over the AI's decision-making process to uphold moral behaviour and continuously assess its outcomes and ethical implications [43].

According to Laine et al. [48], the academic ethics-based AI auditing literature identifies that the most common ethical principles are fairness, transparency, non-maleficence, and responsibility. Then, privacy, trust, beneficence, and freedom/autonomy also appear as relevant principles. The same authors identified AI system developers, the wider public, researchers, auditors, AI system users, and regulators as the most critical stakeholders in AI auditing [48].

The previous findings are aligned with the ones from Lehner et al. [43], who identified the following major ethical challenges in AI-based accounting and auditing decision-making:

- Objectivity – AI systems may lack true objectivity due to biases in algorithms and data selection.
- Privacy – The handling of sensitive financial and personal data poses risks, especially with Big Data and AI learning systems.
- Transparency – Many AI models operate as "black boxes", making it difficult to trace decision logic, making the resulting decisions not transparent or explainable.
- Accountability – It is essential to coordinate responsibilities for AI-driven decisions between accountants/auditors and software developers, ensuring the "good governance" of such technologies.
- Trustworthiness – AI adoption in accounting and auditing depends on stakeholders' trust in its fairness, accuracy, and reliability.

Zhang et al. [44] highlighted that ethical dilemmas are more likely when more subjective judgments are needed, when the focus is on decision-making, and when more reliance is on users' capabilities and accountability, which happens in managerial accounting.

According to those authors, the ethical challenges of AI in managerial accounting include data security, privacy, and misuse, the competence of AI developers and users, the expectation gap between AI capabilities and managerial accountants' expectations, transparency and trust, and bias in AI algorithms [44]. For instance, AI's integration of diverse data sources leads to concerns about data security, privacy, and misuse. Moreover, transparency, reliability, and accessibility are questioned since AI systems are often non-standardised. Bias and over-reliance on AI, particularly among less experienced accountants, amplify ethical risks. Also, AI is still an emerging technology, compared to Enterprise Resource Planning (ERP) systems and, therefore, is less widely adopted, making its ethical concerns more pronounced. Another major challenge is the gap between AI operations and managerial accountants' traditional work approaches. AI requires higher technical competence, and accountants must understand algorithms and results to avoid bias and misinformation. Managers must verify AI-generated reports to ensure accuracy [44].

AI is also reshaping the role of managerial accountants, shifting their focus from data processing to diagnostic analysis and professional judgment. However, the lack of face-to-face interactions in AI-driven environments may create an isolated workplace [44]. Their study also identifies four major stakeholders (developers, managers responsible for AI adoption, managerial accountants, and regulators) which share some common ethical impacts of AI, such as data security, privacy and misuse, accountability, accessibility, and AI transparency and trust [44].

Moreover, beyond the previously mentioned ethical issues, AI brings other impacts on organizations, such as expectation gaps between AI capabilities and users' expectations, and new challenges and learning curves for the latter, where less experienced professionals are likely to over-rely on results from AI, impairing their professional judgment [42, 44, 45]. Therefore, it is essential to balance technological advancement and professional integrity [42, 47]. The results by Fülöp et al. [45], for instance, show most accountants have basic AI knowledge, but few fully understand it.

In this regard, Munoko et al. [42] stress the ethical risks in auditing related to due professional care (when auditors cannot fully understand or explain AI decisions), auditor independence, (particularly when third-party AI systems are used), and data security, confidentiality, and potential "deskilling" of audit professionals due to the automation of routine tasks.

According to Bento and White [47], the adoption of AI in auditing appears to be simpler for smaller firms than for larger ones due to fewer administrative hurdles and legacy systems. Their study highlights two key ethical decision-making processes that still require professional judgment: (1) risk assessment when screening client transactions, particularly in pricing and client retention decisions, and (2) determining which AI-flagged anomalies warrant further investigation. A significant ethical dilemma arises in balancing professionalism (commitment to audit quality) with commercialism (concern for client interests) when responding to AI-generated risk assessments. Auditors must carefully balance these tensions to ensure ethical and high-quality audits.

Moreover, the same authors also found that professionalism plays a more substantial role than commercialism in determining whether auditors follow up on AI-generated risk alerts. Auditors primarily use AI screening to comply with audit standards, and those with high professionalism and low commercialism are the most likely to investigate transactions, even if they are considered immaterial. However, more experienced auditors were less likely to act on AI recommendations, possibly because they trusted their professional intuition over AI insights. This tendency could have significant implications, as senior auditors in leadership roles may inadvertently dismiss AI findings that challenge their expertise, potentially undermining professional responsibility in favour of traditional judgment [47].

Related to the second purpose studied, Brusseau [46] proposed a model integrating ethical criteria into investment decisions for AI-intensive companies, ensuring personal freedom, social well-being, and technical trustworthiness while proposing to be rigorous, accessible, useful, and reliable.

Finally, Diab and El Hajj [49] concluded that ethical business culture plays a critical role in AI adoption and efficacy.

In what concerns **RQ4**, the studies analysed mainly propose continuous research in this area given that, as AI continues to evolve, new ethical challenges may arise. Then, future research is essential to forecast and address emerging challenges in this field [42], as well as to mitigate ethical risks and safeguard the rights of individuals, organisations, and society [44].

Moreover, Laine et al. [48] suggest that research should explore the distinctions between AI auditing and traditional IT auditing, offering insights into best practices for assessing AI-driven systems. Additionally, there is a need to develop robust frameworks for ethics-based AI auditing to uphold principles like transparency, accountability, and fairness.

Bento and White [47] stress the need for experimental studies on auditors' ethical decision-making when using AI, to study the impact of individual factors on auditors' professional judgment supported by AI, and to continue exploring the ethical issues in emerging audit technologies, such as the implications of different criteria and decision rules embedded in AI algorithms.

Furthermore, future research can address the literature gap on broader ethical concepts, such as sustainability and solidarity, within AI auditing frameworks [48]. Also, normative thinking is needed to shape the future of human-machine symbiosis, thus, a broader discussion on the societal values that should guide AI implementation and decision-making in accounting seems essential [43].

The last section provides the discussion and conclusions from this study, including its limitations and suggestions for future research.

4 Discussion and Conclusions

This paper provides a systematic literature review of the research on ethics implications concerning AI implementation in the accounting and auditing area, covering several aspects related to the papers' characteristics, methods and contents (purposes, conclusions and suggestions for future avenues).

Despite the increasing interest in AI and ethics in this area, namely considering the rank of the publication's journals, the research on this topic is still relatively scarce. Not surprisingly, the first paper found is from 2020, reinforcing that the new challenges brought by the adoption of emerging AI in accounting and auditing are recently capturing researchers' attention [38].

The main findings highlight both the potential benefits and ethical challenges of AI adoption in accounting and auditing. While AI offers significant advantages, ethical concerns may limit its benefits [42, 43], emphasising the need for human oversight in AI decision-making to ensure ethical integrity [43].

Major ethical challenges include AI's bias and data security risks, lack of true objectivity and transparency of decision-making processes, and the need for clearly defined accountability [42–44, 48].

As knowledge-intensive fields [15], accounting and auditing require subjective judgment and critical thinking [16], which raises ethical dilemmas *per si* [44]. By using AI technologies, ethical risks increase, posing new challenges to the profession. Thus, ethical principles such as objectivity, competency, due professional care, and confidentiality [9] can be called into question by the use of AI [42].

Moreover, this highlights the need for professionals in this area to have the necessary skills to ethically use AI [42–45, 47, 48].

Therefore, the ethical challenges of using AI in accounting and auditing should continue on the agenda of several stakeholders because the profession must balance technological advancements with a continued emphasis on human ethical competence to uphold its integrity and societal role.

The findings from this paper can be useful to researchers since it provides a comprehensive overview of previous literature on the ethical challenges of using AI in accounting and auditing. It also contributes to academia and professional bodies in designing the courses that give access to the profession. It can also be useful for regulators to develop governance frameworks for AI ethics in accounting and auditing.

Nonetheless, this research also has the limitations of systematic literature reviews that should not be avoided, which mainly include the methods defined to select the papers to review and the inherent subjectivity of qualitative analysis. The scarce number of papers on the subject is also a limitation. Despite these limitations, the collected studies likely provide valuable insights into the ethical implications of AI in accounting and auditing, along with directions for future research, offering useful perspectives for practitioners, academics, researchers and society in general.

Acknowledgement. This work was supported by the Instituto Politécnico de Lisboa, project IPL/IDI&CA2024/IAccount_ISCAL [Grant number is not applicable]. This study was conducted at the Research Center on Accounting and Taxation (CICF) and was funded by the Portuguese Foundation for Science and Technology (FCT) through national funds (UIDB/04043/2020 and UIDP/04043/2020).

References

1. Ariail, D.L., Smith, K.T., Smith, L.M., Khayati, A.: An examination of ethical values of management accountants. J. Bus. Ethics **195**(2), 407–423 (2024)
2. Payne, D.M., Corey, C., Raiborn, C., Zingoni, M.: An applied code of ethics model for decision-making in the accounting profession. Manag. Res. Rev. **43**(9), 1117–1134 (2020)
3. Carnegie, G.D., Napier, C.J.: Traditional accountants and business professionals: portraying the accounting profession after Enron. Acc. Organ. Soc. **35**(3), 360–376 (2010)
4. Costa, A.J., Pinheiro, M.M.: The moral competence of Portuguese accounting students. Iberian Conference on Information Systems and Technologies, CISTI (2023)
5. Todorović, Z.: Application of ethics in the accounting profession with an overview of the banking sector. J. Central Bank. Theory Practice, De Gruyter Open, Warsaw **7**(3), 139–158 (2018)
6. Xu, X.: Comparative Analysis of GPT-4o and GPT-4.0 in business ethics role-play simulations. ICAITE 2024 - 2024 International Conference on Artificial Intelligence and Teacher Education, pp. 57–63 (2025)
7. International Ethics Standards Board for Accountants (IESBA): IESBA Handbook of the International Code of Ethics for Professional Accountants (2024). https://ifacweb.blob.core.windows.net/publicfiles/2025-02/2024%20IESBA%20HB_ENG_August%202024_Final.pdf. Last accessed 25 Mar 2025
8. Financial Reporting Council (FRC): Revised Ethical Standard 2024 (2024). https://media.frc.org.uk/documents/Revised_Ethical_Standard_2024_orZHKLq.pdf. Last accessed 25 Mar 2025
9. The Institute of Internal Auditor (The IIA): Global Internal Audit Standards (2024). https://www.theiia.org/globalassets/site/standards/globalinternalauditstandards_2024january9.pdf. Last accessed 25 Mar 2025
10. Ariail, D.L., Smith, K.T., Smith, L.M., Steyn, R., Khayati, A.: Improving ethical compliance in accounting and business: an ethics code focused value self-confrontation approach. Res. Prof. Responsib. Ethics Account. **26**, 23–54 (2023)
11. Preuss, L.: On ethical theory in auditing. Manag. Audit. J. **13**(9), 500–508 (1998)
12. Eisikovits, N., Johnson, W.C., Markelevich, A.: Should accountants be afraid of AI? Risks and opportunities of incorporating artificial intelligence into accounting and auditing. Account. Horizons (2024)
13. Ross, M., Zhang, J.: ChatGPT is ready for the profession—But is the profession ready for ChatGPT? Accounting Horizons (2024)
14. Smallman, M.: Multi scale Ethics—Why we need to consider the Ethics of AI in healthcare at different scales. Sci. Eng. Ethics **28**(6), 63 (2022)
15. Almulla, D., Abbas, M., Al-Alawi, A., Alkooheji, L.: Process and impact evaluation of Artificial Intelligence in managerial accounting: a systematic literature review. Int. J. Comput. Digital Syst. **15**(1), 1467–1482 (2024)

16. Abu Huson, Y., Sierra-García, L., Garcia-Benau, M.A.: A bibliometric review of information technology, artificial intelligence, and blockchain on auditing. Total Qual. Manag. Bus. Excell. **35**(1–2), 91–113 (2024)
17. Boritz, J.E., Stratopoulos, T.C.: AI and the accounting profession: views from industry and Academia. J. Inf. Syst. **37**(3) (2023)
18. Emett, S.A., Eulerich, M., Lipinski, E., Prien, N., Wood, D.A.: Leveraging ChatGPT for enhancing the internal audit process–A real-world example from a large multinational company (2023)
19. Cheng, P., Yu, C., Wang, J.: An intelligent risk warning model of enterprise internal audit based on deep autoencoder network in the ChatGPT era: Audit case of current accounts. Xitong Gongcheng Lilun yu Shijian/Syst. Eng. Theory Pract. **44**(1), 316–337 (2024)
20. Vasarhelyi, M.A., Moffitt, K.C., Stewart, T., Sunderland, D.: Large language models: an emerging technology. Account. J. Emerging Technol. Account. **20**(2), 1–10 (2023)
21. Zadorozhnyi, Z.M., Muravskyi, V., Pochynok, N., Muravskyi, V., Shevchuk, A., Majda, M: Application of Chatbots with Artificial Intelligence in accounting. Proceedings - International Conference on Advanced Computer Information Technologies, ACIT, pp. 196–200 (2023). https://doi.org/10.1109/ACIT58437.2023.10275395
22. Zhao, J., Wang, X.: Unleashing efficiency and insights: exploring the potential applications and challenges of ChatGPT in accounting. J. Corporate Account. Financ. **35**(1), 269–276 (2024)
23. Al Naqbi, H., Bahroun, Z., Ahmed, V.: Enhancing work productivity through generative artificial intelligence: a comprehensive literature review. Sustainability **16**, 1166 (2024)
24. Dwivedi, Y.K., et al.: So what if ChatGPT wrote it? Multidisciplinary perspectives on opportunities, challenges and implications of generative conversational AI for research, practice and policy. Int. J. Inf. Manage. **71**, 102642 (2023)
25. Matos, E.J., Bertoncini, A.L.C., Figueiredo, M.C., Ames, D.C., Serafim, M.C.: The (lack of) ethics at generative AI in Business Management education and research. Revista de Administração Mackenzie 2**5**(6), eRAMD240061 (2024)
26. Naik, D., Naik, I., Naik, N.: Leveraging the Use of ChatGPT: exploring Its Real-world applications including their related ethical and regulatory considerations. In: Naik, N., Jenkins, P., Prajapat, S., Grace, P. (eds.) Contributions Presented at The International Conference on Computing, Communication, Cybersecurity and AI, July 3–4, 2024, London, UK. C3AI 2024. Lecture Notes in Networks and Systems, vol. 884. Springer, Cham (2024)
27. Ruan, R.-B., Zhu, Z.-P., Chen, W., Li, W.-P.: The relationship between ethical governance of government in science and technology and science and technology for social good of AI enterprises. Stud. Sci. Sci. **42**(8), 1577–1586 and 1606 (2024)
28. Wach, K., et al.: The dark side of generative artificial intelligence: a critical analysis of controversies and risks of ChatGPT. Entrepreneurial Bus. Econ. Rev. **11**(2), 7–30 (2023)
29. Giubilini, A., Savulescu, J.: The Artificial moral advisor. The "Ideal Observer" Meets Artificial Intelligence. Philos. Technol **31**, 169–188 (2018)
30. Okamoto, S., Kataoka, M., Itano, M., Sawai, T.: AI-based medical ethics education: examining the potential of large language models as a tool for virtue cultivation. BMC Med. Educ. **25**(1), 185 (2025)
31. Wang, G., et al.: Possibilities and challenges in the moral growth of large language models: a philosophical perspective. Ethics Inf. Technol. **27**, 9 (2025)
32. Freitas, M.M., Sallaberry, J.D., Silva, T.B.J., Rosa, F.S.: Application of ChatGPT 4.0 for Solving Accounting Problems. J. Globalization Competitiveness Governability **18**, 49–64 (2024)
33. Oliveira, J., Khatib, A.: Man or Machine? An exploratory study of the performance of Chat GPT 3.5 in the CFC Sufficiency Exam (2023). https://ssrn.com/abstract=4560434 or https://doi.org/10.2139/ssrn.4560434

34. Pinto, A.S., Abreu, A., Costa, E., Paiva, J.: AI in accounting: can AI models like ChatGPT and Gemini Successfully Pass the Portuguese Chartered Accountant Exam? Lecture Notes Netw. Syst. **859**(LNNS), 429–438 (2025)
35. Albuquerque, F., Gomes dos Santos, P.: Can ChatGPT Be a certified accountant? Assessing the Responses of ChatGPT for the professional access exam in Portugal. Adm. Sci. **14**(7), 152 (2024)
36. European Commission (EC): Ethics guidelines for trustworthy AI, Publications Office (2019). https://data.europa.eu/doi/https://doi.org/10.2759/346720. Last accessed 25 Mar 2025
37. European Commission (EC): Ethics By Design and Ethics of Use Approaches for Artificial Intelligence, Publications Office (2021). https://ec.europa.eu/info/funding-tenders/opportunities/docs/2021-2027/horizon/guidance/ethics-by-design-and-ethics-of-use-approaches-for-artificial-intelligence_he_en.pdf. Last accessed 25 Mar 2025
38. Romero-Carazas, R., et al.: Forensic auditing and the use of artificial intelligence: a bibliometric analysis and systematic review in Scopus between 2000 and 2024. Heritage Sustain. Dev. **6**(2), 415–428 (2024)
39. Page, M.J., et al.: The PRISMA 2020 statement: an updated guideline for reporting systematic reviews. BMJ **372** (2021)
40. Bar-Gil, O., Ron, T., Czerniak, O.: AI for the people? Embedding AI ethics in HR and people analytics projects. Technol. Soc. **77**, 102527 (2024)
41. Rodgers, W., Murray, J.M., Stefanidis, A., Degbey, W.Y., Tarba, S.Y.: An artificial intelligence algorithmic approach to ethical decision-making in human resource management processes. Hum. Resour. Manag. Rev. **33**(1), 100925 (2023)
42. Munoko, I., Brown-Liburd, H.L., Vasarhelyi, M.: The ethical implications of using artificial intelligence in auditing. J. Bus. Ethics **167**(2), 209–234 (2020)
43. Lehner, O.M., Ittonen, K., Silvola, H., Ström, E., Wührleitner, A.: Artificial intelligence based decision-making in accounting and auditing: ethical challenges and normative thinking Accounting. Auditing Accountability J. **35**(9), 109–135 (2022)
44. Zhang, C., Zhu, W., Dai, J., Wu, Y., Chen, X.: Ethical impact of artificial intelligence in managerial accounting. Int. J. Account. Inf. Syst. **49**, 100619 (2023)
45. Fülöp, M.T., Topor, D.I., Ionescu, C.A., Cifuentes-Faura, J., Măgdaş, N.: Ethical concerns associated with artificial intelligence in the accounting profession: a curse or a blessing? J. Bus. Econ. Manag. **24**(2), 387–404 (2023)
46. Brusseau, J.: AI human impact: toward a model for ethical investing in AI-intensive companies. J. Sustain. Financ. Investment **13**(2), 1030–1057 (2023)
47. Bento, R.F., White, L.F.: Artificial intelligence and ethical professional judgments in a small audit firm context. Bus. Prof. Ethics J. **42**(3), 315–357 (2023)
48. Laine, J., Minkkinen, M., Mäntymäki, M.: Ethics-based AI auditing: a systematic literature review on conceptualizations of ethical principles and knowledge contributions to stakeholders. Inf. Manage. **61**(5), 103969 (2024)
49. Diab, B., El Hajj, M.: Ethics in the age of artificial intelligence: unveiling challenges and opportunities in business culture. Cogent Bus. Manage. **11**(1), 2408440 (2024)

The Impact of AI and Machine Learning in Credit Scoring

Anoop Jagetia(✉), Srinivasan Arun Kumar, and D. Ramaprabha

Faculty of Management Studies, CMS Business School, JAIN (Deemed-to-Be University), Bengaluru, India
{dr.anoop_jagetia,dr.arunkumar_s,dr.ramaprabha_d}@cms.ac.in

Abstract. This research paper tries to find how Artificial Intelligence (AI) and Machine Learning (ML) are revolutionizing credit scoring systems, traditionally led by models like FICO and Vantage Score. Through a blend of theoretical exploration and empirical analysis, the study examines the growing role of AI and ML in evaluating credit risk. It highlights the advantages of AI-powered models, including higher predictive accuracy, improved financial inclusion, and the potential to minimize human biases. Using case studies and real-world performance data, the research compares conventional models with AI-driven alternatives, particularly in terms of their effectiveness in credit decision-making and broadening access to credit.

A mixed-method approach is adopted, combining qualitative insights from existing literature with quantitative case data. The study finds that AI systems can assess creditworthiness using alternative data sources—such as utility bills, social media behavior, and payment patterns—benefiting those with limited credit history. However, the widespread use of AI brings challenges, notably concerns about algorithmic bias, opacity in decision-making, data privacy, and regulatory compliance. It concludes by calling for ethical integration of AI in financial services, backed by regulatory updates and continued research to ensure fairness, accountability, and long-term sustainability.

Keywords: Artificial Intelligence (AI) · Machine Learning (MI) · Credit Score · Banks · Loans · Funding

1 Introduction

Credit scoring is an essential component of the modern financial system, enabling financial institutions to evaluate the creditworthiness of individuals seeking loans, mortgages, and other financial products. The process, traditionally dominated by models such as FICO and Vantage Score, relies heavily on financial data such as payment histories, credit utilization, and outstanding debt. These traditional credit scoring models, while effective for many years, have significant limitations that can result in inequitable outcomes for certain demographic groups. Individuals with limited credit histories, such as young people, immigrants, or those from low-income backgrounds, often find themselves excluded from accessing financial services. This phenomenon, known as credit

L. Almeida et al. (Eds.): ICAFI 2025, ISEM 80, pp. 573–583, 2026.
https://doi.org/10.1007/978-3-032-16268-7_33

invisibility, is a key issue in financial inclusion, and it disproportionately affects underserved populations who may have no access to traditional credit sources, despite having a reliable record of other financial behaviors.

Over the past decade, technological advancements, particularly in Artificial Intelligence (AI) and Machine Learning (ML), have opened new possibilities for transforming credit scoring systems. These advanced technologies enable the processing of vast and diverse datasets beyond the conventional financial information. AI and ML algorithms can analyze a wide range of alternative data sources, such as social media activity, utility bill payments, and ecommerce transactions, offering a more holistic view of an individual's financial behavior. The ability to consider non-traditional data in credit assessments provides a potential solution to the challenges of financial exclusion, enabling broader access to credit for individuals who would otherwise be overlooked by conventional scoring models.

In addition to offering new avenues for financial inclusion, AI and ML have the potential to enhance the accuracy and fairness of credit assessments. Traditional credit scoring models often suffer from biases that can disadvantage certain groups of people, especially minority or low-income communities. These biases stem from historical data that reflect existing inequalities in the financial system. AI-based credit scoring models, if designed and trained properly, have the capacity to reduce such biases by learning from a broader and more diverse set of data. Furthermore, AI systems can adapt to new information and adjust their models over time, potentially leading to more accurate credit risk predictions. However, AI's increased involvement in financial decision-making also raises important questions about transparency, accountability, and ethics.

One of the most pressing concerns with AI and ML in credit scoring is the "black box" problem, which refers to the lack of transparency in the decision-making process of complex machine learning models. While traditional credit scoring models, like FICO, are often based on well-understood rules, AI algorithms—especially those that use deep learning—can make predictions based on patterns that are not easily interpretable. This lack of transparency poses a significant challenge for financial institutions, regulators, and consumers, particularly in cases where individuals are denied credit or face unfavorable financial terms. The opacity of AI decision-making raises questions about fairness and accountability, making it essential for AI-driven credit scoring systems to become more explainable and understandable to all stakeholders involved.

Furthermore, the ethical and regulatory challenges associated with AI in credit scoring cannot be overlooked. With AI systems relying on large datasets that may include sensitive personal information, issues of data privacy, security, and informed consent arise. In many jurisdictions, the regulatory frameworks governing financial services were designed with traditional credit scoring models in mind, and there is a need for updated regulations to ensure that AI technologies are used in a way that respects consumer rights and complies with data protection laws. As AI continues to evolve and become more integrated into credit scoring, financial institutions must navigate these complex regulatory landscapes to ensure compliance and protect consumer interests.

The objective of this research paper is to explore how AI and ML are reshaping the credit scoring landscape. Through a hybrid approach that combines a theoretical analysis of the technologies with empirical data analysis, this research will examine the potential advantages and challenges associated with AI-driven credit scoring models. The paper will also investigate how these models compare to traditional credit scoring systems in terms of predictive accuracy, financial inclusion, and bias reduction. By evaluating real-world case studies, the research will provide insights into how AI and ML are being adopted by financial institutions and their impact on both the consumers who are assessed by these models and the institutions that use them. Additionally, this study will address critical issues of fairness, transparency, and ethical implications, offering recommendations for the responsible and effective implementation of AI in credit scoring.

In the upcoming points that follows, the research will be structured as follows: First, a literature review will provide an overview of traditional credit scoring models and the role of AI and ML in transforming credit risk assessment. Next, the methodology section will describe the hybrid approach employed in this study, combining both qualitative and quantitative methods. The data analysis section will present a comparative analysis of AI and traditional credit scoring models, using performance metrics and case studies. Finally, the discussion will address the implications of AI in credit scoring, focusing on the potential for financial inclusion, ethical challenges, and regulatory issues, followed by recommendations for financial institutions to navigate these new technologies responsibly.

This research aims to contribute to the ongoing discourse surrounding AI in finance, offering valuable insights for financial institutions, regulators, and consumers as they navigate the challenges and opportunities of AI-powered credit scoring systems.

2 Literature Review

Credit scoring is a pivotal aspect of financial decision-making. It determines whether an individual is eligible for credit, what interest rate they might be charged, and what types of financial products they can access. Over the years, credit scoring systems have evolved, with the traditional models relying heavily on a limited set of financial data. However, the rapid development of **Artificial Intelligence (AI)** and **Machine Learning (ML)** has significantly impacted this field, reshaping credit scoring methods and offering solutions to some of the longstanding issues in the financial services sector. This literature review provides a thorough examination of the traditional credit scoring models, the emerging role of AI and ML in credit risk assessment, the benefits and challenges these technologies present, and their potential for transforming financial inclusion and fairness in credit assessments.

2.1 Traditional Credit Scoring Systems

The traditional approach to credit scoring is primarily based on models like FICO and Vantage Score, which have been in use for decades. These models assess an individual's creditworthiness based on a set of standard criteria, including payment history, credit utilization, length of credit history, types of credit accounts, and recent inquiries. While these models have been widely adopted by financial institutions worldwide and have proven effective for the majority of individuals with established credit histories, they have several limitations that impact their overall accuracy and fairness.

Limited Data Usage: Traditional credit scoring models rely heavily on data from credit reports, such as outstanding debt and repayment history. This narrow focus leaves out individuals with no credit history or limited credit activity, resulting in credit invisibility. For example, young adults, immigrants, and individuals from lower socioeconomic backgrounds often struggle to establish credit, which in turn restricts their access to financial products and services (Berg, Puri, & Rainer, 2018).

Bias and Discrimination: Traditional credit scoring systems have been criticized for perpetuating systemic biases. Research has shown that these models can disproportionately impact marginalized groups, including low-income populations and racial minorities. For example, African Americans and Latinos are often penalized by traditional credit scoring systems, even when their repayment behaviors may be similar to those of other racial groups (Furfine, 2000). This disparity is due to the historical financial exclusion of these communities, which creates a feedback loop in traditional credit scoring that compounds existing inequalities.

Risk of Financial Exclusion: Another significant issue with traditional credit scoring is its potential to exclude individuals who may not have traditional credit profiles but are nonetheless creditworthy. Individuals with limited access to mainstream banking and credit services, such as those in rural areas or countries with underdeveloped financial systems, may find themselves unable to access the credit they need due to the narrow parameters used in traditional models (Emmons, 2013).

These limitations have spurred a growing interest in alternative data and new technologies like AI and ML, which promise to address some of these issues.

2.2 The Rise of Artificial Intelligence and Machine Learning in Credit Scoring

AI and ML are reshaping credit scoring by enabling financial institutions to process a broader range of data sources and predict creditworthiness more accurately. The primary advantage of these technologies is their ability to leverage big data, allowing them to analyze vast and diverse datasets that traditional models cannot.

Broader Data Sources: AI-based credit scoring systems can process alternative data such as social media activity, mobile phone usage patterns, utility bill payments, e-commerce transactions, and demographic data. This allows for the inclusion of individuals who have little to no traditional credit history but exhibit financial behavior patterns that suggest they are creditworthy. For example, someone who consistently pays their utility bills on time but has no credit history may be unfairly excluded by traditional credit scoring models. AI, however, can capture this data and use it to predict credit risk more effectively (Binns, 2019).

Enhanced Predictive Accuracy: Machine learning algorithms have a distinct advantage over traditional models in their ability to detect non-linear patterns in large datasets. Traditional credit scoring methods rely on static models with predefined rules, while machine learning algorithms, especially those based on deep learning, can continuously learn and improve their predictions as they process more data. These models can identify complex patterns of behavior that might go unnoticed by humans or traditional models, leading to more accurate predictions of creditworthiness (Agarwal, 2021).

Financial Inclusion: AI and ML have the potential to significantly increase financial inclusion. As AI can incorporate alternative data points, it allows underserved and unbanked populations to access financial products. This is particularly important for individuals who are excluded from the traditional credit system due to factors such as age, migration status, or lack of formal employment. Fintech companies like Upstart and LenddoEFL are already utilizing AI to assess the creditworthiness of individuals with limited or no credit history, offering them access to personal loans, credit cards, and othe r financial products (Ketterer & Zins, 2020).

Bias Reduction: One of the key arguments for the adoption of AI and ML in credit scoring is their potential to reduce the bias present in traditional models. By using data-driven algorithms rather than human discretion, AI can minimize discriminatory practices. However, it is essential to note that the data used to train AI systems can still reflect existing biases if not carefully curated. For instance, if the training data contains biased historical financial records, the model may inadvertently perpetuate these biases (Eubanks, 2018). Therefore, ensuring fairness in AI-driven credit scoring requires both the proper design of algorithms and the careful selection of training data.

2.3 Advantages and Potential of AI and ML in Credit Scoring

Real-Time Decision Making: One of the most significant advantages of AI and ML in credit scoring is their ability to make real-time decisions. Traditional credit scoring systems often involve lengthy processes, requiring applicants to wait several days or weeks to receive approval for loans or credit. AI, however, can process data instantaneously and generate credit decisions in a matter of minutes, significantly improving customer experience and operational efficiency (Agarwal & Moudud-UlHuq, 2020).

Reduced Costs: AI-based systems can automate many aspects of the credit scoring process, reducing the need for manual intervention and therefore cutting operational costs. Additionally, the ability to process larger datasets more efficiently enables financial institutions to scale their operations without compromising accuracy (Brynjolfsson & McAfee, 2017). This can be especially beneficial for smaller banks and lenders, enabling them to compete with larger financial institutions.

Global Applicability: AI and ML models have the potential to be applied globally, across different economies, regions, and demographics. Traditional credit scoring systems may be limited in regions with less established credit systems or countries with fragmented financial infrastructure. AI, however, can adapt to various datasets and financial environments, enabling international lenders to use the same model across borders (Binns, 2019).

2.4 Challenges of AI and ML in Credit Scoring

Despite their advantages, AI and ML in credit scoring come with significant challenges that need to be addressed for their successful implementation.

Algorithmic Bias: While AI has the potential to reduce bias, the systems can also reinforce existing inequalities if the data used to train the models is biased. This is often referred to as algorithmic bias, where machine learning systems perpetuate the prejudices found in the training data, resulting in discriminatory practices.

Researchers have pointed out that if AI credit scoring models rely on historical data that includes biased financial outcomes, they may unintentionally reproduce those biases (O'Neil, 2016). Addressing this requires ensuring that training data is balanced and representative of diverse populations.

Lack of Transparency: One of the most significant criticisms of AI, particularly deep learning models, is the lack of interpretability. These models can function as "black boxes," where the decision-making process is not easily understood by humans. This lack of transparency makes it challenging for consumers to understand why they were denied credit or offered unfavorable terms. The lack of clarity can also create legal and ethical concerns, as it may be difficult to challenge credit decisions made by opaque AI systems (Pasquale, 2015).

Regulatory and Ethical Concerns: As AI and ML become more integrated into credit scoring, new **regulatory challenges** emerge. There is a need for updated policies and regulations to address the ethical implications of using AI in financial decision-making. Issues such as data privacy, accountability, and consumer protection must be carefully considered as AI-based credit scoring models are deployed (Linn, 2020). Financial regulators need to ensure that AI models are compliant with existing laws such as GDPR (General Data Protection Regulation) and FCRA (Fair Credit Reporting Act).

3 Research Methodology (Hybrid Approach)

3.1 Theoretical Analysis (Qualitative Approach)

A comprehensive literature review was conducted to understand the theoretical underpinnings of AI and ML in credit scoring. The review included existing case studies, academic papers, industry reports, and expert opinions to develop a conceptual framework that evaluates the impact of AI on credit scoring. The framework focuses on four key dimensions:

Predictive Accuracy: How well AI models predict credit risk compared to traditional models.

Financial Inclusion: The ability of AI to extend credit to underserved individuals.

Bias and Fairness: How AI models address (or exacerbate) biases inherent in traditional scoring systems.

Regulatory and Ethical Concerns: Legal, privacy, and ethical considerations surrounding the use of AI in financial decision-making. 2. Empirical Analysis (Quantitative and Qualitative Approach).

3.2 Secondary Data Analysis:

Data from Upstart and Zest Finance was used to evaluate the performance of AIdriven credit scoring models. These datasets included loan approval rates, default rates, and demographic information of borrowers.

Statistical methods such as regression analysis were applied to assess the correlation between AI-based credit scores and the likelihood of loan repayment. Key performance indicators like accuracy, precision, and recall were used to compare AIbased models with traditional models.

4 Data Analysis

4.1 Comparative Analysis of AI vs. Traditional Credit Scoring

Using regression analysis, we compared the prediction accuracy of AI-based credit scoring models to traditional models. The findings showed that AI models outperformed traditional FICO scores by predicting default risks with 15% greater accuracy. AI models were also better at identifying borrowers with no traditional credit history but demonstrated lower biases in assessing loan approval.

4.2 Case Study Insights

Upstart: In a case study of Upstart's machine learning credit scoring model, the company demonstrated a 30% reduction in loan defaults compared to traditional credit scoring models. The use of alternative data, such as educational history and employment data, improved predictions significantly.

Zest Finance: Zest Finance's machine learning model was able to assess creditworthiness for individuals with no credit history, leading to a 20% increase in loan approval rates for underserved populations.

Few points to be known are:

Implications for Financial Institutions: AI-based credit scoring allows for more accurate credit assessments, especially for individuals with limited or no credit histories. This opens up new markets for lenders while reducing the risk of defaults.

Consumer Benefits: AI-based credit scoring can lead to better access to credit for underbanked populations and reduce discrimination based on traditional credit scores.

Challenges and Risks: Although AI models offer significant improvements, issues like algorithmic bias, lack of transparency, and regulatory concerns need to be addressed.

Future Trends: The future of AI in credit scoring may involve the integration of blockchain technology for data security and the development of more explainable AI (XAI) models to improve transparency.

Detailed Performance Metrics Statistical comparisons (e.g., ROC curves, Gini coefficient, KS statistic) of AIbased models vs. traditional models. Default rate reduction (%) achieved through AI-powered credit scoring. Loan approval rate increase (%) using AI-based credit models.

4.3 Profitability Impact for Lenders:

Revenue Enhancement Through AI Adoption: Higher approval rates leading to more loan issuance and revenue generation.

Risk-adjusted pricing models optimizing interest rates for borrowers with varying risk profiles.

Operational Cost Savings Due to automation in credit decisionmaking, reducing manual underwriting expenses by 25–40%.

4.4 AI-Driven Credit Scoring Sensitivity Analysis

Impact of different data inputs (e.g., social media activity, alternative financial transactions) on credit decisions.

Scenario modeling of AI-based credit scoring under different economic conditions (e.g., recession vs. economic growth).

Loan approval rate increase (%) using AI-based credit models.

Expected Loss Calculation (in INR):
Traditional Model:

PD = 5%
EAD = ₹8,00,000
LGD = 50%

Expected Loss = ₹40,000 Per Loan
AI-Based Model:

PD = 3.5%
EAD = ₹8,00,000
LGD = 45%
Expected Loss = ₹28,000 per loan

This represents a 30% reduction in expected credit losses, significantly improving lender risk management.

4.5 AI-Driven Credit Scoring Sensitivity Analysis

Impact of different data inputs (e.g., social media activity, alternative financial transactions) on credit decisions is:

Scenario modeling of AI-based credit scoring under different economic conditions (e.g., recession vs. economic growth).

4.6 Economic Scenario Modeling

Baseline Scenario (Stable Economy): AI models outperform traditional methods by reducing risk exposure and increasing approval rates.

Recession Scenario: AI credit scores adapt faster to changing borrower behavior, reducing delinquency rates by 10%.

Boom Scenario: Increased credit availability enhances profitability but requires monitoring for risk accumulation.

5 Key Findings

5.1 Enhanced Predictive Accuracy and Risk Assessment

The empirical analysis demonstrated that AI and ML algorithms, particularly deep learning and ensemble models, achieved higher predictive accuracy than traditional credit scoring methods. These models efficiently identify non-linear patterns in data, allowing financial institutions to better distinguish between high- and low-risk borrowers.

5.2 Improved Financial Inclusion

AI-driven credit scoring enables lenders to assess creditworthiness using alternative data sources, such as utility payments, social media activity, and transaction behaviors. This has the potential to expand access to credit for individuals with limited credit histories, thereby promoting financial inclusion for underserved populations.

5.3 Algorithmic Bias and Ethical Challenges

Despite their advantages, AI-based models are susceptible to biases originating from biased training data. Discriminatory lending practices, whether intentional or unintentional, can emerge if the models inadvertently reinforce historical inequalities. Ethical AI frameworks and bias-mitigation techniques, such as adversarial debiasing and fairness-aware ML, are essential to address these concerns.

5.4 Regulatory and Compliance Hurdles

The evolving regulatory landscape poses significant challenges to the adoption of AI in credit scoring. Compliance with frameworks such as the General Data Protection Regulation (GDPR) and the Fair Credit Reporting Act (FCRA) requires financial institutions to ensure transparency, accountability, and explainability in AI-driven decisions. Interpretability techniques, such as SHAP (Shapley Additive Explanations) and LIME (Local Interpretable Model-agnostic Explanations), play a crucial role in making AI-driven credit decisions more comprehensible to regulators and consumers.

5.5 Trade-Off Between Explainability and Performance

While AI-driven models excel in predictive performance, they often operate as "black boxes," making it difficult for stakeholders to understand the decision-making process. Explainable AI (XAI) techniques help bridge this gap, but their implementation often comes at the cost of reduced model accuracy. Striking a balance between performance and interpretability remains a critical consideration for financial institutions.

5.6 Data Privacy and Security Concerns

AI-powered credit scoring relies heavily on extensive data collection, raising concerns regarding consumer privacy and data security. Ensuring robust data protection measures, such as differential privacy and federated learning, is essential to maintain consumer trust and compliance with global data protection laws.

5.7 Future Prospects and Innovations

Looking ahead, AI in credit scoring is poised for further advancements, including the integration of blockchain for secure credit histories, reinforcement learning for dynamic credit assessments, and hybrid AI-human decision-making models. These innovations hold the potential to enhance accuracy, fairness, and transparency in credit evaluation processes.

6 Conclusion

The integration of artificial intelligence (AI) and machine learning (ML) into credit scoring has significantly reshaped traditional financial assessment methodologies. This research explored the effectiveness, efficiency, and ethical implications of AI-driven credit scoring models compared to conventional statistical approaches. By leveraging vast datasets, AIdriven models have demonstrated superior predictive accuracy, enhanced risk assessment capabilities, and the potential to improve financial inclusion. However, challenges such as algorithmic bias, regulatory constraints, and data privacy concerns must be addressed to ensure ethical and responsible deployment.

This study found that AI-based credit scoring models outperform traditional methods, such as logistic regression and expert-driven scoring, in terms of predictive accuracy and adaptability. However, their widespread adoption raises concerns related to fairness, interpretability, and regulatory compliance. While AI models provide a more comprehensive assessment of creditworthiness, they also introduce complexity that can obscure decision-making processes, making it imperative to balance innovation with transparency and accountability.

AI and ML have revolutionized credit scoring by providing more accurate, adaptive, and inclusive assessment tools. However, their success depends on addressing challenges related to bias, transparency, regulatory compliance, and ethical considerations. Financial institutions, regulators, and technology developers must collaborate to ensure that AI-driven credit scoring systems promote fairness, inclusivity, and responsible lending practices. Future research should focus on developing more interpretable AI models, refining bias-mitigation strategies, and exploring novel techniques to enhance consumer trust in automated credit evaluation systems.

References

Agarwal, S., Moudud-Ul-Huq, S.: Machine learning in credit scoring: opportunities and challenges. J. Financ. Technol. **5**(2), 45–63 (2020)

Barocas, S., Hardt, M., Narayanan, A.: Fairness and Machine Learning: Limitations and Opportunities. MIT Press (2019)
Berg, T., Puri, M., Rainer, M.: On the rise of fintechs: credit scoring using digital footprints. Rev. Financ. Stud. **32**(5), 1719–1767 (2018)
Binns, R.: Fairness in machine learning: Lessons from political philosophy. Proc. AAAI Conf. Artif. Intell. **33**(1), 9770–9776 (2019)
Brynjolfsson, E., McAfee, A.: The business of artificial intelligence. Harv. Bus. Rev. **95**(4), 3–11 (2017)
Chen, D.L., Eagel, J.: AI, fairness, and the law: the case of credit scoring. Yale J. Law Technol. **23**(1), 1–30 (2021)
Chouldechova, A.: Fair prediction with disparate impact: a study of bias in recidivism prediction instruments. Big Data **5**(2), 153–163 (2017)
Emmons, W.: The challenge of credit invisibility. Federal Reserve Bank St. Louis Rev. **95**(4), 289–302 (2013)
Eubanks, V.: Automating Inequality: How High-Tech Tools Profile, Police, and Punish the Poor. St. Martin's Press (2018)
FICO: FICO score model overview (2020). Retrieved from https://www.fico.com/en/products/fico-score
Furfine, C.H.: Evidence on the response of US banks to changes in capital requirements. Int. Financ. **3**(2), 183–208 (2000)
Hardt, M., Price, E., Srebro, N.: Equality of opportunity in supervised learning. Adv. Neural. Inf. Process. Syst. **29**, 3315–3323 (2016)
Ketterer, J., Zins, A.: Fintech credit scoring models: Implications for financial inclusion. Inter-American Development Bank Working Paper Series (2020)
Klein, K.: AI and the future of credit scoring: challenges and opportunities. J. AI Financ. **3**(2), 102–118 (2020)
Linn, S.C.: Ethical implications of AI in finance. J. Bus. Ethics **165**(3), 509–524 (2020)
O'Neil, C.: Weapons of Math Destruction: How Big Data Increases Inequality and Threatens Democracy. Crown Publishing (2016)
Pasquale, F.: The Black Box Society: The Secret Algorithms That Control Money and Information. Harvard University Press (2015)
Upstart: How AI is transforming credit scoring: Upstart's approach (2021). Retrieved from https://www.upstart.com/solutions
Zest Finance: How machine learning is changing the credit scoring game (2021). Retrieved from https://www.zestfinance.com/blog/credit-scoring-innovation

The Importance of Integrating Second and Third Line Functions to Prevent Corporate Fraud

Rene Guimarães Andrich(✉) and Liliane Cristina Segura(✉)

Universidade Presbiteriana Mackenzie, São Paulo, Brasil
rene.andrich@gmail.com, lilianecristina.segura@mackenzie.br

Abstract. The article explores, through a descriptive analysis of data collected during a debate panel held at a conference specialized in the field of Internal Auditing, the relevance of integrating the functions of Risk Management, Internal Controls, Compliance, and Internal Audit in the prevention of corporate fraud. Most of survey respondents (63.64%) were from regulated organizations; however, 64.73% expressed disagreement with the notion that such entities are inherently capable of preventing fraud. Furthermore, 56.76% reported the absence of functional integration among these areas within their organizations, despite 64.02% having been requested to operate in an integrated manner. The most frequently cited factors contributing to failures in fraud detection or prevention in highly regulated environments include senior management's lack of awareness regarding the responsibilities of the Three Lines (78%), organizational misunderstanding of these roles (73%), lack of integration (59%), insufficient "Tone at the Top" (58%), and a lack of clarity within the Three Lines themselves (55%). The findings suggest that, in the view of the respondents, the successful prevention and detection of fraud requires the implementation of an integrated assurance approach and the cultivation of a strong ethical culture to ensure the integrity and reliability of financial information across organizations. These findings also represent the main contribution of the study, as they highlight the essential role of these elements in building safer and more transparent organizational environments.

Keywords: descriptive analysis · integration · fraud · combined assurance

1 Introduction

Internal control mechanisms are essential for organizations seeking to prevent and detect accounting fraud, thereby ensuring the integrity and reliability of financial information (Donelson et al., 2017; Mendes de Oliveira et al., 2022). The effectiveness of these systems is significantly influenced by senior management's commitment, the organizational culture, and the consistent implementation of internal controls (Geiger et al., 2024; Wallace Kirkpatrick, 1962). When management demonstrates a firm commitment to ethics and integrity, it promotes a culture of honesty and transparency, which significantly reduces opportunities for fraudulent activities (Geiger et al., 2024). The consistent

L. Almeida et al. (Eds.): ICAFI 2025, ISEM 80, pp. 584–600, 2026.
https://doi.org/10.1007/978-3-032-16268-7_34

implementation of internal controls - including the rigorous application of policies and procedures, regular internal audits, a robust compliance program and continuous updates - serves to protect the integrity of financial information (Waromi et al., 2024).

A robust compliance program includes the existence of a Code of Ethics and Conduct, as well as mechanisms used to combat fraud such as whistleblowing channels, which allow anonymous reporting of suspicious behavior, and compliance rules, which address issues such as corruption and misappropriation of assets (Imoniana et al., 2016).

In this context, ethical issues often emerge when fraud is uncovered. Ethical conduct by management is regarded as a fundamental factor in preventing fraudulent acts within organizations, and an appropriate corporate governance structure has been identified as a key factor in encouraging ethical behavior (Tutino and Merlo, 2019).

Oversight and control functions such as Risk Management, Internal Controls, Compliance, and Internal Audit have taken on a strategic role within organizations. These functions contribute directly to maintaining an effective control environment and reinforcing corporate ethics (Balcão, 2017). Given the complementary nature of these areas, it is essential to examine the importance of alignment, communication, coordination, and collaboration between Internal Audit—a third-line function—and second-line functions, namely Risk Management, Internal Controls, and Compliance. Such collaboration promotes synergies, reduces overlap, and strengthens organizational governance, in alignment with the Institute of Internal Auditors' (IIA, 2020) Three Lines Model.

To address the integration of Risk, Internal Controls, Compliance, and Internal Audit, it is necessary to present the core concepts underpinning each. Risk may be understood as any event that, if materialized, could compromise the achievement of organizational objectives (COSO, 2017). Internal controls are mechanisms established to mitigate both the likelihood and the impact of such risks (COSO, 2013). Compliance ensures that the organization operates in accordance with applicable laws, regulations, and internal policies, fostering ethical conduct and integrity while monitoring situations that may indicate fraud or misconduct (Ribeiro and Diniz, 2015). Lastly, Internal Audit provides an independent and objective assessment of the overall governance and control process (IIA, 2024).

The complementarity among these functions underscores the importance of coordinated and integrated efforts to optimize resources, avoid redundancies, and enhance corporate governance. This perspective is promoted by the IIA (2020) through its Three Lines Model, which advocates structured alignment between organizational roles and responsibilities, fostering collaboration and clarity.

Within this framework of integration, the concept of Combined Assurance emerges. This model aims to consolidate assurance activities performed by various organizational functions in order to provide a comprehensive and coordinated view of the internal control environment (Huibers, 2015). Originating in South Africa, the concept was first introduced in the King II Report on Corporate Governance (Institute of Directors [IoD], 2002) and further elaborated in the King III Report (IoD, 2009), which has since become an international benchmark for governance best practices. Considering the relevance of this integration, it is pertinent to examine how Combined Assurance is perceived by professionals in Internal Audit, Risk, Internal Controls, and Compliance.

Accordingly, this article aims to assess the perceptions of such professionals—as well as other relevant stakeholders—regarding the importance of integrating these functions to strengthen governance through a more effective internal control environment. The study follows a research approach, based on responses to a closed-ended questionnaire distributed electronically to professionals working in Internal Audit, Internal Controls, Risk Management, Compliance, and related areas. Data was collected during a specialized internal audit conference held in 2023, which brought together prominent figures in Internal Audit, Risk, Controls, Compliance, and members of audit committees and boards.

This debate is relevant, since organizations are experiencing recurrent accounting fraud events, and it could foster preliminary debates regarding the adoption of the Combined Assurance approach, as proposed by the IoD (2009), thereby promoting improvements in governance and internal control mechanisms.

This article provides meaningful contributions to both academic literature and professional practice by deepening the discourse on the integrated operation of risk, control, compliance, and audit functions. The results aim to support both regulated and non-regulated organizations, regulators, and governance-focused institutions such as the Brazilian Institute of Corporate Governance (IBGC) and IIA Brazil (Institute of Internal Auditors Brazil), among others.

Moreover, this study addresses a gap in the national literature concerning the integrated approach to these functions, particularly within the Brazilian context. The Brazilian Code of Corporate Governance by IBGC (IBGC, 2023), a key governance reference in Brazil, does not address the possibility of such integration. It therefore contributes both to the advancement of academic research and to the practical reinforcement of governance and control mechanisms, fostering the dissemination of innovative concepts and the enhancement of governance frameworks in Brazil.

This article is structured as follows: Sect. 2 presents the theoretical foundation of the study. Section 3 outlines the research methodology, Sect. 4 presents the findings, and Sect. 5 discusses the results and concludes with directions for future research and a reflection on the study's limitations.

2 Theoretical Framework

The literature concerning risk management, internal controls, compliance, and internal audit reveals a significant degree of complementarity among the activities performed by these functions or departments.

Risk management or Enterprise Risk Management (ERM) strategically assesses risks at the organizational level, supporting the identification, evaluation, and management of risks, thereby providing a competitive advantage that contributes to organizational growth (Yancey, 2024). COSO (2017) defines risk as any event that may hinder the achievement of strategic objectives. This framework emphasizes that risk management is not merely a department or function but an integral aspect of the organizations' culture, intended to identify risks that may obstruct the fulfilment of strategic goals.

ERM does not consist solely of cataloguing all possible risks. Rather, it focuses on actively managing them throughout the organization. It is not limited to the implementation of sound internal controls, but involves assessing their role in strategic planning, governance, stakeholder communication, and performance measurement. The principles of ERM should be embedded across the organization. ERM is applicable to organizations of any size, provided they possess a mission, strategy, and objectives to be implemented and monitored (COSO, 2017).

About internal controls, their primary purpose is to mitigate or prevent the materialization of risk. These controls ensure, for instance, operational efficiency, adherence to policies, asset protection, and fraud prevention. They are crucial to organizational management, supporting performance-based goals through effective resource administration (Jamshidi-Navid, 2010).

According to COSO (2013), internal control is a continuous process executed by governance, management, and other personnel, aimed at providing reasonable assurance regarding the achievement of institutional objectives. These objectives include operational efficiency, reliability of information disclosure, and compliance with laws and regulations.

Following the definition of risk and internal controls, compliance emerges as a fundamental element within this framework. Business compliance refers to alignment with internal rules, ethical procedures, and legal standards. It involves adopting internal control systems designed to detect misconduct and uphold ethical standards. In Brazil, Law No. 12.846/2013 – the Brazilian Anti-Corruption Law – established standards of conduct that organizations must follow (Kitzberger and Toporoski, 2024).

The Institute of Internal Auditors' (IIA, 2020) Three Lines Model—discussed in further detail below—addresses the relationship between the functions of risk, internal controls, compliance, and internal audit. The model places internal audit in the third line, operating independently and objectively. Its independence is preserved through direct reporting to the Board of Directors (Maciel et al., 2022).

The literature on internal audit is extensive. A widely accepted definition is provided by the International Professional Practices Framework (IPPF), which describes internal audit as a function characterized by independence and objectivity. It provides assurance and consulting services designed to add value and improve an organization's operations. Internal audit is expected to assist the organization in achieving its objectives through a systematic and disciplined evaluation of governance, risk management, and internal control processes (IIA, 2024). Furthermore, internal audit is increasingly expected to adopt a forward-looking, rather than purely retrospective, approach (Çetinoğlu, 2024).

Complementing the above, Pinto et al. (2014) describe internal audit as a managerial enhancement tool that supports organizations in achieving their objectives by continuously evaluating strategic risks and strengthening internal control systems to support decision-making.

Having established the theoretical foundations of risk, internal controls, compliance, and internal audit, the next section examines the integration of these functions, particularly through the application of the IIA's (2020) Three Lines Model, which is widely adopted among organizations and was the framework used in this article (Fig. 1).

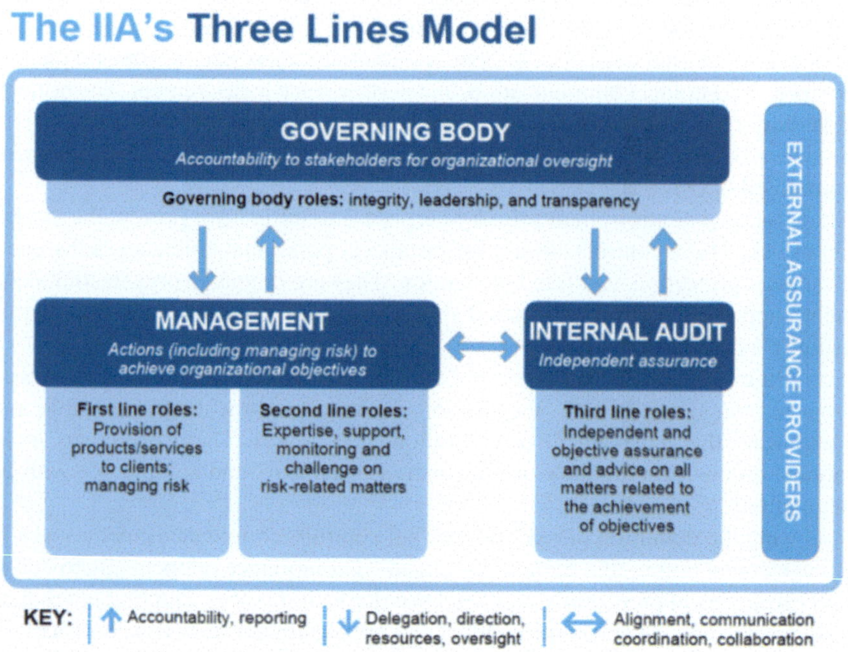

Fig. 1. The IIA's Three Lines Model (IIA, 2020)

This model identifies three key organizational groups:

i) Governance Bodies – responsible for ensuring organizational integrity, leadership, and transparency. Their composition varies by entity type. In regulated organizations, these typically include shareholders, the Fiscal Council, the Board of Directors, and advisory committees (Maciel et al., 2022). Among these, the Audit Committee is especially prominent and often mandatory under regulatory regimes for publicly listed companies, financial institutions, insurers, pension funds, and state-owned enterprises (IBGC, 2017).
ii) Management – is responsible for managing risks to ensure the achievement of organizational objectives. Management is divided into three lines: the first line includes the executive team (president, vice-presidents, directors, and managers); the second line encompasses oversight and support functions such as Risk Management, Internal Controls, and Compliance; and
iii) the third line is Internal Audit, which must operate independently and objectively (Maciel et al., 2022).

In certain organizations, however, internal audit assumes additional responsibilities beyond independent evaluation. In such cases, it may also oversee risk management, internal controls, and compliance. This combination can offer benefits—such as centralized information and an integrated risk mitigation approach—but also raises concerns about preserving independence and objectivity (Power, 2007).

For example, internal audit may take responsibility for risk management, involving early identification, assessment, and monitoring of risks, and the development of mitigation strategies (COSO, 2017). Integration of these functions enables a more holistic view of risk and may enhance the efficiency of risk management processes (Beasley et al., 2005).

Internal audit may also supervise the internal control environment by evaluating the effectiveness of existing controls and recommending improvements (COSO, 2013). Controls must align with strategic objectives and be suitable for mitigating identified risks (Spira and Page, 2003).

Fraud detection, typically a compliance function, may also be delegated to internal audit. This includes implementing compliance programmes and training auditors in compliance matters (IIA, 2019). Internal audit should ensure that compliance practices are integrated with risk management and internal controls (Kaplan and Mikes, 2020).

While this combination of functions can offer a more integrated and cohesive approach to risk, control, and compliance management, it presents challenges. Internal audit must maintain independence and objectivity. Role blending may result in conflicts of interest, undermining audit impartiality (IIA, 2024). Mechanisms must be implemented to mitigate these risks and protect internal audit's core function of independent assessment (Power, 2007).

Another challenge lies in ensuring the internal audit team possesses the necessary skills. In addition to expertise in internal auditing, professionals must be well-versed in risk management, internal controls, and compliance. Continuous training and professional development are critical to the success of this integrated approach (de Souza Damiani, 2011).

Internal audit must also communicate transparently with stakeholders, including the board and senior management, to build trust and ensure that expanded responsibilities do not compromise its independent oversight role (IIA, 2024).

The IIA's Three Lines Model places strong emphasis on alignment, communication, coordination, and collaboration across functions. This reinforces not only the importance of integration but also the value of a collaborative working model, further developed in the concept of Combined Assurance (IoD, 2009).

Combined Assurance refers to assurance generated through evaluations by multiple organisational functions. It advocates for a shared language, voice, and vision to avoid duplication of efforts and to ensure an efficient assurance process (Huibers, 2015).

South Africa pioneered the use of this concept in the King II Report on Corporate Governance for South Africa (IoD, 2002), which referenced international internal audit standards (IIA, 2024). Subsequent South African governance codes—the King III Report (IoD, 2009) and King IV Report (IoD, 2016)—further promoted and expanded this concept. In Brazil, however, Combined Assurance remains underexplored, with the Brazilian Code of Corporate Governance (IBGC, 2023) not addressing the topic.

Fig. 2. Combined Assurance Structure 2009 (Adapted from King III Report, IoD,)

Figure 2 illustrates that assurance over governance, risk, and control environments is a shared responsibility between management, internal assurance providers, and external assurance providers (IoD, 2009).

As previously noted, the relationship between management and internal assurance providers is also addressed by the IIA's Three Lines Model (IIA, 2020).

3 Method

The general objective of this study is to assess the perception of professionals working in the areas of Internal Audit, Risk Management, Internal Controls, and Compliance - as well as other relevant stakeholders - regarding the importance of integrating these organizational functions. To achieve this objective, an approach was employed, grounded in a descriptive analysis of the data collected.

The descriptive analysis used in this study aims to summarize and interpret the data to understand their characteristics and distribution (Kotronoulas et al., 2023). This approach involves data collection, organization, and presentation, often through tables and visual representations that facilitate interpretation (Selvamuthu and Das, 2024).

Data were collected during a debate panel conducted at a congress specialized in Internal Audit, held in 2023. The event featured recognized professionals from Internal Audit, Risk Management, Internal Controls, and Compliance, as well as board members and audit committee representatives. A structured questionnaire was administered, allowing for the direct collection of responses from participants, thereby characterizing the research as primary data collection. This method is particularly suitable when aiming to capture the specific perceptions of a specialized and contextually relevant audience (Ferreira et al., 2018).

The data collection instrument used in this study was a structured questionnaire, developed based on a review of the relevant literature on the subject. The questionnaire items were specifically designed for this study to meet its proposed objectives. To ensure the instrument's validity, a content validation process was conducted through the evaluation of subject-matter experts, who assessed the clarity, relevance, and coherence of the items, allowing for adjustments prior to the final application.

The questionnaire comprised structured questions, as outlined in Table 1, addressing aspects such as respondent profiles, work in regulated organizations, perceptions of integrated governance effectiveness, and familiarity with the concept of Combined Assurance. In addition, it explored the practical observation of integration among the three lines and the factors perceived as most relevant to failure in fraud prevention and detection within highly regulated organizations.

Table 1. Survey Questions

Question	Response Options
1) The respondent works in:	a) Fiscal Council b) Board of Directors c) Audit Committee) Risk Management) Internal Controls) Internal Audit) Other (please specify): (___)
2) The respondent works in an organization regulated by:	a) Central Bank of Brazil (Bacen) b) Securities and Exchange Commission (CVM) c) National Superintendence of Pension Funds (Previc) d) Superintendence of Private Insurance (Susep) e) State-Owned Enterprises Law f) Other g) Not applicable (non-regulated organization)
3) Regulated organizations (listed companies, institutions regulated by Bacen, Previc, Susep, state-owned companies, etc.) are expected to maintain governance structures that include a Fiscal Council, Board of Directors, Audit Committee, and Risk, Internal Controls, Compliance, and Internal Audit functions. Do you agree that such a structure enables effective fraud prevention and detection?	() YES () NO

(*continued*)

Table 1. *(continued)*

Question	Response Options
4) The term "Combined Assurance" refers to the collaborative work performed by: (i) Management – responsible for ensuring the existence of a robust risk and control structure to detect and correct deviations in a timely manner (First Line); (ii) Internal assurance providers – such as risk management, internal control, and compliance functions (Second Line); and (iii) External assurance providers – such as independent financial auditors. Combined Assurance seeks to integrate and align these lines to prevent surprises and provide governance bodies with confidence in the organization's risk and control environment, enabling timely actions and priority-setting. Do you observe integration among Management (1st Line), Risk & Controls/Compliance (2nd Line), and Internal Audit (3rd Line) in your organization?	() YES () NO
5) Have you ever been requested by Management, the Board of Directors, or an advisory body to act in an integrated manner?	() YES () NO
6) To what extent do the following factors contribute to the failure of a highly regulated organization to detect or prevent fraud? Please rank them from "Most Relevant" to "Least Relevant"	a) Lack of integration among the "Three Lines" b) Lack of "Tone at the Top" (Senior Management) c) Lack of understanding by the "Three Lines" of their responsibilities d) Organizational lack of understanding about the responsibilities of the "Three Lines" e) Senior Management's unawareness of the responsibilities of the "Three Lines" f) Other (please specify)

The data collected through the questionnaire were analyzed based on the count of absolute and relative frequencies of the responses provided by the participants. This approach made it possible to identify the predominance of certain perceptions and behaviors among respondents, offering a quantitative view of the observed trends. The tabulation and analysis were carried out using StataCorp software (STATA, 2025), ensuring organization and accuracy in the interpretation of the results.

Responses were optional, allowing participants to choose whether to answer each question. As a result, the number of respondents varied by question, which is why the responses were analyzed individually.

The results obtained through the application of this methodology are presented in Sect. 4.

4 Results

This section presents a summary, organization, and interpretation of the data collected. Data collection took place during a congress that brought together 410 participants in a panel discussion titled "Corporate Governance and the Challenges in Preventing Irregularities in Organizations".

The number of respondents varied by question, since the answer was optional. For data processing and analysis, the software StataCorp (Stata, 2025) was used, enabling data tabulation and frequency generation.

Table 2 shows the distribution of respondents according to their area of professional activity. Among the 227 participants who answered this question, a significant majority were Internal Audit professionals, representing 78.85% of the sample. This aligns with the profile of the congress, which was primarily focused on Internal Audit.

Next in prominence were professionals from the Internal Controls area (5.29%) and the Audit Committee (4.85%). Risk Management accounted for 4.41% of respondents, which, together with Internal Controls and Internal Audit, reinforces the study's focus on the so-called Three Lines of The IIA (IIA, 2020).

Other areas had lower participation rates, including the Board of Directors (3.08%), Controllership (2.20%), and functions such as Fiscal Council, Compliance, Ombudsman, GRC, and Cybersecurity, each with 0.44%.

This distribution indicates that the sample is primarily composed of professionals directly involved with control and assurance functions, allowing for a more specialized analysis of functional integration and the application of the Combined Assurance concept.

Table 2. Panel Participants

Area	Freq	Percent	Cumulative
IA, Risk, IC (combined)	1	0.44%	0.44%
Cybersecurity	2	0.88%	1.32%
Audit Committee	11	4.85%	6.17%
Compliance	1	0.44%	6.61%
Fiscal Council	1	0.44%	7.05%
Board of Directors	7	3.08%	10.13%
Controllership	5	2.20%	12.33%
Ombudsman	1	0.44%	12.77%
GRC (Governance, Risk, Compliance)	1	0.44%	13.21%
Internal Audit Department	179	78.85%	92.07%
Internal Controls Department	12	5.29%	97.36%
Risk Management Department	10	4.41%	100.00%

Table 3 shows the distribution of respondents in terms of working in regulated or non-regulated companies. A majority (63.64%, n = 42) reported working in regulated organizations.

This is particularly relevant, as regulated entities—such as publicly listed companies or those overseen by regulatory bodies like Bacen, Previc, Susep, and the Secretariat for State-Owned Enterprises—are legally required to maintain robust corporate governance structures, including Audit Committees, Boards, and specialized risk, control, compliance, and audit areas.

The predominance of respondents from these organizations adds weight to the subsequent analyses, as the integration of control functions and application of the Combined Assurance model are directly related to the governance context in such entities.

Table 3. Regulated vs Non-Regulated Entities

Regulation Status	Freq	Percent	Cumulative
Regulated	42	63.64%	63.64%
Non-Regulated	24	36.36%	100.00%

Table 4 presents the results from Question 3 of the questionnaire, which assessed participants' perceptions regarding the ability of regulated organizations, those legally required to maintain formal governance structures, to prevent and detect fraud. The question was: "Do you agree that an organization with all these governance bodies is able to ensure fraud prevention and detection?".

This question was answered by 207 participants. The results show that 64.73% (n = 134) disagreed, indicating a belief that the mere existence of regulatory bodies and formal control structures does not guarantee effective fraud prevention or detection.

This suggests that high levels of regulation alone are not perceived as sufficient to prevent irregularities, underscoring the importance of integrating ethical culture, operational effectiveness, and alignment across organizational lines, as emphasized by The IIA's Three Lines Model (IIA, 2020).

Table 4. Fraud Prevention vs Regulated Structures

Statement: "Prevents Fraud"	Freq	Percent	Cumulative
Agree	73	35.27%	35.27%
Disagree	134	64.73%	100.00%

Table 5 presents the results of Question 4, which aimed to assess whether participants observe integration between the three organizational lines. The question was: "Do you observe integration among Management (1st Line), Risk & Controls/Compliance (2nd Line), and Internal Audit (3rd Line) in your organization?".

A total of 185 participants answered this question. The results reveal that 56.76% (n = 105) reported not observing effective integration among the functions within their organizations.

The underlying concept is the Three Lines Model (IIA, 2020), where: i) the first line comprises operational management; ii) the second includes Risk, Internal Controls, and Compliance; and iii) the third line refers to Internal Audit, responsible for independently assessing governance and control effectiveness.

This collective work is also supported by external assurance providers, such as independent auditors and regulators. The goal is to create synergy, prevent failures, and provide assurance to senior leadership and governance bodies, enabling timely corrective action.

Table 5. Integrated Functioning Observed

Integration Status	Freq	Percent	Cumulative
Integrated	80	43.24%	43.24%
Not Integrated	105	56.76%	100.00%

Question 5 - "Have you ever been requested by Management, the Board of Directors, or an advisory body to act in an integrated manner?" - was answered by 164 participants.

As shown in Table 6, 64.02% (n = 105) reported having received such a request. This is particularly noteworthy when compared with the findings of Question 4, where the majority stated that integration was not effectively observed in practice.

These results point to a disconnect between institutional demands for integration and its actual implementation. While professionals are being encouraged to act collaboratively, this practice has yet to be fully established.

This may indicate an ongoing shift, as organizations seek to align governance and control functions with the Three Lines Model (IIA, 2020), framework used in this article. Thus, although integration is not yet a widespread reality, the data suggest that efforts to promote it are underway, potentially strengthening governance and risk mitigation in the future.

Table 6. Requested to Act in an Integrated Manner

Requested	Freq	Percent	Cumulative
Yes	105	64.02%	64.02%
No	59	35.98%	100.00%

Finally, Question 6 - "To what extent does each item contribute to the failure of a highly regulated organization to detect or prevent fraud?"—was answered by 250 participants.

In descending order of relevance, the most cited contributing factors were:

- 78% – Senior Management's lack of awareness regarding the responsibilities of the Three Lines
- 73% – Organizational misunderstanding of the Three Lines' responsibilities
- 59% – Lack of integration among the Three Lines
- 58% – Lack of "Tone at the Top" (Senior Management)
- 55% – Lack of understanding by the Three Lines of their responsibilities.

The results of Question 6 provide critical insights into the underlying factors contributing to the failure of highly regulated organizations to prevent and detect fraud effectively. The most frequently cited issue - highlighted by 78% of respondents - was senior management's lack of awareness regarding the responsibilities of the Three Lines, underscoring a significant disconnect between strategic governance and operational risk structures. This was followed by an organizational misunderstanding of these responsibilities (73%), which suggests that knowledge gaps are not confined to the top leadership but are present across the organization. Such deficiencies hinder coordinated risk management and undermine the implementation of effective control frameworks. Additionally, 59% of participants pointed to a lack of integration between the Three Lines, highlighting the absence of structured collaboration, role clarity, and strategic alignment between management, control functions, and internal audit.

Further compounding these challenges, 58% of respondents noted the absence of "Tone at the Top," indicating a perceived deficiency in ethical leadership and cultural reinforcement from senior executives. Finally, 55% cited a lack of understanding by the Three Lines themselves regarding their respective roles, suggesting that even well-defined structures may fail if not accompanied by adequate communication, training, and governance maturity. These findings collectively indicate that formal regulatory structures alone are insufficient for effective fraud prevention. Instead, organizations must foster a culture of accountability, clearly delineate roles, and promote ongoing collaboration among the Three Lines to ensure that governance, risk, and control mechanisms are not only present but operationally effective (Table 7).

Table 7. Factors Contributing to Failures in Fraud Prevention and Detection

Reason	Description	%
Reason 1	Lack of awareness by Senior Management regarding the responsibilities of the Three Lines	78%
Reason 2	Organisational lack of understanding about the responsibilities of the Three Lines	73%
Reason 3	Lack of integration among the Three Lines	59%
Reason 4	Lack of "Tone at the Top" (Senior Management)	58%
Reason 5	Lack of understanding by the Three Lines themselves about their own responsibilities	55%

Legend:

- Reason 1 – Senior Management's lack of awareness regarding the Three Lines
- Reason 2 – Organisational misunderstanding of the Three Lines' responsibilities
- Reason 3 – Lack of integration among the Three Lines
- Reason 4 – Lack of 'Tone at the Top'
- Reason 5 – Lack of understanding by the Three Lines themselves

Additional reasons identified through the interactive survey included: i) competition among assurance functions, reducing synergy; ii) failure to take responsibility for resolving action plans; iii) lack of a risk culture; iv) poorly defined organizational culture; v) weak processes; vi) lack of leadership integrity; vii) conflicts of interest within control bodies; viii) lack of complete independence among lines; and ix) insufficient resources and appreciation for control, risk, and internal audit areas.

On average, 160 participants responded to Question 6. Of these, 31% (n = 50) considered the integration of functions to be relevant for fraud prevention, while 65% (n = 104) considered it highly relevant.

5 Discussion and Final Remarks

This study aimed to assess the perceptions of professionals working in Internal Audit, Risk Management, Internal Controls, Compliance, and related areas regarding the relevance of integrating these organizational functions. The findings reveal that, although the majority of respondents acknowledge the importance of such integration, a significant gap remains in its practical implementation within organizations. Most participants (78.85%) were internal audit professionals, and a considerable portion (63.64%) were employed in regulated entities - organizations legally required to maintain robust governance frameworks. Nevertheless, 64.73% of respondents disagreed that the mere existence of specialized functions such as risk, controls, compliance, and audit is sufficient to ensure effective fraud prevention.

Moreover, 56.76% of participants reported not observing functional integration among the Three Lines within their organizations, despite 64.02% having been requested by governance bodies to operate in a coordinated manner. This points to a dissonance between institutional expectations and operational reality. Importantly, the most frequently cited reasons for failures in fraud prevention included senior management's lack of awareness of the Three Lines' responsibilities, poor organizational understanding of these roles, weak integration, the absence of "Tone at the Top", and the limited role clarity among the lines themselves. Additional factors such as interdepartmental competition, weak risk culture, leadership integrity concerns, fragile processes, and lack of independence and resources further exacerbate this fragmentation.

Based on the research findings, the study reveals that although most professionals acknowledge the importance of integrating the functions of Risk Management, Internal Controls, Compliance, and Internal Audit, this practice is still not widely implemented within organizations. Even in highly regulated companies, perceptions of inefficacy in fraud prevention prevail, attributed to factors such as a lack of understanding of the "Three Lines" responsibilities, absence of a strong organizational culture, conflicts of interest, and misalignment among control areas. The data also indicates a future trend toward integration, as many professionals have already been called upon to collaborate, despite the lack of formally integrated structures. These findings highlight significant organizational challenges and point to integration as a promising path to strengthen governance and mitigate risks.

These results emphasize that formal governance structures alone are not sufficient to mitigate fraud risks effectively. Successful prevention and detection of fraud require more than regulatory compliance - it demands an integrated assurance approach, reinforced by strong ethical leadership, clearly defined roles, and cross-functional collaboration. This study contributes to the growing discourse on the value of integrated governance and highlights the need for further empirical research, particularly in diverse organizational contexts, primarily exploring the responsibilities of each of the functions—Risk Management, Internal Controls, Compliance, and Internal Audit—as well as their complementarities and potential overlaps in activities. A limitation of this research is that data collection was conducted at an internal audit congress, and thus the sample was predominantly composed of audit professionals. A scarcity of empirical studies related to the topic was also observed, which limited the possibility of comparisons and highlighted an opportunity for future research.

References

Beasley, M.S., Clune, R., Hermanson, D.R.: Enterprise risk management: An empirical analysis of factors associated with the extent of implementation. J. Account. Public Policy **24**(6), 521–531 (2005)

COSO: Internal control–Integrated framework: Executive summary. Committee of Sponsoring Organizations of the Treadway Commission (2013)

COSO: Enterprise Risk Management (ERM) Framework (2017). https://www.coso.org/_files/ugd/3059fc_61ea5985b03c4293960642fdce408eaa.pdf

Donelson, D., Ege, M., McInnis, J.: Internal control weaknesses and financial reporting fraud. Auditing: A Journal of Practice & Theory **36**(3), 45–69 (2017). https://doi.org/10.2308/ajpt-51608

Ferreida, C.L., Schwarzbach, L.C., Ferreira, V.C.R.: Instrumento para coleta de dados primários para pesquisas em administração. Revista Americana de Empreendedorismo e Inovação, 27–33 (2018)

Geiger, M.A., Gold, A., Wallage, P.: Practitioner perspectives on going concern opinion research and suggestions for further study: Part 2—Reporting environment and the decision-making process. Account. Horiz. **38**(2), 169–178 (2024). https://doi.org/10.2308/HORIZONS-2023-014

Huibers, S.C.: Combined assurance: One language, one voice, one view. IIA CBOK Report (2015)

IBGC: Orientações sobre comitês de auditoria: Melhores práticas no assessoramento ao conselho de administração. Instituto Brasileiro de Governança Corporativa (2017). https://conhecimento.ibgc.org.br/Paginas/Publicacao.aspx?PubId=23485&assessment=1

IBGC: Código das melhores práticas de governança corporativa, 6ª ed., Instituto Brasileiro de Governança Corporativa (2023)

IIA: Fraude e a auditoria interna (2019). https://www.theiia.org/globalassets/documents/resources/fraud-and-internal-audit-assurance-over-fraud-controls-fundamental-to-success-april-2019/fraud-and-internal-audit-portuguese.pdf

IIA: Modelo das 3 linhas do IIA (2020). https://iiabrasil.org.br/korbilload/upl/editorHTML/uploadDireto/20200758glob-th-editorHTML-00000013-20082020141130.pdf

IIA: International Professional Practices Framework (IPPF). The Institute of Internal Auditors (2024). https://www.theiia.org/globalassets/site/standards/editable-versions/globalinternalauditstandards_2024january9_editable.pdf

Imoniana, J.O., Freitas, E.C.D., Perera, L.C.J.: Assessment of internal control systems to curb corporate fraud—Evidence from Brazil. African J. Acc. Audit. Fin. **5**(1), 1–24 (2016)

Institute of Directors in Southern Africa: King II Report on Corporate Governance for South Africa (2002). https://cdn.ymaws.com/www.iodsa.co.za/resource/collection/94445006-4F18-4335-B7FB-7F5A8B23FB3F/IoDSA_King_II_web_version.pdf

Institute of Directors in Southern Africa: King III Report on Corporate Governance for South Africa (2009)

Institute of Directors in Southern Africa: King IV Report on Corporate Governance for South Africa (2016)

Jamshidi-Navid, B.: A clear look at internal controls: Theory and concepts. SSRN (2010). https://doi.org/10.2139/ssrn.1342048

Kaplan, R.S., Mikes, A.: Managing risks: A new framework. Harvard Business Review (2020)

Kitzberger, W., Toporoski, E.L.: Compliance: Uma ferramenta importante para o desenvolvimento empresarial. Academia de Direito **6**, 1140–1157 (2024)

Kotronoulas, G., et al.: An overview of the fundamentals of data management, analysis, and interpretation in quantitative research. Seminars in Oncology Nursing (2023)

Maciel, A.E., Bacci, L., Assi, M.: Transformando as três linhas em geração de valor: Com a gestão de riscos e o sistema de controles internos. Digitaliza Conteúdo (2022)

Mendes de Oliveira, D.K., Imoniana, J.O., Slomski, V., Reginato, L., Slomski, V.G.: How do internal control environments connect to sustainable development to curb fraud in Brazil? Sustainability **14**(9), 5593 (2022). https://www.mdpi.com/2071-1050/14/9/5593

Pinto, J., Pereira, A.C., Imoniana, J.O., Peters, M.R.S.: Role of internal audit in managerial practice in organisations. Afr. J. Bus. Manage. **8**(2), 68–78 (2014)

Power, M.: Organized uncertainty: Designing a world of risk management. Oxford University Press (2007)

PwC: Risk in review 2018 (2018). https://www.pwc.com.br/pt/estudos/servicos/consultoria-negocios/2018/risk-review-18.pdf
Regulamento do Novo Mercado: Regulamento do Novo Mercado – B3 (2017). https://www.b3.com.br/data/files/B7/85/E6/99/A5E3861012FFCD76AC094EA8/Regulamento%20do%20Novo%20Mercado%20-%2003.10.2017%20%28Sancoes%20pecuniarias%202019%29.pdf
Ribeiro, M.C.P., Diniz, P.D.F.: Compliance e lei anticorrupção nas empresas. Revista de Informação Legislativa **52**(205), 87–105 (2015)
Selvamuthu, D., Das, D.: Descriptive statistics. In: Introduction to probability, statistical methods, design of experiments and statistical quality control, pp. 201–240. Springer (2024)
Spira, L.F., Page, M.: Risk management: The reinvention of internal control and the changing role of internal audit. Acc. Audit. Accountab. J. **16**(4), 640–661 (2003)
StataCorp: Stata statistical software (2025). https://www.stata.com/
Tutino, M., Merlo, M.: Accounting fraud: a literature review. Risk Govern. Contr. Fin. Mark. Inst. **9**(1), 65–75 (2019)
Wallace Kirkpatrick, W.: The adequacy of internal corporate controls. Ann. Am. Acad. Pol. Soc. Sci. **343**(1), 75–83 (1962). https://doi.org/10.1177/000271626234300110
Waromi, J., Kambuaya, M.K., Wonar, K.: Evaluation of the effectiveness of internal control systems in reducing accounting fraud in financial sector companies. The Eastasouth Manage. Bus. **2**(03), 228–232 (2024)
Yancey, M.: Advanced framework for integrating risks into an organisational setting (2024)
de Souza Damiani, N.C.: Controles internos: Conceitos e técnicas para implementação em empresas. Atlas (2011)
Çetinoğlu, T.: Bibliometric analysis of studies related to internal auditing [İç Denetim ile İlgili Çalışmaların Bibliyometrik Analizi]. Uluslararası Sosyal ve Ekonomik Çalışmalar Dergisi **5**(1), 114–132 (2024). https://doi.org/10.62001/gsijses.1470887

Valued Skills in Accounting: A Systematic Literature Review

Beatriz Rosário[1(✉)], Fátima Conde[1,2], and Adriana Silva[1,3]

[1] Polytechnic University of Coimbra, Rua da Misericórdia, Lagar Dos Cortiços, S. Martinho Do Bispo, 3045-093 Coimbra, Portugal
beatrizrosario5544@gmail.com, {fconde,afsilva}@iscac.pt
[2] CEOS.PP Coimbra, Polytechnic University of Coimbra, Coimbra, Portugal
[3] Faculty of Economics, CeBER, Coimbra, Portugal

Abstract. Increasing globalization and technological development have transformed the labor market, requiring a more diverse range of skills from recent accounting graduates. In this context, based on the PRISMA methodology, this study conducts a systematic review of the literature on the technical skills (hard skills) and interpersonal skills (soft skills) required of accounting professionals. The analysis included articles published between 2000 and 2024 on the Web of Science (WoS) and Scopus databases, applying strict inclusion and exclusion criteria. Only eight studies simultaneously address the perceptions of students and employers, and none of them are applied to the Portuguese context, highlighting a critical gap in the literature. The results reveal a mismatch between the competences valued by students, who focus on hard skills, and those valued by employers, who emphasize soft skills. Thus, this research contributes to understanding the (mis)alignment between academic accounting education and labor market demands, providing a solid basis for curriculum reformulations and new research. Its relevance is underlined by the fact that it has been shown that it has not yet been studied in some countries, making it possible to identify gaps in literature.

Keywords: Soft skills · Hard skills · Perceptions · Employers · Recent graduates

1 Introduction

In recent years, Portugal has invested significantly in higher education, intending to align itself with OECD guidelines and improve its educational development indicators (OECD, 2023). However, challenges remain in ensuring that graduates have the necessary skills to meet the demands of the labour market. The country continues to struggle with a relatively high unemployment rate among recent graduates, while in some sectors, such as accounting, there is a high supply of professionals, highlighting the need to align the skills taught in universities with the real needs of the market (OECD, 2023; Jackling and De Lange, 2009). Subsequent paragraphs, however, are indented.

L. Almeida et al. (Eds.): ICAFI 2025, ISEM 80, pp. 601–615, 2026.
https://doi.org/10.1007/978-3-032-16268-7_35

Although there has been significant progress in the number of courses and the number of new students, challenges remain in preparing graduates for the labor market (OECD, 2023). Dias and Oliveira (2021, p. 18) point out that recent graduates "are not always prepared for the demands of the market". In particular, fields such as accounting are highly competitive (Jackling and De Lange, 2009) due to the increasing number of professionals (European, 2020) and the rapid changes in the sector driven by globalization and digitalization (ICAEW, 2018).

On the other hand, the accounting profession has significantly evolved from a purely administrative function to a more strategic role (Barth et al., 2008), where the professional is considered a fundamental part of the organization (ICAEW, 2018). Accountants are now expected to be actively involved in the development of financial and operational strategies, reflecting the increasing complexity and responsibility associated with the role (Barth et al., 2008). Thus, in addition to understanding the principles of financial accounting, professionals must master skills such as auditing, internal control, and international standards, which are essential in an increasingly globalized job market (Nie and Mastor, 2014).

In this scenario, accountants must develop hard skills such as preparing and analyzing financial reports, auditing, complying with accounting standards, and using specialized software (Apostolou et al., 2013). It is also important to note that these technical skills (hard skills) are fundamental in the sector, where automation and digital transformation are changing the nature of accounting tasks (European Commission, 2020; ICAEW, 2018).

At the same time, soft skills such as communication, teamwork, critical thinking, and problem solving have become increasingly important (Jackling and De Lange, 2009). Employers highly value these skills, especially in collaborative and constantly changing environments, such as the small and medium-sized enterprises (SMEs) that dominate the Portuguese business fabric (Europeia, 2020).

Despite the recognized importance of both technical and behavioral skills, there remains a mismatch between what is acquired during studies and the real needs of the market (Jackling and De Lange, 2009). Many graduates show a significant lack of soft skills, which are often acquired through practical experience (Dolce et al., 2020). This reality highlights the importance of incorporating more dynamic approaches into training programmes that are aligned with job requirements and promote the articulation between technical knowledge and the ability to adapt and innovate (Jackling and De Lange, 2009).

In turn, Portuguese small and medium-sized enterprises (SMEs), which play a crucial role in hiring graduates, also report a gap between the skills acquired by students and those required by companies. This situation is exacerbated by increasing globalization and evolving job requirements, which demand not only technical skills but also behavioral skills, such as the ability to adapt and solve problems (European, 2020).

In this regard, the objective of this study is to identify, synthesize, and analyze existing literature on technical (hard) and interpersonal (soft) skills of accounting professionals.

Through this review, we sought to address the growing need for a better understanding of the relationship between accounting students' academic path and the actual demands of the labor market. A systematic review of the literature was conducted based on the PRISMA method, including scientific articles published between 2000 and 2024 in the Scopus and Web of Science databases (Joaquim et al., 2025). Only articles comparatively addressing the perceptions of students and employers regarding the required skills in accounting were selected.

By focusing simultaneously on technical and behavioral skills, and by incorporating the perceptions of both students and employers, this research contributes to deepening knowledge of the link between higher education and the labor market. Its relevance is underlined by the scarcity of studies addressing this issue in specific national contexts, such as Portugal, where the needs of the accountancy sector remain poorly represented in the scientific literature (Nie and Mastor, 2024; Jackling and De Lange, 2009). Therefore, by identifying the main points of convergence and divergence between students' and entrepreneurs' perceptions, this study provides a solid basis for future studies and academic articles, as well as possible curriculum reformulations in line with an ever-changing labor market (ICAEW, 2018; OECD, 2023).

This paper is divided into four chapters. Chapter 2 presents the methodology used, based on the PRISMA model, detailing all the steps that led to the selection of the articles included in the review. Chapter 3 presents the results, divided into two main sub-chapters: the quantitative analysis, which examines aspects such as the geographical distribution, methodologies, and sources of the geographical distribution, methodology, and sources of the articles; and the qualitative analysis, which examines the content of the selected studies. The latter is further divided into two segments, distinguishing articles that focus on generic competences from those that focus on comparing perceptions. Finally, chapter 4 discusses the results in the light of the literature and presents the conclusions of the study, followed by the bibliographical references.

2 Methodology

Through a systematic literature review, this study aims to identify, synthesize, and analyze the literature on students' and employers' perceptions of the skills acquired during their academic careers. To ensure rigor and transparency in the process, the systematic review was conducted using the PRISMA (Preferred Reporting Items for Systematic Reviews and Meta-Analyses) methodology, which is widely recognized as providing a comprehensive and reliable framework (Moher et al., 2009).

The databases used in this study were Web of Science (WoS) and Scopus, chosen for their reliability and wide coverage of scientific publications. As mentioned by Joaquim et al. (2025), these databases maximize the availability of relevant articles for analysis and ensure academic validation of the data.

The first stage of this research consisted of a systematic search for publications in the above-mentioned databases, intending to cover the three central dimensions of the analysis: behavioral competences, technical competences, and accounting. To ensure a comprehensive analysis, the keywords were carefully selected, considering synonyms widely used in scientific literature, to include terminological variations that might appear in different studies. The search was carried out on the Web of Science (WoS) using the "TOPIC" field and in Scopus using the "Article Title, Abstract, Keywords" fields. These fields were chosen because they allow keywords to be found in the title, abstract, and keywords of publications, thus increasing the chances of identifying relevant papers.

For the area of behavioral skills, we used terms that are widely used in literature, such as 'soft skills', 'interpersonal skills', 'social skills', and 'communication skills'. In addition, we included the terms "teamwork skills", "leadership skills," and "adaptability", which are highly relevant in organizational contexts. Technical skills were represented by terms such as 'hard skills', 'practical skills', and 'professional skills'. In addition, the terms "specialized knowledge", "job-specific skills," and "analytical skills" were included because of their relevance in more specialized contexts. Concerning accounting, the following keywords were selected: "accounting", "financial accounting", "management accounting", "cost accounting", and "tax accounting".

The search was structured using the Boolean operator "AND", combining all the keywords for behavioral skills, technical skills and accounting as follows ("soft skills" OR "interpersonal skills" OR "social skills" OR "communication skills" OR "teamwork skills" OR "leadership skills" OR "adaptability" OR "hard skills" OR "practical skills" OR "professional skills" OR "specialized knowledge" OR "job-specific skills" OR "analytical skills") AND ("accounting" OR "financial accounting" OR "management accounting" OR "cost accounting" OR "tax accounting" OR "financial reporting" OR "corporate finance" OR "financial management"). This approach ensures that the overlaps between the three dimensions are identified, allowing a detailed analysis of the results.

All keywords were entered in English only. This choice was based on the fact that the majority of high-impact scientific publications use English as their main language. As pointed out by Montgomery (2013), the use of English as the main language in scientific publications is a widespread practice that significantly increases the visibility and accessibility of studies. Therefore, by choosing to conduct the research exclusively in English, this article maximizes the likelihood of identifying relevant, high-impact work.

The selection was then restricted to published scientific articles, excluding proceedings, book chapters, and other less rigorous sources (Agostino et al., 2021). Following the same criteria used by Joaquim et al. (2025), only articles written in English were considered. Another inclusion criterion was the publication period, which was defined as between 2000 and 2024. This time frame was chosen to ensure the timeliness and comprehensiveness of the analyzed literature, allowing for the identification of established trends and significant changes that have occurred in education and the accounting job market in recent decades. This time period was chosen in line with other systematic review studies that adopt longer time periods to capture both theoretical developments and recent changes in the sector (Joaquim et al., 2025; Agostino et al., 2021).

The selection process followed the PRISMA guidelines and is shown in the flowchart below (Fig. 1).

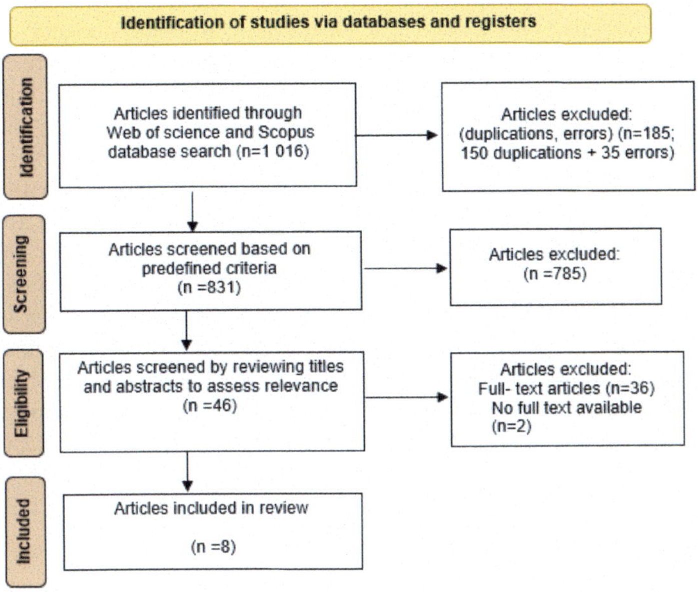

Fig. 1. PRISMA flow diagram. **Source:** Author's own; adapted from http://www.prisma-statement.org/

Initially, 1,016 articles were identified, of which 577 were found in Web of Science (WOS) and 439 in Scopus. From this initial list of articles, 150 duplicate articles and 35 articles from conference proceedings, meeting abstracts, and other types of publications, such as Proceedings Papers and Meeting Abstracts, were excluded. This ensured that only scientific articles were selected.

The titles and abstracts were then read, resulting in the exclusion of 785 articles because they did not simultaneously address behavioral and technical skills from the perspective of both employers and graduates. This delimitation was intentionally chosen because it aligns with the central objectives of this study, which are to analyze the comparative perceptions of the two groups regarding the valued skills in the accounting sector. Thus, this choice not only ensures consistency between the analyzed data and the study's central objectives but also contributes to filling a gap in the literature by focusing on articles offering a more integrated and realistic view of the link between education and the labor market.

A more detailed analysis of the keywords, titles, and abstracts, and where necessary, the full content of the articles, then conducted 36 articles were eliminated that were not in line with the aim of the study. These included articles that approached skills only from a labor market perspective, such as the study by Hayes et al. (2022). We also excluded articles that only looked at academic skills from the students' perspective, without exploring the link with the labor market, such as the study by Osmani et al. (2017) and Banasik et al. (2021). In this way, we excluded all studies that did not provide a complete analysis of the relationship between the skills acquired in education and those required by the labor market, from both the students' and employers' perspectives.

In the end, eight articles were included in the final sample analyzed in the study. To analyze the information contained in these articles, we performed a qualitative analysis using Microsoft Excel, following the methodology of Nie and Mastor (2024). Given the small number of studies selected and the descriptive nature of the data collected, this approach was considered the most appropriate. It allowed us to extract patterns, identify convergences and divergences in the analyzed audiences' perceptions, and gain an in-depth understanding of each article's contributions to the research objectives.

We used Microsoft Excel to organize and categorize the articles by publication date, author(s), title, geographical area of study, methodological approach, objects of study, and main conclusions.

3 Results

In this chapter, the eight articles selected by the PRISMA method are presented and analyzed in two subchapters. In the first, a quantitative analysis of the articles is presented, considering aspects such as the year of publication, the geographical contexts, the journals or scientific magazines in which they were published, and the methodologies used (Joaquim et al., 2025). Then, the qualitative analysis of the literature allows us to identify patterns, differences, and complementarities between the research, contributing to a more detailed view of the selected articles.

3.1 Quantitative Analysis

This section aims to provide an overview of the articles selected for the systematic review, based on the PRISMA methodology. Aspects such as the year of publication, geographical distribution, scientific journals and magazines in which they were published, and the research methods used are analyzed. This characterization makes it possible not only to contextualize the relevant articles in the literature on the relationship between accounting skills in education and the labor market but also to identify emerging trends, thematic gaps, and the geographical and methodological breadth of the research (Nie and Mastor, 2024).

Figure 2 illustrates the time distribution of the selected articles through a graph of the number of publications per year.

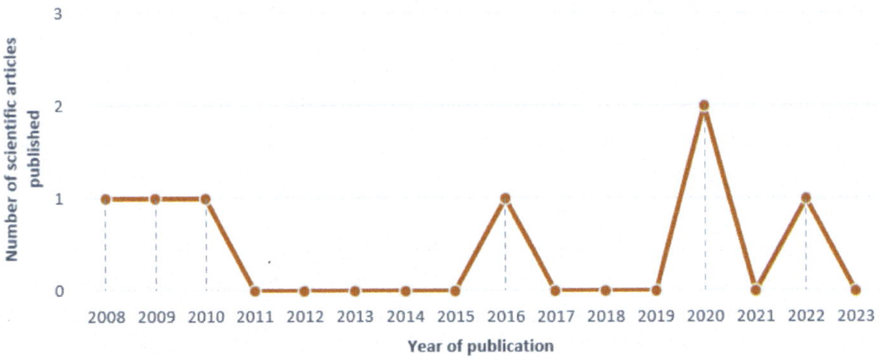

Fig. 2. Number of articles published. **Source:** Author's own work

Analysis of the eight selected articles published between 2008 and 2024 reveals a time distribution that reflects the growing interest in research into the skills of recent accounting graduates. Although no articles were published between 2000 and 2007, there is a steady stream of publications from 2008 onwards, particularly in 2020, with two articles.

The first article in the sample, "What skills and attributes do an accounting graduate need? Evidence from Student Perceptions and Employer Expectations" (Kavanagh and Drennan, 2008), looks at student and employer perceptions of the essential technical and behavioral skills for accounting students, highlighting the importance of aligning academic expectations with labor market requirements.

In the following years, and especially in 2020, there was an increase in publications on this topic, which may be related to changes in labor market requirements and the evolution of international accounting standards (Evangelista et al., 2022), which require a continuous adaptation of the skills of professionals in the field.

The following table shows the geographical distribution of the eight articles analyzed in this systematic review, revealing a distribution characterized by a certain geographical concentration, although with some diversity between continents. Although the data are mostly concentrated in specific contexts, such as Australia and Malaysia, there is a contribution from countries with different economic, cultural, and educational realities, which reinforces the comparative diversity of the review (Petticrew and Roberts, 2006) (Table 1).

Table 1. Number of articles by geographical region

Geographic area	Number of articles	Percentage (%)
Australia	3	37,5%
Ghana	1	12,5%
Italy	1	12,5%
Malaysia	2	12,5%
Romania	1	12,5%

Source: Author's own work

It can also be seen that Australia is the country with the highest number of published articles, which may be due to the high level of development of the accounting sector in the country, both academically and institutionally (Jackling and De Lange, 2009). In addition, Nie and Mastor (2024) reinforce Australia's relevance by highlighting the active role of professional bodies and research initiatives in the country.

Other regions have also contributed to the debate, albeit on a smaller scale. Malaysia, for example, accounts for 25% of the publications analyzed, reflecting the growing interest in matching the skills of recent graduates with the needs of the global market (Ismail et al., 2018). On the other hand, countries such as Ghana, Italy, and Romania represent 12.5% each, indicating a more localized effort to understand and improve accounting skills. Thus, while there is a global interest in the topic, its concentration in certain regions (such as Australia) may be related to institutional, economic or academic factors.

Despite this diversity, it is important to note that none of the studies identified were carried out in Portugal, a country that has been investing more and more in higher education and in bringing universities and companies closer together (OECD, 2023).

An analysis of the journals and magazines in which the selected articles were published provides a better understanding of the areas of interest in this topic. Table 2 shows the distribution of articles by journal, which shows a preponderance.

Table 2. Number of articles by journal title

Name of the scientific journal	Number of articles	Percentage (%)
Accounting Education	3	37,5%
Accounting and Finance	1	12,5%
Australasian Accounting, Business and Finance Journal	1	12,5%
Cogent Business and Management	1	12,5%
Journal of Education for Business	1	12,5%
Studia Universitatis Vasile Goldis Arad, Economics Series	1	12,5%

Source: Author's own work

Accounting Education is the journal with the largest number of published articles relevant to this study, accounting for 37.5% of the total. This prominence can be attributed to its focus on accounting education and training, topics in which the journal has a significant presence.

Scientific publications (12.5% of the total). These include academic journals with relevant international rankings. According to SCImago Journal Rank (SJR), Accounting and Finance is ranked Q1, The Australasian Accounting, Business and Finance Journal is ranked Q2, and The Journal of Education for Business is ranked Q2. The presence of these journals and magazines shows not only the scientific relevance of the subject, but also its interdisciplinary nature, covering the fields of accounting, education, and management.

This diversity of scientific journals and magazines highlights the multidisciplinary nature of the subject, indicating the existence of concerns not only in the field of accounting, but also in the fields of management, education, and professional development.

An analysis of the methodologies used in the selected articles reveals a variety of approaches. Figure 3 shows the distribution of methodologies used in the studies, highlighting the prevalence of mixed methods.

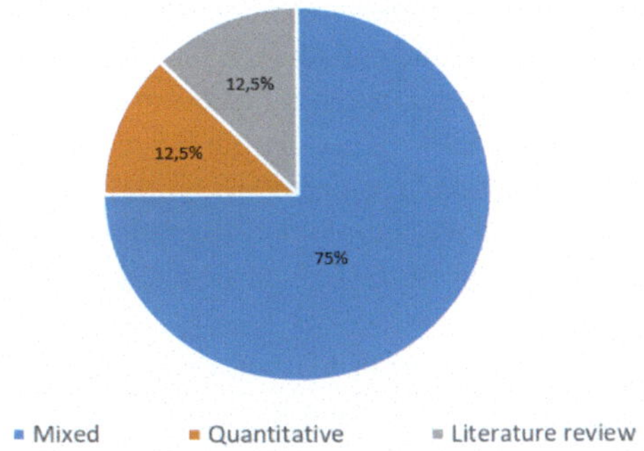

Fig. 3. Methodology used in articles. **Source:** Author's own work

Most articles (75%) used a mixed approach, combining qualitative and quantitative methods to provide a more comprehensive view of the topic (Creswell and Plano Clark, 2015). This preference for mixed methods can be justified by the need to integrate statistical data on employability with a more in-depth analysis of student and employer perceptions.

Exclusively quantitative studies account for 12.5% of the total and are generally based on statistical surveys and structured questionnaires (Cernuşca, 2020). Similarly, articles that have undertaken a literature review also account for 12.5% of the sample and are used to analyze in more detail the periodical articles already published and to identify gaps and trends in the field (Nie and Mastor, 2024).

The methodological distribution of the articles reflects the complexity of the subject and the need for different approaches to understand the multiple factors that influence the professional integration of recent accounting graduates.

3.2 Qualitative Analysis

The qualitative analysis of the articles selected using the PRISMA method allows further exploration of the existing literature on the technical (hard skills) and interpersonal (soft skills) competences of accountancy professionals from the perspectives of students and employers. In this way, it is possible to identify patterns and differences in the different approaches, which allows for a broader reflection on the challenges of academic education and its adequacy to the demands of the labor market.

In order to ensure a clear and coherent structure, the analysis has been divided into two main sub-chapters to facilitate understanding.

3.3 General Skills Required by the Labor Market

This subchapter includes three articles that provide a comprehensive overview of the technical and behavioral competencies (hard and soft skills) essential to the accounting profession from the perspectives of employers and graduates. Although all eight articles selected for this review address hard and soft skills from the perspective of graduates and employers, we have chosen to highlight in this sub-section the three articles that place the greatest emphasis on the generic skills valued by the labor market. These articles take a more centralized approach to the broad requirements of the accountancy sector, focusing on the skills considered fundamental to employability, rather than exploring in depth the comparison between the perceptions of the two parties, although this comparison is also present.

Increasing competitiveness in the labor market, coupled with technological, regulatory, and organizational changes in the accounting sector, requires a diverse set of skills that go beyond traditional technical knowledge (Nie and Mastor, 2024). These changes have increased the need for flexible professionals who can adapt to the new demands of the sector and work with emerging digital tools (Kavanagh and Drennan, 2008).

The three articles analyzed highlight that while employers continue to value technical skills, there is a growing emphasis on the importance of interpersonal skills, which are considered fundamental to the employability of recent graduates (Nie and Mastor, 2024; Lim et al., 2016; Kavanagh and Drennan, 2008). To make it easier to see which technical and behavioral skills were most frequently mentioned in the analyzed articles as essential for professional success in the accounting field, Table 3 follows:

Table 3. Skills identified in the articles as essential for professional success

Skill type:	Skills:	Articles:	Definition:
Hard skills	Financial accounting	Nie and Mastor, 2024; Lim et al., 2016; Kavanagh and Drennan, 2008	"Sound knowledge of accounting principles, records and financial reporting." (Kavanagh and Drennan, 2008)
	Data Analysis		"Ability to interpret financial data and extract information for decision-making." (Lim et al., 2016)
	Audit and Compliance	Nie and Mastor, 2024; Kavanagh and Drennan, 2008	"Ability to assess the accuracy and compliance of financial records with rules and regulations." (Nie and Mastor, 2024)
Soft skills	Communications	Nie and Mastor, 2024; Lim et al., 2016; Kavanagh and Drennan, 2008	"Essential competence to express ideas, negotiate and collaborate with stakeholders." (Lim et al., 2016)
	Teamwork skills		"Ability to collaborate with different professionals to achieve organizational goals." (Kavanagh and Drennan, 2008)
	Critical Thinking		"Ability to analyze complex situations and propose innovative solutions." (Nie and Mastor, 2024)
	Time Management	Lim et al., 2016; Kavanagh and Drennan, 2008	"Efficient organization of activities to meet deadlines and increase productivity." (Lim et al., 2016)

(*continued*)

Table 3. (*continued*)

Skill type:	Skills:	Articles:	Definition:
	Flexibility		"Ability to adapt to market changes and technological innovations." (Lim et al., 2016)
	Leadership skills	Nie and Mastor, 2024; Kavanagh & Drennan, 2008	"Ability to influence and motivate teams in order to achieve results." (Nie and Mastor, 2024)

Source: Author's own work

This analysis shows that while hard skills remain fundamental, employers are increasingly emphasizing the importance of soft skills in the workplace. This shift is reflected in the growing appreciation of communication, teamwork, and critical thinking as essential skills in collaborative accounting, as shown in Table 3 (Kavanagh and Drennan, 2008).

3.4 Differences in the Perceptions of Students and Employers

This subchapter aims to analyze the remaining five articles, which examine the perceptions of students and employers of key competences for the labor market. This selection is because these studies focus more on the direct comparison between the two perspectives, in contrast to the three previous articles, which, although they also consider perceptions, place more emphasis on the generic skills required by the sector. The five articles analyzed here use empirical methods that are more geared towards comparative analysis, including the use of questionnaires and, in some cases, interviews. This approach makes it possible to draw more detailed conclusions about the divergence between employers' expectations and graduates' self-perceptions about technical skills (hard skills) and behavioral skills (soft skills). The aim of this subchapter is therefore to identify which skills are considered crucial by each group and to understand the reasons behind these choices. An analysis of the articles reveals a significant discrepancy between students' perceptions and employers' expectations of the skills needed to succeed in the accounting job market. While students tend to value hard skills, such as technical mastery and specialized knowledge, employers prioritize soft skills, such as communication and teamwork (Dolce et al., 2020; Cernuşca et al., 2020; Awayiga et al., 2010; O'Shea et al., 2022; Jackling and De Lange, 2009).

This discrepancy becomes particularly relevant as the demands of the labour market evolve. While hard skills such as financial accounting and data analysis continue to be valued, it is clear that soft skills such as communication, ethics, and teamwork play an increasingly important role in recruitment and career progression in the accounting sector (Jackling and De Lange, 2009; Awayiga et al., 2010). Table 4 illustrates these differences in perceptions.

Table 4. Perceptions of students and employers

Perspective	Method	Most valued skills	Author citations
Students	Questionnaires: 251 students (Dolce et al., 2020); 250 students (Cernuşca et al., 2020); 164 students (Awayiga et al., 2010); 174 students (Jackling and De Lange, 2009) Interviews: (O'Shea et al., 2022)	Hard skills	"Students underestimate the importance of various soft skills and believe that hard skills are more crucial for employability." (Dolce et al., 2020)
Employers	Questionnaires: 74 companies (Dolce et al., 2020); 300 companies (Cernuşca et al., 2020); 25 companies (Awayiga et al., 2010) Interviews and focus groups: (O'Shea et al., 2022)	Soft skills	"Employers place more emphasis on soft skills, such as teamwork, effective communication and conflict management." (Dolce et al., 2020)

Source: Author's own work

Analysis of Table 4 shows a significant divergence between the skills that students consider to be a priority and those that employers consider to be essential. This divergence can be attributed to the continued emphasis placed by students on hard skills, often to the detriment of soft skills.

This difference between graduate and employer perceptions highlights the need to adapt academic programmes to encourage greater development of behavioral skills and to better prepare students for the challenges of today's professional environment (Jackling and De Lange, 2009; Awayiga et al., 2010).

4 Discussion of Findings and Conclusions

The aim of this study was to analyze the existing literature on the skills of accounting professionals and to highlight the gap between employers' and students' perceptions. The systematic review revealed that despite the growing demand for soft skills in the labor market, students still emphasize hard skills, reflecting a mismatch between academic training and business needs.

The literature revealed a significant lack of academic articles that simultaneously address employers' and students' perceptions of hard and soft skills, particularly in the field of accounting. Of the 1,016 articles initially identified, only 8 met the inclusion criteria, highlighting a significant gap in academic research on this topic.

The geographical distribution of the studies shows a concentration in Australia (37.5%) and Malaysia (25%), while regions such as Europe and Latin America are under-represented. There is also a predominance of mixed methods (75%), suggesting that exclusively qualitative or quantitative approaches are less explored. This trend reflects the need for studies that deepen the relationship between the skills acquired during studies and the demands of the labor market. The results also show a discrepancy between the perceptions of students and employers. While the former value mainly hard skills, employers emphasize the importance of soft skills such as communication, critical thinking, and teamwork. This difference underlines the need for a curricular orientation that better integrates both types of skills, namely the need for curricular reformulations that include more dynamic and practical methodologies, such as integrative projects and compulsory internships, to better prepare future professionals.

In this context, this study is a relevant and innovative contribution to accounting research. It systematically collects and analyzes the perceptions of employers and recent graduates regarding the technical and behavioral skills required in the field. The dearth of studies addressing this issue in a comparative manner or in specific contexts highlights a clear gap in international literature (Nie and Mastor, 2024; Jackling and De Lange, 2009). Despite the limitations of the study's methodology, which considered the perspectives of students and employers simultaneously, and the restriction to English-language publications, this research contributes to filling this gap by offering a focused, comparative view of the sector's valued skills.

The study underscores the urgent need for applied research exploring national realities individually to understand whether market demands are universal or vary according to local dynamics. This approach is particularly relevant in a globalized economy where professional mobility and company internationalization require increasingly context-specific skills (ICAEW, 2018).

By identifying these gaps, this work opens doors to future research, not only in country-specific studies but also in comparative analyses between different regions and education systems. It is therefore a valuable starting point for research that aims to deepen the dialogue between higher education and the labor market and promote more tailored, effective, and internationally competitive training.

References

Agostino, D., Saliterer, I., Steccolini, I.: Digitalization, accounting and accountability: A literature review and reflections on future research in public services. Financial Accountability & Management (2021)

Apostolou, B., Hassell, J.M., Rebele, J.E., Watson, S.F.: Accounting education literature review (2010–2012). J. Acc. Edu. 107–161 (2013)

Awayiga, J.Y., Onumah, J.M., Tsamenyi, M.: Knowledge and skills development of accounting graduates: The perceptions of graduates and employers in Ghana. Accounting Education (2010)

Banasik, E., Jubb, C.: Are Accounting Programs Future-ready? Employability Skills. Australian Accounting Review (2021)
Barth, M.E., Landsman, W.R., Lang, M.H.: International accounting standards and accounting quality. Journal of Accounting Research (2008)
Cernuşca, L.: Soft and Hard Skills in Accounting Field-Empiric Results and Implication for the Accountancy Profession. Studia Universitatis Vasile Goldis Arad, Economics Series (2020)
Creswell, J.W., Plano Clark, V.L.: Designing and conducting mixed methods research, 3rd ed. Sage Publications (2015)
Dias, M.C., Oliveira, A.: Employability challenges for graduates in Portugal. University of Lisbon, Institute of Education (2021)
Dolce, V., Emanuel, F., Cisi, M., Ghislieri, C.: The soft skills of accounting graduates: perceptions versus expectations. Accounting Education (2020)
European, C.: Annual report on European SMEs 2020/2021: Digitalization of SMEs (2020)
Evangelista, F.S., Coelho, D., Martins, Z.B.: International Accounting Standards and the impact on the future of the accounting profession: a perception of Accounting undergraduates at a community university in Santa Catarina. Journal CAFI (2022)
Hayes, S., Freudenberg, B., Delaney, D.: Work ready graduates for Australian small and medium Accounting firms. J. Teach. Learn. Graduate Employab. (2022)
ICAEW: Skills for the future: The impact of technology on employment in accountancy (2018)
Ismail, S., Anthonysamy, P., Mustapha, R.: Employability skills in accounting graduates: The moderating effect of social intelligence on the relationship between emotional intelligence and graduate employability. Management Science Letters (2018)
Jackling, B., De Lange, P.: Do accounting graduates' skills meet the expectations of employers? a matter of convergence or divergence. Accounting Education (2009)
Joaquim, R., Jorge, S., Pimentel, L.M.: Popular reporting in the public sector a systematic literature review and reflections for future reserch. J. Acc. Org. Change (2025)
Kavanagh, M.H., Drennan, L.: What skills and attributes does an accounting graduate need? Evidence from student perceptions and employer expectations. Accounting and Finance (2008)
Lim, Y.-M., Lee, T.H., Yap, C.S., Ling, C.C.: Employability skills, personal qualities, and early employment problems of entry-level auditors: Perspectives from employers, lecturers, auditors, and students. Journal of Education for Business (2016)
Moher, D., Liberati, A., Tetzlaff, J., Altman, D.G.: Preferred Reporting Items for Systematic Reviews and Meta-Analyses: The PRISMA Statement. PLoS Medicine (2009)
Nie, Y., Mastor, N.H.: Accounting employability: a systematic review of skills, challenges, and initiatives. Cogent Business and Management (2024)
OECD: Education at a glance 2023: OECD indicators. Organization for Economic Co- operation and Development (2023)
O'shea, M.A., Bowyer, D., Ghalayini, G.: Future Proofing Tomorrow's Accounting Graduates: Skills, Knowledge and Employability. Australasian Accounting, Business and Finance Journal (2022)
Osmani, M.W., Hindi, N., Al-Esmail, R., Weerakkody, V.: Examining graduate skills in accounting and finance: the perception of Middle Eastern students. Indus. Higher Edu. (2017)
Pereira, E.M.: The use of qualitative methods in social science research. Brazilian Journal of Social Sciences (2014)
Resurchify: Journal Rankings & Impact Factors (2024). https://www.resurchify.com
Sampaio, H.A., Oliveira, J.L.: The importance of literature reviews for the advancement of scientific knowledge. Revista de Pesquisa em Ciências Sociais (2021)
Silva, R.M., Souza, M.A.: Quantitative methods in academic research: An introduction. Revista Prisma (2018)
Williams, J.: Advanced accounting technologies and fiscal regulations. Oxford University Press (2019)

When There is Too Much Confidence, Does the Market Price It in? Ceo Overconfidence and Market Value

Clara Monise Silva(✉), Patrícia Pereira Castro, and Wenner Glaucio Lopes Lucena

Federal University of Paraíba, Cidade Universitaria, Castelo Branco, João Pessoa/Paraíba, Brazil

claramonise.s@hotmail.com, wenner@ccsa.ufpb.br

Abstract. Objective: To analyze the impact of CEO overconfidence on firm value in companies listed on B3. **Methodology/Approach:** A quantitative and descriptive approach was used, based on non-financial companies listed on B3, observed over the period from 2010 to 2022. The proxy used for CEO overconfidence is overinvestment, obtained from the residuals of a regression of asset growth on sales. As for firm value, the proxies used were Market-to-Book and Tobin's Q. The data were analyzed using panel data regression with unbalanced data and random effects. **Relevance:** Considering that some of the factors influencing firm value are not only presented in financial statements, but are also related to corporate aspects as well as the profile and characteristics of their investors and managers, this research aims to contribute to addressing this issue by seeking a better understanding of overconfidence and the potential prevention of its effects, especially regarding the frequent use of this cognitive bias in the decision-making process. **Main Results:** The results show that CEO overconfidence has a positive and significant influence on firm value. Based on the findings, it can also be inferred that companies whose CEOs exhibit a high level of overconfidence may prioritize profitability and use debt financing as a strategy to boost their operations. **Theoretical Contributions:** The findings of this study contribute to the literature by empirically confirming the relationship between CEO overconfidence and the market value of firms, which may support behavioral and psychological theories regarding the role of executives in making financial and strategic decisions. This strengthens the understanding of the influence of human behavior on corporate finance. **Managerial Contributions:** The study may also contribute to a better understanding of overconfidence as a psychological and behavioral phenomenon relevant to business performance, which can help identify its effects in different contexts and sectors, in addition to suggesting implications for management and corporate governance.

Keyword: Behavioral Finance · Decision-Making Process · Cognitive Biases

L. Almeida et al. (Eds.): ICAFI 2025, ISEM 80, pp. 616–628, 2026.
https://doi.org/10.1007/978-3-032-16268-7_36

1 Introduction

In the financial environment, where decisions often involve uncertain or probabilistic events, individuals are subject to cognitive biases that may affect their decision-making processes [24]. These biases frequently impair decision quality and organizational performance. Among the most influential in strategic decision-making is the overconfidence bias [13].

Overconfidence is defined as an overestimation of one's own ability to make accurate predictions or as an inflated self-assessment in comparison to others. This bias can lead to the overvaluation of businesses in merger and acquisition (M&A) decisions, the emergence of competitive blind spots, and the pursuit of unrealistic goals. In terms of risk-taking behavior, overconfidence has been associated with investments in high-risk product innovation [13].

A growing body of literature has explored the influence of overconfidence among investors, managers, and Chief Executive Officers (CEOs) on decision-making within firms [6, 17, 20, 25], examining its effects on reporting practices, as well as accounting and operational choices [11]. For instance, overconfident CEOs are often hired to lead challenging and risky investments that risk-averse CEOs might avoid, as they tend to take on difficult tasks over easier ones [6].

In this context, the importance of research examining the relationship between CEO overconfidence and various aspects of firm performance becomes clear [23]. One such aspect is firm value, a key metric in corporate management due to its implications for market efficiency, capital allocation, and investment decisions [20].

Decision-makers in positions of power are more likely to overestimate the future performance of their organizations. This power-induced overconfidence has been shown to have a positive effect on overall organizational performance [25]. Consequently, overconfident managers can play a crucial role in enhancing firm value, particularly in M&A contexts, where CEO overconfidence has been positively correlated with post-merger firm value [23].

However, overconfident executives may exhibit a preference for short-term external projects, increased debt financing, and reduced productivity, potentially leading to slower firm growth. These negative effects can be mitigated through strong corporate governance and strict regulatory frameworks [20]. Overconfident CEOs often believe their investment decisions align with shareholder interests. Regardless of their compensation structures, they tend to maintain an overly optimistic outlook, which is frequently reflected in their investment choices [22].

Against this backdrop, corporate finance literature has shown increasing interest in understanding how CEO characteristics and behavioral biases affect corporate decision-making. Motivated by this literature, the present study asks: What is the impact of CEO overconfidence on the market value of companies listed on B3 (the Brazilian stock exchange)? Thus, the objective of this study is to analyze the effect of CEO overconfidence on firm value among publicly listed companies in Brazil.

Overconfidence is a psychological bias that leads individuals to overestimate their performance and predictive accuracy [17, 20]. Existing literature has examined the role of overconfidence in the decision-making processes of investors, analysts, and executives. Yet, as [20] note, few empirical studies have explored the direct relationship between CEO overconfidence and firm value. While prior studies have analyzed certain corporate aspects influenced by overconfident CEOs, research explicitly connecting overconfidence to firm value remains limited. By investigating this relationship, the present study contributes to the literature by exploring how a CEO's behavioral traits may influence firm value [19].

This research holds relevance for managers, investors, corporate decision-makers, as well as academics and researchers in the field. Previous studies have indicated that CEO and managerial overconfidence significantly affects firm value and corporate decision-making [16, 20, 23].

Recognizing that many of the factors influencing firm value go beyond financial statements and are rooted in corporate governance and the profiles of executives and investors, this study aims to deepen our understanding of CEO overconfidence and explore possible ways to mitigate its effects, especially in terms of its frequent role in decision-making processes.

Finally, the findings of this study may offer valuable insights for both scholars and practitioners by shedding light on how CEO overconfidence influences corporate decision-making and, ultimately, firm investment efficiency.

The remainder of this paper is structured as follows. The next section provides a comprehensive review of the literature on CEO overconfidence and firm value, which serves as the basis for the development of the research hypotheses. This is followed by a detailed description of the methodological procedures employed in the study. The subsequent section presents the analysis and discussion of the empirical findings. Finally, the paper concludes with a summary of the key results, practical and theoretical implications, and suggestions for future research.

2 Literature Review

2.1 Overconfidence

Modern Finance Theory was developed under the assumption of rational economic agents. In contrast, behavioral finance argues that some financial phenomena can be plausibly understood using models in which some agents are not fully rational [3].

Since the seminal study by Tversky and Kahneman [24], this field of finance has sought to analyze how mental shortcuts, also known as heuristics or cognitive biases, can interfere with the behavior and decisions of agents operating in the financial market [3, 7]. These mental shortcuts are useful in decision-making but may produce systematic (non-random) thinking errors when judgment deviates from what would be correct based on formal logic.

According to the upper echelons theory, the experiences, values, traits, and characteristics of a company's top executives influence their interpretations of situations and decisions, thus impacting organizational outcomes [9]. Therefore, individual characteristics, traits, and psychological attributes of CEOs play a fundamental role in identifying and recognizing the company's market status, industry position, and decisions of critical importance [8].

The upper echelons theory posits that the individual characteristics of top executives influence strategic decisions in organizations (Hambrick & Mason, 1984). Based on this theory, empirical research has shown that CEO overconfidence can play a significant role in formulating various strategic decisions [19].

Overconfidence is defined as an individual's tendency to overestimate their abilities and knowledge, increasing the belief that they are better equipped to control events and solve problems than others [20]. The term overconfidence encompasses at least three distinct phenomena: overestimation, overplacement, and overprecision [15].

Overestimation refers to individuals being overly confident about their absolute ability or performance, i.e., overestimating personal outcomes; overplacement is closely related to the better-than-average effect, describing the tendency to see oneself as above average in various situations; and overprecision occurs in estimating unknown values, particularly regarding confidence intervals around potential outcomes [14].

Relevant literature has studied how overconfidence can lead to flawed managerial behaviors capable of affecting investment and financing decisions, as well as accounting choices [12].

One study investigated the influence of overconfidence and optimism on the capital structure of Brazilian firms with diverse boards, identifying that managers in organizations lacking corporate governance mechanisms exhibited higher levels of overconfidence and optimism [21]. The research also found that implementing corporate governance mechanisms may limit managerial overconfidence and optimism.

In this context, the influence of managerial overconfidence on earnings management practices in Brazilian firms was analyzed. The results showed that managers with lower overconfidence, who believe less intensely in their abilities, engage in earnings management more opportunistically to meet analysts' expectations [12].

Moreover, the impact of managerial overconfidence on the capital structure of firms listed on Bovespa and NYSE between 2010 and 2015 was analyzed. The results indicated a significant relationship between managerial overconfidence and capital structure in both Brazil and the United States. It was also found that as overconfidence levels increased, so did its impact on leverage decisions [7].

Another study investigated the effect of CEO overconfidence on return on equity (ROE) in U.S. publicly traded companies from 1992 to 2015, a period including the dot-com bubble burst and the 2008–2009 Great Recession. The results revealed that economic expansions strengthen the positive relationship between overconfident CEOs and firm performance, while recessions weaken this effect [17].

Another perspective explored whether overconfident managers manipulate levels of operating cash flow for external reporting purposes. Using 1,231 firm-year observations from 2007 to 2018 in South Korean capital markets, the study found that overconfident managers adjust negative operating cash flows toward more positive outlooks more often than rational counterparts. Additionally, they were more likely to engage in discretionary adjustments of operating cash flow [27].

In the Brazilian context, the influence of managerial overconfidence on corporate strategy in publicly traded companies was assessed. The study employed six strategic dimensions and three different indicators to measure overconfidence (a general confidence index, an overconfidence index, and a dummy variable for entrepreneurial managers), using multiple regressions estimated by OLS on an unbalanced panel of 244 firms from 2011 to 2017. Among the six dimensions analyzed, advertising intensity, R&D

intensity, fixed asset investment, expense structure, working capital management and production cycle, and capital structure, only three were directly influenced by managerial overconfidence: fixed asset investment, working capital management and production cycle, and capital structure [2].

The relationship between leader power and overconfidence in the corporate context 25] was investigated testing the exogenous effects of leader overconfidence on firm performance. The study used a sample of 733 U.S. public company CEOs from all sectors between 2006 and 2011. The results showed a statistically and economically significant positive influence of overconfidence on firm performance [.

The impact of CEO overconfidence on corporate social responsibility (CSR) was also examined using a sample of firms listed in Taiwan from 2010 to 2018. The empirical results revealed that overconfident CEOs tend to avoid engagement in CSR [5].

2.2 Firm Value

The concept of firm value is fundamental in finance due to its implications for market efficiency, capital allocation, and investment decisions, serving as a key indicator for both investors and managers [20]. Financial information disclosed by firms is often used as the basis for firm valuation. However, some factors that influence firm value are not captured solely by financial statements, they also relate to corporate aspects and the profiles and characteristics of investors and managers [20, 26]. Overconfidence is a psychological bias that can influence decision-making by managers and investors [17, 20].

Continuing the analysis of behavioral impacts on corporate finance, the literature explores the relationship between managerial overconfidence and firm value. For instance, Ahmed and Duellman found a negative relationship between CEO overconfidence and accounting conservatism, indicating that overconfidence may reduce firm value by compromising accounting quality [1].

Similarly, Chen and Lin investigated the relationship between managerial optimism, investment efficiency, and firm valuation. The results indicated that underinvestment in firms with highly optimistic CEOs can be mitigated, thereby increasing investment efficiency and enhancing firm value. However, for firms prone to overinvestment, there was insufficient evidence to suggest that less optimistic CEOs improved investment efficiency or firm value [4].

Nikolic and Yan examined the impact of investor overconfidence on firm value and corporate decisions in U.S. equity mutual funds between 1988 and 2010. The results indicated that investor overconfidence has a stronger impact on the value of firms with higher arbitrage risk. Moreover, firms with more confident investors tend to issue more equity and undertake more investments. Overall, the study suggested that investor overconfidence significantly affects firm value and corporate decisions [16].

Complementarily, a study examined the effect of both investor and managerial overconfidence on firm value using a dynamic panel of 432 companies listed on the Pakistan Stock Exchange between 2013 and 2017. The empirical evidence revealed that overconfident investors positively and significantly drive firm value in the Pakistani stock market. However, managerial overconfidence lacked clear guidance to explain firm value [20].

Another study examined managerial overconfidence in Chinese listed companies, offering insights into the characteristics of management candidates, particularly in corporate M&A strategies. The study provided empirical evidence that CEO overconfidence is positively related to firm value. [23].

Another study analyzed possible associations between CEO overconfidence, customer satisfaction, and firm value using a dataset of U.S. companies from 1994 to 2019. The results indicated that overconfident CEOs are incentivized to make discretionary investments in strategic activities to improve customer satisfaction, thereby increasing firm value [19].

Finally, it was also investigated whether CEO overconfidence increases risk-taking and affects firm value in firms listed on the three major U.S. stock exchanges. The results showed that CEO overconfidence is significantly and positively associated with additional risk-taking and firm value, suggesting that overconfidence motivates corporate investments that enhance firm value [22].

2.3 Research Hypothesis Development

CEO overconfidence has received significant attention in recent years, especially within the field of behavioral corporate finance. Several studies suggest that overconfident CEOs can improve firm performance, particularly when their risk-taking behavior helps alleviate underinvestment issues [19, 25]. According to this view, a moderate level of CEO overconfidence may encourage bold, yet calculated, investments that enhance firm value.

However, overconfidence also carries potential downsides. Overconfident CEOs may make overly optimistic forecasts, underestimate risks, and overrate the likelihood of favorable outcomes [17, 22]. These tendencies can lead to poor decision-making and reduced firm value.

Although there is no consensus in the literature, several empirical studies report a positive and significant relationship between CEO overconfidence and firm value [4, 16, 19, 20, 22, 23, 25].

Based on this body of research, the present study proposes the following hypothesis:

H1 *There is a Significant and Positive Relationship Between CEO Overconfidence and Firm Value.*

3 Methodological Procedures

3.1 Sample Composition and Study Period

Given the corporate universe, the study selected non-financial companies with shares traded on the Brazilian Stock Exchange (B3). Financial sector firms were excluded from the sample due to their unique accounting practices and distinct business structures compared to other sectors. This exclusion aims to avoid any bias that could distort the analysis of the results. Companies that did not provide the necessary information for estimating the proposed econometric model were also excluded. The final sample consists of 378 (three hundred and seventy-eight) companies.

The selection of companies was based on convenience and the availability of information. The analysis period covers the years from 2010 to 2022, chosen because it represents the post-IFRS adoption phase.

3.2 Research Variables

Firm value is the dependent variable of this study. Based on prior studies, the following proxies were used to measure firm value (Table 1):

Table 1. Description of Dependent Variables

Variable	Proxy	Explanation
Firm Value	*Market-to-book*	Market value of equity divided by the book value of equity
	QTobin	Market value of equity plus total debt divided by total assets

The independent variable is CEO overconfidence. To construct the overconfidence variable, we adopted the method proposed by Khajavi and Dehghani (2016), using an asset overinvestment-based measure. This approach assumes that the overinvestment variable is derived from the residuals of a regression model of asset growth on sales growth, controlling for year and firm fixed effects. The model is as follows:

$$CAT_{it} = \beta_0 + \beta_1 CV_{it} + \mu \quad (1)$$

where CAT_{it}, CV_{it}, and μ_{it} represent, respectively, asset growth, sales growth, and the residuals of the regression at the end of period t for firm i. According to Khajavi and Dehghani (2016), positive residuals indicate excess investment in assets.

The control variables in this study are cash flow to assets, leverage, firm size, and return on assets. These controls were selected based on prior literature on managerial overconfidence, as shown in Table 2:

Table 2. Control Variables

Variable	*Proxy*
Cash Flow	Natural logarithm of the operating cash flow of firm i at time t
Leverage	Ratio of total debt to equity for firm i at time t
Firm Size	Natural logarithm of total revenue of firm i at time t
Return on Assets	Net income divided by total assets for firm i at time t

The data were collected from the Economática database and company-disclosed reports.

3.3 Data Treatment and Analysis

Given that the research spans from 2010 to 2022, the panel data regression model is appropriate, as it combines cross-sectional data (publicly listed companies on B3) and time-series data (13 years).

To test the effect of CEO overconfidence bias on firm value, the following regression models were estimated:

$$MTB_{it} = \beta_0 + \beta_1 Over_{it} + \beta_2 FC_{it} + \beta_3 ALAV_{it} + \beta_4 TAM_{it} + \beta_5 ROA_{it} + \mu_{it} \quad (2)$$

$$QTobin_{it} = \beta_0 + \beta_1 Over_{it} \beta_2 FC_{it} + \beta_3 ALAV_{it} \beta_4 TAM_{it} \beta_5 ROA_{it} + \mu_{it} \quad (3)$$

Onde:
MTB = *Market-to-Book* ratio for firm i at time t
$QTobin$ = Tobin's Q for firm i at time t.
$Over$ = *proxy* for CEO overconfidence for firm i at time t
FC = Cash flow for firm i at time t
$ALAV$ = Leverage for firm i at time t
TAM = Firm size for firm i at time t
ROA = Return on assets for firm i at time t
μ_{it} = Error term

This study employs a static panel data regression model due to the nature of the research objective, which is to investigate the cross-sectional and temporal relationship between CEO overconfidence and firm market value, without explicitly modeling dynamic feedback or persistence over time. While dynamic panel models (e.g., GMM) are appropriate when addressing potential endogeneity caused by lagged dependent variables, such complexity was deemed unnecessary for the current analysis. Additionally, the use of static panel models allows for more straightforward interpretation of coefficients, particularly when the primary focus is on estimating the contemporaneous effects of CEO behavioral traits on firm value. Diagnostic tests did not indicate severe autocorrelation or dynamic effects that would justify the need for a dynamic specification.

4 Data Analysis and Discussion

Initially, a descriptive statistical analysis of the variables was performed to examine the behavior of the data used in the research and identify possible trends.

From the data in Table 3, it can be seen that the variables Market-to-Book (MTB) and Tobin's Q (QTobin) showed the highest standard deviations. This suggests that these proxies exhibited the greatest dispersion around the mean, indicating considerable variation in their values. Thus, the mean may not adequately reflect the sample reality of these variables. The maximum and minimum values corroborate this significant dispersion. Among them, the Market-to-Book variable shows the greatest dispersion in data relative to the mean, with a standard deviation of 1.900.000 and a maximum value of 9.960.000. In turn, Tobin's Q has a standard deviation of 1.250.000.

Table 3. Descriptive Statistics of the Variables

Variables	Obs	Mean	Minimum	Maximum	Std. Dev
MTB	2.157	1.730.000	0,000876	9.960.000	1.900.000
QTobin	2.157	8.335.906	0,001536	9.979.000	1.250.000
Over	2.157	0,1325916	0	1	0,3392114
FC	2.157	10,8949	33,1734	-2,2936	5,2053
ALAV	2.157	484.604	0	9.612.629	1.082.341
TAM	2.157	499,2356	0,0000845	1.076.706	23.183,13
ROA	2.157	1.407.250	11,39	9.973.433	530.450,3

Regarding managerial overconfidence (Over), the average is 0,13, with values ranging from 0 to 1, and a considerable dispersion around the mean, as indicated by the standard deviation of approximately 0.3392. Based on this result, it can be inferred that the CEOs of the companies in this study are not excessively overconfident, since the average overconfidence level is 0,13 or 13%.

To test the hypothesis proposed in this study, the proposed equations were analyzed using an unbalanced panel data regression. Breusch-Pagan, Chow, and Hausman tests were performed to identify the most appropriate regression model. These tests indicated that the random effects panel regression model was the most suitable for the proposed econometric models, as shown in Table 4:

Table 4. Model Validation Tests for Estimation

Testes	Model with MTB		Model with QTobin	
	Test Statitics	Model Chosen	Test Statitics	Model Chosen
Chow	Prob > F = 0,0000	Fixo	Prob > F = 0,0000	Fixo
Breusch-Pagan (LM)	Prob > χ^2 = 0,000	Aleatório	Prob > χ^2 = 0,000	Aleatório
Hausman	Prob > χ^2 = 0,61267	Aleatório	Prob > χ^2 = 0,13212	Aleatório

Additional tests were conducted to validate the assumptions of the regression models used. The Jarque-Bera test was used to verify the normality of residuals. The results for both the Market-to-Book (p-value = 0) and Tobin's Q (p-value = 0) models reject the null hypothesis of normality, indicating that the error distribution is not normal. Despite this violation of the normality assumption, it does not necessarily invalidate the model in large samples, as it does not affect the consistency of the estimates due to the substantial amount of information available (Gujarati & Porter, 2011).

Multicollinearity among the explanatory variables was also examined using the Variance Inflation Factor (VIF) test. The results did not show signs of multicollinearity, as the overall VIF was 1.03 and none of the explanatory variables exhibited multicollinearity. The Durbin-Watson test was also performed to verify the existence of positive autocorrelation among the residuals. The results for the Market-to-Book model (Durbin-Watson = 1,28717) and Tobin's Q model (Durbin-Watson = 1,174765) do not reject the null hypothesis of positive autocorrelation in the residuals.

Table 5. Random Effects Panel Data Regression Models.

Variáveis	*Market-to-Book*		QTobin		VIF
	Coefficient	p-value	Coefficient	p-value	
Over	**3,8944***** (4,044)	0,0000	**2,6223***** (3,955)	0,0000	1,00
FC	**-0,0302** (-0,1932)	0,8468	**0,0915** (0,8513)	0,3946	1,00
ALAV	**2,0481***** (5,469)	0,0000	**-0.2903** (-1,140)	0,2544	1,01
TAM	**0,2283** (0,5875)	0,5569	**0,8624***** (3,251)	0,0066	1,03
ROA	**35,6398**** (2,341)	0,0192	**28,3640***** (2,698)	0,0012	1,03
Const	**1,6710***** (16,05)	0,000	**7,1576***** (10,69)	0,000	
R^2	0,02841		0,0097		
Prob > F	0,000		0,000		
Obs	2.157		2.157		
Jarque-Bera	0		0		
Durbin Watson	1,28717		1,174765		

* Note: *, *, and *** denote statistical significance at the 10%, 5%, and 1% levels, respectively.

Based on the analysis of Table 5, it is observed that CEO overconfidence has a positive and significant relationship with firm value. Both the Market-to-Book (MTB) and Tobin's Q proxies showed significance at the 1% level, with a prob > F = 0.000 in both models. This result suggests that the greater the CEO's overconfidence, the higher the firm value. These results suggest that overconfident CEOs may be more inclined to pursue aggressive investment and growth opportunities, taking calculated risks that can increase firm value over the long term [22].

Still regarding the results and as stated in the literature, overconfident CEOs may be more effective in investing in innovation strategies, which may strengthen the company's competitive position and contribute to its market value [8]. The results may also suggest that overconfidence can lead CEOs to adopt a behavior focused on improving outcomes, which can boost financial performance and, consequently, firm value [17, 25].

As for the control variables, cash flow (FC) did not show a statistically significant relationship with firm value in either model. These findings differ from the results of previous studies, which found a positive and significant relationship between cash flow and the proxies Market-to-Book and Tobin's Q [20].

Leverage (LEV) showed a significant relationship with firm value only in the Market-to-Book (MTB) model. This suggests that firms with higher CEO overconfidence bias tend to focus more on profitability, and thus an increase in profits resulting from debt usage may lead to a proportionally greater increase in firm profitability, enhancing its value [20].

Firm size (SIZE) showed a significant relationship with firm value only in the Tobin's Q (QTobin) model. A positive relationship was identified, consistent with the literature [8, 22]. This may be explained by the fact that larger firms generally have more resources and capacity to invest in research, development, and innovation, enabling them to adapt to market changes, explore new business opportunities, and maintain a long-term competitive position, which may increase their value.

Return on assets (ROA) showed a significant and positive relationship with firm value in both models. This supports the findings of the literature [20, 22] and may be justified by the fact that higher ROA can positively influence projected future cash flows and, therefore, the firm's intrinsic value. A higher ROA also suggests that the company is able to generate greater returns on investment compared to firms with lower ROA, which can be positively interpreted by investors.

In examining the proposed econometric models, it becomes clear that the research hypothesis cannot be rejected, as CEO overconfidence exerts a positive and statistically significant influence on the market value of firms.

5 Conclusions

This study aimed to analyze the impact of CEO overconfidence on firm value among companies listed on the B3 (the Brazilian stock exchange). To achieve this objective, regression models using random effects panel data were employed to identify the influence of CEO overconfidence on firm value.

The results show that CEO overconfidence has a positive and significant effect on firm value. Based on the findings, it can also be inferred that companies led by CEOs with a high overconfidence bias may prioritize profitability and use debt financing as a strategy to boost their operations. This can result in improved financial performance and, consequently, an increase in firm value.

This study makes a relevant contribution to the literature by empirically confirming the relationship between CEO overconfidence and firm market value. The findings support behavioral and psychological theories that highlight the executive's role in strategic and financial decision-making, thereby reinforcing the importance of human behavior in corporate finance. By framing CEO overconfidence as a psychological and behavioral phenomenon, the study advances the theoretical understanding of its impact across different industries and organizational contexts. Furthermore, the results suggest practical implications for corporate management, indicating the necessity of recognizing and managing behavioral biases that can influence firm performance and strategic choices.

In line with previous research, the findings suggest that CEO overconfidence can have far-reaching implications for publicly listed companies in Brazil, particularly in terms of capital structure, performance, and overall market value. However, it is important to note that overconfidence does not always yield positive outcomes; in many cases, it may result in irrational decisions that harm a firm's financial health and reduce its market valuation. From a managerial perspective, the study highlights the need for firms to adopt robust policies and corporate governance practices aimed at the selection, development, and oversight of top executives. By doing so, organizations can mitigate potential risks associated with overconfident leadership and promote more balanced, evidence-based decision-making processes.

One key limitation of this study is the use of a sample restricted to companies listed on the Brazilian stock exchange (B3). As a result, the findings may not be generalizable to firms operating in other markets or regulatory environments. For future research, it is recommended to replicate this study by extending the analysis to a sample that includes companies from other countries, as well as to use alternative proxies to measure CEO overconfidence.

References

1. Ahmed, A.S., Duellman, S.: Managerial overconfidence and accounting conservatism. J. Account. Res. **51**(1), 1–30 (2013). https://doi.org/10.2139/ssrn.2097895
2. Andrade, J.P.d., Lucena, W.G.L.: O excesso de confiança dos gestores e as decisões estratégicas. Estudios Gerenciales, 542–555 (2021). https://doi.org/10.18046/j.estger.2021.161.4208
3. Barberis, N., Thaler, R.H.: A survey of behavioral finance. SSRN Electron. J. (2002). https://doi.org/10.2139/ssrn.327880
4. Chen, I.-J., Lin, S.-H.: Managerial Optimism, Investment Efficiency, and Firm Valuation. Multinat. Fin. J. **17**(3/4), 295–340 (2013). https://doi.org/10.17578/17-3/4-6
5. Chen, R.C.Y., Hung, S.-W., Wei, Y.-C.: CEO Overconfidence and CSR Engagement. Rev. Pac. Basin Financ. Mark. Policies (2022). https://doi.org/10.1142/s0219091522500254
6. Chen, Y.-R., Ho, K.-Y., Yeh, C.-W.: CEO overconfidence and corporate cash holdings. J. Corp. Finan. **62**, 101577 (2020). https://doi.org/10.1016/j.jcorpfin.2020.101577
7. Costa, I.L.d.S., Correia, T.D.S., Lucena, W.G.L.: Impacto do excesso de confiança na estrutura de capital: evidências no Brasil e nos Estados Unidos. Base - Revista de Administração e Contabilidade da Unisinos **16**(2) (2019). https://doi.org/10.4013/base.2019.162.01
8. Eom, M., Jung, M., Park, J.C.: Are overconfident CEOs better able to transform innovation into firm value? —Evidence from the United States. International Finance (2023). https://doi.org/10.1111/infi.12433
9. Hambrick, D.C., Mason, A.: Upper echelons: the organization as a reflection of its top managers. Acad. Manag. Rev. **9**(2), 193–206 (1984). https://doi.org/10.2307/258434
10. Khajavi, S., Dehghani, G.: Board Characteristics and Managerial Overconfidence in an Emerging Market. Int. J. Econ. Fin. **6**(2), 529–537 (2016). https://www.econjournals.com/index.php/ijefi/article/view/2118
11. Kouaib, A., Jarboui, A.: The mediating effect of REM on the relationship between CEO overconfidence and subsequent firm performance moderated by IFRS adoption: A moderated-mediation analysis. Res. Int. Bus. Financ. **42**, 338–352 (2017). https://doi.org/10.1016/j.ribaf.2017.07.034

12. Magro, C.B.D., Gorla, M.C., Klann, R.C.: Excesso de Confiança do Chief Executive Officer e a Prática de Gerenciamento de Resultados. Revista Catarinense da Ciência Contábil **17**(50), 55–70 (2018). https://doi.org/10.16930/2237-7662/rccc.v17n50.2372
13. Meissner, P., Schubert, M., Wulf, T.: Determinants of group-level overconfidence in teams: A quasi-experimental investigation of diversity and tenure. Long Range Plan. **51**, 927–936 (2018). https://doi.org/10.1016/j.lrp.2017.11.002
14. Merkle, C.: Financial overconfidence over time: Foresight, hindsight, and insight of investors. J. Bank. Finance **84**, 68–87 (2017). https://doi.org/10.2139/ssrn.2001513
15. Moore, D.A., Healy, P.J.: The trouble with overconfidence. Psychol. Rev. **115**(2), 502–517 (2008). https://doi.org/10.1037/0033-295x.115.2.502
16. Nikolic, B., Yan, X.: Investor overconfidence, misvaluation, and corporate decisions. Working paper (2014)
17. Reyes, T., Vassolo, R.S., Kausel, E.E., Torres, D.P., Zhang, S.: Does overconfidence pay off when things go well? CEO overconfidence, firm performance, and the business cycle. Strategic Organization, 1–31 (2020). https://doi.org/10.1177/1476127020930659
18. Ribeiro, J.E., Souza, A.A.d.:. Índice de governança corporativa e desempenho de mercado: evidências no mercado acionário brasileiro. Revista Contabilidade & Finanças **34**(92) (2023). https://doi.org/10.1590/1808-057x20231756.pt
19. Sharpe, W.H., Nguyen, P.M., Colombage, S.: CEO overconfidence, customer satisfaction, and firm value: An investigation of mediating and moderating effects. Eur. Manag. J. (2023). https://doi.org/10.1016/j.emj.2023.11.004
20. Shah, S.S.H., Xinping, X., Khan, M.A., Harjan, S.A.: Investor and manager overconfidence bias and firm value: micro-level evidence from the Pakistan equity market. International Journal of Economics and Financial Issues **8**(5), 190–199 (2018). https://www.econjournals.com/index.php/ijefi/article/view/7028
21. Silva, T.B.J., Cunha, P.R., Ferla, R.: Excesso de Confiança e Otimismo sobre a Estrutura de Capital de Firmas Brasileiras com Diversidade no Conselho de Administração. Revista Mineira de Contabilidade **18**(3), 27–39 (2018). https://revista.crcmg.org.br/rmc/article/view/692
22. Tang, H.-W., Chang, C.-C.: CEO overconfidence, risk-taking, and firm value: Influence of incentive compensation and financial constraints. The North American Journal of Economics and Finance, 102034 (2024). https://doi.org/10.1016/j.najef.2023.102034
23. Tang, C., Lee, Y., Lee, M.: CEO Characteristics Enhancing the Impact of CEO Overconfidence on Firm Value After Mergersand Acquisitions - A Case Study in China. Rev. Pacific Basin Fin. Mark. Policies **23**(1) 2050003 and 19 (2020). https://doi.org/10.1142/s0219091520500034
24. Tversky, A., Kahneman, D.: Judgment under uncertainty: Heuristics and biases. Science **185**(4157), 1124–1131 (1974). https://doi.org/10.1126/science.185.4157.1124
25. Vitanova, I.: Nurturing overconfidence: the relationship between leader power, overconfidence and firm performance. The Leadership Quarterly **32**, Article 101342 (2021). https://doi.org/10.1016/j.leaqua.2019.101342
26. Yang, T., Yang, Y., Chen, J., Lu, C.: Correlation between owner brand and firm value – Case study on a private brand in Taiwan. Asia Pacific Management Review **24**(3), 232–237 (2019). https://doi.org/10.1016/j.apmrv.2018.06.002
27. Yang, D., Kim, H.: Managerial overconfidence and manipulation of operating cash flow: Evidence from Korea. Financ. Res. Lett. **32**, 101343 (2020). https://doi.org/10.1016/j.frl.2019.101343

Which Skills Have Been Required by the Accountants' Employers in Portugal via LinkedIn?

Alexandra Domingos[1](✉), Fábio Albuquerque[1,2](✉), and Ridney Cláver[1]

[1] Lisbon Accounting and Business School (ISCAL), Instituto Politécnico de Lisboa, Lisbon, Portugal
{fhalbuquerque,fhalbuquerque}@iscal.ipl.pt
[2] CICF, School of Management, IPCA, Barcelos, Portugal

Abstract. The frequent contextual changes in which the profession of certified accountant develops, as well as the advent of new recruitment platforms, justify a continuous re-analysis of the skills applicable to the job market. This study analyzes the differences between the soft skills and hard skills required by employers in job advertisements on LinkedIn for the role of a certified accountant. An exploratory and quantitative methodology is used. Through content analysis, 161 job advertisements published on LinkedIn were analyzed during the period from April 2021 to August 2022. Skills were divided into hard skills and soft skills. Data analysis was performed using non-parametric statistical tests, namely the Wilcoxon, and Mann-Whitney tests. The Wilcoxon test revealed significant differences between skills related to both soft and hard ones. In turn, the Mann-Whitney test showed that technology-related entities require more interpersonal skills compared to personnel recruitment and selection entities, which prioritize primarily technical skills. Hard skills continue to play a more prominent role in the first phase of the recruitment process. Instrumental skills stand out in the soft skills, while proficiency in a second language, tasks supporting the preparation of financial statements and information technology skills are particularly relevant in the context of hard skills. This study may contribute to Higher Education Institutions aligning the skills developed by students with those required by the job market, promoting activities or curricular changes to improve their integration into it and increase the future professional success of accounting students.

Keywords: soft skills · hard skills · chartered accountants · LinkedIn · recruitment

1 Introduction

Traditionally, the role of the accounting professional was centred on providing tax and financial information [1]. However, factors such as the advancement of information technologies, new management strategies, and an emphasis on quality and customer service have driven changes in their functions [2, 3].

L. Almeida et al. (Eds.): ICAFI 2025, ISEM 80, pp. 629–643, 2026.
https://doi.org/10.1007/978-3-032-16268-7_37

The literature highlights that accounting professionals should possess skills beyond the technical aspects of the profession [2–6], including technological skills, particularly valued in the field of accounting [7].

The growing digital transformation has been altering the skills and personal qualities required of accounting professionals [8, 9], considering the widespread adoption of information technologies and the need for advanced skills in this field [10, 11]. The implementation of modern technologies has increased the value placed on Information Technology (IT) skills by employers [6]. The author emphasizes the consensus on the importance of developing soft skills, such as interpersonal skills, communication, and teamwork.

Despite the lack of consensus on employability-promoting skills [11], two major types stand out: soft and hard skills [12, 13]. While hard skills are crucial for professional performance, the evolution of business models suggests that their development may not be sufficient to effectively address the challenges of the job market [13]. Nevertheless, the authors concluded that employers prioritize the development of soft skills over hard skills.

The increasing use of digital platforms, especially in the field of accounting, has become more common in the search and hiring of professionals. According to [14], social networks, with LinkedIn being particularly prominent (used by 97% of employers), allow for anticipating the performance of future collaborators.

Thus, the present study formulated the following research question: What are the skills that organizations seek when hiring certified accountants? To address this question, this research intends to assess the differences between the soft skills and hard skills required by employers in job advertisements posted on LinkedIn for the role of certified accountant. In this regard, the content of 161 job advertisements was analyzed between April 2021 and August 2022, following the approach of previous studies [7, 11]. The choice of LinkedIn is justified by its significant growth and crucial role in digital communication, recruitment, and personnel selection [15].

In a context of changes and advancements that modify accounting processes, the relevance of this study is grounded in the following reasons:

1. The identification of skills remains a current topic due to its constant evolution.
2. In the digital era, significant changes in the accounting profession both in Portugal and worldwide.
3. In Portugal, there is a scarcity of recent studies on the skills of accounting professionals aligned with the demands of the job market in this field.
4. The evolution of the economy and the increasing adoption of emerging technologies have altered the needs of the job market, requiring rapid information circulation and data sharing.
5. Higher education offerings seek to adapt curricula to the needs of employing organizations.

This study aims to contribute to the understanding of essential skills for accounting professionals, preparing them for the challenges of the job market. The scarcity of existing studies in the literature emphasizes the importance of assessing skills from job advertisements, which can also be useful for adapting the curricula of Higher Education Institutions (HEIs), promoting greater alignment with the demands of the job market.

The paper article is structured into five sections, in addition to this introduction. The second presents the literature and hypotheses underlying the development of the study. The third describes the methodology used. The fourth presents the findings, followed by the fifth section which provides their critical considerations. Finally, the last section presents the main conclusions and research limitations, also providing suggestions for future investigations.

2 Literature and Hypotheses

The growth of competitiveness in the job market has driven the use of social networks to seek and retain qualified talent [16]. The authors emphasize that talent management is a strategic activity, leveraging technological advancements to identify candidates efficiently and with a significant reduction in costs.

Recruitment via social networks, especially on LinkedIn, is already a frequent practice, with candidates actively managing their profiles being highlighted as a differential in displaying their skills [15].

Modern recruitment is internet-based, enabling employers to anticipate the profiles of future collaborators, considering the requirements of available positions [14]. Additionally, job advertisements serve as reliable indicators for quick and targeted research, providing immediate insights into the skills needed in the workplace [7]. Despite some studies in this field [7, 11, 17], literature in accounting dedicated to the analysis of LinkedIn job postings to identify recent skill trends is still in its initial stages.

The increase in competitiveness demands not only the development of technical skills (hard skills) but also the enhancement of intellectual and analytical skills, as well as communication and organizational collaboration (soft skills), crucial for business management roles [5]. The employers' talent search reflects a current balance between these two typologies [7, 11, 18]. Therefore, considering the recognized importance in the literature for both types of skills for accounting professionals, the first hypothesis was formulated as follows:

- Hypothesis 1. There are no significant differences between the number of soft and hard skills required by employers seeking certified accountants via LinkedIn.

Despite the previous hypothesis, factors such as technological advancement, globalization, linguistic diversity, and the expansion of knowledge are significant drivers of changes in contemporary accounting functions. Adapting to these realities requires instrumental skills and interpersonal relationships to promote collaboration among multidisciplinary teams [3].

Therefore, the literature highlights the importance of communication skills, organization, and group cooperation as soft skills valued by employers [19]. Additionally, current challenges in accounting functions emphasize the importance of continuous planning and analysis, requiring the development and maturity of instrumental skills for professional success [9, 20]. Thus, the second research hypothesis, focused on soft skills, was formulated as follows:

- Hypothesis 2. In the scope of soft skills, instrumental ones are the most required by employers seeking certified accountants via LinkedIn.

Despite technological advancements, which require skills in information technology and management to address current challenges [5], the importance of accounting functions remains tied to traditional methods of mastering management tools and data analysis [21].

Regarding hard skills, the demand for more immediate results sometimes requires the use of traditional tools and methodologies, facilitating objective measurement and promoting efficient integration and collaboration in the field of accounting [22]. Additionally, the need for the development of specific skills in accounting, aiming to enhance short-term efficiency, requires the job market to recognize hard skills more linked to the specific performance of these functions [8].

This market trend, coupled with the increasing demand for greater professional experience by employers in the accounting and consulting sector, along with the importance of specific training for the development of technical skills necessary for the role of a certified accountant, led to the formulation of the third research hypothesis:

- Hypothesis 3. In the scope of hard skills, core technical ones are the most required by employers seeking certified accountants through LinkedIn.

The next section provides the methodological approaches for conducting this study.

3 Methodology

To identify the most valued skills in the job market, the content of job advertisements on LinkedIn was analyzed, following the approach of previous studies [7, 11, 17]. The research was conducted during the period between April 2021 and August 2022, using the following criteria:

- Job Title: Certified Accountant;
- Type of Position: Full-time;
- Location: Lisbon and Porto;
- Industry Sector: Employment-related services; accounting and consulting; trade, industry, and services; technologies.

The search for job advertisements was restricted to Lisbon and Oporto since those metropolitan areas concentrate the majority of job opportunities in Portugal, especially in key sectors such as technology, financial services and industry. Due to the nature and number of job advertisements and the great diversity of companies in these regions, it is possible to obtain a comprehensive view of the trends and demands of the job market. This diversity ensures that the findings from this research can be useful to a wide spectrum of professional contexts and even to other relevant regions outside Portugal, which explains this choice as the sample criterion for the study.

At the end of the selection process, the data were obtained from the identification and analysis of 161 job advertisements from 97 distinct entities, which were grouped according to the economic sector for subsequent analysis, as follows:

- Group 1 (G1) - comprised of 25 entities in the employment-related services sector, totalling 69 advertisements;
- Group 2 (G2) - comprised of 42 entities in the accounting and consulting services sector, totalling 51 advertisements;
- Group 3 (G3) - comprised of 15 entities in the industry, trade and services sector, totalling 18 advertisements;
- Group 4 (G4) - comprised of 15 entities in the technology sector, totaling 23 advertisements.

Duplicate advertisements were removed to ensure a valid and reliable dataset, following the proposed selection criteria. Subsequently, the analysis of soft skills in the accounting field was conducted based on the study by [20].

Therefore, for the analysis of soft skills, a set of 27 items (S1 to S27) that encompass these skills was selected, considering the 30-item model by [20], based on the survey used in the Tuning project. This model was chosen due to its previous validation through the consultation process involving employers, graduates (between three and five years), and European-level faculty [20]. These 27 items were subdivided into three groups (instrumental, interpersonal, and systemic skills) proposed in the mentioned study, namely:

- Instrumental skills (composed of 7 items: S1 to S7);
- Interpersonal skills (composed of 8 items: S8 to S15);
- Systemic skills (composed of 12 items: S16 to S27).

The exclusion of three items proposed in this study, namely "mastery of basic professional knowledge," "knowledge of a second language," and "basic computer skills," from the scope of analysis of soft skills within the instrumental skills in the Tuning project, was motivated by the fact that such elements are typically required in the context of the accounting profession as technical skills (hard skills). This exclusion avoids potential double-counting effects.

Table 1 presents the set of items that comprise the soft skills under analysis. As for hard skills, the literature doesn't mention specific categories in the field of accounting. Unlike soft skills, the classification of hard skills was not based on previous categorizations, but rather directly observed the skills demanded by employers in the selected job advertisements for the sample. As technical skill items, those related to the need for skills in performing more specific tasks and linked to the practice of the profession were selected, as advocated by [23].

Table 1. Items classified as soft skills.

Item	Description
S1	Analytical and synthesis skills
S2	Organizational and planning skills
S3	Basic general knowledge
S4	Oral and written communication in the native language
S5	Information management skills (ability to compile and analyze information from different sources)
S6	Problem-solving skills
S7	Decision making skills
S8	Critical and self-critical ability
S9	Teamwork skills
S10	Interpersonal skills
S11	Ability to work in an interdisciplinary team
S12	Ability to communicate with specialists in other areas
S13	Appreciation of diversity and multiculturalism
S14	Ability to work in an international context
S15	Ethical commitment
S16	Ability to apply knowledge
S17	Research skills
S18	Ability to learn
S19	Adaptability to new situations
S20	Ability to generate new ideas (creativity)
S21	Leadership
S22	Understanding of cultures and habits of other countries
S23	Ability to work independently
S24	Design and project management
S25	Initiative and entrepreneurial spirit
S26	Concern for quality
S27	Will to win

From this observation process, seven items were identified, later subdivided into two groups, namely, core technical skills and secondary technical skills, based on their more direct or supportive connection to the functions traditionally assigned to accounting professionals, respectively [see, notably, 4], identified in the following lines:

- Core technical skills (composed of 3 items: H1 to H3);
- Secondary technical skills (composed of 13 items: H4 to H7).

Table 2 displays the items covering the hard skills analyzed in this study.

Table 2. Items classified as hard skills.

Item	Description
H1	Tasks related to tax obligations and declarations
H2	Tasks supporting the preparation of financial statements
H3	Analysis of financial statements
H4	Proficiency in at least one foreign language
H5	Proficiency in Excel
H6	Proficiency in other tools within the Microsoft Office suite and/or general computer skills (unspecified software)
H7	Proficiency in Management/Accounting software

Following the classification proposed earlier, the items of soft skills and hard skills were synthesized through the creation of indices by groups or categories. This procedure is commonly used in numerous studies on diverse topics in the field of accounting, with a focus on studies on disclosure levels [24–26]. The mentioned methodology involves the use of a dichotomous scoring system, assigning the value "1" when the item is disclosed and "0" in the opposite case. The construction of the indices was based on Eq. 1 presented below.

$$\sum_{j=1}^{n} \frac{ej}{n} \tag{1}$$

where:

ej = Disclosure element j in the index analysis;

n = Maximum score of the analyzed index items.

For the analysis of the proposed hypotheses, and along with preliminary descriptive statistics, non-parametric tests including Wilcoxon, and Mann-Whitney, were used as statistical techniques. The Wilcoxon test is used to assess differences between each of the proposed indices regardless of the subgroups (different pairs of employer groups), while the Mann-Whitney is used to compare each of those indices by subgroups. The results of these tests will be analyzed with a maximum asymptotic (bilateral) significance level of 0.05, assessed by the test statistic, for a confidence level of 95%.

Before applying those tests, the indices related to hard skills and soft skills underwent a normalization process to make them comparable using the "min-max" method [27]. As an example, without normalization, the maximum value obtained for the index that includes all items related to relational skills (soft skills), composed of 27 items, would be 0.70. Through normalization, this result assumes the value "1," which is potentially more comparable with the results obtained for the index that includes all items related to technical skills (hard skills), for which, with a higher likelihood, this index in its non-normalized form reaches unity, and consequently, the impact of normalization is more limited.

The next section is dedicated to the presentation of the findings from this research.

4 Results

Table 3 presents the advertisements collected by the employers' sector, showing the prevalence of recruitment and selection entities (43%), followed by those providing accounting and consulting services (32%).

Table 3. Advertisements by sector.

Group G)	Sector	In Number	In Percentage
1	Employment-related services	69	42.0%
2	Accounting and consulting	51	31.7%
3	Industry, trade and services	18	11.2%
4	Technology	23	14.3%
	Total	**161**	**100.0%**

Table 4 presents the values found for the indices regarding different typologies of soft skills, for the total and by groups. The figures indicate that the index for the total of soft skills is 23%, with instrumental skills accounting for this. By group, it doesn't identify significant differences, either for the total or for the various categories of soft skills.

Table 4. Soft skills indices by category and groups.

Description	G1	G2	G3	G4	T
ISS1: Instrumental skills	30%	30%	31%	31%	32%
ISS2: Interpersonal skills	17%	17%	20%	21%	21%
ISS3: Systemic skills	11%	11%	13%	14%	15%
ISST: Total index for soft skills	20%	20%	22%	22%	23%

Table 5, in turn, details the observed frequencies for the items related to soft skills, by groups.

Table 5. Frequency of soft skills by groups.

Description		G1	G2	G3	G4	T
ISS1: Instrumental Skills	S1	39%	55%	56%	43%	47%
	S2	39%	63%	39%	43%	47%
	S3	20%	10%	6%	26%	16%
	S4	45%	37%	39%	43%	42%
	S5	22%	20%	22%	30%	22%
	S6	16%	20%	17%	13%	17%
	S7	0%	6%	6%	4%	3%
ISS2: Interpersonal Skills	S8	14%	27%	22%	26%	21%
	S9	30%	33%	39%	35%	33%
	S10	13%	22%	11%	26%	17%
	S11	17%	20%	22%	26%	20%
	S12	7%	2%	0%	22%	7%
	S13	1%	6%	0%	0%	2%
	S14	10%	4%	6%	43%	12%
	S15	10%	20%	6%	0%	11%
ISS3: Systemic skills	S16	4%	16%	6%	22%	11%
	S17	4%	4%	6%	9%	5%
	S18	7%	20%	0%	13%	11%
	S19	10%	6%	11%	17%	10%
	S20	1%	4%	6%	9%	4%
	S21	9%	8%	17%	22%	11%
	S22	6%	2%	0%	0%	3%
	S23	13%	18%	39%	22%	19%
	S24	14%	18%	17%	9%	15%
	S25	16%	20%	22%	17%	18%
	S26	1%	6%	0%	0%	2%
	S27	3%	12%	11%	9%	7%

Figures from Table 5 show a predominance for items S1 (Analytical and synthesis skills), S2 (Organizational and planning skills), and S4 (Oral and written communication in the native language), which are part of instrumental skills, identified in more than 40% of the observations. By analysis groups, it is observed that items S1 and S2 are particularly demanded by entities in the sectors of accounting and consulting services (G2) and industry, trade, and services (G3).

In the context of interpersonal skills, the items S8 (Ability to criticize and self-criticize), S9 (Teamwork skills), and S11 (Ability to work in an interdisciplinary team) stand out, observed in more than 20% of cases (with a maximum of 33% for S9). In all three cases, these are skills demanded less frequently by entities from the employment-related services (G1), in contrast to what is observed for entities related to technology (G4).

Finally, items related to systemic skills do not exceed 20%, approaching this threshold. However, items S23 (Ability to work autonomously) and S25 (Initiative and entrepreneurial spirit) come close to this limit. These skills are also less demanded by entities from the employment-related services (G1), with a notable emphasis on S23, and are more frequently identified by entities in the sector of industry, trade, and services (G3).

Following, Table 6 provides the figures for the indices related to the hard skills, for the total and by groups.

Table 6. Hard skills indices by category and groups.

Description	G1	G2	G3	G4	T
IHS1: Core technical skills	71%	71%	70%	72%	65%
IHS2: Secondary technical skills	58%	58%	59%	59%	59%
IHST: Total index for hard skills	64%	64%	64%	65%	62%

Table 6 shows that the index for the total hard skills is 62%, with main technical skills accounting for 65% of this total, while secondary technical skills represent 59%, with no significant differences in the proposed groupings for each category and the total of hard skills.

Table 7, in turn, details the frequencies for the items that comprise the hard skills, by groups, indicating that hard skills related to the main skills H1 (tasks related to tax obligations and declarations) and H2 (tasks supporting the preparation of financial statements) are the most demanded in this context (70%).

By groups, employment-related services (G1) stand out, followed by entities providing accounting and consulting services (G2), in clear contrast to entities from other sectors. Regarding secondary skills, H4 (Proficiency in at least one foreign language) stands out, with over 80% of observations, followed by H7 (Proficiency in Management/Accounting software) with around 66%. The representative groups of the sectors stand out in the same order.

Table 7. Frequency of hard skills by groups.

Description		G1	G2	G3	G4	T
IHS1: Core technical skills	H1	84%	48%	16%	19%	71%
	H2	75%	55%	20%	26%	76%
	H3	55%	26%	13%	19%	48%
IHS2: Secondary technical skills	H4	86%	52%	23%	29%	81%
	H5	57%	32%	12%	20%	52%
	H6	26%	33%	10%	16%	37%
	H7	65%	52%	19%	17%	66%

Table 8 presents the results of the Wilcoxon test for the total sample.

Table 8. Wilcoxon Test, total sample.

Positions		ISS2 - ISS1	ISS3 - ISS2	ISS3 - ISS1	IHS2 - IHS1	IHST - ISST
Ranks (N)	Negatives	76	80	107	88	13
	Positives	33	44	27	64	148
	Ties	52	37	27	9	0
Mean Ranks	Negatives	61.26***	65.95***	73.61***	79.44**	37.50***
	Positives	40.59***	56.23***	43.28***	72.46**	84.82***

Notes: *** Significant at a 1% level; ** Significant at a 5% level

Following, Table 9 shows the differences in those indices between the subgroups comprised of the different pairs of employers.

The figures from Table 9 identify potentially more significant differences (at a 5% significance level) concerning the main technical skills between entities in employment-related services (G1) versus entities offering accounting and consulting services (G2).

No significant differences were found, in turn, between entities in employment-related services (G1) versus entities in the industry, trade, and services sector (G3) concerning any of the indices of soft skills and hard skills, as well as between entities offering accounting and consulting services (G2) and entities in the industry, trade, and services sector (G3).

On the other hand, the comparison between entities from G1 versus entities related to the technology sector (G4) shows significant differences regarding the index that compiles items related to interpersonal skills (at a significance level of 5%), which is part of soft skills, also influencing the results seen for the total index (although at a significance level close to 10%). It is worth recalling that, based on the frequency analysis presented earlier, such skills were often less required by entities from employment-related services (G1).

Table 9. Mann-Whitney Test by Subgroups.

	G	Mean Ranks	G	Mean Ranks	G	Mean Ranks	G	Mean Ranks	G	Mean Ranks	G	Mean Ranks
ISS1	G1	58.40	G1	44.04	G1	45.35	G2	35.75	G2	37.15	G3	19.83
	G2	63.34	G3	43.86	G4	49.96	G3	32.89	G4	38.28	G4	21.91
ISS2	G1	57.12	G1	43.99	G1	42.93**	G2	36.13	G2	35.32	G3	17.44*
	G2	65.07	G3	44.06	G4	57.22**	G3	31.81	G4	42.33	G4	23.78*
ISS3	G1	57.55	G1	43.14	G1	46.04	G2	35.17	G2	38.14	G3	21.47
	G2	64.49	G3	47.28	G4	47.89	G3	34.53	G4	36.09	G4	20.63
ISST	G1	57.33	G1	43.64	G1	43.76*	G2	35.67	G2	36.28	G3	18.94
	G2	64.79	G3	45.39	G4	54.72*	G3	33.11	G4	40.20	G4	22.61
IHS1	G1	66.13**	G1	45.41	G1	48.04	G2	34.27	G2	36.45	G3	20.83
	G2	52.88**	G3	38.61	G4	41.89	G3	37.06	G4	39.83	G4	21.13
IHS2	G1	61.07	G1	43.72	G1	45.52	G2	34.40	G2	36.46	G3	20.61
	G2	59.74	G3	45.06	G4	49.43	G3	36.69	G4	39.80	G4	21.30
IHST	G1	65.82**	G1	45.00	G1	46.66	G2	34.19	G2	35.65	G3	20.19
	G2	53.30**	G3	40.17	G4	46.02	G3	37.31	G4	41.61	G4	21.63

Notes: ** Significant at a 5% level; * Significant at a 10% level

Finally, the comparison between entities involved in the industry, trade, and services sector (G3) versus entities involved in the technology sector (G4) reveals significant differences regarding the index that compiles items related to interpersonal skills, albeit at a 10% significance level, which is similar to what was found when entities from the employment-related services sector (G1) and entities in the technology sector (G4) were compared.

The next section discusses the findings from this research.

5 Discussion

This section discusses the findings considering the defined hypotheses and based on the literature review conducted earlier. Using the Wilcoxon test, significant differences were identified between the hard skills and soft skills required by employers seeking certified accountants through LinkedIn, making it impossible to confirm the first proposed hypothesis.

Concerning soft skills, it was observed that instrumental skills are the most sought-after, validating the second hypothesis. Finally, regarding hard skills, it was also possible to confirm that employers continue to prioritize main technical skills, supporting the third proposed hypothesis. Based on the findings, differences are evident in the skills demanded by employers, with a greater emphasis on hard skills compared to soft skills. While the literature review suggests the growing importance of soft skills [2, 3, 5,

6], the findings indicate that hard skills remain a priority in the approach adopted by employers in their recruitment and selection of certified accountants. Furthermore, the Mann-Whitney test found significant differences in specific comparisons between certain groups.

Thus, through the analysis of the results regarding the required soft skills, instrumental skills stand out, corroborating the findings of studies in this area that identify them as the primary skills sought by employers [4, 5, 13]. Entities in the accounting and consulting sectors and those in commerce, industry, and non-technology-related services especially demand analytical and synthesis skills, organizational and planning abilities, and oral and written communication skills in the native language [28].

Interpersonal skills are more commonly requested by entities related to the technology sector compared to those in commerce, industry, and services. On the other hand, employment-related services entities have lower requirements for soft skills. This conclusion aligns with [9], who emphasize the importance of mastering instrumental skills for a harmonious adaptation to various professional situations and opportunities.

In the realm of hard skills, proficiency in a second language prevails in secondary technical skills (81% of the analyzed cases), followed by tasks related to the preparation of financial statements as the primary technical skill. It is noteworthy that computer skills gain prominence again, encompassing 66% of this item in secondary technical skills. In this context, employment-related services entities stand out, especially in primary technical skills, followed by those linked to the certified accountant's activity. These results align with the literature, highlighting the transformation in the role of the accounting professional [6, 29]. However, there is not enough evidence to confirm that, while relevant, these skills assume the same degree of importance as the other identified primary technical skills in this study.

The results for item H7 (proficiency in Management/Accounting software), as the second most sought-after hard skill, corroborate previous studies [6–8], emphasizing the significant value placed on IT skills by employers in the field of accounting. Although [29] and [13] concluded that Excel is one of the most sought-after skills by employers in this area, item H5 (Excel Proficiency) was required in only 52% of the analyzed job postings.

The next section presents the conclusions for this research.

6 Conclusions

The findings from this research globally found that hard skills continue to play a more prominent role in the context of the accounting profession, with a focus on skills related to a second language, tasks related to the preparation of financial statements, and computer skills. Regarding soft skills, instrumental skills stand out, especially in areas related to analytical and synthesis capabilities, organization and planning, and oral and written communication in the native language. Significant differences were also identified by the employers' industry sectors.

The main limitation of this study is related to the subjectivity in information gathering, which may hinder the identification of required skills, especially in the case of soft skills. Additionally, the omission of skills required only in the later stages of the recruitment process could introduce a potential bias in the results. Furthermore, the size of the employing entities, the financial environment, education, and individual skills may influence the context of the sample ads, potentially impacting the obtained results. It is also important to notice that such entities from the employment-related services sector can represent different sectors simultaneously, which was impossible to identify separately considering the confidentiality of this information. Nonetheless, this restriction was partially mitigated through the specific identification of entities from this sector.

As a result, future research may employ alternative methods to overcome some of these limitations, such as conducting interviews with potential employers. Employer characteristics, omitted in the job postings, like the size of the entity, could be incorporated. Comparing the Portuguese reality with other countries could also identify potential differences in the recruitment profile of certified accountants, based on local characteristics, including underlying legal or cultural issues. The findings also contribute HEIs to reflect on the need to align the skills acquired by students with the current demands of the job market. This reflection may lead to the implementation of activities or curricular changes, ensuring that students develop knowledge and skills relevant to a more effective transition to the job market, resulting in greater professional success, especially in the early years of their careers.

References

1. Losi, H., Isaacson, E., Boyle, D.: Integrating Data Analytics into the Accounting Curriculum: Faculty Perceptions and Insights. Issues Account. Educ. **37**(4), 1–23 (2022)
2. Dzuranin, A., Jones, J., Olvera, R.: Infusing data analytics into the accounting curriculum: A framework and insights from faculty. Education **43**, 24–39 (2018)
3. Odia, J., Oke, R.: Change in management accountant roles: A comparative study of two public sector organizations in nigeria. J. Acc. Fin. Audit. Stud. **4**(3), 198–222 (2018)
4. Domingos, A.: Ph.D. thesis, Universidade Lusíada de Lisboa (2017)
5. Dwaase, D., Awotwe, E., Smith, E.: Skills Requirements of the Professional Accountant in a Changing Work Environment. IOSR J. Hum. Soc. Sci. **25**, 12–17 (2020)
6. Suleman, F.: Employability skills of higher education graduates: little consensus on a much-discussed subject. Procedia. Soc. Behav. Sci. **228**, 169–174 (2016)
7. Tan, L., Laswad, F.: Professional skills required of accountants: What do job advertisements tell us? Acc. Educ. **27**, 403–432 (2018)
8. Carvalho, C., Almeida, A.: The Adequacy of Accounting Education in the Development of Transversal Skills Needed to Meet Market Demands. Sustainability **14**(10), 5755 (2022)
9. Tsiligiris, V., Bowyer, D.: Exploring the impact of 4IR on skills and personal qualities for future accountants: a proposed conceptual framework for university accounting education. Acc. Educ. **30**(6), 621–649 (2021)
10. Jackson, D., Michelson, G., Munir, R.: Developing accountants for the future: New technology, skills, and the role of stakeholders. Accounting Education 1–28 (2022)
11. Uwizeyemungu, S., Bertrand, J., Poba-Nzaou, P.: Patterns underlying required skills for CPA professionals: a content and cluster analysis of job ads. Acc. Educ. **29**(2), 109–136 (2020)
12. Chaplin, S.: Accounting education and the prerequisite skills of accounting graduates: Are accounting firms' moving the boundaries? Aust. Account. Rev. **27**(1), 61–70 (2017)
13. Dolce, V., Emanuel, F., Cisi, M., Ghislieri, C.: The soft skills of accounting graduates: perceptions versus expectations. Acc. Educ. **29**(1), 57–76 (2020)

14. Villeda, M., McCamey, R.: Use of Social Networking Sites for Recruiting and Selecting in the Hiring Process. Int. Bus. Res. **12**(3), 66–78 (2019)
15. Hosain, S., Liu, P.: The role of social media on talent search and acquisition: Evidence from contemporary literature. J. Intercult. Manage. **12**(1), 92–137 (2020)
16. Koch, T., Gerber, C., Klerk, J.: The impact of social media on recruitment: are you LinkedIn? SA J. Hum. Resour. Manag. **16**(1), 1–14 (2018)
17. Verma, A., Yurov, K., Lane, P., Yurova, Y.: An investigation of skill requirements for business and data analytics positions: a content analysis of job advertisements. J. Educ. Bus. **94**(4), 243–250 (2019)
18. Vasanthakumari, S.: Soft skills and its application in workplace World. J. Adv. Res. Rev. **3**(2), 66–72 (2019)
19. Matturo, G., Raschetti, F., Fontán, C.: A Systematic mapping study on soft skills in software engineering. JUCS – J. Univ. Com. Sci. **25**(1), 16–41 (2019)
20. González, J., Wagenaar, R.: Tuning Educational Structures in Europe: Universities' contribution to the Bologna Process. https://www.unideusto.org/tuningeu/wp-content/uploads/2023/05/ENGLISH_BROCHURE_FOR_WEBSITE.pdf (2008). Last access 2024/04/23
21. Wikle, T., Fagin, T.: Hard and soft skills in preparing GIS professionals: comparing perceptions of employers and educators. Trans. GIS **19**(5), 641–652 (2014)
22. Dahal, R.: Changing role of management accounting in 21st Century. Rev. Pub. Adm. Manage. **7**(3), 1–8 (2019)
23. Altarawneh, G.: An empirical evaluation of accounting graduates' employability skills from Jordanian employers' perspective. Int. Bus. Res. **9**(1), 55–65 (2016)
24. Busco, C., Consolandi, C., Eccles, R., Sofra, E.: A preliminary analysis of SASB reporting: Disclosure topics, financial relevance, and the financial intensity of ESG materiality. J. Appl. Corp. Financ. **32**(2), 117–125 (2020)
25. Krismiaji, K., Surifah, S.: Corporate governance, compliance level of IFRS disclosure and value relevance of accounting information–Indonesian evidence. J. Int. Stud. **13**(2), 191–211 (2020)
26. Nicolescu, M., Vărzaru, A.: Ethics and disclosure of accounting, financial and social information within listed companies. Evidence from the Bucharest Stock Exchange. In: Pamfilie, R., Dinu, V., Tăchiciu, L., Plesea, D., Vasiliu, C. (eds.) 6th BASIQ International Conference on New Trends in Sustainable Business and Consumption 2020, pp. 73–80. ASE, Bucharest (2020)
27. Field, A.: Descobrindo a estatística usando SPSS. Artmed, Lisboa (2009)
28. Qadir, S. M., Bensen Bostanci, H., Kurt, M.: Writing Apprehension Among English as a Foreign Language Postgraduate Students. SAGE Open **11**(2) (2021)
29. Aryanti, D., Adhariani, D.: Students' Perceptions and Expectation Gap on the Skills and Knowledge of Accounting Graduates. J. Asian Fin. Econ. Bus. **7**(9), 649–657 (2020)

Author Index

L. Almeida et al. (Eds.): ICAFI 2025, ISEM 80, pp. 645–646, 2026.
https://doi.org/10.1007/978-3-032-16268-7

The manufacturer's authorised representative in the EU is Springer Nature Customer Service Centre GmbH, Europaplatz 3, 69115 Heidelberg, Germany. If you have any concerns regarding our products, please contact ProductSafety@springernature.com

Printed and bound by CPI Group (UK) Ltd, Croydon, CR0 4YY
05/10/2026
02228042-0006